16th Asia Pacific Bioinformatics Conference (APBC 2018)

Yokohama, Japan
15 – 17 January 2018

Editors:

Yi-Ping Phoebe Chen
Yasubumi Sakakibara
Yoshihiro Yamanishi

ISBN: 978-1-5108-6019-3

Printed from e-media with permission by:

Curran Associates, Inc.
57 Morehouse Lane
Red Hook, NY 12571

proceedings.com

Some format issues inherent in the e-media version may also appear in this print version.

Printed by Curran Associates, Inc. (2018)

For further information, please contact BioMed Central Ltd
at the address below. The collection is available online at
https://bmcgenomics.biomedcentral.com/articles/supplements/volume-19-supplement-2

BioMed Central Ltd
236 Gray's Inn Road
London WC1X 8HB
UK

Phone: +44 (0)20 3192 2000
Fax: +44 (0)20 3192 2010

Email: info@biomedcentral.com
Web: www.biomedcentral.com

Additional copies of this publication are available from:

Curran Associates, Inc.
57 Morehouse Lane
Red Hook, NY 12571 USA
Phone: 845-758-0400
Fax: 845-758-2633
Email: curran@proceedings.com
Web: www.proceedings.com

TABLE OF CONTENTS

Mukherjee *et al. BMC Systems Biology* 2018, **12**(Suppl 1):6
https://doi.org/10.1186/s12918-018-0533-6

BMC Systems Biology

RESEARCH

Open Access

Counting motifs in dynamic networks

Kingshuk Mukherjee*, Md Mahmudul Hasan, Christina Boucher and Tamer Kahveci

From The Sixteenth Asia Pacific Bioinformatics Conference
Yokohama, Japan. 15-17 January 2018

Abstract

Background: A network motif is a sub-network that occurs frequently in a given network. Detection of such motifs is important since they uncover functions and local properties of the given biological network. Finding motifs is however a computationally challenging task as it requires solving the costly subgraph isomorphism problem. Moreover, the topology of biological networks change over time. These changing networks are called dynamic biological networks. As the network evolves, frequency of each motif in the network also changes. Computing the frequency of a given motif from scratch in a dynamic network as the network topology evolves is infeasible, particularly for large and fast evolving networks.

Results: In this article, we design and develop a scalable method for counting the number of motifs in a dynamic biological network. Our method incrementally updates the frequency of each motif as the underlying network's topology evolves. Our experiments demonstrate that our method can update the frequency of each motif in orders of magnitude faster than counting the motif embeddings every time the network changes. If the network evolves more frequently, the margin with which our method outperforms the existing static methods, increases.

Conclusions: We evaluated our method extensively using synthetic and real datasets, and show that our method is highly accurate($\geq$ 96%) and that it can be scaled to large dense networks. The results on real data demonstrate the utility of our method in revealing interesting insights on the evolution of biological processes.

Keywords: Biological networks, Dynamic networks, Motif finding

Background

Biological networks capture complex interactions among molecules (e.g., genes, proteins, etc.) which perform various cellular functions [1]. These networks can be represented as graphs where the set of nodes represents the set of molecules and the set of edges represents the set of known interactions among these molecules. Depending on what the nodes and edges represent (e.g., proteins and signalling events), we classify a biological network as either a protein-protein interaction network, a gene regulatory network, or a signalling network—just to name a few. Interactions among these nodes can be undirected (e.g., protein-protein interaction network) or directed (e.g., gene regulatory network). To understand how cells perform diverse functions, we need to study the networks that model these interactions. Analysing biological networks has proved to be useful in many applications, such as identifying functionally similar pathways in multiple species [2, 3], reconstructing metabolic networks from newly sequenced genome [4] or identifying drug targets [5].

One of the fundamental problems in analyzing biological networks is identification of network motifs. A *network motif* is a small subnetwork that occurs frequently in a given network [6, 7]. These motifs can be viewed as the basic building block of a biological network [6] and thus, uncover functions and local properties of it [8]. Finding network motifs is a computationally hard problem [9]. One way to identify the topological structure of a motif of n nodes is to generate all possible subnetwork topologies of n nodes and search these topologies in the given target network. This problem becomes intractable as the value of n increases since the number of possible topologies grows exponentially with this value. Furthermore, given

*Correspondence: kingdgp@ufl.edu
Department of Computer and Information Science and Engineering, University of Florida, Gainesville, FL, USA

a motif topology, counting the number of embeddings of this topology is identical to the subgraph isomorphism problem, which is NP-complete [10].

One common formulation to count the number of embeddings of a given motif in a given network is to allow overlap between the subnetworks (i.e. share nodes or edges). Most existing methods for motif counting use this overlap assumption [11–16]. An alternative formulation is to count only disjoint embeddings of each motif—i.e., no two embeddings of the same motif share an edge—in the target network [17]. A third and more restrictive formulation requires that no two embeddings of the same motif share a node in the target network. Counting non overlapping embeddings in a given network requires solving the maximum independent set problem which is NP-complete [9]. The complexities of the motif counting methods also grow rapidly as the number of nodes in the motif and the underlying network increases. Since all these methods try to solve the subgraph isomorphism problem, scaling these methods to large networks remains to be a difficult task.

The motif counting problem, when applied to biological networks, introduces a subtle, yet massive challenge, which is often ignored by most existing studies. This challenge arises due to evolving nature of biological networks. The topology of biological networks change over time. For instance, human embryonic stem cell differentiates into hematopoietic stem cell, then to various other cell types such as liver, kidney, etc. during the development process. Even without cell differentiation, as the chromosomes' chromatin structures change through folding and unfolding events, different sets of genes get exposed for transcription and thus for interaction. As the network evolves, frequency of each motif in the current network topology can also change. Thus, even if we know the count of a given motif prior to the topological alteration of the network, this number becomes invalid after the alteration. To resolve this issue, we need to update the frequency of each motif such that it effectively mirrors the current snapshot of the network. Methods that compute frequency of motifs assume that the topology of the network remains unchanged once the computation is performed. One trivial way to adopt these methods to dynamically evolving network topologies is to re-compute the frequency of motifs from scratch each time the network evolves. This strategy however renders to be very expensive and impractical particularly for large and highly evolving networks. We need new strategies that quickly adapt to the topological changes as the network evolves.

Our Contributions. In this article, we design and develop a scalable method for counting the number of embeddings of given motifs in a dynamic biological network. Briefly, our method first computes the number of embeddings of each motif in the initial network topology. For each motif, we store its embeddings in a list. As the topology of the network evolves through network edit operations (i.e., node (edge) insertion (deletion)), we dynamically process the list of embeddings and update the count of motifs. Lastly, we demonstrate that our method can dynamically update the frequency of each motif in orders of magnitude faster than existing methods, and that it has very high accuracy (e.g., $\geq 96\%$).

Preliminaries

Basic definitions and notation

We restrict our interest to biological networks that can be modeled as a undirected graphs. That said, the methods we develop here can be applied to directed graphs as well. We denote a graph as $G = (V, E)$ with a set of nodes V connected by a set of edges E. A graph $G' = (V', E')$ is a subgraph of $G = (V, E)$ if $V' \subseteq V$ and E' is a subset of edges of E connecting the nodes in V' (i.e., $E' \subseteq E \cap V \times V$).

A *path* is a sequence of edges in a graph that constitutes an ordered sequence of distinct nodes. For example, the sequence of nodes $[a, b, e, f]$ is a path that is shown in Fig. 1a. We define the *length* of a path as the number of edges on that path. We define the *distance* between two nodes in a graph as the length of the shortest path which connects them.

We next define the concept of connectedness of nodes and graphs. Nodes u and v of graph G are connected if there is a path between u and v in G. Thus, a graph $G = (V, E)$ is connected if all node pairs $u, v \in V$ in G are connected. Figure 1a depicts a connected graph. This paper only considers connected subgraphs as motifs. Therefore, in the rest of this article, the term subgraph implies that it is a connected subgraph.

In a given graph $G = (V, E)$, we define the *k-neighborhood* of an edge $e = (u, v)$ as a subgraph $G' = (V', E')$, where $E' \subseteq E$ contains all edges which have at least one incident node whose distance to either u or v is at most k. The set of nodes, $V' \subseteq V$ contains all nodes connected by E'.

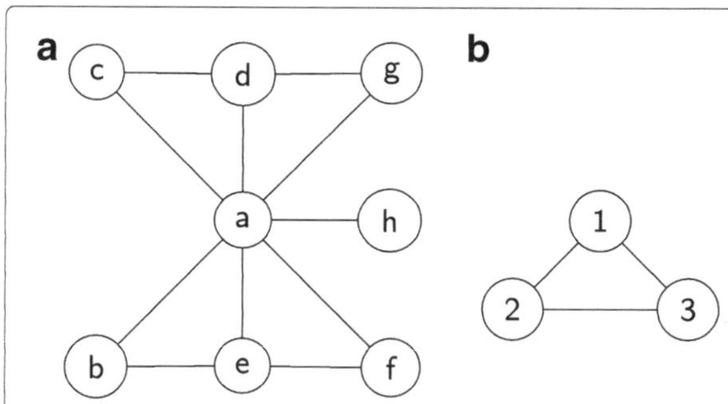

Fig. 1 An example to demonstrate pattern and its frequencies

Graph isomorphism

We say that two graphs $G = (V_G, E_G)$ and $H = (V_H, E_H)$ are isomorphic if there is a bijection function $\pi : V_G \rightarrow V_H$ such that edge $(u, v) \in E_G \iff (\pi(u), \pi(v)) \in E_H$. Let us denote the graph in Fig. 1a with G. Figure 1b presents a graph that forms a triangle. Let us denote this with P. There are four subgraphs in G which are isomorphic to P. These subgraphs consist of the following sets of edges $\{(a, b), (a, e), (b, e)\}$, $\{(a, e), (a, f), (e, f)\}$, $\{(a, c), (a, d), (c, d)\}$, and $\{(a, d), (a, g), (d, g)\}$. We denote these subgraphs with S_1, S_2, S_3 and S_4 respectively. We denote the set of all subgraphs in G which are isomorphic to P with $\mathcal{S} = \{S_1, S_2, S_3, S_4\}$

Isomorphism is a transitive relationship. This indicates that all subgraphs in $\mathcal{S}$ are isomorphic to each other. Thus $\mathcal{S}$ defines a topological equivalence class of subgraphs. We represent subgraphs in an equivalence class with a pattern. For example, the triangle pattern P represents the equivalence class $\mathcal{S}$. Notice that the nodes of a pattern are unlabeled and it only defines the topology of the subgraph in an equivalence class.

We say that two subgraphs of a given graph *overlap* if they share at least one edge. Otherwise, we call them *disjoint*. For example, in Fig. 1a subgraphs S_1 and S_2 overlap since they share the edge (a, e). Subgraphs S_1 and S_4 on the other hand are disjoint.

The *frequency* of a pattern P in a graph G is a function of the set of subgraphs of G which are isomorphic to P. We call these subgraphs as the *embeddings* of P in G. The fact that these subgraphs may overlap with each other has led to alternative definitions to the frequency function. The classic definition of frequency ignores overlaps between isomorphic subgraphs. It thus reports the frequency of a pattern as the cardinality of its equivalence class. This frequency function is denoted as the $F1$ count of a pattern [18]. For example, each subgraph in $\mathcal{S}$ is an embedding of the pattern P in G. Therefore, the $F1$ count of pattern P in G is four. An alternative measure counts only the disjoint embeddings of a pattern. This frequency is denoted as the $F2$ count. For example, in Fig. 1a the $F2$ count of P in G is two (e.g., $\{S_1, S3\}$). A more restrictive version of frequency definition of a pattern does not allow two embeddings to have a common node. This is denoted as the $F3$ count of a pattern. For example, $F3$ count of pattern P in G is one since node a is common to all embeddings in $\mathcal{S}$.

Dynamic biological networks

Biological networks are inherently dynamic structures. Their topologies change with time through network edit operations. There are four possible network edit operations that perturb the topology of a network, namely (i) node insertion, (ii) node deletion, (iii) edge insertion, and (iv) edge deletion. It is easy to represent node insertion (deletion) using edge insertion (deletion). For example, a node insertion can be represented as a set of edge insertions between existing nodes and the new node. Similarly, a node deletion is equivalent to deleting all edges connected to the deleted node. Formally we define a dynamic network as follows: a sequence of edge deletions and additions $e_{\pi_1}, e_{\pi_2}, e_{\pi_3} ..., e_{\pi_i}$ $\forall e_{\pi_i} \in V \times V$ are performed on $G = (V, E)$ such that

$$G_i = \begin{cases} (V, E_i = E_{i-1} \cup \{e_{\pi_i}\}), & \text{if } e_{\pi_i} \notin E_{i-1} \\ (V, E_i = E_{i-1} - \{e_{\pi_i}\}), & \text{otherwise} \end{cases}$$

where G_i represents the topology of the network after the ith edge insertion or deletion and G_0=G

Related work

We classify existing methods that are related to our work in two categories: methods for static networks and methods for dynamic networks. The methods on static networks can be broadly divided into two categories: (1) those that count motifs in a single network, and (2) those that count motifs in a set of networks. SUBDUE [19] and GREW [20], for example, find the most frequent subgraphs in a given large network. Complete subgraph finding algorithms are those which are guaranteed to find all subgraphs that satisfy some constraints in a given network. Most subgraph finding algorithms are complete; however, their efficiency falls when they operate on large dense graphs. Algorithms such as SUBDUE and GREW are scalable to an extent but they find only a small number of subgraphs than those discovered by complete algorithms.

The meaning of frequency changes when we consider methods finding frequent subgraphs in a set of networks. If a subgraph exists in a network, the frequency of that subgraph increases by one, irrespective of the number of embedding found in that particular network. FSG [21], gSpan [22] and SPIN [23] fit in this category. FSG generates possible subgraphs by growing them one edge at a time. The number of graphs that contains a certain subgraph is referred to as the support of that subgraph. Frequent subgraphs are reported which exceeds a certain set support threshold. The method becomes expensive as it computes the canonical labels of many redundant subgraphs. gSpan performs a DFS(Depth First Search) and finds out the minimum representation of vertex orderings and then organizes the canonical labels into a hierarchical spanning tree. The frequent subgraphs are discovered by traversing the tree and finding subgraphs which exceed the minimum support. In this way gSpan avoids generating canonical labels for redundant subgraphs. SPIN improves on gSpan by reducing the mining area as it only looks at maximal frequent subgraphs which are not part of larger frequent subgraphs. As mentioned, these methods do not count the number of occurrences of a subgraph in a network but rather check is the subgraph appears at least once in a given network.

CODENSE [24] finds coherent, dense subgraphs in large biological networks. GraphSig [25] mines significant and frequent subgraphs in which graphs are represented as feature vectors. SiS [26] finds subgraphs with the largest probability to appear in a set of biological networks. Most of these methods only focus on the $F1$ count of the patterns (e.g., [11–16]). Others such as MAVISTO [18] computes both $F2$ and $F3$ counts of motifs.

Elhesha et al. [17] finds the $F2$ count of large motifs in a biological network by first computing the $F2$ counts of 4 basic motifs and then joining those basic motifs iteratively to generate larger motifs. Since this paper is one of the most recent papers which computes the $F2$ count of a given motif in a static network, we have used this method for the initial computation of the $F2$ counts and have also compared the running time and accuracy of our method against this method. We explain this in more detail in "Evaluation criteria" section.

There are few methods which focus on dynamic networks and none aim to solve motif counting. Wackersreuther et al. [27] discover frequent subgraphs in dynamic networks. From a time series of graphs they generate a union graph which they call dynamic graph. Each edge in the dynamic graph is denoted with an *edge existence string* which contains the label of that edge at different points of the time series. From the dynamic graph they find subgraphs which share the same common dynamic pattern i.e. they appear and disappear together from the time series of graphs. Qin et al. [28] solve a similar problem but they label the edges 1 or 0 depending on whether that edge exists in a given time point or not. The algorithm finds significant substructures whose edge labels show some fixed properties (patterns of appearance and disappearance from the network) and are within a user defined distance of each other in the dynamic graph.

These methods require the networks at different time points as input and they do all their processing on a union graph constructed from this sequence of networks. Also, these methods only look at the $F1$ counts and do not consider the $F2$ frequency of subgraphs.

Methods

We begin by describing a method that computes the count of $F1$ and $F2$ for a given motif in a static network. Next, we describe possible network operations that change the topology of the networks, and discuss how to dynamically update the count of $F1$ and $F2$ for each of these operations.

Motif counting in static networks

Assume that we are given a motif topology denoted with P. Given a graph G, we want to compute the count of $F1$ and $F2$ of pattern P in G. Let us denote the set of all embedding of P in G with S. We denote the cardinality of the set S (i.e., $F1$ count of P) with $|S|$. Recall that the $F2$ count of pattern P is the cardinality of the maximal set of embedding where two embeddings do not share edges. We denote such set with S'. To compute the $F2$ count of P, we introduce the concept of an *overlap graph*, which is unique to P and G. Let us denote the overlap graph with $G^o = (V^o, E^o)$. Here, each node in V^o corresponds to an embedding of P listed in S. Let us denote the relationship between the nodes in V^o and the embeddings in S with a bijection function $\phi : V^o \rightarrow S$. Each edge $(u, v) \in E^o$ indicates that the two embeddings $\phi(u)$ and $\phi(v)$ share at least one edge.

We use the overlap graph to generate the maximal, non-overlapping embedding set S' in an iterative fashion. First, we find the node $u \in V^o$ with the smallest degree. If there are multiple nodes with the same smallest degree, we randomly select one of them. We insert the corresponding embedding $\phi(u)$ into S'. Since S' only contains non-overlapping embeddings, we remove node u from V^o along with all the nodes $v \in V^o$, such that $(u, v) \in E^o$. We repeat this process to populate S' until V^o becomes the empty set.

Motif counting in dynamic networks

Let us denote the given network with $G = (V, E)$. Also, let us denote the topology of the network after the ith edge insertion or deletion with $G_i = (V, E_i)$. Thus, we have G_0=G and $\forall\ i \geq 0$, $|E_i - E_{i-1}| = 1$. Given a motif topology denoted with pattern M, we compute the $F1$ and $F2$ counts of M in the initial network G_0 by using the method described in "Motif counting in static networks" subsection. As the network G evolves (i.e., new edges are added and/or deleted), the count of $F1$ and $F2$ of M can change. Next, we will show an algorithm for efficiently updating the $F1$ and $F2$ counts as the network evolves from G_i to $G_{i+1}\ \forall\ i \geq 0$. By repeatedly applying our algorithm, after each network edit operation, the motif count is updated for arbitrarily large sequence of network updates.

Updating the F1 count

We now describe our method for updating the count of $F1$ of M as G_i evolves into G_{i+1}. We assume that $F1$ for G_i is known. Our algorithm for updating $F1$ relies on initially constructing and maintaining an auxiliary data structure that allows for the embeddings containing an edge to be efficiently queried. Thus, at the beginning of our algorithm, we find all embeddings of a given motif M in the initial network G_0. After finding these embeddings, we create a list of embeddings for each edge $e \in E$, denoted as D_e, which stores all embeddings that contain e. That is, for a motif M, let m be an embedding in a given network. Then $m \in D_e$ if $e \in m$. This data structure, which we refer to as the *edge compressed bitmap*, is updated each time an edge is either added or deleted. The $F1$ is then updated based on the *edge compressed bitmap*.

Suppose that as the network G_i evolves to G_{i+1} the $e \in E_i$ is deleted. This reduces the $F1$ count of motif M, if the deleted edge is a part of embeddings of M. From the *edge compressed bitmap*, we find the set of embeddings of M which contain e. We remove this set (D_e) from the edge compressed bitmap and reduce the $F1$ count of M by the cardinality of this set.

Next, assume that an edge $e \notin E_i$ is added to G_i. Unlike the edge deletion, prior to this update, we do not know whether e is a part of an embedding of M in G_{i+1}. We locate such embeddings of M in G_{i+1} as follows. Let us denote the diameter of M with k. We search the *k-neighborhood* of e in G_{i+1}. The set of embeddings of M which contains e can be formed with its neighboring edges. We add this set to the edge compressed bitmap and increase the $F1$ count of M by the cardinality of the set of new embeddings.

Updating the F2 count

After updating the $F1$ count, we proceed to update the $F2$ count. Updating the $F2$ count is more challenging than updating $F1$ because computing the count of $F2$ is NP-complete [9] and the methods used are heuristics. As a result, the $F2$ count we compute even for a single static network may deviate from the optimal result. We would like to minimize the additional errors introduced by dynamic updates.

First, we assume that we have already computed the $F1$ and $F2$ counts of the given motif M in G_i and the $F1$ count of M in G_{i+1}. Next, we describe how we update the $F2$ count for G_{i+1}. There are following two possible scenarios: (1) an edge has been deleted from G_i, and (2) and edge has been added to G_i. In the first scenario, the removal of an edge e from G_i will cause the $F2$ count to either remain the same or decrease by one. The former case occurs when none of the embeddings in the set D_e contribute to the $F2$ count in G_i. The latter case occurs when one of the embeddings from the set D_e, contributes to the $F2$ count of G_i. Let us denote that embedding with X ($X \in D_e$). After removing e, the embedding X does not exist in G_{i+1}. This reduces the $F2$ count of M by one. However, it is possible that there is another embedding (say Y), which can be included in the $F2$ count for G_{i+1} to replace X. For this to happen, Y must satisfy two conditions: (i) Y overlaps with X, and (ii) Y does not overlap with any other embedding included in the $F2$ count of M in G_i. If such an embedding Y exists, we include it in the $F2$ set. Thus the $F2$ count remains unaltered. Otherwise the $F2$ count decreases by one.

In order to identify any embedding Y that satisfies the two condition above, we explore the *neighbors* of X in the overlap graph. Recall that the *neighbors* of an embedding in the overlap graph are those embeddings of M which share at least one common edge with that embedding. If say, X consists of edges e_1, e_2 and e_3 then the *neighbors* of X will be the union of sets D_{e_1}, D_{e_2} and D_{e_3}.

From the set of *neighbors* of X, we consider each embedding and check if they can be included in the updated $F2$ count. If an embedding Y in that set, has all of its edges free then we include it in the $F2$ set for G_{i+1}. Therefore, if such an embedding Y exists, the $F2$ count remains unaltered as the inclusion of Y compensates for the deletion of X. Otherwise we decrease the $F2$ count by one.

Assume that an edge e, where $E_{i+1} - E_i = \{e\}$ is added G_i.This addition will either increase the $F2$ count of M in G_{i+1} by one or has no influence. The new edge can form new embeddings of M in G_{i+1}. We explain how we obtain such new embeddings in "Updating the $F1$ count" section. We then check if any of these new embeddings can be included in the updated $F2$ count. To do this, we consider each new embedding, and check if all of its edges are uninvolved in the $F2$ count (they could be involved in the $F2$ count with other embeddings). If such an embedding exists, we include it in the $F2$ set and increase the $F2$ count by one.

Results

Datasets

We use synthetic and real datasets to evaluate our method. First, we describe our synthetic dataset. We use synthetic networks to observe the performance of our method by varying a broad suite of parameters, namely the number of nodes, the average number of edges per node in the network and the topology model. We call the number of nodes in the networks and average number of edges per node as network *size* and *degree* respectively. We generate undirected random networks using three generative models, namely (i) Erdös-Rényi (ER) [29], ii) Watts-Strogatz (WS) [30] and (iii) Barabási-Albert (BA) [31]. We experiment with networks of different sizes and degree values. We generate networks with sizes ranging from 1000 to 5000 at increments of 1000. For each network size, we set the network degree to ten, fifteen and twenty. For each parameter combination, we randomly generate 10 undirected synthetic networks for each model. In total, we generated 450 (i.e., $3 \times 5 \times 3 \times 10$) synthetic networks.

Next, we describe our secondary dataset. We generate gene co-expression network using the transcription data from Rivera-Mulia et al. [32]. This dataset contains the gene expressions for 19 human cell lines obtained at various stages of development from embriyonic stem cell to various cells. Table 1 lists these cell lines. From these cell lines, we construct co-expression networks at three different stages of development (i) *Post stem cell* uses all the cell lines except the stem cells. (ii) *Pre-pancreatic cell* includes all cell lines but the pancreas cells. (iii) *Pre-liver cell* contains all cell lines except liver cells. We use the datasets (i) and (ii) above to build a pathway of two networks for

Table 1 Cell lines used in our experiments

Cell line	Cell type
ESC_Cyt49	ESC
ESC_H9	ESC
NC_Cyt49	Neural crest
NC_H9	Neural crest
MSC_H9	MSCs
NPC_H9	NPCs
LPM_H9	Lateral plate mesodem
Splanc_Cyt49	Splanchnic mesoderm
Splanc_H9	Splanchnic mesoderm
Epic_Cyt49	Mesothelium
Epic_H9	Mesothelium
SM_H9	Smooth muscle
DE_Cyt49	Definitive endoderm
Liver_d5_Cyt49	Immature hepatic
Liver_d8_Cyt49	Hepatoblast
Liver_d16_Cyt49	Liver
Panc_d5_Cyt49	Primitive gut
Panc_d8_Cyt49	Posterior foregut
Panc_d12_Cyt49	Pancreas

pancreas cell development. We use (i) and (iii) to build a second pathway for liver development. For the first pathway above we construct two networks as follows. We compute the Pearson's correlation between the transcriptions of all gene pairs in the two datasets ((i) and (ii)). We use two correlation thresholds; *stringent cutoff* set to 0.95, and a *relaxed cutoff* set to 0.75. Assume that the post stem cell dataset is the primary and the Pre-pancreatic cell dataset is the secondary dataset. Given a pair of genes u and v we include an edge between them in the primary dataset network if one of the two conditions hold: (1) correlation between u and v within the primary dataset is above the stringent cutoff, or (2) correlation between u and v within the secondary dataset is above the stringent cutoff while that within the primary dataset is above the relaxed cutoff. We then construct the network for the Pre-pancreatic dataset similarly by assuming it to be the primary and the post stem cell dataset as the secondary one. We repeat the same process to construct another sequence of two networks for the liver pathway. The details of the three simulated networks and two pathways are summarized in Fig. 2.

Evaluation criteria

We simulate dynamic changes in
as follows. For each network, we perform a sequence of 1000 edit operations (i.e., edge insertion and deletion). We conduct the topological perturbations using the degree preserving edge shuffling method [33]. Each shuffle takes two edges (a, b) and (u, v), such that they share no nodes. It then replaces these edges with (a, v) and (u, b). Thus, one edge shuffling introduces two edge insertions and two edge deletions in a random order. Such edge shuffling ensures that all the nodes in the network preserves their degrees.

In our experiments, we use a small set of motifs that have been used frequently in literature. Figure 3 shows these motifs. Note that these are all possible topologies with two and three edges. We call these topologies the *basic* motifs. Having said that, our method is generic and can be employed to count any motif topology. Given a sequence of network edit operations, we first compute the number of embeddings for each motif in Fig. 3 before applying these edit operations. We do this using the method presented in Elhesha et al. [17] as it is one of the most recent papers which compute the $F2$ frequency of motifs in a static network. We then apply each edit operation in the order it was given to update the network topology. After each edit operation, we dynamically update the count of each motif using our method.

We compare our method against Elhesha et al. [17] as it also computes the $F2$ count of the four basic motifs. We evaluate our method in terms of its accuracy and running time. All experiments were performed on a server with 3 GB RAM and AMD Opteron dual core processors (2.2 GHz) running Linux. Notice that unlike our method, Elhesha et al. computes the $F2$ count from scratch when the network topology changes. Thus, at any point in the evolution, we consider their count as the ground truth. We report the accuracy of our method as the ratio of the $F2$ count of our method to the $F2$ count returned by their method at that instance of the network. We measure the running time of the static method for the initial network topology and denote it with *static time*. We also measure the running time of our method after each edit operation and denote it with *dynamic time*.

Running time on the synthetic dataset

Effect of network model

We first investigate the effects of different network models on the running time of our algorithm. We set the network size to 5000 nodes and the average node degree to 15. We

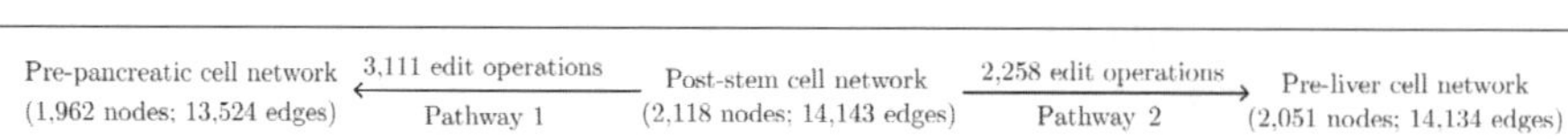

Fig. 2 Summary of simulated networks from real data

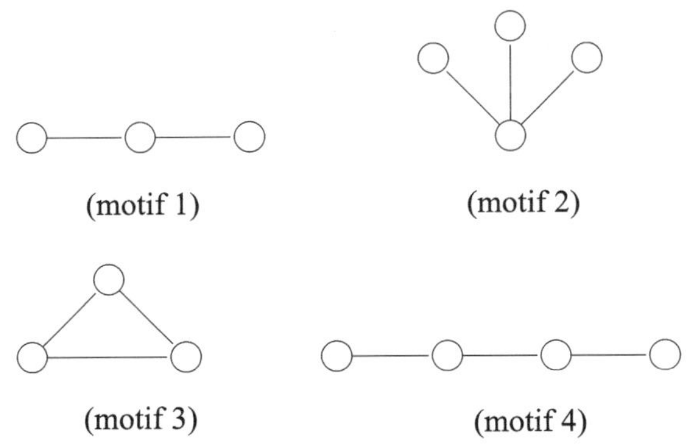

Fig. 3 Motifs used in the experiment

perform 1000 random edit operations and dynamically update the motif counts for all motif topologies in Fig. 3. Figure 4 presents the results.

We observe that the running time of our method for all three models are significantly less than that of the static method. The static method needs to recompute the $F2$ count from scratch every time the network topology changes. To illustrate this, we plot the cost of running the static algorithm twice, i.e., one for initial topology and the other for an arbitrary intermediate topology as the network evolves. We observe that, despite the additional cost our method incurs for initialization, the total cost of our method for 1000 dynamic topologies is less than that of the static method for only two topologies. More specifically our method updates the motif count for 1000 evolving topologies more than four times faster than the static method on a single network topology.

We observe that both static and dynamic methods incur the largest running time for the BA model. This is because the degree distribution in the BA model forces motif counting algorithms to evaluate many possibilities not encountered in other models, hence the increase in the running time. We observe much faster running time for the other two models.

Effect of network Size

In this experiment, we fix the average degree of the networks to 15, and vary the number of nodes from 1000 to 4000 with increment of 1000. We perform the experiments for each motif and random network model combination. We process 1000 random edit operations and dynamically update the motif counts. We compute the total update time as the processing time starting from the first edit operation until the end of the last edit operation. We compute the average update time as the total update time divided by the number of edit operations. Figure 5 shows the results.

We observe that the average update time of our method is very small even for the largest network size. This shows the effectiveness of using dynamic update in the motif counting problem.

Effect of average degree

In this experiment, we fix the network size to 4000 nodes and set the average node degree to 10, 15 and 20 respectively. We perform experiments for each motif and random network model. We process 1000 random edit operations and dynamically update the motif counts. We report the average update time for the edit operations. Figure 6 presents the results.

We observe that the average update time of our method is very small for all degree values. Average update time increases as the node degrees increase. Among the three models, we observe that the BA model takes longer time to update and the other two models show similar performance.

These experiments support our conjecture that dynamic update time per edit operation is very small for all parameter combinations. It scales well to networks with large sizes and node degrees. The potency of our method is important in evolving biological networks where running

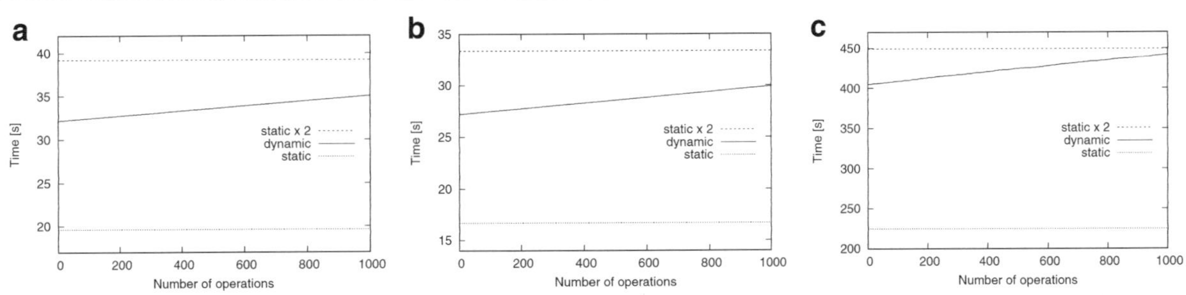

Fig. 4 Effect of network models on the running time. The three models are (i) Erdös-Rényi (ER), ii) Watts-Strogatz (WS) and (iii) Barabási-Albert (BA). The lower dotted line represents the running time for the static algorithm and the upper dotted line represents the twice the running time for the static algorithm. The solid line shows the running time for our dynamic algorithm as it computes the counts of the initial network and then updates them as the network evolves from G_0 to G_{1000}. **a** ER, **b** WS, **c** BA

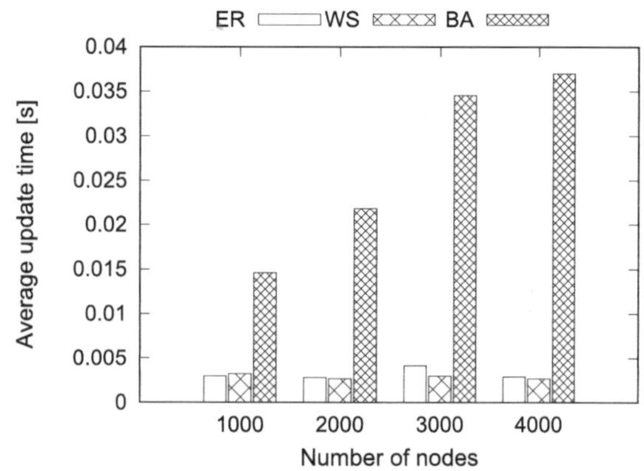

Fig. 5 Effect of network size on average update time for different random network models

static method after each topological change is not practical.

Accuracy on the synthetic dataset

In the previous section, we show that our method dynamically updates the motif count very fast in all parameter combinations. Recall however that it is possible that our dynamic method may yield fewer $F2$ count as compared to the static method. Faster running time is desirable only if our method produces accurate results. Here, we evaluate the accuracy of our method as compared to the static one. We run our experiments with network size and degree set to 5000 and 15 respectively. To compare with static method, we also run static method [17] after every 100 edit operations. Figure 7 presents the results.

We observe that our method is highly accurate for all motifs and random network models. The $F2$ count reported by our method is very close to that of the static method. As the number of edit operation increases, the gap between the two $F2$ counts gradually grow. This is expected since the static method computes the number of embeddings from scratch, whereas our method dynamically updates the count from the initial computation. That said, the gap remains to be negligible even after applying 1000 edit operations.

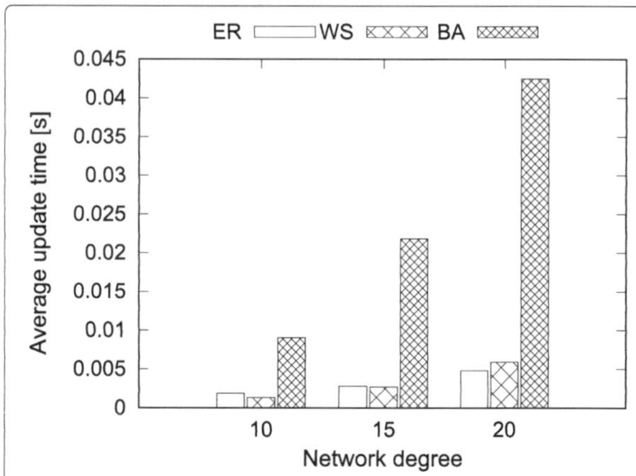

Fig. 6 Effect of node degree on average update time

Effect of network size

In this experiment, we fix the average degree of the networks to 15, and vary the network sizes from 2000 to 5000 with increment of 1000. In this experiment, we dynamically compute the motif counts after each of the 1000 edit operations. In the final topology, we also compute the results using the static method. We report the accuracy of our method for each motif and random model combinations. Figure 8 shows the results.

Our results demonstrate that our method is highly accurate for all network sizes, motifs, and random network models. The accuracy of our method is over 95% in all of our experiments. This shows the robustness of our method.

Effect of average degree

In this experiment, we fix the network size to 5000 and vary the degrees to 10, 15 and 20. We dynamically compute the motif counts after each of the 1000 edit operations. In the final topology, we also compute the results using the static method. Here, we report the accuracy of our method in different random models. Figure 9 shows the results.

Our method is highly accurate for all degree values. For instance, accuracy values of motifs one, two and four are over 97% for each network size. For all degree and random network combinations, our method counts the motifs with more than 95% accuracy.

In summary, our method yields highly accurate results for all network sizes, density values, motif topologies, and random network models. These results suggest that our method is robust to a broad spectrum of parameter settings.

Evaluation on the real dataset

We perform experiments on two pathways using the gene co-expression dataset described earlier in "Datasets" section. The first one is from post stem cell network to Pre pancreatic cell network. The second one is from post-stem cell network to pre-liver cell network. We consider the first network in each pathway as the *starting network* and the second one as the *final network* in the dynamic evolution. The difference in the set of edges between the starting and the final networks is the set of edit operations that the starting network goes through to evolve into the final network. As the order in which these edit operations take place are not known, we rank

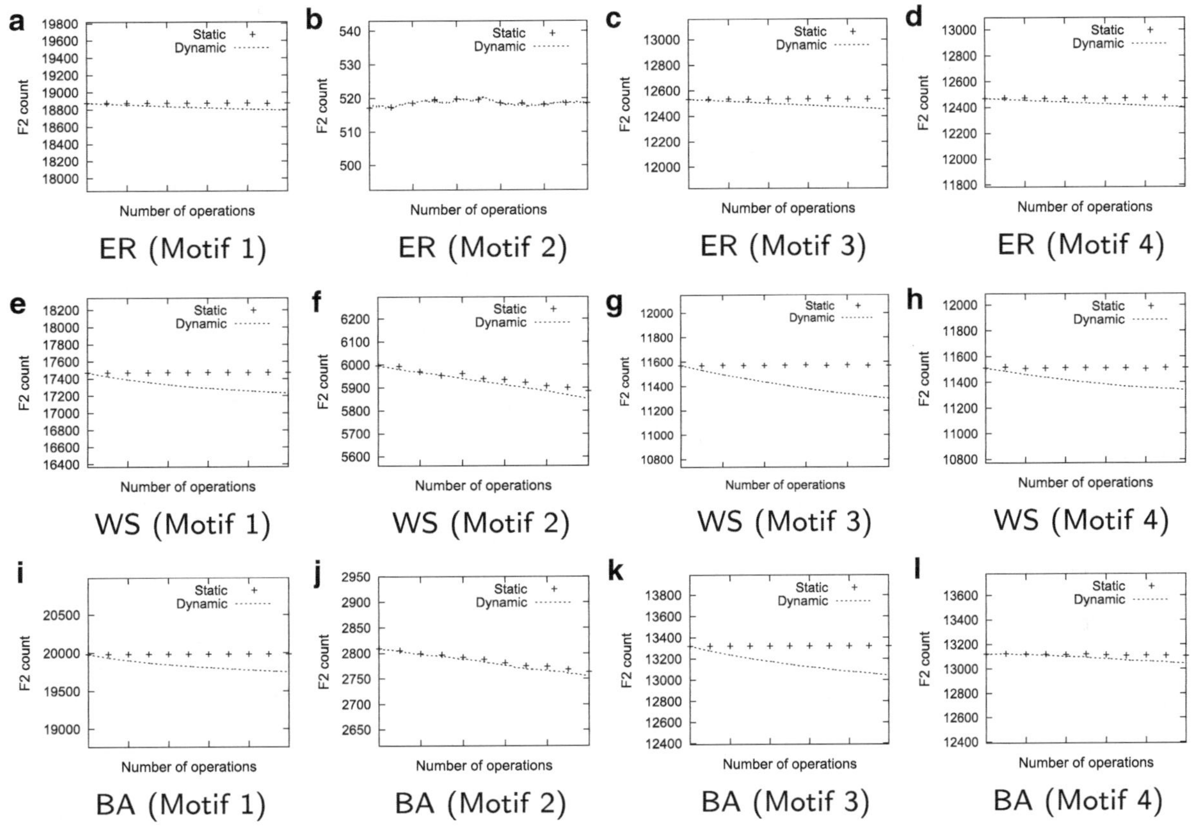

Fig. 7 Number of embeddings found by our dynamic method and static method as the network evolves through 1000 edit operations. **a** ER (Motif 1), **b** ER (Motif 2), **c** ER (Motif 3), **d** ER (Motif 4), **e** WS (Motif 1), **f** WS (Motif 2), **g** WS (Motif 3), **h** WS (Motif 4), **i** BA (Motif 1), **j** BA (Motif 2), **k** BA (Motif 3), **l** BA (Motif 4)

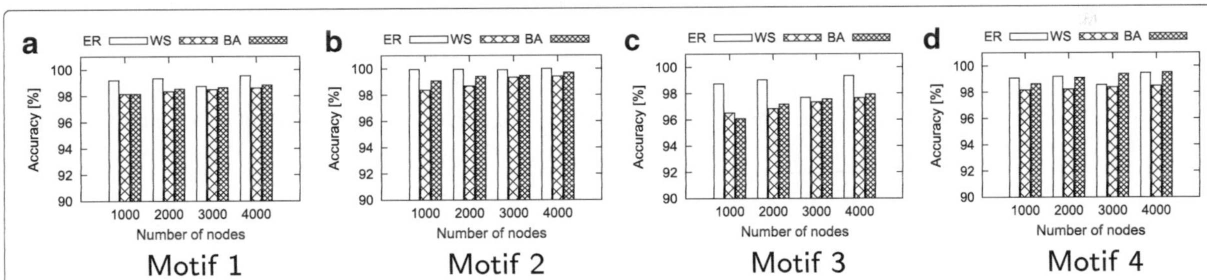

Fig. 8 Accuracy found by our dynamic method and static method. **a** Motif 1, **b** Motif 2, **c** Motif 3, **d** Motif 4

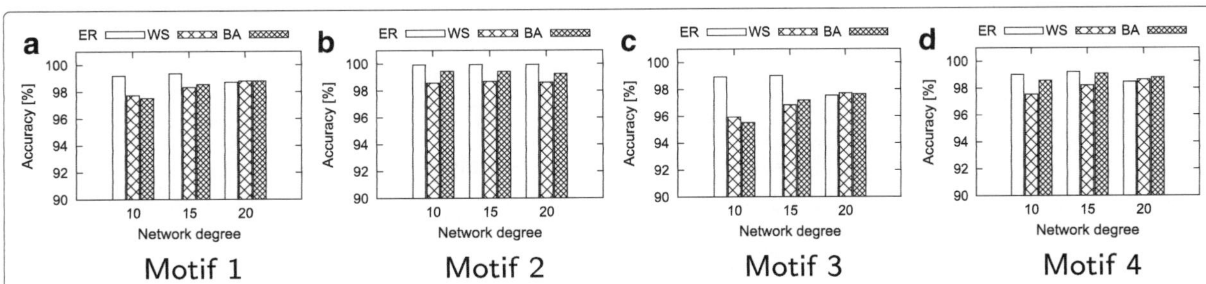

Fig. 9 Accuracy found by our dynamic method and static method. **a** Motif 1, **b** Motif 2, **c** Motif 3, **d** Motif 4

them randomly. We then use the static algorithm on the starting network to find the motif embeddings for the same motifs used in "Evaluation criteria" section. We then apply the edit operations in the given order and update the motif counts and embeddings. We also find all the motif embeddings and the F2 count of the motif in the final network using the static algorithm for comparison.

Evaluation of running time

Figure 10 presents the running time results for our method on the two pathways. For comparison, it also shows the running time of the static algorithm when it is applied to only the starting network, and both starting and final networks once. Our results demonstrate that our dynamic algorithm takes less time to update the *F*2 counts for over thousands of edit operations as compared to running the static algorithm once on two network instances only (the starting and final networks). This is a dramatic improvement as our method computes the motif count for not only those two networks but also thousands of network topologies in between these two states of networks. In other words our algorithm can identify motif counts for thousands of networks while the static algorithm fails to do that without running it from scratch. We observe that while the static algorithm takes around 60 s per network to count motifs, our dynamic algorithm updates the motif count in only a few milliseconds per intermediate network.

Evaluation of accuracy

The performance improvement is desirable only if the algorithm remains accurate. Here, we evaluate whether our method correctly computes motif counts. We focus on the final network of each pathway for this purpose. We run our algorithm to dynamically update the *F*2 motif count until we reach to the final network. Let us denote this count with $c_{dynamic}$. We then compute the same count using the static algorithm when it is applied on the final network only. Let us denote this count with c_{static}. We report the accuracy as the percentage of the motif count our method finds over the static one (i,e., accuracy = $100 \times c_{dynamic}/c_{static}$. Figure 11 presents the results.

Similar to the synthetic datasets, we observe that our dynamic algorithm is highly accurate on the real datasets as well. For all the motifs tested, the accuracy is over 97% for both pathways. We observe higher accuracy for the second pathway. This is possibly because the number of network edit operations for the second patway is lesser than that of the first one (see Fig. 10). The results are very promising for our method as both pathways are outcomes of a large number of edit operations.

Discussion

From the results on the simulated dataset we observe that the running time of our method for all three models are dramatically less than that of the static method. For the initial network topology, the dynamic method spends a little more time than the static method to compute the *F*2 count. This is because it initializes the relevant data structures which are needed to quickly update the *F*2 count as the network evolves. Once the data structures are set up, however, the cost of dynamic update becomes negligible. This demonstrates the key advantage of our dynamic method over the static one. The average update time grows very slowly as the network size increases. This suggests that our method scales to large networks and very large number of network edit operations. Among the three models, we observe that relative to the other two models, the BA model takes longer time to update. This is because the networks in BA model yields significantly more motif instances. The potency of our method is important in evolving biological

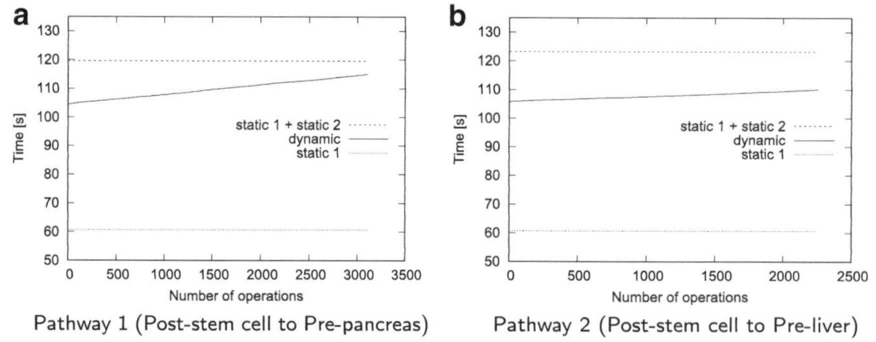

Fig. 10 Time analysis for real experiments. The lower dotted line represents the time taken by the Static algorithm to compute the counts in the *starting network* in the pathway. The upper dotted line represents the time taken by the static algorithm to compute the counts in the starting network plus the time taken by the static algorithm to compute the counts in the *final network*. The solid line represents the time taken by our dynamic algorithm to update the counts from the *starting network* to the *final network* as it goes through the network edit operations (plotted in X-axis). **a** Pathway 1 (Post-stem cell to Pre-pancreas), **b** Pathway 2 (Post-stem cell to Pre-liver)

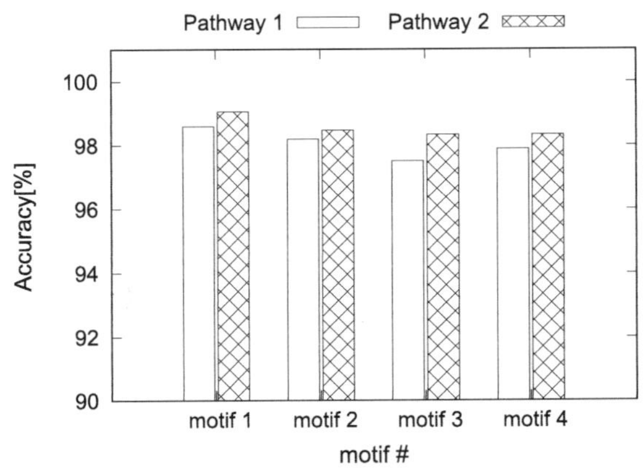

Fig. 11 Accuracy of the dynamic algorithm for the real networks

networks where running static method after each topological change is not practical.

We also observe that our method is highly accurate for all motifs and random network models. The accuracy of our method remains stable with increasing network size while the accuracy values for BA model increase as the network size grows. Since the BA model represents the biological networks well, these results promise the applicability of our method in real networks. We also observe that our method is stable for growing node degree values.

Our experiments with the real dataset shows that our methods have similar efficacy on real data as well. Therefore, it can be used on biological dynamic networks to uncover interesting observations on the evolution of biological processes.

Conclusions

Several approaches exist to compute the number of embeddings for a given motif in static network; however, no such method exists for dynamic networks. In this article, we address this problem and describe a method that incrementally updates the motif count as the network changes its topology. We evaluated our method extensively using synthetic and real datasets, and show that our method is highly accurate and that it can be scaled to large dense networks. The results on real data demonstrate the utility of our method in revealing interesting insights on the evolution of biological processes.

Funding
This work was supported partially by US National Science Foundation (NSF) under grant 1262451. The publication of this article was also sponsored by this grant.

Availability of data and materials
Data from the research of Rivera-Mulia et al. is publicly available at http://www.replicationdomain.org.

About this supplement
This article has been published as part of *BMC Systems Biology* Volume 12 Supplement 1, 2018: Selected articles from the 16th Asia Pacific Bioinformatics Conference (APBC 2018): systems biology. The full contents of the supplement are available online at https://bmcsystbiol.biomedcentral.com/articles/supplements/volume-12-supplement-1.

Authors' contributions
KM, MMH and TK worked on the algorithmic development. KM implemented the method, compiled the datasets and ran all experiments. KM, MMH, TK and CB worked on writing the manuscript. All authors edited the manuscript. All authors read and approved the final manuscript.

Ethics approval and consent to participate
Not applicable.

Consent for publication
Not applicable.

Competing interests
The authors declare that they have no competing interests.

Publisher's Note

Springer Nature remains neutral with regard to jurisdictional claims in published maps and institutional affiliations.

Published: 11 April 2018

References

1. Hartwell LH, Hopfield JJ, Leibler S, Murray AW. From molecular to modular cell biology. Nature. 1999;402:47–52.
2. Dost B, et al. QNet: A tool for querying protein interaction networks. J Comp Biol. 2008;15(7):913–25.
3. Hasan MM, Kahveci T. Indexing a protein-protein interaction network expedites network alignment. BMC Bioinformatics. 2015;16(1):326.
4. Francke C, Siezen RJ, Teusink B. Reconstructing the metabolic network of a bacterium from its genome. Trends Microbiol. 2005;13(11):550–8.
5. Sridhar P, Kahveci T, Ranka S. An iterative algorithm for metabolic network-based drug target identification. Pac Symp Biocomput. 2007;88–99.
6. Milo R, et al. Network motifs: simple building blocks of complex networks. Science. 2002;298(5594):824–7.
7. Wernicke S. A faster algorithm for detecting network motifs. In: Workshop on Algorithms in Bioinformatics. Berlin, Heidelberg: Springer; 2005. p. 165–77.
8. Milenković T, Lai J, Pržulj N. GraphCrunch: a tool for large network analyses. BMC Bioinformatics. 2008;9(1):70.
9. Garey MR, Johnson DS. Computers and Intractability: A Guide to the Theory of NP-Completeness. New York: W. H. Freeman & Co.; 1979.
10. Cook SA. The complexity of theorem-proving procedures. In: Proceedings of the Third Annual ACM Symposium on Theory of Computing. STOC '71. New York: ACM; 1971. p. 151–8.
11. Grochow JA, Kellis M. Network Motif Discovery Using Subgraph Enumeration and Symmetry-Breaking. In: Research in Computational Molecular Biology. Berlin, Heidelberg: Springer; 2007. p. 92–106.
12. Kashtan N, Itzkovitz S, Milo R, Alon U. Efficient sampling algorithm for estimating subgraph concentrations and detecting network motifs. Bioinformatics. 2004;20(11):1746–1758.
13. Omidi S, Schreiber F, Masoudi-Nejad A. Moda: An efficient algorithm for network motif discovery in biological networks. Genes Genet Syst. 2009;84(5):385–95. http://doi.org/10.1266/ggs.84.385.
14. Wernicke S. Efficient detection of network motifs. IEEE/ACM Trans Comput Biol Bioinfo. 2006;3(4):347–59.
15. Chen J, Hsu W, Lee ML, Ng S. NeMoFinder: Dissecting genome-wide protein-protein interactions with meso-scale network motifs. In: Proceedings of the ACM SIGKDD International Conference on Knowledge Discovery and Data Mining, vol. 2006; 2006. p. 106–15.
16. Kashani ZRM, et al. Kavosh: a new algorithm for finding network motifs. BMC Bioinformatics. 2009;10(1):318.
17. Elhesha R, Kahveci T. Identification of large disjoint motifs in biological networks. BMC Bioinformatics. 2016;17(1):408. http://doi.org/10.1186/s12859-016-1271-7.
18. Schreiber F, Schwöbbermeyer H. Frequency concepts and pattern detection for the analysis of motifs in networks. In: Transactions on

Computational Systems Biology III. Berlin, Heidelberg: Springer; 2005. p. 89–104.
19. Cook DJ, Holder LB. Substructure discovery using minimum description length and background knowledge. J Artif Int Res. 1994;1(1):231–55.
20. Kuramochi M, Karypis G. GREW - a scalable frequent subgraph discovery algorithm. In: Fourth IEEE International Conference on Data Mining, 2004. ICDM '04; 2004. p. 439–42.
21. Kuramochi M, Karypis G. An efficient algorithm for discovering frequent subgraphs. IEEE Trans Knowl Data Eng. 2004;16:1038–51.
22. Yan X, Han J. gSpan: Graph-based substructure pattern mining. In: IEEE International Conference on Data Mining, 2002. ICDM '02; 2002. p. 721–4.
23. Huan J, Wang W, Prins J. SPIN: mining maximal frequent subgraphs from graph databases. In: Proceedings of the Tenth ACM SIGKDD International Conference on Knowledge Discovery and Data Mining. KDD '04. New York: ACM; 2004. p. 581–6.
24. Hu H, et al. Mining coherent dense subgraphs across massive biological networks for discovery. Bioinformatics. 2005;21:213–21.
25. Ranu S, Singh AK. GraphSig: A scalable approach to mining significant subgraphs in large graph databases. In: 2009 IEEE 25th International Conference on Data Engineering; 2009. p. 844–55.
26. Hasan MM, Kahveci T. Color distribution can accelerate network alignment. In: ACM BCB. New York: ACM; 2013. p. 52–61.
27. Wackersreuther B, et al. Frequent subgraph discovery in dynamic networks. In: Proceedings of the eighth workshop on mining and learning with graphs. MLG '10. New York: ACM; 2010. p. 155–62.
28. Qin G, Gao L, Yang J. Significant substructure discovery in dynamic networks. Curr Bioinform. 2013;8(1):46–55.
29. Erdős P, Rényi A. On random graphs. I. Publ Math Debrecen. 1959;6:290–7.
30. Watts DJ, Strogatz SH. Collective dynamics of'small-world'networks. Nature. 1998;393:409–10.
31. Barabási AL, Albert R. Emergence of scaling in random networks. Science. 1999;286(5439):509–12. https://doi.org/10.1126/science.286.5439.509.
32. Rivera-Mulia JC, et al. Dynamic changes in replication timing and gene expression during human development. Genome Res. 2015;25(8): 1091–103.
33. Milo R, et al. On the uniform generation of random graphs with prescribed degree sequences. 2003. Arxiv preprint cond-mat/0312028.

Buiga *et al. BMC Systems Biology* 2018, **12**(Suppl 1):11
https://doi.org/10.1186/s12918-018-0534-5

BMC Systems Biology

RESEARCH

Open Access

Regulation of dual specificity phosphatases in breast cancer during initial treatment with Herceptin: a Boolean model analysis

Petronela Buiga[1,2], Ari Elson[1], Lydia Tabernero[2] and Jean-Marc Schwartz[2*]

From The Sixteenth Asia Pacific Bioinformatics Conference
Yokohama, Japan. 15-17 January 2018

Abstract

Background: 25% of breast cancer patients suffer from aggressive HER2-positive tumours that are characterised by overexpression of the HER2 protein or by its increased tyrosine kinase activity. Herceptin is a major drug used to treat HER2 positive breast cancer. Understanding the molecular events that occur when breast cancer cells are exposed to Herceptin is therefore of significant importance. Dual specificity phosphatases (DUSPs) are central regulators of cell signalling that function downstream of HER2, but their role in the cellular response to Herceptin is mostly unknown. This study aims to model the initial effects of Herceptin exposure on DUSPs in HER2-positive breast cancer cells using Boolean modelling.

Results: We experimentally measured expression time courses of 21 different *DUSPs* between 0 and 24 h following Herceptin treatment of human MDA-MB-453 HER2-positive breast cancer cells. We clustered these time courses into patterns of similar dynamics over time. In parallel, we built a series of Boolean models representing the known regulatory mechanisms of DUSPs and then demonstrated that the dynamics predicted by the models is in agreement with the experimental data. Furthermore, we used the models to predict regulatory mechanisms of DUSPs, where these mechanisms were partially known.

Conclusions: Boolean modelling is a powerful technique to investigate and understand signalling pathways. We obtained an understanding of different regulatory pathways in breast cancer and new insights on how these signalling pathways are activated. This method can be generalized to other drugs and longer time courses to better understand how resistance to drugs develops in cancer cells over time.

Keywords: Boolean model, Herceptin, Breast cancer, DUSPs

Background

Breast cancer represents a major public health issue, affecting one in eight women in the Western world [1]. Approximately 25% of human breast cancer cases are HER2-positive, due to overexpression of the HER2 protein or amplification of the *HER2* gene, or due to activating mutations that increase HER2 activity. HER2-positive tumours are associated with poor prognosis [2, 3]. HER2 is a 185 kDa protein that belongs to the family of tyrosine kinase epidermal growth factor receptors, and is also known as Neu or ErbB2. HER2 has no known ligand and is activated by dimerization or by incorporation into heterodimers with its related family members HER1, HER3 or HER4 [4]. Increased membrane receptor concentration of HER2 in HER2-overexpressing cells can also activate HER2 by ligand-independent homo-dimerization, leading to activation of the HER2 signalling pathway [5]. Activated HER2 auto-phosphorylates and associates with other molecules, activating several signalling pathways and mediators, such as the Mitogen-Activated Protein Kinases (MAPKs), which promote cell growth, proliferation, and survival [6].

* Correspondence: Jean-marc.Schwartz@manchester.ac.uk
[2]School of Biological Sciences, Faculty of Biology, Medicine and Health, University of Manchester, Manchester, UK
Full list of author information is available at the end of the article

A major drug for treating HER2-positive breast cancer is Herceptin, a humanized monoclonal antibody that binds the extracellular Domain IV of HER2. Herceptin reduces HER2 dimerization leading to, among other effects, inhibition of MAPK signalling [6]. Herceptin has high efficacy in patients with HER2-positive breast cancer, but resistance to its effects often develops and poses a significant therapeutic challenge [7, 8].

MAPKs, such as ERK1, ERK2, p38 and JNK, are major regulators of HER2 signalling and of other signalling events that take place in cancer cells. MAPKs are activated by phosphorylation of specific threonine and tyrosine residues present in a TXY motif by upstream kinases. Dephosphorylation of either or both the threonine and tyrosine residues of this motif downregulates MAPK activity. Dephosphorylation of both residues is carried out by specialized enzymes, the dual-specificity phosphatases (DUSPs) that are part of the larger tyrosine phosphatase superfamily [9]. DUSPs comprise a subfamily of 10 MAPK phosphatases (MKPs) that are key players in inactivating MAPKs, leading to down-regulation of MAPK signalling [10, 11]. A second subfamily of DUSPs, the atypical DUSPs, are also involved in regulating the same pathways [12]. Each DUSP has different subcellular localization and substrate specificity [13]. DUSPs are of particular interest in the context of cancer, since they play putative roles in tumour progression and in resistance to therapy [14]. Deeper understanding of the mechanisms that regulate DUSP expression in HER2 signalling is crucial to clarify their contribution in cancer promotion and in development of resistance of Herceptin. A good experimental model to study HER2-positive breast cancer is MDA-MB-453, a human breast cancer cell line that overexpresses endogenous HER2 protein [15, 16].

One of the most useful methods to predict the outcome of signalling pathways is a systems biology approach through logical modelling [17]. Logical models such as Boolean models are powerful tools to represent dynamic molecular processes in complex cellular systems, where construction of detailed kinetic models would be prohibitively expensive and time-consuming, or where complete knowledge of molecular interactions or quantitative details of biological processes is lacking [18]. In a Boolean model, nodes represent biological components of interest and are connected by edges describing the nature of the interaction occurring between them. The activity of biological components is represented by discrete values (states), generally 0 (OFF, inactive) or 1 (ON, active), and actions are represented by activation (positive regulation) or inhibition (negative regulation) [18].

Several studies of qualitative and quantitative modelling related to HER signalling have been performed. Chen et al. presented a quantitative model of HER signalling during the immediate-early phase of ligand-stimulated cell signalling [19]. Samaga and colleagues built a comprehensive logical model of the signalling network of EGFR in liver cancer, which was based on the comprehensive EGFR network map published by Oda and colleagues [20, 21]. Von der Hyde et al., constructed a Boolean model of HER2 signalling to identify individual drug response patterns and their resistance in HER2-positive breast cancer [22]. However, none of these studies addressed DUSPs specifically, leaving a major gap in our understanding of how Herceptin affects these important regulators of cell signalling.

Our work analyses Herceptin action on DUSP expression in HER2-positive breast cancer cells, in order to investigate and in some cases predict regulatory mechanisms for *DUSP*s in this context. The current study examines DUSPs during the initial phase of Herceptin treatment, in which some changes in gene expression occur, but which is still too early for compensatory mutations to arise or to be selected for. Initially, *DUSP* mRNA expression in HER2-positive breast cancer cells was determined experimentally during the period between 0 to 24 h following cell exposure to Herceptin. In parallel, we constructed a series of models that collectively accounted for the response of each *DUSP* to the drug and for the variations in responses among DUSPs, and enabled prediction of new regulatory mechanisms that were unknown.

Methods

Cell line

The human breast cancer cell line MDA-MB-453 (American Type Culture Collection, Manassas, VA, USA) was cultured in RPMI 1640 medium (Gibco, Carlsbad, CA, USA) containing 50 U/ml penicillin, 50 mg/ml streptomycin and 10% heat-inactivated foetal bovine serum (FBS - Gibco) at 37 °C in an atmosphere of 5% CO_2.

Herceptin treatment

Cells were plated in 60 mm plates in duplicate for each time point, and cultured for two days until they reached 75% confluency. Cells were then treated with 50 μM Herceptin (Roche, Switzerland) continuously for 2, 4, 12 or 24 h and then collected. Plates prepared and grown in parallel but not exposed to Herceptin were used as time 0. Total RNA was extracted from cells using the RNeasy Mini Kit (QIAGEN, Germany) and was treated with DNase I at room temperature for 15 min. One microgram of total cellular RNA was reverse transcribed in a final volume of 20 μl using the qScript cDNA synthesis kit (Quanta Biosciences, Beverly, MA, USA) according to the manufacturer's protocol. 0.5 μl (25 ng) cDNA was used to perform quantitative PCR (RT-qPCR) using KAPA SYBR Fast qPCR Master Mix (2X) ABI Prism (Kapa Biosystems, Wilmington, MA, USA) and target-specific forward and reverse primers (Additional file 1) in an AB StepOnePlus instrument (Applied Biosystems, Foster City, CA, USA).

The amplification conditions consisted of an initial activation step of 95 °C for 20 s, followed by 40 cycles of 95 °C for 3 s and 60 °C for 30 s, and ending with one run of 95 °C for 15 s, 60 °C for 60 s and 95 °C for 15 s.

Standard curves were used to assess primer efficiency, and average change in threshold cycle (ΔCT) values were determined for each of the samples relative to endogenous Actinb and GAPDH (internal control) levels using the ΔΔCT method, where the fold change of target gene expression is determined relative to a reference sample, normalized to the reference genes (Actinb and GAPDH) [23]. Experiments were performed in triplicates to determine mean along with standard error, having two biological repeats. DUSP-specific forward and reverse primers were designed using the Primer3 software [24, 25].

Levels of Actinb and GAPDH mRNA transcripts measured as described above were used to normalize the amount of input RNA amplified in each reaction. Each sample was amplified in duplicate and the resultant mean values were used for analysis. Student T-test statistical analysis was performed in GraphPad Prism version 7.0a for Mac OS X (GraphPad Software, La Jolla California, USA).

Correlations of *DUSP* expression

DUSPs time courses were further analysed by computing Pearson correlation coefficients between all time series; the $\log_{10}$ of raw intensity values was taken, then correlations were computed using the *corr* function in Python 2.7. Since we wanted clusters to be composed of only positively correlated time courses, we set all negative correlations to zero then computed clusters using the *clustermap* function from the *seaborn* library, with the default average linkage method; clusters were plotted using *matplotlib*.

Construction of Boolean models

Boolean models were manually constructed following an extensive literature review. Previously reported network maps of HER signalling were mined to find the essential proteins and interactions for HER2 downstream signalling. Relevant proteins and interactions found to be involved in downstream signalling of MAPK - ERK as described by Chen et al., and other components of MAPK downstream signalling, JNK and p38, in models developed by Oda et al. and Samaga et al., were included. All the selected interactions have been checked for cellular context and verified using multiple literature sources [19–21].

In normal cells, the basal level of the downstream MAP kinases (ERK, JNK and p38) leads to proliferation; increased activity of both JNK and p38 leads to apoptosis and inhibition of cellular growth and proliferation [26–29]. However, cancer cells often do not undergo apoptosis because of the inhibitory action of DUSPs on both JNK and p38. Detailed interactions of DUSPs with their cognate kinases were not included in any of the previously reported maps or models. These were added to our models, based on literature reports of substrate specificity [12, 13] and the necessity to investigate their contribution.

Graphical maps of all the models were created using the yWorks yEd graph editor (version 3.14.4). All the interacting components of the model were represented as nodes (rectangular boxes), while the interactions were represented as edges (sharp arrow for activation and blunt arrow for inhibition). Circles represent the AND function which combines several interactions. Model construction and time course simulation of MAPK signalling pathways in response to Herceptin treatment was done using the BooleanNet toolkit under Python 2.7 [30]. Both synchronous and asynchronous simulations were performed. In synchronous updating rules, all variables are updated at the same time, while in asynchronous simulations a single random variable is updated at a time. The asynchronous simulations confirmed the results obtained from the synchronous simulations, hence we only present the results of the latter. Heatmaps were generated and exported using the Python tool matplotlib [31].

Herceptin was selected to be a user defined input, while Survival was chosen as a readout. Each node in the model can be either in an active state (ON) or in an inactive state (OFF), referring to 1 and 0 respectively in a binary domain $n \in \mathbb{N}$; $t \in \mathbb{Z}+$: $n(t) = 0$ or 1. In this model interactions were represented as either activation or inhibition, combined using AND, OR and NOT Boolean functions, which are sufficient to represent any logical relationship. All of these reactions were considered to have equal time scale $t = 1$. A set of initial conditions were assumed considering at least one node to be active, then other nodes either become inactive (1 ➔ 0), active (0 ➔ 1) or maintain their state (0 ➔ 0 or 1 ➔ 1) after each time increment. These changes are dependent on the upstream nodes and interaction conditions. Where difficulties arose in determining the use of AND or OR functions, this was solved by reviewing the specific literature. The AND function was used when there was an explicit action of two or more proteins acting collectively on a node, otherwise the OR function was applied [32].

Results

DUSP expression measured by qPCR

We measured gene expression of 21 *DUSPs* (10 MKPs and 11 atypical *DUSPs*) by RT-qPCR in the human HER2-positive MDA-MB-453 breast cancer cell line at 0, 2, 4, 12 and 24 h following Herceptin exposure. Data from time point 0 (non-treated samples) was the reference against which expression levels at the other time points were compared for each *DUSP*. We limited our

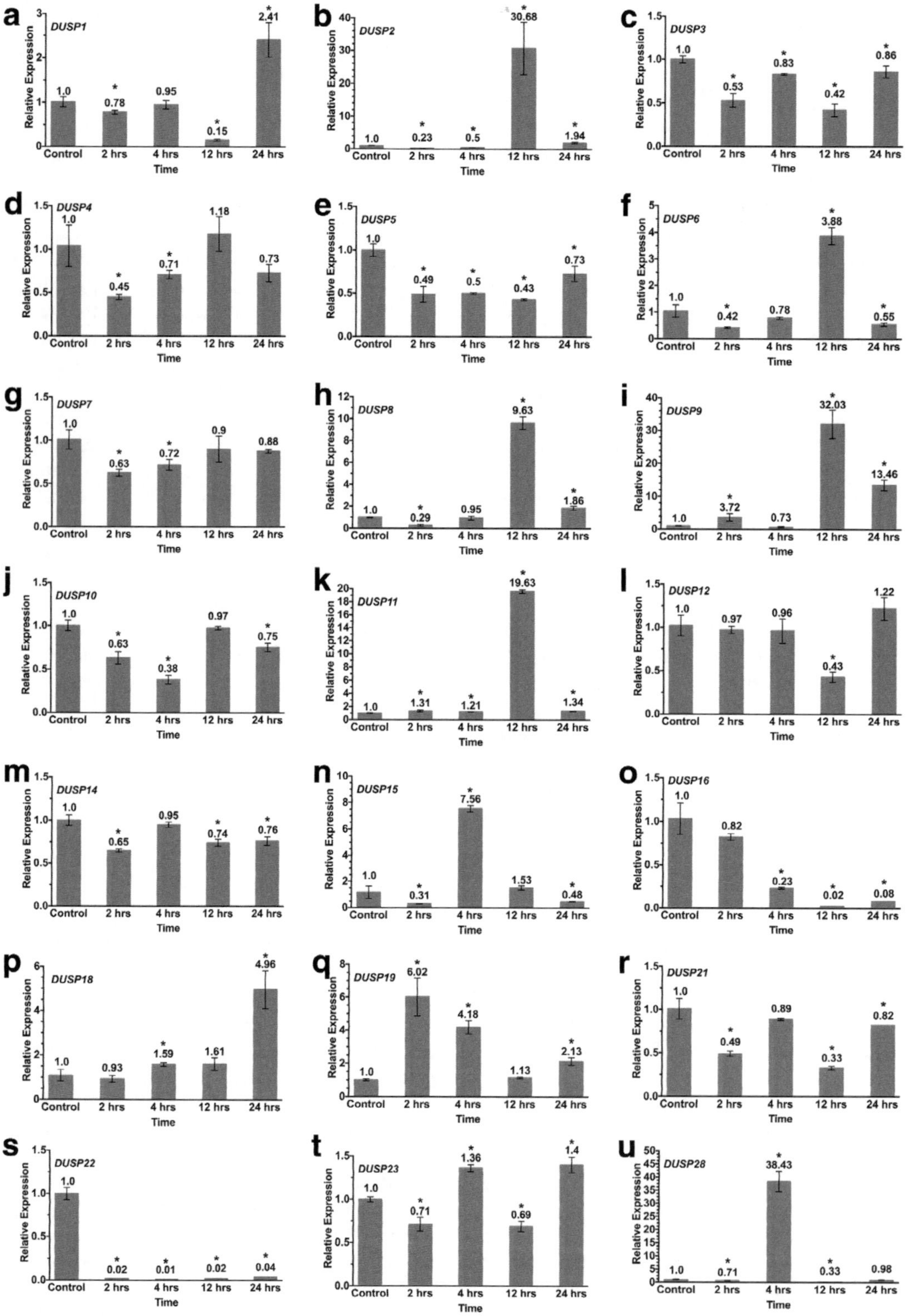

Fig. 1 *DUSP* mRNA expression measured by RT-qPCR. (**a**-**u**) *DUSP* expression measured by RT-qPCR, comparing expression of the indicated *DUSPs* in control (non-treated, time 0) samples with cells at 2, 4, 12 and 24 h following exposure to Herceptin. Bars represent mean ± SE. Two biological repeats (each in triplicate) were performed. An asterisk indicates a statistically-significant difference of $p < 0.05$ from the value at $t = 0$

analysis to 24 h in order to focus on the initial response in *DUSP* gene expression and cell signalling to Herceptin. RT-qPCR expression levels of the 21 selected *DUSPs* are shown in Fig. 1(a-u).

Expression of individual *DUSPs* was variable among the different time points following Herceptin treatment. Herceptin treatment led to a decrease in expression of most *DUSPs*: expression of *DUSPs* 3, 5, 7, 10, 14, 16, 21 and 22 was decreased during the entire exposure to Herceptin, while expression of *DUSPs* 1 and 12 was decreased until 12 h. Some of the *DUSPs* showed no clear trend or no strong change, while other *DUSPs* decreased at the beginning, but rose strongly at later time points (e.g. *DUSP9*). Expression of a number of *DUSPs* increased immediately, but subsequently decreased: *DUSP19* increased up to 2 h and decreased later. *DUSPs* 2, 6 and 8 showed a peak at 12 h, *DUSP28* and 15 showed a peak at 4 h. *DUSP22* expression was abolished over the entire time series, while in *DUSP16* expression showed a gradual decrease. In contrast, *DUSP18* showed a gradual increase in expression throughout the entire time course of this study.

Correlation of *DUSP* expression and clustering

In order to characterise these different expression patterns in a more rigorous manner, we calculated correlations between expression time courses of all individual *DUSPs*. Figure 2 shows an overview of matching *DUSP* expression patterns, which revealed seven clusters that were selected for further modelling and analysis.

From the correlation matrix and visual inspection of the expression time courses we defined the following clusters: cluster 1 is composed of *DUSPs* 2, 6 and 8; cluster 2 includes *DUSPs* 9 and 11; *DUSPs* 4 and 7 belong to cluster 3; cluster 4 consists of *DUSPs* 23 and 28; *DUSPs* 5 and 22 are part of cluster 5; *DUSPs* 3 and 21 belong to cluster 6, while cluster 7 includes *DUSPs* 1 and 12. There were six remaining *DUSPs* which did not fit into any cluster, these are: *DUSPs* 10, 14, 15, 16, 18 and 19.

The distinct expression patterns over time for each cluster are shown in Fig. 3, where expression of all *DUSPs* in a given cluster is plotted on logarithmic scale. Generally, an immediate decrease in *DUSP* expression can be seen in the clusters. For cluster 1, the immediate decrease is followed by a sharp increase after 2 h then an abrupt reduction after 12 h. Cluster 2 shows little change for the first 4 h, followed by a marked increase until 12 h, ending with a slight decrease until 24 h. In cluster 3, an initial reduction for the first 4 h is followed by an increase in *DUSP* expression until 12 h. Cluster 4 shows an immediate decrease then abrupt increase at 4 h, followed by a sharp decrease until 12 h, ending with

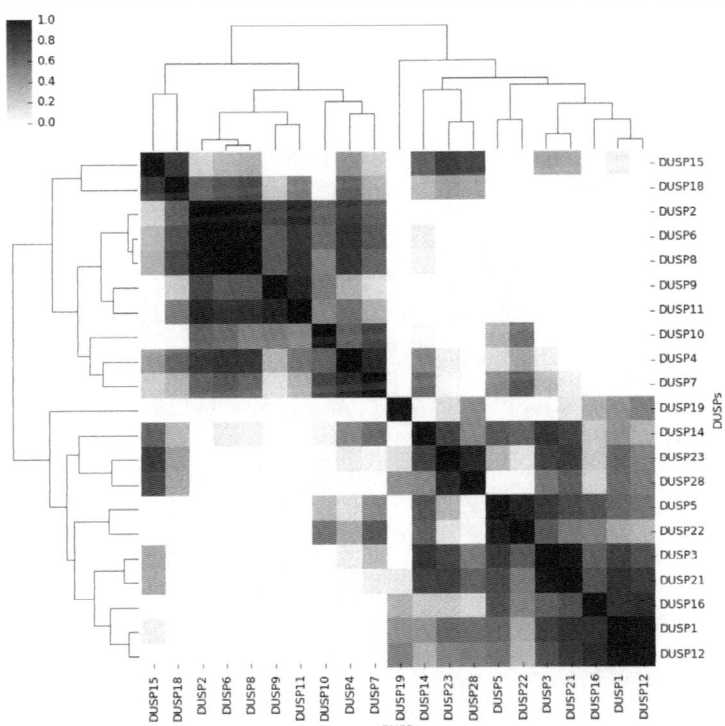

Fig. 2 Clustering of *DUSP* expression time courses. Clustering of *DUSP* expression time courses between 0 and 24 h after exposure to Herceptin. Colours represent the Pearson correlation coefficient between each pair of *DUSPs*, with negative values clipped to zero

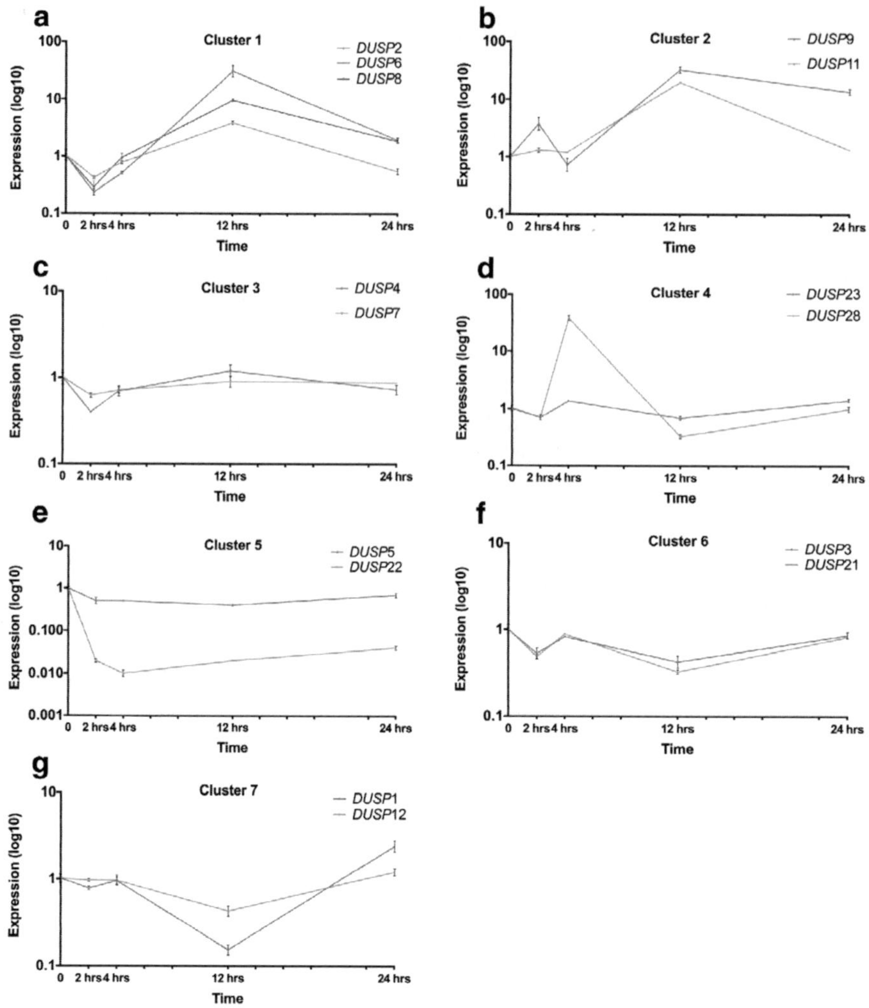

Fig. 3 Time series of *DUSP* expression exposed to Herceptin grouped by selected clusters. The horizontal axis represents time, the vertical axis represents measured expression expressed in logarithmic scale. Expression levels were normalised with respect to the control (1). Data represents mean ± SE (two biological repeats, each in triplicate) as noted in Fig. 1. **a** Cluster 1; **b** Cluster 2; **c** Cluster 3; **d** Cluster 4; **e** Cluster 5; **f** Cluster 6; **g** Cluster 7

an increase until 24 h. In cluster 5 the expression shows a moderate oscillating pattern, the expression varies within a small range. In cluster 6 a reduction in *DUSP* expression is observed until 12 h followed by an increase until 24 h. Finally, cluster 7 shows an overall pattern of a slight increase followed by a sharp decrease.

Boolean modelling of the selected clusters

In order to interpret the biological findings, a series of models were created to simulate and represent the possible combinations of *DUSP* regulation (including induction of expression and substrate specificity). Each analysed cluster has been matched with a model. Figures 4-7 show the model and simulation results for one representative *DUSP* from each cluster for which regulatory mechanisms were known: *DUSP*2 for cluster 1, *DUSP*4 for cluster 3, *DUSP*5 for cluster 5 and *DUSP*1 for cluster 7.

The representative of cluster 1, DUSP2, is induced by ERK1/2 and JNK1/2; it inhibits ERK1/2 and p38 [13]. When exposed to Herceptin over time, *DUSP2* shows an oscillating pattern in the Boolean model that agrees with the quantified gene expression data by RT-qPCR (see Figs. 3a and 4).

The regulation of DUSP4, the chosen representative of cluster 3, is known: it is induced by ERK1/2 and it inhibits ERK1/2 and JNK1/2 [13]. When exposed to Herceptin over time, *DUSP4* shows oscillating expression in

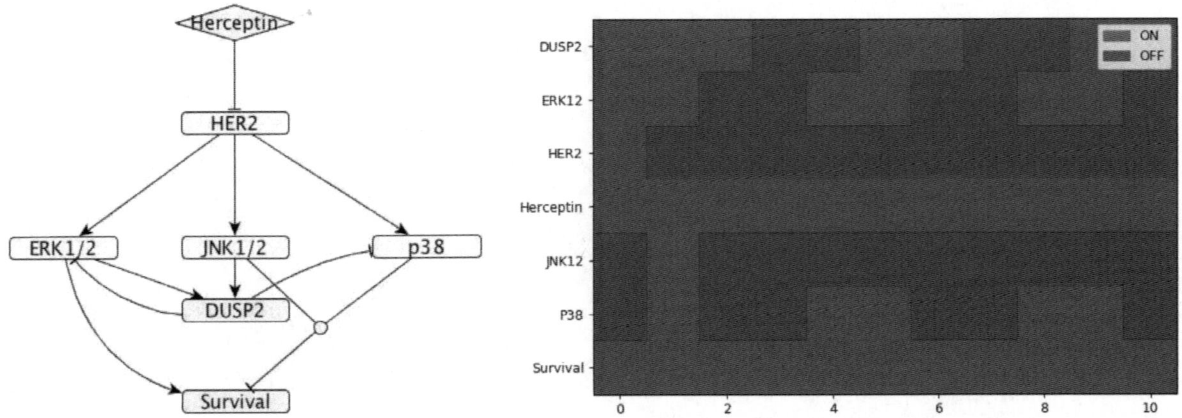

Fig. 4 *DUSP2* model. (Left). Network map of *DUSP2* regulation (cluster 1) in MDA-MB-453 HER2-positive breast cancer cells exposed to Herceptin. (Right). Corresponding Boolean time series represented as a heatmap showing gene expression. Green and red indicate nodes in the ON and OFF states, respectively. Note that time units in Boolean simulations are arbitrary and do not necessarily match the 24 h period of the study, but the overall expression pattern can be compared to the overall Boolean time series

the Boolean model that agrees with the quantified gene expression data by RT-PCR (see Figs. 3c and 5).

The representative of cluster 5, DUSP5, is induced by ERK1/2 and it inhibits ERK1/2 [13]. When exposed to Herceptin over time, DUSP5 shows oscillating behaviour in the Boolean model that agrees with the quantified gene expression data by RT-PCR (see Figs. 3e and 6).

DUSP1, the representative for cluster 7, is induced by ERK1/2 and p38, and inhibits JNK1/2 [13]. When exposed to Herceptin, Boolean modelling indicates that *DUSP1* expression is initially stable and then decreases, in accordance with the quantified gene expression data by RT-PCR up to 12 h (see Fig. 3g and 7).

Figures 8-10 show representative models matching expression trends with their corresponding predicted heatmaps for *DUSPs* whose regulation is only partially known. Each model and heatmap shows one representative *DUSP* per cluster: *DUSP9* (cluster 2), *DUSP23* (cluster 4) and *DUSP3* (cluster 6). For all these *DUSPs*, it is currently unknown by which MAPKs they are induced.

In cluster 2, *DUSP9* was chosen as a representative and its regulation is partially known: its substrate is ERK, but it is unknown by which of the MAP kinases it is induced. Therefore, we simulated the possible combinations of all regulators and we found that the only solution that showed a convincing agreement between Boolean model simulation and experimental time course was the one where *DUSP9* is induced by JNK. The agreement between Boolean model (Fig. 8) and *DUSP9* expression time course enables us to predict that *DUSP9* might be regulated by JNK. Since the two MAP kinases JNK and p38 are symmetrical

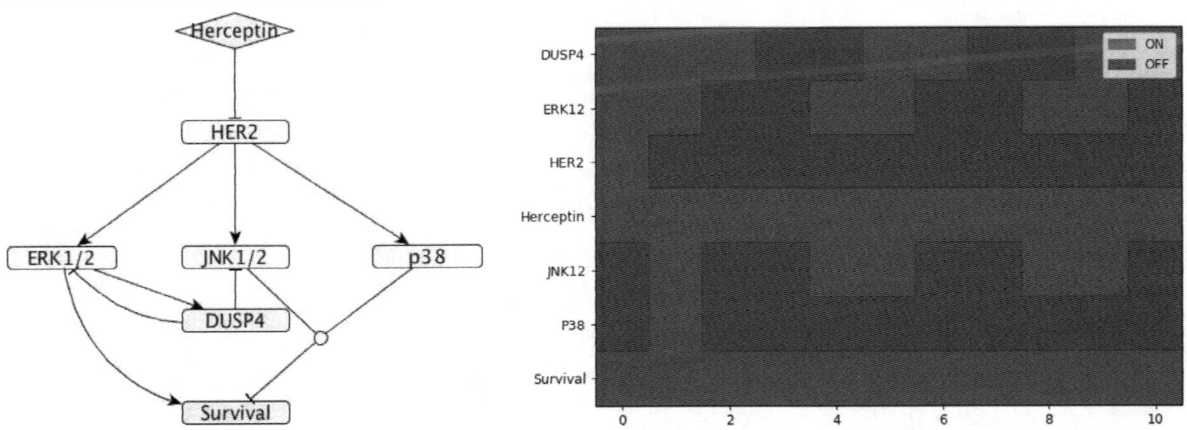

Fig. 5 *DUSP4* model. (Left). Network map of *DUSP4* (cluster 3) regulation in MDA-MB-453 HER2-positive breast cancer cells exposed to Herceptin. (Right). Corresponding Boolean time series represented as a heatmap showing gene expression as described in Fig. 4

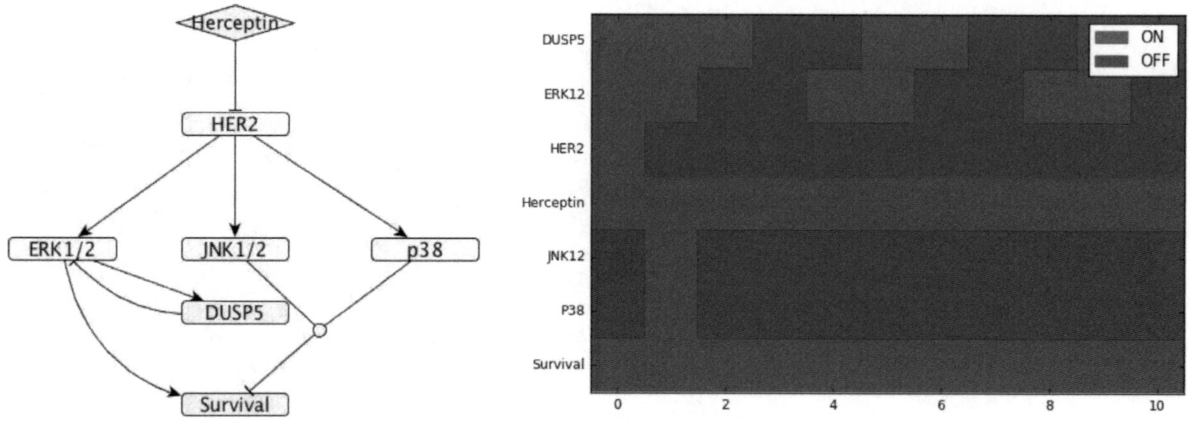

Fig. 6 *DUSP5* model. (Left). Network map of *DUSP5* (cluster 5) regulation in MDA-MB-453 HER2-positive breast cancer cells exposed to Herceptin. (Right). Corresponding Boolean time series represented as a heatmap showing gene expression as described in Fig. 4

in the model, the same outcome can be achieved if *DUSP9* is induced by p38.

In cluster 4, *DUSP23* is the smallest active phosphatase, whose regulation is only partially known: its substrates are ERK [33], JNK1/2 and p38 [34], but it is unknown by which of the MAP kinases it is induced. The Boolean model shown in Fig. 9 agrees with the experimental time course and enables us to predict that *DUSP23* might be regulated by JNK1/2. Since the two MAP kinases JNK and p38 are symmetrical in the model, the same outcome can be achieved if *DUSP23* is induced by p38.

In cluster 6, *DUSP3* is an atypical phosphatase that has a similar catalytic domain as the typical MKPs. Therefore, *DUSP3* was suggested to dephosphorylate MAP kinases in the same manner as MKPs [12]: JNK, ERK [35] and also p38 [36]. The Boolean simulation shown in Fig. 10 agrees with the experimental time course, enabling us to predict that *DUSP3* might be regulated by JNK1/2.

DUSP16 has an inhibitory action on JNK and p38 [13], but it is unknown by which kinase it is induced. The model presented in Fig. 11 is in agreement with the measured *DUSP16* expression time course, therefore we can predict that *DUSP16* might be induced by ERK. It is worth noting that all models presented above show survival to be ON, but in the case of the *DUSP16* model, survival switches to OFF after some time.

Discussion

Previous studies of Herceptin treatment in HER2 positive breast cancer and development of resistance to Herceptin focused on its effects on various signalling pathways and on cell survival [22, 37–39]. Györffy and colleagues correlated Herceptin resistant tumours from patients with DUSPs involved in development of Herceptin resistance

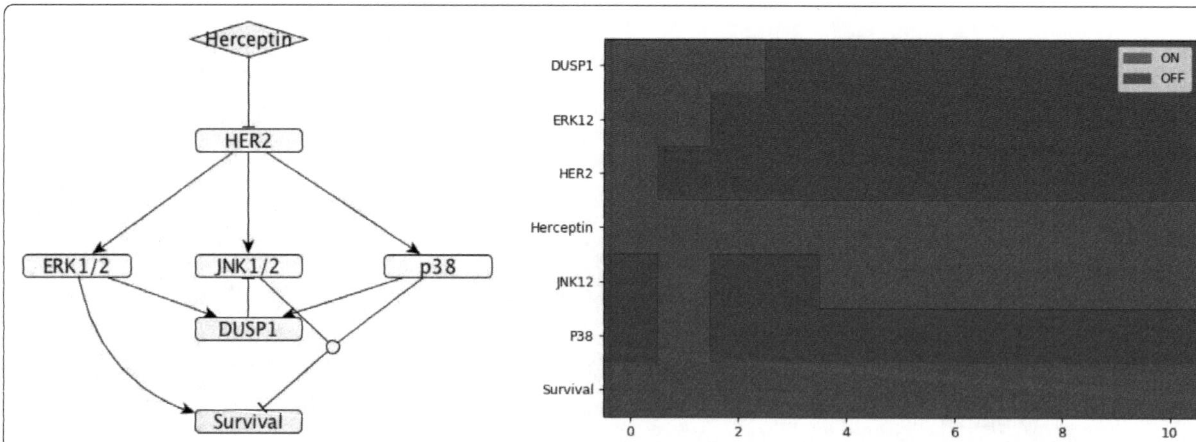

Fig. 7 *DUSP1* model. (Left). Network map of *DUSP1* (cluster 7) regulation in MDA-MB-453 HER2-positive breast cancer cells exposed to Herceptin. (Right). Corresponding Boolean time series represented as a heatmap showing gene expression as described in Fig. 4

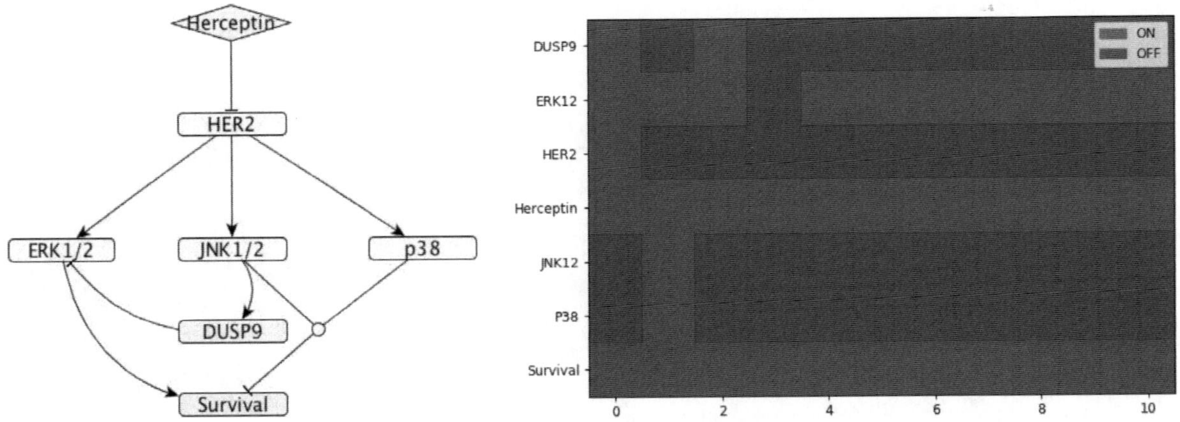

Fig. 8 *DUSP9* model. (Left). Network map of *DUSP9* (cluster 2) regulation in MDA-MB-453 HER2-positive breast cancer cells exposed to Herceptin. (Right). Corresponding Boolean time series represented as a heatmap showing gene expression as described in Fig. 4

in HER2 positive breast cancer, and suggested that DUSP4 is involved in this process [40]. Nonetheless, there are significant gaps in our understanding of whether and how DUSPs participate in development of Herceptin resistance in breast tumours.

Here we presented a comprehensive analysis of the early-stage responses of 21 selected *DUSPs* to Herceptin treatment in human breast cancer cells. The cell line MDA-MB-453 is reported in the literature in some cases as being Herceptin sensitive [41–43] and in others as Herceptin resistant [44–46]. Our study focuses on short-term effects of exposure to Herceptin in order to examine the initial response of HER2-expressing tumour cells to Herceptin. We expect that this response is based on the pre-existing cell signalling pathways – possibly with some rapid re-configuration induced by Herceptin – and would shed light on the structure of these pathways. In contrast, the time frame of this study effectively prevents mutations in genes from affecting the cellular response. *DUSPs* 1, 3, 4, 7, 10, 12, and 21 showed a decrease in expression compared to the control, non-treated group during the entire time series, while the other *DUSPs* showed a pattern of increasing expression over time. Herceptin reduces mitogenic and survival signals from HER2. It is then possible that the initial decrease in *DUSP* expression aims to counter this effect by removing DUSPs that down-regulate MAPK activity and thus also counter mitogenic and survival signalling. At later times additional roles of *DUSPs* become evident, and their expression patterns diverge. For example, at 24 h expression of *DUSPs* 1, 12, 23, 28, 3, 21 increases, possibly indicating the beginning of emerging resistance. Alternatively, these *DUSPs* may perform additional roles upon which cell survival depends and which limit the length of time the cells can withstand their reduced expression.

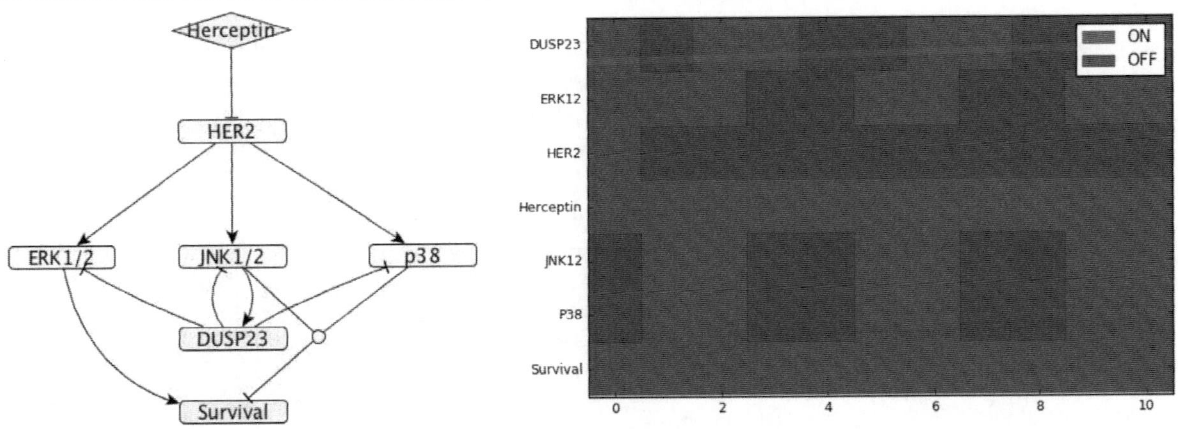

Fig. 9 *DUSP23* model. (Left) Network map of *DUSP23* (cluster 4) regulation in MDA-MB-453 HER2-positive breast cancer cells exposed to Herceptin. (Right). Corresponding Boolean time series represented as a heatmap showing gene expression as described in Fig. 4

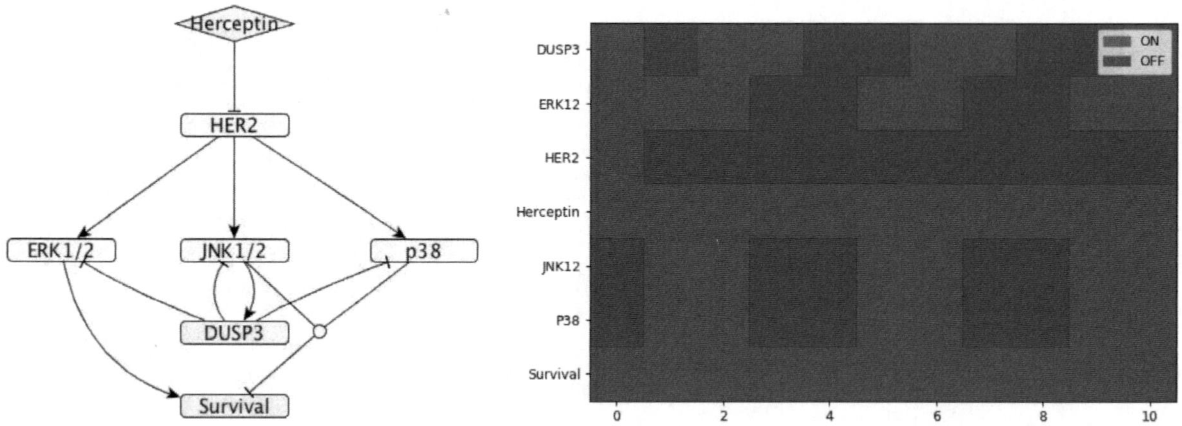

Fig. 10 *DUSP3* model. (Left). Network map of *DUSP3* (cluster 6) regulation in MDA-MB-453 HER2-positive breast cancer cells exposed to Herceptin. (Right). Corresponding Boolean time series represented as a heatmap showing gene expression as described in Fig. 4

Uniquely for *DUSP16*, whose expression is decreased dramatically following treatment with Herceptin, the model shows that Survival switches OFF after some time. Although cells are not dying in the time frame of 24 h, reduced expression of phosphatase might play an important role in driving apoptosis of cells treated with Herceptin, and thus prevent resistance to the drug from arising. Additional experimental and modelling studies are therefore required to understand cell behaviour at later time points.

Previous studies have utilized Boolean modelling to understand cellular responses in cancer. Fumia and Martins constructed and described theoretical models which included the main signalling pathways in cancer in order to interpret mutational events and cancer cell response [47]; Grieco et al. performed in silico simulations to understand the role of deregulations in MAPK signalling on cell fate decision in cancer [48]. Von der Heyde et al. used reverse and forward engineering techniques in order to reveal individual mechanisms of drug response or resistance in breast cancer [22]. Sahin et al. constructed a literature-based Boolean model to model the regulation of pRB through ERBB-receptor signalling in native Herceptin resistant cell lines. These authors integrated the experimental data into their model and overcame discrepancies with additional experimental validation [49]. Our approach is distinct in that we combine pre-existing biological knowledge about *DUSP* regulation to construct mechanistic Boolean models, together with independent experimental analysis to challenge and validate these models. The models presented here were not trained to fit experimental data but we showed that the predicted dynamics of *DUSP* expression is in accordance with experimentally observed time courses, which gives stronger confidence that

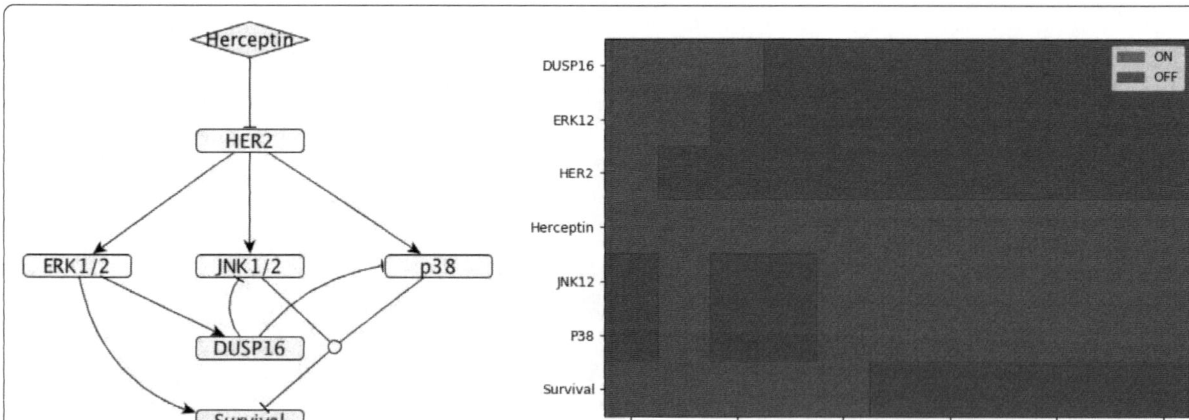

Fig. 11 *DUSP16* model. (Left). Network map of *DUSP16* regulation in MDA-MB-453 HER2-positive breast cancer cells exposed to Herceptin. (Right). Corresponding Boolean time series represented as a heatmap showing gene expression as described in Fig. 4

the models are mechanistically correct. Moreover, this approach allowed us to predict new regulatory mechanisms for DUSPs with partially known regulation, such as DUSPs 9, 23, 3 and 16. In the future, this method can be extended by measuring and simulating longer time series, as well as by including additional signalling pathways in order to better understand the dynamics of development of resistance to Herceptin in breast cancer.

Additional file

Additional file 1: List of primers used for qPCR analyses of the expression of the *DUSP* mRNAs examined in this study. Description: All sequences are of human origin except for beta actin, which is from mouse. (DOCX 198 kb)

Acknowledgements
We thank Dr. Maayan Barnea, Weizmann Institute, for advice on the qPCR experiments.

Funding
This study was supported by a studentship to PB funded by Weizmann UK and by the Faculty of Biology, Medicine and Health, University of Manchester (UoM), a "Making-Connections" grant from Weizmann-UK (P17231) to LT, AE and JMS, and a grant from the Lord Alliance "Get Connected" program between UoM and WIS to LT, AE and JMS. AE is incumbent of the Marshall and Renette Ezralow Professorial Chair. Funding for the publication of this article was provided by grant P17231 from Weizmann-UK.

Availability of data and materials
All data generated or analysed during this study are included in this published article and its Additional file.

About this supplement
This article has been published as part of *BMC Systems Biology* Volume 12 Supplement 1, 2018: Selected articles from the 16th Asia Pacific Bioinformatics Conference (APBC 2018): systems biology. The full contents of the supplement are available online at https://bmcsystbiol.biomedcentral.com/articles/supplements/volume-12-supplement-1.

Authors' contributions
JMS, AE and LT designed the study and experiments. PB carried out the experiments, PB and AE analysed the experiments. PB and JMS constructed the models and analysed the simulations. PB wrote the manuscript, JMS, AE and LT edited the manuscript. All the authors read and approved the final manuscript.

Ethics approval and consent to participate
Not applicable.

Consent for publication
Not applicable.

Competing interests
The authors declare that they have no competing interests.

Publisher's Note
Springer Nature remains neutral with regard to jurisdictional claims in published maps and institutional affiliations.

Author details
[1]Department of Molecular Genetics, The Weizmann Institute of Science, Rehovot, Israel. [2]School of Biological Sciences, Faculty of Biology, Medicine and Health, University of Manchester, Manchester, UK.

Published: 11 April 2018

References
1. Spanhol F, Oliveira L, Petitjean C, Heutte L. A dataset for breast cancer histopathological image classification. IEEE Trans Biomed Eng. 2016;63(7):1455–62.
2. Slamon D, Godolphin W, Jones L, Holt J, Wong S, Keith D, Levin W, Stuart S, Udove J, Ullrich A, et al. Studies of the HER-2/neu proto-oncogene in human breast and ovarian cancer. Science. 1989;244(4905):707–12.
3. Dittrich A, Gautrey H, Browell D, Tyson-Capper A. The HER2 signaling network in breast cancer—like a spider in its web. J Mammary Gland Biol Neoplasia. 2014;19(3–4):253–70.
4. Hurvitz SA, Hu Y, O'Brien N, Finn RS. Current approaches and future directions in the treatment of HER2-positive breast cancer. Cancer Treat Rev. 2013;39(3):219–29.
5. Schulz R, Streller F, Scheel AH, Rüschoff J, Reinert MC, Dobbelstein M, Marchenko ND, Moll UM. HER2/ErbB2 activates HSF1 and thereby controls HSP90 clients including MIF in HER2-overexpressing breast cancer. Cell Death Dis. 2014;5(1):e980.
6. Vu T, Claret FX. Trastuzumab: updated mechanisms of action and resistance in breast cancer. Front Oncol. 2012;2:62.
7. Wang Y-C, Morrison G, Gillihan R, Guo J, Ward RM, Fu X, Botero MF, Healy NA, Hilsenbeck SG, Phillips G, et al. Different mechanisms for resistance to trastuzumab versus lapatinib in HER2- positive breast cancers – role of estrogen receptor and HER2 reactivation. Breast Cancer Res. 2011;13(6):R121.
8. Blackwell KL, Pegram MD, Tan-Chiu E, Schwartzberg LS, Arbushites MC, Maltzman JD, Forster JK, Rubin SD, Stein SH, Burstein HJ. Single-agent lapatinib for HER2-overexpressing advanced or metastatic breast cancer that progressed on first- or second-line trastuzumab-containing regimens. Ann Oncol. 2009;20(6):1026–31.
9. Alonso A, Sasin J, Bottini N, Friedberg I, Friedberg I, Osterman A, Godzik A, Hunter T, Dixon J, Mustelin T. Protein tyrosine Phosphatases in the human genome. Cell. 2004;117(6):699–711.
10. Owens DM, Keyse SM. Differential regulation of MAP kinase signalling by dual-specificity protein phosphatases. Oncogene. 2007;26(22):3203–13.
11. Kondoh K, Nishida E. Regulation of MAP kinases by MAP kinase phosphatases. Biochimica et Biophysica Acta (BBA) - Molecular Cell Research. 2007;1773(8):1227–37.
12. Bayón Y, Alonso A. Emerging signaling pathways in tumor biology. Transworld Research Network: India; 2010.
13. Kidger AM, Keyse SM. The regulation of oncogenic Ras/ERK signalling by dual-specificity mitogen activated protein kinase phosphatases (MKPs). Semin Cell Dev Biol. 2016;50:125–32.
14. Hoeger B, Diether M, Ballester PJ, Kohn M. Biochemical evaluation of virtual screening methods reveals a cell-active inhibitor of the cancer-promoting phosphatases of regenerating liver. Eur J Med Chem. 2014;88:89–100.
15. Seo H-S, Ku JM, Choi H-S, Woo J-K, Jang B-H, Shin YC, Ko S-G. Induction of caspase-dependent apoptosis by apigenin by inhibiting STAT3 signaling in HER2-overexpressing MDA-MB-453 breast cancer cells. Anticancer Res. 2014;34(6):2869–82.
16. Mishra SK, Bhowmick SS, Chua H, Zhang F, Zheng J. Computational cell fate modelling for discovery of rewiring in apoptotic network for enhanced cancer drug sensitivity. BMC Syst Biol. 2015;9(Suppl 1):S4.
17. Wang R-S, Saadatpour A, Albert R. Boolean modeling in systems biology: an overview of methodology and applications. Phys Biol. 2012;9(5):055001.
18. Tian K, Rajendran R, Doddananjaiah M, Krstic-Demonacos M, Schwartz J-M. Dynamics of DNA damage induced pathways to cancer. PLoS One. 2013;8(9):e72303.
19. Chen WW, Schoeberl B, Jasper PJ, Niepel M, Nielsen UB, Lauffenburger DA, Sorger PK. Input–output behavior of ErbB signaling pathways as revealed by a mass action model trained against dynamic data. Mol Syst Biol. 2009;5(1):239.
20. Samaga R, Saez-Rodriguez J, Alexopoulos LG, Sorger PK, Klamt S. The logic of EGFR/ErbB signaling: theoretical properties and analysis of high-throughput data. PLoS Comput Biol. 2009;5(8):e1000438.
21. Oda K, Matsuoka Y, Funahashi A, Kitano H. A comprehensive pathway map of epidermal growth factor receptor signaling. Mol Syst Biol. 2005;1(1):2005.0010.
22. Von der Heyde S, Bender C, Henjes F, Sonntag J, Korf U, Beißbarth T. Boolean ErbB network reconstructions and perturbation simulations reveal

individual drug response in different breast cancer cell lines. BMC Syst Biol. 2014;8(1):75.
23. Rao X, Huang X, Zhou Z, Lin X. An improvement of the 2^(−delta delta CT) method for quantitative real-time polymerase chain reaction data analysis. Biostat Bioinforma Biomath. 2013;3(3):71–85.
24. Untergasser A, Cutcutache I, Koressaar T, Ye J, Faircloth B, Remm M, Rozen S: Primer3 - new capabilities and interfaces. Nucleic Acids Res 2012;40(15): e115.
25. Koressaar T, Remm M. Enhancements and modifications of primer design program Primer3. Bioinformatics. 2007;23(10):1289–91.
26. Woodgett J, Avruch J, Kyriakis J. The stress activated protein kinase pathway. Cancer Surv. 1996;27:127–38.
27. Loda M, Capodieci P, Mishra R, Yao H, Corless C, Grigioni W, Wang Y, Magi-Galluzzi C, Stork P. Expression of mitogen-activated protein kinase phosphatase-1 in the early phases of human epithelial carcinogenesis. Am J Pathol. 1996;149(5):1553.
28. Wang H-y. Cheng Z, Malbon CC: Overexpression of mitogen-activated protein kinase phosphatases MKP1, MKP2 in human breast cancer. Cancer Lett. 2003;191(2):229–37.
29. Haagenson KK, Wu GS. The role of MAP kinases and MAP kinase phosphatase-1 in resistance to breast cancer treatment. Cancer Metastasis Rev. 2010;29(1):143–9.
30. Albert I, Thakar J, Li S, Zhang R, Albert R. Boolean network simulations for life scientists. Source Code Biol Med. 2008;3:16.
31. Hunter JD. Matplotlib: a 2D graphics environment. Computing In Science & Engineering. 2007;9(3):90–5.
32. Hetmanski JH, Zindy E, Schwartz JM, Caswell PT. A MAPK-driven feedback loop suppresses Rac activity to promote RhoA-driven cancer cell invasion. PLoS Comput Biol. 2016;12(5):e1004909.
33. Wu Q, Li Y, Gu S, Li N, Zheng D, Li D, Zheng Z, Ji C, Xie Y, Mao Y. Molecular cloning and characterization of a novel dual-specificity phosphatase 23 gene from human fetal brain. Int J Biochem Cell Biol. 2004;36(8):1542–53.
34. Takagaki K, Satoh T, Tanuma N, Masuda K, Takekawa M, Shima H, Kikuchi K. Characterization of a novel low-molecular-mass dual-specificity phosphatase-3 (LDP-3) that enhances activation of JNK and p38. Biochem J. 2004;383(3):447–55.
35. Alonso A, Fau SM, Williams S, Mustelin T. Inhibitory role for dual specificity phosphatase VHR in T cell antigen receptor and CD28-induced Erk and Jnk activation. J Biol Chem. 2001;276(7):4766–71.
36. Schumacher MA, Todd JL, Rice AE, Tanner KG, Denu JM. Structural basis for the recognition of a Bisphosphorylated MAP Kinase peptide by human VHR protein Phosphatase. Biochemistry. 2002;41(9):3009–17.
37. Chakrabarty A, Bhola NE, Cammie S, Ghosh R, Kuba MG, Dave B, Chang JC, Arteaga CL. Trastuzumab-resistant cells rely on a HER2-PI3K-FoxO-survivin axis and are sensitive to PI3K inhibitors. Cancer Res. 2013;73(3):1190–200.
38. Crawford A, Nahta R. Targeting Bcl-2 in herceptin-resistant breast cancer cell lines. Current Pharmacogenomics and Personalized Medicine (Formerly Current Pharmacogenomics). 2011;9(3):184–90.
39. Nahta R, Takahashi T, Ueno NT, Hung M-C, Esteva FJ. P27kip1 down-regulation is associated with trastuzumab resistance in breast cancer cells. Cancer Res. 2004;64(11):3981–6.
40. Menyhart O, Budczies J, Munkácsy G, Esteva FJ, Szabó A, Miquel TP, Győrffy B. DUSP4 is associated with increased resistance against anti-HER2 therapy in breast cancer. Oncotarget. 2017;8(44):77207–18.
41. Sharieh EA, Awidi AS, Ahram M, Zihlif MA. Alteration of gene expression in MDA-MB-453 breast cancer cell line in response to continuous exposure to Trastuzumab. Gene. 2016;575(2 Pt 2):415–20.
42. Gong C, Yao Y, Wang Y, Liu B, Wu W, Chen J, Su F, Yao H, Song E. Up-regulation of miR-21 mediates resistance to trastuzumab therapy for breast cancer. J Biol Chem. 2011;286(21):19127–37.
43. Li X, Xu J, Tang X, Liu Y, Yu X, Wang Z, Liu W. Anthocyanins inhibit trastuzumab-resistant breast cancer in vitro and in vivo. Mol Med Rep. 2016; 13(5):4007–13.
44. Ginestier C, Adelaide J, Goncalves A, Repellini L, Sircoulomb F, Letessier A, Finetti P, Geneix J, Charafe-Jauffret E, Bertucci F, et al. ERBB2 phosphorylation and trastuzumab sensitivity of breast cancer cell lines. Oncogene. 2007;26(50):7163–9.
45. Lin XL, Wang XL, Ma B, Jia J, Yan Y, Di LJ, Yuan YH, Wan FL, Lu YL, Liang X, et al. HER2-specific T lymphocytes kill both trastuzumab-resistant and trastuzumab-sensitive breast cell lines in vitro. Chin J Cancer Res. 2012;24(2):143–50.
46. Canonici A, Gijsen M, Mullooly M, Bennett R, Bouguern N, Pedersen K, O'Brien NA, Roxanis I, Li J-L, Bridge E. Neratinib overcomes trastuzumab resistance in HER2 amplified breast cancer. Oncotarget. 2013;4(10):1592.
47. Fumia HF, Martins ML. Boolean network model for cancer pathways: predicting carcinogenesis and targeted therapy outcomes. PLoS One. 2013; 8(7):e69008.
48. Grieco L, Calzone L, Bernard-Pierrot I. Radvanyi Fo, Kahn-Perlès B, Thieffry D: Integrative modelling of the influence of MAPK network on cancer cell fate decision. PLoS Comput Biol. 2013;9(10):e1003286.
49. Sahin Ö, Fröhlich H, Löbke C, Korf U, Burmester S, Majety M, Mattern J, Schupp I, Chaouiya C, Thieffry D, et al. Modeling ERBB receptor-regulated G1/S transition to find novel targets for de novo trastuzumab resistance. BMC Syst Biol. 2009;3(1):1.

Teraguchi and Kumagai *BMC Systems Biology* 2018, **12**(Suppl 1):15
https://doi.org/10.1186/s12918-018-0526-5

BMC Systems Biology

RESEARCH Open Access

Estimation of diffusion constants from single molecular measurement without explicit tracking

Shunsuke Teraguchi[1,2*] and Yutaro Kumagai[2*]

From The Sixteenth Asia Pacific Bioinformatics Conference
Yokohama, Japan. 15-17 January 2018

Abstract

Background: Time course measurement of single molecules on a cell surface provides detailed information about the dynamics of the molecules that would otherwise be inaccessible. To extract the quantitative information, single particle tracking (SPT) is typically performed. However, trajectories extracted by SPT inevitably have linking errors when the diffusion speed of single molecules is high compared to the scale of the particle density.

Methods: To circumvent this problem, we develop an algorithm to estimate diffusion constants without relying on SPT. The proposed algorithm is based on a probabilistic model of the distance to the nearest point in subsequent frames. This probabilistic model generalizes the model of single particle Brownian motion under an isolated environment into the one surrounded by indistinguishable multiple particles, with a mean field approximation.

Results: We demonstrate that the proposed algorithm provides reasonable estimation of diffusion constants, even when other methods suffer due to high particle density or inhomogeneous particle distribution. In addition, our algorithm can be used for visualization of time course data from single molecular measurements.

Conclusions: The proposed algorithm based on the probabilistic model of indistinguishable Brownian particles provide accurate estimation of diffusion constants even in the regime where the traditional SPT methods underestimate them due to linking errors.

Keywords: Brownian motion, Diffusion constants, Expectation maximization algorithm, Probabilistic model, Single molecular measurement

Background

Sensing the extracellular environment is crucial for cells to properly respond and function. The information from the environment is typically encoded in microscopic molecular signals that are recognized by cell surface receptors. The signaling of cell surface receptors involves several physical processes, including ligation to their ligands, oligomerization, and subsequent binding to the downstream signaling components in cytosol. Although many details of these processes have been inferred from biochemical, genetic, and molecular or cell biological studies, their physical and dynamical aspects at the microscopic level are still largely unknown [1].

Recent development of techniques for single molecular measurement such as total internal reflection fluorescence (TIRF) microscopy [2] provides a chance to directly observe the dynamics of these processes from time course images of fluorescently-labeled single molecules on cell surfaces [3, 4]. A typical workflow for such data is single particle tracking (SPT) [5]. In SPT, the positions of particles in each time frame are first detected. With the help of the sophisticated detection algorithms, the spatial resolution of the detected position could be of

* Correspondence: teraguchi@megabank.tohoku.ac.jp; ykumagai@biken.osaka-u.ac.jp
[1]Tohoku Medical Megabank Organization, Tohoku University, 2-1 Seiryo-machi, Aoba-ku, Sendai 980-8573, Japan
[2]Quantitative Immunology Research Unit, Immunology Frontier Research Center, Osaka University, 3-1 Yamada-oka, Suita, Osaka 565-0871, Japan

sub-pixel order [6]. The next step is linking, where the trajectory of each molecule is inferred by connecting seemingly identical particles in subsequent frames. Usually, the nearest particle in the subsequent frame with global consistency is identified as the same particle [7, 8].

The identified trajectories of particles must be further analyzed quantitatively to find biologically relevant physical parameters. The diffusion constant, which characterizes the diffusion speed of the particles, is one such parameter, and has been the target for subsequent analyses [9–11]. It has been shown that the diffusion constants of membrane proteins such as cell surface receptors can change along with biophysical events such as binding to their ligand or cytosolic adaptor molecules. For example, the diffusion constants of the epidermal growth factor receptor (EGFR), which belongs to a family of receptor tyrosine kinase, have been found to decrease after binding to EGF, and to transduce signals via subsequent binding with its adaptor Grb2 protein [12, 13]. It has also been shown that intracellular signaling proteins functioning on the membrane have multiple states, each of which have different diffusion constants [14, 15].

Although SPT methods are widely used, they encounter difficulties when the density of particles is higher. When the particle density becomes comparable to the scale of diffusion in the time resolution of the measurement, the expected area of diffusion of a particle tends to contain several irrelevant particles purely by chance. Since, in typical experiments, visualized molecules are indistinguishable from fluorescent signals, linking errors of SPT are inevitable. Then, trajectories from such erroneous SPT lead to underestimation of diffusion constants, and incorrect biological interpretations. Note that this problem of linking error may occur even in the regime where the detection error coming from the diffraction limit of a microscope is negligible.

In this paper, we address this problem of linking error in diffusion constant estimation. As we have seen, the problem arises from the impossibility of perfect hard linking of identical particles in SPT. Here instead of linking the nearest particles in subsequent frames, we only assign a probability of such possible identification with respect to the particle density around the position, and directly estimate the diffusion constant without specifying concrete trajectories. For this purpose, we derive a probabilistic model of the distance to the nearest neighbor by generalizing the canonical theory of single Brownian motion into multiple indistinguishable particles. The resultant algorithm successfully estimates diffusion constants even under high particle density conditions where SPT based methods underestimate them. The proposed algorithm shows some resemblance to another SPT free diffusion constant estimation method, namely particle image correlation spectroscopy (PICS) [16], which was inspired by image correlation microscopy [17–20]. The advantages of our algorithm over PICS include lower variances of estimated diffusion constants, lower numbers of hyperparameters to be determined before the analysis, and the applicability to cases with inhomogeneous particle distributions, whereas PICS assumes a homogeneous distribution.

In this paper, we first introduce the probabilistic model of the positions of the nearest neighbors of a diffusing particle surrounded by indistinguishable particles and then formulate the inference of diffusion constants in terms of maximum likelihood estimation based on this model. In a simple setting with a homogeneous particle distribution, our algorithm can be considered to be a natural generalization of the canonical diffusion constant estimation from the mean square displacement (MSD) to the case of finite density of surrounding particles. Our algorithm is further generalized to allow multiple states with different diffusion constants with the help of the expectation maximization (EM) algorithm [21]. Comparison of the performance of our proposed method based on simulated artificial diffusion data with other diffusion constant-estimation methods indicates the advantage of the proposed algorithm. Finally, we demonstrate that the algorithm can be used to infer the state of each molecule and visualize the single molecular data with such information.

Theory

A probabilistic model of a diffusing particle surrounded by indistinguishable particles

To develop the probabilistic model for estimating the 2D lateral diffusion constants under high particle density, we focus on a single Brownian particle in a time frame (Fig. 1). Without loss of generality, we take the position of the particle as the origin of our polar coordinates. As is well known, the probability of finding the same Brownian particle at a position with a radial distance greater than Δr after a time-lag Δt is given by [22].

$$P_{\text{dif}}(r > \Delta r|D) = e^{-\frac{\Delta r^2}{4D\Delta t}}, \tag{1}$$

where the parameter $_D$ is the diffusion constant of the particle.

In typical time-lapse single molecular imaging of cells, particles are indistinguishable from one another. By assuming the independence of the dynamics of each particle, we can model the distribution of such indistinguishable surrounding particles by a local uniform density,ρ, which is a sort of mean field approximation of the surrounding particles. In this approximation, we can derive the probability of having the nearest surrounding particle at a distance greater than Δr as follows. We begin with a finite case where there are, on average, N surrounding particles in the disk with a radius R around

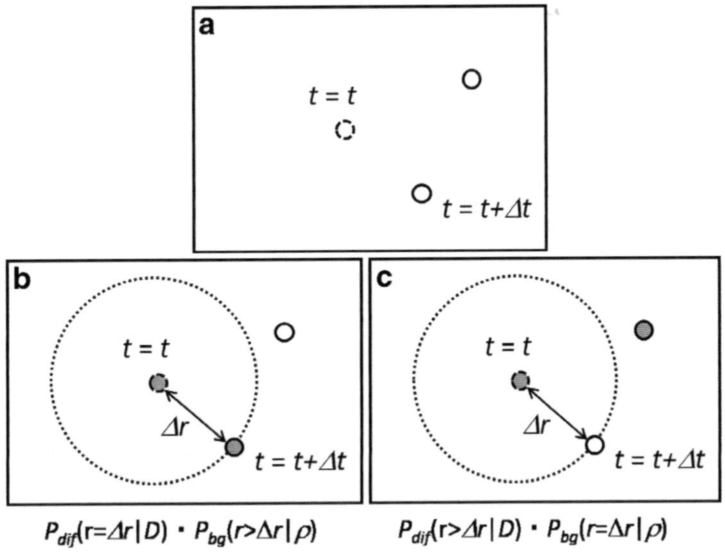

Fig. 1 Schematic of the probabilistic model. **a** a typical distribution of particles at $t + \Delta t$ (thick circles) with an indication of the position of a representative particle at t (dashed circle). **b** the case where the nearest particle is the original particle. **c** the case where the nearest particle is a surrounding particle. Gray color indicates the identification of the original particle. The large dotted circles indicate the distance to the nearest particle. The distance to the nearest neighbor of the origin at the subsequent time frame is modeled by the probabilistic model with respect to the diffusion constant of the original particle and the particle density at the origin

a point. We assume that the surrounding particles are uniformly distributed within the disk. If we consider a smaller disk with a radius Δr inside the disk, the probability of a single surrounding particle being found outside of the smaller disk is $1 - a/A$, where $a = \pi\Delta r^2$ and $A = \pi R^2$ correspond to the areas of the smaller and bigger disks, respectively. Then, the probability that all the N surrounding particles are also found outside of the smaller disk is $(1 - a/A)^N$. Assuming that a is much smaller than A, this probability can be approximated as

$$\left(1-\frac{a}{A}\right)^N = \exp\left(N \log\left(1-\frac{a}{A}\right)\right) \cong \exp\left(-\frac{aN}{A}\right) = \exp\left(-\rho\pi\Delta r^2\right),$$

where $\rho = N/A$ is the local particle density. Thus, the probability of having the nearest surrounding particle at a distance greater than Δr is given by

$$P_{\mathrm{bg}}(r > \Delta r|\rho) = e^{-\rho\pi\Delta r^2}. \quad (2)$$

By combining the above results together, the probability of detecting the nearest particle at a distance greater than Δr would be given by

$$P_{\mathrm{nn}}(r > \Delta r|\rho, D) = P_{\mathrm{dif}}(r > \Delta r|D) P_{\mathrm{bg}}(r > \Delta r|\rho) = e^{-\rho\pi\Delta r^2 - \frac{\Delta r^2}{4D\Delta t}}. \quad (3)$$

This is the fundamental probabilistic model upon which we develop the estimation algorithm of the diffusion constant in this paper (Fig. 1). This probabilistic model generalizes the theory of Brownian motion of a single isolated particle into that of a single particle surrounded by indistinguishable particles.

The indication of the model becomes more manifest if we calculate the expected mean square displacement to the nearest particle (MSDN) as

$$\mathrm{MSDN} = E(\Delta r^2) \equiv \int_0^\infty \Delta r^2 \left(-\frac{d}{d(\Delta r)} P_{\mathrm{nn}}(r > \Delta r|\rho, D)\right) d(\Delta r) = \frac{4D\Delta t}{1 + 4\rho\pi D\Delta t}. \quad (4)$$

This is a natural generalization of the well-known relationship between the MSD of a single diffusing particle and the diffusion constant [22],

$$\mathrm{MSD} = 4D\Delta t. \quad (5)$$

As expected, MSDN goes back to the original MSD in the limit of ρ being zero, (i.e., where there are no surrounding particles). Due to the additional term in the denominator, the MSDN is, in general, smaller than MSD. This is because the nearest particle can be the original particle diffused from the origin as in MSD, or even a nearer surrounding particle.

This relationship can be easily solved with respect to D, allowing it to be estimated as

$$D = \frac{\text{MSDN}}{4\Delta t(1 - \rho\pi\text{MSDN})}. \tag{6}$$

Compared to the standard estimation from MSD,

$$D = \frac{\text{MSD}}{4\Delta t}, \tag{7}$$

the estimated diffusion constant acquires a fold increase of $1/(1 - \rho\pi\text{MSDN})$, which compensates for the apparent reduction of the displacement compared to MSD. In Fig. 2, we show the MSDN for simulated data. As Δt increases, the points deviate from the line $4D\Delta t$ and obey the above theoretical prediction as expected. Note that the time course of MSDN is conceptually different from that of MSD in a trajectory after SPT. In the case of SPT, the identification of the same particle is consecutively performed using all measured time points during Δt. On the other hand, in MSDN, the nearest point after time duration Δt was chosen without referring to the measured time points before Δt.

Maximum likelihood estimation of diffusion constants for local particle density

Though the above relationship between the diffusion constant and MSDN allows us to estimate diffusion constants for the case of a uniform particle distribution, it is difficult to generalize it into an inhomogeneous particle distribution, which is a less ideal but much more relevant situation. In such a case, a constant particle density ρ alone cannot capture the underlying particle distribution.

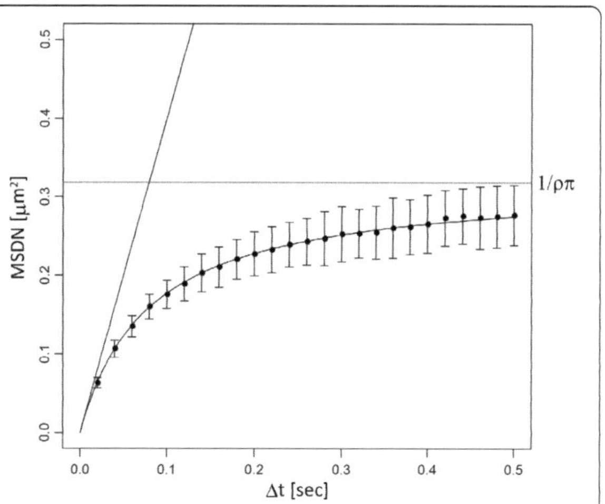

Fig. 2 Mean square displacement to the nearest particle. A comparison of MSDN and MSD. The black straight line corresponds to the expected MSD, while the black curve is the expected MSDN, with D=1 μm^2/s and ρ=1 particles/μm^2. The points are the mean MSDN directly calculated from corresponding simulated data. The error bars indicate the standard deviation from one thousand independent simulations. The red line indicates the asymptotic value of the expected MSDN at $\Delta t \to \infty$

Here, we formulate a more general estimation algorithm of diffusion constants using a maximum likelihood estimation based on the above probabilistic model. The log-likelihood of an observed dataset is given by

$$\begin{aligned} l &= \log \prod_{i=1}^{N} P_{\text{nn}}(r = \Delta r_i | \rho_i, D) \\ &= \sum_{i=1}^{N} \left[\log\left(2\rho_i\pi + \frac{1}{2D\Delta t}\right) + \log(\Delta r_i) - \rho_i\pi\Delta r_i^2 - \frac{\Delta r_i^2}{4D\Delta t} \right]. \end{aligned} \tag{8}$$

Here, the index i represents each particle in the preceding time frame, Δr_i is the distance to the nearest particle in the subsequent time frame, and ρ_i is the local particle density around particle i. If we assume a uniform distribution (i.e., that all ρ_i are the same), this maximum likelihood estimation of D is analytically tractable and reduces to the same relation between the diffusion constant and MSDN described above.

In the case of general ρ_i, it is convenient to utilize the EM algorithm [21, 23]. For this purpose, we introduce a latent variable $q_i \in \{0, 1\}$, which takes the value of zero if the nearest point comes from the surrounding particles, but becomes one if it is the original particle diffused from the origin. Then the complete-data log-likelihood with the information of the latent variable is given by

$$l' = \log \prod_{i=1}^{N} p(\Delta r_i, q_i | \rho_i, D). \tag{9}$$

Here, the joint probability distribution is defined as

$$p(\Delta r_i, q_i | \rho_i, D) = \begin{cases} 2\rho_i\pi\Delta r_i e^{-\rho_i\pi\Delta r_i^2 - \frac{\Delta r_i^2}{4D\Delta t}} & \text{for } q_i = 0 \\ \frac{\Delta r_i}{2D\Delta t} e^{-\rho_i\pi\Delta r_i^2 - \frac{\Delta r_i^2}{4D\Delta t}} & \text{for } q_i = 1 \end{cases} \tag{10}$$

In the EM algorithm, instead of maximizing the log-likelihood directly, a quantity $Q(D, D^l)$ is maximized with respect to D by iteration:

$$Q(D, D^l) = \sum_{i=1}^{N} \sum_{q \in \{0,1\}} \log(p(\Delta r_i, q | \rho_i, D)) p(q | \Delta r_i, \rho_i, D^l). \tag{11}$$

Here, D^l is the estimation of the diffusion constant D at the l-th iteration. The conditional probability based on D^l is calculated from the above joint probability as

$$p(q = 0 | \Delta r_i, \rho_i, D^l) = \frac{4\rho_i\pi D^l\Delta t}{4\rho_i\pi D^l\Delta t + 1}, \tag{12}$$

$$p(q=1|\Delta r_i,\rho_i,D^l)=\frac{1}{4\rho_i\pi D^l\Delta t+1}. \tag{13}$$

Taking the derivative of Q with respect to D and equating it to zero,

$$\frac{dQ}{dD}=\sum_{i=1}^{N}\left[\frac{\Delta r_i^2}{4D^2\Delta t}p(q=0|\Delta r_i,\rho_i,D^l)+\left(\frac{\Delta r_i^2}{4D^2\Delta t}-\frac{1}{D}\right)p(q=1|\Delta r_i,\rho_i,D^l)\right]=0, \tag{14}$$

we obtain the update rule

$$D^{l+1}=\frac{\langle\Delta r^2\rangle}{4\Delta t\langle P(q=1)\rangle_{D^l}}, \tag{15}$$

where we have defined the expected fraction of data points with $q = 1$ as

$$\langle P(q=1)\rangle_{D^l}\equiv\frac{1}{N}\sum_{i=1}^{N}p(q=1|\Delta r_i,\rho_i,D^l). \tag{16}$$

Now, the correction from the original MSD relation is neatly summarized by this expected fraction of the data whose nearest points come from the original particle diffused from the origin.

Generalization to models with multiple diffusive states

In this subsection, we further generalize the maximum likelihood estimation of diffusion constants into the case where particles take multiple states with different diffusion constants. It has been revealed that some membrane proteins change their physical properties upon binding to other molecules or spontaneous change of their conformation, and that these changes can be inferred from the change of the diffusion constant in some cases [14, 15]. Here we consider this type of change of diffusion constants, which we shall refer as to the change of their states. In this paper, we only provide the solution for relatively simpler situations of the dynamics with multiple diffusive states where the interconversion of different states can be ignored in the time resolution [10].

This simple generalization is practically quite useful, even when there is no biological reason to expect the existence of such multiple states of the target molecule. In a real experiment, many fluorescently-dyed surface molecules disappear for several reasons, such as internalization of the particle, breaching of the fluorescent dye, and so on. Such disappearance of particles can be modeled in the above framework by adding an additional state whose diffusion constant is infinitely large. In addition, some accidental peaks of fluorescent intensity may be wrongly detected as particles due to the low signal-to-noise ratio of the original images (false detections). Those spurious particles also tend to disappear in the subsequent time frame. Thus, we can reduce the effects of such false detections by introducing such a state in advance. We will address this issue again in Result section.

The derivation of the corresponding EM algorithm is largely parallel to the one in the previous subsection. In addition to the latent variable q_i, which specifies whether or not the nearest particles are the original particle itself, we introduce an additional latent variable specifying states of the particle i, $s_i \in \{1,\cdots,M\}$, where M is the number of possible states.

The joint probability distribution of this model is given by

$$p(\Delta r_i,q_i,s_i|\rho_i,D_{s_i},\alpha_{s_i})=\begin{cases}2\rho_i\pi\alpha_{s_i}\Delta r_i e^{-\rho_i\pi\Delta r_i^2-\frac{\Delta r_i^2}{4D_{s_i}\Delta t}} & \text{for } q_i=0\\ \frac{\alpha_{s_i}\Delta r_i}{2D_{s_i}\Delta t}e^{-\rho_i\pi\Delta r_i^2-\frac{\Delta r_i^2}{4D_{s_i}\Delta t}} & \text{for } q_i=1\end{cases}, \tag{17}$$

where D_{S_i} is the diffusion constant of the state s_i, and α_{s_i} is the probability of being the state s_i.

The quantity Q for deriving the update rule of the EM algorithm is similarly defined by

$$Q(D,D^l)=\sum_{i=1}^{N}\sum_{s=1}^{M}\sum_{q\in\{0,1\}}\log(p(\Delta r_i,q,s|\rho_i,\theta))\,p(q,s|\Delta r_i,\rho_i,\theta^l). \tag{18}$$

Here, θ collectively denotes all of the parameters to be estimated, namely, $\theta = \{D_1,\cdots,D_M,\alpha_1,\cdots\alpha_M\}$. The conditional probability is calculated from the joint probability as follows:

$$p(q=0,s|\Delta r_i,\rho_i,\theta^l)=\frac{2\rho_i\pi\alpha_s^l e^{-\frac{\Delta r_i^2}{4D_s\Delta t}}}{\sum_{s'=1}^{M}\left(2\rho_i\pi+\frac{1}{2D_{s'}{}^l\Delta t}\right)\alpha_{s'}^l e^{-\frac{\Delta r_i^2}{4D_{s'}\Delta t}}}, \tag{19}$$

$$p(q=1,s|\Delta r_i,\rho_i,\theta^l)=\frac{\frac{\alpha_s^l}{2D_{s'}{}^l\Delta t}e^{-\frac{\Delta r_i^2}{4D_s\Delta t}}}{\sum_{s'=1}^{M}\left(2\rho_i\pi+\frac{1}{2D_{s'}{}^l\Delta t}\right)\alpha_{s'}^l e^{-\frac{\Delta r_i^2}{4D_{s'}\Delta t}}}. \tag{20}$$

Compared to the single state case, here, the joint probability also depends upon the displacement, Δr_i.

By maximizing Q under the restriction of conservation of probability, $\sum_s \alpha_s = 1$, we obtain

$$\alpha_s^{l+1} = \frac{1}{N}\sum_{i=1}^{N}\sum_{q\in\{0,1\}} p\left(q, s|\Delta r_i, \rho_i, \theta^l\right), \tag{21}$$

$$D_s^{l+1} = \frac{\sum_{i=1}^{N}\sum_{q\in\{0,1\}} \Delta r_i^2 p\left(q, s|\Delta r_i, \rho_i, \theta^l\right)}{4\Delta t \sum_{i=1}^{N} p\left(q = 1, s|\Delta r_i, \rho_i, \theta^l\right)}. \tag{22}$$

This is our final update rule for maximum likelihood estimation for the multi state model.

Methods

Monte Carlo simulation

To compare the performance of the proposed and existing methods, we generate artificial data of single molecular particle diffusion with Monte Carlo simulation. Depending on the purpose of simulation, we generate simulated data in two different ways.

Pairwise simulation

In [16], to evaluate the performance of PICS algorithm, the authors utilized simulated data generated as pairs of time frames, rather than a single time course of diffusing particles. Since it allows precise controls of the distribution of the simulated data, it makes subsequent comparison among algorithms and interpretation of observed performance easier. Thus, we follow the same strategy to simulate diffusion dynamics in some of our simulations in Result section.

First we draw a fixed number of positions of particles from the corresponding probability distribution of particles for the preceding time frame. In the case of uniform particle distribution, we sample the particles over a much larger area than the area of interest, in order to keep the same distribution after the diffusion steps. Next, we generate the subsequent frame by adding a displacement drawn from the two-dimensional normal distribution with a variance of $2D\Delta t$ to each position. When needed, another fixed number of particles are drawn from the same particle distribution, and added independently to both the preceding and subsequent frames to represent the existence of false detections, which typically occur in detection from low signal-to-noise ratio image data. In the simulation with false detections, we set the fraction of false detections to 20%. Each estimation of diffusion constants is performed against 10 pairs of time frames. The simulation is repeated 100 times for each condition. All simulations are performed using R (http://www.r-project.org/).

Image based time course simulation

In the above simulation method, positions of detected points were directly generated by Monte Carlo simulation. Thus, no particular bias coming from detecting particle positions from image data is taken into account. In order to take account of such uncontrollable effects, we further examine diffusion constant estimation algorithms by artificially generated time course image data of single molecular measurements. For this purpose, we utilize the image data generator provided as a plugin "ISBI Challenge Track Generator" [24] of an open platform software "ICY" [25] for bioimage analysis.

We set the parameters of the plugin software as follows; SNR = 4, sequence length = 10, particle density = 100, 500 and 1000, sigma = 1, 2, 3, 5, 7 and 10 in the particle motion with creator type "BROWNIAN_UNIFORM". The image size is 512 pixels × 512 pixels. The other parameters (except for seeds) are set to default, which means the extinction rate of each particle is 0.05. The particles in generated image data are detected by another plugin "Spot Detector" of the ICY software. The detection of bright spots by Spot Detector plugin is performed with default parameters. The simulation is repeated 3 times for each condition with different seed values.

Other algorithms to estimate diffusion constants

To evaluate the performance of our proposed method, we compare it with existing algorithms. To make the comparison make sense, we examine algorithms that are applicable to the same type of the data, namely, the time series of the location of detected points. For example, some of algorithms utilized to estimate the diffusion constants under higher density or higher diffusion speed cannot be compared because they require specially designed data set for the algorithms [26, 27]. As a result, our comparison is made mainly with PICS algorithm which is particularly designed for estimating diffusion constants under higher particle density, in addition to SPT based methods.

PICS

We implement the PICS algorithm in R to enable automatic parameter estimation from the Monte Carlo simulation data. A minor difference from the original implementation described in [16] is that we fit the whole cumulative correlation function at once to simplify the automation instead of separately fitting the linear and non-linear parts of the cumulative correlation function to the data. In our experience, this implementation of PICS provides comparative or even better performance compared to the original one (data not shown).

Local SPT

As an example of the most naïve approach, we make trajectories by simply associating each particle to the nearest particle in the subsequent frame without considering global consistency. Unlike the case of global SPT described below, in this approach, a particle in a subsequent time frame might be associated with several particles in the preceding time frame.

Global SPT

As a representative of SPT method, we implement the global linking algorithm based on a greedy hill-climbing optimization with topological constraints following the literature [24, 28]. This algorithm was used in one of the best performance groups in the international competition of particle tracking methods [24]. In this algorithm, there is no conflict between the associations of each particle. We set the maximum distance parameter for limiting the association of subsequent particles to a large enough value to link all particles. For the pairwise simulation, this procedure provided sensible estimation of diffusion constants independent of the details of the exact value of the maximum distance parameter, as far as the particle density is not very high (data not shown).

After obtaining the distribution of diffusion step sizes with local or global SPT, we estimate the diffusion constant with a maximal likelihood estimation based on the assumption that each single particle exhibits Brownian motion.

Particle density estimation for the simulated data

To apply our algorithm, we have to estimate the (local) particle density. In the case of a uniform distribution, we estimate the density by simply dividing the total particle number in the frame by the area of interest. In an inhomogeneous case, it is difficult to accurately estimate the local particle density based on just a single time frame. Therefore, we estimate the local probabilistic density by a k nearest-neighbor algorithm after merging all subsequent frames in the dataset except for the one in the frame of interest. Then, the particle density at the point is obtained by weighting the probabilistic density with the number of particles in the frame of interest. The value of k from the k nearest neighbor density estimation in the merged data is chosen to be the number of time frames utilized, which corresponds to the length scale of $k = 1$ in a time frame.

Estimation of diffusion constants for the real data

HeLa cells grown on glass coverslips (Matsunami) in a 6-well plate were transfected with Lyn_{11}-Halotag using Lipofectamine 2000 (Invitrogen). Afters 4 h, the culture medium was replaced with DMEM and the cells were incubated at 37 °C for 24 h. The culture medium was exchanged with Opti-MEM (Gibco), and the cells were incubated at 37 °C. After 2 h, the cells were washed once with OPTI-MEM and incubated with 0.03 nM of Halotag TMR ligand (Promega) in Opti-MEM for 30 min in a CO2 incubator. The cells were then washed three times with Opti-MEM and single-molecule imaging was performed using a TIRF microscope. Single particle detection and estimation of diffusion constants were done using ICY and PNN algorithm, respectively.

Results and Discussion

Dependence of estimated diffusion constants on particle density

Both PICS and our estimation algorithm, hereafter called the probabilistic nearest neighbor (PNN) estimation, have been designed to accurately estimate diffusion constants under the condition of high particle density. We first compare these methods to SPT-based methods with and without global optimization of linking (referred to as global SPT and local SPT, respectively) with pairwise simulated data (see Method section for details).

First, we examine the effect of particle density under the ideal condition of a homogeneous distribution (Additional file 1: Figure S1 and Fig. 3). We vary the particle density from 0.1 to 10 particles/μm^2, fixing the diffusion constant to be 1μm^2/s. The time resolution, Δt, of the data acquisition is assumed to be 20 ms [16]. Note that, in this ideal situation of Brownian motion, only the ratio of the scales of the diffusion constant and the particle density is the relevant parameter. Thus, the effects of changing the particle density with a fixed diffusion constant are effectively equivalent to the ones of changing the diffusion constant with a fixing particle density.

As expected, the change in particle density significantly affects the diffusion constants estimated by the simplest method, local SPT (Fig. 3). In this method, each pair of nearest neighbor points in the subsequent time frame is simply identified as the same physical particle without consideration of the behaviors of other particles. With this simple method, even with one-order lower particle density, the estimation accuracy is low due to the bias caused by the linking error (Additional file 1: Figure S1).

After global optimization (global SPT) of the linking, the estimation accuracy of SPT method is improved. In particular, under lower particle density conditions, it reproduces the true diffusion constants to great accuracy (Additional file 1: Figure S1). However, in the condition with higher particle density ($\rho \geq 2$), this method also underestimates the diffusion constants. This value of the particle density roughly corresponds to that where $4\rho\pi D\Delta t$ becomes comparable to 1 in Eq. 1. This result suggests the limitation in SPT methods under high particle density conditions.

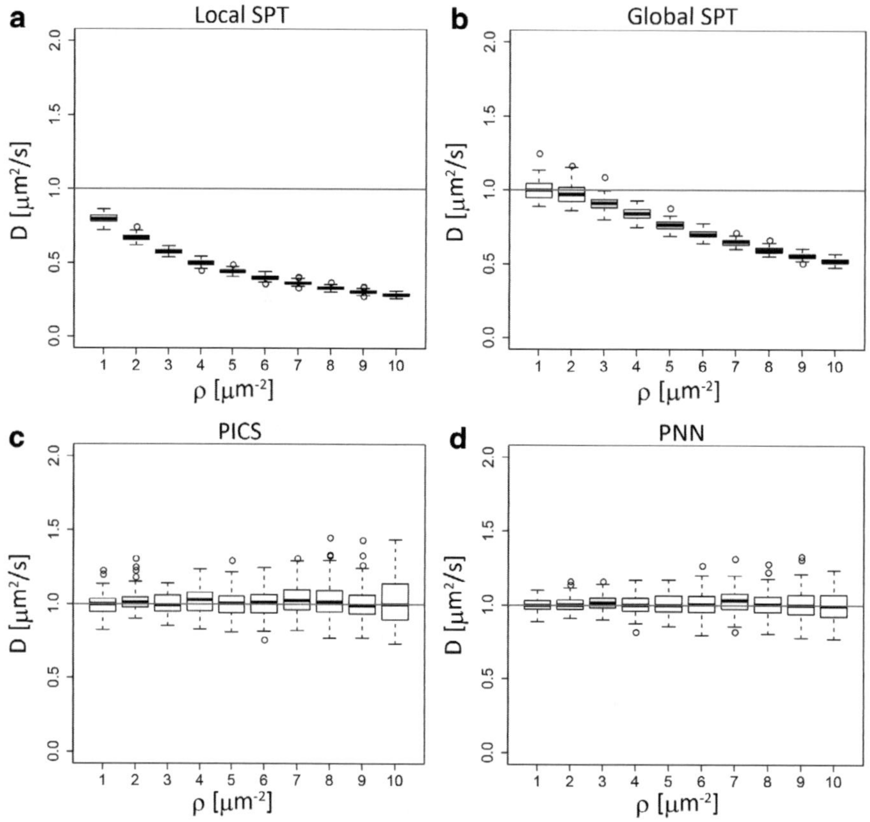

Fig. 3 Comparison of the performance of different algorithms in a uniform distribution. Box plots summarizing a comparison of the algorithms. The x axis is the particle density and the y axis is the estimated diffusion constant. The red line indicates the true diffusion constant. **a** local SPT. **b** global SPT. **c** PICS and **d** PNN

On the other hand, the two SPT-free methods PICS and PNN, which take the effects of surrounding particles explicitly into account, estimate the diffusion constants quite well over the whole range of particle densities under consideration (Fig. 3 and Additional file 1: Figure S1). Though the standard deviations among independent simulations tend to increase along with the increase of particle density, these could be reduced if more data in the same condition became available [16].

Thus, the estimation of diffusion constants using PICS or PNN leads to similar performance with SPT-based methods under lower particle density and outperforms them under higher particle density. Therefore, we focus on these two methods in the following discussion.

Effect of false detections

By comparing PNN and PICS from the above results, one might conclude that the accuracy of PNN is slightly better than that of PICS because the standard deviation of the estimated results is smaller in the former than the latter. However, the above comparison was performed based on simulation in a quite ideal condition: particles distributed uniformly without any false detection. On the other hand, real single molecular measurements tend to be performed under less ideal conditions with a lower signal-to-noise ratio. This affects the accuracy of the detection of peak positions from raw images, leading to spurious particles that are wrongly detected in such noisy images.

In order to mimic such a situation, we artificially introduce additional particles independently drawn from the same distribution in each time frame. We simply refer to these additional particles as false detections. The existence of false detections significantly degrades the estimation accuracy (Fig. 4, left panels) of both PNN and PICS. The effects of false detections in the diffusion constant estimation are two-fold. One effect is to increase the apparent density of surrounding particles in the subsequent time frames, and the other is the addition of spurious particles in the preceding time frames that immediately disappear from the scope. The former effect is, by design, treated both in PICS and PNN since the particle density is estimated with both physical particles and false detections. On the other hand, the spurious

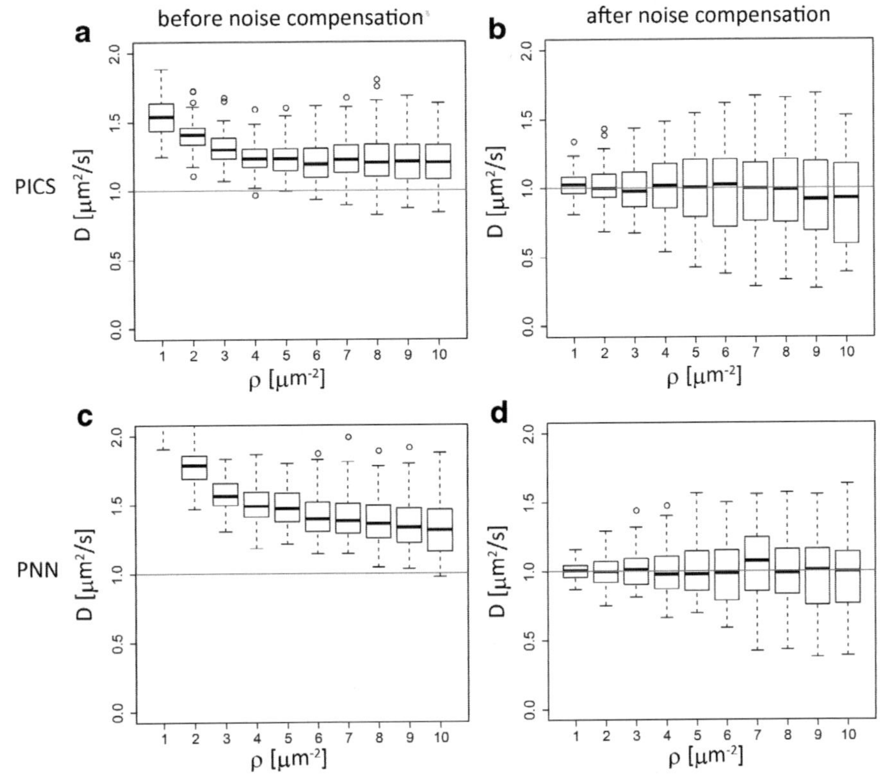

Fig. 4 Comparison of the performance of PICS and PNN in a uniform distribution with false detections. Box plots summarizing the comparison of PICS (**a** and **b**) and PNN (**c** and **d**). The top row is for PICS and the bottom row is for PNN. The first column is the result before introducing the state corresponding to the false detections. The second column is the result after introducing the state for false detection compensation. The x axis is the particle density and the y axis is the estimated diffusion constant. The red line indicates the true diffusion constant

particles coming from false detections in the preceding time frames behave like particles with an infinitely high diffusion constant. Therefore, the addition of false detections biases the estimated diffusion constants towards higher values. Note also that similar effects may occur when actual particles disappear by internalization or dissociation of surface protein from the membrane, bleaching of fluorescent dye and so on.

Fortunately, as commented in Theory section, this effect of false detections can be addressed by generalizing the probabilistic model both in PICS and PNN by introducing an additional state for false detections with an infinitely large diffusion constant. With this generalization, both PICS and PNN improve their prediction accuracy (Fig. 4, right panels) with a cost of larger standard deviation, which originates from the increase of the number of the parameters to be estimated, namely the fraction of false detections.

Estimation with an inhomogeneous distribution

As mentioned above, another idealization in the above simulation was the assumption of a uniform distribution of the particles. In fact, this is one of the key assumptions in the PICS algorithm. On the other hand, we have designed PNN to be applicable beyond this assumption. Here, we compare the performance of these two methods under three inhomogeneous distributions: Gaussian, circular and Gaussian mixture.

Figure 5, Additional file 2: Figure S2 and Additional file 3: Figure S3 show the results of estimation of diffusion constants under three classes of inhomogeneous distributions, a Gaussian distribution, a circular distribution forming an annulus and Gaussian mixture distributions, respectively. Panel B of each figure shows the results of PICS, where the estimated diffusion constants are biased, especially for the higher particle density. This result is more or less expected, since this type of inhomogeneous condition is beyond the original scope of PICS.

Panel C of each figure is the result of the PNN estimation with the known theoretical distribution utilized to generate simulated data. In this case, the estimated diffusion constants are much closer to their true values. Of course, in a real situation, we cannot access to the true underlying distribution of the particles. Thus, we have to estimate the distribution from the data, and the accuracy of the diffusion constant estimation depends

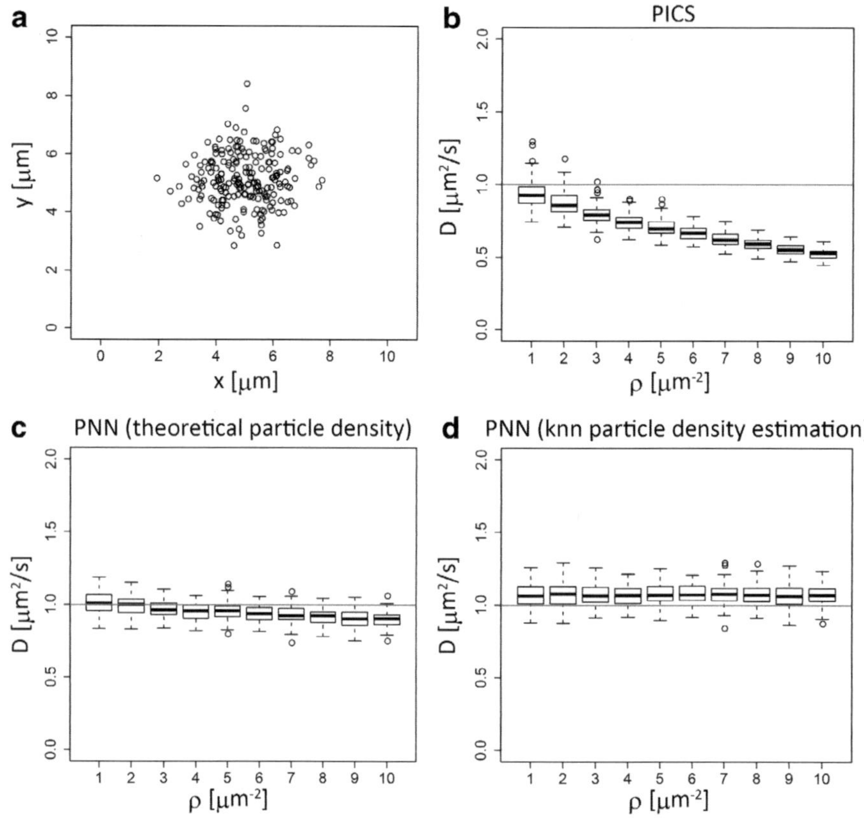

Fig. 5 Comparison of the performance of PICS and PNN in a Gaussian distribution. **a** a representative snapshot of the particle distribution. **b**, **c**, and **d** box plots summarizing the comparison between PICS and PNN under a Gaussian distribution. **b** PICS. **c** PNN, where the known particle density distribution for the simulation is used for the diffusion constant estimation. **d** PNN where the particle density distribution is estimated from the data using a k nearest neighbor algorithm. The x axis is the mean particle density over the area of interest, and the y axis is the estimated diffusion constant. The red line indicates the true diffusion constant

upon the accuracy of the density estimation. However, the results here demonstrate that as far as the particle density is estimated accurately enough, PNN should work reasonably well.

Panels D of Fig. 5, Additional file 2: Figure S2 and Additional file 3: Figure S3 show the results of PNN with a particle density estimated from the data itself. Here, in order to estimate the particle density, we use k nearest neighbor estimation. In general, there is a tradeoff between spatial resolution and statistical error in density estimation. Since our algorithm of PNN relies on the (first) nearest neighbor, smaller *k* values with high spatial resolution would be preferable. However, density estimation based on a smaller *k* tends to have a larger variance. In order to circumvent this problem, we estimate the particle density using all the post frames in the dataset except for the one in the frame of interest while keeping the effective *k* value equal to one (see Method section for details). The accuracy of the resultant diffusion constant is comparable to the accuracy using theoretical distributions. Our result here demonstrates that, with a suitable choice of density estimation methods, our algorithm can be utilized to estimate the diffusion constant, even under an inhomogeneous particle distribution.

Image based simulation

To mimic a realistic situation of diffusion constant estimation from typical single molecular measurements, we further examine our algorithm and others using artificial image data generator for an open competition of SPT organized in 2012 [24]. The image data generator is provided as a plugin "ISBI Challenge Track Generator" of an open platform "ICY" for bioimage analysis. We generate image data of diffusion dynamics as triplicates for each condition. We set the parameters of the simulator to be relatively low signal-to-noise ratio, and short sequence length, to increase the difficulty of the estimation in the category of "BROWNIAN_UNIFORM" (see Method section for details). Note also that the particles in this simulation disappear with an extinction rate of

0.05. A representative movie and images of this simulation are in Additional file 4: Movie S1 and in Additional file 5: Figure S4, respectively. The detection of the particles from image data was made by another plugin "Spot Detector" of the ICY software.

The results of the estimation of diffusion constants are summarized in Fig. 6. Here, we show the diffusion constants estimated by PNN, PICS and Local SPT. Though we have also applied global SPT to the same data, it showed very strong dependence on the maximum distance parameter and we could not obtain sensible estimation from the analysis (data not shown). Thus, the results of global SPT are omitted. We observe very similar tendency as in the previous simulations. PNN provides the most accurate results over the range of simulated conditions. Local SPT shows very strong bias depending on the particle density (number) and true diffusion constants. PICS does not show particular bias but tends to have higher variances. Visual inspection of the fitted curves of PICS clearly indicated poor fitting due to the effect of diffraction, as discussed in the original paper of PICS [16]. In the paper, they discussed how to mitigate the effect of diffraction in an iterative algorithm. Here, instead of implementing their iterative algorithm, we apply PICS to the corresponding ground truth data provided by the simulator, which are free from all the effects of diffraction (Panel D). Though this additional favor improves the fitting and the performance of PICS, PNN still seems to outperform the ground truth based PICS (Fig. 6). These results indicate the advantage of PNN in the application to the real image data from single molecular measurement of living cells.

3D visualization of particle states

The key of the proposed algorithm is that it assigns a probability of taking each possible state to each particle detected without specifying a trajectory. This property of the algorithm can be utilized to visualize time course data itself. The data shown in the upper panel of Fig. 7 consists of particles taking three different states, namely slower diffusion (0.2 $\mu m^2/s$), faster diffusion (2 $\mu m^2/s$), and false detections. The particle density including all of the three states is 1 particles/μm^2. The lower left panel is the same data in color (red: slower particle, cyan:

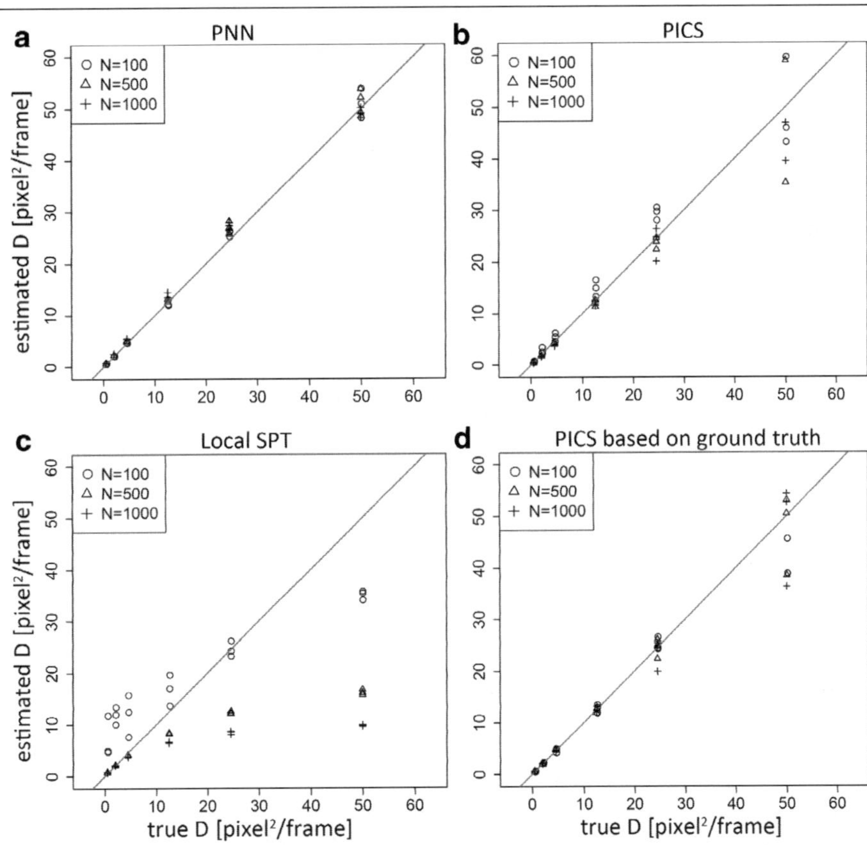

Fig. 6 The performance of different algorithms in image based simulations. Scatter plots summarizing the performance of the algorithms. The x axis is the true diffusion constant used for the simulation and the y axis is the estimated diffusion constant. The red line indicates the diagonal line corresponding to the successful estimation. **a** PNN. **b** PICS. **c** local SPT and **d** PICS applied to the corresponding ground truth data

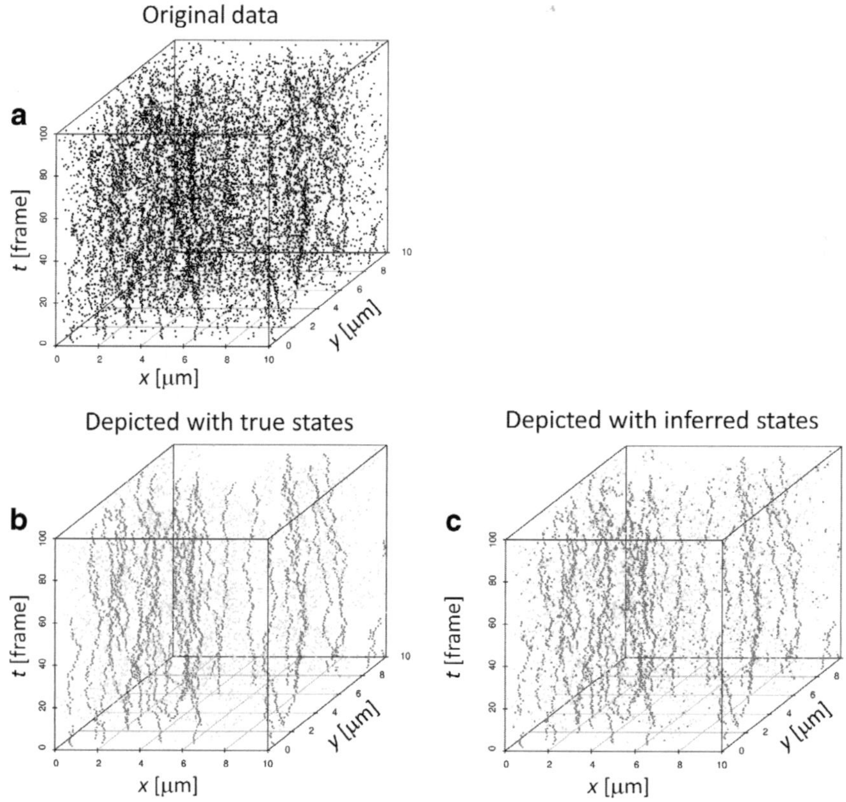

Fig. 7 3D visualization of particle positions and states. 3D representation of the time course simulated data of diffusing particles. The z axis corresponds to time while the other two axes correspond to the x- and y-axes of the original data. **a** the original data. **b**, the same data depicted in color (red: slower particle (0.2 μm^2/s), cyan: faster particle (2 μm^2/s)) after removing the false detection. **c** the same data depicted in colors based on the particle states inferred by PNN

faster particle) after removing the false detections. We apply the PNN algorithm to the data and infer the state of each particle by choosing the most probable one among the assigned probabilities. As shown in the lower right panel, the resultant figure bears a strong resemblance to the original data, giving another support for the validity of this algorithm. Unlike canonical SPT methods attempting to determine a hard-wired trajectory, our algorithm keeps several possibilities at the same time. This application of PNN to a visualization purpose would be useful, particularly when one is interested in identifying rare events like interactions between pairs of particles.

Application to real data

We applied the PNN algorithm to a real data. Lyn_{11}-Halotag construct, which is localized on cytoplasmic membrane, was expressed in HeLa cells and single-molecule imaging was performed (Additional file 6: Movie S2). Particle detection data was generated by using ICY and subjected to estimation of diffusion constant by PNN. PNN with *k* nearest neighbor particle density estimation resulted in diffusion constant of 2.81×10^{-2} μm^2/s under the assumption where molecules take only one state and those of 5.20×10^{-3} μm^2/s and 6.15×10^{-2} μm^2/s under the assumption where molecules take two states. We could also calculate AIC under these assumptions and the result supported the latter. In the case, fractions of false detection, slow and fast states were estimated to be 36%, 24% and 40%, respectively. A previous report [29] suggested that Lyn, origin of the Lyn_{11} tag, exhibits two states, namely lateral diffusion and transient confinement in a lipid region through lipid-lipid interaction. This supports our result where Lyn_{11}-Halotag has two states with slow and fast diffusion. Together, this result suggests that the proposed algorithm works well for real data and helps to understand dynamics of molecules.

Conclusions

In this paper, we proposed a novel diffusion constant estimation algorithm based on a probabilistic model of the nearest point without explicitly performing SPT. Though conventional SPT methods try to link pairs of particles in the subsequent frames in a hard manner, such hard linking inevitably leads to erroneous pairing if no other information to distinguish particles is available. We have derived a probabilistic model by explicitly considering a Brownian particle surrounded by indistinguishable particles in a mean field approximation. Since our probabilistic model allows us to estimate diffusion constants without relying on particular hard-linked trajectories, it performs well even in the cases with higher particle density or higher diffusion speed, where standard SPT methods underestimate the diffusion constant. Since particle density is difficult to control in real experiments, this is advantageous in practical usage.

We have also provided a generalization of our algorithm to multiple diffusive states. This generalization was the key to address the case with false detections, since disappearing particles behave like particles with the additional diffusive state whose diffusion constant is infinity. Thus, in practice, one is recommended to examine both models with and without a fraction of disappearing particles, and select a model by comparing a statistical indicator like the Akaike Information Criterion [30].

In addition to high prediction accuracy, one of the advantages of PNN is its applicability beyond a uniform particle distribution. This has been the limitation on PICS, another existing SPT-free algorithm. We have demonstrated that, with or without knowledge of the underlying distribution, our algorithm accurately estimates diffusion constants even for the cases where PICS cannot be properly applied. In general, without prior knowledge of the underlying particle distribution, the actual performance of diffusion constant estimation also depends upon the accuracy of the estimation of the underlying particle distribution from the data, though the investigation of optimal density estimation itself is beyond the scope of this paper.

Since PNN considers each particle separately, it allows us to obtain detailed information about each particle. With the help of the EM algorithm, PNN estimates the probability that each particle is in each state. This kind of information, combined with their spatial distribution, can be used for providing further insights into the underlying biology, as briefly demonstrated in Fig. 7.

Another advantage of the proposed method, which is not apparent from the above benchmark results, is the small number of hyperparameters to be determined before analyses. For example, SPT based methods typically have a hyperparameter corresponding to the maximum distance parameter, which specifies the possible maximum displacement of diffusing particles to avoid connections of completely irrelevant particles. As mentioned in Result section, the estimated diffusion constants tend to largely depend on the choice of such a hyperparameter especially when particle density or diffusion speed is higher. PICS also requires several number of hyperparameters to perform a fitting to the empirical cumulative correlation functions [16], including the bin size and the range of consideration, whose optimal values may depend on data. On the other hand, PNN under a homogeneous particle distribution effectively has only a single hyperparameter, the margin to define the range of interest of the preceding time frames compared to the subsequent time frames, which also needed for PICS in addition to the ones mentioned above. We have confirmed that PNN has very weak dependency on the margin parameter as expected from the construction of the algorithm (data not shown). In the case of PNN under inhomogeneous particle distribution, the number of hyperparameters may vary depending on the chosen method of particle distribution estimation. In fact, in combination with the k nearest neighbor estimation of particle distribution we utilized in this paper, no hyperparameter, even the margin parameter, is required. This nature of small number of hyperparameters in PNN is very convenient in practice, since, otherwise, many trials and errors are needed to optimize hyperparameters. In particular, when the absolute value of estimated parameters is of concern, it is not a trivial matter to choose such hyperparameters objectively.

Finally, we would like to emphasize the complementary role of diffusion constant estimation methods. First of all, all of the methods we examined in this paper are based on the assumption that identification of single molecules from the image data is more or less possible. If the particle density is too high, the resultant images cannot have the resolution of single molecular imaging. In this extreme case, other methods without relying on particle detection at all, like image correlation microscopy [17–20], would be preferable. Or, if one can use specially designed experimental equipments, some other choices of methods to finely estimate diffusion constants like [26, 27] are available. In the case of the typical time lapse images of single molecules [3, 4, 12–15] which we considered in this paper, we have demonstrated that PNN is favorable than PICS, in terms of the accuracy, the applicability to inhomogeneous distribution and the convenience of the analysis due to the small number of hyperparameters. However, PICS analysis also has an advantage that the analysis is more graphical than PNN and one might be able to address the validity of the model by a visual inspection as far as the underlying spatial distribution of the particle is uniform. In turn, though canonical SPT methods tend to underestimate the diffusion constant and largely depend on hyperparameters under higher particle distribution, they allow one to analyze individual trajectories which may provide otherwise inaccessible information of each trajectory. In this sense, these methods can be utilized in combination. For example, one may first apply PNN to robustly estimate diffusion constants. This information of diffusion constants in turn might be utilized to determine the hyperparameters of SPT methods to minimize the linking error of SPT. Then, the resultant trajectories may be utilized, not to estimate the diffusion constants any more, but to extract other biologically interesting parameters which PNN cannot infer. Thus, having different diffusion estimation algorithms enlarges our freedom to analyze data, and increases the chance of obtaining biologically meaningful information from various single molecular time course datasets. In this regard, our algorithm opens a new window for accessing diffusion constants, in particular, in the regime where the particle density becomes comparable to the effective scale of diffusion.

Additional files

Additional file 1: Figure S1. Comparison of the performance of different algorithms in uniform distributions with lower particle densities. Box plots summarizing the comparison of the algorithms as in Fig. 3. The x axis is the particle density and the y axis is the estimated diffusion constant. The red line indicates the true diffusion constant. A, local SPT. B, global SPT. C, PICS and D, PNN. (PNG 105 kb)

Additional file 2: Figure S2. Comparison of the performance of PICS and PNN in a circular distribution. A, a representative snapshot of the particle distribution. B, C, and D, box plots summarizing the comparison between PICS and PNN under a circular distribution. B, PICS. C, PNN, where the known particle density distribution for the simulation is used for the diffusion constant estimation. D, PNN where the particle density distribution is estimated from the data using the k nearest neighbor algorithm. The x axis is the mean particle density over the area of interest, and the y axis is the estimated diffusion constant. The red line indicates the true diffusion constant. (PNG 171 kb)

Additional file 3: Figure S3. Comparison of the performance of PICS and PNN in Gaussian mixture distributions. A, a representative snapshot of the particle distribution. The red crosses represent centers of three Gaussian distributions. B, C, and D, box plots summarizing the comparison between PICS and PNN under a Gaussian mixture distribution. B, PICS. C, PNN, where the known particle density distribution for the simulation is used for the diffusion constant estimation. D, PNN where the particle density distribution is estimated from the data using the k nearest neighbor algorithm. The x axis is the mean particle density over the area of interest, and the y axis is the estimated diffusion constant. The red line indicates the true diffusion constant. (PNG 306 kb)

Additional file 4: Movie S1. A representative movie of the image based simulation. A representative movie generated by the plugin, ISBI Challenge Track Generator, of an open platform software ICY. Seed = 123,456, SNR = 4, sequence length = 10, particle density = 1000, sigma = 10 in the particle motion with creator type "BROWNIAN_UNIFORM". The other parameters are set to default, which means the extinction rate of each particle is 0.05. (TIFF 2561 kb)

Additional file 5: Figure S4. Representative images of the image based simulation. Representative images generated by the plugin, ISBI Challenge Track Generator, of an open platform software ICY. Seed = 123,456, SNR = 4, sequence length = 10, sigma = 10 in the particle motion with creator type "BROWNIAN_UNIFORM". Particle densities are 100, 500 and 1000, respectively. The other parameters are set to default. (PNG 517 kb)

Additional file 6: Movie S2. TIRF microscopic single molecule video image of Lyn_{11}-Halotag in HeLa cells. Membrane-localized single Lyn_{11}-Halotag protein molecules in a HeLa cell were observed by a TIRF microscope as described in Methods. (AVI 2796 kb)

Acknowledgements
The authors thank M. Ueda and J. Kozuka for help in single-molecule imaging and valuable feedback for this manuscript, R. Mishima for help in preparation of plasmids, A. Yoshimura, E. Kurumatani, and Y. Kimura for help in cell culture, M. Ogawa for secretarial assistance. S. T. thanks Y. Miyanaga for letting us know about the work by Semrau and Schmidt. The authors would like to thank Enago (https://www.enago.com/) for the English language review.

Funding
This work has been partially supported by a grant-in-aid from the SENTAN program of the Japan Agency for Medical Research and Development (AMED), a combined research grant provided by IFReC, a research grant from The Uehara Memorial Foundation, and JSPS Grant-in-Aid for Young Scientists (B) (25870396). Funding for the publication of this article was provided by a research grant from The Uehara Memorial Foundation.

Availability of data and materials
All data generated or analyzed during this study, all the codes used, and all the experimental materials used are available upon request.

About this supplement
This article has been published as part of *BMC Systems Biology* Volume 12 Supplement 1, 2018: Selected articles from the 16th Asia Pacific Bioinformatics Conference (APBC 2018): systems biology. The full contents of the supplement are available online at https://bmcsystbiol.biomedcentral.com/articles/supplements/volume-12-supplement-1 .

Authors' contributions
ST and YK conceived and designed the study. ST developed and implemented the algorithm. YK performed the imaging experiment. ST and YK wrote the paper. Both of the authors have read and approve the final manuscript.

Ethics approval and consent to participate
Not applicable.

Consent for publication
Not applicable.

Competing interests
The authors declare that they have no competing interests.

Publisher's Note

Published: 11 April 2018

References

1. Liu Z, Lavis LD, Betzig E. Imaging live-cell dynamics and structure at the single-molecule level. Mol Cell Elsevier Inc. 2015;58:644–59.
2. Axelrod D. Total internal reflection fluorescence microscopy in cell biology. Traffic. 2001;2:764–74.
3. Sako Y, Minoghchi S, Yanagida T. Single-molecule imaging of EGFR signalling on the surface of living cells. Nat Cell Biol. 2000;2:168–72.
4. Ueda M, Sako Y, Tanaka T, Devreotes P, Yanagida T. Single-molecule analysis of chemotactic signaling in Dictyostelium cells. Science. 2001;294:864–7.
5. Saxton MJ, Jacobson K. Single-particle tracking: applications to membrane dynamics. Annu Rev Biophys Biomol Struct. 1997;26:373–99.
6. Gelles J, Schnapp BJ, Sheetz MP. Tracking kinesin-driven movements with nanometre-scale precision. Nature. 1988;331:450–3.
7. Reid DB. An algorithm for tracking multiple targets. Autom. Control. IEEE. Trans. 1979;24:843–54.
8. Jaqaman K, Loerke D, Mettlen M, Kuwata H, Grinstein S, Schmid SL, et al. Robust single-particle tracking in live-cell time-lapse sequences. Nat Methods. 2008;5:695–702.
9. Selvin P, Ha T. Single-molecule techniques: a laboratory manual. Selvin P. & Ha T, editor. Cold Spring Harbor Laboratory Press; 2008.
10. Sako Y, Ueda M. Cell signaling reactions: single-molecular kinetic analysis. In: Sako Y, Ueda M, editors. Signal Transduct: Springer; 2011. http://www.springer.com/la/book/9789048198634.

11. Vestergaard CL, Blainey PC, Flyvbjerg H. Optimal estimation of diffusion coefficients from single-particle trajectories. Phys Rev E - Stat Nonlinear, Soft Matter Phys. 2014;89:022726.
12. Morimatsu M, Takagi H, Ota KG, Iwamoto R, Yanagida T, Sako Y. Multiple-state reactions between the epidermal growth factor receptor and Grb2 as observed by using single-molecule analysis. Proc Natl Acad Sci U S A. 2007;104:18013–8.
13. Low-Nam ST, Lidke KA, Cutler PJ, Roovers RC, van PMP B e H, Wilson BS, et al. ErbB1 dimerization is promoted by domain co-confinement and stabilized by ligand binding. Nat Struct Mol Biol Nature Publishing Group. 2011;18:1244–9.
14. Hibino K, Watanabe TM, Kozuka J, Iwane AH, Okada T, Kataoka T, et al. Single- and multiple-molecule dynamics of the signaling from H-Ras to cRaf-1 visualized on the plasma membrane of living cells. ChemPhysChem. 2003;4:748–53.
15. Matsuoka S, Shibata T, Ueda M. Asymmetric PTEN distribution regulated by spatial heterogeneity in membrane-binding state transitions. PLoS Comput Biol. 2013;9:e1002862.
16. Semrau S, Schmidt T. Particle image correlation spectroscopy (PICS): retrieving nanometer-scale correlations from high-density single-molecule position data. Biophys. J. Elsevier. 2007;92:613–21.
17. Hebert B, Costantino S, Wiseman PW. Spatiotemporal image correlation spectroscopy (STICS) theory, verification, and application to protein velocity mapping in living CHO cells. Biophys. J. Elsevier. 2005;88:3601–14.
18. Kolin DL, Costantino S, Wiseman PW. Sampling effects, noise, and photobleaching in temporal image correlation spectroscopy. Biophys J Elsevier. 2006;90:628–39.
19. Kolin DL, Wiseman PW. Advances in image correlation spectroscopy: measuring number densities, aggregation states, and dynamics of fluorescently labeled macromolecules in cells. Cell Biochem Biophys. 2007;49:141–64.
20. Pandžić E, Rossy J, Gaus K. Tracking molecular dynamics without tracking: image correlation of photo-activation microscopy. Methods Appl Fluoresc. 2015;3:14006.
21. Dempster APA, Laird NMN, DDB R. Maximum likelihood from incomplete data via the EM algorithm. J R Stat Soc Ser B Methodol. 1977;39:1–38.
22. Van Kampen NG. Stochastic processes in physics and chemistry. Third ed: North-holl. Pers. Libr. Elsevier. 2007. https://www.elsevier.com/books/stochastic-processes-in-physics-and-chemistry/van-kampen/978-0-444-52965-7.
23. Bilmes JA. A gentle tutorial of the EM algorithm and its application to parameter estimation for Gaussian mixture and hidden Markov models. 1997;
24. Chenouard N, Smal I, de Chaumont F, Maška M, Sbalzarini IF, Gong Y, et al. Objective comparison of particle tracking methods. Nat Methods. 2014;11:281–9.
25. de Chaumont F, Dallongeville S, Chenouard N, Hervé N, Pop S, Provoost T, et al. Icy: an open bioimage informatics platform for extended reproducible research. Nat Methods. 2012;9:690–6.
26. Di Rienzo C, Piazza V, Gratton E, Beltram F, Cardarelli F. Probing short-range protein Brownian motion in the cytoplasm of living cells. Nat Commun Nature Publishing Group. 2014;5:5891.
27. Digman MA, Gratton E. Imaging barriers to diffusion by pair correlation functions. Biophys. J. Biophysical. Society. 2009;97:665–73.
28. Sbalzarini I, Koumoutsakos P. Feature point tracking and trajectory analysis for video imaging in cell biology. J Struct Biol. 2005;151:182–95.
29. Suzuki KGN, Fujiwara TK, Sanematsu F, Iino R, Edidin M, Kusumi A. GPI-anchored receptor clusters transiently recruit Lyn and G alpha for temporary cluster immobilization and Lyn activation: single-molecule tracking study 1. J Cell Biol. 2007;177:717–30.
30. Akaike H. Information theory and an extension of themaximum likelihood principle. 2nd Int Symp Inf Theory, Akademinai Kiado. 1973:267–81.

Mori et al. BMC Systems Biology 2018, 12(Suppl 1):37
https://doi.org/10.1186/s12918-018-0527-4

BMC Systems Biology

RESEARCH Open Access

ncRNA-disease association prediction based on sequence information and tripartite network

Takuya Mori[1†], Hayliang Ngouv[2†], Morihiro Hayashida[3], Tatsuya Akutsu[2] and Jose C. Nacher[1*]

From The Sixteenth Asia Pacific Bioinformatics Conference
Yokohama, Japan. 15-17 January 2018

Abstract

Background: Current technology has demonstrated that mutation and deregulation of non-coding RNAs (ncRNAs) are associated with diverse human diseases and important biological processes. Therefore, developing a novel computational method for predicting potential ncRNA-disease associations could benefit pathologists in understanding the correlation between ncRNAs and disease diagnosis, treatment, and prevention. However, only a few studies have investigated these associations in pathogenesis.

Results: This study utilizes a disease-target-ncRNA tripartite network, and computes prediction scores between each disease-ncRNA pair by integrating biological information derived from pairwise similarity based upon sequence expressions with weights obtained from a multi-layer resource allocation technique. Our proposed algorithm was evaluated based on a 5-fold-cross-validation with optimal kernel parameter tuning. In addition, we achieved an average AUC that varies from 0.75 without link cut to 0.57 with link cut methods, which outperforms a previous method using the same evaluation methodology. Furthermore, the algorithm predicted 23 ncRNA-disease associations supported by other independent biological experimental studies.

Conclusions: Taken together, these results demonstrate the capability and accuracy of predicting further biological significant associations between ncRNAs and diseases and highlight the importance of adding biological sequence information to enhance predictions.

Keywords: ncRNA-disease association predictions, Tripartite network, Resource allocation

Background

Recent studies have investigated the biological functions, transcriptome, and regulation of non-coding RNAs (ncRNAs) of all sizes in a wide range of organisms [1]. siRNA (short interfering RNA), miRNA (microRNA) and piRNA (piwi-interacting RNA) are the three main types of short ncRNAs (less than 30 nucleotides). They play an important role in histone modification, gene silencing, heterochromatin formation, and DNA methylation, targeting at the transcriptional and post-transcriptional levels. Long non-coding RNAs (lncRNAs), which are greater than 200 nucleotides, are relevant in fundamental processes of gene regulation, such as chromatin modification and transcriptional regulation [2]. Studies indicate that lncRNAs can be categorized in one or more of the four archetypes, which include signal archetype (molecular signal or transcriptional activity indicator), decoy archetype (measure and adjust the balance of RNA regulation), guide archetype (directs the localization of ribonucleoprotein complexes to their targets), and scaffold archetype (a structural role for relevant proteins or RNAs to resemble). However, the main functions, structures, and mechanisms of lncRNAs remain unknown [3].

Because of their categorization into the four archetypes, ncRNAs have been proposed to have strong

* Correspondence: nacher@is.sci.toho-u.ac.jp
†Equal contributors
[1]Department of Information Science, Toho University, Miyama 2-2-1, Funabashi, Chiba 274-8510, Japan
Full list of author information is available at the end of the article

connections with the development and pathophysiology of diseases. Computational and experimental studies have demonstrated that alteration and deregulation of both short ncRNAs and lncRNAs at the structural and expression levels, can cause various types of cancer, such as breast cancer, leukemia, hepatocellular, and colon cancer, as well as neurodegenerative disorders, cardiovascular diseases, and immune-mediated diseases [4].

X-inactive specific transcript (Xist) is a lncRNA located on the X chromosome of the placental mammals that plays an important role in the X inactivation process. Lee et al. [5] demonstrated that Xist is a potent suppressor of hematological cancer in mice. Xist is required for hematopoietic stem cell survival and function. Therefore, Xist deletion results in leukemia, marrow fibrosis, and histiocytic sarcoma.

Yang et al. [6] also demonstrated a strong correlation between lncRNAs in tumor tissues and hepatocellular carcinoma (HCC). According to their experimental results, Cox regression analysis showed that both lncRNAs H19 and UCA1 were serious risk factors for HCC. Furthermore, logistic regression also indicated that H19 was overexpressed in hepatitis B virus-infected individuals.

A study on lncRNA PRINS (Psoriasis susceptibility-related RNA Gene Induced by Stress) provides further evidence on an association between ncRNAs and diseases. PRINS is transcribed by RNA polymerase II and expressed at various levels of human tissues [7]. PRINS is believed to play a role in susceptibility to psoriasis and has been linked to autoimmune diseases [8].

These findings are representative of the small number of ncRNA-disease connections that have been functionally established. This poses a major obstacle for biophysicians in their quest to formulate new hypotheses for molecular mechanisms underlying complexes diseases, and to enhance the efficacy and efficiency of disease diagnosis and treatment. For the purpose of determining an association, bio-physicians must partition patients into appropriate groups and accurately investigate them. This method indeed establishes the correlation and expands the acknowledgement of the association in various phenomena. Although it is necessary to predict and infer ncRNA-disease associations, it is an incredible cost and time burden for bio-physicians.

To address this problem, some computational models have been proposed for ncRNA-disease association inference. Chen et al. [9] introduced Laplacian Regularized Least Squares for LncRNA-Disease Association (LRLSLDA), a semi-supervised learning method based on the framework of Laplacian Regularized Least Square and the assumption that ncRNAs with similar functions tend to interact with similar diseases. LRLSLDA can predict ncRNA-disease associations reliably. Yet, there are obstacles with parameter selection and classifier combinations. Another method called RWRLncD has been proposed by Sun et al. [10] that infers ncRNA-disease associations by integrating ncRNAs functional similarity network, disease similarity network, and known ncRNA-disease associations. RWRLncD cannot be employed without any known ncRNA-disease associations. Li et al. [11] introduced a genomic location-based computational method for ncRNA-disease association prediction. However, the approach has not been evaluated with statistical tests, and not all ncRNAs were associated with their neighbor genes, a major limitation of the method. Yang et al. [12] introduced a method employing a bipartite network with resource-allocation technique to infer new ncRNA-disease connections. Their experiment to validate the performance of their method mainly focused on only one dataset collected from Chen et al. (2013), with roughly 1028 interactions between 322 ncRNAs and 221 diseases. They also included additional interactions via deep literature mining. Alaimo et al. then proposed a new ncRNA-disease association prediction method called ncPred, which was shown to outperform Yang et al.'s method [13]. ncPred is a resource-propagation-based method applied on an ncRNA-target-disease tripartite network. The tripartite network, formed up by ncRNA-target and target-disease interaction bipartite networks, guided the resource transferring process to infer new associations. Targets refer to a group of genes, microRNAs or proteins that are related to particular ncRNAs in terms of expression regulation and binding activities etc. With inclusion of the targets, ncPred was experimentally shown to infer more biological information with higher reliability than Yang et al.'s method. Yet, there is still some room for improvement, particularly with regards to establishing the biological significance of identified associations. Chen et al. [14] developed hypergeometric distribution for lncRNA-disease association inference (HGLDA) to predict lncRNA-disease associations by integrating miRNA-disease interactions and lncRNA-miRNA associations. HGLDA applied p-value matrices obtained from interaction networks, with false discovery rate (FDR) correction. lncRNA-disease pairs with FDR less than 0.05 were predicted to be potential lncRNA-disease associations. However, HGLDA provided the least biological information, which led to weaker biological significance reliability.

Here, we introduce a method that improves on Alaimo et al.'s ncPred. The proposed method integrates a resource-allocation technique in the ncRNA-target-disease tripartite network, with pairwise similarity information obtained from sequence expressions. The prior knowledge combined with the biological information is carried and transferred throughout the network, and finally utilized to infer new associations.

To validate the performance of our proposed method, we conducted a 5-fold-cross-validation procedure to demonstrate its reliability, accuracy and efficiency, on a dataset reconstructed from Chen et al. (2013) [15]. In addition, by using a database of experimentally confirmed interactions between ncRNAs and miRNAs shown in Helwak et al. [16], we performed another validation of our proposed method. Our results demonstrate that our proposed method outperforms ncPred.

Methods

Data sources

In order to evaluate the performance of the approach, we prepared two kinds of data, ncRNA-target interaction matrix (LncRNADisease database (2015)) and target-disease interaction matrix (DisGeNET database) which are the same database sources used in Chen et al. (2013) [15], to form the tripartite network, as well as the sequence expression of those targets and ncRNAs (extracted from Uniprot and LncRNADisease databases, respectively). Since there are ncRNA sequence expressions and targets that are still unknown, we could only collect 76 ncRNAs, 109 targets, and 514 diseases (see Table 1). In Fig. 1, the degree distribution of the resulting network is shown. The results indicate that the entire network may follow a power law distribution. Moreover, we collected experimentally confirmed interactions between ncRNAs and miRNAs from the database shown in [16] and from Alaimo [13] Supplementary Information files and reconstructed another dataset composed on 151 ncNRAs, 179 targets and 134 diseases (see Helwak dataset in Table 1).

Computational approach

In our approach, disease-target and target-ncRNA interaction matrices are required to construct the tripartite network (See Fig. 2). Sequence expressions of both targets and ncRNAs are also needed to generate string-kernel features for the prediction method. The overall algorithm of our method is shown in Fig. 3.

Let $D = \{d_1, d_2, ..., d_m\}$ be a set of diseases, let $T = \{t_1, t_2, ..., t_n\}$ be a set of targets which refer to genes or microRNA, and let $R = \{r_1, r_2, ..., r_p\}$ be a set of non-coding RNAs. Let $SR^l = \{sr_1^l, sr_2^l \quad , sr_p^l\}$ and let $ST^l = \{st_1^l, st_2^l, ..., st_n^l\}$ be sets of the sequence expressions of the targets and ncRNAs respectively.

Table 1 Description of the datasets

Type of Data	Chen et al	Helwak et al.
Diseases	514	134
Targets	109	179
ncRNAs	76	151
Disease-target interactions	580	1572
Target-ncRNA interactions	111	610
Average degree	1.977	9.405

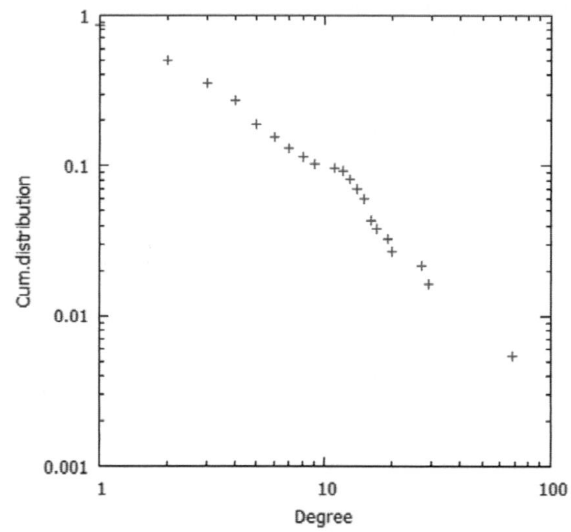

Fig. 1 Cumulative degree distribution of the tripartite ncRNA-target disease network for Chen et al. dataset

L-gram-string kernel feature extraction and standardization

In our approach, the features of sequence expressions were extracted via the *l-gram-string* kernel method [17], in which each string is transformed into a vector consisting of the number of occurrences of each substring of length l. In bioinformatics, string kernel is considered a function that measures the similarity of a pair of sequence expressions (strings) with finite length, for the purpose of generating real-value feature vectors. In our case, the method was evaluated experimentally using length $l = 1, 2, 3$ and 4.

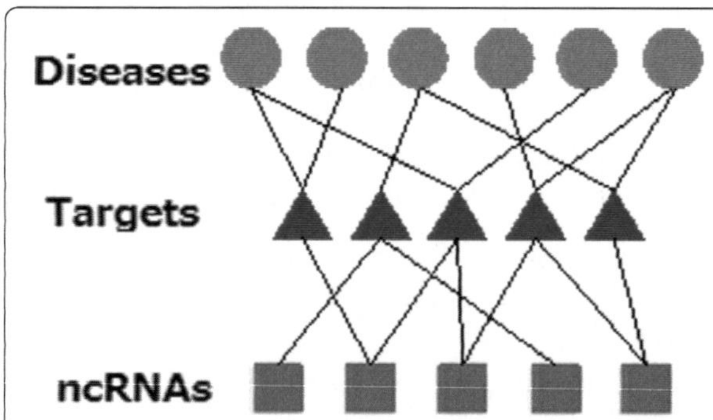

Fig. 2 Example of a tripartite network used in this work. Targets integrate information from ncRNAs and diseases

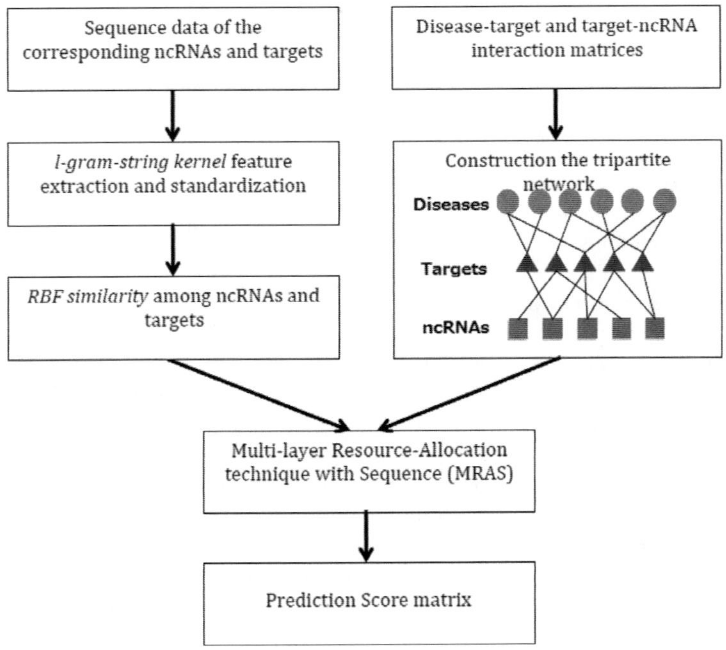

Fig. 3 Illustration of the proposed computational method. Sequence information analysis is combined with a tripartite network structure. A multi-layer resource-allocation technique integrates the information and predicts associations between ncRNAs and human diseases

In order to improve the quality and reduce the redundancy of the computed features from the previous step, the standardization method below was applied [18].

$$x' = \frac{x - \overline{x}}{\sigma},$$

where x is the original feature vector, $\overline{x}$ is the mean and σ is the standard deviation.

Thus, let $SST^l = \{sst_1^l, sst_2^l, ..., sst_n^l\}$ and $SSR^l = \{ssr_1^l, ssr_2^l, ..., ssr_p^l\}$ be sets of the normalized real-value feature vectors computed from the sequence expressions of the targets and ncRNAs respectively.

RBF kernel similarity among ncRNAs and targets

To effectively compute the similarity among each pair of ncRNA-ncRNA and target-target, the RBF kernel similarity technique was employed. In statistical learning, RBF kernel can be viewed as a common kernel function to measure the similarity [19].

Let $k\_rna\ (i,j)$ and $k\_target\ (i,j)$ be the RBF kernel similarity function of pair ncRNA-ncRNA and target-target respectively.

Thus, the RBF similarity of ith ncRNA and j^{th} ncRNA can be computed as

$$k\_rna\ (i,j) = \exp\left(-\gamma_1 \left\| ssr_i^l - ssr_j^l \right\|^2\right),$$

where $\gamma_1 = \frac{1}{2\sigma_1^2}$, σ_1 is the parameter $0 < \sigma_1 < 1$.

The RBF similarity of ith target and j^{th} target can be computed as

$$k\_target\ (i,j) = \exp\left(-\gamma_2 \left\| sst_i^l - sst_j^l \right\|^2\right),$$

where $\gamma_2 = \frac{1}{2\sigma_2^2}$, σ_2 is the parameter $0 < \sigma_2 < 1$.

Constructing the tripartite network disease-target-ncRNA

G (D, T, R, E), where *E* denotes a set of edges, is a tripartite network (Fig. 2) representing the disease-target interactions and target-ncRNA interactions. $A^{DT} = \{a_{ij}^{DT}\}_{m \times n}$ denotes the adjacency matrix of the disease-target network, and $A^{TR} = \{a_{ij}^{TR}\}_{n \times p}$ represents the adjacency matrix of target-ncRNA network.

Targets represent a group of biomolecules functionalized by non-coding RNAs. Targets can be genes, proteins, microRNAs, etc., whose activities include expression regulation, binding, and complex formation. In our method, targets function as a bridge allowing us to extract further biological information, for enhancing the accuracy of disease-ncRNA association inference.

Our method integrates the resource-allocation algorithm in the network with pairwise similarity information derived from sequence expressions to produce scores showing the level of certainty of the interaction. Essentially, the resource-allocation algorithm carries prior understanding of the bipartite network, which can be employed to predict the interactions.

Multi-layer resource-allocation technique with sequence (MRAS) information

Since the tripartite network consists of two bipartite networks, two-layer resource-allocation techniques were applied. For the first-layer (disease-target bipartite network), the resource will be transferred from the nodes in T (targets) to the nodes in D (diseases) integrating with pairwise similarity information between each pair of the diseases, then move back to the nodes in T. For the second-layer (target-ncRNA bipartite network), the resource integrated with pairwise similarity information between each pair of the ncRNAs, is allocated from the nodes in R (ncRNAs) to the nodes in T, and then combined with the resource from the previous layer. Finally, the weights computed within the two layers were merged into one, called combined weight (W_C), indicating the likelihood that in case a disease associates with a target t_i, it then possibly interacts with ncRNA r_j. Prediction scores (P) can then be computed from the combined weight W_C, and the higher the score, the greater the certainty that the ncRNA will associate with a particular disease.

Let deg(x) be the degree of node x in the disease-target network, and deg′(y) be the degree of node y in the target-ncRNA network.

- Layer 1: disease-target bipartite network
  Let $W^T = \{w_{ij}^T\}_{n\times n}$ be the probability that the ith target interacts with the jth target when both of them interact with the same disease:

$$w_{ij}^T = k\_target(i,j) \cdot \sum_{l=1}^{m} \frac{a_{li}^{DT} a_{lj}^{DT}}{deg(d_l)}$$

- Layer 2: target-ncRNA bipartite network
  Let $W^R = \{w_{ij}^R\}_{p\times p}$ be the probability that the ith ncRNA interacts with the jth ncRNA when both of them interact with the same target:

$$w_{ij}^R = k\_rna(i,j) \cdot \sum_{l=1}^{n} \frac{a_{li}^{TR} a_{lj}^{TR}}{\deg(t_l)}$$

Hence, the combined weight $W_C = \{w_{ij}^C\}_{n\times p}$ of the two bipartite networks together with the similar neighborhood of both targets and ncRNAs can be obtained as follows. The formula assigned more weight to the path with higher frequency.

$$w_{ij}^C = \sum_{t=1}^{n} \left[w_{it}^T \sum_{r=1}^{p} \left(a_{tr}^{TR} \cdot w_{rj}^R \right) \right]$$

Finally, the prediction score matrix can be computed based on the following formula: $P = \{p_{ij}\}_{m\times p} = A^{DT} \cdot W_C$

Based on the above procedures, the prediction score matrix was produced via the integration of biological similarities of sequence expressions and the resource-allocation technique in the tripartite network. The score matrix thus predicted associations between ncRNAs and diseases. The higher the score, the higher the certainty of ncRNA-disease connectivity.

Results

Algorithm prediction performance and evaluation

As mentioned, the performance of our method was measured by applying a 5-fold-cross-validation procedure as well as other computational operations. The evaluation algorithm is described as follows:

1) First, we consider all pairs of ncRNA and disease (N pairs) exiting in the tripartite network. These pairs are expected to be predicted by the algorithm in the score matrix P.
2) Fix a pair of values for σ_1 and σ_2 .
3) Consider k-fold cross validation with $k = 5$. Therefore, all N pairs are divided into 5 groups.
4) Each group of $N/5$ elements becomes a test data one time. The rest four groups are training data
5) From each test data group, we scan those pairs that are connected through targets and disconnect them (see Fig. 4b)
6) Run the algorithm and make predictions only for the pairs in the test group P_{test}.
7) Construct a ROC curve for the test group based on $N/5$ scores for P_{test}.
8) Repeat 3~ 7 for different folds. We average all five ROC curves.
9) Repeat 3~ 8 for randomization of test and training groups five more times and obtained an averaged ROC curve.
10) Repeat 2~ 9 for each σ_1 [0.1,0.2,...,0.9] x σ_2 [0.1,0.2,..,0.9] and pick up the best AUROC among all results.

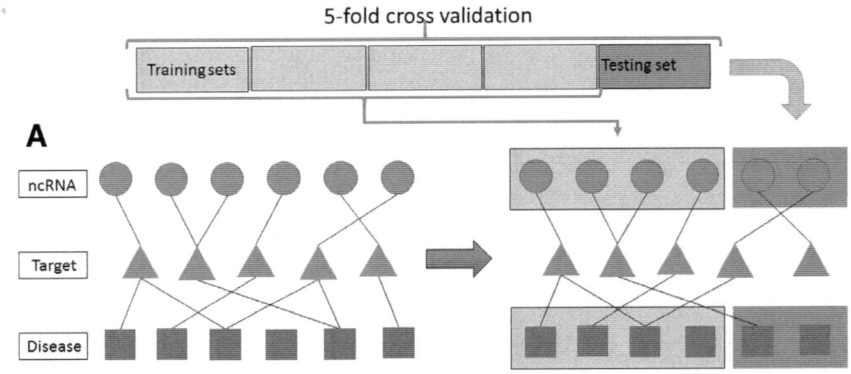

Fig. 4 Algorithm performance evaluation methodology using 5-fold cross validation. From the constructed network as shown in (**a**), we divide all ncRNA-disease pairs into 5 groups. Each time one group performs as a testing dataset. To evaluate whether the algorithm is able to predict the interacting pairs without memory of existing interactions, we delete links that connect ncRNAs with disease if they belong to the testing set as shown in (**b**)

Evaluation results

We took into account the area under receiver operating characteristic (ROC) curve (AUC) values to assess the reliability, credibility, and accuracy of the prediction method. Table 2 demonstrates that our proposed method clearly outperforms ncPred in terms of the average values of AUC for the analysed datasets as shown in Table 1. These results emphasize the improved reliability and accuracy of predicting new ncRNA-disease associations or biological relevance using our approach, compared to that of Alaimo et al. which uses ncPred.

In our approach, the RBF kernel technique was selected to conduct the similarity measure. In order to prove that RBF kernel can interpret the similarity of biological significance more appropriately, we conducted comparison experiments on the dataset listed in Table 1 with three representative kernel functions: linear kernel, polynomial (degree = 2) kernel and RBF kernel. Table 3 clearly illustrates the prediction performance of our approach underlying each kernel function. The results demonstrate that RBF kernel successfully outperforms the other two kernel techniques, showing the highest average AUC values (see Fig. 5b).

The *l-gram-string kernel* technique was applied to extract biological features from the sequence data of ncRNAs and targets in the proposed method. In order to investigate which length of l can interpret and extract better information from the sequence data, we validated the performances underlying different lengths of l. We also compared the results of the proposed method under two assumptions. By using the cut link method described in algorithm performance evaluation section and without cut link. Table 4 shows the comparison for both methods and for *l-gram-string kernel* (when $l = 1, 2, 3$ and 4) in terms of AUC values using the dataset described in Table 1. The results show that link cut notably decreases the prediction of the algorithm as expected from 0.75 to 0.57 (see Fig. 6a and b). However, even though we perform this strict procedure, the algorithm is able to predict correctly interactions with an AUC above 0.5. The results also show that performance slightly improved at $l = 3$. Thus, $l = 3$ is the optimal parameter for the method (see Table 4 and Fig. 6b). Note that for Helwak dataset the optimal length is $l = 1$.

For parameter tuning of RBF kernel similarity function σ_1 and σ_2 we selected the optimal parameters as explained in the evaluation method procedure. Table 5 shows the optimal values identified in our computation.

Case studies

In order to demonstrate the credibility and functionality of the proposed method, we compared predicted disease-ncRNA pairs using Chen et al. dataset with recent experimental studies, as shown in Table 6, on six diseases, namely lung cancer, liver cancer, breast cancer,

Table 2 Comparison of the proposed method and ncPred through averaged area under ROC curve (ROC)

AUC		
	Proposed method	ncPred
Chen et al.(2013)	*0.57135*	0.50242
Helwak et al.(2013)	*0.77444*	0.50295

Italic numbers indicate the best performance

Table 3 Comparison of performance using three representative kernel functions

Dataset	AUC		
	Linear Kernel	Polynomial Kernel (d = 2)	RBF Kernel
Chen et al. (2013)	0.51561	0.54175	**0.57135**

Performance results for the following three kernel functions: linear kernel, polynomial (degree = 2) kernel and RBF kernel. The length of *l-gram* is set to $l = 3$ in this experiment

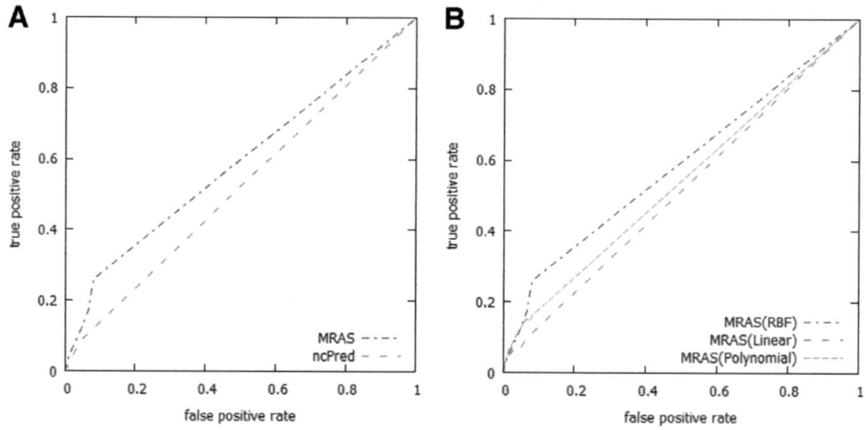

Fig. 5 Comparison between the proposed method (MRAS) and ncPred. **a** shows an ROC curves comparison between the proposed method (MRAS) and ncPred with evaluation of the Chen et al. dataset. The ROC curves were drawn up based on the average results of the simulation, and repeated to ensure reliable estimates as described in the evaluation method procedure. The results also convey that the proposed method produced a higher true positive rate, which demonstrates its superior performance over its competitor. **b** shows the results of the proposed method using different kernel functions

urinary bladder neoplasms, prostatic neoplasms and stomach neoplasms.

Lung cancer

Lung Cancer (LC) is one of the most deadly diseases affecting both men and women worldwide. There are three types of lung cancer: non-small cell lung cancer, small cell lung cancer, and lung carcinoid tumor. In 2015 lung and bronchus cancer accounted for 13.3% of all cancer cases in the US. The proposed method predicted that seven ncRNAs, H19, HOTAIR, MALAT1, PVT1, WRAP53, XIST, CDKN2B-AS1 are associated with LC. Kondo et al. reported that the alteration and deregulation of H19 impacts lung cancer cell growth [20]. HOX transcript antisense RNA (HOTAIR) prevents gene expression via collection of chromatin modifiers. Loewen et al. demonstrated that HOTAIR plays an important role in the intervention of lung cancer [21]. lncRNA metastasis associated lung adenocarcinoma transcript 1 (MALAT1) can impair in vitro cell motility of lung cancer cells and simultaneously influence a number of genes (Tano et al. & Tseng et al. [22, 23]). Yang et al. reported that an increased expression of the lncRNA PVT1 promotes tumorigenesis in non-small cell lung cancer [24]. Tantai et al. suggested a combined identification of long non-coding RNA XIST and HIF1A-AS1 in serum as an effective screening for non-small cell lung cancer [25]. Park et al., observed evidences for lung cancer with two variants located in cancer pleiotropic regions, namely *TERT* and risk of lung adenocarcinoma and *CDKN2BAS1* with risk of lung squamous cell carcinoma [26].

Table 4 Comparison of *l-gram-string kernel* in terms of average Area Under the Curve (AUC) values

		l = 1	*l* = 2	*l* = 3	*l* = 4
Chen et al.(2013)	**AUC**	0.7507	0.7506	*0.7513*	0.7433
	Cut link AUC	0.5706	0.5644	*0.5713*	0.5476
Helwak et al.(2013)	**AUC**	*0.8543*	0.7855	0.7338	0.7080
	Cut link AUC	*0.7744*	0.7098	0.6430	0.6441

The performance of the *l-gram-string kernel* for *l* = 1, 2, 3 and 4 was examined using the AUC metric. The computation was done using the RBF kernel. Table also shows the prediction performance when the links between a disease and their associated ncRNAs are deleted (Cut link UAC). Italic numbers indicate the best performance

Liver cancer

Liver cancer is the third most deadly cancer worldwide [27]. Very few patients receive curative treatments, while the majority do not recover since they are diagnosed at later stages. Our method predicted that ncRNAs H19, PCNA-AS1 and WRAP51 are correlated with liver cancer. H19 ncRNA expression was shown to result in high H19 protein expression in liver cancer whenever there is a loss of imprinting [28]. Iizuka et al. investigated further the epigenetic abnormalities in the insulin-like growth factor 2 (IGF2) and H19 genes observed in hepatocellular carcinoma (HCC) [29]. Yuan et al. reported that antisense long non-coding RNA PCNA-AS1 promotes tumor growth in hepatocellular carcinoma [30]. Several studies have also linked WRAP51 with HCC [31].

Breast cancer

Breast cancer is a ductal carcinoma beginning in the cells of lobules, ducts and other tissues of the breast. In the United States, breast cancer is the second most common cancer, just after skin cancer. Both women and

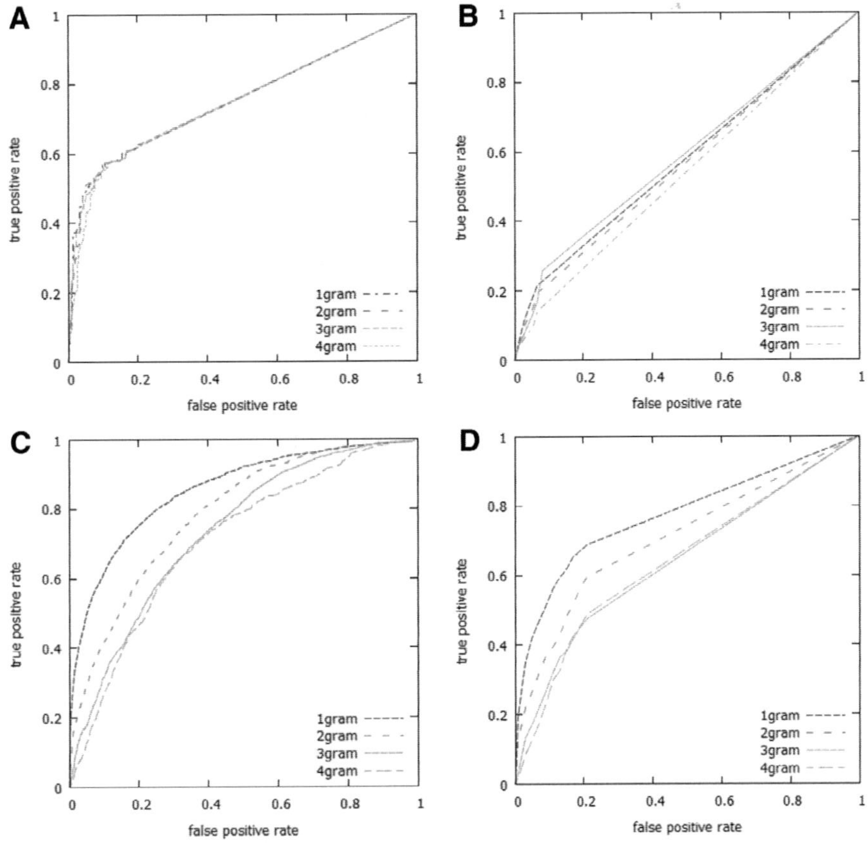

Fig. 6 Illustration of the ROC curves for the proposed algorithm. ROC curves for the proposed algorithm MRAS using different values of *l*-gram string (l = 1, 2, 3 and 4) and using RBF kernel (**a**) without link cut (AUC = 0.75) and (**b**) with link cut (AUC = 0.57) for the Chen dataset. The results for the Helwak dataset using RBF kernel (**c**) without link cut (AUC = 0.85) and (**d**) with link cut (AUC = 0.77)

men can suffer from this severe disease. Our method predicted that four ncRNAs, H19, PVT1, SRA1 and WRAP53 may be associated with this disease. Berteaux et al. found out that H19 transcript antisense RNA stabilize breast cancer cells and is overexpressed in breast tumors [32]. Others evidences were found across literature [33, 34]. Zhang et al. reported that the non-protein coding plasmacytoma variant translocation 1 (PVT1) has been implicated in human cancers [35, 36]. In addition, Hube et al. reported that alternative splicing of SRA1 could lead to the generation of coding and non-coding RNA isoforms in breast cancer cell lines [37]. Cao et al. recently reported about the association between the WRAP53 gene rs2287499 C > G polymorphism and cancer risk [38].

Table 5 Optimal values of σ_1 and σ_2 parameters

	σ_1	σ_2
Chen et al.(2013)	0.6	0.7
Helwak et al.(2013)	0.9	0.1

Urinary bladder Neoplasms

Urinary bladder neoplasms are the result of abnormal growth of bladder cells, and are considered as one of the common cancers. Men are at higher risk for this disease than women. The predictive results derived from our method inferred that ncRNAs H19 and WRAP53 might be associated with urinary bladder cancer. Luo et al. demonstrated that the abnormal expression of H19, particularly its up-regulation, contributed to cell proliferation in bladder neoplasms [39]. Wrap53 is a multi-functional gene which is capable of regulating p53 levels in both normal and cancer cell lines [40].

Prostatic Neoplasms

Prostatic neoplasm is caused by uncontrolled growth of cells located in the prostate causing tumors. Based upon our proposed method, prostate cancer was predicted to be associated with H19, MALAT1, CDKN2B-AS1, HOTAIR, PCAT1, SRA1, XIST. Zhu et al. demonstrated that H19 was significantly downregulated in the metastatic prostatic tumor cell line M12 [41]. Perez et al.

Table 6 The predicted ncRNAs related to the six diseases examined in this study

Disease	ncRNA	Evidence (PMID)
Breast Neoplasms	H19	15985428;27540977;9811352;24780616
Breast Neoplasms	PVT1	17908964;10485452;23907597
Breast Neoplasms	SRA1	17710122;26460974
Breast Neoplasms	WRAP53	27525856;15736456
Carcinoma, Hepatocellular	H19	9570364
Carcinoma, Hepatocellular	PCNA-AS1	24704293
Carcinoma, Hepatocellular	WRAP53	23836507
Lung Neoplasms	HOTAIR	25010625;22088988
Lung Neoplasms	MALAT1	23243023;26490983
Lung Neoplasms	PVT1	25400777
Lung Neoplasms	XIST	26339353
Prostatic Neoplasms	CDKN2B-AS1	27197191;24988946
Prostatic Neoplasms	H19	19513555;25895025
Prostatic Neoplasms	HOTAIR	26411689
Prostatic Neoplasms	MALAT1	23845456
Prostatic Neoplasms	PCAT1	21804560
Prostatic Neoplasms	SRA1	16607388
Prostatic Neoplasms	XIST	16261845
Stomach Neoplasms	CDKN2B-AS1	24810364;26187665
Stomach Neoplasms	HOTAIR	25063030;23898077
Stomach Neoplasms	MALAT1	24857172
Urinary Bladder Neoplasms	H19	23354591
Urinary Bladder Neoplasms	WRAP53	27912092

The predicted ncRNAs related to lung cancer, liver cancer, breast cancer, urinary bladder neoplasms, prostatic neoplasms and stomach neoplasms based upon our method. Each research article reporting on a specific ncRNA is shown with a unique ID (*PMID*) by PubMed

showed that the antisense intronic transcript of MALAT1 is correlated with tumor differentiation in prostate cancer [42]. Fehringer reported the involvement of CDKBN2B-AS1in Cross-cancer genome-wide analysis of lung, ovary, breast, prostate and colorectal cancer using a cross-cancer genome-wide analysis [43]. Zhang et al. discovered that LncRNA HOTAIR enhances the Androgen-Receptor-Mediated Transcriptional Program and Drives Castration-Resistant Prostate Cancer [44]. Ren et al. suggested the Long noncoding RNA MALAT-1 as a potential therapeutic target for castration resistant prostate cancer [45]. Presner et al. results implied that rhe Long Non-Coding RNA PCAT-1 Promotes Prostate Cancer Cell Proliferation through cMyc [46]. The same research group used a transcriptome sequencing across a prostate cancer cohort to identify PCAT-1 as an unannotated lincRNA implicated in disease progression [47]. Several works also reported on the involvement of SRA1 [48] and XIST [49] in prostatic neoplasm.

Stomach Neoplams

Stomach neoplams that occurs as a consequence of abnormal grothw of stomach cells have also been associated to ncRNAs in several works. Our method predicted CDKN2B-AS1, HOTAIR and MALAT1 as main correlated molecules with stomach neoplasms. Zhang et al., reported that ANRIL (CDKN2B-AS1) which recruits and binds to PRC2 is usually observed upregulated in human gastric cancer (GC) tissues [50]. Lee et al. found that long non-coding RNA HOTAIR tends to promote not only carcinogenesis but also invasion of gastric adenocarcinoma [51]. Wang et al. also observed that MALAT1 promotes cell proliferation in gastric cancer by recruiting SF2/ASF [52].

Discussion

Our work has improved the prediction quality and performance of Alaimo et al.'s method ncPred by integrating a biological feature information derived from sequence data with weight calculated by the multi-layer resource allocation technique, throughout the whole disease-target-ncRNA tripartite network.

The proposed method is unique because involves three types of biological information, namely ncRNAs sequences, target sequences and diseases. This rich biological information is also organized in a complex tripartite network in which targets integrate information from ncRNAs and diseases. Our study extended Alaimo's approach [13] because we integrated biological sequence information. The uniqueness of the datasets and the complexity of the networks structure make it difficult to perform a straightforward comparison of our predictions with other approaches besides ncPred algorithm [13]. Indeed, most of the previous works have been done using a more simple approach involving a bipartite network. For example, Yang el al [12] constructed a bipartite network composed of only lncRNAs and diseases. On this network, a propagation algorithm navigated to infer lncRNAs implicated in diseases. Alaimo et al. [13] have already shown that their method outperformed the method by Yang et al. [12].

Our computational analyses indicate that our approach could result in more reliable and biologically efficient disease-ncRNA associations prediction than ncPred (See Tables 2 and 6). The prediction scores were obtained based on significant biological information, which provided useful suggestions of which ncRNA-disease associations have stronger interactions. This paves an easy path for pathologists to analyze and interpret ncRNA-disease associations. However, in order to accurately determine and confirm the associations, suitable patients and document cases are still needed.

The proposed method still has limitations that need to be considered. For instance, it lacks ncRNA-target interaction data and their particular sequence data. Those sequence data are required to extract the feature vector, which could extend the computational time. Furthermore, the lengths of ncRNAs and targets vary, ranging from less than 100 nucleotides to more than 10,000 nucleotides. This impacts the ranking of predicted ncRNA-disease associations. Accordingly, more related datasets are required to improve and expand the reliability and quality of ncRNA-disease association inference.

Acknowledgments
JCN was partially supported by JSPS KAKENHI Grant Number JP25330351. MH was partially supported by JSPS KAKENHI Grant Number 16 K00392. TA was partially supported by JSPS KAKENHI Grant Numbers 26240034 and 26540125. This research was also partially supported by the Collaborative Research Program of the Institute for Chemical Research, Kyoto University.

Funding
Publication of this article was funded by Toho University's research budget.

Availability of data and materials
The datasets used in our study were downloaded from the publicly available databases mentioned in the text. The source code is available from the corresponding author on reasonable request.

About this supplement
This article has been published as part of *BMC Systems Biology* Volume 12 Supplement 1, 2018: Selected articles from the 16th Asia Pacific Bioinformatics Conference (APBC 2018): systems biology. The full contents of the supplement are available online at https://bmcsystbiol.biomedcentral.com/articles/supplements/volume-12-supplement-1 .

Authors' contributions
TM and JCN made available computer programs and performed rigorous computational experiments by designing the algorithmic evaluation. HN designed the method and performed preliminary computational experiments. TM, JCN and MH contributed to the data acquisition, data analysis and interpretation of data. TA supervised HN's work with giving various advices and comments. TM, HN and JCN wrote the draft. All of the authors have read and approved the final manuscript.

Ethics approval and consent to participate
Not applicable

Consent for publication
Not applicable

Competing interests
The authors declare that they have no competing interests.

Publisher's Note
Springer Nature remains neutral with regard to jurisdictional claims in published maps and institutional affiliations.

Author details
[1]Department of Information Science, Toho University, Miyama 2-2-1, Funabashi, Chiba 274-8510, Japan. [2]Bioinformatics Center, Institute for Chemical Research, Kyoto University, Gokasho, Uji, Kyoto 611-0011, Japan. [3]Department of Electrical Engineering, Matsue College of Technology, Matsue 690-8518, Japan.

Published: 11 April 2018

References
1. Mercer TR, Dinger ME, Mattick JS. Long non-coding RNAs: insights into functions. Nat Rev Genet. 2009;10:155–9.
2. Wang KC, Chang HY. Molecular mechanisms of long noncoding RNAs. Mol Cell. 2011;43(6):904–14.
3. Derrien T, Johnson R, Bussotti G, Tanzer A, Djebali S, Tilgner H, et al. The GENCODE v7 catalog of human long noncoding RNAs: analysis of their gene structure, evolution, and expression. Genome Res. 2012;22(9):1775–89.
4. Wapinski O, Chang HY. Long noncoding RNAs and human disease. Trends Cell Biol. 2011;21(6):354–61.
5. Lee JT, Bartolomei MS. X-inactivation, imprinting and long noncoding RNAs in health and disease. Cell. 2013;152:1308–23.
6. Yang Z, Lu Y, Xu Q, Tang B, Park C-K, Chen X. HULC and H19 played different roles in overall and disease-free survival from hepatocellular carcinoma after curative hepatectomy: a preliminary analysis from gene expression omnibus. Di Markers. 2015;2015:191029.
7. Scaria V, Pasha A. Long non-coding RNAs in infection biology. Front Genet. 2012;3:308.8.
8. Li J, Xuan Z, Liu C. Long non-coding RNAs and complex human diseases. Int J Mol Sci. 2013;14(9):18790–808. https://doi.org/10.3390/ijms140918790.
9. Chen X, Yan GY. Novel human lncRNA-disease association inference based onlncRNA expression profiles. Bioinformatics. 2013;29:2617–24.
10. Sun J, et al. Inferring novel lncRNA–disease associations based on a random walk model of a lncRNA functional similarity network. Mol BioSyst. 2014;10:2074–81.
11. Li J, et al. A bioinformatics method for predicting long noncoding RNAs associated with vascular disease. Sci China Life Sci. 2014;57:852–7.
12. Yang X, Gao L, Guo X, et al. A network based method for analysis of lncRNA-disease associations and prediction of lncRNAs implicated in diseases. PLoS One. 2014;9(1):e87797.
13. Alaimo S, Giugno R, Pulvirenti A. ncPred: ncRNA-disease association prediction through tripartite network-based inference. Front Bioeng Biotechnol. 2014;2:71.
14. Chen X. Predicting lncRNA-disease associations and constructing lncRNA functional similarity network based on the information of miRNA. Sci Rep. 2015;5:13186.
15. Chen G, Wang Z, Wang D, Qiu C, Liu M, Chen X, et al. LncRNADisease: a database for long-non-coding RNA-associated diseases. Nucleic Acids Res. 2013;41:D983–6.
16. Helwak A, Kudla G, Dudnakova T, Tollervey D. Mapping the human miRNA interactome by CLASH reveals frequent noncanonical binding. Cell. 2013; 153(3):654–65.
17. Sören S, Gunnar R, Konrad R, Bottou L, Chapelle O, DeCoste D, Weston J, editors. In: Large-Scale Kernel Machines, chap. 4. Cambridge: MIT Press; 2007. p. 73–104.
18. Ismail B, Dauda U. Standardization and its effects on k-means clustering algorithm. Res J Appl Sci Eng Technol. 2013;6(17):3299–303.
19. Vert JP, Tsuda K, Scholkopf B. A primer on kernel methods. Kernel Methods in Comput Biol. 2004;35–70:7.
20. Kondo M, Suzuki H, Ueda R, Osada H, Takagi K, Takahashi T, Takahashi T. Frequent loss of imprinting of the H19 gene is often associated with its overexpression in human lung cancers. Oncogene. 1995;10(6):1193–8.
21. Loewen G, Jayawickramarajah J, Zhuo Y, Shan B. Functions of lncRNA HOTAIR in lung cancer. J Hematol Oncol. 2014;7:90.
22. Tano K, Akimitsu N. Long non-coding RNAs in cancer progression. Front Genet. 2012;3:219.

23. Tseng JJ, Hsieh YT, Hsu SL, Chou MM. Metastasis associated lung adenocarcinoma transcript 1 is up-regulated in placenta previa increta/percreta and strongly associated with trophoblast-like cell invasion in vitro. Mol Hum Reprod. 2009;15(11):725–31.
24. Yang YR, Zang SZ, Zhong CL, Li YX, Zhao SS, Feng XJ. Increased expression of the lncRNA PVT1 promotes tumorigenesis in non-small cell lung cancer. Int J Clin Exp Pathol. 2014;7(10):6929–35.
25. Tantai J, Hu D, Yang Y, Geng J. Combined identification of long non-coding RNA XIST and HIF1A-AS1 in serum as an effective screening for non-small cell lung cancer. Int J Clin Exp Pathol. 2015;8(7):7887–95.
26. Park SL, et al. Pleiotropic associations of risk variants identified for other cancers with lung cancer risk: the PAGE and TRICL consortia. J Natl Cancer Inst. 2014;106(4):dju061.
27. Llovet JM, Lok A. Hepatitis B virus genotype and mutants: risk factors for hepatocellular carcinoma. JNCI J Natl Cancer Ins. 2008;100(16):1121–3.
28. Qi P, Du X. The long non-coding RNAs, a new cancer diagnostic and therapeutic gold mine. Mod Pathol. 2013;26:155–65.
29. Iizuka N, Oka M, Tamesa T, Hamamoto Y, Yamada-Okabe H. Imbalance in expression levels of insulin-like growth factor 2 and H19 transcripts linked to progression of hepatocellular carcinoma. Anticancer Res. 2004;24(6):4085–9.
30. Yuan SX, Tao QF, Wang J, Yang F, Liu L, Wang LL, Zhang J, Yang Y, Liu H, Wang F, Sun SH, Zhou WP. Antisense long non-coding RNA PCNA-AS1 promotes tumor growth by regulating proliferating cell nuclear antigen in hepatocellular carcinoma. Cancer Lett. 2014;349(1):87–94.
31. Ortiz-Cuaran S, Cox D, Villar S, Friesen MD, Durand G, Chabrier A, Khuhaprema T, Sangrajrang S, Ognjanovic S, Groopman JD, Hainaut P, Le Calvez-Kelm F. Association between TP53 R249S mutation and polymorphisms in TP53 intron 1 in hepatocellular carcinoma. Genes Chromosomes Cancer. 2013;52(10):912–9.
32. Berteaux N, Lottin S, Monté D, Pinte S, Quatannens B, Coll J, Hondermarck H, Curgy JJ, Dugimont T, Adriaenssens E. H19 mRNA-like noncoding RNA promotes breast cancer cell proliferation through positive control by E2F1. J Biol Chem. 2005;280(33):29625–36.
33. Zhang K, Luo Z, Zhang Y, Zhang L, Wu L, Liu L, Yang J, Song X, Liu J. Circulating lncRNA H19 in plasma as a novel biomarker for breast cancer. Cancer Biomark. 2016;17(2):187–94.
34. Adriaenssens E, Dumont L, Lottin S, Bolle D, Leprêtre A, Delobelle A, Bouali F, Dugimont T, Coll J, Curgy JJ. H19 overexpression in breast adenocarcinoma stromal cells is associated with tumor values and steroid receptor status but independent of p53 and Ki-67 expression. Am J Pathol. 1998;153(5):1597–607.
35. Zhang Z, Zhu Z, Zhang B, Li W, Li X, Wu X, Wang L, Fu L, Fu L, Dong JT. Frequent mutation of rs13281615 and its association with PVT1 expression and cell proliferation in breast cancer. J Genet Genomics. 2014;41(4):187–95.
36. Guan Y, Kuo WL, Stilwell JL, Takano H, Lapuk AV, Fridlyand J, Mao JH, Yu M, Miller MA, Santos JL, Kalloger SE, Carlson JW, Ginzinger DG, Celniker SE, Mills GB, Huntsman DG, Gray JW. Amplification of PVT1 contributes to the pathophysiology of ovarian and breast cancer. Clin Cancer Res. 2007;13(19):5745–55.
37. Hube F, Guo J, Chooniedass-Kothari S, Cooper C, Hamedani MK, Dibrov AA. Alternative splicing of the first intron of the steroid receptor RNA activator (SRA) participates in the generation of coding and noncoding RNA isoforms in breast cancer cell lines. DNA Cell Biol. 2006;25:418–28.
38. Cao HY, Wang S, Zhang ZY, Lou JY. Association between the WRAP53 gene rs2287499 C>G polymorphism and cancer risk:a meta-analysis. Genet Mol Res. 2016;15(3):25.
39. Luo M, Li Z, Wang W, Zeng Y, Liu Z, Qiu J. Upregulated H19 contributes to bladder cancer cell proliferation by regulating ID2 expression. FEBS J. 2013;280(7):1709–16.
40. Mahmoudi S, Henriksson S, Corcoran M, Méndez-Vidal C, Wiman KG, Farnebo M. Wrap53, a natural p53 antisense transcript required for p53 induction upon DNA damage. Mol Cell. 2009;33(4):462–71.
41. Zhu M, Chen Q, Liu X, Sun Q, Zhao X, Deng R, Wang Y, Huang J, Xu M, Yan J, et al. LncRNA h19/miR-675 axis represses prostate cancer metastasis by targeting TGFBI. FEBS J. 2014;281:3766–75.
42. Perez DS, Hoage TR, Pritchett JR, Ducharme-Smith AL, Halling ML, Ganapathiraju SC, Streng PS, Smith DI. Long, abundantly expressed non-coding transcripts are altered in cancer. Hum Mol Genet. 2008;17:642–55.
43. Fehringer G, et al. Cross-cancer genome-wide analysis of lung, ovary, breast, prostate and colorectal cancer reveals novel pleiotropic associations. Cancer Res. 2016;76(17):5103–14.
44. Zhang A, Zhao JC, Kim J, Fong KW, Yang YA, Chakravarti D, Mo YY, Yu J. LncRNA HOTAIR enhances the androgen-receptor-mediated transcriptional program and drives castration-resistant prostate cancer. Cell Rep. 2015;13(1):209–21.
45. Ren S, Liu Y, Xu W, Sun Y, Lu J, Wang F, Wei M, Shen J, Hou J, Gao X, Xu C, Huang J, Zhao Y, Sun Y. Long noncoding RNA MALAT-1 is a new potential therapeutic target for castration resistant prostate cancer. J Urol. 2013;190(6):2278–87.
46. Prensner JR, Chen W, Han S, Iyer MK, Cao Q, Kothari V, Evans JR, Knudsen KE, Paulsen MT, Ljungman M, Lawrence TS, Chinnaiyan AM, Feng FY. The long non-coding RNA PCAT-1 promotes prostate cancer cell proliferation through cMyc. Neoplasia. 2014;16(11):900–8.
47. Prensner JR, Iyer MK, Balbin OA, Dhanasekaran SM, Cao Q, Brenner JC, Laxman B, Asangani IA, Grasso CS, Kominsky HD, Cao X, Jing X, Wang X, Siddiqui J, Wei JT, Robinson D, Iyer HK, Palanisamy N, Maher CA, Chinnaiyan AM. Transcriptome sequencing across a prostate cancer cohort identifies PCAT-1, an unannotated lincRNA implicated in disease progression. Nat Biotechnol. 2011;29(8):742–9.
48. Kurisu T, Tanaka T, Ishii J, Matsumura K, Sugimura K, Nakatani T, Kawashima H. Expression and function of human steroid receptor RNA activator in prostate cancer cells: role of endogenous hSRA protein in androgen receptor-mediated transcription. Prostate Cancer Prostatic Dis. 2006;9(2):173–8.
49. Laner T, Schulz WA, Engers R, Müller M, Florl AR. Hypomethylation of the XIST gene promoter in prostate cancer. Oncol Res. 2005;15(5):257–64.
50. Zhang EB, Kong R, Yin DD, You LH, Sun M, Han L, Xu TP, Xia R, Yang JS, De W, Jf C. Long noncoding RNA ANRIL indicates a poor prognosis of gastric cancer and promotes tumor growth by epigenetically silencing of miR-99a/miR-449a. Oncotarget. 2014;5(8):2276–92.
51. Lee NK, Lee JH, Park CH, Yu D, Lee YC, Cheong JH, Noh SH, Lee SK. Long non-coding RNA HOTAIR promotes carcinogenesis and invasion of gastric adenocarcinoma. Biochem Biophys Res Commun. 2014;451(2):171–8.
52. Wang J, Su L, Chen X, Li P, Cai Q, Yu B, Liu B, Wu W, Zhu Z. MALAT1 promotes cell proliferation in gastric cancer by recruiting SF2/ASF. Biomed Pharmacother. 2014;68(5):557–64.

Qiu et al. BMC Systems Biology 2018, **12**(Suppl 1):7
https://doi.org/10.1186/s12918-018-0528-3

BMC Systems Biology

METHODOLOGY

Open Access

Discovery of Boolean metabolic networks: integer linear programming based approach

Yushan Qiu[1], Hao Jiang[2*], Wai-Ki Ching[3] and Xiaoqing Cheng[4]

From The Sixteenth Asia Pacific Bioinformatics Conference
Yokohama, Japan. 15-17 January 2018

Abstract

Background: Traditional drug discovery methods focused on the efficacy of drugs rather than their toxicity. However, toxicity and/or lack of efficacy are produced when unintended targets are affected in metabolic networks. Thus, identification of biological targets which can be manipulated to produce the desired effect with minimum side-effects has become an important and challenging topic. Efficient computational methods are required to identify the drug targets while incurring minimal side-effects.

Results: In this paper, we propose a graph-based computational damage model that summarizes the impact of enzymes on compounds in metabolic networks. An efficient method based on Integer Linear Programming formalism is then developed to identify the optimal enzyme-combination so as to minimize the side-effects. The identified target enzymes for known successful drugs are then verified by comparing the results with those in the existing literature.

Conclusions: Side-effects reduction plays a crucial role in the study of drug development. A graph-based computational damage model is proposed and the theoretical analysis states the captured problem is NP-completeness. The proposed approaches can therefore contribute to the discovery of drug targets. Our developed software is available at "http://hkumath.hku.hk/~wkc/APBC2018-metabolic-network.zip".

Keywords: Metabolic network, Integer linear programming, Boolean model

Background

One of the most important biological processes in organism is metabolism and its dysregulation can contribute to many human diseases. The manipulation of metabolism has been extensively studied in the field of metabolic engineering as many of metabolic process produce commodity and specialty chemicals. In particular, production of industrially valuable products using a microbial host with recombinant technology becomes one of the most successful applications of metabolic engineering [1–3]. Metabolic network is used to model the behavior of reactions and chemicals [4]. Authors in [5] have proposed a new mathematical model for large metabolic network and demonstrated that the outcome of network expansion is robust against of single or few reactions. Indeed, for the analysis of metabolic networks, many studies have been conducted so as to develop Boolean models. For example, Smart et al. [6] studied the effect of deletion of each enzyme in metabolic network of a Boolean model, and Lemke et al. [7] considered almost the same problem from the viewpoint of the Boolean aspect. Tamura et al. [8] developed an integer linear programming method for Boolean reaction cut (BRC) problem. Furthermore, in [9], the computational complexity of BRC was analyzed. In the Boolean model of metabolic networks, compounds and reactions can be represented by "OR" and "AND" nodes, respectively. Li et al. [10] have developed methods for finding a set of enzymes whose inhibition stops the production of the target compound with a minimum elimination of non-target compounds. Furthermore, Takemoto et al. [11] and Lee et al. [12] estimated the size distribution

*Correspondence: jiangh@ruc.edu.cn
[2]Department of Mathematics, School of Information, Renmin University of China, No.59 Zhong Guan Cun Avenue, Hai Dian District, 100872 Beijing, China
Full list of author information is available at the end of the article

of the deletion effect of enzymes using a branching process. Enumerating all the possible enzyme combination is infeasible since the number of combination increases exponentially with the number of enzymes. Lu et al. [13] developed an integer linear programming method for designing synthetic metabolic networks by considering minimum reaction insertion in a Boolean model. The proposed method can appropriately solve minimum reaction insertion in a Boolean network.

In pharmaceutics, the development of drugs mainly focus on target identification and lead inhibitor identification [14]. Furthermore, many researchers have done a lot of work on the efficacy of drugs regardless of their side-effects (toxicity). However, toxicity or lack of efficacy may occur when the compounds which are not the intended targets in the metabolic network. Thus how to effectively reduce the side-effect of drugs has become an important and challenging issue. The current works aim to identify the biological targets (could be enzyme or protein) for drugs which can be manipulated so as to produce the desired effect (i.e., curing a disease) with minimum disruptive side-effects [15, 16]. Drug targets, in particular enzymes, are chosen to reduce abnormal metabolites by formulating an optimal combination problem in enzyme combination in metabolic networks [10, 17]. It is known that the function of enzymes is to catalyze reactions and produces metabolites (compounds) in metabolic networks. Human diseases, especially metabolic diseases, may be directly caused by the accumulation of certain compounds due to the malfunctions of enzymes. We denote such compounds as *target compounds* while the others as *non-target compounds*. For example, the malfunction of enzyme phenylalanine hydroxylase results in accumulation of the amino acid phenylalanine which causes the disease phenylketonuria [18]. Therefore, there is a great need to identify a set of enzymes which are manipulated by drugs to prevent the buildup of target compounds with minimal damage. Here *damage* corresponds to the number of non-target compounds whose production are forced to stop by the inhibition of that enzyme (or enzyme set).

Then the problem becomes to identify the optimal enzyme set whose inhibition eliminates the target compounds and at the same time, incurs minimum damage based on the given metabolic network and a set of target compounds. We denote such problem as enzyme combination identification (ECI). Sridhar et al. [19] developed a branch-and-bound algorithm to dynamically explore the search space. Furthermore, two filtering strategies are proposed to prune the search space to guarantee an optimal solution. However, their algorithm is complicated and impractical for us to use it. In this paper, we develop an efficient method based on Integer Linear Programming [20–22] which has a wide application in solving NP-hard problems. Furthermore, all instances of the original problem are needed to convert into integer programming formalization so as to apply the existing free solver called CPLEX [23].

This paper has two main contributions. First, we formulate a new biological problem based on Boolean metabolic network to identify the drug target with minimum damage. Second, we propose an effective and efficient method, Integer Linear Programming, (needs $O(m+n)$ variables) to solve the captured problem and it is easy to implement. By integrating the human metabolic networks, we have shown that the proposed approach can accurately identify the target enzyme set for known drugs in the literature. Furthermore, experiments have shown that our proposed method is extremely efficient which can effectively solve the problem in seconds.

The remainder of the paper is organized as follows. In "Problem formulation" section, we formulate the captured problem. "Methods" section presents the algorithm ILP-ECI. In "Results and discussions" section, we discuss the experimental results and give a theoretical illustration. Finally, conclusions are given in the last section.

Problem formulation

A metabolic network can be represented by a directed graph $G=(V,A)$ which captures the interaction between reactions, compound, and enzymes. Let C, R, and E denote the set of compounds, reactions, and enzymes, respectively. Then V_C, V_R and V_E denote the set of vertices from C, R, and E, where $V_C=\{v_{c_1},v_{c_2},\cdots,v_{c_m}\}$ denotes a set of *compound nodes*, $V_R=\{v_{r_1},v_{r_2},\cdots,v_{r_n}\}$ represents a set of *reaction nodes* and $V_E=\{v_{e_1},v_{e_2},\cdots,v_{e_l}\}$ defines a set of *enzyme nodes*. The subscript m,n,l denote the number of *compound nodes*, *reaction nodes* and *enzyme nodes*, respectively. Here $V=V_C\cup V_R\cup V_E$ and $V_C\cap V_R=\{\}, V_C\cap V_E=\{\}, V_R\cap V_E=\{\}$ hold. Here $\{\}$ is the empty set. Let $V_s\subseteq V_C$ and $V_t\subseteq V_C$ be sets of *source nodes* and *target nodes*, respectively. Source nodes are those nodes having no incoming edges and they are the seed compounds. A metabolic network can be defined as follows:

- For each edge $(u,v)\in A$, either $(u\in V_C)\wedge(v\in V_R)$ or $(u\in V_R)\wedge(v\in V_C|v\in V_E)$ holds and "$\wedge$" represents "AND" function;
- Each source node has no incoming edges;
- For nodes $v\notin V_s$ and $v\in V_C$ have at least one incoming edge.

The main problem **Enzyme Combination Identification (ECI)** in a Boolean network is first described with a simple example and then followed by its mathematical formalization.

A small hypothetical metabolic network is shown in Fig. 1, where circles, rectangles and triangles represent compounds, reactions and enzymes, respectively. For instance, R_1 is a reaction, its substrates are C_1 and E_2 and its product is C_2. Here C_5 is the target compound in this figure which shall be stopped. To stop the production of C_5, reaction R_2 must be prevented from taking place. One of the possible solution is to disrupt one of its catalyzing enzymes (E_1 for instance). Another is to stop the production of its reactant compounds (i.e., C_2 or C_3). If C_2 is stopped, then two possible ways can achieve this effect (i.e., enzyme E_2 or E_1). It is noted that C_2, R_2, C_7 and R_1 forms a cycle, the inhibition of E_1 will result in inactivating reaction R_2 which will further makes C_2 inproducible. The other way is inhibiting enzyme E_2 which makes R_1 inactive and further stop the production of compound C_2. Therefore, the inhibition of enzyme E_1 or E_2 can result in stopping the production of the target compound.

We define the **Enzyme Combination Identification** as follows.

- **Input:** A metabolic network and a set of target compounds $T(T \in C)$, C is the set of compounds.
- **Output:** Find a set of enzymes $X(X \in E)$ with minimum damage, whose inhibition stops the production of all the compounds in T.

For simplicity, we assume there are no external inputs to all reactions and all the input compounds related to reactions are shown in the network. We note that different compounds and enzymes may have different levels of importance in the metabolic networks. Here we assume that all the compounds and enzymes are of equal importance. We assign binary value (i.e., 0 or 1) to each node V in the Boolean model. "0" means that the corresponding compound is not producible or the corresponding reaction is inactive while "1" means that the corresponding compound is producible and the corresponding reaction is active. For the related enzyme catalyzing the reaction, "1" means that it is active and "0" means that it is inhibited. Let G be such an assignment (G is a function from V to $\{0, 1\}$). For each node $v \in V$, we write $v = 1$ (resp., $v = 0$) if 1 (resp., 0) is assigned to v. Then G can be regarded as a *valid assignment* if the following conditions are satisfied:

(i) For each $v \in V_s, v = 1$.
(ii) For each $v \notin V_s, v = 1$ if and only if there is u such that $(u, v) \in E$ and $u = 1$.
(iii) For each $v \in V_r, v = 1$ if and only if $u = 1$ holds for all u such that $(u, v) \in E$.

Condition (ii) implies that compound nodes correspond to "OR" nodes. Condition (iii) implies that reaction nodes correspond to "AND" nodes, which means that the output is forced to 0 when a node is inactivated.

From Fig. 1, we can see that inhibition of E_1 results in the knock out of compounds C_4, $C7$, C_8 and C_9 in addition to the target compound C_5. Then we denote the number of non-target compounds knocked out as the *damage*, which is caused by the manipulating the enzyme set in the metabolic network. It can be seen that the damage of inhibiting E_2 is 2 (i.e., C_2 and C_4). Compound C_7 is still producible because it can be produced by R_3 even after the inhibition of E_2. The damage effect of inhibition of E_1 is 4 (i.e., C_4, $C7$, C_8 and C_9). Both E_1 and E_2 are potential drug targets since they can achieve the effect of disrupting the target compound C_5. However, E_2 is a better drug target than E_1 owing to the fact that it causes less damage.

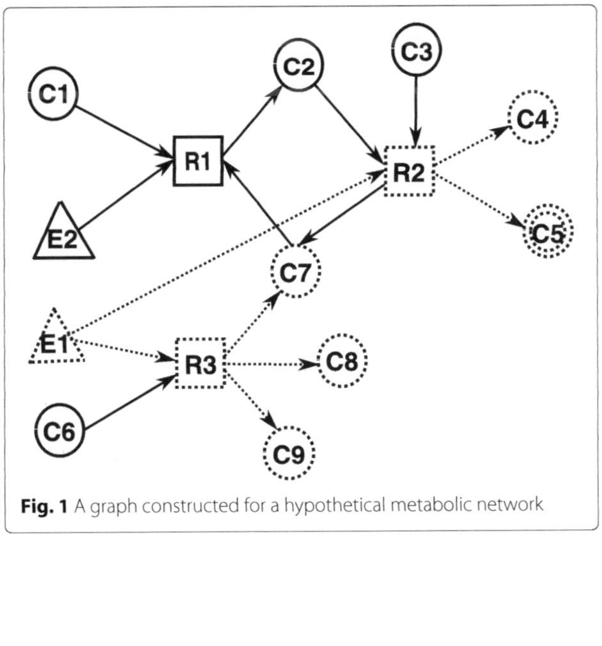

Fig. 1 A graph constructed for a hypothetical metabolic network

Methods

In this section, we introduce integer programming-based methods for ECI. Integer programming, in particular, Integer Linear Programming (ILP) is set to minimize (or maximize) a linear objective function under linear constraints with all the variables taking integer values. In the following, each variable takes including the binary value (i.e., 0 or 1), representing the Boolean values. We apply ILP to ECI since ILP is widely used for solving NP-hard problems.

The ILP formulation for the network in Fig. 1 is as follows:

ILP-ECI

$$\min \left\{ \sum_{i=1}^{m} TCi \right\} \tag{1}$$

subject to

$$FC5 = 1 \tag{2}$$

$$\begin{aligned} &TR1 + FC1 + FC7 + FE2 \geq 1,\\ &FR1 + TC1 \geq 1,\\ &FR1 + TC7 \geq 1,\\ &FR1 + TE2 \geq 1, \end{aligned} \tag{3}$$

$$\begin{aligned} &TR2 + FC2 + FC3 + FE1 \geq 1,\\ &FR2 + TC2 \geq 1,\\ &FR2 + TC3 \geq 1,\\ &FR2 + TE1 \geq 1, \end{aligned} \tag{4}$$

$$\begin{aligned} &TR3 + FC6 + FE1 \geq 1,\\ &FR3 + TC6 \geq 1,\\ &FR3 + TE1 \geq 1 \end{aligned} \tag{5}$$

$$TC2 = TR1 \tag{6}$$

$$TC4 = TR2 \tag{7}$$

$$TC5 = TR2 \tag{8}$$

$$\begin{aligned} &FC7 + TR2 + TR3 \geq 1,\\ &TC7 + FR2 \geq 1,\\ &TC7 + FR3 \geq 1 \end{aligned} \tag{9}$$

$$TC8 = TR3 \tag{10}$$

$$TC9 = TR3 \tag{11}$$

$$TC1 = 1, \qquad TC3 = 1, \qquad TC6 = 1 \tag{12}$$

$$TX + FX = 1 \text{for any X} \tag{13}$$

We denote the above formalization as **ILP-ECI**. Here all variables including the value of reaction compound and enzyme nodes take either 1 or 0. Thus, v_{r_i} can be either 0 or 1, and v_{c_i} and v_{e_i} also take 0 or 1. In this example, $v_{r_i} = 0$ (resp. $v_{r_i} = 1$) indicating that the value of reaction i takes 0 is represented by $FRi = 1$ (resp. $TRi = 1$) which implies that the reaction is inactivated (otherwise, the $TRi = 1$ implies the reaction is activated). Therefore, $TRi = 0$ (equivalent to $FRi = 1$) means the corresponding value for true reaction takes 0, which implies the reaction is inactive. And $FRi = 0$ (equivalent to $TRi = 1$) indicates the corresponding value for false reaction takes 0, which implies the reaction is active. Thus, $TRi + FRi = 1$ holds for any node i in the network. Similarly, TCi and FCi are used to represent the values of compound nodes. For instance, $TC2 = 1$ means that $v_{c_2} = 1$ and in other words, $FC2 = 0$ since $TCi + FCi = 1$. Furthermore, v_{r_i} corresponds to "AND" node which implies that if $v_{e_i} = 0$ will inactivate v_{r_i}.

The objective function (1) means that the damage should be minimized. $FCi = 1$ (or $TCi = 0$) means that a compound v_{c_i} is not producible. Equation (2) means that the target compound v_{c_5} should be 0 after the 0-1 assignment converges. Equation (3) represents the Boolean relation $v_{r_1} = v_{c_1} \wedge v_{c_7} \wedge v_{e_2}$. Note that the Boolean relations such as "$\vee$" or "$\wedge$" cannot be used in ILP formulation, we need to convert them into linear equations and/or inequations. Actually, "$\vee$" indicates "AND" function and "$\wedge$"represents "OR" function. Since $x_1 = x_2 \wedge x_3 \wedge \cdots \wedge x_n$ can be represented by

$$(x_1 \vee \overline{x_2} \vee \cdots \vee \overline{x_n}) \wedge (\overline{x_1} \vee x_2) \wedge (\overline{x_1} \vee x_3) \wedge \cdots \wedge (\overline{x_1} \vee x_n) = 1,$$

the constraint $v_{r_1} = v_{c_1} \wedge v_{c_7} \wedge v_{e_2}$ can be converted into

$$\left(v_{r_1} \vee \overline{v_{c_1}} \vee \overline{v_{c_7}} \vee \overline{v_{e_2}}\right) \wedge \left(\overline{v_{r_1}} \vee v_{c_1}\right) \wedge \left(\overline{v_{r_1}} \vee v_{c_7}\right) \wedge \left(\overline{v_{r_1}} \vee v_{e_2}\right) = 1.$$

Thus Eq. (3) is obtained. Similarly, Eqs. (4)-(5) represent the constraints of v_{r_2} and v_{r_3}, respectively.

For a compound node with indegree is 1 which indicates the node has only one incoming edge, the value of the predecessor is just copied. For instance, since v_{c_2} has only one predecessor v_{r_1}, v_{c_2} is just copied from v_{r_1} as shown in Eq. (6). Similarly, v_{c_4} is just copied from v_{r_2} which is shown in Eq. (7).

However, for a compound node with indegree more than 1, it is necessary to convert the "$\vee$" relation into linear equation or inequations. Equation (9) represents the Boolean relation $v_{c_7} = v_{r_2} \vee v_{r_3}$. Since $x_1 = x_2 \vee x_3 \vee \cdots \vee x_n$ can be represented by

$$(\overline{x_1} \vee x_2 \vee \cdots \vee x_n) \wedge (x_1 \vee \overline{x_2}) \wedge (x_1 \vee \overline{x_3}) \wedge \cdots \wedge (x_1 \vee \overline{x_n}) = 1,$$

$v_{c_7} = v_{r_2} \vee v_{r_3}$ can be turned into $\left(\overline{v_{c_7}} \vee v_{r_2} \vee v_{r_3}\right) \wedge \left(v_{c_7} \vee \overline{v_{r_2}}\right) \wedge \left(v_{c_7} \vee \overline{v_{r_3}}\right)$. Thus, Eq. (9) can be obtained.

Equation (6) - Eq. (12) represent the constraints of $v_{c_1}, \cdots, v_{c_9}$ respectively. Equation (12) means that v_{c_1}, v_{c_3} and v_{c_6} are 1 since their indegrees is 0.

Equation (13) means that "T" and "F" correspond to "true (1)" and "false (0)", respectively, and complement each other. X in Eq. (13) means any component or reaction in the metabolic network.

The above formalization can clearly solve ECI and obtain the correct solution $\{E_2\}$. Besides, the number of variables is $O(m + n)$ in the above formalization where m and n are the number of compounds and reactions, respectively.

It is noted that solving ILP is NP-complete, however, a problem that can be formalized as ILP is not always NP-complete. Thus in the following, we prove that ECI is NP-complete.

Theorem *ECI is NP-complete problem with the maximum indegree and outdegree being bounded by 2.*

Proof Obviously, the problem is in NP, it suffices to show that it is NP-hardness. The proof is by a polynomial time reduction from minimum edge cover (MEC), which is a problem for a given graph to find the minimum number of edges so that each node is incident to at least one of the selected edges. For instance, $E_1 = \{e_2, e_3, e_6\}$ is one of optimal solutions of MEC for graph shown in Fig. 2. Let $G = (V, E)$ be an instance of MEC,

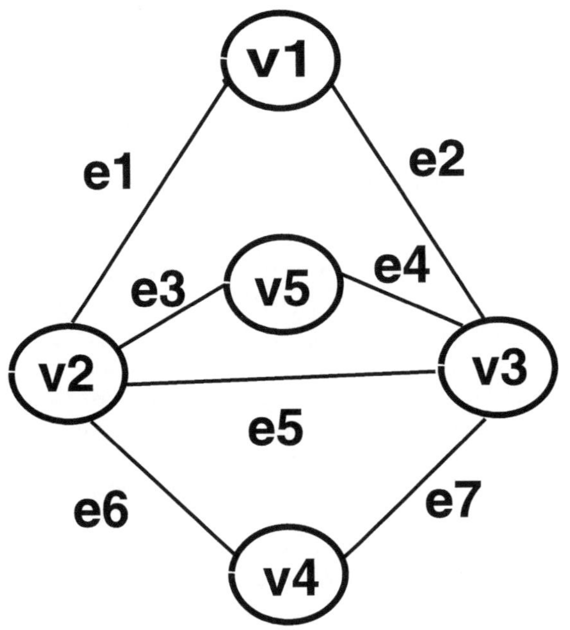

Fig. 2 An instance of minimum edge cover (MEC)

where $V = \{v_1, v_2, \cdots, v_n\}$ and $E = \{e_1, e_2, \cdots, e_m\}$. We then construct the corresponding ECI as below. The metabolite network $G = (V_c \cup V_r \cup V_e, E)$ is given by

$$V_c = \{c_1, c_2, \cdots, c_m\} \cup c_t, \quad V_r = \{r_1, r_2, \cdots, r_n\},$$

$$E = \left\{\{c_i, r_j\} | i = 1, \cdots, m, j = 1, \cdots, n, \text{ if } v_j \text{ is an end point of } e_i.\right\} \cup \left\{\{r_j, c_t\} | j = 1, \cdots, n\right\}$$

It is noted that the minimum *damage* is determined uniquely by the inhibition of enzyme set. Furthermore, our objective is to minimize the "damage" (side-effects). Then V_e can be regarded as virtual nodes and denoted as an empty set in this case. The ECI problem can be converted into the problem of identifying the minimum set of non-target compounds. Thus the graph for MEC shown in Fig. 2 is converted into ECI shown in Fig. 3. It is clear that this conversion can be done in polynomial time. Then we show that MEC for G has a solution of size z if and only if ECI has a minimum damage of size z. To guarantee that the target compound c_t is stopped (i.e., $c_t = 0$), it implies that all r_j ($j = 1, \cdots, n$) takes the value 0. If G has an edge cover of size z, then it follows that the minimum number of c_i taking 0 should be z. On the other hand, if the minimum damage for ECI is z, then each r_j must be 0 so as to satisfy $c_t = 0$, we have at least predecessor of each r_j must be included in the minimum damage set. Since there is an edge between c_i and r_j if and only if v_j is incident to e_i. Thus $\{c_i | c_i \in \text{minimum damage set}\}$ is an edge cover of size z. □

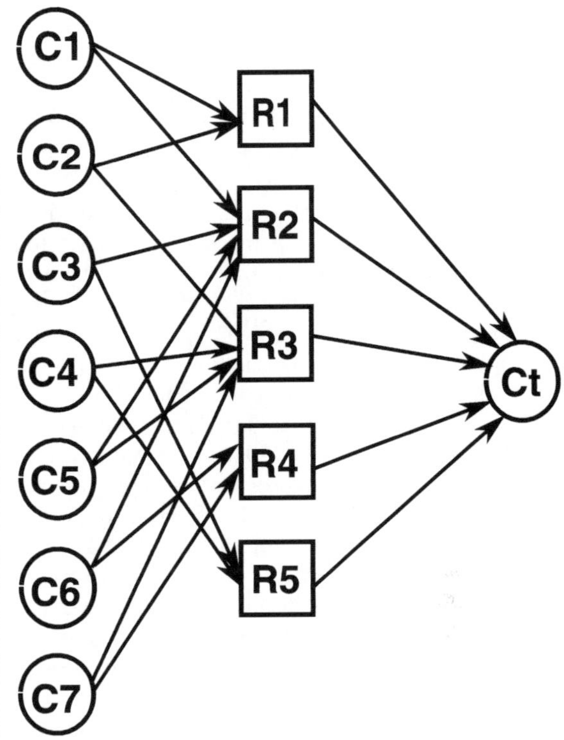

Fig. 3 The polynomial time reduction from minimum edge cover (MEC) problem to minimum damage identification problem

Results and discussions

In this section, we verify the biological validity of our proposed method by using known drugs. Besides, the performance of the ILP-ECI algorithm is evaluated by using the execution time which indicates the total time taken by the method. For the experimental data, we extracted the metabolic network information from the Kyoto Encyclopedia of Genes and Genomes (KEGG) database [24]. KEGG is database which provides known drug molecules along the the enzymes they inhibit and their therapeutic category. Then we use drugs at this database as our benchmarks and we report two of them due to the space limitation. And the value in parenthesis that starts with letter "C", "D", and "E" (e.g., E1.13.11.34) is the unique identifier assigned to the corresponding compound, drug, and enzyme respectively in KEGG.

The drug we used in this paper is *Benoxaprofen* (D03080) which inhibits arachidonate 5-lipoxygenase (E1.13.11.34) [25]. This enzyme appears in several networks including arachidonic acid metabolism network (hsa00590). In Pharmacology, the inhibition of 5-lipoxygenase will decrease the biosynthesis of LTB4 (C02165), cysteinylcontaining leukotrienes LTC4 (C02166), LTD4 (C05951) and LTE4 (C05952) (see, for instance, Fig. 4) [26]. According to our graph model, the removal of E1.13.11.34 will eliminate four compounds (LTB4, LTC4, LTD4 and LTE4) in arachidonic acid metabolism network. Furthermore, these four compounds play an important role in the mechanisms of toxic brain damage in acute methanol poisoning in humans [27]. Thus, we chose them as the target compounds. Apart from that, inhibiting this enzyme also eliminates four more compounds (i.e., 5(S)-HPETE(C05356), 5-HETE(C04805), LTA4(C00909), and 20-OH-LTB4(C04853)) [28]. These compounds can be regarded as "damage" in our model. Furthermore, when applying ILP-ECI with LTB4, LTC4, LTD4 and LTE4 as the target compounds, we find LTA4H (E3.3.2.6) and LTC4 synthase (E4.4.1.20) as the optimal enzyme set since the inhibition of these two enzymes eliminates only one non-target compound namely 20-OH-LTB4 (C04853). Thus, it is shown that ILP-ECI potentially finds a better solution in this experiment than the existing drugs since the same target compounds are eliminated by the existing drug in addition to other four compounds. Indeed, recent research validated our model since the anti-inflammatory effect of the levels of LTA4H [29] and LTC4 [30] has been observed. The computational time in this experiment takes only 5.19 s.

Another experiment we conducted is the histidine metabolism network (hsa00340). The enzyme amine oxidase (E.1.4.3.4) appeared in hsa00340 can be inhibited by drug (Rasagiline (D02562)). According to our graph model, the removal of enzyme E.1.4.3.4 will result in eliminating two compounds Methylimidazoleacetic acid (C05828) and Methylimidazole acetaldehyde (C05827). It should be noted that the level of pros-methylimidazoleacetic acid is closely related to severity of Parkinson disease in patients [31, 32]. Running ILP-ECI with Methylimidazoleacetic acid and Methylimidazole acetaldehyde as the target compounds finds amine oxidase as the optimal enzyme. It takes only 3.26 s to run ILP-ECI. This experiment verifies that Rasgiline targets the optimal enzyme. An important advantage of our Boolean model is its capability of detecting the lack of substrates where the connectivity-based methods fail to handle this. Another advantage of this model is its capability of handling branches and cycles in a pathway from the source compound to the target compound. However, there are still have some limitations in this method. One of the major drawbacks is the assumption that all the compounds and enzymes are of equal importance. However, different compounds and enzymes

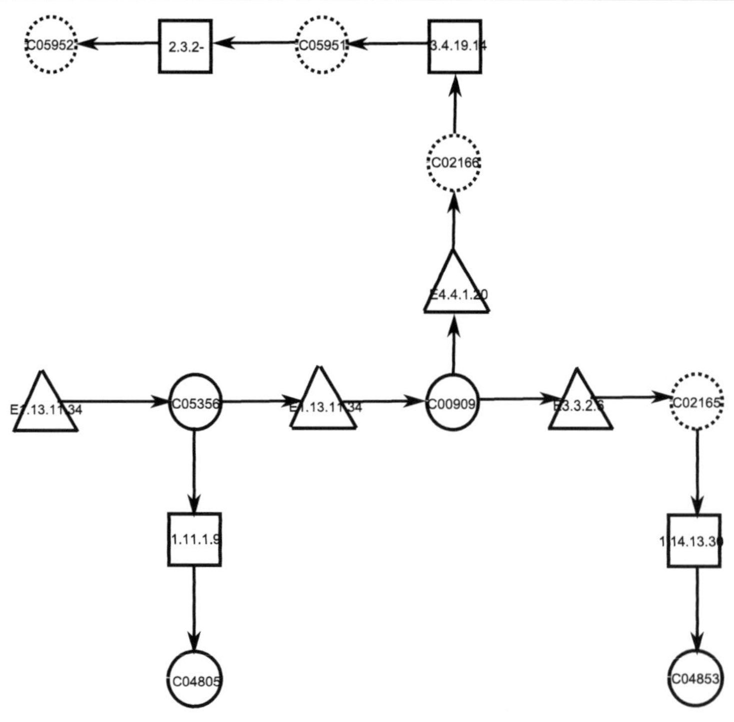

Fig. 4 Subgraph of arachidonic acid metabolism

may have different levels of importance in the metabolic networks. Our future work will focus on developing other models which include the weight of different nodes.

Conclusions

In this paper, we formulate the optimal enzyme combination identification (ECI) problem as an optimization problem in Boolean metabolic networks. We have proven that ECI in the Boolean model is NP-complete and the target enzyme set is uniquely determined when the target compounds are given. Furthermore, considering that an exhaustive search cannot be used to solve ECI when the network is large, we developed an ILP-based algorithm for ECI. Considering that the computational time of IP-based method is exponential to the number of variables, to improve the scalability of the developed method, it is vital to reduce the number of variables appearing in IP formalization. And our proposed IP-based method needs $O(m + n)$ variables.

The efficiency and effectiveness are validated by the computational experiments in which datasets were downloaded from the KEGG database. The results demonstrate that the proposed model can accurately identify the target enzymes for known successful drugs in the literature. Specifically, ILP-ECI has found a different enzyme set for the target compounds of Benoxaprofen which indicates that our method has a great potential to be better than Benoxaprofen. The reason is that the solution of our algorithm damages only one non-target compound while Benxaprofen damages 4 non-target compounds including the compound damaged by ILP-ECI's solution. Besides, ILP-ECI has found that the same target enzyme like Rasagiline when its target compounds are given in advance. It is to be noted that our problem needs only $O(m + n)$ variables in the IP formalization. The experiments also show that ILP-ECI can solve the problem in a short time which confirms the efficiency of our algorithm.

Acknowledgements
The authors would like to thank anonymous reviewers for your helpful and constructive comments.

Funding
This research is supported in part of Natural Science Foundation of SZU (Grant No. 2017058) and National Natural Science Foundation of China NSFC (Grant Nos. 91730301, 11626229 and 11671158). The publication costs are funded by National Natural Science Foundation of China NSFC (Grant Nos. 91730301).

Availability of data and materials
All the data sets are publicly available and can be accessed from the KEGG databases.

About this supplement
This article has been published as part of *BMC Systems Biology* Volume 12 Supplement 1, 2018: Selected articles from the 16th Asia Pacific Bioinformatics Conference (APBC 2018): systems biology. The full contents of the supplement are available online at https://bmcsystbiol.biomedcentral.com/articles/supplements/volume-12-supplement-1.

Authors' contributions
YQ designed the research. YQ and WKC proposed the methods and did theoretical analysis. YQ and HJ collected the data. YQ, HJ and XC conducted the experiments and analyze the results. YQ, HJ, WC and XC wrote the manuscript. All authors read and approved the final manuscript.

Ethics approval and consent to participate
Not applicable.

Consent for publication
Not applicable.

Competing interests
The authors declare that they have no competing interests.

Author details
[1]College of Mathematics and Statistics, Shenzhen University, Nanhai Avenue 3688, 518060 Shenzhen, China. [2]Department of Mathematics, School of Information, Renmin University of China, No.59 Zhong Guan Cun Avenue, Hai Dian District, 100872 Beijing, China. [3]Department of Mathematics, The University of Hong Kong, Pokfulam Road, Hong Kong, Hong Kong. [4]School of Mathematics and Statistics, Xi'An Jiaotong University, No.28 West Xianning Road, 710049 Xi'An, China.

Published: 11 April 2018

References

1. Bro C, Regenberg B, Förster J, Nielsen J. In silico aided metabolic engineering of saccharomyces cerevisiae for improved biothanol production. Metab Eng. 2006;8(2):102–11.
2. Lee SK, Chou H, Ham TS, Lee TS, Keasling JD. Metabolic engineering of microorganisms for biofuels production: from bugs to synthetic biology to fuels. Curr Opin Biotechnol. 2008;19:556–63.
3. Alper H, Jin YS, Moxley JF, Stephanopoulos G. Identifying gene tarets for the metebolic engineering of lycopene biosynthesis in escherichia coli. Metab Eng. 2005;7:155–64.
4. Soh KC, Hatzimanikatis V. Dreams of metabolism. Trends Biotechnol. 2010;28:501–8.
5. Handorf T, Ebenhöh O, Heinrich R. Expanding metabolic networks: scopes of compounds, robustness, and evolution. J Mol Evol. 2005;61: 498–512.
6. Smart AG, Amaral LA, Ottino JM. Cascading failure and robustness in metabolic networks. Proc Natl Acad Sci. 2008;105:13223–8.
7. Lemke N, Herédia F, Barcellos CK, Dos Reis AN, Mombach JC. Essentiality and damage in metabolic networks. Bioinformatics. 2004;20:115–9.
8. Tamura T, Takemoto K, Akutsu T. Finding minimum reation cuts of metabolic networks under a boolean model using integer programming and feedback vertex sets. Int J Knowl Discov Bioinformatics (IJKDB). 2010;1:14–31.
9. Tamura T, Akutsu T. Exact algorithms for fidning a minimum reaction cut under a boolean model of metabolic networks. IEICE Trans Fundam Electron. 2010;93:1497–1507.
10. Li Z, Wang RS, Zhang XS, Chen L. Detecting drug targets with minimum side effects in metabolic networks. IET Syst Biol. 2009;3:523–33.
11. Takemoto K, Tamura T, Akutsu T. Theoretical estimation of metabolic network robustness against multiple reaction knockouts using branching process approximation. Physica A Stat Mech Appl. 2003;392:5525–535.
12. Lee D, Goh KI, Kahng B. Branching process approach for boolean bipartite networks of metabolic reations. Phys Rev E. 2012;86:027101.
13. Lu W, Tamura T, Song J, Akutsu T. Integer programming-based method for designing synthetic metabolic networks by minimum reation insertion in a boolean model. Plos ONE. 2014;9:92637.
14. Drews J. Drug discovery: A historical perpective. Science. 2000;287:1960–4.
15. Smith C. Hitting the target. Nature. 2003;422:341–7.
16. Takenaka T. Classical vs reverse pharmacology in drug discovery. BJU Int. 2001;88:7–10.
17. Li Z, Wang RS, Zhang XS. Two-stage flux balance analysis of metabolic networks for drug target identification. BMC Syst Biol. 2011;5:11.
18. Surtees R, Blau N. The neurochemistry of phenylketonuria. Eur J Pediatr. 2000;159:109–13.
19. Sridhar P, Song B, Kahveci T, Ranka S. Mining metabolic networks for optimal drug targets. Pac Symp Biocomput. 2008;13:291–302.

20. Schrijver A. Theory of Linear and Integer Programming. New York: John Wiley & Sons, Inc; 1986.
21. Zhenping L, Zhang S, Wang Y, Zhang XS, Chen L. Alignment of molecular networks by integer quadratic programming. Bioinformatics. 2007;23:1631–9.
22. Cheng X, Qiu Y, Hou W, Ching WK. Integer programming-based method for observability of singleton attractors in boolean networks. IET Syst Biol. 2017;11:30–5.
23. IBM 2010, IBM ILOG CPLEX Optimizer. http://www-01.ibm.com/software/commerce/optimization/cplex-optimizer/index.html. Accessed 2 Mar 2017.
24. Kanehisa M, Goto S. Kegg: Kyoto encyclopedia of genes and genomes. Nucleic Acids Res. 2000;28:27–30.
25. Fogh K, Herlin T, Kragballe K. In vitro inhibition of leukotriene b4 formation by exogeneous 5-lipoxygenase inhibitors is associated with enhanced generation of 15-hydroxy-eicosatetraenoic acid (15-hete) by human neutrophils. Arch Dermatol Res. 1988;280:430–6.
26. Lötzer K, Funk CD, Habenicht AJ. The 5-lipoxygenase pathway in arterial wall biology and atherosclerosis. Biochim Biophys Acta. 2005;1736:30–7.
27. Zakharov S, Kotikova K, Nurieva O, Hlusicka J, Kacer P, Urban P, Vaneckova M, Seidl Z, Diblik P, Kuthan P, Navratil T, Pelclova D. Leukotriene-mediated neuroinflammation, toxic brain damage, and neurodegeneration in acute methanol poisoning. lin Toxicol (Phila). 2017;55:249–59.
28. Sweeney FJ, Eskra JD, Carty TJ. Development of a system for evaluating 5-lipoxygenase inhibitors using human whole blood. Prostaglandins Leukot Med. 1987;28:73–93.
29. Rao NL, Dunford PJ, Xue X, Jiang X, Lundeen KA, Coles F, Riley JP, Williams KN, Grice CA, Edwards JP, Karlsson L, Fourie AM. Anti-inflammatory activity of a potent, selective leukotriene a4 hydrolase inhibitor in comparison with the 5-lipoxygenase inhibitor zileuton. J Pharmacol Exp Ther. 2007;321:1154–60.
30. Torres-Galván MJ1, Ortega N, Sánchez-García F, Blanco C, Carrillo T, Quiralte J. Ltc4-synthase a-444c polymorphism: lack of association with nsaid-induced isolated periorbital angioedema in a spanish population. Ann Allergy Asthma Immunol. 2001;87:506–10.
31. Blandina P, Cherici G, Moroni F, Prell GD, Green JP. Release of glutamate from striatum of freely moving rats by pros-methylimidazoleacetic acid. J Neurochem. 1995;64:788–93.
32. Prell GD, Khandelwal JK, Burns RS, Blandina P, Morrishow AM, Green JP. Levels of pros-methylimidazoleacetic acid: Correlation with severity of parkinson's disease in csf of patients and with the depletion of striatal dopamine and its metabolites in mptp-treated mice. J Neural Transm. 1991;3:1435–63.

Ranjan *et al. BMC Systems Biology* 2018, **12**(Suppl 1):10
https://doi.org/10.1186/s12918-018-0529-2

BMC Systems Biology

RESEARCH Open Access

Composite mathematical modeling of calcium signaling behind neuronal cell death in Alzheimer's disease

Bobby Ranjan[1], Ket Hing Chong[1] and Jie Zheng[1,2,3*]

From The Sixteenth Asia Pacific Bioinformatics Conference
Yokohama, Japan. 15-17 January 2018

Abstract

Background: Alzheimer's disease (AD) is a progressive neurological disorder, recognized as the most common cause of dementia affecting people aged 65 and above. AD is characterized by an increase in amyloid metabolism, and by the misfolding and deposition of β-amyloid oligomers in and around neurons in the brain. These processes remodel the calcium signaling mechanism in neurons, leading to cell death via apoptosis. Despite accumulating knowledge about the biological processes underlying AD, mathematical models to date are restricted to depicting only a small portion of the pathology.

Results: Here, we integrated multiple mathematical models to analyze and understand the relationship among amyloid depositions, calcium signaling and mitochondrial permeability transition pore (PTP) related cell apoptosis in AD. The model was used to simulate calcium dynamics in the absence and presence of AD. In the absence of AD, i.e. without β-amyloid deposition, mitochondrial and cytosolic calcium level remains in the low resting concentration. However, our *in silico* simulation of the presence of AD with the β-amyloid deposition, shows an increase in the entry of calcium ions into the cell and dysregulation of Ca^{2+} channel receptors on the Endoplasmic Reticulum. This composite model enabled us to make simulation that is not possible to measure experimentally.

Conclusions: Our mathematical model depicting the mechanisms affecting calcium signaling in neurons can help understand AD at the systems level and has potential for diagnostic and therapeutic applications.

Keywords: Calcium signaling, Neuronal cell death, Alzheimer's disease, Mathematical modeling

Background

Alzheimer's disease (AD) is characterized by the deposition of β-amyloid (Aβ) oligomers in and around neurons in the brain accompanied by dysfunctional neuronal calcium homeostasis. Autophagy is generally an efficient mechanism for removing amyloids. During the onset of AD, autophagy is increased but the transfer of autophagic vesicles to the lysosomes is blocked [1]. This may contribute to the accumulation of amyloids. There is increasing evidence to support the hypothesis that Aβ induces an up-regulation of intracellular Ca^{2+} and leads to AD. Multiple studies on AD mouse models have shown that Ca^{2+} dysregulation leads to increased Ca^{2+} entry into the cytoplasm resulting in neuronal cell death and AD [2, 3].

The observations about the effect of A-β oligomers on neuronal calcium signaling led to the formulation of the calcium hypothesis of AD [4]. The basic argument behind the hypothesis is that the activation of the amyloidogenic pathway results in a remodeling of the neuronal calcium signaling pathway. The up-regulation of Ca^{2+} distorts the normal neuronal Ca^{2+} signaling by increasing the amount

*Correspondence: zhengjie@ntu.edu.sg
[1]Biomedical Informatics Lab, School of Computer Science and Engineering, Nanyang Technological University, 50 Nanyang Avenue, 639798 Singapore, Singapore
[2]Genome Institute of Singapore, Biopolis, 138672 Singapore, Singapore
Full list of author information is available at the end of the article

of Ca^{2+} being taken up by the mitochondria. A sustained increase in the mitochondrial Ca^{2+} may activate the mitochondria to initiate the intrinsic pathway of Ca^{2+}-induced apoptosis, as described in the calcium hypothesis of Alzheimer's disease [4].

According to Berridge, the increased output of Ca^{2+} due to the hypersensitivity of the Ca^{2+} signaling system may activate the mitochondria to initiate the intrinsic pathway of Ca^{2+}-induced apoptosis by opening up the mitochondrial permeability transition pore (PTP), causing collapse of the mitochondrial membrane potential and releasing cytochrome c and other factors that activate the caspase cascade responsible for apoptosis.

Ichas and Mazat [5] demonstrated that the mitochondrial PTP operates at the crossroads of 2 distinct physiological pathways i.e. the Ca^{2+} signaling network during the life of the cell and the effector phase of the apoptotic cascade during Ca^{2+}-dependent cell death. It has 2 open conformations correspondingly. The low-conductance state, which allows the diffusion of small ions like Ca^{2+}, is pH-operated, promoting spontaneous closure of the channel. A high-conductance state, which allows the unselective diffusion of big molecules, stabilizes the channel in open conformation [5].

Mitochondria in open high-conductance state can no longer maintain a proton gradient, and thus cannot sustain oxidative phosphorylation, resulting in an arrest of aerobic ATP synthesis (necrotic cell death). This also results in an oncotic imbalance in the mitochondria causing it to swell up. The cristae formed by the inner membrane unfold, leading to rupture of the outer membrane that brings into direct communication the former intermembrane space and the extra-mitochondrial medium. Soluble components like cytochrome c and Apoptosis Inducing Factor (AIF), which are normally trapped in the intermembrane space, are released into the cytosol, thereby inducing cell apoptosis [5].

Thul [6] described the use of Ordinary Differential Equations (ODEs) to model intracellular Ca^{2+} oscillations, assuming intracellular Ca^{2+} concentration to be spatially homogeneous. The use of ODEs is widespread among modelers because: (a) the study of ODEs is computationally well-supported, with a large body of techniques available to investigate ODEs in great detail, and (b) lack of sufficient experimental data to develop a spatially extended model.

In this paper the processes mentioned above have been modeled mathematically using ODEs to allow for quantitative understanding of the dynamics of neuronal cell death in AD. We achieved this by integrating three models: Fall-Keizer Model [7], Mitochondrial PTP Model [8], and Amyloid Metabolism Model [9]. These models are explained in detail in the next section. The results obtained are qualitatively consistent with all the three papers. To demonstrate the validity of our composite model, we tested two hypotheses proposed by these individual models:

(1) When there is no abnormality in β-amyloid folding and deposition, the initiation of an action potential does not lead to long-term sustained oscillations (an extension of the result detailed in [7]). This will be explained further in the Results section.
(2) When β-amyloid misfolding affects calcium signaling within the neuron, the action potential is prevented from dying down immediately (a more complex expression of the qualitative trend represented by [9]). As β-amyloid deposition increases with time, the rate of entry of calcium ions into the cell increases, thus allowing the calcium ion oscillations to continue by maintaining high calcium ion levels in the cytosol. As this rate of entry continues to increase, the cytosol and mitochondria attempt to release excess calcium ions to one another more frequently and in smaller amounts, thus resulting in smaller but more rapid oscillations.

We used time-course simulation to relate the occurrence of events to biological processes, thereby verifying our model. We also quantified certain findings from the time-course simulations, with special emphasis on the time taken for the PTP to open in high conductance state.

Methods

The process of building a composite model from individual models is depicted in the flow chart in Fig. 1. Our objective is that the composite model should not only satisfy the properties of the individual models but also help lead biologists using this model to additional insights regarding the processes described here. Specifically, for an in-depth study of calcium signaling we integrate key components from the three models to investigate the dynamics of calcium oscillations to neuronal cell death. The three models are explained in the following subsections.

As depicted in the model schematic diagram (Fig. 2), the composite model consists of four main molecular species, namely Cytosolic Ca^{2+} (CAC), Mitochondrial Ca^{2+} (CAM), Endoplasmic Reticulum Ca^{2+} (CAER) and beta-amyloid (Aβ). In addition, a node is included to represent the mitochondria in high-conductance state (PTPh).

The Fall-Keizer model

The Fall-Keizer model is an integrated model depicting mitochondrial Ca^{2+} handling and metabolic function. It integrates the Magnus-Keizer model [10, 11] and the De Young-Keizer model [12]. The Magnus-Keizer model is a comprehensive mitochondrial model with six proton transfer mechanisms that affect Ca^{2+} signaling.

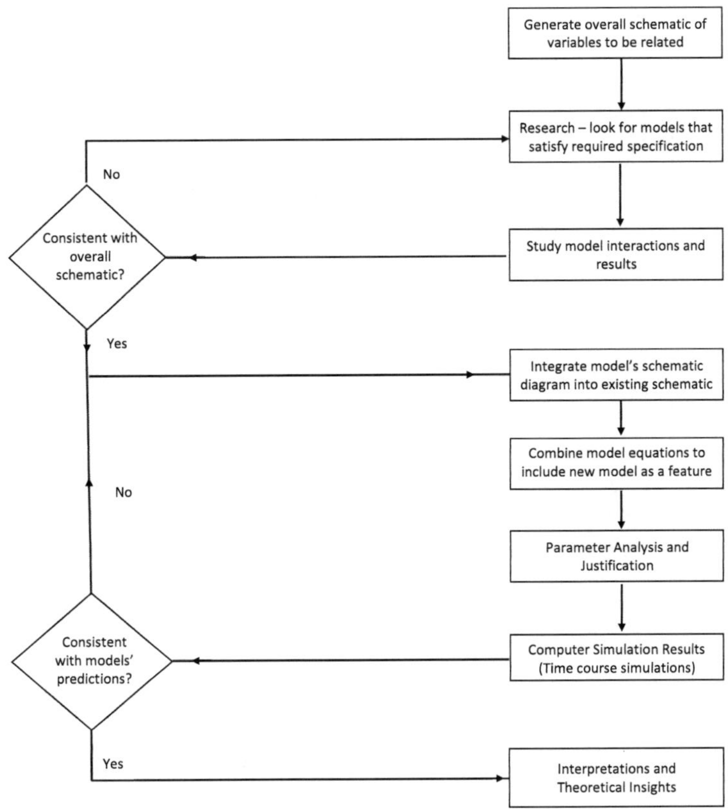

Fig. 1 Flow chart of analyzing existing models and constructing one composite model

A key motivation for using the Fall-Keizer model was to improve the original Magnus-Keizer model, by modifications to the Ca^{2+} uniporter so that the prediction of Ca^{2+} signaling can be more accurate.

The inclusion of the De Young-Keizer model for inositol-1,4,5-triphosphate (IP3)-mediated Ca^{2+} release along with appropriate scaling and provisions for accommodating different cell types allows a modeler to easily shape this model to his or her objective. Furthermore, the comprehensive nature and the modularity of the model that have been maintained by Fall and Keizer make this model an ideal choice for the purposes of this paper.

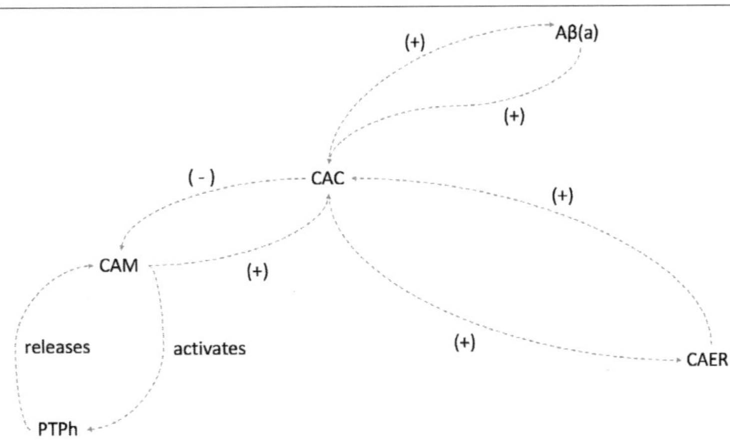

Fig. 2 Composite Model Schematic relating Cytosolic Ca^{2+} concentration (CAC), Mitochondrial Ca^{2+} concentration (CAM), Endoplasmic Reticulum Ca^{2+} concentration (CAER), β-amyloid (a) concentration and PTP high-conductance activation state (PTPh)

The model has over 12 variables interacting with each other to depict the Ca^{2+} signaling in a cell. The basic mitochondrial function has been taken from the Magnus-Keizer model, with a modified mechanism for Ca^{2+} uptake introduced by Fall and Keizer [7]. Furthermore, the modular nature of this model allows for addition of the PTP characteristics which were omitted in the Fall-Keizer model.

$$\frac{dCAC}{dt} = \frac{f_i}{V_c \tau_{min}} \left[M \left(J^{Ca}_{Na,ex} - J^{Ca}_{uni} - J^{Ca}_{PTP} \right) - E \left(J_{serca} - J_{er,out} \right) \right] \tag{1}$$

Table 1 List of parameters used in our model with their values and biological significance [7–9]

Parameters	Value	Biological significance
V	1 ml	Total volume
$uMmM$	1000	Converts μM to mM
τ_{min}	60	Converts minutes to seconds
$p_{cytosol}$	0.5	Proportion of volume occupied my cytosol
$d_{cytosol}$	75 mg/ml	Density of cytosolic protein mg/ml
p_{mito}	0.05	Proportion of volume occupied my mitochondria
d_{mito}	1000 mg/ml	Density of mitochondrial protein mg/ml
p_{er}	0.1	Proportion of volume occupied my ER
d_{er}	1000 mg/ml	Density of ER protein mg/ml
c_{mito}	0.0725 $\frac{nmol}{mV*mg}$	Mitochondrial calcium concentration
ρ_{uni}	300 $\frac{nmol}{mg*min}$	Maximum rate of transport through mitochondrial uniporter
ρ^{Ca}_{Na}	3 $\frac{nmol}{mg*min}$	Maximum rate of transport through Na^+/Ca^{2+} exchanger
ρ_{res}	0.4	Mitochondrial respiration co-efficient
ρ_F1	0.7 $\frac{nmol}{mg*min}$	Mitochondria - Fo/F1 ATPase
$[P_i]_m$	20 mM	Concentration of free phosphates
ρ_{leak}	0.2 $\frac{nmol}{mg*min}$	Mitochondrial membrane proton leak
$J_{red,basal}$	20 $\frac{nmol}{mg*min}$	NADH reduction rate
$J_{max,ANT}$	900 $\frac{nmol}{mg*min}$	ATP/ADP antiport flux
glc	1 mM	Glucose concentration in cytosol
$J_{hyd,max}$	30.1 mM	Cytosol hydrolysis of ATP
V_{IP3}	3000 μM	IP_3 receptor volume
V_{leak}	0.1	proportion of leakage from IP_3 receptor
J_{leak}	0.1	ER leak
d_{IP3}	0.25 μM	IP_3 receptor sensitivity
d_{ACT}	1 μM	IP_3 receptor Ca^{2+} activation constant
d_{INH}	1.4 μM	IP_3 receptor Ca^{2+} inhibition constant
τ	4s	IP_3 receptor inhibitory time constant
V_{serca}	110 $\frac{nmol}{mg*min}$	SERCA pump flux
k_{serca}	0.4μM	SERCA pump Ca^{2+} sensitivity
V_1	0.0065 nM/s	Constant rate of β-amyloid synthesis
V_α	0.05 nM/s	Maximal rate of β-amyloid synthesis
K_α	120 nM	Half-saturation constant
k_β	0.2 nM^3/s	Rate constant of increased Ca^{2+} entry
K_1	0.01 s^{-1}	Rate constant of β-amyloid elimination
n	2	Hill coefficient for activation of β-amyloid synthesis
m	4	Cooperativity coefficient
f_m	0.0003	Mitochondrial Ca^{2+} buffering coefficient
f_i	0.01	Cytosolic Ca^{2+} buffering coefficient

Table 2 List of parameters used in IP_3 step function for IP_3-mediated calcium oscillations [7–9]

Parameters	Value	Biological significance
baseline	0.3 μM	Base concentration
amplitude	0.3 μM	Oscillation amplitude
init	10 *ms*	Initial time
duration	100 *ms*	Duration of oscillations

Equation (1) was obtained from the Fall-Keizer model [7]. CAC is directly proportional to the total volume of mitochondria in the cell (M) multiplied by the rate of transfer of calcium ions into the cytosol from the mitochondria, which is dependent on the sodium-calcium ion exchanger, the calcium uniporter and the PTP. CAC is also directly proportional to the total volume of ER in the cell (E) multiplied by the rate of transfer of calcium ions into the cytosol from the ER, which is affected by the SERCA pump and the leakage of calcium ions from the ER.

Since our focus is on mitochondria-induced cell apoptosis, this model provides us with the appropriate foundation to build upon.

The Mitochondrial PTP model

The model proposed by Oster et al. [8] provides a representation of the mitochondrial permeability transition pore (PTP) behavior. In our model, we have focused on the high conductance state of PTP. In its high-conductance conformation, PTP opening induces unselective solute fluxes that dissipate the concentration gradients of relatively big molecules. However, since most proteins remain trapped in the matrix, the resulting oncotic imbalance (at least in vitro) causes high amplitude swelling of the organelle. The subsequent unfolding of the inner membrane causes rupture of the outer membrane, which results in the release of soluble components (mainly cytochrome c and Apoptosis Inducing Factor (AIF)) that are normally located in the intermembrane space [13–15].

The transition of the pore to a high-conductance state requires prolonged levels of high mitochondrial Ca^{2+}. Once the pore opens to this state, it remains open, leading to cell death. The model proposed by Oster et al. [8] assumes that whether or not the pore enters the high-conductance state depends on a secondary slow process, which in turn depends on the overall mitochondrial Ca^{2+} load.

The Amyloid metabolism model

It is widely known that amyloids perturb Ca^{2+} homeostasis, and β-amyloids perturb the balance between Ca^{2+} entry in and extrusion out of the cytoplasm. In healthy neurons, these processes equilibrate, leading to a basal Ca^{2+} level in the range of 50-100 nM [16]. Studies on the cortical neurons of AD stricken animals found a basal Ca^{2+} level of around 250 nM, i.e. around twice that found in controls [17].

Based on these observations, we believe that the model proposed by Caluwe and Dupont [9] provides an accurate representation of the relationship between β-amyloid protein concentration and cytoplasmic calcium concentration. Their depiction of a positive feedback loop supports the increased basal Ca^{2+} levels observed in cortical neurons with the accumulation of amyloids. Furthermore, research conducted on the production of the toxic oligomers found that Ca^{2+} ions actually promote the synthesis of β-amyloids [18]. These results support the existence of a bistable state switch in the neurons due to the positive feedback loop between the amyloid concentration and the calcium ion concentration. Equation (2), describing the change in intracellular

Table 3 List of parameters used in PTP Integration for PTPh and PTPl states [7–9]

Parameters	Value	Biological significance
CAM^*	$4\mu M$	Threshold Mitochondrial Ca^{2+}
y^*	0.8	Secondary process threshold
f_{H_M}	1.28×10^{-6}	Fast buffering constant for protons in mitochondria
p_1	0.022	Parameter p_1
p_2	0.0001	Parameter p_2
p_3	0.0231	Parameter p_3
p_4	0.0001	Parameter p_4
amp_τ	26000	Amplitude for time constant
p_6	0.001	Parameter p_6
$perm_l^H$	3.0	PTP Permeability to protons
$perm_{Ca}$	0.4	PTP Permeability to calcium ions
postptp	2	PTP opening indication constant

Table 4 List of model ODEs from [7–9]

Equation	Biological significance
$dNADH_m/dt = (J_{red} - J_o) * M / (uMmM * V_m * minute)$	ODE for mitochondrial NADH concentration change (mM/s)
$dADP_m/dt = \left(J_{ANT} - J_{p,TCA} - J_{p,F1}\right) * M / (uMmM * V_m * minute)$	ODE for mitochondrial ADP concentration change (mM/s)
$dADP_i/dt = \left(-J_{ANT} * M + \left(J_{hyd} - J_{p,gly}\right) * C\right) / (uMmM * V_c * minute)$	ODE for cytosolic ADP concentration change (mM/s)
$dPSI/dt = -\left(-J^H_{res} + J_{H,F1} + J_{ANT} + J^H_{PTP} + J^H_L + 2 * J_{uni} + 2 * J^{Ca}_{PTP}\right) * M / (c_{mito} * minute)$	ODE for mitochondrial inner membrane voltage
$dh/dt = (d_{INH} - (CAC + d_{INH}) * h) / \tau$	ODE for change in percentage of closed channels

calcium concentration, has been taken from the Caluwe and Dupont model [9].

$$\frac{dCAC}{dt} = V_2 + K_\beta a^m - k_2 * CAC \qquad (2)$$

This equation shows the effect of amyloids on intracellular calcium concentration. It was formulated based on the assumption that amyloids increase intracellular calcium concentration by increasing the permeability of plasma membranes. This assumption is also made in our model. Here, the rate at which calcium enters the cytoplasm and the rate of eliminating calcium ions from the cytoplasm are assumed to be constant. Furthermore, calcium ion concentration in the cytoplasm is assumed to have first order kinetics.

Composite model

We used XPPAUT [19], an .ode model file as in Additional file 1 and MATLAB to plot our time-course simulations. Details of the composite model constructed based on the afore-mentioned three models are given below. A schematic diagram of the composite model is illustrated in Fig. 2. The parameters used in our composite model were obtained from the individual models.

The model equations for these molecular species are given below and the model parameters are given in Table 1. For the meaning of the rate constants and other parameters readers are to refer to Tables 1, 2 and 3. In an attempt to extend the models described previously and make them more comprehensive, we have combined components and effects that act on the same ion concentration. For instance, here, we used the components affecting CAC in Eqs. (1) and (2) to create the composite equation shown in Eq. (3):

$$\frac{dCAC}{dt} = \frac{f_i}{V_c \tau_{min}} \left[M\left(J^{Ca}_{Na,ex} - J^{Ca}_{uni} - J^{Ca}_{PTP}\right) - E\left(J_{serca} - J_{er,out}\right)\right] + k_\beta a^m \qquad (3)$$

This equation shows the effect of amyloid concentration inducing calcium ion entry into the cell by taking into account its effect of increasing plasma membrane permeability. The effect of amyloid deposition on cytosolic calcium concentration is quantified as the rate of increase

Table 5 List of initial conditions [7–9]

Parameters	Value	Biological significance
$PSI(0)$	164 mV	Base potential at time 0
$CAM(0)$	0.05 μM	Mitochondrial Ca^{2+} concentration at time 0
$CAC(0)$	0.05 μM	Cytosolic Ca^{2+} concentration at time 0
$CAER(0)$	11 μM	ER Ca^{2+} concentration at time 0
$ADP_m(0)$	4.46 mM	Mitochondrial ADP concentration at time 0
$ADP_i(0)$	0.028 mM	Cytosolic ADP concentration at time 0
$NADH_m(0)$	0.16 mM	Mitochondrial NADH concentration at time 0
$h(0)$	95%	Percentage of closed channels at time 0
$a(0)$	0 mM	β-amyloid concentration at time 0
$y(0)$	0	Secondary slow process involved in PTP opening at time 0
$PTP_h(0)$	0	PTP closed at time 0
$PTP_l(0)$	0	PTP closed at time 0

Table 6 List of model equations used in calculation of protein amounts [7–9]

Equation	Biological significance
$M = V * p_{mito} * d_{mito}$	Calculation of mitochondrial protein amount
$C = V * p_{cytosol} * d_{cytosol}$	Calculation of cytosolic protein amount
$E = V * p_{er} * d_{er}$	Calculation of ER protein amount

of the Ca^{2+} entry multiplied by the amyloid concentration raised to its cooperativity coefficient.

$$\frac{dCAM}{dt} = \frac{f_m M}{V_m \tau_{min}} \left(J_{uni}^{Ca} - J_{Na,ex}^{Ca} + J_{PTP}^{Ca} \right) \tag{4}$$

The dynamics of mitochondrial calcium ion concentration is shown in Eq. (4), obtained from Oster et al. [8]. It increases with the calcium ion influx through the uniporter and PTP, and it decreases with the calcium ion efflux through the Na^+-Ca^{2+} ion exchanger.

$$\frac{dCAER}{dt} = \frac{f_i E}{V_e \tau_{min}} \left(J_{serca} - J_{er,out} \right) \tag{5}$$

ER calcium ion concentration varies with influx through the SERCA (Sarcoplasmic/Endoplasmic Reticulum Calcium ATPase) pump and efflux by leakage of calcium ions, obtained from Fall and Keizer (2001) [7].

$$\frac{da}{dt} = V_1 + V\alpha \left(\frac{1}{1 + \frac{K_\alpha^n}{CAC^n}} \right) - K_1 a \tag{6}$$

Equation (6) focuses on the rate of change of amyloid concentration (a) in the cell. Like Caluwe and Dupont [9], we picked amyloid concentration as a generic quantity, encompassing intracellular and extracellular amyloid concentrations, along with amyloid compounds of different lengths and oligomerization states. The rate of amyloid synthesis is assumed to be a constant. The increase of the cytosolic calcium concentration increases the amyloid concentration as well, according to Caluwe and Dupont [9]. Moreover, the increased amyloid concentration negatively affects its own increase in gradient, allowing some form of regulation of amyloids in spite of cytosolic calcium ion concentration.

$$\frac{dPTP_h}{dt} = \frac{PTP_h^\infty - PTP_h}{\tau_h} \tag{7}$$

Equation (7) is derived from Oster et al. [8]. It depicts the mitochondrial PTP in its high conductance state as a binary state controlled by a Heaviside function. This depends on a secondary slow process, which is also a Heaviside function function that depends on a mitochondrial calcium ion concentration crossing a particular threshold. This is the time constant used to provide a time lag in simulating the actual switching of the PTP state. Tables 4, 5, 6, 7, 8, 9, 10, 11, 12, 13, 14, 15, 16 and 17 contain the equations, parameters and initial conditions from other models that have been used in this model.

Table 7 List of model equations used in calculation of compartment volumes [7–9]

Equation	Biological significance
$V_m = (V * p_{mito})$	Calculation of mitochondrial compartment volumes
$V_c = (V * p_{cytosol})$	Calculation of cytosolic compartment volumes
$V_e = (V * p_{er})$	Calculation of ER compartment volumes

Table 8 List of model equations used in nucleotide conversion/conservation relations (obtained from [10, 11])

Equation	Biological significance
$ATP_m = (12 * d_{mito}/uMmM) - ADP_m$	Mitochondrial ATP concentration
$NAD = (8 * d_{mito}/uMmM) - NADHM$	Mitochondrial NAD concentration

Results

In the absence of pathology

Figures 3 and 4 show the time-course simulation of CAC, CAM and PTPh in a neuron stimulated by an action potential but devoid of deposition of misfolded β-amyloid oligomers. The β-amyloid concentration has been mathematically made null to depict the normal response of these parameters in a neuron to such a stimulus. In this scenario, there is an initial spike in cytosolic calcium level due to the action potential, which is subsequently followed by oscillations of lower amplitude in the ER, cytosol and mitochondria. The oscillations during such a stimulus are consistent with those observed in the Fall-Keizer model [7]. On adding mitochondria to IP_3-mediated calcium ion release in isolation, long-term oscillations are observed (as detailed in [7]). However in our model, on incorporation of uniporter and PTP transport, there is transport of calcium ions outside this oscillatory mechanism, because of which the oscillations cannot be maintained and will stop. These oscillations are temporary, and die down after around 110 ms, restoring the calcium level in the ER, mitochondria and cytosol to their original values. There is naturally no stimulation of the mitochondrial PTP to open in the high conductance state due to the low resting concentration of Ca^{2+} in the mitochondria.

In this case, the neuron responds as it would normally do to a stimulus, with the Ca^{2+} ions being released rapidly from the ER via the IP_3 channels to maintain calcium homeostasis across the cell. This sudden spurt results in the onset of oscillations of Ca^{2+} ion concentration in the

Table 9 List of model equations used in calculation of proportion of free nucleotides (obtained from [11])

Equation	Biological significance
$ADP_{mf} = 0.8 * ADP_m$	Unbound mitochondrial ADP concentration
$ADP_{if} = 0.3 * ADP_i$	Unbound cytosolic ADP concentration

Table 10 List of model equations used in calculation of proportion of charged, free nucleotides (obtained from [11])

Equation	Biological significance
$[ADP^{3-}]_m = 0.45 * ADP_{mf}$	Unbound, 3- charged mitochondrial ADP concentration
$[ADP^{3-}]_i = 0.45 * ADP_{if}$	Unbound, 3- charged cytosolic ADP concentration
$[MgADP^{-}]_i = 0.55 * ADP_{if}$	Unbound, 1- charged cytosolic $MgADP^{-}$ concentration
$[ATP^{4-}]_i = 0.05 * ATP_i$	Unbound, 4- charged cytosolic ATP concentration
$[ATP^{4-}]_m = 0.05 * ATP_m$	Unbound, 4- charged mitochondrial ATP concentration

Table 11 List of model equations used in mitochondrial Ca^{2+} handling (obtained from [10])

Equation	Biological significance
Mitochondrial uniporter equations	
$MWC_{num} = (CAC/6) * ((1 + (CAC/6))^3)$	MWC numerator
$MWC_{denom} = ((1 + (CAC/6))^4) + (50/((1 + (CAC/0.38))^{2.8}))$	MWC denominator
$MWC = MWC_{num}/MWC_{denom}$	MWC fraction value
$V^D_{uni} = (PSI - 91)/13.35$	Uniporter potential exponent
$J_{uni} = (\rho_{uni} * V^D_{uni} * (MWC - CAM * exp(-V^D_{uni}))/(1 - exp(-V^D_{uni}))) * (1 - PTP_h)$	Rate of transport through uniporter considering PTP in high conductance state
Na^+/Ca^{2+} exchanger equations	
$V^D_{nc} = exp((PSI - 91)/53.4)$	Na^+/Ca^{2+} exchanger potential generated
$J_{nc} = (\rho_{nc} * V^D_{nc} * (1/(1 + (9.4/30) * *2)) * (1/(1 + (0.003 * Dmito/CAM)))) * (1 - PTP_h)$	Rate of Na^+/Ca^{2+} exchange

Table 12 List of model equations used in calculation of mitochondrial respiration equations (obtained from [11])

Equation	Biological significance
$A_{res} = (1.35e18) * NADHM^{0.5}/(NAD)^{0.5}$	A_{res} = affinity bracketed expression
$V^D_{res} = exp(0.191 * PSI)$	Respiration potential generated
Proton pump equations	
$r_1 = 7e - 7$	Variable r_1
$r_2 = (2.54e - 3) * A_{res}$	Variable r_2
$r_3 = 0.639 * V^D_{res}$	Variable r_3
$r_4 = 7.58e13 + (1.57e - 4) * A_{res}$	Variable r_4
$r_5 = (1.73 + A_{res} * 1.06e - 17) * V^D_{res}$	Variable r_5
$J^H_{res} = 360 * \rho_{res} * ((r1 + r2 - r3)/(r4 + r5))$	Rate of transport through proton pump during respiration
Oxygen consumption rate equations	
$o_1 = A_{res} * 2.55e - 3$	Variable o_1
$o_2 = A_{res} * 2.00e - 5$	Variable o_2
$o_3 = 0.639 * (V^D_{res})$	Variable o_3
$o_4 = (V^D_{res}) * A_{res} * 8.63e - 18$	Variable o_4
$o_5 = (1 + A_{res} * 2.08e - 18) * 7.54e13$	Variable o_5
$o_6 = (1.73 + 1.06e - 17 * A_{res}) * V^D_{res}$	Variable o_6
$J_o = 30 * \rho_{res} * (o1 + o2 - o3 + o4)/(o5 + o6)$	Rate of oxygen consumption during respiration

Table 13 List of model equations used in calculation of mitochondrial Fo/F1-ATPase equations (obtained from [11])

Equation	Biological significance
$A_{F1} = (1.71e9) * (ATP_m)/(ADP_{mf} * pim)$	A_{F1} = affinity bracketed expression
$V_{F1}^D = exp(0.112 * PSI)$	ATPase potential generated
F_0/F_1 ATPase phosphorylation of ADP_m	
$f_1 = 10.5 * A_{F1}$	Variable f_1
$f_2 = 166 * V_{F1}^D$	Variable f_2
$f_3 = (4.85e - 12) * A_{F1} * V_{F1}^D$	Variable f_3
$f_4 = (1e7 + 0.135 * A_{F1}) * 275$	Variable f_4
$f_5 = (7.74 + (6.65e - 8) * A_{F1}) * V_{F1}^D$	Variable f_5
$J_{p,F1} = -60 * \rho_{F1} * ((f_1 - f_2 + f_3) / (f_4 + f_5))$	Rate of F_0/F_1 ATPase phosphorylation
$J_{H,F1} = -180 * \rho_{F1} * (0.213 + f_1 - 169 * V_{F1}^D) / (f_4 + f_5)$	Proton flux due to ATPase
$J_{H,leak} = \rho_{leak} * (PSI + 24.6)$	Mitochondrial membrane proton leak
$f_{PDH} = 1/(1 + (1.1 * (1 + (15/(1 + (CAM/0.05))^2))))$	Fraction of activated pyruvate
$J_{red} = J_{red,basal} + 6.3944 * f_{PDH} * J_{gly,total}$	NADH reduction rate
ATP/ADP antiport flux	
$ant_1 = ([ATP^{4-}]_i / [ADP^{3-}]_i) * ([ADP^{3-}]_m / [ATP^{4-}]_m) * exp(-PSI/26.7)$	Variable ant_1
$ant_2 = 1 + ([ATP^{4-}]_i / [ADP^{3-}]_i) * exp(-PSI/53.4)$	Variable ant_2
$ant_3 = 1 + ([ADP^{3-}]_m / [ATP^{4-}]_m)$	Variable ant_3
$J_{ANT} = J_{max,ANT} * ((1 - ant_1) / (ant_2 * ant_3))$	Rate of Adenine Nucleotide Translocator (ANT) activity
Phosphorylation of ADP_m from TCA cycle	
$J_{p,TCA} = (J_{red,basal}/3) + 0.84 * f_{PDH} * J_{gly,total}$	Unbound, 3- charged mitochondrial ADP concentration

Table 14 List of model equations used in calculation of cytosolic components (obtained from [11])

Equation	Biological significance
Glycolytic rate based on hexokinase	
$gly_{num} = (123.3 * (1 + 1.66 * glc) * (glc * ATP_i)) * 0.0249$	Glycolytic rate numerator
$gly_{denom} = 1 + (4 * ATP_i) + ((1 + 2.83 * ATP_i) * 1.3 * glc) + ((1 + 2.66 * ATP_i) * 0.16 * glc^2)$	Glycolytic rate denominator
$J_{gly,total} = gly_{num}/gly_{denom}$	Glycolytic rate
$J_{p,gly} = 2 * J_{gly,total}$	Phosphorylation of ADPi from glycolysis
$J_{hyd} = 41 * (ATP_i) + J_{hyd,max}/(1 + (8.7/glc)^{2.7})$	Cytosolic hydrolysis of ATP

Table 15 List of model equations used in calculation of ER Ca^{2+} handling (modified from [21])

Equation	Biological significance
$J_{er,out} = (V_{IP3} * ((IP_3/(IP_3 + d_{IP3}))^3) * ((CAC/(CAC + d_{ACT}))^3) * (h^3) + v_{leak}) * (CAER - CAC)$	IP_3 receptor and leak
$J_{serca} = V_{serca} * CAC^2/(k_{serca}^2 + CAC^2)$	SERCA pump

Table 16 List of model equations used in IP_3 step function (modified from [21])

Equation	Biological significance
$stepupf = heav(t - init)$	Heaviside step up function
$stepdownf = heav(t - (init + duration))$	Heaviside step down function
$IP_3 = baseline + amplitude * (stepupf - stepdownf)$	IP_3 step function

mitochondria, cytosol and ER. There is no further activity, and the concentration of Ca^{2+} is sustained at a low equilibrium value.

In the presence of pathology

In Fig. 5, the neuron is affected by β-amyloid deposition, causing an increase in the entry of calcium ions into the cell and dysregulation of Ca^{2+} channel receptors on the ER. At the start of the time-course simulation, an action potential is simulated causing the calcium levels in the cytosol to jump to over 1.4 μM. Then, owing to Calcium-Induced Calcium Release (CICR), the process of sequestering and subsequent release of Ca^{2+} ions, with the ER and mitochondria, the calcium level continues to oscillate for a while until it dies down and returns to its original level of around 0.05-0.1 μM. As the amyloid concentration grows in the process, there is a sudden jump in the cytosolic and correspondingly mitochondrial calcium ion concentration due to the sudden release of a large volume of calcium ions from the ER (not shown in this graph). Amyloid metabolism affects the ryanodine and IP_3 receptors located on the surface of the ER that regulate calcium release from the ER [20]. The sudden rise in calcium levels can be attributed to the ER sequestering a large proportion of the calcium that was initially in the cytosol and mitochondria.

As this continues, oscillation of calcium levels resumes, except this time the mitochondrial calcium ion concentration continuously increases, indicating an increase in sequestration of calcium ions by the mitochondria. As this constant rise is maintained for a period of time, as detailed in Oster et al. [8], a slower secondary process is activated, which results in the PTP opening in the high-conductance state. The opening of PTP in high-conductance state detailed by Oster et al. [8] follows exactly the same pattern, with the high, sustained levels of mitochondrial calcium ion concentration resulting in the activation of a slower, secondary process. This secondary slow process, on completion, results in the PTP opening in high conductance state.

The β-amyloid concentration rises slowly with the growth in cytosolic calcium, and this observation can be attributed to the relatively slow nature of amyloid processing to produce β-amyloids as compared to transfer of Ca^{2+} ions (Fig. 6).

Figure 7 shows that the PTPh starts opening at around 742 ms and takes 123 ms to reach the high-conductance state at around 865 ms. At the point of opening of the PTP in high conductance state (near 741 ms), mitochondrial Ca^{2+} rushes out through the PTP and into the cytosol, which in turn is sequestered by the ER. The spurted release results in an oscillation of Ca^{2+} ions between the

Table 17 List of model equations used in PTP Integration (modified from [21])

Equation	Biological significance
$\tau_y = 1000 * ((1000/cosh\,(CAM/0.1)) + 0.1)$	Time constant for secondary slow process
$\tau_h = \tau_y/8$	Time constant for PTP high conductance state
$PTP_h^{\infty} = heav\,(y - y^*)$	Heaviside step function for PTP max value
$y^{\infty} = heav\,(CAM - CAM^*)$	Heaviside step function for y threshold value
$dy/dt = (y^{\infty} - y)\,/\tau_y$	Secondary slow process involved in opening of PTP high conductance state
$dPTP_h/dt = \left(PTP_h^{\infty} - PTP_h\right)/\tau_h$	PTP high conductance state dynamics
$J_{PTP}^H = perm_l^H * PTP_l * PSI * (H_M - 0.0000000398 * exp\,(-37.434 * PSI)\,/\,(1 - exp\,(-37.434 * PSI)))$	Proton flux through PTP in high conductance state
$J_{PTP}^{Ca} = perm_{Ca} * PTP_l * J_{uni} * (1 - postptp * PTP_h)$	Rate of Ca^{2+} ion transport across PTP
$dH_M/dt = \left(f_{H_M}/\tau_h\right) * \left(J_L^H + J_{F1}^H - J_{res}^H + J_{PTP}^H\right)$	Change in mitochondrial proton concentration
$\tau_l = p_6 + amp_\tau/cosh\,((H_M - p_3)\,/p_4)$	Time constant for PTP low conductance state
$PTP_l^{\infty} = 0.5 * (1 + tanh\,((p1 - H_M)\,/p_2))$	Rate of change of polling function
$dPTP_l/dt = \left(PTP_l^{\infty} - PTP_l\right)/\tau_l$	PTP low conductance state dynamics

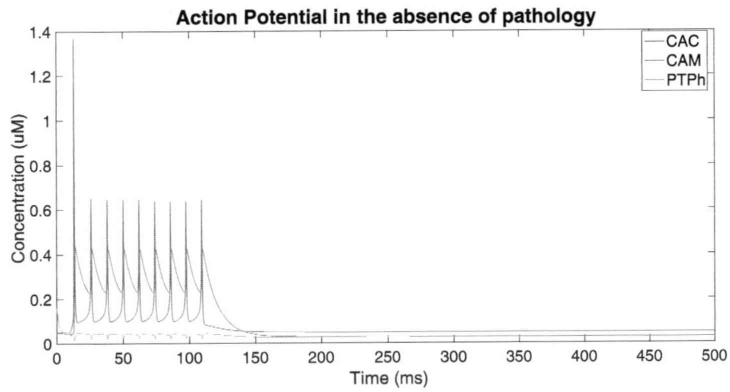

Fig. 3 Overall time-course simulation showing cytosolic (CAC) and mitochondrial (CAM) Ca^{2+} ion concentration along with tendency for opening of PTP in high-conductance state (PTPh) over a time of 500ms in the absence of β-amyloid deposition (by setting $V_1 = 0$, $V_\alpha = 0$ and $K_1 = 0$)

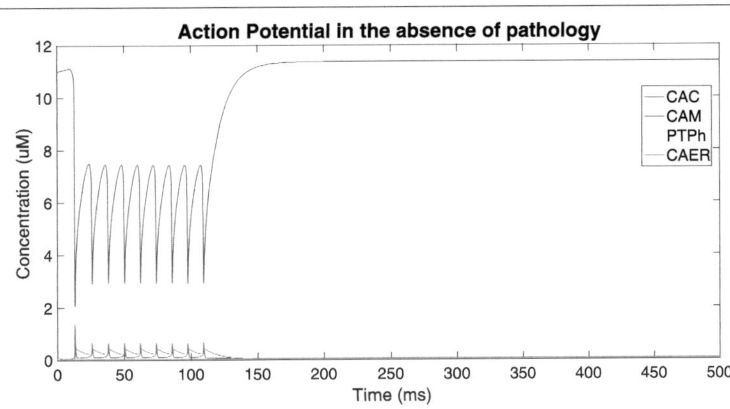

Fig. 4 Addition of ER Ca^{2+} ion concentration (CAER) to Fig. 3 and a visual representation of the scale of difference between the resting ER Ca^{2+} level and cytosolic and mitochondrial resting Ca^{2+} levels

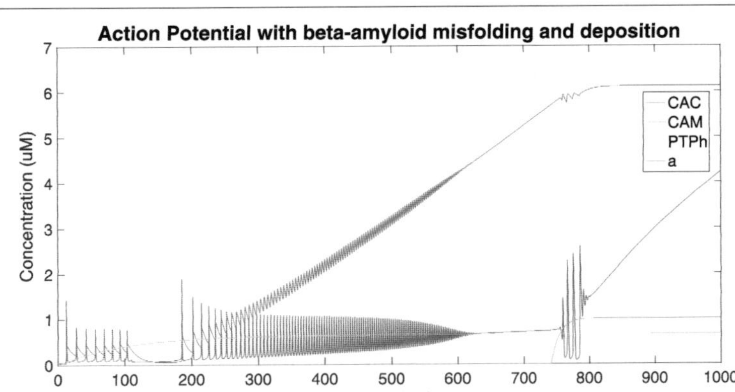

Fig. 5 Graph depicting cytosolic (CAC) and mitochondrial (CAM) Ca^{2+} ion concentration, β-amyloid concentration (a) and tendency for opening of PTP in high-conductance state (PTPh) over a time of 1000ms - in the presence of pathology (using the provided equations, parameter values and initial conditions in the XPP .ode file)

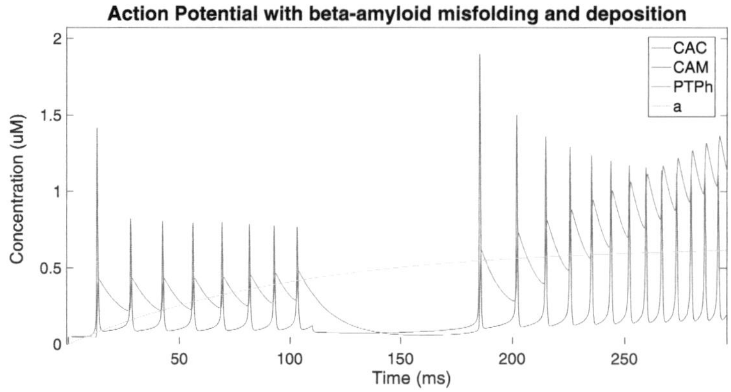

Fig. 6 Zooming into 0-250 ms of Fig. 5 to visualise the oscillations and growth of amyloid concentration

ER and the cytosol, which dies down as the mitochondrial calcium level stagnates, indicating that the mitochondria is no longer functional.

When the PTPh opens in the high conductance state, calcium ions rush out of the mitochondria causing the CAM to level off. As the mitochondria stop functioning due to the PTP opening in high conductance, its calcium ion concentration is frozen, with CAC and CAER skyrocketing due to uncontrollable amounts of calcium ions flowing into the cell. This may be because the amyloid concentration has also increased over time (although in a much more regulated fashion), causing the plasma membrane to become more permeable and allowing more calcium ions into the neuron. Consequently, a jump occurs in cytosolic calcium ion concentration, which oscillates for a while and once stabilized begins to increase at an accelerated rate.

Conclusion and discussion

This paper presents a mathematical model of the biological processes involved in the deposition of β-amyloids in and around the neurons and its effect on neuronal calcium signaling homeostasis. Using a detailed mitochondrial model of calcium signaling [7], the paper relates this change in calcium signaling in the cytosol to the calcium level in the mitochondrial matrix. Moreover, the introduction of the permeability transition pore and its characteristics allowed the model to depict the irreversible onset of apoptosis. By combining the models and features regarding β-amyloid deposition, mitochondrial calcium signaling and permeability transition pore activity, we simulated the opening of the permeability transition pore using amyloid deposition as the trigger. We found that, using the parameters and model equations above, high-conductance state is reached at around 865 ms after amyloid deposition begins.

The lack of comprehensive models of subcellular dynamics in Alzheimer's disease poses a challenge for computational scientists to explore. We envisage a need for integrated modeling, i.e. selecting individual and specialized models and finding ways to combine them, to formulate a composite model of a neuron's cell fate in AD.

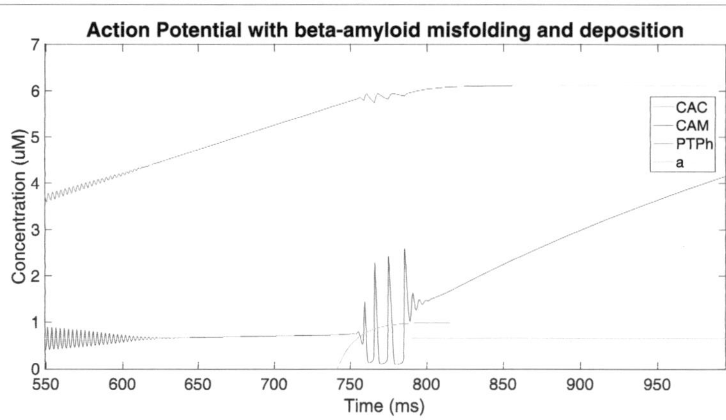

Fig. 7 Zooming into 550-1000 ms of Fig. 5 to observe changes on opening of the PTP

In this paper, we have modeled only a single neuron undergoing mitochondrial PTP-related apoptosis, which is one of the mechanisms of cell death in AD. We have not studied necrosis due to Ca^{2+} excitotoxicity, or looked at the effect of β-amyloid deposition and misfolding on cognitive functions, both of which can substantially extend the scope of this research. Furthermore, the effect of the neuron on its neighboring neurons in such a scenario has not been considered in this paper, as we assumed the neuron to exist in isolation. These issues are pertinent, and extending our composite model to address them could significantly improve our understanding of the disease.

In summary, we constructed a composite model by integrating three individual models to recapitulate a sequence of events and their repercussions that eventlead to neuronal death. Through this model, we are able to shed new light on a single sequence of processes starting from the deposition and misfolding of β-amyloid in and around the neuron all the way to neuronal apoptosis. Compared with existing models, this model provides a more comprehensive view of these molecular processes occurring in Alzheimer's disease. It represents a step towards building more realistic models to facilitate the diagnosis and treatment of this aging-related disease.

Additional file

Additional file 1: This file contains the parameters, initial conditions and equations used to generate the results in this paper. It can be run using XPP [19]. (ODE 8.18 kb)

Acknowledgements
We would like to thank Dr. Andrew Oster for his support in this project in answering our numerous questions and providing supplementary material for his work. We also wish to acknowledge the funding support for this project from Nanyang Technological University under the Undergraduate Research Experience on CAmpus (URECA) programme.

Funding
We wish to acknowledge the funding support for this project from Nanyang Technological University under the Undergraduate Research Experience on CAmpus (URECA) programme. This work was also supported by the MOE AcRF Tier 1 grant (2015-T1-002-094), Ministry of Education, Singapore. The funding for publication of this article was provided by the MOE AcRF Tier 1 grant (2015-T1-002-094), Ministry of Education, Singapore.

Availability of data and materials
Not applicable.

About this supplement
This article has been published as part of *BMC Systems Biology* Volume 12 Supplement 1, 2018: Selected articles from the 16th Asia Pacific Bioinformatics Conference (APBC 2018): systems biology. The full contents of the supplement are available online at https://bmcsystbiol.biomedcentral.com/articles/supplements/volume-12-supplement-1.

Authors' contributions
Experimental concept and design: BR, CKH, ZJ; Sample collection and data contribution: BR; Data analysis and interpretation: BR, CKH, ZJ; Computational support: BR, CKH; Manuscript writing and figures: BR, ZJ; Manuscript review: All authors read and approved the final manuscript.

Ethics approval and consent to participate
Not applicable.

Consent for publication
Not applicable.

Competing interests
The authors declare that they have no competing interests.

Author details
[1]Biomedical Informatics Lab, School of Computer Science and Engineering, Nanyang Technological University, 50 Nanyang Avenue, 639798 Singapore, Singapore. [2]Genome Institute of Singapore, Biopolis, 138672 Singapore, Singapore. [3]Complexity Institute, Nanyang Technological University, Nanyang Drive, 637723 Singapore, Singapore.

Published: 11 April 2018

References

1. Nixon RA. Autophagy, amyloidogenesis and alzheimer disease. J Cell Sci. 2007;120(23):4081–91.
2. Demuro A, Mina E, Kayed R, Milton SC, Parker I, Glabe CG. Calcium dysregulation and membrane disruption as a ubiquitous neurotoxic mechanism of soluble amyloid oligomers. J Biol Chem. 2005;280(17): 17294–300.
3. Demuro A, Parker I, Stutzmann GE. Calcium signaling and amyloid toxicity in alzheimer disease. J Biol Chem. 2010;285(17):12463–8.
4. Berridge MJ. Calcium hypothesis of alzheimer's disease. Pflügers Archiv-European J Physiol. 2010;459(3):441–9.
5. Ichas F, Mazat JP. From calcium signaling to cell death: two conformations for the mitochondrial permeability transition pore. switching from low-to high-conductance state. Biochim Biophys Acta Bioenergetics. 1998;1366(1):33–50.
6. Thul R. Translating intracellular calcium signaling into models. Cold Spring Harb Protoc. 2014;2014(5):066266.
7. Fall CP, Keizer JE. Mitochondrial modulation of intracellular ca 2+ signaling. J Theor Biol. 2001;210(2):151–65.
8. Oster AM, Thomas B, Terman D, Fall CP. The low conductance mitochondrial permeability transition pore confers excitability and cicr wave propagation in a computational model. J Theor Biol. 2011;273(1): 216–31.
9. De Caluwé J, Dupont G. The progression towards alzheimer's disease described as a bistable switch arising from the positive loop between amyloids and ca 2+. J Theor Biol. 2013;331:12–18.
10. Magnus G, Keizer J. Minimal model of beta-cell mitochondrial ca2+ handling. Am J Physiol Cell Physiol. 1997;273(2):717–33.
11. Magnus G, Keizer J. Model of β-cell mitochondrial calcium handling and electrical activity. i. cytoplasmic variables. Am J Physiol-Cell Physiol. 1998;274(4):1158–73.
12. Keizer J, De Young G. Simplification of a realistic model of ip3-induced ca2+ oscillations. J Theor Biol. 1994;166(4):431–42.
13. Kantrow SP, Piantadosi CA. Release of cytochrome c from liver mitochondria during permeability transition. Biochem Biophys Res Commun. 1997;232(3):669–71.
14. Scarlett JL, Murphy MP. Release of apoptogenic proteins from the mitochondrial intermembrane space during the mitochondrial permeability transition. FEBS Lett. 1997;418(3):282–6.
15. Zamzami N, Hirsch T, Dallaporta B, Petit PX, Kroemer G. Mitochondrial implication in accidental and programmed cell death: apoptosis and necrosis. J Bioenerg Biomembr. 1997;29(2):185–93.
16. Shi C, Wu F, Xu J. H 2 o 2 and paf mediate aβ1-42-induced ca 2+ dyshomeostasis that is blocked by egb761. Neurochem Int. 2010;56(8): 893–905.
17. Lopez JR, Lyckman A, Oddo S, LaFerla FM, Querfurth HW, Shtifman A. Increased intraneuronal resting [Ca^{2+}] in adult alzheimer's disease mice. J Neurochem. 2008;105(1):262–71.
18. Itkin A, Dupres V, Dufrêne YF, Bechinger B, Ruysschaert JM, Raussens V. Calcium ions promote formation of amyloid β-peptide (1–40) oligomers causally implicated in neuronal toxicity of alzheimer's disease. PLoS ONE. 2011;6(3):18250.

19. Ermentrout B. Simulating, Analyzing, and Animating Dynamical Systems: a Guide to XPPAUT for Researchers and Students. Philadelphia: SIAM; 2002.
20. Area-Gomez E, Castillo MDCL, Tambini MD, Guardia-Laguarta C, De Groof AJ, Madra M, Ikenouchi J, Umeda M, Bird TD, Sturley SL, et al. Upregulated function of mitochondria-associated er membranes in alzheimer disease. EMBO J. 2012;31(21):4106–23.
21. Li YX, Rinzel J. Equations for insp3 receptor-mediated [ca2+] i oscillations derived from a detailed kinetic model: a hodgkin-huxley like formalism. J Theor Biol. 1994;166(4):461–73.

Zhang *BMC Systems Biology* 2018, **12**(Suppl 1):8
https://doi.org/10.1186/s12918-018-0530-9

BMC Systems Biology

RESEARCH

Open Access

Comparisons of gene coexpression network modules in breast cancer and ovarian cancer

Shuqin Zhang

From The Sixteenth Asia Pacific Bioinformatics Conference
Yokohama, Japan. 15-17 January 2018

Abstract

Background: Breast cancer and ovarian cancer are hormone driven and are known to have some predisposition genes in common such as the two well known cancer genes BRCA1 and BRCA2. The objective of this study is to compare the coexpression network modules of both cancers, so as to infer the potential cancer-related modules.

Methods: We applied the eigen-decomposition to the matrix that integrates the gene coexpression networks of both breast cancer and ovarian cancer. With hierarchical clustering of the related eigenvectors, we obtained the network modules of both cancers simultaneously. Enrichment analysis on Gene Ontology (GO), KEGG pathway, Disease Ontology (DO), and Gene Set Enrichment Analysis (GSEA) in the identified modules was performed.

Results: We identified 43 modules that are enriched by at least one of the four types of enrichments. 31, 25, and 18 modules are enriched by GO terms, KEGG pathways, and DO terms, respectively. The structure of 29 modules in both cancers is significantly different with *p*-values less than 0.05, of which 25 modules have larger densities in ovarian cancer. One module was found to be significantly enriched by the terms related to breast cancer from GO, KEGG and DO enrichment. One module was found to be significantly enriched by ovarian cancer related terms.

Conclusion: Breast cancer and ovarian cancer share some common properties on the module level. Integration of both cancers helps identifying the potential cancer associated modules.

Keywords: breast cancer, Ovarian cancer, Module comparison

Background

Despite of decades of intensive study and substantial progress in understanding breast cancer and ovarian cancer, these two diseases remain the most prevalent malignancy in women, and the important causes of death in women. Among solid gynaecological tumors, breast cancer is the most often diagnosed tumor while ovarian cancer is the most deadly gynaecological neoplasia. Both breast and ovarian cancer are hormone driven and are known to have some predisposition genes in common. The major genes associated with susceptibility to breast and ovarian cancer are the two well-known high-penetrance cancer genes: BRCA1 and BRCA2 [1–3]. However, mutations in these genes account for only a very small percent of all breast cancers and ovarian cancers. Other genes such as TP53, PTEN, and STK11/LKB1, are even less common causes of breast and ovarian cancer [4]. Despite tremendous efforts to conquer such malignant diseases, research on studying the mechanism of cancer development and developing effective preventive measures is still a hot topic.

The high speed development of high-throughput technologies such as next generation sequencing of the human genome, gene expression microarrays, identification of the changes of copy numbers has dramatically accelerated

Correspondence: zhangs@fudan.edu.cn
Center for Computational Systems Biology, Shanghai Key Laboratory for Contemporary Applied Mathematics, School of Mathematical Sciences, Fudan University, No.220 Handan Road, 200433 Shanghai, China

the study aiming at predicting and curing such diseases. Many works have been published to address the topics on associated susceptibilities, potential biomarkers, cancer predictions and so on [5–19]. Several of them put breast cancer and ovarian cancer together in their studies [5–10]. These works either borrowed information from each other with the assumption that both cancers have similar etiologies [5, 7, 9], or conducted research on the differences between the pathogenic mechanisms [6, 8]. Some review papers also analyzed the related research progresses of both cancers together [4, 10]. In this work, we also put both cancers together to study the complex gene coexpression patterns with the network tools.

Network has been widely applied to study the complex interactions between genes, proteins, and other small molecules. It is also a popular tool for studying breast cancers and ovarian cancers [11–16, 20–22]. When using networks to study the complex interactions, one typical concept is module, which is the densely connected subnetworks. With the network module analysis, we may infer the susceptibility genes in cancer [16], identify the biomarkers [14], and predict the prognosis of the cancer patients [11]. Current studies on cancer-related network modules are mainly based on one network, which is also true for breast cancer [14] and ovarian cancer [16]. Even in the study of more than one network, the module identification process is one network by one network, and then comparisons between different networks are performed [16]. Recently, several module identification methods on multiple networks are proposed [22–24]. Among them, the method proposed in [22] introduced an algorithm to find the differential modules in different networks. It mainly concentrates on the differential part. While in the paper [24], the method not only can find the modules in each network, but also can align the modules at the same time. Thus both the common and the differential parts can be detected. In the following, we compared the modules that were identified from the gene coexpression networks of breast cancer and ovarian cancer using the method in [24]. We analyzed the basic properties of the modules including density, average degree, distribution difference etc., and we did enrichment analysis of Gene Ontology (GO), KEGG pathway, Disease Ontology (DO) and Gene Set Enrichment Analysis (GSEA). By comparing the modules, both the common properties and the differences between the two cancers are detected.

Methods

Data sets

The level 3 gene expression data for breast cancer (BRCA) and ovarian cancer (OV) were downloaded from The Cancer Genome Atlas (TCGA). The gene expression data were generated with UNC AgilentG4502A. We chose the samples from the solid tissues only. There are 526 and 572 samples for both BRCA and OV, respectively. The expression value of 17,814 genes was measured. The missing data for each specific gene was imputed with the mean value of the known samples. We also downloaded the most updated protein-protein interaction (PPI) data for humans from BioGrid *https://thebiogrid.org*. We chose the genes having PPIs in the gene expression data, and 9603 genes were selected.

To choose the differential expressed genes for BRCA and OV, *t*-test and Kolmogorov-Smirnov test were applied. The genes having *p*-values less 0.01 in both tests were chosen, where the *p*-values were adjusted by controlling the false discovery rate. We then got 7742 genes.

Coexpression network construction

We first computed the pairwise Pearson correlation coefficient r_{ij} to measure the coexpression levels between gene i and gene j. To make the correlations across the two networks comparable, we 'normalized' the Pearson correlation coefficients with the same method as shown in [25]. Fisher's z transformation score for each r_{ij} was first computed as $z_{ij} = 0.5 \log \frac{1+r_{ij}}{1-r_{ij}}$. The z-scores were then normalized to make z_{ij} follow normal distribution. After this, the 'normalized' correlations were obtained by transforming back to r_{ij}. With these steps, the correlations in the two networks will be on the same level. We then did hard thresholding to make the network be unweighted. We took 99.5% quantile of the transformed correlation coefficients as the threshold. If the absolute value of the correlation coefficient is greater than the threshold, we assigned an edge between the corresponding genes, otherwise, there is no edge. With this method, the average degree is about 37 in both networks.

Module identification in BRCA and OV coexpression networks

Before we did module identification, we first removed the genes having no links with any other genes in both networks. The left gene number became 6779. Consider the constructed gene coexpression network G_1 (BRCA), G_2 (OV) consisting of 6779 genes. We let the adjacency matrices for both networks be A_1, A_2, where $A_k(i,j) = 1$ represents there is an edge between gene i and gene j. We first applied the model proposed in [24] to cluster the genes in lower subspace for both networks. This model aims at finding the clusters in multiple networks and aligning the clusters at the same time. The main idea is to use spectral clustering to find the cluster in each network, and align them by maximizing the cluster similarity of multiple networks. Here, we only have two networks.

We assume the putative number of clusters M in both networks is given first.

Let S^k be the assignment of the 6779 vertices into M clusters for the network G_k, where $S^k_{im} = 1$, if $i \in G_k$ belongs to the m-th cluster, otherwise $S^k_{im} = 0$ for $i = 1, 2, \cdots, 6779, m = 1, 2, \cdots, M, k = 1, 2$. The optimization model is formulated as:

$$\begin{aligned} \min \quad & \sum_{k=1,2}\sum_{m=1}^{M} \frac{{S^k_{,m}}^T (D_k - A_k) S^k_{,m}}{{S^k_{,m}}^T S^k_{,m}} - \beta \sum_{k,l=1,2}\sum_{m=1}^{M} \frac{{S^k_{,m}}^T S^l_{,m}}{\|S^k_{,m}\|_2 \|S^l_{,m}\|_2} \\ \text{s.t.} \quad & S^k_{i,m} \in \{0,1\}, i = 1, 2, \cdots, 6779; m = 1, 2, \cdots, M; k = 1, 2; \\ & \sum_{m=1}^{M} S^k_{,m} = \mathbf{1}, \text{for } k = 1, 2. \end{aligned} \tag{1}$$

The first term in the objective function is to do clustering in both networks separately, and the second term defines the similarity of the clusters in both networks measured by cosine function. β balances the contributions from inter-network and intra-network.

To solve the optimization problem (1), the variable S^k_{im} is relaxed. Using the same technique as spectral clustering, the optimization problem is transformed to:

$$\min \ \tilde{\Psi}(\tilde{S}) = \text{Tr}(\tilde{S}^T C \tilde{S}), \text{ s.t. } \tilde{S}^T \tilde{S} = 2I_M,$$

where $C = \begin{pmatrix} L_1 & \mathbf{0} \\ \mathbf{0} & L_2 \end{pmatrix} - \beta \begin{pmatrix} \mathbf{0} & I_n \\ I_n & 0 \end{pmatrix}, \tilde{S} = \begin{pmatrix} \tilde{S}^1 \\ \tilde{S}^2 \end{pmatrix}, \tilde{S}^k_{,m} = \frac{S^k_{,m}}{\|S^k_{,m}\|_2}$,
and $L_k = D_k - A_k$, $\mathbf{0}$ is an $n \times n$ matrix with all entries being zero.

By computing the eigenvectors corresponding to the M smallest eigenvalues of matrix C, the original vertices in the networks are projected to a space of dimension M. To get the clusters, we may use k-means clustering to cluster the data points similar to spectral clustering. Due to the large size of the network, k-means does not work well. Instead, we applied hierarchical clustering with complete linkage to cluster the vertices. The distance is chosen to be the spearman distance. This is because when the size of matrix C is large, the range of the eigenvector entries is large, but their order is comparatively stable. The algorithm is summarized in '**Algorithm**'.

Selection of parameter β and M The parameter β controls the connections between the vertices in both networks. When $\beta = 0$, it is equivalent to finding the clusters in two networks separately. When β becomes larger, the corresponding vertices in both networks tend to belong to the same cluster. We note that even when a group of vertices are densely connected in the first network, while their corresponding parts are isolated in the second network, the method will put all the isolated vertices in the same cluster as in the first network. Here, since both networks were controlled to have a close number of total connections, we directly set $\beta = 1$, which means the connection weight between two networks is the same as that within both networks. We note that when $\beta > 1$ and it is within a reasonable range, the results do not change much.

Algorithm:
Input: Adjacency matrix $A_k, k = 1, 2$, and M, which is the putative number of clusters.

1. Construct the matrix C;
2. Compute the M eigenvectors $\mathbf{u}_1, \mathbf{u}_2, \cdots, \mathbf{u}_M$ corresponding to the M smallest eigenvalues of matrix C;
3. Construct a new matrix $T \in R^{2 \cdot 6779 \times M}$ with columns $\mathbf{u}_1, \mathbf{u}_2, \cdots, \mathbf{u}_M$;
4. Cluster the points constructed from each row of matrix T into M clusters using hierarchical clustering with spearman distance.

Output: Cluster label in both networks.

The number of clusters M was chosen according to the eigenvalues of matrix C. M corresponds to the first big eigengap [26]. We note that here M is not the number of clusters in either of the two networks because by choosing β, the isolated vertices in one network can also be clustered together depending on the other network. It should be the number of the union clusters in both networks. Thus the method can find both the consistent clusters and the differential clusters having quite different connection probabilities.

After the above clustering procedures, we can get a cluster label for each gene. The corresponding clusters in both networks may include different genes. We take the union of the genes as the cluster. Due to the large number of genes, the clustering method may have some bias. Some unconnected subnetworks may be clustered together. Before going to further analysis, we need to check each cluster such that it cannot include unconnected subnetworks. These resulted subnetworks are defined as the modules we identified.

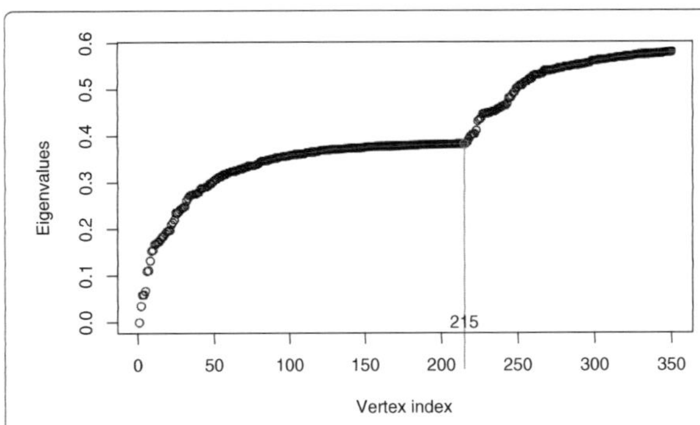

Fig. 1 Eigenvalues of matrix C

Table 1 Statistics of the identified modules

No.	Size	$\bar{d}_{BRCA}$	$\bar{d}_{OV}$	D_{BRCA}	D_{OV}	*p*-value	N_{GO}	N_{KEGG}	N_{DO}	N_{GSEA}
1	66	1.36	7.97	0.02	0.12	1.06E-37	2			
2	56	9.79	7.14	0.18	0.13	2.67E-4	1	1		
3	111	3.59	20.29	0.03	0.18	6.88E-160	6	1		
4	6	0.67	1.67	0.13	0.33	0.39	133	1		
5	202	3.65	16.95	0.02	0.08	2.20E-200	53	1		
6	7	3.43	2.57	0.57	0.43	0.54	42		3	
7	92	1.87	2.15	0.02	0.02	0.37	26	5	21	
8	10	0	3.8	0	0.42	3.33E-06	2	1		
9	71	3.38	10.39	0.05	0.15	3.42E-32		2		
10	50	2.28	11.4	0.05	0.23	5.65E-40			1	
11	14	0.43	2.71	0.03	0.21	6.48E-4	22			
12	8	0.5	1.5	0.07	0.21	0.25	9			
13	11	0.36	2	0.04	0.2	0.02	6			
14	25	1.12	5.84	0.05	0.24	1.76E-11	286	27	53	17
15	16	0.88	2.88	0.06	0.19	3.41E-3	4	9	15	
16	29	1.10	4.97	0.04	0.18	5.33E-10		1	3	
17	6	1.33	4	0.27	0.8	0.01	131	7	1	
18	22	1.27	1.36	0.06	0.06	1	16			
19	22	1	2.82	0.05	0.13	2.11E-3	2	2	4	
20	6	2.33	0.67	0.47	0.13	0.11			15	
21	6	2	3.33	0.4	0.67	0.27	15	2	4	
22	13	0.15	2.31	0.01	0.19	6.02E-4	8			
23	40	1.25	6.85	0.03	0.18	3.19E-20	180	8		
24	22	0.64	2.64	0.03	0.13	2.68E-4		1	5	
25	7	0.86	1.71	0.14	0.29	0.45	1			
26	9	0.22	2.44	0.03	0.31	4.42E-3		1		
27	127	3.40	14.87	0.03	0.12	7.67E-109	24			
28	14	0.57	3.71	0.04	0.29	2.72E-05		4	2	
29	14	5.14	5.43	0.40	0.42	0.88	74	2	17	
30	22	3.91	8.82	0.19	0.42	8.08E-08		2		
31	15	0.67	3.73	0.05	0.27	3.03E-05	1	2		
32	7	2.57	0.29	0.43	0.05	0.01	138	2		
33	213	0.94	49.94	0.004	0.24	0	130	3	16	
34	7	0.86	2	0.14	0.33	0.28	40	2	16	
35	12	1.67	4.17	0.15	0.38	5.77E-3	3		1	
36	6	3.33	2.33	0.67	0.47	0.46	38	6		
37	9	2	4	0.25	0.5	0.05	1	1	8	
38	14	2.14	2.86	0.16	0.22	0.45	36		3	
39	6	2	3.33	0.4	0.67	0.27	37			
40	102	4.16	16.35	0.04	0.16	3.08E-91	8			
41	17	1.65	3.88	0.10	0.24	3.89E-3	7			
42	46	2.78	10.04	0.06	0.22	1.68E-25	2			
43	125	10.54	9.28	0.09	0.07	0.02	7			
Total							31	25	18	1

Table 2 Enriched GO terms with p-value< 10^{-5} in all the modules

No.	GO ID	Description	p-value
5	GO:0016259	Selenocysteine metabolic process	4.86E-59
	GO:0006614	SRP-dependent cotranslational protein targeting to membrane	2.03E-56
	GO:0006613	Cotranslational protein targeting to membrane	3.84E-56
	GO:0045047	Protein targeting to ER	7.94E-56
	GO:0072599	Establishment of protein localization to endoplasmic reticulum	4.45E-55
	GO:0070972	Protein localization to endoplasmic reticulum	4.45E-55
	GO:0001887	Selenium compound metabolic process	1.37E-53
	GO:0000184	Nuclear-transcribed mRNA catabolic process, nonsense-mediated decay	1.64E-52
	GO:0019080	Viral gene expression	2.59E-47
	GO:0006415	Translational termination	3.29E-47
	GO:0019083	Translational elongation	7.19E-47
	GO:0044033	Multi-organism metabolic process	2.32E-46
	GO:0000956	Nuclear-transcribed mRNA catabolic process	6.16E-44
	GO:0006612	Protein targeting to membrane	6.16E-44
	GO:0006402	mRNA catabolic process	2.23E-42
	GO:0006401	RNA catabolic process	1.09E-39
	GO:0043624	Cellular protein complex disassembly	7.83E-39
	GO:0006413	Translational initiation	1.70E-38
	GO:0043241	Protein complex disassembly	6.44E-38
	GO:0032984	Macromolecular complex disassembly	3.32E-37
	GO:0006575	Cellular modified amino acid metabolic process	3.50E-34
	GO:1901605	Alpha-amino acid metabolic process	1.89E-33
	GO:0090150	Establishment of protein localization to membrane	1.44E-32
	GO:0034655	Nucleobase-containing compound catabolic process	3.53E-32
	GO:0019058	Viral life cycle	1.09E-30
	GO:0044270	Cellular nitrogen compound catabolic process	4.69E-30
	GO:0046700	Heterocycle catabolic process	4.69E-30
	GO:0019439	Aromatic compound catabolic process	1.27E-29
	GO:1901361	Organic cyclic compound catabolic process	2.56E-28
	GO:0072657	Protein localization to membrane	5.23E-28
	GO:0042254	Ribosome biogenesis	3.59E-14
	GO:0022613	Ribonucleoprotein complex biogenesis	5.19E-14
	GO:0042255	Ribosome assembly	1.56E-12
	GO:0042273	Ribosomal large subunit biogenesis	3.61E-12
	GO:0000027	Ribosomal large subunit assembly	9.99E-11
	GO:0022618	Ribonucleoprotein complex assembly	1.23E-09
	GO:0071826	Ribonucleoprotein complex subunit organization	2.74E-09
	GO:0042274	Ribosomal small subunit biogenesis	5.64E-06
6	GO:0009952	Anterior/posterior pattern specification	1.91E-10
	GO:0003002	Regionalization	1.90E-09
	GO:0007389	Pattern specification process	7.21E-09
	GO:0001501	Skeletal system development	1.71E-06
14	GO:0032496	Response to lipopolysaccharide	1.02E-08
	GO:0002237	Response to molecule of bacterial origin	1.02E-08
	GO:0050727	Regulation of inflammatory response	5.43E-08
	GO:1903034	Regulation of response to wounding	8.40E-07
	GO:0030595	Leukocyte chemotaxis	1.24E-06
	GO:0050900	Leukocyte migration	4.99E-06
	GO:0060326	Cell chemotaxis	5.27E-06
27	GO:0044782	Cilium organization	3.90E-10
	GO:0060271	Cilium morphogenesis	7.14E-09
	GO:0042384	Cilium assembly	7.14E-09
	GO:0007018	Microtubule-based movement	1.80E-08
	GO:0010927	Cellular component assembly involved in morphogenesis	7.24E-08
	GO:0030031	Cell projection assembly	3.37E-07
	GO:0042073	Intraciliary transport	4.27E-06

Table 2 Enriched GO terms with p-value$< 10^{-5}$ in all the modules *(Continued)*

No.	GO ID	Description	p-value
	GO:0098840	Protein transport along microtubule	4.27E-06
36	GO:1901685	Glutathione derivative metabolic process	8.14E-12
	GO:1901687	Glutathione derivative biosynthetic process	8.14E-12
	GO:0006749	Glutathione metabolic process	1.03E-10
	GO:0042537	Benzene-containing compound metabolic process	1.49E-09
	GO:0006805	Xenobiotic metabolic process	2.90E-08
	GO:0071466	Cellular response to xenobiotic stimulus	2.90E-08
	GO:0009410	Response to xenobiotic stimulus	2.90E-08
	GO:0044272	Sulfur compound biosynthetic process	3.74E-08
	GO:0006575	Cellular modified amino acid metabolic process	2.18E-07
	GO:0006790	Sulfur compound metabolic process	5.04E-07
39	GO:0048706	Embryonic skeletal system development	1.28E-08
	GO:0009952	Anterior/posterior pattern specification	7.49E-08
	GO:0048704	Embryonic skeletal system morphogenesis	4.26E-07
	GO:0003002	Regionalization	4.47E-07
	GO:0007389	Pattern specification process	1.51E-06
	GO:0001501	Skeletal system development	1.86E-06
	GO:0048705	Skeletal system morphogenesis	4.37E-06

Enrichment analysis

To see the associations between the identified gene modules and gene functions, we did Gene Ontology (GO)[27], KEGG pathway [28], and Disease Ontology (DO) [29] enrichment analysis for each module. GO annotates genes to biological processes (BP), molecular functions (MF), and cellular components (CC) in a directed acyclic graph structure. We only considered BP terms here. KEGG annotates genes to pathways, and Disease Ontology (DO) annotates genes with human disease associations. To see whether the identified modules show statistically significant, concordant differences between the two diseases, we also did Gene Set Enrichment Analysis (GSEA) [30]. GSEA was also done for the three types of enrichment analysis: GO, KEGG, and DO. We implemented the enrichment analysis with 'clusterProfiler' [31]. For all the cases, we let the cutoff be the Benjamini-Hochberg adjusted p-value 0.05, and recorded all the enriched terms with p-value less than 0.05.

Results

We did module identification in the gene coexpression networks of both breast cancer and ovarian cancer simultaneously. Figure 1 shows the first 350 eigenvalues of the matrix C. We chose M to be 215. After we clustered the genes into 215 clusters with hierarchical clustering, we removed those with size less than 5 and greater than 800. With the method described, finally, we got 62 modules.

To look at the module structures in both networks, we first computed the average degree ($\bar{d}_{BRCA}$, $\bar{d}_{OV}$) and density(D_{BRCA}, D_{OV}) for each module. Besides, we did statistical test for each module to see whether the modules in the two networks have the same connection distribution. We assume the connection between any two vertices is randomly generated following Bernoulli distribution with a given probability. We applied t-test to see whether the probability in the two networks is the same. The p-values are recorded. For all the 62 modules, after we did all the enrichment analysis, we removed those modules that have no enriched terms. Finally, 43 modules were found to have at least one type of enrichment. We put the module size, the average degree, the density, the t-test p-value, and the number of enriched terms for GO, KEGG pathway, DO, GSEA in Table 1. For GSEA, we only found GSEA GO enrichment terms.

From the t-test p-values of the modules in both coexpression networks, the structures of 29 modules are significantly different with p-values less than 0.05, of which 25 modules have larger densities in OV. Thirty one modules are enriched by GO terms, 25 modules are enriched by KEGG pathways, 18 modules are enriched by DO terms, and one module is enriched by GSEA GO terms. One module (module 14) is enriched by all the four terms. Nine modules (module 7,15,17,19,21,29,33,34,37) are enriched by GO, KEGG, and DO. We checked the details of each module and the enriched terms. All the enrichment results are put in the Additional files 1, 2, 3, 4 and 5.

GO enrichment analysis

We listed the modules that have an enriched p-value less than 10^{-5} in Table 2. Three modules including module 6, 36, and 39 have the same connection distribution in both cancers. These modules have small sizes, and very significant enrichments. Six among the 7 genes in module 6 are involved in the regionalization and pattern specification process. Module 36 is mainly related to glutathione metabolic process and xenobiotic metabolic process. Five of the 6 genes are involved in these processes. Module 39 is mainly related to skeletal system development. Five among 6 genes are involved in this process. In other three modules 5, 14, and 27, the connections in the OV coexpression network are much denser. There are several isolated genes in the BRCA coexpresson network. These three modules are involved in many complex biological processes significantly. One typical example is the enriched term 'GO:0016259 selenocysteine metabolic process'. There are 89 genes in the background 16655 genes involved in this process. In module 5, 44 among the 195 (overlapping with the background) genes are involved in this process. For the term 'GO:0006614 SRP-dependent cotranslational protein targeting to membrane', 45 genes among the 195 genes are involved in the process compared to the background 108 genes of 16655 genes in this process. These show the high correlations among the genes involved in the same process. Compared to OV, the genes involving in the same biological processes in BRCA have much less correlations, which leads to the sparser module structures.

KEGG enrichment analysis

The KEGG pathways that enrich the modules having a p-value less than 10^{-4} are listed in Table 3. Module 5 is enriched by the pathway 'hsa03010 Ribosome'. Forty-six out of 130 (overlapping with the background) genes in this module belong to this pathway compared to 154 out of 7274 in the background genes. From the GO enriched terms of this module, it is clear that this module is mainly involved in the translation process. The pathways related to module 14 are mainly related to diseases. 'IL-17 signaling pathway' [32], 'TNF signaling pathway' [33], 'Rheumatoid arthritis' [34], and 'MAPK signaling pathway' [35] were shown to have relations with BRCA. For these enriched terms, no existing literatures addressed their associations with OV to the best of our knowledge. Module 23 is also related to cancers. Eight genes in this module belong to 'Pathways in cancer'. It is also enriched

Table 3 Enriched KEGG terms with p-value< 10^{-4} in all the modules

No.	KEGG ID	Description	p-value
5	hsa03010	Ribosome	1.02E-43
14	hsa04668	TNF signaling pathway	3.80E-09
	hsa04657	IL-17 signaling pathway	1.97E-06
	hsa05323	Rheumatoid arthritis	4.57E-05
	hsa04380	Osteoclast differentiation	1.96E-04
	hsa04010	MAPK signaling pathway	3.00E-04
23	hsa05217	Basal cell carcinoma	2.59E-05
	hsa05200	Pathways in cancer	9.26E-05
33	hsa04610	Complement and coagulation cascades	2.68E-06
36	hsa00480	Glutathione metabolism	1.12E-10
	hsa00982	Drug metabolism-cytochrome P450	1.43E-10
	hsa01524	Platinum drug resistance	1.43E-10
	hsa00980	Metabolism of xenobiotics by cytochrome P450	1.43E-10
	hsa05204	Chemical carcinogenesis	1.93E-10
	hsa05418	Fluid shear stress and atherosclerosis	2.37E-09

Table 4 Enriched DO terms with p-value< 0.01 in all the modules

No.	DO ID	Description	p-value
14	DOID:3770	Pulmonary fibrosis	7.40E-04
	DOID:1602	Lymphadenitis	7.40E-04
	DOID:9942	Lymph node disease	7.40E-04
	DOID:1936	Atherosclerosis	7.40E-04
	DOID:1036	Chronic leukemia	7.40E-04
	DOID:2348	Arteriosclerotic cardiovascular disease	7.40E-04
	DOID:2349	Arteriosclerosis	7.79E-04
	DOID:3459	Breast carcinoma	1.10E-03
	DOID:3082	Interstitial lung disease	1.94E-03
	DOID:0070004	Myeloma	3.21E-03
	DOID:75	Lymphatic system disease	3.21E-03
	DOID:4960	Bone marrow cancer	3.21E-03
	DOID:865	Vasculitis	5.68E-03
	DOID:13378	Kawasaki disease	7.82E-03
	DOID:9538	Multiple myeloma	9.27E-03
15	DOID:2730	Epidermolysis bullosa	3.27E-04
	DOID:2731	Vesiculobullous skin disease	3.27E-04
	DOID:299	Adenocarcinoma	6.72E-03
	DOID:4766	Embryoma	9.26E-03
	DOID:688	Embryonal cancer	9.71E-03
17	DOID:700	Mitochondrial metabolism disease	1.77E-03
33	DOID:1882	Atrial heart septal defect	2.16E-03
	DOID:9477	Pulmonary embolism	2.16E-03
	DOID:1681	Heart septal defect	8.06E-03
	DOID:1682	Congenital heart disease	8.06E-03
	DOID:2214	Inherited blood coagulation disease	8.06E-03
	DOID:780	Placenta disease	8.13E-03
34	DOID:2893	Cervix carcinoma	1.62E-03
	DOID:4362	Cervical cancer	1.62E-03
	DOID:120	Female reproductive organ cancer	4.60E-03
37	DOID:10652	Alzheimer's disease	8.20E-03
	DOID:680	Tauopathy	8.20E-03
38	DOID:3113	Papillary carcinoma	3.66E-03
	DOID:2394	Ovarian cancer	4.14E-02

by 'breast cancer' with a p-value 0.049, with 3 genes associated with breast cancer in this module. This module has no enrichments related to OV. The pathway 'hsa05418 Chemical carcinogenesis' that enriches module 36 is also related to cancers. Chemical carcinogens may contribute significantly to the causation of a sizable fraction, perhaps a majority, of human cancers [36].

DO enrichment analysis

Table 4 lists the DO enriched terms in all the modules with p-value less than 0.01, and one term that is related to OV. Module 14 is enriched by BRCA, with 7 among 20 (overlapping with the background) genes in this module associated with BRCA. Module 38 is enriched by OV with 4 among 11 (overlapping with the background) genes in this module being associated with OV. There are no OV related enriched terms in module 14, and no BRCA related terms in module 38. This is mainly due to the different known genes associated with the two cancers. We note that module 34 is also enriched by the female organ cancer.

From all the above analysis, we found that most modules are the general modules that may not be associated with BRCA and OV. The validated enriched terms related to BRCA are much more than that of OV. One reason may be there are more researches conducted on BRCA. Among all the enriched modules, module 14 is enriched by BRCA, and module 38 is enriched by OV. Module 36 is not enriched by these two diseases, but the enriched terms are related to cancer treatment. Among these three modules, the structure of module 14 is shown to be different with a p-value 1.76E-11 in these two cancers. In the following, we give some details of these three modules.

Enrichment analysis for module 14

Module 14 consists of 25 genes. The module structure in both networks is significantly different with a p-value 1.76E-11. Figure 2 shows the module structures. It is much denser in OV compared to that in BRCA. Figure 3 shows the dotplot of GO, KEGG, DO enrichment results. For GO and DO, we plotted the first 30 enriched terms with the smallest p-values. We plotted all the 27 enriched terms for KEGG. Figure 4 shows the associations between the genes and the enriched terms. We selected the most enriched 12 GO terms with p-value less than 10^{-4}, 10 KEGG terms with p-value less than 0.01, and 15 DO terms with p-value less than 0.01.

The enriched GO terms are mainly related to different responses, such at inflammatory response, response to molecule of bacterial origin, responses to wounding, immune response, etc.. Several responses have been shown to be associated with cancers [37, 38]. For example, the inflammation as the seventh hallmark of cancer plays important roles in cancer development. Inflammatory cells may facilitate angiogenesis and promote the growth, invasion, and metastasis of tumor cells, which may change the genetic instability in cancer cells. Controlling the regulation of inflammatory response has a potential in both prevention and treatment of cancer [37]. There are 9 genes included in this module associated with the enriched term 'regulation of inflammatory response', which achieves a p-value of 5.43E-08. This module also includes 7 genes that are associated with the enriched term 'adaptive immune response', which achieves a p-value of 8.78E-05. The immune response can result in the proliferation of antigen-specific lymphocytes. When antibodies and T-cell receptors are expressed and up-regulated, immunity is acquired. Then the immune systems will initiate antigenic responses against carcinomas. A new approach to the treatment of cancer is immunotherapy, which aims to up-regulate the immune system in order that it may better control carcinogenesis [38]. Figure 4a shows the relations between the 12 GO terms with p-value less than 10^{-4} and the 25 genes in this module. The genes 'CXCL1', 'CCL20', 'ANXA1', 'S100A9', 'S100A12', 'TNF', 'DUSP10', 'TNFAIP2' are included in more than 6 enriched terms. There are 10 genes associated with the most enriched terms 'response to lipopolysaccharide'

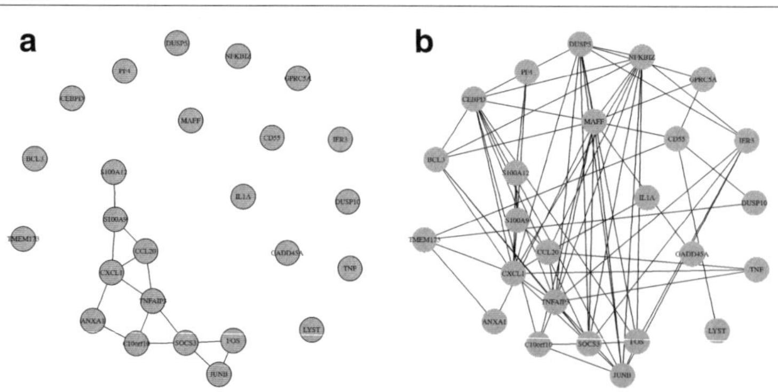

Fig. 2 Network structure of module 14. **a** BRCA; **b** OV

and 'response to molecule of bacterial origin'. For these two terms, we have not found the related literature that addresses their relations to cancers.

In the KEGG enriched terms, 'TNF signaling pathway' [33],'Rheumatoid arthritis' [34], 'MAPK signaling pathway' [35], and 'IL-17 signaling pathway' [32] have shown to be associated with BRCA. TNF is a major inflammatory cytokine shown to be highly expressed in breast carcinomas. It induces a wide range of intracellular signal pathways including apoptosis and cell survival as well as inflammation and immunity [33]. 'MARK signaling pathway' is involved in various cellular functions, including

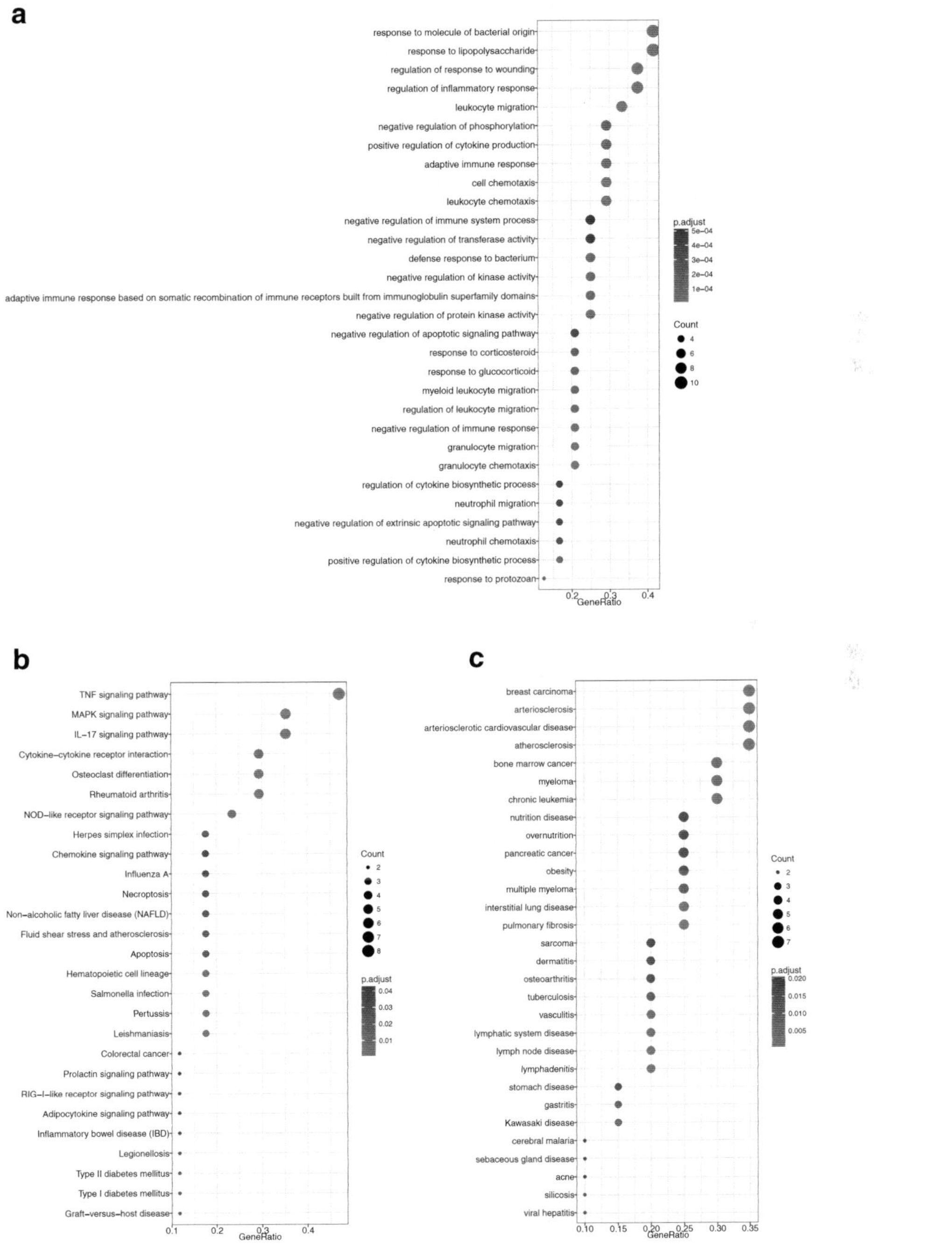

Fig. 3 Enrichment results for module 14. **a** GO, 30 terms with minimal *p*-values; **b** KEGG, all enriched 27 terms; **c** DO, 30 terms with minimal *p*-values

cell proliferation, differentiation and migration. Research on signaling pathway switch in breast cancer shows that in a large proportion of breast cancer, MARK signaling pathway is repressed, while another important pathway is activated. This mechanism may have impacts on the balance between self-renewal, proliferation, and differentiation of the tumor-initiating cells [35]. IL-17 plays crucial roles in both acute and chronic inflammatory responses. It is shown to have a direct association with breast cancer invasion in human breast tumors. IL-17 directly induced breast cancer cell invasion. There should be a potential mechanism for breast cancer invasion and tumor progression [32]. 'Rheumatoid arthritis' is mainly related to immune systems. Research shows that the risk of breast cancer is increased in non-Caucasians patients with rheumatoid arthritis while it decreased in Caucasian population [34]. Figure 4b shows the associations between the enriched KEGG terms and the genes. Four genes including 'CXCL1', 'TNF', 'FOS', and 'IL1A' connect to at least 6 of the 10 terms. There are at least 6 genes associated with 'TNF signaling pathway', 'MAPK signaling pathway', and 'IL-17 signaling pathway'. For these enriched pathways, we have not found their associations with OV.

In the DO enriched terms, 'breast carcinoma' reaches the p-value 0.001. Seven genes including 'S100A9', 'TNF', 'SOC53', 'CD55', 'IL1A', 'ANXA1', 'GADD45A' among the 25 genes in this module are associated with BRCA. Several other diseases also enrich module 14, including arteriosclerosis disease, nutrition disease, etc.. However, OV is not on this list. Figure 4c shows the associations between the genes and the diseases. 'S100A9', 'TNF', 'SOCS3', and 'IL1A' connect to at least 10 diseases among the 15 enriched diseases.

In the GSEA study, we ordered the genes according to the t-test p-value between the two diseases and did the analysis. Finally, 17 GO terms enrich this module. Table 5 shows the enriched terms. The 8 sequential genes having the largest t-test p-values are all in the enriched biological processes. They are 'TNFAIP3', 'S100A9', 'BCL3', 'MAFF', 'TMEM173', 'JUNB', 'CEBPD', 'NFKBIZ'. Figure 5 shows the patterns for the running enrichment score. All the enriched terms have a similar pattern for these 8 genes.

By comparison of this module structure between BRCA and OV, we found module 14 is closely related to BRCA from the above enrichment analysis. From the gene coexpression network of BRCA (Fig. 2), we see some breast cancer associated genes are isolated, such as 'TNF', 'CD55', 'IL1A'. In OV network, these genes are highly correlated and clustered into one module. This may show that due to the tumor, the correlations of some cancer related genes decrease in BRCA.

Enrichment analysis for module 36

This module includes 6 densely connected genes in both diseases, which has a p-value 0.46 in t-test for the structure difference. Of the 6 genes, 5 genes GSTM1, GSTM2, GSTM3, GSTM4, GSTM5 encode the glutathione S-transferase that belongs to the mu class. These genes

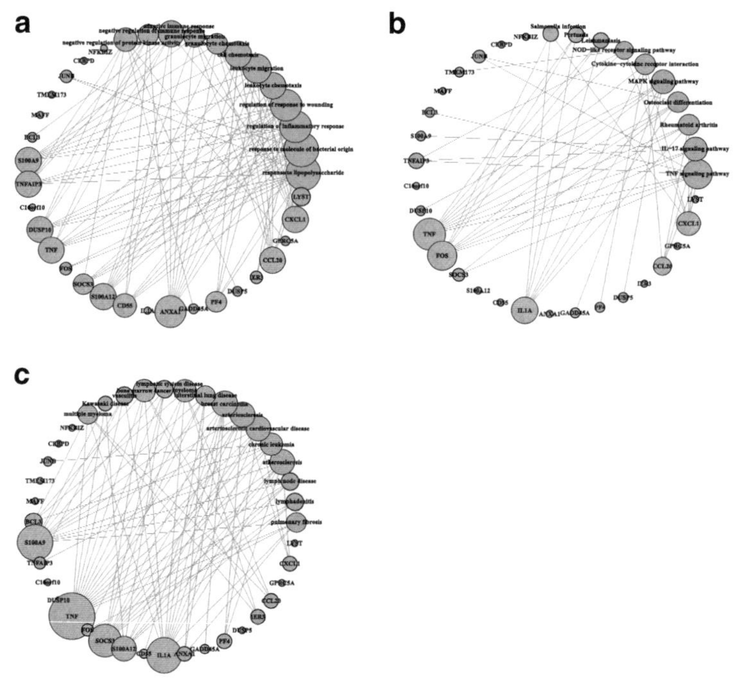

Fig. 4 Enrichment results for module 14. **a** GO terms with p-value $< 10^{-4}$; **b** KEGG terms with p-value < 0.01; **c** DO terms with p-value < 0.01

function in the detoxification of electrophilic compounds such as carcinogens, therapeutic drugs, environmental toxins and products of oxidative stress, by conjugation with glutathione. Thus this module is enriched by the related biological processes such as: 'glutathione derivative metabolic process', 'xenobiotic metabolic process', 'sulfur compound biosynthetic process', etc.. It is also enriched by the related pathways such as 'glutathione metabolism', 'drug metabolism-cytochrome P450', 'chemical carcinogenesis', and so on. Another gene is BCAR3, which is associated with estrogen resistance and breast cancer. It is translated to the breast cancer anti-estrogen resistance protein 3. Although this module is not enriched by BRCA and OV in DO significantly, it is related to the treatment of breast cancer [39].

Enrichment analysis for module 38

Module 38 includes 11 genes. Figure 6 shows the module structure in both cancers. The connection probability in these two networks is statistically the same with t-test, although the detailed connections are different.

Table 5 Enriched GO terms with GSEA for module 14

GO ID	Description	Set Size	*p*-value
GO:0010467	Gene expression	14	0.04
GO:0010468	Regulation of gene expression	14	0.04
GO:0034645	Cellular macromolecule biosynthetic process	14	0.04
GO:0006351	Transcription, DNA-templated	13	0.04
GO:0006355	Regulation of transcription, DNA-templated	13	0.04
GO:0016070	RNA metabolic process	13	0.04
GO:0018130	Heterocycle biosynthetic process	13	0.04
GO:0019438	Aromatic compound biosynthetic process	13	0.04
GO:0032774	RNA biosynthetic process	13	0.04
GO:0034654	Nucleobase-containing compound biosynthetic process	13	0.04
GO:0044271	Cellular nitrogen compound biosynthetic process	13	0.04
GO:0051252	Regulation of RNA metabolic process	13	0.04
GO:0097659	Nucleic acid-templated transcription	13	0.04
GO:1901362	Organic cyclic compound biosynthetic process	13	0.04
GO:1903506	Regulation of nucleic acid-templated transcription	13	0.04
GO:2000112	Regulation of cellular macromolecule biosynthetic process	13	0.04
GO:2001141	Regulation of RNA biosynthetic process	13	0.04

This module is enriched by 36 GO terms with adjusted p-value less than 0.05. The gene number involved in the related biological processes is at most 4. It is enriched by 3 diseases including ovarian cancer. Four of the 14 genes are associated with OV using DO enrichment. One typical gene is 'WT1', which connects to several other genes in the OV network, while it has no connections in the BRCA network. This gene is necessary for the development of the ovaries in females, and thus is associated with ovarian cancer. The 'WT1' protein has been found to bind a host of cellular factors such as p53. It has been ranked as the No.1 target for cancer immunotherapy by the National Cancer Institute. However, it has no associations with BRCA to the best of our knowledge. A densely connected subnetwork in this module is Kallikrein-related peptidases (KLK5, KLK6, KLK7, KLK8, KLK10). This gene family can be taken as the novel cancer biomarkers as shown in [40]. The potential of KLKs as diagnostic, prognostic, and treatment monitoring biomarkers for many types of malignancies has been extensively investigated including breast cancer and ovarian cancer. Overall, this module is closely related to cancers including breast cancer and ovarian cancer, and is more associated with ovarian cancer.

Discussion and conclusion

Breast cancer and ovarian cancer are important causes of death for women. Both of them are harmone driven and are known to have some common susceptible genes such as BRCA1 and BRCA2. Several published works have studied both cancers together with the assumption that they have the same etiologies. Coexpression network modules for both breast cancer and ovarian cancer have

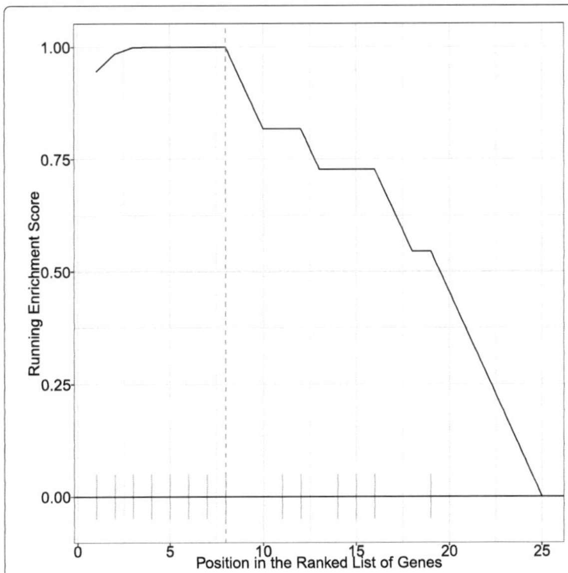

Fig. 5 GSEA GO enrichment for module 14

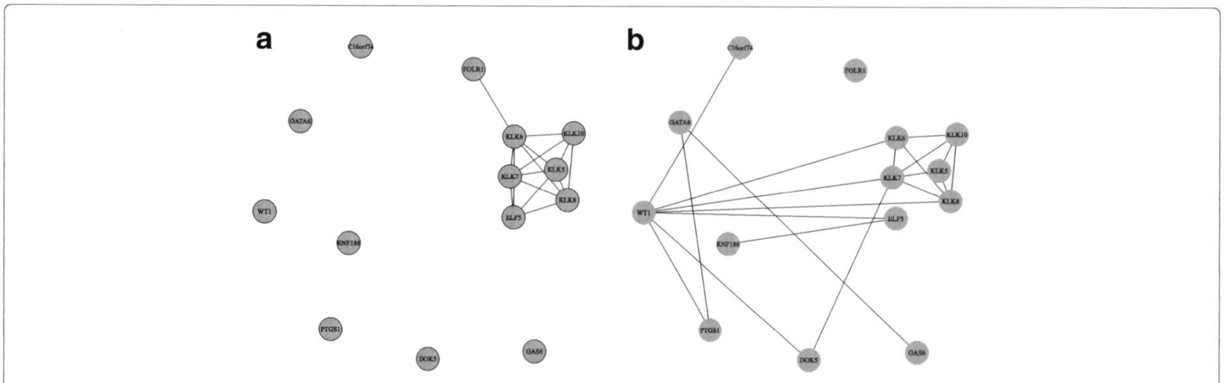

Fig. 6 Network structure of module 38. **a** BRCA; **b** OV

been studied separately by several researchers. However, there are no comparisions between the coexpression network modules between breast cancer and ovarian cancer. By comparing the modules in both cancers, we aim at finding more relations between both cancers including both the similarities and the differences.

In this work, we compared the coexpression network modules of both cancers by simultaneously identifying the modules of both cancers. By projecting the genes into a lower space representation, we did hierarchical clustering of all the genes with complete linkage of spearman distance to get the modules. 43 modules were identified to be enriched by at least one of GO, KEGG pathway, and DO terms. In most of these modules, density in OV network is larger than that in BRCA network. There are 31, 25 and 18 modules enriched by GO, KEGG pathway, and DO terms, respectively. By checking the details of each enriched term, we found that one module (module 14) is enriched by GO, KEGG and DO terms related to breast cancer. Among all GO and KEGG enriched terms, several have been validated. And there are 7 genes associated with breast cancer in this module. However, there is no enrichment information related to ovarian cancer. Another module 38, which is enriched by ovarian cancer in DO, is closely related to several cancers including breast cancer and ovarian cancer, and is more associated with ovarian cancer. From the analysis of the genes in module 36, we found that it is related to the treatment of breast cancer, although it is not enriched by breast cancer terms in DO.

Comparison of the identified modules shows the differences of breast cancer and ovarian cancer on the module level. Different modules are enriched by the two cancers significantly. It also shows some common properties of both cancers such as the KLKs family in module 36. More importantly, by simultaneous clustering of both cancers, the potential cancer related modules can be identified. For example, module 14 is significantly enriched by breast cancer related terms, but much connection information is borrowed from the ovarian cancer network. This may imply the associations between this module and breast cancer.

Additional files

Additional file 1: The identified modules. We listed all the identified modules. The second column in each sheet lists the gene names. (XLSX 337 kb)

Additional file 2: Statistics of the modules. The module size, average degree, density, number of enriched GO, KEGG, DO terms are listed. The module names are the sheet name in Additional file 1. There is a map between the module names and the module number in the paper. (CSV 5 kb)

Additional file 3: GO enrichment results. This file lists all the GO-BP enriched terms for each module. 'geneID' is the Entrez gene id. (XLSX 193 kb)

Additional file 4: KEGG enrichment results. This file lists all the KEGG enriched terms for each module. 'geneID' is the Entrez gene id. (XLSX 59 kb)

Additional file 5: DO enrichment results. This file lists all the DO enriched terms for each module. 'geneID' is the Entrez gene id. (XLSX 54 kb)

Acknowledgements

S. Zhang's research is supported in part by NSFC grant No.11471082, Science and Technology Commission of Shanghai Municipality 16JC1402600.

Funding

The publication of this work is supported by NSFC grant No. 11471082.

Availability of data and materials

The level 3 gene expression data for breast cancer (BRCA) and ovarian cancer (OV) were downloaded from The Cancer Genome Atlas (TCGA). The gene expression data are generated with UNC AgilentG4502A. The most updated protein-protein interaction (PPI) data for humans were downloaded from BioGrid https://thebiogrid.org.

About this supplement

This article has been published as part of *BMC Systems Biology* Volume 12 Supplement 1, 2018: Selected articles from the 16th Asia Pacific Bioinformatics Conference (APBC 2018): systems biology. The full contents of the supplement are available online at https://bmcsystbiol.biomedcentral.com/articles/supplements/volume-12-supplement-1.

Authors' contributions

Not applicable.

Ethics approval and consent to participate
Not applicable.

Consent for publication
Not applicable.

Competing interests
The author declares that she has no competing interests.

Publisher's Note

Springer Nature remains neutral with regard to jurisdictional claims in published maps and institutional affiliations.

Published: 11 April 2018

References

1. Miki Y, Swensen J, Shattuckeidens D, Futreal PA, Harshman K, Tavtigian SV, Liu Q, Cochran C, Bennett L, Ding W, et al. A strong candidate for the breast and ovarian cancer susceptibility gene brca1. Science. 1994;266(5182):66–71.
2. Wooster R, Neuhausen SL, Mangion J, Quirk Y, Ford D, Collins N, Nguyen K, Seal S, Tran T, Averill D, et al. Localization of a breast cancer susceptibility gene, brca2, to chromosome 13q12-13. Science. 1994;265(5181):2088–90.
3. Claus EB, Schildkraut JM, Thompson WD, Risch N. The genetic attributable risk of breast and ovarian cancer. Cancer. 1996;77(11): 2318–24.
4. Wooster R, Weber BL. Breast and ovarian cancer. N Engl J Med. 2009;348(23):2339–47.
5. M'a'att'a KM, Nurminen R, Kankuri-Tammilehto M, Kallioniemi A, Laasanen S-L, Johanna S. Germline emsy sequence alterations in hereditary breast cancer and ovarian cancer families. BMC Cancer. 2017. 17(496).
6. Pearce CL, Hirschhorn JN, Wu AH, Burtt NP, Stram DO, Young S, Kolonel LN, Henderson BE, Altshuler D, Pike MC. Clarifying the progins allele association in ovarian and breast cancer risk: A haplotype-based analysis. J Natl Cancer Inst. 2005;97(1):51–9.
7. Song H, Ramus SJ, Kjaer SK, Dicioccio RA, Chenevixtrench G, Pearce CL, Hogdall E, Whittemore AS, Mcguire V, Hogdall C, et al. Association between invasive ovarian cancer susceptibility and 11 best candidate snps from breast cancer genome-wide association study. Hum Mol Genet. 2009;18(12):2297–304.
8. Eckstein N. Platinum resistance in breast and ovarian cancer cell lines. J Exp Clin Cancer Res. 2011;30(1):91.
9. Thompson ER, Dragovic RL, Stephenson S, Eccles D, Campbell IG, Dobrovic A. A novel duplication polymorphism in the fanca promoter and its association with breast and ovarian cancer. BMC Cancer. 2005;5(1):43.
10. Newman B, Millikan RC, King M. Genetic epidemiology of breast and ovarian cancers. Epidemiol Rev. 1997;19(1):69–79.
11. Wang X, Wang S, Zhou L, Yu L, Zhang L. A network-pathway based module identification for predicting the prognosis of ovarian cancer patients. J Ovarian Res. 2016;9(1):73.
12. Wu G, Stein L. A network module-based method for identifying cancer prognostic signatures. Genome Biol. 2012;13(12):1–14.
13. Hofree M, Shen JP, Carter H, Gross AM, Ideker T. Network-based stratification of tumor mutations. Nat Methods. 2013;10(11):1108–15.
14. Yang R, Daigle BJ, Petzold LR, Doyle FJ. Core module biomarker identification with network exploration for breast cancer metastasis. BMC Bioinformatics. 2012;13(1):12.
15. Chuang H, Lee E, Liu Y, Lee D, Ideker T. Network-based classification of breast cancer metastasis. Mol Syst Biol. 2007;3(1):140.
16. Hong S, Dong H, Jin L, Xiong M. Gene co-expression network and functional module analysis of ovarian cancer. Int J Comput Biol Drug Des. 2011;4(2):147.
17. Lee J-H, Zhao X-M, Yoon I, et al. Integrative analysis of mutational and transcriptional profiles reveals driver mutations of metastatic breast cancers. Cell Discov. 2016;2:16025.
18. Liu K-Q, Liu Z-P, Hao J-K, et al. Identifying dysregulated pathways in cancers from pathway interaction networks. BMC Bioinformatics. 2012;13(126).
19. Zhao X-M, Liu K-Q, Zhu G, et al. Identifying cancer-related micrornas based on gene expression data. Bioinformatics. 2015;31(8):1226–34.
20. Song W-MS, Bin Z. Multiscale embedded gene co-expression network analysis. PLoS Comput Biol. 2015;11(11):1004574.
21. Gao C, McDowell IC, Zhao S, Brown CD, Engelhardt BE. Context specific and differential gene co-expression networks via bayesian biclustering. PLoS Comput Biol. 2016;12(7):1004791.
22. Zhu L, Ding Y, Chen C-Y, et al. Metadcn: meta-analysis framework for differential co-expression network detection with an application in breast cancer. Bioinformatics. 2017;33(8):1121–9.
23. Huang H, Chuang Y, Chen C. Affinity aggregation for spectral clustering. In: IEEE conference on Computer Vision and Pattern Recognition (CVPR) 2012; 2012. p. 773–80.
24. Zhang S, Zhao H, Ng MK. Functional module analysis for gene coexpression networks with network integration. IEEE/ACM Trans Comput Biol Bioinforma. 2015;12(5):1146–60.
25. Li W, Liu C, Zhang T, Li H, Waterman MS, Zhou XJ. Integrative analysis of many weighted co-expression networks using tensor computation. PLoS Comput Biol. 2011;7(6).
26. Von Luxburg U. A tutorial on spectral clustering. Stat Comput. 2007;17(4): 395–416.
27. Ashburner M, Ball CA, Blake JA, Botstein D, Butler H, Cherry JM, Davis AP, Dolinski K, Dwight SS, Eppig JT, et al. Gene ontology: tool for the unification of biology. Nat Genet. 2000;25(1):25–9.
28. Kanehisa M, Goto S, Furumichi M, Tanabe M, Hirakawa M. Kegg for representation and analysis of molecular networks involving diseases and drugs. Nucleic Acids Res. 2010;38:D355–D360.
29. Osborne JD, Flatow J, Holko M, Lin S, Kibbe WA, Zhu LJ, Danila M, Feng G, Chisholm RL. Annotating the human genome with disease ontology. BMC Genomics. 2009;10(1):1–8.
30. Subramanian A, Tamayo P, Mootha VK, Mukherjee S, Ebert BL, Gillette MA, Paulovich AG, Pomeroy SL, Golub TR, Lander ES, et al. Gene set enrichment analysis: A knowledge-based approach for interpreting genome-wide expression profiles. Proc Natl Acad Sci U S A. 2005;102(43):15545–50.
31. Yu G, Wang L, Han Y, He Q. clusterprofiler: an r package for comparing biological themes among gene clusters. Omics A J Int Biol. 2012;16(5): 284–7.
32. Zhu X, Mulcahy LA, Mohammed R, Lee AHS, Franks HA, Kilpatrick L, Yilmazer A, Paish EC, Ellis IO, Patel PM, et al. Il-17 expression by breast-cancer-associated macrophages: Il-17 promotes invasiveness of breast cancer cell lines. Breast Cancer Res. 2008;10(6):1–11.
33. Kamel MW, Shouman S, Elmerzebany M, Kilic GS, Veenstra TD, Saeed M, Wagih ME, Diazarrastia CR, Patel D, Salama SA. Effect of tumour necrosis factor-alpha on estrogen metabolic pathways in breast cancer cells. J Cancer. 2012;3(1):310–21.
34. Tian G, Liang J, Wang Z, Zhou D. Breast cancer risk in rheumatoid arthritis: An update meta-analysis. BioMed Res Int. 2014;2014:453012.
35. Guille A, Chaffanet M, Birnbaum D. Signaling pathway switch in breast cancer. Cancer Cell Int. 2013;13(1):66.
36. Luch A. Nature and nurture lessons from chemical carcinogenesis. Nat Rev Cancer. 2005;5(2):113–25.
37. Colotta F, Allavena P, Sica A, Garlanda C, Mantovani A. Cancer-related inflammation, the seventh hallmark of cancer: links to genetic instability. Carcinogenesis. 2009;30(7):1073–81.
38. Adam JK, Odhav B, Bhoola KD. Immune responses in cancer. Pharmacol Ther. 2003;99(1):113–32.
39. Van Agthoven T, Van Agthoven TLA, Dekker A, Der Spek PJV, Vreede L, Dorssers LCJ. Identification of bcar3 by a random search for genes involved in antiestrogen resistance of human breast cancer cells. EMBO J. 1998;17(10):2799–808.
40. Kontos CK, Scorilas A. Kallikrein-related peptidases (klks): a gene family of novel cancer biomarkers. Clin Chem Lab Med. 2012;50(11):1877–91.

Lyu *et al. BMC Systems Biology* 2018, **12**(Suppl 1):5
https://doi.org/10.1186/s12918-018-0531-8

BMC Systems Biology

RESEARCH

Open Access

KDiamend: a package for detecting key drivers in a molecular ecological network of disease

Mengxuan Lyu†, Jiaxing Chen†, Yiqi Jiang, Wei Dong, Zhou Fang and Shuaicheng Li*

From The Sixteenth Asia Pacific Bioinformatics Conference
Yokohama, Japan. 15-17 January 2018

Abstract

Background: Microbial abundance profiles are applied widely to understand diseases from the aspect of microbial communities. By investigating the abundance associations of species or genes, we can construct molecular ecological networks (MENs). The MENs are often constructed by calculating the Pearson correlation coefficient (PCC) between genes. In this work, we also applied multimodal mutual information (MMI) to construct MENs. The members which drive the concerned MENs are referred to as key drivers.

Results: We proposed a novel method to detect the key drivers. First, we partitioned the MEN into subnetworks. Then we identified the most pertinent subnetworks to the disease by measuring the correlation between the abundance pattern and the delegated phenotype—the variable representing the disease phenotypes. Last, for each identified subnetwork, we detected the key driver by PageRank. We developed a package named KDiamend and applied it to the gut and oral microbial data to detect key drivers for Type 2 diabetes (T2D) and Rheumatoid Arthritis (RA). We detected six T2D-relevant subnetworks and three key drivers of them are related to the carbohydrate metabolic process. In addition, we detected nine subnetworks related to RA, a disease caused by compromised immune systems. The extracted subnetworks include InterPro matches (IPRs) concerned with immunoglobulin, Sporulation, biofilm, Flaviviruses, bacteriophage, *etc.*, while the development of biofilms is regarded as one of the drivers of persistent infections.

Conclusion: KDiamend is feasible to detect key drivers and offers insights to uncover the development of diseases. The package is freely available at http://www.deepomics.org/pipelines/3DCD6955FEF2E64A/.

Keywords: Molecular ecological network, Microbiome, Key driver, Delegated phenotype, Disease

Background

Assessment and characterization of microbiota are prevalent in human disease studies [1–3]. When the species within the microbial community interact with each other in equilibrium, serving as co-adapted colonists and providing beneficial goods and services, disruption of such alliances may induce health issue [4]. For example, the imbalance in the community could lead to bacterial overgrowth and the development of respiratory infections [5]. In this case, network analysis, for instance, differential network analysis, which identifies biomarker candidates by detecting changes in the correlation relationships between different experimental conditions [6], provides a better understanding towards disease. Thereafter, in microbiome area, molecular ecological networks (MENs) [7] can be constructed to perform network analysis for different types of actors within the microbial community, for examples, species, taxons, or phylogenetic gene markers, and they are referred to as phylogenetic molecular ecological networks (pMENs) where phylogenetic gene markers

*Correspondence: shuaicli@gmail.com
†Equal contributors
Department of Computer Science, City University of Hong Kong, Tat Chee Avenue, Hong Kong, China

serve as the actors [8]. Similar to co-expression networks, Deng et al. proved that the MENs are scale-free and small world [7].

In a MEN, the removal of some species could be disproportionately deleterious. These species are referred to as *keystone species*. Keystone species are topologically important molecules in the MEN. Berry et al. has studied the detection of keystone species in MENs thoroughly [9]. They applied a brute-force leave-one-out strategy to evaluate the keystoneness of a species in a given MEN, and demonstrated the impact of the keystone species on species richness. They also classified keystone species according to their topological properties using linear discriminant analysis. Deng et al. proposed a method to detect keystone species from the MENs by integrating phenotype information [10]. They identified keystone species by calculating the correlation between a phenotype variable and the abundance pattern of species clusters. Researchers also considered species connected to many others in MENs as keystone species (also referred to as hub nodes) [11].

Key drivers, which are major components that drive the disease concerned MENs, provide hints to understand the mechanisms of disease and are intensively studied with RNA data. There are various of methods to identify the key drivers in a co-expression network. One method is to incorporate the annotation of genes and pathways of diseases in order to locate the key drivers by considering enrichment of statistic of genes neighborhood [12, 13]. Another category of method distinguishes important MENs by calculating associations between gene modules with meta information like phenotype and GWAS analysis [14, 15], and then detects the key drivers by measuring the genes topology effect. For example, MEGENA [16] did multiscale hub analysis and Zhang et al. examined the number of N-hob downstream nodes [17]. Those methods on detecting key drivers in RNA data analysis can be adopted to detect key drivers in MENs. Even though Portune et al. locates important microbial species and genes with the assistance of gene annotation to study the MENs [18], the annotation for microbial genes and species yet demands intensive efforts and the pathways of diseases are incomplete.

The distinction between keystone species and key drivers is that the keystone species are only topologically important, while key drivers motivate disease associated networks. MENs of diseases can be different compared to those from healthy individuals. By analyzing the factors driving the differences, we can uncover the development of the disease.

Inspired by key drivers analysis with RNA data and keystone species studies in MENs, we proposed a method to perform key drivers analysis without the availability of annotation information. Given a microbial abundance profile, we first construct the MEN, in which the nodes represent the microbial species or phylogenetic gene markers and the edges capture the associations between their respective nodes. Then we divide the MEN into multiple subnetworks and extract the subnetworks that are most relevant to the disease by calculating the associations between subnetworks and phenotype variables. A single phenotype variable could be insufficient to capture the changes in disease networks from healthy networks and it can be biased. To address this issue, we applied principal component analysis to extract delegated phenotype, which is more robust. Last, our method detects the key driver based on PageRank, which utilizes node topological properties within each extracted subnetwork. It captures the global link structure of subnetworks thus outperforms statistical algorithms that only use local information.

There are multiple ways to calculate inference of MENs, of which two of the most popular ways are Pearson correlation coefficient (PCC) and mutual information (MI) [19]. A review of correlation detection strategies in MENs [20] suggests that although some methods outperform others, the inference calculating method still needs further improvement. To reduce the effect of the high proportion of zero counts, Paulson et al. applied a mixture model that implemented a zero-inflated Gaussian (ZIG) distribution of mean group abundance for each taxonomic feature to do differential abundance analysis. Experiments show the improvement of mixture model compared to other models, for instance, DESeq, edgeR and Kruskal-Wallis test [21, 22]. Inspired by the above trials of solving rare microbes issues with mixture models, we proposed to construct the network by multimodal mutual information (MMI) [23] under the assumption of the Gaussian mixture model. In KDiamend, we implemented both PCC and MMI to infer the associations between nodes in the MENs. However, correlation-based methods, like PCC, have their limitations. To be more specific, it is hard to distinguish correlation with causation [24]. There are many other arbitrary methods to construct networks, like Bayesian network [24] and WGCNA [25], which apply topology overlaps to measure the similarities between nodes. These various methods can be implemented to construct the network as potential options in our framework. Nevertheless, it is out of the range of this work.

Our main contribution is that we refined the framework of key driver detection, and proposed delegated phenotype to capture the changes in disease networks from healthy networks. To validate our method, we performed experiment based on simulated data. Then, we tested KDiamend with two real microbiome datasets. We conducted key drivers analysis on Type 2 diabetes (T2D) and Rheumatoid Arthritis (RA), whose data are from gut microbiome and oral microbiome respectively. For each disease, we also compared experiment using PCC and

MMI as two different inference methods, and acquired both consensus and divergence. Experiments of the two inference methods identified multiple identical phylogenetic gene markers and identified consensus pattern of disease-associated networks, indicating the robustness of our framework. On the other hand, the two different inference methods also led to specific findings, providing us with various aspects to study the mechanisms of diseases. We detected six T2D-relevant subnetworks and identified key drivers for each of them correspondingly. The identified key drivers include IPR006047, IPR018485 and IPR003385 related to the carbohydrate metabolic process, while the carbohydrate metabolic process is an important issue during the development of T2D [26]. In addition, we also detected key drivers for RA. Both PCC and MMI experiments located multiple InterPro matches (IPRs) which are related membrane and infection. Six subnetworks were extracted by PCC, containing IPRs concerned with immunoglobulin, Sporulation. Three subnetworks were detected by MMI, with IPRs about biofilm, Flaviviruses, bacteriophage, etc. The result is inspiring since the development of biofilms is regarded as one of the drivers of persistent infections [27] and some biofilms-growing bacterias contribute to RA [28].

Methods

Our method is to detect the key drivers which drive the diseases related networks in the microbial community. The key drivers can be microbial species or phylogenetic gene markers. For simplicity, we present our method with nodes as genes in the subsequent descriptions.

The detection of key drivers consists of following steps (see Fig. 1). First, we construct a MEN to represent the relationship between genes based on microbial abundance profiles and infer the weight of each edge. Second, we cluster the genes and partition the MEN into multiple subnetworks. Third, we analyze the phenotype variables and extract the delegated phenotype. By computing the associations between subnetworks and delegated phenotype, we obtain subnetworks that are most related to the disease. Last, based on PageRank, we identify actors with top influence over others in each subnetwork as key drivers.

Inference method

In KDiamend, we provide two ways to compute distances between genes. The first one is PCC, which is the most popular way to capture similarities between genes. In addition, inspired from the inference of gene regulatory network in RNA analysis and mixture models in the microbiome analysis, we adopted the MMI and normalization processes in Context Likelihood of Relatedness(CLR) [29]. MI, which uses the mutual dependency and common uncertainty as for the measurement of connection between genes, does not assume linear, or continuous dependence like correlation [19, 30], so it can detect interactions which might be missed by PCC. MMI, under the assumption of the Gaussian mixture model, is for dealing with the high proportion of zero counts issue. The adopted CLR, which considers the context of the whole network and eliminates noises from the background, makes MMI more tolerant for noises when measuring the interactions.

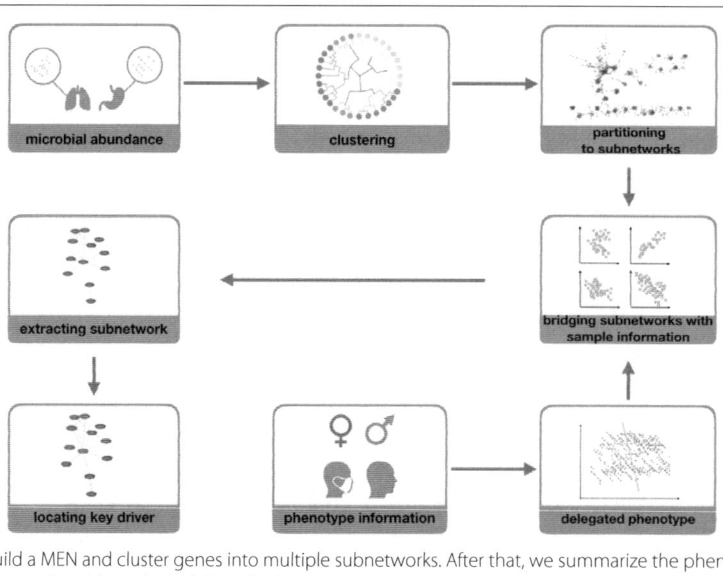

Fig. 1 Flowchart. First, we build a MEN and cluster genes into multiple subnetworks. After that, we summarize the phenotype variables and connect it to subnetworks. Then, we locate key drivers through PageRank

At the beginning, we have an abundance matrix E, which contains abundance value of n genes in m samples. For each gene i, we have a vector of $X_i = (E_{i1}, E_{i2}, \ldots, E_{im})$. The strength of the relationship between gene i and gene j can be measured by PCC:

$$PCC(i,j) = \frac{cov(X_i, X_j)}{\sigma_i, \sigma_j}, \quad (1)$$

where $cov(X_i, X_j)$is the covariance of X_i and X_j, and σ_i is the standard deviation of X_i. The adjacency matrix A of network can be generated from $A_{ij} = PCC(i,j)$. Then the distance between gene i and gene j can be interpreted as $D_{ij} = 1 - |PCC(i,j)|$.

Apart from PCC, we also implemented MMI [23]. First, we decomposed X_i into c_i bins as $X_{i,1}, \ldots, X_{i,c_i}$. In this case, X_i was distributed according to the following function:

$$f_{X_i}(x) = \sum_{k=1}^{c_i} \pi_{i,k} g_{X_{i,k}}(x), \quad (2)$$

where $g_{X_{i,k}}(x)$ $(1 \leq k \leq c_i)$ denotes the density function for $C_{i,k}$, and $\pi_{i,k}$ is the proportion for each sample in $C_{i,k}$. As proved in former work [23], assuming that $X_{i,k}$ fits in a Gaussian distribution, we can estimate the mutual information between X_i and X_j as:

$$MMI(X_i, X_j) = MMI^O(X_i, X_j) + MMI^I(X_i, X_j). \quad (3)$$

The "outer" MI, $MMI^O(X_i, X_j)$, captures discretized dependency, while the "inner" MI, $MMI^I(X_i, X_j)$, refers to the weighted aggregation of MI for each bin.

After computing MMI between all the genes and get a matrix M, we normalized the distance between gene i and gene j by: $CLR(Z_i, Z_j) = \sqrt{\left(Z_i^2, Z_j^2\right)}$ where Z_i and Z_j are z-scores of M_{ij} taking M_i and M_j as background respectively [29]. Then we applied hierarchical clustering and partitioned the MEN into multiple subnetworks according to the distance between genes.

Delegated phenotype

To best capture phenotype change in disease networks from healthy networks, we generated delegated phenotype by rotating Coordinates in the PC space of phenotype variable matrix S. Each row in S represents a sample while each column refers to a phenotype variable, such as gender, age, disease state, etc. If the properties are non-numerical, we converted data into numbers before further analysis. Suppose one column v in S indicates whether each sample is collected from a person with or without this disease. That is $v = (v_1, v_2, \ldots, v_k, \ldots, v_m), v_k \in (Y, N), 1 < k < m$ where m is the number of samples, Y indicates that this sample is collected from a person with the disease, and N means not.

We applied principal component analysis (PCA) to conclude S. We consider the first two principal components (PCs) to be enough for explaining disease variability, since the number of phenotype variables is relatively small in our test data. When phenotype data are more complicated, we may need extra analysis to decide the number of PCs we use to conclude delegated phenotype. We investigated the first two PCs and plotted samples in the coordinate of PC1 and PC2, regarding every sample as a point. Consequently, we got m points and each point refers to one sample. The coordinates for point k, is expressed as (x_k, y_k). We rotated PC to PC' and make sure it has the largest correlation with v upon rotation. PC' can best explain variability related to the disease. The angle of rotation is the θ that maximize $f(\theta)$.

$$f(\theta) = \sum_{k, v_k \in N} (x_k cos\theta_k + y_i sin\theta_k) - \sum_{k, v_k \in Y} (x_k cos\theta_k + y_k sin\theta_k) \quad (4)$$

That is to say, the angle between $PC1$ and $PC1'$ is:

$$\theta = arctan \frac{\sum_{k, v_k \in N} x_k - \sum_{k, v_k \in Y} x_k}{\sum_{k, v_k \in N} y_k - \sum_{k, v_k \in Y} y_k} \quad (5)$$

For example (see delegated phenotype in Fig. 1), blue points represent $v_k = Y$ and red points represent $v_k = N$. By calculating θ, we got the line which implies the direction most correlated to the disease state.

We acquired delegated phenotype PC', which has the largest correlation with the disease state and outperforms other single variables on explaining the variability of phenotype information at the meantime. For every subnetwork, we concluded the abundance pattern of genes as eigengenes [31]. Then we bridged the subnetworks to phenotype information by calculating the correlation between eigengenes and PC'. Subnetworks which have strong relationships to PC' are extracted as disease-relevant subnetworks.

Identifying Key driver

Last, in every extracted disease relevant subnetwork, we applied network topology analysis and assigned every gene a PageRank score. PageRank (PR) ranks the nodes in a graph according to the structures of links with others and is used by Google's search engine to compute rankings of websites. In this algorithm, the score for one node can be affected by its neighbors [32] and if one's neighbors have high scores, its score increases iteratively [33].

As stated in [34], letting F_u be the nodes linking to the node, B_u be the nodes linked from it, and $N_u = |F_u|$ be the magnitude of F_u. Besides, considering there might be other factors towards the ranking, let $E(u)$ be the vector concerned with some of the rank. Then, the PageRank of the node is defined as.

$$R(u) = c \sum_{\upsilon \in B_u} \frac{R(\upsilon)}{N_\upsilon} + cE(u) \quad (6)$$

Results

Gut microbiome

We tested our method with real microbiome datasets and compared PCC with MMI in this framework. First, In order to detect key drivers for T2D, we downloaded processed InterPro matches (IPR) abundance data from EBI (SRP008047), which is gut metagenome (microbiome) data from Chinese samples. InterPro [35] provides a functional analysis of protein sequences by classifying them into families and predicting the presence of domains and important sites. The phenotype information of the dataset is provided in related paper [1]. We used the 145 samples from stage one.

We first trimmed IPRs with low abundance in relative abundance matrix and then applied quantile normalization [36, 37]. By computing PCC and MMI between pairs of genes, we reconstructed a MEN and conducted clustering to partition the MEN into multiple subnetworks. For all subnetworks, eigengenes were calculated by selecting the first PC of abundance. The eigengene was used for summarizing the abundance pattern in each subnetwork and to bridge it with phenotype information. On the other hand, we digitalized the phenotype matrix and applied PCA to it. We generated delegated phenotype by rotating coordinates in the PC space to best capture the phenotype change in disease networks from healthy networks. We extracted three subnetworks for PCC experiment and three subnetworks for MMI according to the correlation between eigengenes and delegated phenotype and the corresponding *p*-value (see Fig. 2). The *p*-value for correlation of each subnetwork was calculated by permuting the same number of genes from the dataset and calculating the correlation between the permuted eigengene and the delegated phenotype. After repeating 1000 times, the rank of the real correlation for the subnetwork is regarded as *p*-value.

PCC experiment and MMI experiment detected 11 consensus IPRs which scattered in three subnetworks for MMI and two subnetworks for PCC. Consequently, the interaction generated from two types of inference connects these five extracted subnetworks together and merges them into one large community (see Fig. 3). Most of the consensus IPRs in the merged community are associated with the metabolic process and the catalytic activity which implies that the process is relevant to the disease. More specifically, two experiments both identified IPR018485 which participates in the carbohydrate metabolic process with the phosphotransferase activity and is active in carrying out ATP-dependent phosphorylation [38]. The extracted disease-relevant subnetworks in T2D are about the carbohydrate metabolic process and phosphorelay.

In addition, for PCC experiment, key drivers in subnetwork 89 and subnetwork 208 are related to the carbohydrate metabolic process, including IPR006047, IPR018485, and IPR003385. The key driver in subnetwork 166 is IPR001789, which plays a role in phosphorelay signal transduction system. MMI also detected IPR018211, IPR005538, IPR003501, and IPR001790 which are related to phosphorelay. PCC and MMI both identified IPRs related to the carbohydrate metabolic process and phosphorelay.

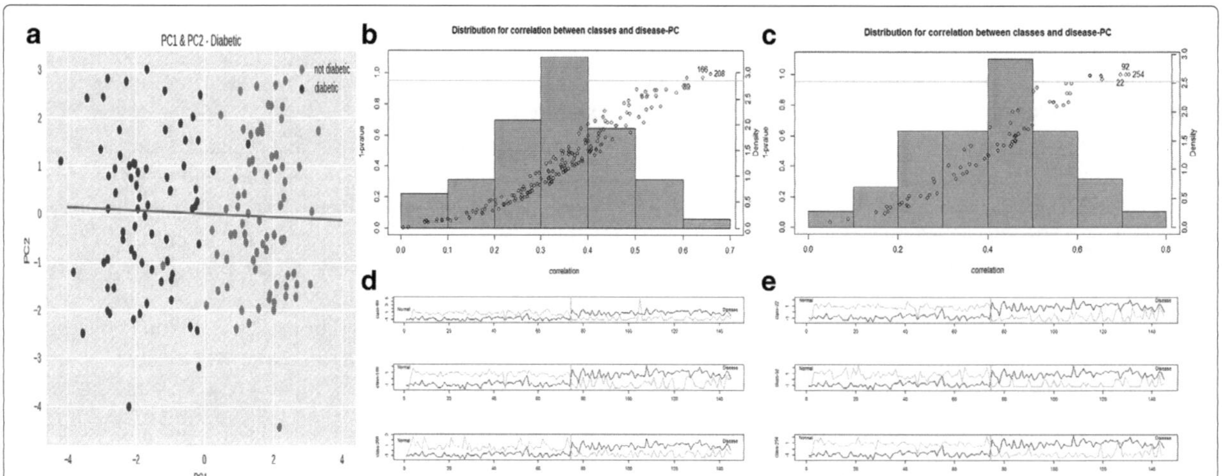

Fig. 2 Experiment results for type 2 diabetic. **a** The delegated phenotype. **b, c** Distribution for correlation between subnetworks and delegated phenotype. Every point refers to a subnetwork. The x-axis of the point refers to the correlation between this subnetwork and the delegated phenotype. The y-axis of the point refers to the 1−*p*-value for the correlation. The y-axis on the right is for the histogram of these correlation. **b** is for PCC experiment and **c** is for MMI experiment. **d, e** Plots for eigengene and delegated phenotype. Blue line refers to the z-score of eigengene for that subnetwork. X-axis refers to samples. **d** is for PCC experiment and E is for MMI experiment

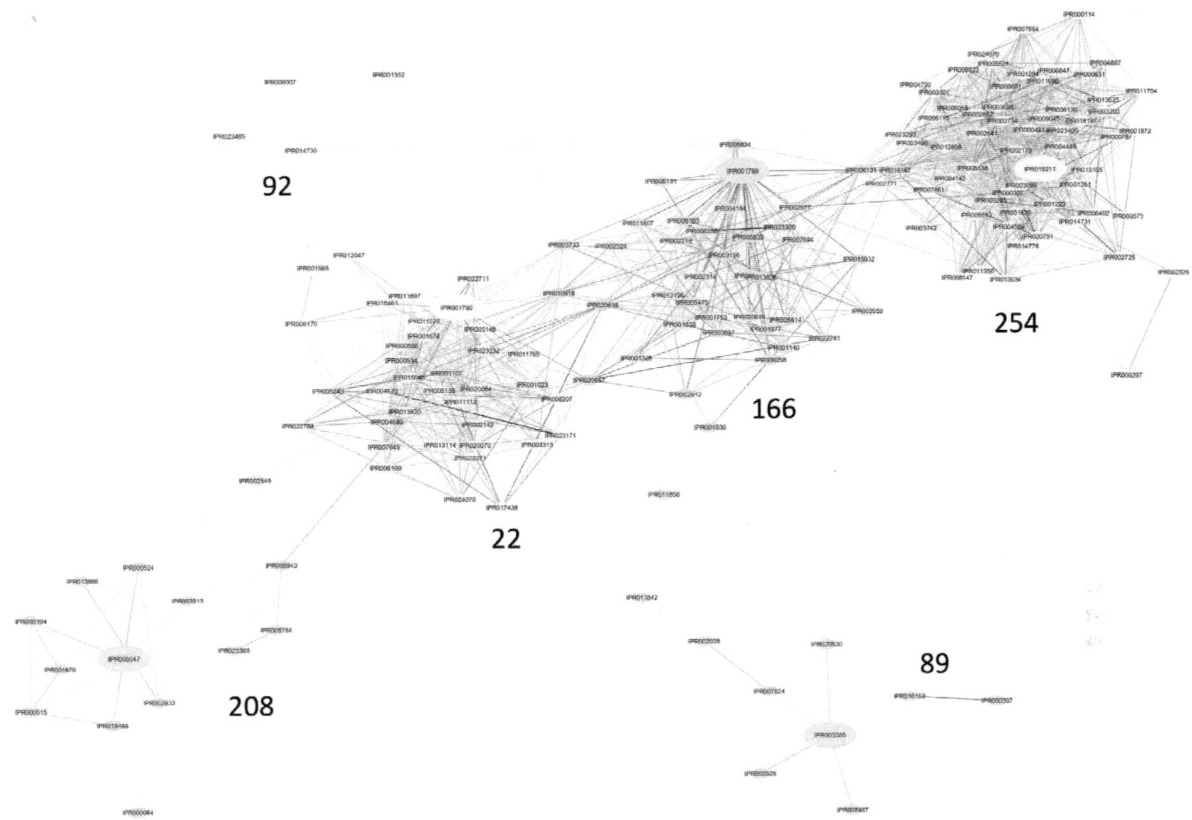

Fig. 3 Extracted networks for T2D. Detected T2D-relevant subnetworks for PCC and MMI experiment. Blue and yellow nodes refer to IPRs identified by PCC and MMI, while green ones refer to IPRs identified by both methods. The weight calculated by PCC and MMI was normalized to the same scale. The corresponding subnetworks are labeled. Nodes with top PageRank in each subnetwork are enlarged

Oral microbiome

We applied our method to oral microbiome to detect the key drivers in microbial community related to dysbiosis in Rheumatoid Arthritis (RA). The abundance data was downloaded from EBI (ERP006678). Information of phenotype variables for different individuals was acquired from published paper [2]. We mapped the samples downloaded from EBI with the individual ID and got 49 oral microbial samples in total. Among them, 27 samples were collected from patients with RA in different disease states, 22 samples, used as the control, were collected from people without RA. 21 of them are saliva samples and 28 are dental samples.

We first conducted filtering and then applied normalization to avoid noises. After that, we constructed the MEN by computing similarities between all pairs of IPRs. Then we partition the MEN into multiple subnetworks by clustering.

Similar to the analysis for T2D, we processed the phenotype matrix and detected subnetworks most related to RA. First, we removed phenotype variables with more than 1/3 missing values. Then, for remaining phenotype variables, we conducted imputation using R package MICE [39]. By computing correlation between delegated phenotype and eigengenes of subnetworks, we extracted six subnetworks most related to RA using PCC and three subnetworks using MMI. Finally, we identified key drivers for detected disease associated subnetworks correspondingly.

We applied key drivers analysis for RA using PCC and MMI as two different inference methods respectively. Both experiments show IPRs, in extracted associated subnetworks, have higher abundance in disease state than in normal state (see Fig. 4). For PCC experiment, annotation shows that most IPRs in subnetwork 335 and 63 are about cell membrane while most IPRs in subnetwork 676, 128, 679 and 680 are about replication and cell growth. Functions for IPRs were inferred according to keywords and Gene Ontology (GO) mentioned in InterPro [35]. Moreover, subnetwork 335 also contains IPR014879(Sporulation initiation factor Spo0A, C-terminal) and IPR013783 (Immunoglobulin-like fold). IPR013783 is about immunoglobulin molecules and T-cell receptor antigen [40, 41], while RA is a disease caused by compromised immune systems [42].

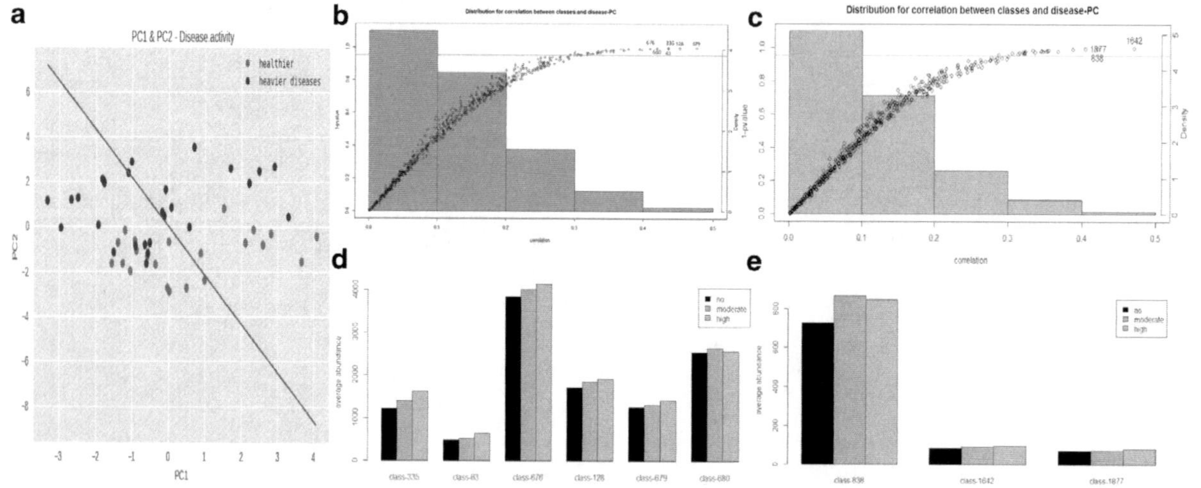

Fig. 4 Experiment results for Oral experiment. **a** The delegated phenotype. **b**, **c** Distribution for correlation between subnetworks and delegated phenotype. Every point refers to a subnetwork. The x-axis of the point refers to the correlation between this subnetwork and the delegated phenotype. The y-axis of the point refers to the $1-p$-value for the correlation. The y-axis on the right is for the histogram of these correlation. **b** is for PCC experiment and **c** is for MMI experiment. **d**, **e** Bar plot for the average abundance of extracted subnetworks. The relative abundance is different in different disease states

For MMI experiment, subnetwork 1642, which has the largest correlation with delegated phenotype, contains multiple IPRs about biofilm: IPR024487, IPR019669, and IPR010344. There are totally 24 IPRs in this subnetwork and top 5 of them are IPR003496, IPR024205, IPR008542, IPR010344, and IPR019669. Specifically, IPR010344 plays a role in biofilm formation and IPR019669 participates in single-species biofilm formation on the inanimate substrate. The development of biofilms is one of the drivers of persistent infections [27]. Some bacteria, when growing in the biofilm, e.g., Porphyromonas gingivalis in dental plaque, can become destructive and may contribute to RA [28]. Besides, subnetwork 1642 also contains IPR013756, associated with Flaviviruses, and IPR009774, related to hypothetical Streptococcus thermophilus bacteriophage, which hints the infection process in this subnetwork.

Discussion

Noise tolerance of delegated phenotype

To test whether our delegated phenotype is robust when phenotypes are deficient, we tried every combination of phenotypes with removing 1,2,3,6,10 of them from the phenotype variables matrix of RA, and generated delegated phenotype for each of them. Then we calculated the correlations between those generated delegated phenotypes and extracted subnetworks. The result is promising and these extracted subnetworks have high correlation values in most cases (see Fig. 5).

Performance of PageRank in searching the key driver

We tested the performance of PageRank on a simulated dataset. At the beginning, we named the driven relationship as sub-gene relationship. We simplified the network by assuming that one gene could only be driven by one gene. Linear function is used to represent the driven relationship. i.e. $y = Ax + n$, where x and y denotes expression levels for gene and its sub-gene and n is the noise following the normal distribution with 0 mean. There are three parameters for the simulation algorithm: the number of sub-genes for each gene, the depth of the network and the noise level. Here, we used the variance to define the noise level and the variance of noise is βx. The structure of the simulated network and three parameters are shown in Fig. 6.

We generated the simulated data with various parameters. For each parameter group, 100 samples were produced. We compared the performance of PageRank with the degree algorithm that locates the key driver with the highest degree. As shown in Fig. 7, the noise level has little effect on prediction precision. The result of the degree algorithm also follows this pattern. To compare these two algorithms, we collected the cases where only one algorithm correctly found the key driver and the result is shown in *C*. Since when the number of sub-genes is large, both algorithms have high prediction precision, we focus more on cases where the sub-gene number is relatively small. In this situation, the PageRank has better performance.

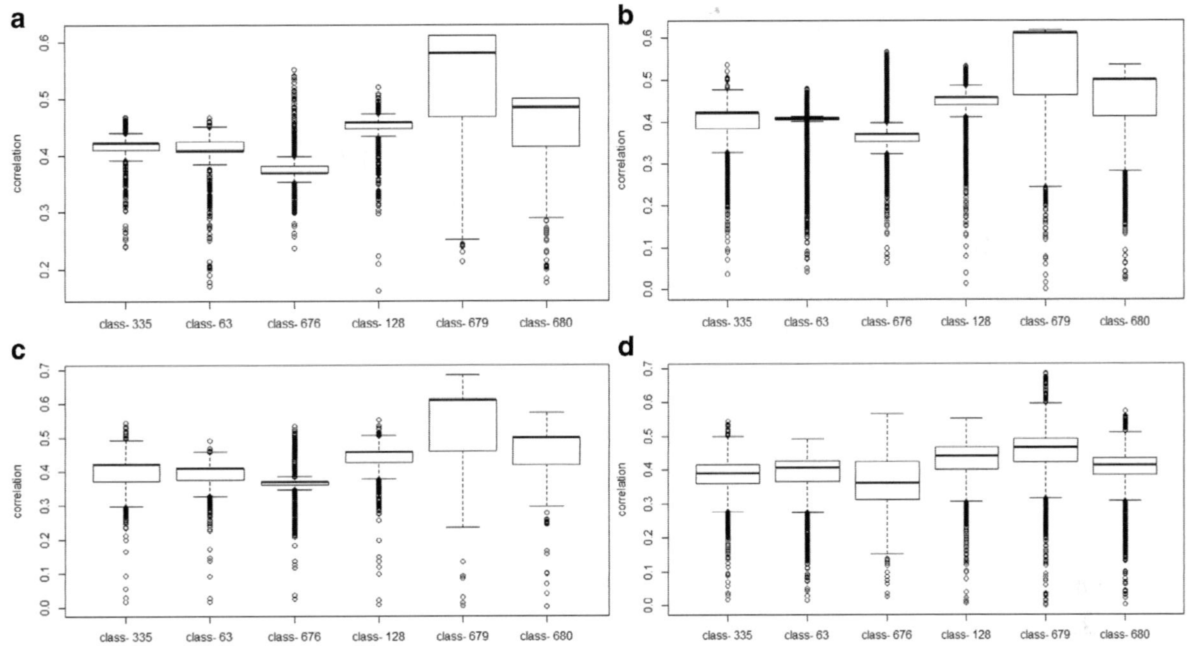

Fig. 5 Noise tolerance of delegated phenotype. **a** The first box plot refers to combination with removal of one, two, or three phenotype variables. **b** The second refers to the removal of six phenotype variables. **c** The third refers to the removal of ten phenotype variables. **d** The last one refers to the overall performance with all these combinations

Application to Alzheimer's disease

To further validate our method is capable of detecting key drivers of disease, we applied KDiamend to Alzheimer's Disease (AD) with analyzing RNA expression profiles, which were downloaded from GEO(GSE44770) [17]. Both of PCC and MMI experiments identified FBXL16 and OLFM1. FBXL16 related pathways are Innate Immune System and Class I MHC mediated antigen processing and presentation, while researches have shown that the activation of the Innate Immune System plays a crucial

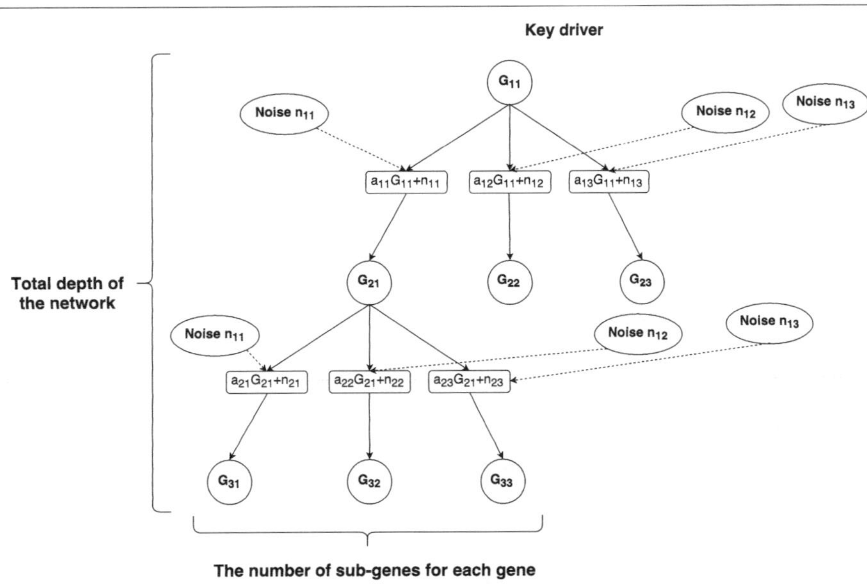

Fig. 6 The structure of simulated network. The simplified network can be transferred into a tree

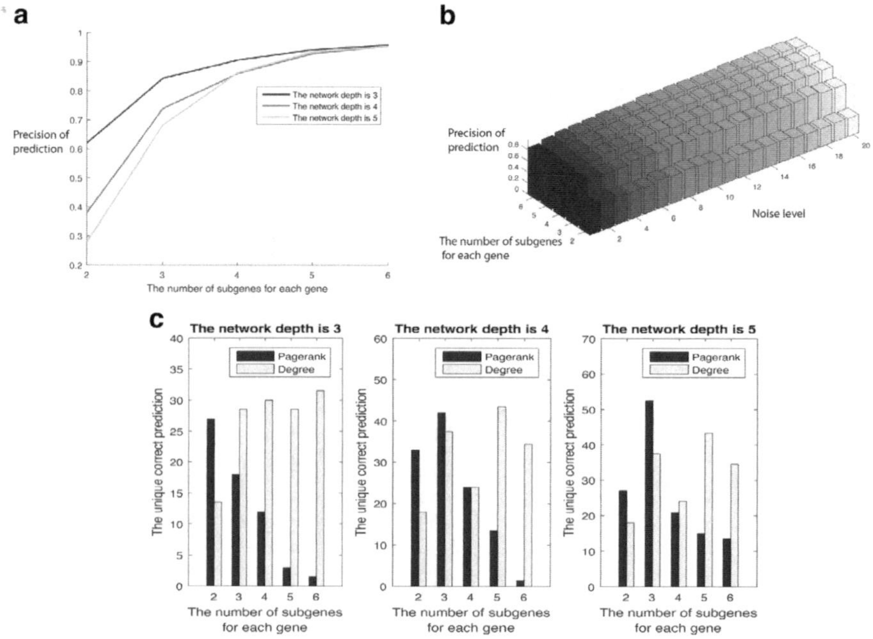

Fig. 7 Testing results for PageRank algorithm to find the key driver. The first two figures show the prediction precision of PageRank under different simulation parameters. **a** is related to the network depth and the number of sub-genes. **b** is for the noise level and the number of sub-genes. The third figure compares the performances of PageRank and degree algorithm. We counted cases where only one algorithm found the key driver correctly

role in promoting AD [43]. OLFM1 is related to nervous system development and Neuroblastoma [44]. Besides, PCC experiment also identified RPS4Y1 and PITPNB as key drivers for extracted disease associated subnetworks. MMI experiment also identified KAZALD1, OR4A47, RNASE11, TXNDC2, 7-Mar, RTN4, TSPAN9, PCNP and PPP2R2C. More specifically, RTN4 is related to Demyelinating Disease [45] and KAZALD1 is related to Lobar Holoprosencephaly [46]. These experiments show a possible application of our method. It is capable of detecting key drivers in the network inferred from not only the microbial abundance profile but also other kinds of abundance data, like RNA expression or proteomics.

Conclusion

We proposed a novel method to detect key actors who drive the disease concerned MENs, which helps to understand microbial factors relevant to the certain disease. We divided the MENs into multiple subnetworks and then, instead of detecting important genes based on pathways or gene annotations, we extracted subnetworks which are most relevant to disease by utilizing the correlation between the patterns of abundance profiles and the delegated phenotype. Lastly, we identified key drivers based on PageRank.

We tested our method with two real microbial datasets. We detected that the disease-relevant subnetworks in T2D are related to the carbohydrate metabolic process and phosphorelay, while RA-relevant subnetworks are related to membrane, cell growth, and infection. The extracted subnetworks for RA include IPRs concerned with immunoglobulin, Sporulation, biofilm, Flaviviruses, bacteriophage, etc. Then we located corresponding key drivers for extracted disease-relevant subnetworks. Besides microbial data, we also tested our method with gene expression profiles to identify key drivers for AD and the outcome was inspiring. Experiments show our method is capable of detecting key drivers and providing hints to understand the mechanisms of diseases.

Acknowledgments
We would like to express sincere gratitude to our colleagues who greatly assisted the research. We thank Tan Bowen for offering assistance during designing the algorithm and Zhao Zicheng, Xu Chang for comments that greatly improved the manuscript.

Funding
This work and the publication of this article were sponsored by the GRF Project from RGC General Research Fund at City University of Hong Kong, Kowloon, HKSAR under Grant 9042181 (CityU 11203115).

Availability of data and materials
Package is available at http://www.deepomics.org/pipelines/3DCD6955FEF2E64A/.

About this supplement
This article has been published as part of *BMC Systems Biology* Volume 12 Supplement 1, 2018: Selected articles from the 16th Asia Pacific Bioinformatics Conference (APBC 2018): systems biology. The full contents of the supplement

are available online at https://bmcsystbiol.biomedcentral.com/articles/supplements/volume-12-supplement-1.

Authors' contributions
LYU was responsible for the implementation of the package. Chen and LYU together designed the algorithm, conducted experiments and completed the manuscript. JIANG provided expertise on metagenomics and provided the data. DONG helped to improve the algorithm, conduct experiments and complete the manuscript. Fang provided assistance for testing and deploying the pipeline on the platform. LI provided insights and also polished the manuscript. All of the authors have read and approved of the final manuscript.

Ethics approval and consent to participate
Not applicable.

Consent for publication
Not applicable.

Competing interests
The authors declare that they have no competing interests.

Publisher's Note

Published: 11 April 2018

References

1. Qin J, Li Y, Cai Z, Li S, Zhu J, Zhang F, Liang S, Zhang W, Guan Y, Shen D, et al. A metagenome-wide association study of gut microbiota in type 2 diabetes. Nature. 2012;490(7418):55–60.
2. Zhang X, Zhang D, Jia H, Feng Q, Wang D, Liang D, Wu X, Li J, Tang L, Li Y, et al. The oral and gut microbiomes are perturbed in rheumatoid arthritis and partly normalized after treatment. Nat Med. 2015;21(8):895.
3. Nakatsu G, Li X, Zhou H, Sheng J, Wong SH, Wu WKK, Ng SC, Tsoi H, Dong Y, Zhang N, et al. Gut mucosal microbiome across stages of colorectal carcinogenesis. Nat Commun. 2015;6:8727.
4. Mack A, Olsen L, Choffnes ER, et al. Microbial Ecology in States of Health and Disease: Workshop Summary. Wasington, DC: National Academies Press; 2014.
5. de Steenhuijsen Piters WA, Sanders EA, Bogaert D. The role of the local microbial ecosystem in respiratory health and disease. Phil Trans R Soc B. 2015;370(1675):20140294.
6. Fukushima A. Diffcorr: an r package to analyze and visualize differential correlations in biological networks. Gene. 2013;518(1):209–14.
7. Deng Y, Jiang YH, Yang Y, He Z, Luo F, Zhou J. Molecular ecological network analyses. BMC Bioinformatics. 2012;13(1):1.
8. Deng Y, Zhou J. Molecular ecological network of microbial communities. Encyclopedia of Metagenomics: Genes, Genomes and Metagenomes: Basics, Methods, Databases and Tools. 2015. p. 504-510.
9. Berry D, Widder S. Deciphering microbial interactions and detecting keystone species with co-occurrence networks. Front Microbiol. 2014;5:219.
10. Deng Y, Zhang P, Qin Y, Tu Q, Yang Y, He Z, Schadt CW, Zhou J. Network succession reveals the importance of competition in response to emulsified vegetable oil amendment for uranium bioremediation. Environ Microbiol. 2016;18(1):205–18.
11. Kurtz ZD, Müller CL, Miraldi ER, Littman DR, Blaser MJ, Bonneau RA. Sparse and compositionally robust inference of microbial ecological networks. PLoS Comput Biol. 2015;11(5):1004226.
12. Zhang B, Zhu J. Identification of key causal regulators in gene networks. In: Proceedings of the World Congress on Engineering, vol. 2. Hong Kong: Newswood Limited; 2013.
13. Huan T, Meng Q, Saleh MA, Norlander AE, Joehanes R, Zhu J, Chen BH, Zhang B, Johnson AD, Ying S, et al. Integrative network analysis reveals molecular mechanisms of blood pressure regulation. Mol Syst Biol. 2015;11(4):799.
14. Filteau M, Pavey SA, St-Cyr J, Bernatchez L. Gene coexpression networks reveal key drivers of phenotypic divergence in lake whitefish. Mol Biol Evol. 2013;30:053.
15. Talukdar HA, Asl HF, Jain RK, Ermel R, Ruusalepp A, Franzén O, Kidd BA, Readhead B, Giannarelli C, Kovacic JC, et al. Cross-tissue regulatory gene networks in coronary artery disease. Cell Syst. 2016;2(3):196–208.
16. Song WM, Zhang B. Multiscale embedded gene co-expression network analysis. PLoS Comput Biol. 2015;11(11):1004574.
17. Zhang B, Gaiteri C, Bodea LG, Wang Z, McElwee J, Podtelezhnikov AA, Zhang C, Xie T, Tran L, Dobrin R, et al. Integrated systems approach identifies genetic nodes and networks in late-onset alzheimer's disease. Cell. 2013;153(3):707–20.
18. Portune KJ, Beaumont M, Davila AM, Tomé D, Blachier F, Sanz Y. Gut microbiota role in dietary protein metabolism and health-related outcomes: The two sides of the coin. Trends Food Sci Technol. 2016;57:213–32.
19. Fraser AM, Swinney HL. Independent coordinates for strange attractors from mutual information. Phys Rev A. 1986;33(2):1134.
20. Weiss S, Van Treuren W, Lozupone C, Faust K, Friedman J, Deng Y, Xia LC, Xu ZZ, Ursell L, Alm EJ, et al. Correlation detection strategies in microbial data sets vary widely in sensitivity and precision. ISME J. 2016;10(7):1669.
21. Paulson JN, Stine OC, Bravo HC, Pop M. Differential abundance analysis for microbial marker-gene surveys. Nat Methods. 2013;10(12):1200–2.
22. McMurdie PJ, Holmes S. Waste not, want not: why rarefying microbiome data is inadmissible. PLoS Comput Biol. 2014;10(4):1003531.
23. Zhang L, Chen JX, Li S. More accurate models for detecting gene-gene interactions from public expression compendia. In: Bioinformatics and Biomedicine (BIBM), 2016 IEEE International Conference On. Red Hook: IEEE; 2016. p. 1871–8.
24. Friedman N. Inferring cellular networks using probabilistic graphical models. Science. 2004;303(5659):799–805.
25. Langfelder P, Horvath S. Wgcna: an r package for weighted correlation network analysis. BMC Bioinformatics. 2008;9(1):559.
26. Meyer KA, Kushi LH, Jacobs DR, Slavin J, Sellers TA, Folsom AR. Carbohydrates, dietary fiber, and incident type 2 diabetes in older women. Am J Clin Nutr. 2000;71(4):921–30.
27. Macia M, Rojo-Molinero E, Oliver A. Antimicrobial susceptibility testing in biofilm-growing bacteria. Clin Microbiol Infect. 2014;20(10):981–90.
28. Marcinkiewicz J, Strus M, Pasich E. Antibiotic resistance: a "dark side" of biofilmassociated chronic infections. Pol Arch Med Wewn. 2013;123(6):309–13.
29. Faith JJ, Hayete B, Thaden JT, Mogno I, Wierzbowski J, Cottarel G, Kasif S, Collins JJ, Gardner TS. Large-scale mapping and validation of escherichia coli transcriptional regulation from a compendium of expression profiles. PLoS Biol. 2007;5(1):8.
30. Roulston MS. Significance testing of information theoretic functionals. Physica D: Nonlinear Phenom. 1997;110(1):62–6.
31. Langfelder P, Horvath S. Eigengene networks for studying the relationships between co-expression modules. BMC Syst Biol. 2007;1(1):54.
32. Manza RR. Computer Vision and Information Technology: Advances and Applications. New Delhi: IK International Pvt Ltd; 2010.
33. Xing W, Ghorbani A. Weighted pagerank algorithm. In: Communication Networks and Services Research, 2004. Proceedings. Second Annual Conference On. Washington, DC: IEEE; 2004. p. 305–14.
34. Page L, Brin S, Motwani R, Winograd T. The PageRank citation ranking: Bringing order to the web. Stanford InfoLab; 1999.
35. Finn RD, Attwood TK, Babbitt PC, Bateman A, Bork P, Bridge AJ, Chang HY, Dosztányi Z, El-Gebali S, Fraser M, et al. Interpro in 2017—beyond protein family and domain annotations. Nucleic Acids Res. 2017;45(D1):190–9.
36. Bullard JH, Purdom E, Hansen KD, Dudoit S. Evaluation of statistical methods for normalization and differential expression in mrna-seq experiments. BMC Bioinformatics. 2010;11(1):1.
37. Qiu X, Wu H, Hu R. The impact of quantile and rank normalization procedures on the testing power of gene differential expression analysis. BMC Bioinformatics. 2013;14(1):1.
38. Zhang Y, Zagnitko O, Rodionova I, Osterman A, Godzik A. The fggy carbohydrate kinase family: insights into the evolution of functional specificities. PLoS Comput Biol. 2011;7(12):1002318.
39. Buuren S, Groothuis-Oudshoorn K. mice: Multivariate imputation by chained equations in r. J Stat Softw. 2011;45(3):1–68.
40. Bork P, Holm L, Sander C. The immunoglobulin fold: structural classification, sequence patterns and common core. J Mol Biol. 1994;242(4):309–20.
41. Halaby D, Poupon A, Mornon JP. The immunoglobulin fold family: sequence analysis and 3d structure comparisons. Protein Eng. 1999;12(7):563–71.

42. Edwards J, Cambridge G, Abrahams V, et al. Do self-perpetuating b lymphocytes drive human autoimmune disease? Immunol Oxford. 1999;97:188–96.
43. Heneka MT, Golenbock DT, Latz E. Innate immunity in alzheimer's disease. Nat Immunol. 2015;16(3):229–36.
44. Yokoyama M, Nishi Y, Yoshii J, Okubo K, Matsubara K. Identification and cloning of neuroblastoma-specific and nerve tissue-specific genes through compiled expression profiles. DNA Res. 1996;3(5):311–20.
45. Murayama KS, Kametani F, Saito S, Kume H, Akiyama H, Araki W. Reticulons rtn3 and rtn4-b/c interact with bace1 and inhibit its ability to produce amyloid β-protein. Eur J NeuroSci. 2006;24(5):1237–44.
46. Peltekova IT, Hurteau-Millar J, Armour CM. Novel interstitial deletion of 10q24. 3–25.1 associated with multiple congenital anomalies including lobar holoprosencephaly, cleft lip and palate, and hypoplastic kidneys. Am J Med Genet A. 2014;164(12):3132–6.

Yu et al. BMC Systems Biology 2018, 12(Suppl 1):14
https://doi.org/10.1186/s12918-018-0532-7

BMC Systems Biology

RESEARCH

Open Access

Predicting and understanding comprehensive drug-drug interactions via semi-nonnegative matrix factorization

Hui Yu[1], Kui-Tao Mao[1], Jian-Yu Shi[2*], Hua Huang[3], Zhi Chen[4], Kai Dong[2] and Siu-Ming Yiu[5]

From The Sixteenth Asia Pacific Bioinformatics Conference
Yokohama, Japan. 15-17 January 2018

Abstract

Background: Drug-drug interactions (DDIs) always cause unexpected and even adverse drug reactions. It is important to identify DDIs before drugs are used in the market. However, preclinical identification of DDIs requires much money and time. Computational approaches have exhibited their abilities to predict potential DDIs on a large scale by utilizing pre-market drug properties (e.g. chemical structure). Nevertheless, none of them can predict two comprehensive types of DDIs, including enhancive and degressive DDIs, which increases and decreases the behaviors of the interacting drugs respectively. There is a lack of systematic analysis on the structural relationship among known DDIs. Revealing such a relationship is very important, because it is able to help understand how DDIs occur. Both the prediction of comprehensive DDIs and the discovery of structural relationship among them play an important guidance when making a co-prescription.

Results: In this work, treating a set of comprehensive DDIs as a signed network, we design a novel model (DDINMF) for the prediction of enhancive and degressive DDIs based on semi-nonnegative matrix factorization. Inspiringly, DDINMF achieves the conventional DDI prediction (AUROC = 0.872 and AUPR = 0.605) and the comprehensive DDI prediction (AUROC = 0.796 and AUPR = 0.579). Compared with two state-of-the-art approaches, DDINMF shows it superiority. Finally, representing DDIs as a binary network and a signed network respectively, an analysis based on NMF reveals crucial knowledge hidden among DDIs.

Conclusions: Our approach is able to predict not only conventional binary DDIs but also comprehensive DDIs. More importantly, it reveals several key points about the DDI network: (1) both binary and signed networks show fairly clear clusters, in which both drug degree and the difference between positive degree and negative degree show significant distribution; (2) the drugs having large degrees tend to have a larger difference between positive degree and negative degree; (3) though the binary DDI network contains no information about enhancive and degressive DDIs at all, it implies some of their relationship in the comprehensive DDI matrix; (4) the occurrence of signs indicating enhancive and degressive DDIs is not random because the comprehensive DDI network is equipped with a structural balance.

Keywords: Drug-drug interaction, Nonnegative matrix factorization, Regression, Network community, Balance theory

* Correspondence: jianyushi@nwpu.edu.cn
[2]School of Life Sciences, Northwestern Polytechnical University, Xi'an, China
Full list of author information is available at the end of the article

Background

When two or more drugs are taken together, they would unexpectedly influence each other in terms of pharmacokinetic or pharmacodynamical behavior [1]. This kind of influence is termed as Drug-Drug Interaction (DDI), which would cause adverse drug reactions (e.g. the reduction in efficacy or the increment on unexpected toxicity among the co-prescribed drugs). As the number of approved drugs increases, the number of unidentified DDIs is rapidly rising, such that more and more adverse effects among the drugs may occur. Unidentified DDIs would lead patients, who are treated with numerous medications, to be in the unsafe treatment of medication errors [2–5]. Understanding DDI is also the first step towards drug combination, which is one of promising strategies for multifactorial complex diseases [6]. Therefore, it is an urgent need to screen and analyze DDIs before clinical medications are administered. However, traditional experimental approaches for DDI identification (e.g. testing cytochrome P450 [7] or transporter-associated interactions [8]) face challenges, such as high cost, long duration, animal welfare considerations [9], the very limited number of participants and the great number of drug combinations under screening in clinical trials. As a result, only a few of DDIs could be identified during drug development (usually the clinical trial phase), and most of them are reported after the drugs are approved.

Computational approaches are promising to discover potential DDIs on a large scale, and they have gained many concerns from both academy and industry recently [10, 11]. Text-mining based computational approaches have been developed for detecting DDIs from different sources [9], including scientific literature [12, 13], electronic medical records [14], and the Adverse Event Reporting System of FDA (http://www.fda.gov). These approaches heavily rely on the clinical evidence in post-market, thus cannot provide alerts of potentially DDIs before clinical medications are administered. In contrast, machine learning-based computational approaches (e.g. naïve similarity-based approach [15], network recommendation-based [9], classification-based [16]) are able to provide such alerts by utilizing pre-marketed drug features or similarities [17], such as, chemical structures [15], targets [18], hierarchical classification codes [16] and side effects [9, 19]. Most of the existing machine learning-based approaches were designed for conventional binary prediction, which only indicates how likely a pair of drugs is a DDI. However, two interacting drugs may increase or decrease their own pharmaceutical behaviors or effects in vivo.

It is more important to know exactly whether the interaction increases or decreases the pharmaceutical behaviors of the drugs when making optimal patient care, establishing the dosage of a drug, designing prophylactic drug therapy, or finding the resistance to therapy with a drug [20]. As one of the important pharmaceutical indices, serum concentration reflects the amount of a drug in the pharmacokinetic circulation [21]. When interacting with other drugs, a drug would increase or decrease the level of its own and its partners' serum concentration. For example, the serum concentration of Dofetilide (whose DrugBank Id is DB00204) decreases when it is taken with Dabrafenib (DB08912) together, whereas its serum concentration increases when taken with Dalfopristin (DB01764). For short, we name the first case of DDI as a degressive DDI and the second case of DDI as an enhancive DDI in the following texts.

To summarize, predicting approaches for comprehensive DDIs helps to uncover the underlying mechanism of how DDIs occur [19]. However, most of current approaches have been developed only for conventional binary DDIs, but not for enhancive and degressive DDIs. Furthermore, there is a lack of systematic analysis on the structural relationship hidden among known DDIs. Revealing such a structural relationship is very important, because it is able to help understand how DDIs occur. A promising solution of these two issues is able to guide medical doctors to make safe co-prescriptions.

In this paper, to address abovementioned issues, we firstly design a novel model (DDINMF) for DDI prediction based on Nonnegative Matrix Factorization (NMF). Representing DDIs as a binary network and a signed network respectively, we then make an attempt to reveal the structural relationship hidden among DDIs by NMF and social balance theory.

Methods

Dataset

We collected 2329 approved drugs from DrugBank [22] and selected 603 drugs, which have DDIs recorded in DrugBank. Among the set of drugs, we also removed those drugs without chemical structures or without the off-label side effects recorded in OFFSIDES [23], and kept the remaining 568 drugs having both of them. The final DDI dataset contains 21,351 DDIs, including 16,757 enhancive DDIs (E-DDI) and 4594 degressive DDIs (D-DDI). Each drug is represented as an 881-dimensional feature vector $\mathbf{f}_{str}$ based on PubChem structure descriptor and also a 9149- dimensional feature vector $\mathbf{f}_{se}$ according to the off-label side effects provided by OFFSIDES [23]. The second feature $\mathbf{f}_{se}$ was proposed firstly in [19]. These two vectors are binary, of which a value of 1 denotes the occurrence of a specific structure fragment in its chemical structure or the observation of a specific side effect in clinic, and 0 if this does not occur or is not observed. By organizing DDIs as a DDI graph, we also investigated the degrees of DDIs, enhancive DDIs and degressive DDIs respectively. The details of our DDI dataset are listed in Table 1.

Problem formulation

Without loss of generality, let $\mathbf{D} = \{d_i\}$, $i = 1, 2, \ldots, m$ be a set of m approved drugs. Their interactions can be organized as an $m \times m$ symmetric interaction matrix $\mathbf{A}_{m \times m} = \{a_{ij}\}$, which can be regarded as the adjacent matrix of DDI network. For the conventional DDI, $a_{ij} = 1$ if d_i interacts with d_j, and $a_{ij} = 0$ otherwise. For the comprehensive DDI, $a_{ij} \in \{-1, 0, +1\}$. In details, if d_i and d_j do not interact with each other, $a_{ij} = 0$. When there is an enhancive DDI or a degressive DDI between d_i and d_j, $a_{ij} = +1$ or $a_{ij} = -1$ respectively. Obviously, as the special case of the comprehensive DDI matrix $\mathbf{A}$, the conventional binary DDI matrix $\mathbf{A}_b$ can be obtained by $\mathbf{A}_b = Binary(\mathbf{A})$.

In addition, each drug d_i in $\mathbf{D}$ is represented as a p-dimensional feature vector $\mathbf{f}_i = [f_1, f_2, \ldots, f_k, \ldots, f_p]$, where $f_k = 1$ denotes the k-th specific structure fragment or observed side effect and $f_k = 0$ otherwise . All the drugs in $\mathbf{D}$ are sequentially organized as an $m \times p$ feature matrix $\mathbf{F}_{m \times p}$.

To obtain a brief description, **a known drug** is referred to as a drug in $\mathbf{D}$ and **a new drug** is defined as the drug having no interaction with any drug in $\mathbf{D}$. A screening scenario for new drugs is considered in this paper. Formally, the task is to predict how likely a newly given drug d_x interacts with one or more known drugs, and how likely a DDI of d_x is an enhancive DDI or a degressive DDI. Fig. 1 illustrates the task.

Prediction model

Since the new drugs have no existing interaction with the drugs in $\mathbf{D}$ and are regarded as isolated nodes in the DDI network (Fig. 1), we cannot utilize their topological information to deduce their potential interactions. This is also the well-known cold start problem in recommendation system. Obviously, we need their additional properties (e.g. chemical structure fragments or side effects), which are called drug features in terms of machine learning. Once the additional features are obtained, we would build a supervised model to perform DDI prediction. The main idea is to build the relationship between the features of known drugs in $\mathbf{D}$ and their DDI network topology. We built the model (DDINMF) for predicting DDI based on nonnegative matrix factorization. It includes two phases (Fig. 2).

(1) In the training phase of DDINMF, the adjacent matrix $\mathbf{A}_{m \times m}$ among m known drugs is first decomposed into two matrices $\mathbf{A}_{m \times m} \approx \mathbf{W}_{m \times r} \times \mathbf{H}_{r \times m}$, where r is the dimension of the latent topological space. Denote $\mathbf{F}_{m \times p}$ as the input feature matrix of those m known drugs (the training drugs). Then the relationship between $\mathbf{F}_{m \times p}$ and $\mathbf{H}_{r \times m}$ can be modeled by a regression $(\mathbf{H}^T)_{m \times r} = \mathbf{F}_{m \times p} \times \mathbf{B}_{p \times r}$, where $\mathbf{B}_{p \times r}$ is the regression coefficient matrix.

(2) In the predicting phase, the learned $\mathbf{B}_{p \times r}$ firstly maps the $n \times p$ input feature matrix of n newly given drugs (denoted as $\mathbf{F}_x$) into the latent topological space by $\mathbf{H}_x^T = \mathbf{F}_x \times \mathbf{B}$. Then the $n \times r$ mapped latent feature matrix $\mathbf{H}_x$ is used to generate the predicted interactions between the new drug and the known drugs by $\mathbf{A}_x = (\mathbf{W}\mathbf{H}_x)^T$.

DDINMF applies the regular nonnegative matrix factorization (NMF) to decompose the DDI adjacent matrix when given a conventional binary adjacent matrix, and applies a variant of NMF, semi-NMF, when given a comprehensive adjacent matrix. A brief introduction of both NMF and semi-NMF can be found in the next section. The regression in DDINMF is solved by Partial Least Square Regression (PLSR) because there is a multi-collinearity between some columns of $\mathbf{F}$ or $\mathbf{H}$. The model of multivariate PLSR can be solved by SIMPLS algorithm [24].

Nonnegative matrix factorization and semi-nonnegative matrix factorization

Given an instance matrix $\mathbf{X}_{m \times n} = [\mathbf{x}_1, \mathbf{x}_2, \ldots, \mathbf{x}_n] \in \mathbb{R}^+$, Nonnegative Matrix Factorization (NMF) aims to find a base matrix $\mathbf{W}_{m \times r} = [\mathbf{w}_{ir}] \in \mathbb{R}^+$ and an encoding matrix $\mathbf{H}_{r \times n} = [\mathbf{h}_{ri}] \in \mathbb{R}^+$, whose product can well approximate the original matrix $\mathbf{X}^+ \approx \mathbf{W}^+\mathbf{H}^+$, where $r < \min(m, n)$. These approximating factors are typically obtained by solving the constrained least square minimization problem [25]:

$$\min_{\substack{W \geq 0 \\ H \geq 0}} \|\mathbf{X}\text{-}\mathbf{W}\mathbf{H}\|_F^2 = \sum_{i,j} \left(\mathbf{x}_{ij} - \sum_{k=1}^{r} \mathbf{w}_{ik}\mathbf{h}_{ki} \right)^2 \tag{1}$$

The algorithm minimizing the objective function is as follows:

Table 1 Details of comprehensive DDI network

Property	Value	Degree	Value	Degree of E-DDI	Value	Degree of D-DDI	Value
#Drug	568	Ave.	75.18	Ave.	59.00	Ave.	16.18
#DDI	21,351	Median	61.50	Median	45.00	Median	8.00
#E-DDI	16,757	Max.	296	Max.	230	Max.	206
#D-DDI	4594	Min.	1	Min.	0	Min.	0

E-DDI: enhancive DDIs, *D-DDI*: degressive DDIs

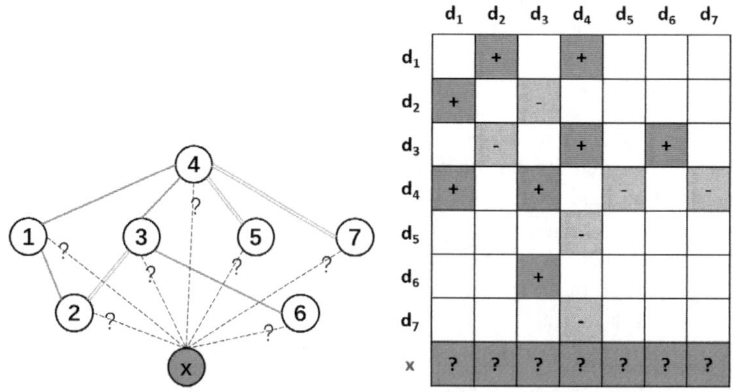

Fig. 1 Illustration of predicting DDIs for a newly given drug. In the left panel, nodes in DDI network represent drugs. The hollow nodes are known drugs (numbered from 1 to 7) and the solid lines between them denote their interactions. Blue lines are enhancive interactions and yellow lines are degressive interactions. The node in red respectively is the newly given drug, tagged as **X**. In the right panel, the adjacent matrix is shown. The cells in it are filled with blue, yellow and red, accounting for the types of DDIs and drug pairs of interest respectively. All the pairwise entries among $\{d_1, d_2 \ldots, d_7\}$ are used to train the model, the entries in the red cells denote the testing entries. Our problem is to determine which known drugs could interact with the new drug **X** and what type these potential interactions are

$$\mathbf{w}_{ik} \leftarrow \mathbf{w}_{ik} \frac{(\mathbf{XH}^T)_{ik}}{(\mathbf{WHH}^T)_{ik}}, \quad \mathbf{h}_{kj} \leftarrow \mathbf{h}_{kj} \frac{(\mathbf{X}^T\mathbf{W})_{kj}}{(\mathbf{H}^T\mathbf{W}^T\mathbf{W})_{kj}} \tag{2}$$

The factor matrices **W** and **H** in the regular NMF are able to provide basic clustering. If we view $\mathbf{W} = [\mathbf{w}_1, \mathbf{w}_2, \ldots, \mathbf{w}_r]$ as the cluster centroids, then $\mathbf{H} = [\mathbf{h}_1, \mathbf{h}_2, \ldots, \mathbf{h}_n]$ can be viewed as the cluster indicators for each datapoint. Especially, when $\mathbf{X}^+$ is a symmetric binary nonnegative matrix (e.g. the adjacent matrix of DDI $\mathbf{A}_b$), **W** denotes the communities in a network while **H** denotes the membership of instances to those communities.

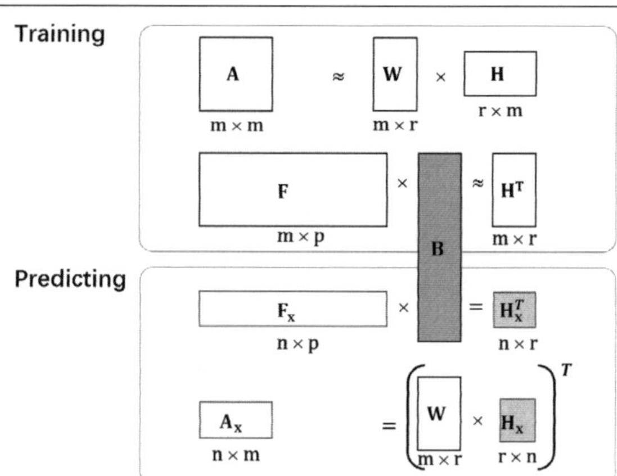

Fig. 2 Overview of DDINMF. DDINMF contains a training phase and a predicting phase. (1) In its training phase, the adjacent matrix **A** is first decomposed into a basis (community) matrix and a latent (encoding) feature matrix by $\mathbf{A} \approx \mathbf{W} \times \mathbf{H}$. Then the relationship between the input feature matrix **F** and the latent feature matrix **H** is modeled by a regression $(\mathbf{H}^T) = \mathbf{F} \times \mathbf{B}$. (2) In the predicting phase, the learned regression coefficient **B** firstly maps the input feature matrix $\mathbf{F}_x$ of *n* newly given drugs into their latent feature matrix by $\mathbf{H}_x^T = \mathbf{F}_x \times \mathbf{B}$. Then the mapped latent feature matrix of $\mathbf{F}_x$ is used to generate the predicted interactions between the new drugs and the known drugs by $\mathbf{A}_x = (\mathbf{W}\mathbf{H}_x)^T$

Though there is a variant of NMF, symmetric NMF (SymNMF) [26], which looks more natural to the symmetric adjacent matrix of binary DDIs, it has disadvantages from a technical point of view. SymNMF has higher complexity than NMF, thus runs slower. Most importantly, it has a larger reconstructed error, especially a higher divergence of the diagonal elements in its reconstructed matrix comparing to all-zero diagonal of the original matrix. Therefore, we apply NMF but not SymNMF to decompose the binary adjacent matrix of DDI.

NMF has a strong constraint of $\mathbf{X} \in \mathbb{R}^+$. When the instance matrix has mixed signs (e.g. the adjacent matrix of comprehensive DDI **A**), we consider Semi-NMF $\mathbf{X}^{\pm} \approx \mathbf{W}^{\pm}\mathbf{H}^+$, in which both $\mathbf{X} \in \mathbf{R}_{\pm}^{m \times n}$ and $\mathbf{W} \in \mathbf{R}_{\pm}^{m \times r}$ have mixed signs while only $\mathbf{H} \in \mathbf{R}_{+}^{r \times n}$ is restricted to comprise strictly nonnegative components [27]. Semi-NMF has the same form of cost/objective function as that of NMF, except for no constraint on **W**. But the algorithm minimizing the objective function is totally different:

$$\mathbf{W} \leftarrow \mathbf{XH}^{\dagger}, \quad \mathbf{H} \leftarrow \mathbf{H} \odot \sqrt{\frac{(\mathbf{W}^T\mathbf{X})^{\text{pos}} + [(\mathbf{W}^T\mathbf{W})^{\text{neg}}\mathbf{H}]_{ik}}{(\mathbf{W}^T\mathbf{X})^{\text{neg}} + [(\mathbf{W}^T\mathbf{W})^{\text{pos}}\mathbf{H}]_{ik}}}, \tag{3}$$

where $\mathbf{H}^{\dagger}$ is the Moore–Penrose pseudo-inverse of **H**, $\mathbf{A}^{\text{pos}}$ is a matrix that has the negative elements of matrix **A** replaced with 0, and $\mathbf{A}^{\text{neg}}$ is the one that has the positive elements of **A** replaced with 0:

$$\forall i, j. \mathbf{A}_{ij}^{\text{pos}} = \frac{|\mathbf{A}_{ij}| + \mathbf{A}_{ij}}{2}, \mathbf{A}_{ij}^{\text{neg}} = \frac{|\mathbf{A}_{ij}| - \mathbf{A}_{ij}}{2}. \quad (4)$$

Though there are other methods of matrix factorization (e.g. Principal Component Analysis, PCA) that can be used to decompose the comprehensive adjacent matrix of DDIs, semi-NMF has the exclusive advantage that both its **W** and **H** show physical meanings, such as the ordinary degree, the difference of positive degree and negative degree, and the balance properties. See also Section *Results and Discussion*.

Finally, NMF and Semi-NMF are adopted to decompose the binary and the comprehensive adjacent matrices of DDIs in our predicting model respectively.

Assessment

K-fold Cross-validation (K-CV) is the well-established approach to validate the power of generalization of algorithms in machine learning. To reflect the fact that new drugs have no interaction and to avoid over-optimistic prediction [17, 28–31], the K-CV scheme should be elaborately designed. For the given drugs having NO known interaction, the K-CV scheme tries to assess the task of predicting new potential interactions between them and those drugs having known interactions. The task corresponding to the K-CV scheme is useful when one tries to extend existing co-prescriptions by adding new drugs.

The generation of both training samples and testing samples is as follows. 1/K drugs are randomly removed out of all the given drugs in **D** as the testing drugs and the remaining drugs are taken as the training drugs. The drug pairs among the training drugs are selected as the training samples. The testing drugs are regarded as new drugs, and only the drug pairs between the testing drugs and the training drugs are selected as the testing samples, which are blind to the training (see also Fig. 1). Finally, the above procedures are repeated K times and the average of predicting performance in all rounds of CV is taken as the final performance. Usually, K = 10.

Two measures are usually adopted to assess the predicting performance, including the area under receiver operating characteristic curve (AUROC) and the area under precision-recall curve (AUPR) [32]. In the prediction of conventional binary DDIs, since positive and negative samples are interactions and non-interactions, both AUROC and AUPR can be calculated by comparing their predicted scores. Considering that the predicted scores of enhancive DDIs tend to be greater than ZERO and the predicted scores of degressive DDIs tend to be less than ZERO, we need to extend the calculation of AUROC and AUPR. In the prediction of comprehensive DDIs, to adopt the same way of calculating AUROC and AUPR, both enhancive and degressive DDIs first are labeled as positive samples while non-interactions are still labeled as negative samples. Then, the union of the predicted scores of enhancive DDIs and the MINUS of the predicted scores of degressive DDIs is compared with those scores of non-interactions by the same way as that in measuring conventional prediction.

Results and Discussion

Feature preprocessing

Because each drug was represented as an 881-dimensional feature vector $\mathbf{f}_{str}$ or as a 9149-dimensional feature vector $\mathbf{f}_{se}$, we reduced the high dimensions to accelerate the calculation. Though the value of the reduced dimension was unknown, we estimated it by performing ordinary PCA on the input feature matrix and counting the obtained PCs from the first one until the one, of which the entries having the values near to zeros. Then, using the number of PCs as the estimated dimension, we applied Kernel PCA [33] to the original input feature matrix and obtained the mapped feature matrix having significantly smaller dimensions. Gaussian function was adopted as the kernel in Kernel PCA with its variance equal to 4 for $\mathbf{f}_{str}$ and 16 for $\mathbf{f}_{se}$. Finally, the reduced dimensions of $\mathbf{f}_{str}$ and $\mathbf{f}_{se}$ were 487 and 557 respectively.

Comparison with the state-of-the-art approaches

Before running prediction, we assigned two parameters in our DDINMF, including the dimension of latent space (r) in matrix factorization and the number of latent factors (k) in PLSR as follows. We fixed the former with $rank(\mathbf{A})/2$ and tuned the latter from the list {10, 20, 30, 40, 50, 60, 70, 80, 90, 100} with the prediction measure of AUROC under 10-CV. Using the structure feature $\mathbf{f}_{str}$ and the side effect feature $\mathbf{f}_{se}$, we run NMF for binary DDI prediction and Semi-NMF for comprehensive DDI prediction respectively. When performing the conventional prediction of binary DDIs, we turned the entries equal to -1 into $+1$ in $\mathbf{A}$ to form the binary DDI adjacent matrix $\mathbf{A}_b$ (see also Section *Problem Formulation*). The best values of the number of latent factors ($k = 40$ and $k = 100$ respectively) were picked up for two kinds of DDI prediction. In addition, we concatenated $\mathbf{f}_{str}$ with $\mathbf{f}_{se}$ and run the prediction again ($k = 100$). The results show that $\mathbf{f}_{str}$ are significantly better than both $\mathbf{f}_{se}$ and their concatenation (Fig. 3). Therefore, only using the chemical structure feature $\mathbf{f}_{str}$, we performed DDI prediction in the following experiments.

First, we compared our approach with two state-of-the-art approaches, including Naïve similarity-based approach [15] and label propagation-based approach [9]. To obtain the robust predicting performance, 10-CV was repeated 50 times under different random seeds. Their final performance was reported by the average performance over 50 repetitions of CV (Table 2). The results show that DDINMF is significantly superior to them in terms of both AUROC

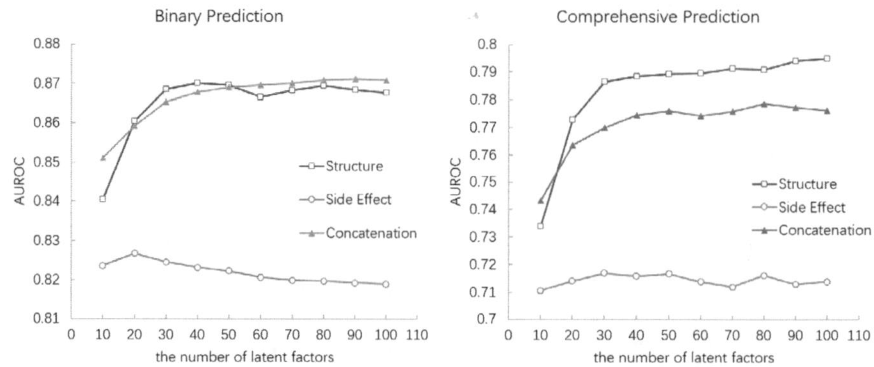

Fig. 3 Illustration of determining the best value for the number of latent factors when given the structure feature and the side effect feature

and AUPR, and also indicate that the comprehensive prediction is more difficult than the conventional prediction.

We also compared NMF with SymNMF in the case of binary DDI prediction. We decomposed our binary DDI adjacent matrix with the dimension of latent space = $rank(\mathbf{A})/2$ by NMF and SymNMF respectively. NMF spends only 4.76 s on the binary adjacent matrix of DDIs while SymNMF spends 97.6 s using a computer equipped with Windows 10 (64-bits), Intel Core i7-4700MQ 2.40G and 16 GB RAM. On the other hand, the reconstructed error of NMF is only 26.1122 while that of SymNMF is up to 94.7374. The mean and standard derivation of the diagonal elements in the reconstructed matrix achieved by NMF are 0.3888 and 0.4003, whereas those achieved SymNMF are 1.4352 and 0.8303 respectively. In addition, the performance (AUROC = 0.870 and AUPR = 0.583) achieved by SymNMF is worse than those achieved by NMF (see also Table 2), especially on AUPR. In summary, NMF is better than SymNMF.

Similarly, we compared semi-NMF with PCA in the case of comprehensive DDI prediction. PCA achieves the prediction performance with AUROC = 0.780 and AUPR = 0.535. Obviously, semi-NMF with AUROC = 0.796 and AUPR = 0.579 outperforms PCA. More importantly, semi-NMF has the advantage that both its **W** and **H** show physical meanings, such as the ordinary degree, the difference of positive degree and negative degree, and the balance properties. See also next section.

Finally, considering the desirable performance of predicting conventional DDIs, we performed a novel prediction. In detail, we randomly collected 50 extra drugs from DrugBank and checked whether they interact with any of the original 568 drugs. According the predicted scores, the top-k drug pairs then were picked out and compared with the records in DrugBank. We counted the records matching the drug pairs of interest. For convenience, the number of such records over k is called hitting ratio. We checked the records from top-10 to top-100 with step 10 (Table 3). The validation demonstrates that our DDINMF is effective to predict DDIs for newly coming drugs.

Properties of DDI network

In this section, two questions are lifted up: (1) how the decomposed factor matrices **W** and **H** reflect the properties of DDI network; (2) whether the binary DDI matrix $\mathbf{A}_b$ implies some relationship about enhancive and degressive DDIs in the comprehensive DDI matrix. Since NMF has the nature of clustering, the analysis based on it would give insights on the corresponding answers.

To describe briefly, we refer to a set of DDIs as a binary DDI network or a signed DDI network according to the type of DDI. For a drug, we also refer to the number of it DDIs, the number of its enhancive DDIs, and the number of its degressive DDIs as *Degree, Positive Degree*, and *Negative Degree* respectively. In this context, we treated the rows of **W** as the community-derived features $\{\mathbf{f}_i^{\mathbf{W}}\}$ and the columns of **H** as the encoding features $\{\mathbf{f}_i^{\mathbf{H}}\}$ of drugs respectively. The former indicates how each drug contributes to comprise the communities, which account for all the columns of **W**, while the latter indicates how

Table 2 Comparison with state-of-the-art methods

	Binary Prediction		Comprehensive Prediction	
Method	AUROC	AUPR	AUROC	AUPR
Naïve Similarity [15]	0.779 ± 0.001	0.342 ± 0.002	0.641 ± 0.002	0.298 ± 0.004
Label Propagation [9]	0.776 ± 0.001	0.327 ± 0.002	0.635 ± 0.004	0.286 ± 0.006
DDINMF	0.872 ± 0.002	0.605 ± 0.006	0.796 ± 0.003	0.579 ± 0.003

Table 3 Novel Prediction

Top-k	10	20	30	40	50	60	70	80	90	100
Hitting ratio(%)	100	100	97	97	92	93	90	88	84	82

likely each drug belongs to those communities. Since their dimensions are high, we applied kernel PCA to them and used the first 3 principal components (PC) of the mapped instances to visualize them in a 3-d space.

We firstly tried to visualize the binary DDI network (Fig. 4). Each data-point in the two spaces is rendered by its drug degrees (Fig. 4a and c) and the difference between its positive degree and negative degree (Fig. 4b and d) respectively. In Fig. 4, Red means the largest degree or the degree difference, blue means the smallest value and other colors transiting from blue to red denote the middle values in an ascending order. Several interesting aspects can be found:

(1) As expected, the mapped spaces show significant clusters. Four significant branches are observed in the mapped space of community-derived features (Fig. 4a) while both two big clusters and one small cluster are found in the mapped space of encoding features (Fig. 4b).

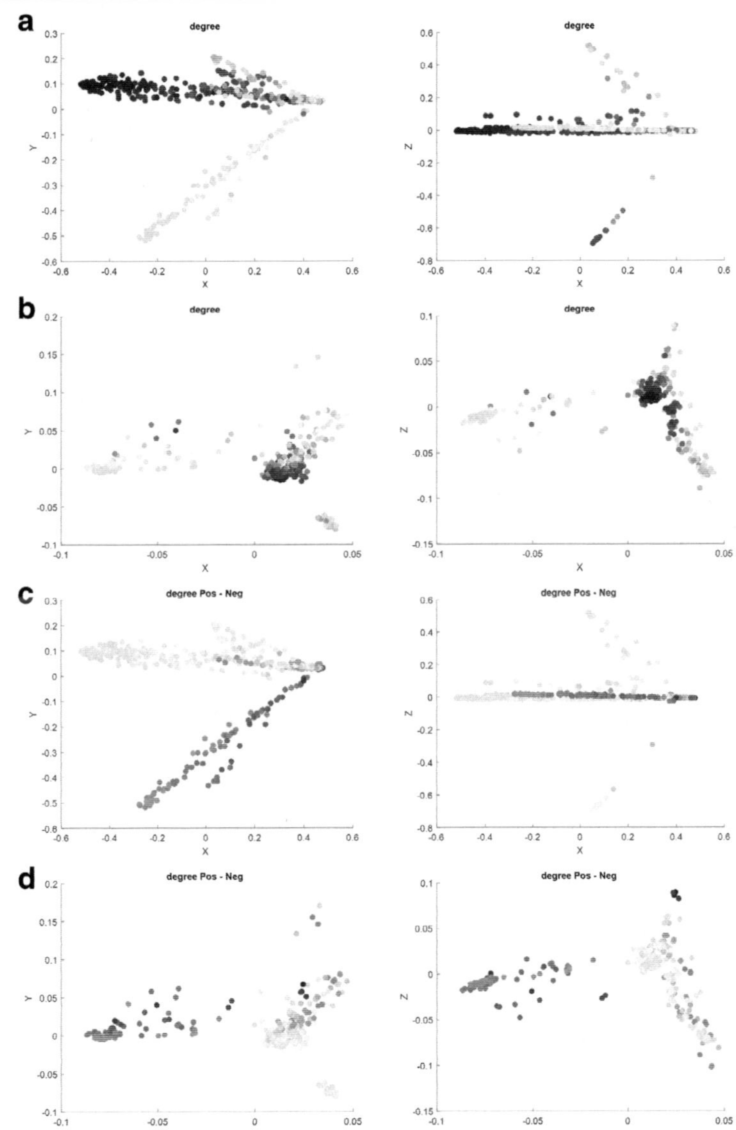

Fig. 4 Illustration of the mapped space of binary DDI network. (**a**) Community-derived features rendered by drug degree; (**b**) encoding features rendered by drug degree; (**c**) community-derived features rendered by the difference between positive degree and negative degree; (**d**) encoding features rendered by the difference between positive degree and negative degree

Degree shows a fairly clear distribution along those branches or in those clusters. For example, the Spearman Correlation between the degrees of the data points in the biggest branch (Fig. 4a) and their first mapped PC (denoted as X) is up to 0.9250.

(2) Surprisingly, the difference between Positive Degree and Negative Degree shows a significant distribution. Especially, both the branch X in Fig. 4c and the left cluster in Fig. 4d contain all the data points, of which their positive degrees are definitely greater than their negative degrees. In addition, the points having large degrees tend to have a large difference between Positive Degree and Negative Degree. The observation on binary DDIs surprisingly implies that the occurrence of signs indicating enhancive and degressive DDIs is not random.

Similarly, we then tried to visualize the comprehensive DDI network (Fig. 5). The mapped space of community-derived features contains four sectors, which show a significant distribution of both degree and positive-negative degree difference in the plane of the first and the third PCs (denoted as X-Z). The mapped space of encoding features also shows that the positive degrees of data points are

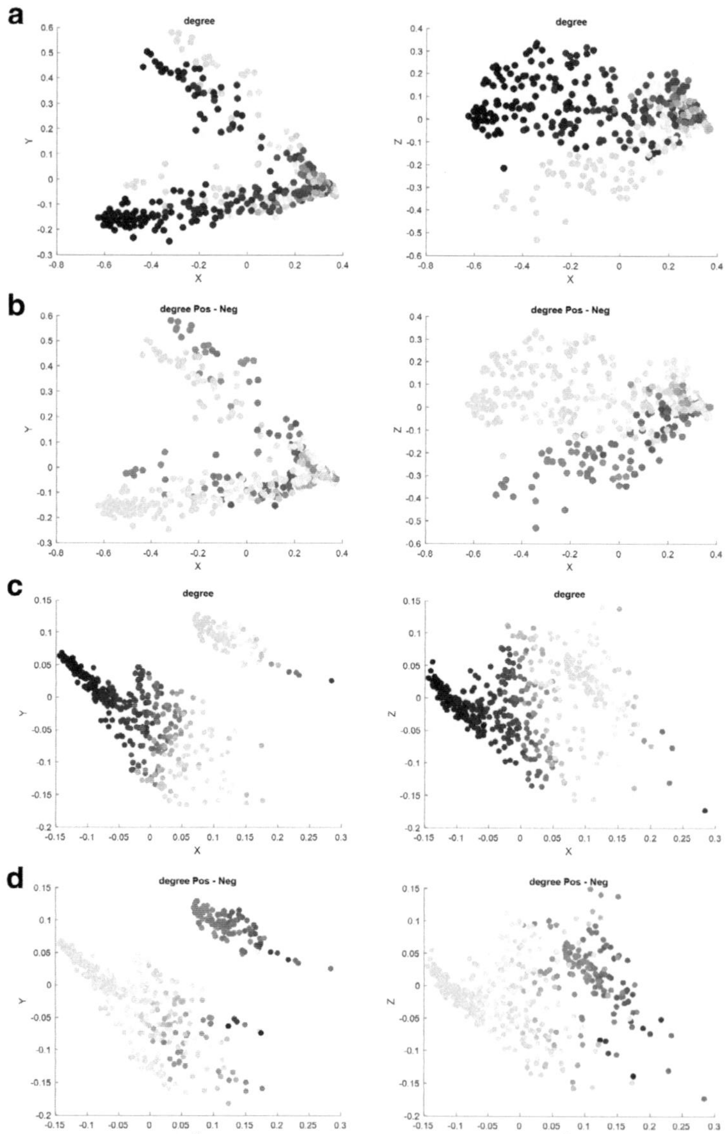

Fig. 5 Illustration of the mapped space of comprehensive DDI network. (**a**) Community-derived features rendered by drug degree; (**b**) community-derived features rendered by the difference between positive degree and negative degree; (**c**) encoding features rendered by drug degree; (**d**) encoding features rendered by the difference between positive degree and negative degree

definitely greater than their negative degrees and that the points having large degrees tend to have a large difference between positive degree and negative degree. Differentially, the mapped space of encoding features presents that its first PC (denoted as X) and the degree are almost completely correlated with each other (Spearman Correlation = 0.9822). In general, the observation is consistent with that in Fig. 4.

In summary, one just keeps it in mind that the binary DDI network contains NO information about enhancive and degressive DDIs at all, but it shows a consistency with the comprehensive DDI network, especially about Degree, Positive Degree and Negative Degree. Therefore, the binary DDI network implies some relationship about enhancive and degressive DDIs in the comprehensive DDI matrix. In other words, the occurrence of signs indicating enhancive and degressive DDIs is not random.

To validate this point further, we checked the signed DDI network to see whether or not it is of a structural balanced network. The relationships in signed networks can be positive ("enhancive", "like", "friend") or negative ("degressive", "dislike", "enemy"). Thus, there are four kinds of triangle relation patterns according to the type of DDI, including (positive, positive, positive), (positive, negative, negative), (negative, positive, positive) and (negative, negative, negative). The first two triangles are called the balanced pattern, the third is called as the unbalanced pattern and the last is named as the weakly balanced pattern. We counted four kinds of triangle patterns in the comprehensive (signed) DDI network and listed their numbers as follows: 374,215, 47,389, 46,478 and 7773. Obviously, ~ 90% triangles are balanced patterns. After checking the balance tendency in the mapped space of community-derived features, we found: (1) all the triangles (54,722) among the drugs having the first PC < 0 are purely balanced, and (2) the interactions of the drugs having the second PC < 0 are purely positive and negative in the two branches respectively. In addition, the small cluster in the mapped space of encoding features contains no weakly balanced triangles but 253,284 balanced triangles as well as 8597 imbalanced triangles. The results demonstrate that the comprehensive DDI network is equipped with a structural balance, such that the occurrence of enhancive and degressive DDIs is not random. The observation helps understand how DDIs gather together.

Conclusions

Existing computational approaches are able to screen potential DDIs on a large scale before drugs are used in the medicine market. However, none of them can predict comprehensive DDIs, including enhancive and degressive DDIs, though it is important to know whether the interaction increases or decreases the behavior of the interacting drugs before making a co-prescription. Moreover, the important structural relationship hidden among DDIs still remains unknown.

To address the abovementioned issues, we have designed a novel approach DDINMF for comprehensive DDI prediction in different scenarios. Treating the set of DDI as a network, it provides a unified framework for predicting conventional binary DDIs as well as comprehensive DDIs. Experiments on the dataset demonstrate that DDINMF is significantly superior to two state-of-the-art approaches in the conventional binary DDI prediction and also shows an acceptable performance in the comprehensive DDI prediction. More importantly, DDINMF enables us to discover crucial knowledge hidden among DDIs, including degree distribution and tendency, as well as the implication of binary network to signed network, especially the structural balance of comprehensive DDI network.

There are still unknown but important factors among DDIs to be uncovered. In the future, we will focus on extracting and integrating more features of drug (e.g. drug targets and pathway) together to achieve improved DDI prediction, building an interpretable mapping between drug features and DDI, and digging out more structural patterns of comprehensive/signed DDI network and even their dynamic properties.

Abbreviations

AUPR: The area under precision-recall curve; AUROC: The area under the receiver operating characteristic curve; CV: Cross-validation; DDI: Drug-drug interaction; NMF: Nonnegative matrix factorization; PCA: Principal Components Analysis; PLSR: Partial Least-Squares Regression

Acknowledgements

The author would like to thank the reviewers for their constructive comments that help make the paper much clearer.

Funding

Research in this work was supported by Aviation Science Fund of China (Grant No.2016ZC53028), the Program of Peak Experience in Northwestern Polytechnical University (2016), the Seed Foundation of Innovation and Creation for Graduate Students in Northwestern Polytechnical University(2018), and China National Training Programs of Innovation and Entrepreneurship for Undergraduates (201710699330). The publication of this article was sponsored by the Fundamental Research Funds for the Central Universities of China (No. 3102017jghk02010).

Availability of data and materials

The dataset used in this work can be download from https://github.com/JustinShi2016/APBC2018.

About this supplement

This article has been published as part of *BMC Systems Biology* Volume 12 Supplement 1, 2018: Selected articles from the 16th Asia Pacific Bioinformatics Conference (APBC 2018): systems biology. The full contents of the supplement are available online at https://bmcsystbiol.biomedcentral.com/articles/supplements/volume-12-supplement-1.

Authors' contributions

HY and JYS conceived and designed the experiments, and draft the manuscript. KTM and HH collected the dataset and performed the experiments. ZC and KD analyzed the results. KTM and HH contributed materials/analysis tools. HY and JYS developed the codes used in the analysis. SMY helped to draft the manuscript. JYS is the corresponding author. All authors read and approved the final manuscript.

Ethics approval and consent to participate
Not applicable.

Consent for publication
Not applicable.

Competing interests
The authors declare that they have no competing interests.

Publisher's Note

Author details
[1]School of Computer Science, Northwestern Polytechnical University, Xi'an, China. [2]School of Life Sciences, Northwestern Polytechnical University, Xi'an, China. [3]School of Software and Microelectronics, Northwestern Polytechnical University, Xi'an, China. [4]Department of Critical Care Medicine, People's Hospital of Jiangxi Province, Nan Chang, China. [5]Department of Computer Science, The University of Hong Kong, Hong Kong, China.

Published: 11 April 2018

References
1. Wienkers LC, Heath TG. Predicting in vivo drug interactions from in vitro drug discovery data. Nat Rev Drug Discov. 2005;4(10):825–33.
2. Leape LL, Bates DW, Cullen DJ, Cooper J, Demonaco HJ, Gallivan T, Hallisey R, Ives J, Laird N, Laffel G, et al. Systems analysis of adverse drug events. ADE prevention study group. JAMA. 1995;274(1):35–43.
3. Businaro R. Why we need an efficient and careful pharmacovigilance. Aust J Pharm. 2013;1:4.
4. Karbownik A, Szałek E, Sobańska K, Grabowski T, Wolc A, Grześkowiak E. Pharmacokinetic drug-drug interaction between erlotinib and paracetamol: a potential risk for clinical practice. Eur J Pharm Sci. 2017;102:55–62.
5. Mulroy E, Highton J, Jordan S. Giant cell arteritis treatment failure resulting from probable steroid/antiepileptic drug-drug interaction. N Z Med J. 2017; 130(1450):102–4.
6. Zhao XM, Iskar M, Zeller G, Kuhn M, van Noort V, Bork P. Prediction of drug combinations by integrating molecular and pharmacological data. PLoS Comput Biol. 2011;7(12):e1002323.
7. Veith H, Southall N, Huang R, James T, Fayne D, Artemenko N, Shen M, Inglese J, Austin CP, Lloyd DG, et al. Comprehensive characterization of cytochrome P450 isozyme selectivity across chemical libraries. Nat Biotechnol. 2009;27(11):1050–5.
8. Huang SM, Temple R, Throckmorton DC, Lesko LJ. Drug interaction studies: study design, data analysis, and implications for dosing and labeling. Clin Pharmacol Ther. 2007;81(2):298–304.
9. Zhang P, Wang F, Hu J, Sorrentino R. Label propagation prediction of drug-drug interactions based on clinical side effects. Sci Rep. 2015;5:12339.
10. Wiśniowska B, Polak S. The role of interaction model in simulation of drug interactions and QT prolongation. Curr Pharmacol Rep. 2016;2(6):339–44.
11. Zhou D, Bui K, Sostek M, Al-Huniti N. Simulation and prediction of the drug-drug interaction potential of Naloxegol by physiologically based pharmacokinetic modeling. CPT Pharmacometrics Syst Pharmacol. 2016;5(5):250–7.
12. Bui QC, Sloot PMA, van Mulligen EM, Kors JA. A novel feature-based approach to extract drug-drug interactions from biomedical text. Bioinformatics. 2014;30(23):3365–71.
13. Zhang Y, Wu HY, Xu J, Wang J, Soysal E, Li L, Xu H. Leveraging syntactic and semantic graph kernels to extract pharmacokinetic drug drug interactions from biomedical literature. BMC Syst Biol. 2016;10 Suppl 3:67.
14. Duke JD, Han X, Wang ZP, Subhadarshini A, Karnik SD, Li XC, Hall SD, Jin Y, Callaghan JT, Overhage MJ, et al. Literature based drug interaction prediction with clinical assessment using electronic medical records: novel Myopathy associated drug interactions. PLoS Comput Biol. 2012;8(8):e1002614.
15. Vilar S, Uriarte E, Santana L, Lorberbaum T, Hripcsak G, Friedman C, Tatonetti NP. Similarity-based modeling in large-scale prediction of drug-drug interactions. Nat Protoc. 2014;9(9):2147–63.
16. Cheng F, Zhao Z. Machine learning-based prediction of drug-drug interactions by integrating drug phenotypic, therapeutic, chemical, and genomic properties. J Am Med Inform Assoc. 2014;21(e2):e278–86.
17. Li J-X, Shi J-Y, Gao K, Lei P, Yiu S-M. Predicting combinative drug pairs via integrating heterogeneous features for both known and new drugs. Springer International Publishing Switzerland: ISBRA: 2016; 2016. p. 297–8.
18. Luo H, Zhang P, Huang H, Huang J, Kao E, Shi L, He L, Yang L. DDI-CPI, a server that predicts drug-drug interactions through implementing the chemical-protein interactome. Nucleic Acids Res. 2014;42(Web Server issue):46–52.
19. Shi J-Y, Huang H, Li J-X, Lei P, Zhang Y-N, Yiu S-M. In: IWBBIO, editor. Predicting comprehensive drug-drug interactions for new drugs via triple matrix factorization, vol. 2017. Spain: Lecture Notes in Computer Science: Bioinformatics and Biomedical Engineering, Springer; 2017. p. 108–17.
20. Koch-Weser J. Serum drug concentrations in clinical perspective. Ther Drug Monit. 1981;3(1):3–16.
21. Bichteler A, Wikoff DS, Loko F, Harris MA. Estimating serum concentrations of dioxin-like compounds in the US population effective 2005-2006 and 2007-2008: a multiple imputation and trending approach incorporating NHANES pooled sample data. Environ Int. 2017;105:112–25.
22. Law V, Knox C, Djoumbou Y, Jewison T, Guo AC, Liu Y, Maciejewski A, Arndt D, Wilson M, Neveu V, et al. DrugBank 4.0: shedding new light on drug metabolism. Nucleic Acids Res. 2014;42(Database issue):D1091–7.
23. Tatonetti NP, Ye PP, Daneshjou R, Altman RB. Data-driven prediction of drug effects and interactions. Sci Transl Med. 2012;4(125):125ra131.
24. Dejong S. Simpls - an alternative approach to partial least-squares regression. Chemometr Intell Lab. 1993;18(3):251–63.
25. Lee DD, Seung HS. Algorithms for non-negative matrix factorization. Adv Neur In. 2001;13:556–62.
26. Kuang D, Ding C, Park H. Symmetric nonnegative matrix factorization for graph clustering. Anaheim: 12th SIAM International Conference on Data Mining; 2012. p. 106–17.
27. Ding C, Li T, Jordan MI. Convex and semi-nonnegative matrix factorizations. IEEE Trans Pattern Anal Mach Intell. 2010;32(1):45–55.
28. Pahikkala T, Airola A, Pietila S, Shakyawar S, Szwajda A, Tang J, Aittokallio T. Toward more realistic drug-target interaction predictions. Brief Bioinform. 2015;16(2):325–37.
29. Shi JY, Yiu SM, Li YM, Leung HCM, Chin FYL. Predicting drug-target interaction for new drugs using enhanced similarity measures and super-target clustering. Methods. 2015;83:98–104.
30. Shi J-Y, Gao K, Shang X-Q, Yiu S-M. LCM-DS: a novel approach of predicting drug-drug interactions for new drugs via Dempster-Shafer theory of evidence: BIBM IEEE; 2016. p. 512–5.
31. Shi J-Y, Li J-X, Lu H-M. Predicting existing targets for new drugs base on strategies for missing interactions. BMC Bioinformatics. 2016;17(Suppl 8):282.
32. Jiao Y, Du P. Performance measures in evaluating machine learning based bioinformatics predictors for classifications. Quant Biol. 2016;4(4):320–30.
33. Scholkopf B, Smola A, Muller KR. Nonlinear component analysis as a kernel eigenvalue problem. Neural Comput. 1998;10(5):1299–319.

Yim et al. BMC Systems Biology 2018, 12(Suppl 1):9
https://doi.org/10.1186/s12918-018-0535-4

BMC Systems Biology

RESEARCH Open Access

Annotating activation/inhibition relationships to protein-protein interactions using gene ontology relations

Soorin Yim[1,2], Hasun Yu[2], Dongjin Jang[1] and Doheon Lee[1,2*]

From The Sixteenth Asia Pacific Bioinformatics Conference
Yokohama, Japan. 15-17 January 2018

Abstract

Background: Signaling pathways can be reconstructed by identifying 'effect types' (i.e. activation/inhibition) of protein-protein interactions (PPIs). Effect types are composed of 'directions' (i.e. upstream/downstream) and 'signs' (i.e. positive/negative), thereby requiring directions as well as signs of PPIs to predict signaling events from PPI networks. Here, we propose a computational method for systemically annotating effect types to PPIs using relations between functional information of proteins.

Results: We used *regulates*, *positively regulates*, and *negatively regulates* relations in Gene Ontology (GO) to predict directions and signs of PPIs. These relations indicate both directions and signs between GO terms so that we can project directions and signs between relevant GO terms to PPIs. Independent test results showed that our method is effective for predicting both directions and signs of PPIs. Moreover, our method outperformed a previous GO-based method that did not consider the relations between GO terms. We annotated effect types to human PPIs and validated several highly confident effect types against literature. The annotated human PPIs are available in Additional file 2 to aid signaling pathway reconstruction and network biology research.

Conclusions: We annotated effect types to PPIs by using *regulates*, *positively regulates*, and *negatively regulates* relations in GO. We demonstrated that those relations are effective for predicting not only signs, but also directions of PPIs. The usefulness of those relations suggests their potential applications to other types of interactions such as protein-DNA interactions.

Keywords: Protein-protein interaction, Activation, Inhibition, Gene ontology

Background

A cell reacts to stimuli through signaling pathways, in which proteins physically interact with each other to transmit signals. Those signals propagate inside a cell, causing various responses such as cell proliferation and differentiation [1–4]. Abnormal signal transduction triggers aberrant biological processes that might result in diseases such as cancer [2–5]. To understand how such signals flow, various high-throughput experiments have been developed to detect protein-protein interactions (PPIs) such as yeast two hybrid and affinity purification-mass spectroscopy [6]. Even though such high-throughput experiments can determine whether two proteins bind to each other or not, they are not sufficient for reconstructing signaling pathways.

To reconstruct signaling pathways from PPI networks, we need to know two aspects of PPIs: 'directions', and 'signs'. Directions of PPIs represent upstream/downstream relationships, indicating the direction of signal flow. Signs of PPIs represent whether the interactions have positive effects or negative effects. By combining directions with signs, we can define activation/inhibition relationships of PPIs, which we call 'effect types'.

* Correspondence: dhlee@biosoft.kaist.ac.kr
[1]Department of Bio and Brain Engineering, KAIST, 291 Daehak-ro, Yuseong-gu, Daejeon 34141, Republic of Korea
[2]Bio-Synergy Research Center, 291 Daehak-ro, Yuseong-gu, Daejeon 34141, Republic of Korea

Effect types are indispensable for not only reconstructing signaling pathways, but also other research areas such as network pharmacology [7, 8]. Without directions, we cannot know causality. This leads to many false positive results that arise from mistaking an effect as a cause [8]. Without signs, we cannot distinguish whether a result is desirable or harmful. For example, when signs are unavailable for drug-disease associations, we cannot differentiate whether a drug cures a disease, or causes a disease as a side effect [9].

Despite effect types are important, no experimental method is available that determines effect types of PPIs in a high-throughput way. Satisfying this need, several computational methods have been proposed to predict signs of PPIs systemically [10–12]. Based on data they used, previous works can be categorized into phenotype-based methods [10, 11] and a Gene Ontology (GO) [13]-based method [12]. Phenotype-based methods used RNA interference (RNAi) screening to identify phenotypes that were affected by a gene knockdown. Then, they predicted signs of PPIs based on the hypothesis that proteins resulting in similar phenotypes would interact positively [10, 11]. Even though they were effective, they have two limitations. Firstly, they ignored directions even though their aim was predicting effect types. Secondly, predicted signs cannot be generically applied to human PPIs. Because conducting RNAi screening for all proteins is experimentally expensive, they applied their method to smaller *Drosophila melanogaster* [10], or HeLa cells [11].

To overcome these limitations, a recent method utilized GO, which is more directly related to proteins [12]. Their hypothesis was that proteins with similar GO annotations would interact positively. They used GO terms as features for representing PPIs, and trained L2-regularized logistic regression model. Even though they improved the performance by using more direct data, they have mainly three limitations. Firstly, still they did not consider direction, leaving the causality between two proteins unknown. Secondly, similar GO annotations not necessarily means two proteins interact positively. In either positive PPIs or negative PPIs, two proteins interact with each other in any case. Therefore, negatively interacting proteins might also participate in the same biological process or have similar molecular function. In fact, the exactly same feature encoding was used for predicting whether two proteins interact or not, treating positive PPIs and negative PPIs equally [13]. Thirdly, they did not consider GO relations. However, GO has *positively regulates* and *negatively regulates* relations, which indicate signs between GO terms. These relations might help to differentiate negative PPIs in which one protein *negatively regulates* a biological process in which the other protein participates. Moreover, those relations indicate directions between GO terms, suggesting their potential use in predicting effect types of molecular interactions.

Here, we propose a method for annotating directions as well as signs to PPIs. We hypothesized that directions and signs between GO terms, represented by *regulates*, *positively regulates*, and *negatively regulates* relations, can be used for predicting directions and signs of PPIs. The rationale behind this hypothesis is as follows. Let us assume that protein p1 and p2 interact with each other, and there is a significant tendency in which GO terms involving p1 *positively regulates* GO terms involving p2. Since protein p1 and p2 interact with each other, the tendency of positive regulation might result from activation of p2 by p1. Based on this hypothesis, we predicted directions of undirected, unsigned PPIs first. Then, we predicted sign for each directed, unsigned PPI. PPIs were represented by features that were generated from *regulates*, *positively regulates*, and *negatively regulates* relations. Then, we trained logistic regression models for predicting directions and signs. Independent test results demonstrated that our method outperforms previous GO-based method, especially for negative PPIs. In addition, we annotated effect types to human PPIs and validated highly confident predictions against literature.

Methods

Method overview

The overall method for annotating effect types to PPIs is illustrated in Fig. 1. The input was an undirected, unsigned PPI network. For each undirected, unsigned PPI, we predicted its direction first. We trained two logistic regression models that predicted whether a signal can flow in left-to-right direction and right-to-left direction, respectively. The two models shared the same feature vectors, which were composed of pairs of GO terms between which *regulates*, *positively regulates*, or *negatively regulates* relation hold. By combining outputs of these two models, we decided final direction of a PPI as one of 'left-to-right', 'right-to-left', and 'bi-directional'. Then, we predicted sign for each directed, unsigned PPI. If the PPI is bi-directional, we predicted each sign for both directions. For predicting signs, we trained two logistic regression models that predicted whether a directed PPI can act as activation and inhibition, respectively. The two models shared an identical feature vector, which was composed of pairs of GO terms between which *positively regulates* or *negatively regulates* relation hold. By combining outputs of these two models, we decided final effect type as one of the followings: 'activation', 'inhibition', 'activation&inhibition', and 'affect'. As a result, we obtained PPIs with effect type.

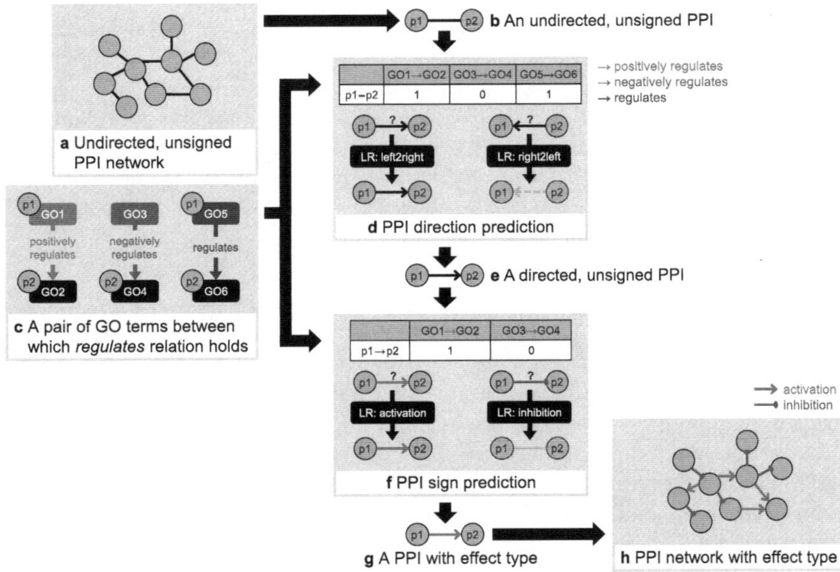

Fig. 1 Method overview. **a** Input was an undirected, unsigned PPI network. **b**, **c**, **d** For each undirected, unsigned PPI, direction was predicted. We trained two logistic regression models that predicted whether the signal can flow in left-to-right direction and right-to-left direction, respectively. Feature vector was composed of pairs of GO terms between which *regulates*, *positively regulates*, or *negatively regulates* relation holds. **e** A directed, unsigned PPI was obtained as a result of direction prediction. **f** A sign of PPI was predicted for each directed PPI. We trained two logistic regression models that predicted whether a directed PPI can act as activation or inhibition, respectively. Feature vector was composed of a pair of GO terms between which *positively regulates* or *negatively regulates* relation holds. **g**, **h** As a result of direction prediction and sign prediction, we annotated effect types to PPI network. Abbreviations: GO, gene ontology; PPI, protein-protein interaction; LR, logistic regression

Dataset

PPI dataset

We collected three PPI datasets: a training set, an independent test set, and a prediction set. To gather reliable datasets, we applied following policies to all datasets: (1) We collected human PPIs only. (2) We removed functional associations and self-interactions. (3) We collected PPIs only when at least one *regulates* relation holds between GO terms to which constituent proteins are annotated. (4) We mapped protein families to their members. (5) We integrated multiple instances of the same PPI to remove redundancy.

For the training set and the independent test set, we determined directions for each protein pair, and then determined sign for each directed PPI. For proteins protein 1 (p1) and protein 2 (p2), if a signal can flow in only one direction, for example from p1 to p2, the direction is 'uni-directional'. On the other hand, if the signal can flow in both directions depending on a context, the direction is 'bi-directional'. Then, we determined sign for each directed PPI. If a PPI is bi-directional, we determined sign for each direction independently. The same directed PPI can act as both activation and inhibition, depending on a context. For example, naked cuticle (NKD) binds to dishevelled segment polarity protein (DVL). This PPI acts as a switch from canonical Wnt signaling pathway to planar cell polarity (PCP) Wnt signaling pathway [14]. This means that NKD inhibits DVL in the aspect of canonical Wnt signaling pathway, whereas NKD activates DVL in the aspect of PCP Wnt signaling pathway. To deal with such context dependency, we categorized effect types of PPIs into four classes: 'activation', 'inhibition', both activation and inhibition are possible depending on a context ('activation&inhibition'), and neither activation nor inhibition ('affect').

We collected PPIs with known effect types as a training set from Kyoto Encyclopedia of Genes and Genomes (KEGG) pathway [15]. KEGG is a manually curated database for pathways, and contains the largest number of PPIs whose effect types are known. Following the policies, we collected 'PPrels' whose subtypes were one of the followings: activation, inhibition, phosphorylation, dephosphorylation, glycosylation, and methylation.

We gathered another set of PPIs with known effect types as an independent test set from Search Tool for the Retrieval of Interacting Genes/Proteins (STRING) [16]. STRING is an integrated database for protein-protein associations, including functional and inferred associations. To secure reliable PPIs, we collected PPIs that were experimentally validated. Moreover, we used PPIs whose scores were higher than 800 out of 1000, which resulted in about 1.28% of the PPIs available in STRING. In addition, we removed PPIs that were in the training set.

We collected prediction set from Biological General Repository for Interaction Datasets (BioGRID), whose effect types were previously unknown and predicted by our method [17]. We collected multi-validated PPIs that were validated in at least two experimental systems or two publications. We removed PPIs that were in the training set or the independent test set. As a result, we obtained 20,192 PPIs as a training set, 3420 PPIs as an independent test set, and 28,742 PPIs as a prediction set as shown in Table 1.

GO dataset

We collected ontologies and GO annotations from GO [18]. We defined a concept of 'regulators' and 'regulatees' for GO terms. A regulator is a GO term that *regulates* another GO term. If it *positively regulates* another GO term, then it is a positive regulator, whereas it is a negative regulator if it *negatively regulates* one. Hereinafter we collectively refer to *regulates*, *positively regulates*, and *negatively regulates* as (*positively/negatively*) *regulates* when any of them are applicable. A regulatee is a GO term that is (positively/negatively) regulated by another GO term. For example, 'chromatin silencing' *negatively regulates* 'transcription, DNA-templated'. Therefore, 'chromatin silencing' is a negative regulator whereas 'transcription, DNA-templated' is a regulatee.

To find all (positive/negative) regulators, we composed GO relations to form a composite relation, such that

$$\text{relation 1} \circ \text{relation 2} \rightarrow \text{composite relation.} \tag{1}$$

For instance, composing *is a* with *positively regulates* becomes *positively regulates*. Since 'actin nucleation' *is a* 'positive regulation of actin filament polymerization', which *positively regulates* 'actin filament polymerization', 'actin nucleation' becomes a positive regulator. This is called 'relation reasoning', and all possible composite relations are listed in Additional file 1: Table S1. We iteratively applied relation reasoning to find all (positive/negative) regulators, thereby increase coverage of our method. Hereinafter, we do not differentiate whether a regulator directly regulates regulatee, or indirectly regulates by a composite relation. The statistics of GO terms are shown in Fig. 2. Two hundred fifty six molecular function terms were regulators among 10,940 molecular function terms. For biological process terms, 11,820 terms were regulators out of 29,584 biological process terms.

Table 1 PPI dataset statistics

Direction					
	Uni-directional	Bi-directional		Total	
KEGG	20,078	114		20,192	
STRING	3110	310		3420	
BioGRID	–	–		28,742	
Effect type					
	Activation	Inhibition	Activation& inhibition	Affect	Total
KEGG	10,603	4548	207	447	15,805
STRING	2732	466	133	13	3344
BioGRID	–	–	–	–	–

Feature generation for representing PPIs

To encode the directions and signs between GO terms, we defined the concept of a p1 → p2 (positive/negative) regulation pair. A p1 → p2 (positive/negative) regulation pair is a pair of a (positive/negative) regulator and a corresponding regulatee, in which protein p1 is annotated to the (positive/negative) regulator and protein p2 is annotated to the regulatee. For example, if protein p1 is annotated to 'chromatin silencing' and protein p2 is annotated to 'transcription, DNA-templated', then 'chromatin silencing' and 'transcription, DNA-templated' constitute a p1 → p2 negative regulation pair as depicted in Fig. 3a.

A p1 → p2 (positive/negative) regulation pair indicates direction and sign between GO terms. We projected such directions and signs between GO terms to PPIs, by using each p1 → p2 (positive/negative) regulation pair as a feature for representing PPIs. We describe feature generation procedure with a toy example illustrated in Fig. 3b, which is a subset of GO terms related to Wnt signaling pathway.

Features were generated by following procedures. Firstly, we collected all GO terms to which proteins were annotated. In our toy example, protein p1 is annotated to four GO terms: 'Wnt signaling pathway', 'positive regulation of Wnt signaling pathway', 'regulation of Wnt signaling pathway', and 'positive regulation of canonical Wnt signaling pathway'. On the other hand, protein p2 is annotated to three GO terms: 'regulation of Wnt signaling pathway', 'canonical Wnt signaling pathway', and 'negative regulation of canonical Wnt signaling pathway'.

Secondly, for all possible pairs of GO1 and GO2 in which p1 is annotated to GO1 and p2 is annotated to GO2, we determined whether GO1 (*positively/negatively*) *regulates* GO2. If it did, then we regarded GO1 and GO2 as a p1 → p2 (positive/negative) regulation pair. However, in many cases, GO1 did not (*positively/negatively*) *regulate* GO2 itself. In such cases, we regarded that if p2 is annotated to GO2, then p2 is also annotated to ancestors of GO2 that have *is a* or *part of* relation with GO2. For example, despite p2 is not directly annotated to 'Wnt signaling pathway', we can say that p2 is related to 'Wnt signaling pathway' because p2 is annotated to

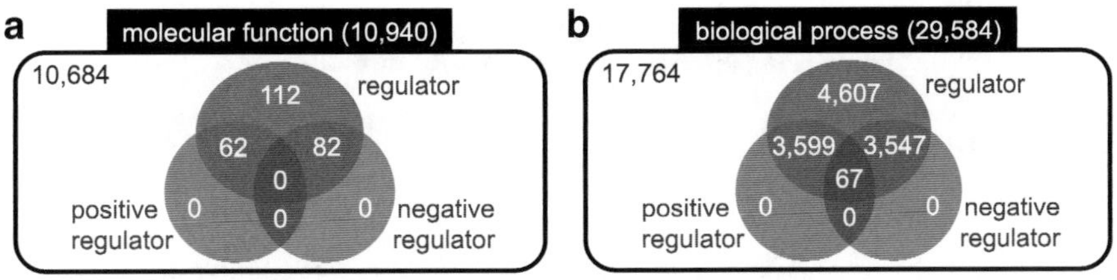

Fig. 2 The statistics of GO terms. **a** Among 10,940 molecular function terms, 256 terms (2.3%) were regulators. **b** There were 11,820 (40.0%) regulators among 29,584 biological process terms, among which 67 terms were both positive and negative regulator

'canonical Wnt signaling pathway'. This kind of extending GO annotations of a protein by using *is a* and *part of* relations in GO is called 'annotation grouping'. To increase our coverage, if GO1 did not (*positively/negatively*) *regulates* GO2 itself, we applied annotating grouping and determined whether GO1 (*positively/negatively*) *regulates* any ancestors of GO2. If it did, then we found the most specific ancestor of GO2 with the highest information content that is regulated by GO1 [19]. Then, we regarded GO1 and the most specific ancestor of GO2 that is regulated by GO1 as a p1 → p2 (positive/negative) regulation pair. Since excessive annotation grouping might result in too high-dimensional feature vectors in which features are highly correlated, we did not applied annotation grouping when GO1 regulates GO2 itself. For the same reason, we used only the most specific ancestor of GO2, not all the ancestors.

For example, since p1 is annotated 'positive regulation of canonical Wnt signaling pathway' and p2 is annotated to 'canonical Wnt signaling pathway', the two GO terms form a p1 → p2 positive regulation pair. On the other hand, 'positive regulation of Wnt signaling pathway' does not regulate 'canonical Wnt signaling pathway'. However, p2 is related to 'Wnt signaling pathway' when we apply annotation grouping. Thus, 'positive regulation of Wnt signaling pathway' and 'Wnt signaling pathway' constitute another p1 → p2 positive regulation pair.

We repeated the same procedure for determining whether GO2 (*positively/negatively*) *regulates* GO1 or its ancestors. As a result, we found six kinds of regulation pairs: p1 → p2 regulation pairs, p1 → p2 positive regulation pairs, p1 → p2 negative regulation pairs, p2 → p1 regulation pairs, p2 → p1 positive regulation pairs, and p2 → p1 negative regulation pairs. The six kinds of regulation pairs were used as features for predicting directions and signs of PPIs.

Feature generation for predicting directions of PPIs

For predicting directions of PPIs, we only considered the directions of (positive/negative) regulation pairs; whether it is from p1 to p2, or from p2 to p1. We did not differentiate regulation pairs, positive regulation pairs, and negative regulation pairs. The value of a (positive/negative) regulation pair GO1-GO2 is defined as:

$$f_{\text{direction}}[\text{GO1-GO2}] = \begin{cases} 1 & \text{if GO1-GO2 is a p1}\rightarrow\text{p2 (positive/negative)} \\ & \text{regulation pair, exclusively} \\ -1 & \text{if GO1-GO2 is a p2}\rightarrow\text{p1 (positive/negative)} \\ & \text{regulation pair, exclusively} \\ 0 & \text{otherwise} \end{cases} \quad (2)$$

In our toy example, since 'positive regulation of canonical Wnt signaling pathway' and 'canonical Wnt signaling pathway' constitute a p1 → p2 positive regulation pair, but not a p2 → p1 positive regulation pair, it has the value of one. On the other hand, 'negative regulation of canonical Wnt signaling pathway' and 'Wnt signaling pathway' form a p2 → p1 negative regulation pair, exclusively. Thus, it has the value of – 1. If the direction of a (positive/negative) regulation pair is both from p1 to p2 and from p2 to p1, then the feature value is zero. In the toy example, since p1 → p2 (positive/negative) regulation pairs outnumber p2 → p1 (positive/negative) regulation pairs, the direction of PPI is more likely to be from p1 to p2. We removed (positive/negative) regulation pairs that were not used in the training set and the number of features for direction was 37,617.

Feature generation for predicting signs of PPIs

Feature generation for predicting signs of PPIs are similar to that for directions; We used each regulation pair as a feature. However, there are also some differences: (1) Since we predicted sign for a directed PPI, we used regulation pairs whose directions were consistent with the direction of the PPI. (2) Since simple regulation pairs are uninformative for predicting signs, we used positive regulation pairs and negative regulation pairs only. (3) We removed regulation pairs that is both positive and negative. For example, 'cell cycle switching, mitotic to meiotic cell cycle' *positively regulates* 'meiotic cell cycle' and *negatively regulates* 'mitotic cell cycle'. Thus, 'cell cycle switching, mitotic to meiotic cell cycle' both *positively* and *negatively regulates* 'cell cycle'. We removed those regulation pairs since they are meaningless. (4) Compared to a feature vector for direction, which

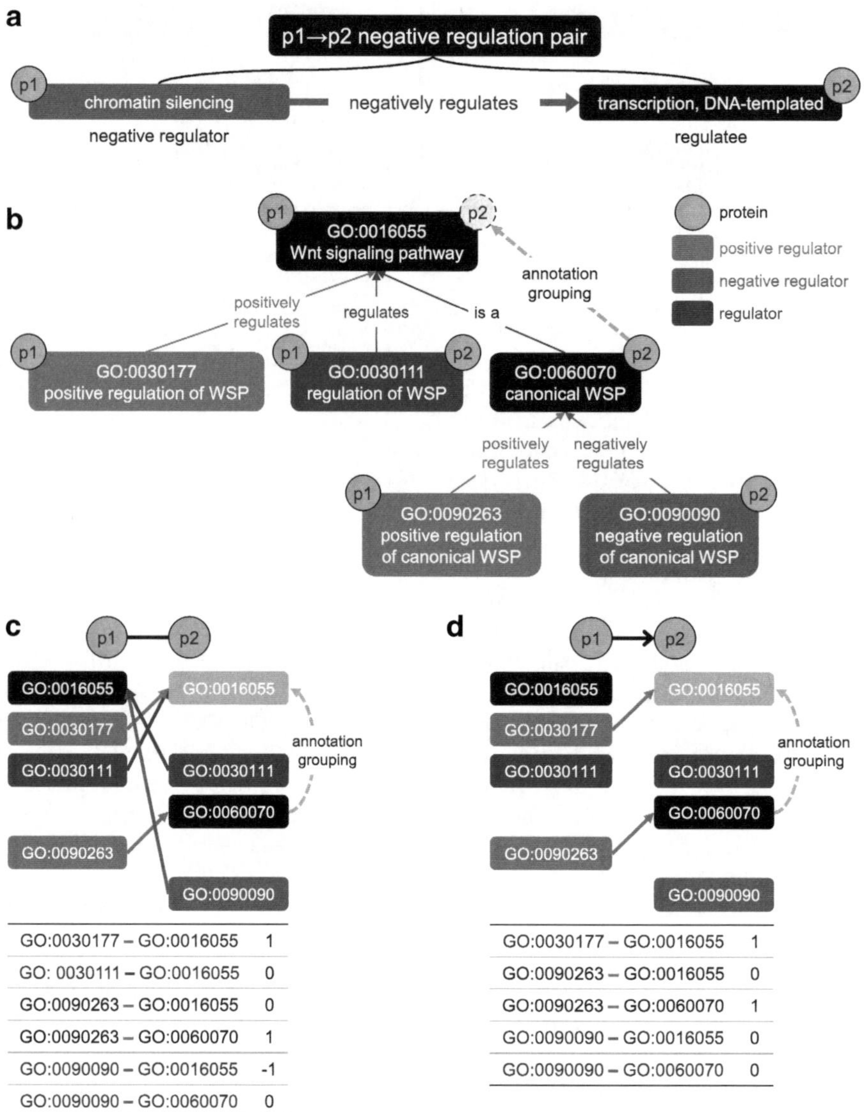

Fig. 3 Feature generation for directions and signs. **a** 'chromatin silencing' *negatively regulates* 'transcription, DNA-templated'. Thus, 'chromatin silencing' is negative regulator whereas 'transcription, DNA-templated' is a regulatee. If protein p1 and p2 is annotated to 'chromatin silencing' and 'transcription, DNA-templated' respectively, the two GO terms compose a p1 → p2 negative regulation pair, which represents direction and sign between two GO terms. **b** Feature generation procedures are explained with a toy example related to Wnt signaling pathway. There are four regulators (GO:0030177, GO:0030111, GO:0090263, GO:0090090), among which two are positive regulators (GO:0030177, GO:0090263) and one is a negative regulator (GO:0090090). Protein p1 is annotated to four GO terms (GO:0016055, GO:0030177, GO:0030111, GO:0090263), whereas protein p2 is annotated to three GO terms (GO:0030111, GO:0060070, GO:0090090). Protein p2 is not directly annotated to 'Wnt signaling pathway', but to 'canonical Wnt signaling pathway'. Nonetheless, since 'canonical Wnt signaling pathway' *is a* 'Wnt signaling pathway', protein p2 is related to 'Wnt signaling pathway'. **c** For GO1 and GO2 to which protein p1 and p2 are annotated respectively, we determined whether GO1 (*positively/negatively*) *regulates* GO2. If it did, GO1 and GO2 became a p1 → p2 (positive/negative) regulation pair. If it did not, we determined whether GO1 (*positively/negatively*) *regulates* any ancestors of GO2. Then, GO1 and the most specific ancestor of GO2 became a p1 → p2 (positive/negative) regulation pair. That way, we found (positive/negative) regulation pairs for p1 → p2, and p2 → p1 direction. To represent PPIs, we used regulation pairs as features. For directions, directions of regulation pairs were encoded as feature values. p1 → p2 (positive/negative) regulation pairs had the value of one, whereas p2 → p1 (positive/negative) regulation pairs had − 1. **d** For signs, signs of regulation pairs were encoded as feature values. p1 → p2 positive regulation pairs had the value of one, whereas p1 → p2 negative regulation pairs had − 1. Abbreviations: GO, gene ontology; WSP: Wnt signaling pathway

signifies the direction between GO terms, a feature vector for sign indicates signs. For predicting sign of a directed PPI p1 → p2, the value of a positive/negative regulation pair GO1-GO2 is defined as:

$$f_{sign}[GO1\text{-}GO2] = \begin{cases} 1 & \text{if GO1-GO2 is a p1}\rightarrow\text{p2 positive regulation pair, exclusively} \\ -1 & \text{if GO1-GO2 is a p1}\rightarrow\text{p2 negative regulation pair, exclusively} \\ 0 & \text{otherwise} \end{cases} \quad (3)$$

In our toy example, since 'positive regulation of canonical Wnt signaling pathway' and 'canonical Wnt signaling pathway' is a p1 → p2 positive regulation pair, it has the value of one. On the other hand, since 'negative regulation of canonical Wnt signaling pathway' and 'Wnt signaling pathway' is a p2 → p1 negative regulation pair, not p1 → p2, it has the value of zero. In the toy example, since p1 → p2 positive regulation pairs outnumber p1 → p2 negative regulation pairs, p1 more likely activates p2 rather than inhibits p2. We removed regulation pairs that were not used in the training set. The number of features for sign was 20,077.

Model generation for performance evaluation

We used L2-regularized logistic regression for predicting directions and signs of PPIs. We used logistic regression because it is interpretable [20]. Moreover, L2-regularization reduces overfitting that might be caused by high-dimensionality of feature vectors. As shown in Table 1, we had much more activating PPIs than inhibiting ones. To overcome this imbalance, we adopted cost-sensitive learning in which class-weight was inversely proportional to the class frequency [21].

Model generation for predicting directions of PPIs

We trained two L2-regularized logistic regression models that shared the same feature vectors for predicting directions of undirected, unsigned PPIs; they predicted whether a signal could flow in left-to-right direction and right-to-left direction, respectively. By combining outputs from the two models, we determined final directions as one of the followings: 'left-to-right', 'right-to-left', or 'bi-directional'. For example, if a signal is predicted to be able to flow in 'left-to-right' direction, but not in 'right-to-left' direction, then the final direction of PPI is 'left-to-right'. If a signal can flow in both direction, then the final direction of PPI is 'bi-directional'. Instead of training one classifier that predicts three outcomes, we trained two classifiers separately because uni-directional PPIs highly outnumbered bi-directional ones as shown in Table 1. During the training and test phase, we randomly divided uni-directional PPIs into two equal-sized sets: left-to-right PPIs, and right-to-left PPIs.

Model generation for predicting signs of PPIs

For each directed, unsigned PPI, we predicted its effect type as one of 'activation', 'inhibition', 'activation&inhibition', and 'affect'. Similar to directions, signs of PPIs were highly imbalanced; the number of 'activation&inhibition' and 'affect' class were very low. Thus, we trained two classifiers, rather than a single classifier that predicts four possible outcomes. The two classifiers shared the same feature vectors and predicted whether a directed PPI can act as activation and inhibition, respectively. Then, final effect types were determined by combining outputs of two classifiers. If a directed PPI can act as 'activation', but not as 'inhibition', then its effect type were determined as 'activation', and vice versa.

Results and discussions

Cross-validation performances

We applied our method to KEGG dataset, and conducted 10-fold cross-validation. In 10-fold cross validation, KEGG dataset is split into ten disjoint subsets. Then, we trained logistic regression models by using nine subsets, and tested the models on the remaining one subset. This procedure was repeated such that the models can be evaluated for each subset. The performance was obtained for each subset, and the mean performance is listed in Table 2. For directions, the performance of left-to-right and right-to-left classifiers were almost identical. F1-score and accuracy were as high as 0.89 for both left-to-right and right-to-left classifiers. Area under receiver operating characteristics (AUROC) and area under precision-recall curve (AUPRC) was 0.95 and 0.94 for both classifiers, respectively [see Additional file 1: Figure S1].

For signs, f1-score of activation and inhibition classifier were 0.91 and 0.80, respectively. F1-score of activation classifier was higher because activating PPIs outnumbered inhibiting ones. Accuracy of two classifiers were identical as 0.88. AUROC were 0.94 and 0.93 for activation and inhibition classifiers, respectively. Also, AUPRC were 0.96 and 0.89 for activation and inhibition classifiers, respectively [see Additional file 1: Figure S1].

Table 2 Performance of classifiers for 10-fold cross validation

Classifier	Precision	Recall	F1-score	Accuracy
Direction				
Left2right	0.894 ± 0.025	0.890 ± 0.029	0.892 ± 0.027	0.892 ± 0.027
Right2left	0.893 ± 0.029	0.892 ± 0.023	0.892 ± 0.025	0.892 ± 0.026
Effect type				
Activation	0.923 ± 0.031	0.902 ± 0.059	0.912 ± 0.040	0.881 ± 0.050
Inhibition	0.791 ± 0.103	0.819 ± 0.088	0.800 ± 0.077	0.876 ± 0.052

The performance is shown as mean ± standard deviation

Independent test performances

To see how well our model generalizes to datasets from different sources, we conducted an independent test. In an independent test, we trained logistic regression models with KEGG dataset, and tested with STRING dataset. The performance for predicting directions and signs are listed in Table 3. The performance of left-to-right classifier and right-to-left classifier were similar. Accuracy of left-to-right classifier and right-to-left classifier were around 0.6 and 0.59, respectively. The AUROC of left-to-right classifier and right-to-left classifier were 0.64 and 0.63, respectively [see Additional file 1: Figure S2].

The performance for predicting sign was higher than predicting directions. The accuracy of activation classifier and inhibition classifier were both 0.69. However, because the dataset was imbalanced, f1-score was much higher in activation classifier than inhibition classifier. AUROC of activation classifier and inhibition classifier were 0.67 and 0.63, respectively [see Additional file 1: Figure S2].

Comparison with the previous GO-based method

To demonstrate the effectiveness of GO relations, we compared the performance of our method to the performance of the previous GO-based method that did not use GO relations. Since the previous work did not consider directions of PPIs, we slightly modified our PPI dataset and feature values for the comparison. For PPI datasets, we originally defined effect types for directed PPIs. Therefore, bi-directional PPIs have two effect types, one effect type for each direction. To compare performance with the previous work, we mapped effect types of directed PPIs to undirected PPIs by using 'OR' operation so that bi-directional PPIs have only one effect type. For an undirected PPI p1-p2, if either directed PPIs p1 → p2 or p2 → p1 can act as activation, then p1-p2 has 'activation' as its effect type. Likewise, if either p1 → p2 or p2 → p1 can act as 'inhibition', then p1-p2 has 'inhibition' as its effect type. The statistics of PPI dataset that were used for the performance comparison is listed in Table 4.

For feature generation, when we predict effect type of a directed PPI p1 → p2, originally we considered p1 → p2

Table 3 Performance of classifiers for independent test

Classifier	Precision	Recall	F1-score	Accuracy
Direction				
Left2right	0.64	0.60	0.62	0.60
Right2left	0.64	0.58	0.61	0.59
Effect type				
Activation	0.91	0.71	0.79	0.69
Inhibition	0.29	0.49	0.37	0.69

Table 4 Statistics of PPI datasets that were used for comparison with previous GO-based method

		Activation	Inhibition	Activation& inhibition	Affect	Total
KEGG	Previous work	14,054	5478	382	552	20,466
	Our work	12,309	4787	369	517	17,982
STRING	Previous work	2849	465	152	15	3481
	Our work	2729	455	152	15	3351

The previous GO-based method covers all PPIs whose constituent proteins are annotated to at least one GO term. However, our method requires at least one positive/negative regulation pair to predict effect type, thereby decreasing our coverage

positive/negative regulation pairs only. p2 → p1 positive/negative regulation pairs were omitted because the direction between GO terms were inconsistent with the direction of the PPI. However, since the directions were not considered in the performance comparison, we used both p1 → p2 positive/negative regulation pairs and p2 → p1 positive/negative regulation pairs.

We conducted independent test for the previous work and our work. In the independent test, logistic regression models were trained with KEGG dataset, and tested with STRING dataset. The independent test results are shown in Fig. 4. For activation classifier, the performance was identical or slightly better in our method. On the other hand, in inhibition classifier, our method outperformed the previous work, especially in terms of recall. These results show that our method solved the second and third limitations of the previous work that we mentioned. The second limitation was that even if one protein inhibits another protein, the two proteins might share the same GO terms because they interact with each other. This may result in similar feature vector between activating PPIs and inhibiting PPIs in the previous work, suggesting that simply considering whether two proteins share the same GO terms are not sufficient for predicting signs of PPIs. We solved this problem by using *positively regulates* and *negatively regulates* relations in GO, not using which was the third limitation. Enhanced performance demonstrates that those relations help to predict signs of PPIs, especially for inhibiting ones.

Even though our method outperformed the previous GO-based method, our method has one drawback; slightly lower coverage. Since we need at least one positive/negative regulation pair for predicting effect types, we covered 89% of PPIs that were covered by the previous method, as shown in Table 4.

Annotation of effect types to human PPIs

We applied our method to the prediction set. We validated top five most confident predictions against

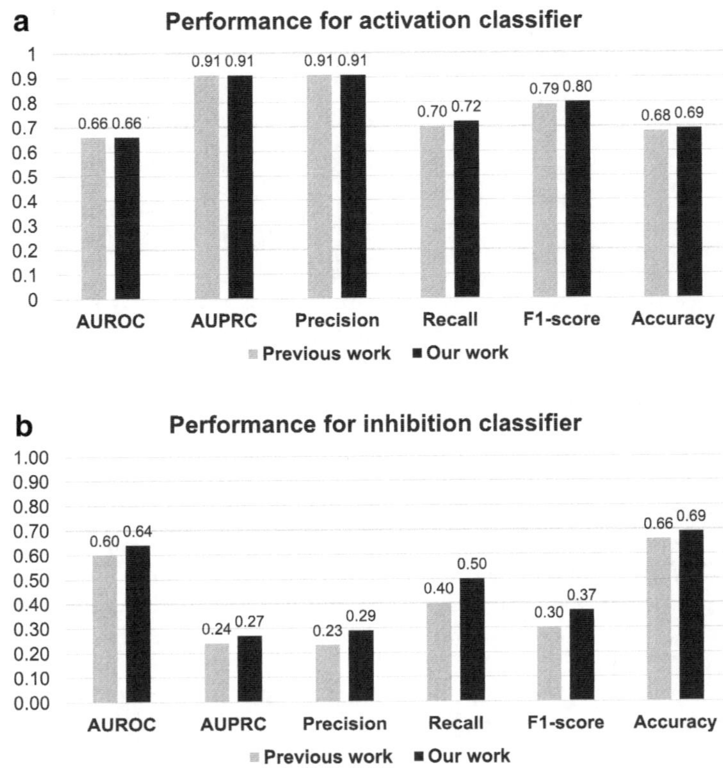

Fig. 4 Performance comparison with the previous GO-based method. We conducted independent test to compare performance with the previous GO-based method that did not consider GO relations. We compared the performance of activation and inhibition classifiers in terms of AUROC, AUPRC, precision, recall, f1-score, and accuracy. Since the previous work did not consider directions, performance for predicting signs was compared. **a** For predicting activation, our work performed equally, or slightly better than the previous work. **b** For predicting inhibition, our work outperformed the previous work by all metrics. These results show that overlap or difference between GO annotations of two proteins are not sufficient discriminating factor for signs of PPIs, given that inhibiting PPIs also participate in the same biological process. Moreover, *positively regulates* and *negatively regulates* relations in GO can be used for enhancing performance for predicting signs of PPIs. Abbreviations: AUROC, area under receiver operating characteristics; AUPRC, area under precision-recall curve

literature, each for activation and inhibition. We were able to find five publications supporting our prediction, as shown in Table 5.

Case study

To demonstrate that our method can predict effect types of PPIs even when one direction is activation and the other direction is inhibition, we conducted a case study. Hras proto-oncogene, GTPase (HRAS) activates mitogen-activated protein kinase 1 (MAPK1) in axon guidance, a process in which axon growth cone migrates in specific direction. On the other hand, MAPK1 inhibits HRAS in neutrophin signaling pathway. In 10-fold cross-validation, our method correctly predicted effect types of two directions, even when one direction is activation and the other is inhibition. This means that since we predict signs for each directed PPI, and use positive/negative regulation pairs only when their directions are consistent with the directions of PPIs, our method is able to predict effect types of PPIs independently for both directions. Regulation pairs used for the prediction is illustrated in Fig. 5. For predicting that HRAS activates MAPK1, 'positive regulation of MAPK cascade' and 'positive regulation of MAP kinase activity' were used. On the other hand, 'negative regulation of cell differentiation' was used for predicting that MAPK1 inhibits HRAS, which is related to the function of neutrophin signaling pathway. This means that some regulation pairs reflect the context in which the PPI occurs.

Conclusions

In this work, we predicted effect types of PPIs by using *regulates*, *positively regulates*, and *negatively regulates* relations in GO. We hypothesized that directions and signs between GO terms can be used to predict directions and signs of PPIs. For an undirected, unsigned PPI, we predicted its direction first. We trained two logistic regression models to predict whether a signal can flow in left-to-right direction and right-to-left direction,

Table 5 Literature validation of top five activating and inhibiting PPIs from effect type-annotated human PPIs

Protein1	Predicted effect type	Protein2	Supporting evidence
KIT	Activation	STAT1	[24]
NRG1	Activation	LIMK1	[25]
ADAM17	Activation	MAPK1	
ABL1	Activation	BRCA1	[26]
FGF2	Activation	RPS19	
AXIN1	Inhibition	GSK3A	
GRB10	Inhibition	IRS1	[27]
MAD7	Inhibition	SMAD4	
GRB10	Inhibition	AKT1	[28]
TLE1	Inhibition	LEF1	

respectively. The directions of (*positively/negatively*) *regulates* relations were encoded as features for representing direction of PPI. Then, we predicted sign for each directed PPI, thereby predicting effect type of PPI. We also trained two logistic regression models for predicting whether a directed PPI can act as activation and inhibition, respectively. We represented a directed PPI with features whose values were signs of *positively/negatively regulates* relations. As a result, we annotated effect types to PPIs, thereby turning PPI network into a directed, signed graph.

Our contribution is two-fold. Firstly, we proposed a concept of p1 → p2 (positive/negative) regulation pair, which is effective for predicting directions, as well as signs of PPIs. This solves the limitation of previous works that were not able to predict directions of PPIs. Secondly, we demonstrated usefulness of (*positively/negatively*) *regulates* relations in GO. Up to date, most of GO-related works have used only *is a* and *part of* relations. In this work, we showed that (*positively/negatively*) *regulates* relations are effective for predicting directions and signs of PPIs, suggesting their extension to other types of interactions. For example, those relations might be used for predicting signs of protein-DNA interactions; whether a transcription factor activates or represses expression of a target gene.

Even though our work improved the performance for predicting signs of PPIs, we have some drawbacks. Since we need at least one positive/negative regulation pairs for predicting signs, our method has lower coverage than the previous GO-based method. We applied relation reasoning and annotation grouping to compensate low coverage of (*positively/negatively*) *regulates* relations, nevertheless, some PPIs were not covered. At this moment, our method did not consider specificity of GO terms. However, more specific GO terms have clearer meaning. This suggests that reflecting specificity of GO terms might improve our method. In addition, the performance of inhibition classifier was much lower than activation classifier since there were much more activating PPIs than inhibiting ones. This imbalance was so significant that cannot be perfectly corrected by cost-sensitive learning. We expect that accumulation of more inhibiting PPIs will enhance the performance in future.

To facilitate signaling pathway reconstruction and network biology research, we provided effect type-annotated human PPIs in Additional file 2. The annotated effect types turned PPI network into a directed signed graph, opening up opportunities for discovering new characteristics of PPI network or signaling pathways. For example, signs of PPIs can be used for measuring stability of PPI

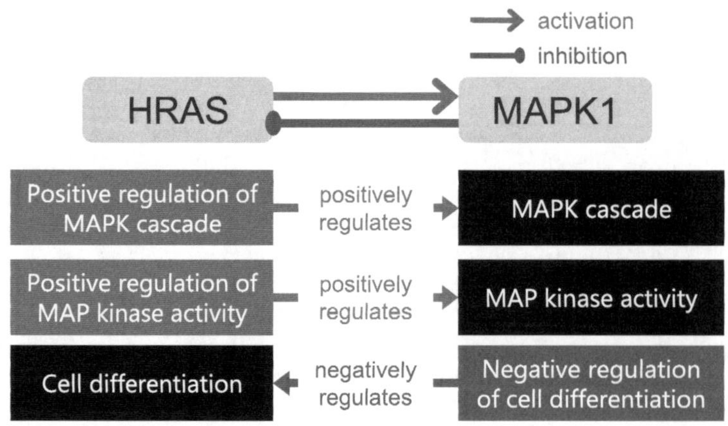

network [10, 22]. In addition, effect types of PPIs can be used for discovering novel regulators of signaling pathways [10, 23], and improving performance for predicting efficacies of drugs [8].

Additional files

Additional file 1: Table S1. Lists all possible combinations of GO relations where relation reasoning can be applied. **Figure S1-S2.** shows ROC and PRC along with their area under the curves obtained from cross-validation, and independent test results. (PDF 799 kb)

Additional file 2: Effect type-annotated human PPIs. This file contains effect type-annotated human PPIs in .xls format. Each row is a triplet of (protein 1, effect type, protein 2), in which protein 1 is a upstream protein whereas protein 2 is a downstream protein. (XLS 1547 kb)

Abbreviations
AUPRC: Area under precision-recall curve; AUROC: Area under receiver operating characteristics; DVL: Dishevelled segment polarity protein; GO: Gene ontology; HRAS: Hras proto-oncogene, GTPase; MAPK1: Mitogen-activated protein kinase 1; NKD: Naked cuticle; PCP: Planar cell polarity; PPI: Protein-protein interaction; RNAi: RNA interference

Acknowledgements
Not applicable

Funding
This work and publication of this article was sponsored by the Bio-Synergy Research Project (NRF-2012M3A9C4048758) of the Ministry of Science and ICT through the National Research Foundation.

Availability of data and materials
The datasets used and/or analyzed during the current study are available from the corresponding author on reasonable request.

About this supplement
This article has been published as part of *BMC Systems Biology* Volume 12 Supplement 1, 2018: Selected articles from the 16th Asia Pacific Bioinformatics Conference (APBC 2018): systems biology. The full contents of the supplement are available online at https://bmcsystbiol.biomedcentral .com/articles/supplements/volume-12-supplement-1 .

Authors' contributions
SY and DL designed this work. SY developed the proposed technique with HY and DJ under the supervision of DL. All authors read, wrote, and approved the manuscript.

Ethics approval and consent to participate
Not applicable

Consent for publication
Not applicable

Competing interests
The authors declare that they have no competing interests.

Publisher's Note
Springer Nature remains neutral with regard to jurisdictional claims in published maps and institutional affiliations.

Published: 11 April 2018

References
1. Anastas JN, Moon RT. WNT signalling pathways as therapeutic targets in cancer. Nat Rev Cancer. 2013;13(1):11–26.
2. Stewart DJ. Wnt signaling pathway in non–small cell lung cancer. Journal of the National Cancer Institute. 2014;106(1):djt356.
3. Takebe N, Miele L, Harris PJ, Jeong W, Bando H, Kahn M, et al. Targeting notch, hedgehog, and Wnt pathways in cancer stem cells: clinical update. Nat Rev Clin Oncol. 2015;12(8):445–64.
4. Yu H, Lee H, Herrmann A, Buettner R, Jove R. Revisiting STAT3 signalling in cancer: new and unexpected biological functions. Nat Rev Cancer. 2014;14(11):736.
5. Baron R, Kneissel M. WNT signaling in bone homeostasis and disease: from human mutations to treatments. Nat Med. 2013;19(2):179–92.
6. Snider J, Kotlyar M, Saraon P, Yao Z, Jurisica I, Stagljar I. Fundamentals of protein interaction network mapping. Mol Syst Biol. 2015;11(12) https://doi.org/10.15252/msb.20156351.
7. Gu H, Ma L, Ren Y, He W, Wang Y, Qiao Y. Exploration of the mechanism of pattern-specific treatments in coronary heart disease with network pharmacology approach. Comput Biol Med. 2014;51:198–204.
8. Yu H, Choo S, Park J, Jung J, Kang Y, Lee D, editors. Prediction of drugs having opposite effects on disease genes in a directed network. BMC systems biology. BioMed Central Ltd. 2016;10:17–25.
9. Yu L, Huang J, Ma Z, Zhang J, Zou Y, Gao L. Inferring drug-disease associations based on known protein complexes. BMC Med Genet. 2015;8(2):S2.
10. Vinayagam A, Zirin J, Roesel C, Hu Y, Yilmazel B, Samsonova AA, et al. Integrating protein-protein interaction networks with phenotypes reveals signs of interactions. Nat Methods. 2014;11(1):94–9.
11. Suratanee A, Schaefer MH, Betts MJ, Soons Z, Mannsperger H, Harder N, et al. Characterizing protein interactions employing a genome-wide siRNA cellular phenotyping screen. PLoS Comput Biol. 2014;10(9):e1003814.
12. Mei S, Zhang K. Multi-label ℓ2-regularized logistic regression for predicting activation/inhibition relationships in human protein-protein interaction networks. Scientific Reports. 2016;6:36453.
13. Mei S. Probability weighted ensemble transfer learning for predicting interactions between HIV-1 and human proteins. PLoS One. 2013;8(11):e79606.
14. Yan D, Wallingford JB, Sun T-Q, Nelson AM, Sakanaka C, Reinhard C, et al. Cell autonomous regulation of multiple Dishevelled-dependent pathways by mammalian Nkd. Proc Natl Acad Sci. 2001;98(7):3802–7.
15. Kanehisa M, Furumichi M, Tanabe M, Sato Y, Morishima K. KEGG: new perspectives on genomes, pathways, diseases and drugs. Nucleic Acids Res. 2017;45(D1):D353–D61.
16. Szklarczyk D, Morris JH, Cook H, Kuhn M, Wyder S, Simonovic M, et al. The STRING database in 2017: quality-controlled protein–protein association networks, made broadly accessible. Nucleic Acids Res. 2017;45(D1):D362–D8.
17. Chatr-aryamontri A, Oughtred R, Boucher L, Rust J, Chang C, Kolas NK, et al. The BioGRID interaction database: 2017 update. Nucleic Acids Res. 2017;45(D1):D369–D79.
18. Consortium GO. Gene ontology consortium: going forward. Nucleic Acids Res. 2015;43(D1):D1049–D56.
19. Lord PW, Stevens RD, Brass A, Goble CA. Investigating semantic similarity measures across the gene ontology: the relationship between sequence and annotation. Bioinformatics. 2003;19(10):1275–83. https://doi.org/10.1093/bioinformatics/btg153.
20. Dreiseitl S, Ohno-Machado L. Logistic regression and artificial neural network classification models: a methodology review. J Biomed Inform. 2002;35(5):352–9.
21. He H, Garcia EA. Learning from imbalanced data. IEEE Trans Knowl Data Eng. 2009;21(9):1263–84.
22. Anchuri P, Magdon-Ismail M, editors. Communities and balance in signed networks: A spectral approach. IEEE/ACM International Conference on Advances in Social Networks Analysis and Mining (ASONAM). ACM; 2012.
23. Vinayagam A, Stelzl U, Foulle R, Plassmann S, Zenkner M, Timm J, et al. A directed protein interaction network for investigating intracellular signal transduction. Sci Signal. 2011;4(189):rs8–rs.
24. Chaix A, Lopez S, Voisset E, Gros L, Dubreuil P, De Sepulveda P. Mechanisms of STAT protein activation by oncogenic KIT mutants in neoplastic mast cells. J Biol Chem. 2011;286(8):5956–66.
25. Yin D-M, Chen Y-J, Lu Y-S, Bean JC, Sathyamurthy A, Shen C, et al. Reversal of behavioral deficits and synaptic dysfunction in mice overexpressing neuregulin 1. Neuron. 2013;78(4):644–57.
26. Maiani E, Diederich M, Gonfloni S. DNA damage response: the emerging role of c-Abl as a regulatory switch? Biochem Pharmacol. 2011;82(10):1269–76.

27. Liu M, Bai J, He S, Villarreal R, Hu D, Zhang C, et al. Grb10 promotes lipolysis and thermogenesis by phosphorylation-dependent feedback inhibition of mTORC1. Cell Metab. 2014;19(6):967–80.
28. Hanson AJ, Wallace HA, Freeman TJ, Beauchamp RD, Lee LA, Lee E. XIAP monoubiquitylates Groucho/TLE to promote canonical Wnt signaling. Mol Cell. 2012;45(5):619–28.

Huang et al. BMC Genomics 2018, 19(Suppl 2):86
DOI 10.1186/s12864-018-4468-5

BMC Genomics

RESEARCH

Open Access

VAReporter: variant reporter for cancer research of massive parallel sequencing

Po-Jung Huang[1,2,4], Chi-Ching Lee[3], Ling-Ya Chiu[4], Kuo-Yang Huang[5], Yuan-Ming Yeh[4], Chia-Yu Yang[4], Cheng-Hsun Chiu[2] and Petrus Tang[2,4,6*]

From The Sixteenth Asia Pacific Bioinformatics Conference 2018
Yokohama, Japan. 15-17 January 2018

Abstract

Background: High throughput sequencing technologies have been an increasingly critical aspect of precision medicine owing to a better identification of disease targets, which contributes to improved health care cost and clinical outcomes. In particular, disease-oriented targeted enrichment sequencing is becoming a widely-accepted application for diagnostic purposes, which can interrogate known diagnostic variants as well as identify novel biomarkers from panels of entire human coding exome or disease-associated genes.

Results: We introduce a workflow named VAReporter to facilitate the management of variant assessment in disease-targeted sequencing, the identification of pathogenic variants, the interpretation of biological effects and the prioritization of clinically actionable targets. State-of-art algorithms that account for mutation phenotypes are used to rank the importance of mutated genes through visual analytic strategies. We established an extensive annotation source by integrating a wide variety of biomedical databases and followed the American College of Medical Genetics and Genomics (ACMG) guidelines for interpretation and reporting of sequence variations.

Conclusions: In summary, VAReporter is the first web server designed to provide a "one-stop" resource for individual's diagnosis and large-scale cohort studies, and is freely available at http://rnd.cgu.edu.tw/vareporter.

Keywords: NGS, Exomes, SNV annotation, TCGA, ICGC

Background

Precision medicine based on massive parallel sequencing technologies is becoming a new trend in the treatment of diseases because it enables improved identification of disease targets, which can reduce health care costs and improve clinical outcomes. This has prompted the move of massive parallel sequencing into the clinic – the U.S. Food and Drug Administration (FDA) approved the first massive parallel sequencer in 2013 for use in clinical setting for searching known diagnostic variants in known disease genes [1]. Many massive parallel sequencing-based multiplexing assays with panels of disease genes have been developed to offer precise molecular diagnoses; these assays comprise nearly all of the Mendelian genes listed in the Online Mendelian Inheritance in Man (OMIM) database [2] and the cancer-associated genes in the Catalogue of Somatic Mutations in Cancer (COSMIC) [3], reflecting the increasing needs of and advances in genetic testing.

While most rare or novel variants are not covered by the currently available disease-targeted sequencing methods, more extensive screening approaches, such as whole-exome sequencing (WES) and whole-genome sequencing (WGS), may assure the most comprehensive collection of variant spectra from individual genomes. Recently, WES has gradually become a dominant genetic test in the diagnostic setting – it decreases the cost of sequencing and has revealed several pathogenic mutations [4–6] and medically actionable targets for subsequent therapeutic research. Despite the potential to

* Correspondence: petang@mail.cgu.edu.tw
[2]Genomic Medicine Research Core Laboratory, Chang Gung Memorial Hospital, Linkou, Taiwan
[4]Molecular Medicine Research Center, Chang Gung University, Taoyuan, Taiwan
Full list of author information is available at the end of the article

provide comprehensive catalogues of genetic profiles, the cost, time and computing resources required to gather all of the genomic information have limited the wide adoption of the WGS assay for clinical applications [7]. Nevertheless, notable accomplishments, such as uncovering important roles of rare genetic variants in common diseases, providing deep characterization of genetic polymorphisms in different human populations, and finalizing the mutation landscapes for the most common cancer types, have still been primarily based on the use of WGS by large-scale genome sequencing centers [8, 9]. However, exploiting such large amounts of data is a substantial challenge for most researchers without bioinformatics support.

To the best of our knowledge, targeted enrichment sequencing is becoming a widely-accepted application for diagnostic purposes and is able to interrogate known diagnostic variants in addition to identifying novel disease markers from panels of entire human coding exomes or disease-associated genes. Although sequencer manufacturers have provided cloud-based solutions for general analysis purposes, these tools are bundled with specific genetic testing products from the relevant manufacturers, which substantially limits their usability. Moreover, most of the existing variant annotation tools [10–12] can perform well on single datasets and are thus suitable for clinical diagnostic tests for individuals. However, they are less likely to meet the requirements for cohort studies because cross-sample analysis is often resource demanding and not readily resolved. Here, we present VAReporter, which is web-based application with an intuitive and friendly environment for prioritizing disease-relevant abnormalities from single patients or study cohorts. VAReporter can provide comprehensive annotation by integrating a wide variety of biomedical databases. Comparison of gene mutation spectra between study cohorts and the Cancer Genome Atlas (TCGA) tumors is feasible with the aid of the visual analytic framework embedded in VAReporter. Moreover, state-of-art algorithms that account for mutation phenotypes are used to rank the importance of mutated genes. Our system also follows the American College of Medical Genetics and Genomics (ACMG) guidelines [13] for nomenclature, interpretation and reporting of sequence variations. In conclusion, VAReporter is designed to meet the requirements of massive parallel sequencing variome studies, ranging from individual diagnostic tests to large-scale cohort studies.

Methods

VAReporter framework

VAReporter provides an intuitive interface and flexible infrastructure for the management and analysis of genetic variants identified from massively parallel sequencing projects (Fig. 1). The system has functionalities that prioritize phenotype-associated variants by annotation, functional prediction, multi-sample comparison, and visual interpretation of the genetic variants. VAReporter has the ability to accept heterogeneous variant call file (VCF) formats from state-of-the-art variant callers, such as GATK [14], VarScan [15], MuTect [16] and VarDict [17], and provides the most comprehensive list of support formats with respect to single and paired samples. A wide variety of biomedical databases, including dbSNP [18], 1000 Genomes [19], COSMIC, the Cancer Gene Census [20], dbNSFP [21], Clinvar [22], OMIM [2], RefSeq [23], UniProt [24], Pfam [25], GO [26], KEGG [27], DrugBank [28], the DGIdb [29] and the Human Gene Mutation Database [30] (HGMD), were compiled as local annotation databases to facilitate the interpretation of biological effects introduced by genetic alterations. A high-performance computing cluster with Sun Grid Engine was used to fulfil

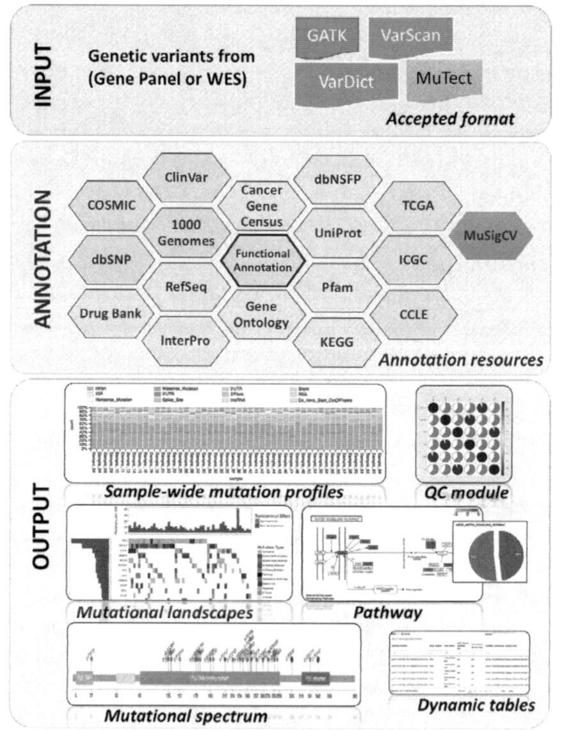

Fig. 1 Framework of VAReporter. VAReporter takes genetic variants from gene panel or whole-exome sequencing as input materials, supporting heterogeneous VCF formats such as GATK, VarScan, MuTect and VarDict. A wide variety of biomedical databases were compiled as local annotation resources to facilitate the interpretation of biological effects introduced by genetic alterations. MutSigCV algorithm was also incorporated into the framework to detect significantly altered genes in study cohorts. Visualization modules are widely used for displaying sample-wide mutation profiles, landscapes, spectra and affected pathways. Dynamic tables with filtering and sorting functionalities are provided to facilitate the prioritization of clinically actionable targets

the computational requirements for measuring variant accuracy and quality, annotating genetic variants, identifying significant mutated genes, and comparing mutation spectra across samples. Dynamic charts, filterable tables and reproducible reports were constructed with Shiny (https://shiny.rstudio.com), which is a web application framework for R, to facilitate data interpretation and target prioritization. For large-scale cohort studies of paired tumor-normal (T/N) samples, lists of single-nucleotide variations, insertions and deletions were subjected to the MutSigCV algorithm to identify the significantly mutated genes from WES or WGS.

Data input

VAReporter begins by uploading a compressed file containing VCF files into the R data structure and storing the genomic features of the individual variants in a sample-specific manner. To make the subsequent quality control and annotation easier and to provide comprehensive support for the data formats of commonly used variant calling tools, such as GATK [14], VarScan [15], MuTect [16] and VarDict [17], the user can assign the corresponding data formats to the relevant VCF files via a drop-down list of variant callers before uploading the compressed file. For cohort studies, sample metadata records can also be uploaded alongside the variant files regardless of whether they are in the VCF format, the International Cancer Genome Consortium (ICGC) TSV, or TCGA MAF formats to perform in-depth comparisons between experimental designs or features. Detailed instructions on the metadata formats can be found on the tutorial page (http://rnd.cgu.edu.tw/vareporter/bookdown-tutorial/_book/intro.html). Timestamps are used as job identifiers and are returned to the user to retrieve the finished jobs.

Quality control and association analysis

The R programming language is used to retrieve the variant allele frequencies from sample pairs according to unique mutation events defined by chromosome, position, reference allele and variant allele. The correlation coefficients between the samples are rendered according to the degree of association between the variables. The R ggplot2 [31] and corrplot [32] packages are used to render the variant frequencies from multiple sample pairs into a grid layout of multiple scatter plots (Fig. 2). Generally, a significant portion of shared mutations from T/N paired samples are distributed closer to the diagonal of the scatter plot with a fair number of sample-specific variants located on the X- or Y-axes. Two additional groups of points located at the top and right axis as indicated by red ovals in Fig. 1b can be easily

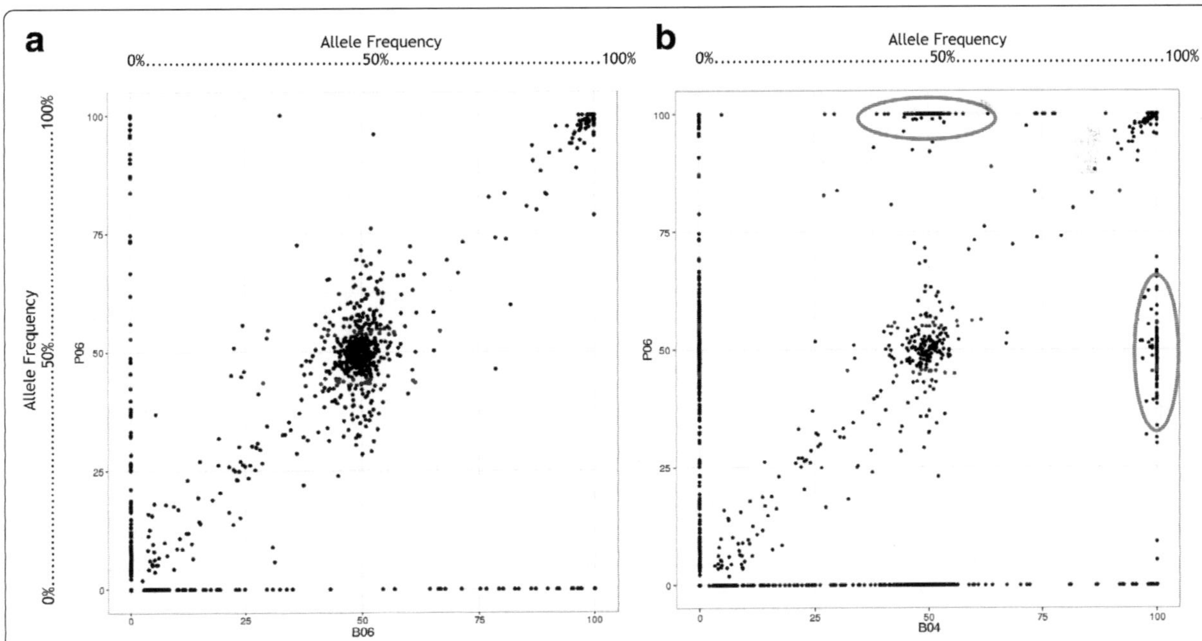

Fig. 2 Identification of mislabeled specimens. The variant allele frequencies are extracted from T/N paired samples according to unique mutation events defined by chromosome, position, reference allele and variant allele, subsequently used to generate a scatter plot between two samples. **a** A significant portion of shared mutations from correct T/N paired samples are distributed closer to the diagonal of the scatter plot with a large majority of heterozygous and homozygous variants located at regions of 50 and 100% allele frequencies. **b** Two additional groups of points located at the top and right axis as indicated by red ovals in the scatter plot can be easily depicted when T/N mismatched samples were used to generate this figure, based on the concept that these variants are less likely to change all their variant frequencies from 50 to 100% through mutation events

depicted when T/N mismatched samples were used to generate this figure, based on the concept that these variants are less likely to change all their variant frequencies from 50 to 100% through mutation events.

Variant annotation and functional effect prediction

As mentioned in a previous study [33], the majority of the existing variant annotation tools, including ANNOVAR [10], SNPeff [11], and Variant Effect Predictor [12], were developed for general non-cancer applications and lack the functionality to automatically select the correct transcript to capture the expected variant annotations in concordance with the existing cancer sequencing studies. The transcript list can be downloaded from Broad Institute through the following link (https://www.broadinstitute.org/~lichtens/oncobeta/tx_exact_uniprot_matches.AKT1_CRLF2_FGFR1.txt) [33], which was constructed from GENCODE version 19, composed of transcripts with 100% sequence identity with UniProt records, followed by manual selection to achieve 100% annotation in concordance with MyCancerGenome [34]. These records were subsequently utilized to determine the consequences on mutations in transcripts and proteins. Functional prediction and conservation scores for coding variants can be retrieved from pre-computed results with algorithms, such as SIFT [35], PolyPhen2 [36], LRT [37], MutationTaster [38], Mutation Assessor [39], FATHMM [40], GERP++ [41] and PhyloP [42] through dbNSFP, which can ease the prioritization of variants based on the functional influences of protein alterations.

Dynamic tables, charts and filters

The JavaScript library DataTables [43] is used to provide features such as filtering, sorting, pagination and saving the table as a PDF. Bar charts are used to present the most frequently mutated genes, highly affected protein domains and detailed variant compositions in individual samples. Flexible filters are provided based on items, such as gene symbol, genomic location, sample name, variant classification, affected protein domain, protein change, and SNVs, in specific ethnic groups in addition to disease information. A highly-integrated framework that seamlessly connects filters, tables, and charts was created with the R Shiny web application (https://shiny.rstudio.com) and is useful in both the exploratory and discovery stages for grasping the global mutational characteristics of a cohort as well as prioritizing candidate targets of interest.

Visual summary of genetic mutations in cancer cohorts (CoMut plot)

CoMut plots are often used in cancer research publications for visual summaries of the genetic mutations in cancer study cohorts [44]. Additionally, the MutSigCV algorithm is used to detect significantly altered genes in cancer cohorts. Because the source code for creating CoMut plots is not currently available, VAReporter uses an in-house script to render significantly altered genes into graphics similar to CoMut plots. The plots are ensembles of multiple simpler plots, such as heat maps and bar graphs, which are aligned and interconnected via common X- or Y-axes and display mutation events in a grid-like form that is particularly suitable for presenting data with intricate and associative natures (Fig. 3). Somatic genome alteration events that affect protein-coding genes within a common signaling pathway exhibit mutual exclusivity among samples, which is a well-known characteristic that is often used to identify driver mutations in cancers. To perform systematic evaluations against all signaling pathways that are plausibly perturbed by somatic mutations, the OncoPrint sorting method [45] is adapted to display genomic alterations in the gene sets of specific signaling pathways in a mutually exclusive manner.

Comparative analysis and visualization of mutation-affected pathways

Pathway component genes are defined as gene sets collected in the Molecular Signatures Database v5.0 (MSigDB) [46] that are curated from KEGG [27], BioCarta [47], Pathway Interaction Database [48], Reactome and Signaling Gateway [49]. VAReporter can assess the mutational events of pathway component genes and display subsets of patients as pie charts and heat maps to identify the most frequently altered pathways in specific TCGA/ICGC tumors and in custom study cohorts. The GenVisR package [50] is integrated into our pipeline to facilitate the identification and visualization of mutually exclusive genetic events in pathway components (Fig. 4a). Mutational events in individual pathway component genes can be retrieved from the mutation profile, which is subsequently mapped to the relevant pathway graph downloaded from KEGG [51]. The R pathview package [52] is used to facilitate pathway-based data integration and visualization (Fig. 4b). However, only KEGG pathway maps are supported by the R pathview package.

Comprehensive mutational spectrum analysis

Over 3 million simple somatic mutations from 66 cancer projects of the TCGA and ICGC were downloaded from the ICGC Data Portal [53] and compiled into our local index databases. Lollipop plots are a simple and widely used graphics for interpreting genetic mutations with protein annotations. An in-house script is used to translate the gene symbols into SwissProt accession numbers that can be used to retrieve protein domains and their corresponding colors from the Pfam database [25]. Diverse mutation types (e.g., missense and nonsense mutations) are denoted by different colors with marker sizes

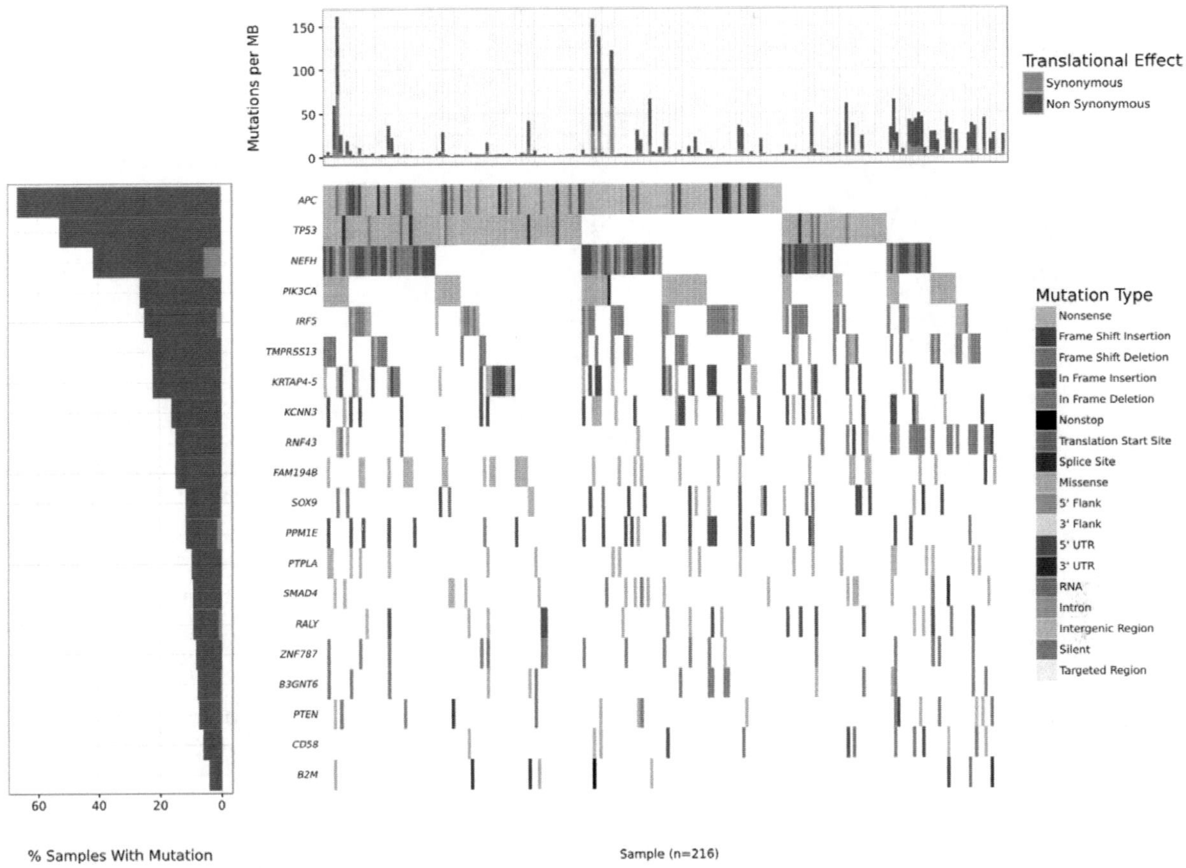

Fig. 3 Displaying mutation landscapes by CoMut plot. In-house script is used to render significantly altered genes and their relevant mutation events into heat maps and bar graphs, which are aligned and interconnected via a common X- or Y-axes, particularly suitable for presenting data with intricate and associative natures. The OncoPrint sorting method is also adapted to display genomic alterations in the gene sets of specific signaling pathways in a mutually exclusive manner and to identify driver mutations in cancers

that are exponentially proportional to the number of affected samples, which provides an intuitive method for inspecting the mutational spectra in individual genes. The lollipop module can display gene-specific mutational spectra from custom cohort studies and TCGA/ ICGA cancer studies at the simultaneously and side by side, which is also a unique feature of VAReporter.

Custom gene panels and reports

As mentioned previously, VAReporter supports almost all of the available disease-targeted gene panels and whole-exome panels ranging from inherited disease genes to cancer-associated genes. Disease testing panels with known diagnostic mutations and their corresponding genes can be selected through a drop-down list provided on the web. In addition to the gene lists defined by existing commercially available gene panels, VAReporter also enables users to create gene lists for their custom-made gene panels through the panel management button.

Results and discussion

Example of use

As a proof-of-principle experiment, we applied VAReporter to perform an in-depth analysis of the ICGC open-access datasets, which contain 216 sets of whole-exome sequencing data from colon cancer patients in United States [54]. In this study cohort, over 796,781 mutation events that correspond to 105,739 unique somatic mutations were identified and recorded in the tab-separated TSV file, which can be downloaded from the ICGC Data Portal through the following link: (https://dcc.icgc.org/api/v1/download?fn=/release_20/Projects/COAD-US/simple_somatic_mutation.open.COAD-US.tsv.gz). A previous study [55] mentioned that calculating statistics for such huge data sets in real-time is a computationally taxing task. To address this issue, the R data structure and Shiny web framework were used to optimize the interactive visualization between the graphs and data sets. Detailed exemplary outputs for 216 mutational profiles of colon tumors can be found on the VAReporter demonstration page at (http://rnd.cgu.edu.tw/vareporter/

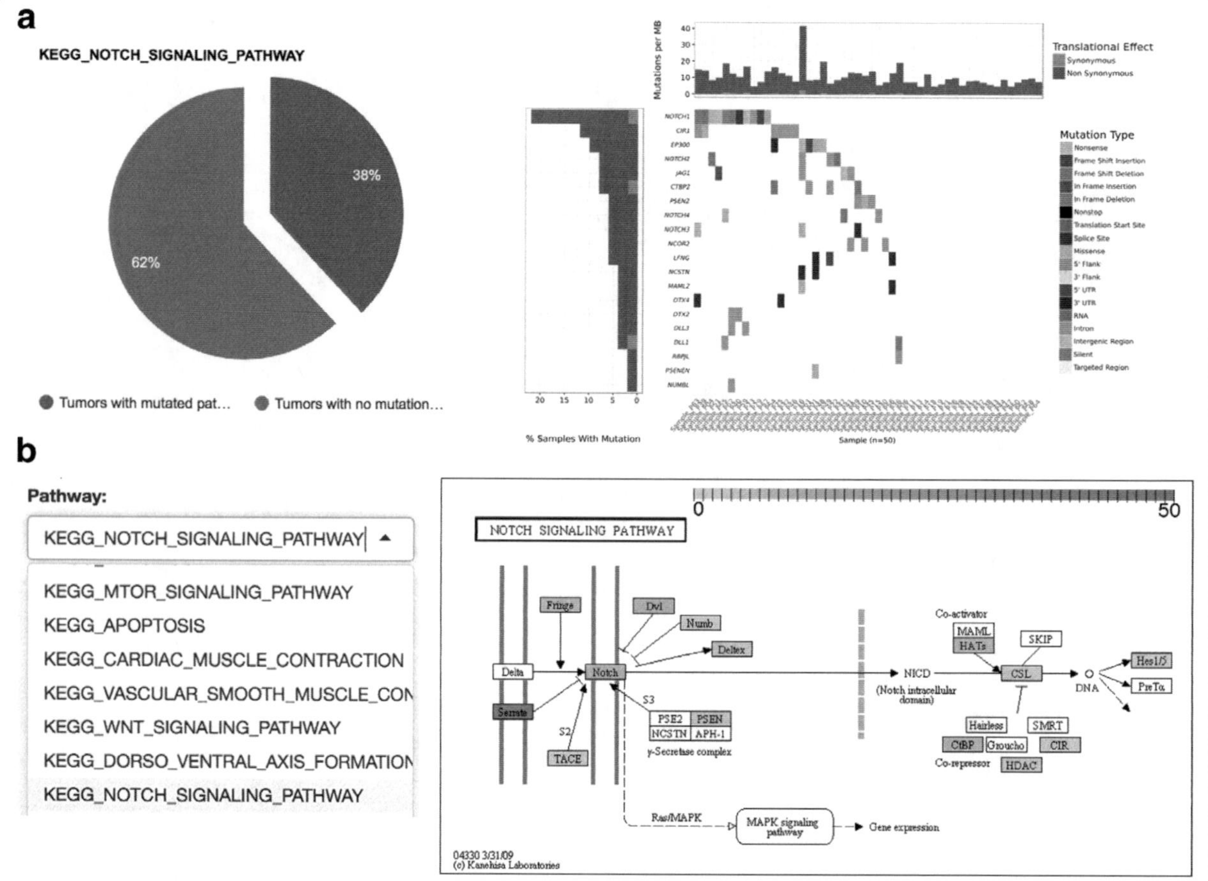

Fig. 4 Pathway visualization. **a** VAReporter can assess the mutational events of pathway component genes and display subsets of patients as pie charts and heat maps to identify the most frequently altered pathways in a study cohort. **b** The R pathview package is used to facilitate pathway-based data integration and visualization based on mutational events identified in the component genes of specific pathway [51]

main.php?jobId=demo.COAD-US). While this demonstration is used to profile somatic mutations, samples related to hereditary diseases may also be analyzed, from which qualitative and quantitative features of causative germline variants could be illustrated with equal efficiency. Detailed instructions can refer to the following link http://rnd.cgu.edu.tw/vareporter/bookdown-tutorial/_book/example-of-use.html .

Output summary

After the successful submission of a job, processing statuses, such as job queuing, file format conversion, variant annotation, significant mutant gene prediction, and report generation, are displayed using a dynamic progress indicator. The output section of tutorial page (http://rnd.cgu.edu.tw/vareporter/tutorial.php) displays the standard output of VAReporter based on the demonstration data sets mentioned above. The standard output consists of eight-page sections.

Global analyses of mutation patterns

The first section summarizes various mutation types as proportions of the total mutations per samples using a stacked bar graph to provide a global view of the mutation patterns in a sample-wide manner. Samples with relatively high or low compositions of specific mutation types can be easily depicted with this graph, which may provide some clues about the sample characteristics and their biological relevance.

The second section uses a bar chart to display the most frequently affected genes and protein domains across the samples. Basic information about each of the mutations in the affected genes and the published rankings of mutated genes for individual TCGA cancer types are also provided to facilitate comparison between the user's data and the published results.

Evaluating mutational consequences from the perspective of the central dogma

The third section provides a series of tables with searching, sorting, and filtering functionalities. The mutational consequences at the DNA, RNA, and protein levels are provided to enable inspection of mutations from the perspective of the central dogma. Human population genetic diversity and disease-related information [22, 30] can also be used as filters to identify ethnic-specific variants and disease-specific germline or somatic variants, respectively. For evaluating the biological consequences of novel candidate variants or mutations that have not been categorized in known variant databases [3, 18], nearly all of the available algorithms are applied to predict the importance and functional effects of each candidate variation.

Mutational landscapes, spectra and significant mutation genes

The fourth and fifth sections employ visual analytic strategies for interactive exploration of multidimensional genomics datasets. CoMut plots are often used in TCGA cancer research publications as visual summaries of genetic variations in study cohorts. OncoPrint is a widely used strategy for identifying cancer-driver genes and pathways and can identify mutations in gene sets of specific pathways that exhibit a pattern of mutually exclusive mutations across a study cohort [56]. However, the source code for generating CoMut plot has not been released to the public and the cBioPortal constrains the OncoPrint module for use with web services only, which make their usability limited to large research institutions with well-established bioinformatics units. Although the GenVisR package [50] provides an alternative method for mimicking the functions of both CoMut plot and OncoPrint, cumbersome steps are required to annotate and render complex genomic alteration events in a cohort into the acceptable format of this visualization package. VAReporter integrates the GenVisR package and has simplified the overall data processing procedures and automated every step, including variant annotation, format conversion and CoMut plot generation. Notably, the resulting CoMut plot can be further focused on specific pathway component genes for the convenience of inspecting the mutually exclusive mutational events in individual pathways, which is a unique and novel feature of VAReporter. With the aim of identifying the dominant altered pathways in a study cohort, VAReporter can assess the mutational events of pathway component genes to identify the most frequently altered pathways in specific tumors. Lollipop plots were first introduced by cBioPortal [45] and are simple and widely used to inspect mutational spectra for individual genes and interpret genetic mutations with protein annotation. The major difference between the VAReporter lollipop module and cBioPortal is that VAReporter can simultaneously display gene-specific mutation spectra from both custom cohort studies and TCGA/ICGA cancer studies, which is also a unique feature of VAReporter. MutSigCV [57] has become a widely-accepted algorithm for distinguishing cancer driver genes from the background of random mutations and incorporates covariate factors, such as patient-specific mutation frequencies, mutation spectra, gene-specific mutation rates, gene expression levels and DNA replication timing, into the evaluation model. This design can substantially reduce false positives in the generated lists of significant genes. To simplify each data processing and preparation step, VAReporter incorporates MutSigCV [57] as a critical component for the identification of cancer driver genes. As illustrated in tutorial page (http://rnd.cgu.edu.tw/vareporter/tutorial.php), not only the significant gene list but also the summary chart of the types of genetic alterations across all samples can be created in an automatic manner.

Mining clinically actionable drug targets

The sixth section provides tables with known information about gene-disease associations to inform clinicians of the reported mutation spectra associated with hereditary disorders or cancers. Because hundreds to thousands of coding variants can be observed in an individual's cancer genome, prioritizing causative variants becomes a major challenge. VAReporter incorporates clinically relevant drug-gene interactions from the Drug Gene Interaction Database [29] (DGIdb) that was assembled through an extensive manual curation effort from 27 sources, including seven resources specifically focused on interactions linked to clinical trials. Users can prioritize clinically actionable drug targets by sorting scores that account for both the number of distinct sources and distinct PubMed IDs. With the potential of directly benefitting the patient, clinically actionable genes are reported alongside their drug recommendations, which may assist physicians in providing the right drug to the right patient.

Experimental validation

The seventh section offers nucleotide sequences that span variant sites for the convenience of subsequent PCR primer design and Sanger validation. Because the Cancer Cell Line Encyclopedia (CCLE) project [58] has conducted a detailed genetic characterization of approximately 1000 human cancer cell lines, and the CCLE recorded variants are generally considered to be known mutations or verified variants, information or validation statuses on cancer-specific somatic mutations are provided to facilitate the prioritization of novel candidate mutations before experimental validations are performed.

Identification of mislabeling errors

The final section was designed to fix the problem of mislabeled specimens in clinical labs. Specimen labeling errors account for a large proportion of identification problems during the sample collection process. Literature reviews have revealed that specimen mislabeling occurs commonly and introduces errors at rates of 0.1 to 6.5% [59]. However, existing quality assessment tools [60] only provide quality checks on sequencing data. These quality assessment methods can only provide information about the per-base sequence quality and laboratory contamination events and thus lack the ability to identify specimens with labeling errors. To provide a quick view of whether there are any problems in the data, the Pearson correlations are calculated between the samples based on the mutant allele frequencies extracted from VCF files and rendered into a correlation matrix like that displayed in Fig. 1. A scatter plot is used for the detailed inspection of variant allele frequencies between samples (Fig. 1). The user can easily identify specimen labeling errors and incorrect T/N paired samples based on the concept that these variants are less likely to change all their frequencies from 50 to 100% through mutation events.

Comparison of the feature of different tools for variant annotation and interpretation

Over the past few years, several packages have been developed to meet the needs of variant annotation and interpretation for massively parallel sequencing. Some tools have been designed to handle the annotation task, while other tools focus mainly on the filtering and interpretation functionalities. Despite the maturation of current tools, there still are to different extent technical weaknesses. VAReporter aims to provide the most comprehensive set of features to address all plausible issues. A more detailed comparison of the current tools is summarized in Table 1.

Benchmarking

The VAReporter web server runs Apache 2.2.15 on a Centos 6.8 Linux machine housing single Intel i7-5820 K 3.30 GHz processor and 64 GB RAM. To test the utility of this pipeline, 200 of variant calling files were randomly selected from WES datasets of colorectal cancer through ICGC Data Portal (https://dcc.icgc.org/api/v1/download?fn=/release_20/Projects/COAD-US/simple_somatic_mutation.open.COAD-US.tsv.gz). For the performance of VAReporter, please refer to http://rnd.cgu.edu.tw/vareporter/tutorial.php?target=benchmark.

Conclusions

In this paper, we proposed the application of VAReporter to meet the requirements for all types of variome studies, including individual diagnostics tests and large-scale cohort studies ranging from single genes to WES/WGS. VAReporter is dedicated to the incorporation of a series of visual analytic and prediction modules for the identification of cancer driver genes, the inspection of mutational landscapes and spectra, the prioritization of clinically actionable genes, and the identification of likely mislabeled samples. Additionally, VAReporter also provides a portal for comparing in-house cancer genomic data with those from TCGA/ICGA to support comprehensive comparisons of the mutational landscapes between cohorts. Overall, VAReporter represents a highly-integrated framework for the in-depth analyses of genetic variants for all types of massively parallel

Table 1 Comparison of features of different tools for massive parallel sequencing annotation and interpretation

Tool	VARepoter	Vanno [61]	Annotate-it [55]	ANNOVAR [10]	Anntools [62]	KGGSeq [63]	SeqAnt [64]	TREAT [65]	Oncotator [33]
Availability	Web	Web	Web	Command line	Command line	Command line	Web	Command line	Both
Tracking mislabeled specimen	✓								
SNPs/1000Genomes/COSMIC	✓	✓	✓	✓	✓	✓	✓	✓	✓
Indels	✓	✓		✓	✓	✓	✓	✓	✓
Cross-sample comparison	✓	✓							
Filters	✓	✓	✓	✓		✓		✓	
Domain information	✓								✓
Dynamic summarized chart	✓	✓							
Gene Ontology	✓	✓	✓						✓
Mutational Landscape	✓								
OMIM	✓		✓					✓	✓
Pathway visualization	✓					✓		✓	✓
dbNSFP	✓	✓	✓			✓			
Sequence retrieval	✓	✓		✓					
ICGC/TCGA comparison	✓								TCGA only

sequencing applications with the aim of translating genomic data into useful clinical insights and moving toward precision medicine.

Availability and requirements

Project name: VAReporter.

Project home page: http://rnd.cgu.edu.tw/vareporter/

Operating system(s): Platform independent.

Programming language(s): R, PHP, Shell Script and JavaScript.

Other requirements: Supported browsers Safari, Google Chrome, Firefox, Internet Explorer 11 and Microsoft Edge.

License: GNU GPL version 3.

Any restrictions to use by non-academics: none

Abbreviations

ACMG: American College of Medical Genetics and Genomics; CCLE: Cancer cell line encyclopedia; COSMIC: Catalogue of somatic mutations in cancer; DGIdb: Drug gene interaction database; GNU: General Public License; HGMD: Human gene mutation database; ICGC: International Cancer Genome Consortium; MAF: Mutation annotation format; OMIM: Online Mendelian Inheritance in Man (OMIM) database; TCGA: The Cancer Genome Atlas; VCF: Variant call format; WES: Whole-exome sequencing; WGS: Whole-genome sequencing

Acknowledgements

We would like to thank Prof. Bertrand Tan (Department of Biomedical Sciences, Chang Gung University) for his critical review of this manuscript.

Funding

This work was supported by the Chang Gung Memorial Hospital, Linkou, Taiwan (CMRPD1G0321/2). The storage and computing facilities were supported by grants from the Ministry of Science and Technology, Taiwan (MOST 104–2321-B-182-007-MY3; 105–2218-E-182-004; 106–2221-E-182-068). Publication of this article was sponsored by MOST 104–2321-B-182-007-MY3 grant.

Availability of data and materials

The demonstration dataset used in this study is available in the ICGC Data Portal.
(https://dcc.icgc.org/api/v1/download?fn=/release_20/Projects/COAD-US/simple_somatic_mutation.open.COAD-US.tsv.gz)

About this supplement

This article has been published as part of *BMC Genomics* Volume 19 Supplement 2, 2018: Selected articles from the 16th Asia Pacific Bioinformatics Conference (APBC 2018): genomics. The full contents of the supplement are available online at https://bmcgenomics.biomedcentral.com/articles/supplements/volume-19-supplement-2.

Authors' contributions

PJH designed the overall framework of this application and prepared the manuscript. CCL maintained the system and prepared the demonstration data sets. LYC, YMY and KYH contributed to the visual interactive interface. CYY and CHC tested the software and prepared the tutorial page. PJH and PT were the main supervisors of the project. All of the authors have read and approve of the final manuscript.

Ethics approval and consent to participate

Not applicable.

Consent for publication

Not applicable.

Competing interests

The authors declare that they have no competing interests.

Publisher's Note

Springer Nature remains neutral with regard to jurisdictional claims in published maps and institutional affiliations.

Author details

[1]Department of Biomedical Sciences, Chang Gung University, Taoyuan, Taiwan. [2]Genomic Medicine Research Core Laboratory, Chang Gung Memorial Hospital, Linkou, Taiwan. [3]Department and Graduate Institute of Computer Science and Information Engineering, Chang Gung University, Taoyuan, Taiwan. [4]Molecular Medicine Research Center, Chang Gung University, Taoyuan, Taiwan. [5]Graduate Institute of Pathology and Parasitology, National Defense Medical Center, Taipei, Taiwan. [6]Graduate Institute of Biomedical Sciences, Chang Gung University, Taoyuan, Taiwan.

Published: 9 May 2018

References

1. Collins FS, Hamburg MA. First FDA authorization for next-generation sequencer. N Engl J Med. 2013;369(25):2369–71.
2. Amberger JS, Bocchini CA, Schiettecatte F, Scott AF, Hamosh A. OMIM.org: online Mendelian inheritance in man (OMIM(R)), an online catalog of human genes and genetic disorders. Nucleic Acids Res. 2015;43(Database issue):D789–98.
3. Forbes SA, Beare D, Gunasekaran P, Leung K, Bindal N, Boutselakis H, Ding M, Bamford S, Cole C, Ward S, et al. COSMIC: exploring the world's knowledge of somatic mutations in human cancer. Nucleic Acids Res. 2015;43(Database issue):D805–11.
4. Saudi Mendeliome G. Comprehensive gene panels provide advantages over clinical exome sequencing for Mendelian diseases. Genome Biol. 2015;16:134.
5. Ng SB, Buckingham KJ, Lee C, Bigham AW, Tabor HK, Dent KM, Huff CD, Shannon PT, Jabs EW, Nickerson DA, et al. Exome sequencing identifies the cause of a mendelian disorder. Nat Genet. 2010;42(1):30–5.
6. Bowles NE, Jou CJ, Arrington CB, Kennedy BJ, Earl A, Matsunami N, Meyers LL, Etheridge SP, Saarel EV, Bleyl SB, et al. Exome analysis of a family with Wolff-Parkinson-white syndrome identifies a novel disease locus. Am J Med Genet A. 2015;167A(12):2975–84.
7. van El CG, Cornel MC, Borry P, Hastings RJ, Fellmann F, Hodgson SV, Howard HC, Cambon-Thomsen A, Knoppers BM, Meijers-Heijboer H, et al. Whole-genome sequencing in health care. Recommendations of the European Society of Human Genetics. Eur J Hum Genet. 2013;21(Suppl 1):S1–5.
8. Wang K, Kim C, Bradfield J, Guo Y, Toskala E, Otieno FG, Hou C, Thomas K, Cardinale C, Lyon GJ, et al. Whole-genome DNA/RNA sequencing identifies truncating mutations in RBCK1 in a novel Mendelian disease with neuromuscular and cardiac involvement. Genome Med. 2013;5(7):67.
9. Kandoth C, McLellan MD, Vandin F, Ye K, Niu B, Lu C, Xie M, Zhang Q, McMichael JF, Wyczalkowski MA, et al. Mutational landscape and significance across 12 major cancer types. Nature. 2013;502(7471):333–9.
10. Wang K, Li M, Hakonarson H. ANNOVAR: functional annotation of genetic variants from high-throughput sequencing data. Nucleic Acids Res. 2010;38(16):e164.
11. Cingolani P, Platts A, Wang le L, Coon M, Nguyen T, Wang L, Land SJ, Lu X, Ruden DM. A program for annotating and predicting the effects of single nucleotide polymorphisms, SnpEff: SNPs in the genome of Drosophila Melanogaster strain w1118; iso-2; iso-3. Fly (Austin). 2012;6(2):80–92.
12. McLaren W, Gil L, Hunt SE, Riat HS, Ritchie GR, Thormann A, Flicek P, Cunningham F. The Ensembl variant effect predictor. Genome Biol. 2016;17(1):122.
13. Richards S, Aziz N, Bale S, Bick D, Das S, Gastier-Foster J, Grody WW, Hegde M, Lyon E, Spector E, et al. Standards and guidelines for the interpretation of sequence variants: a joint consensus recommendation of the American College of Medical Genetics and Genomics and the Association for Molecular Pathology. Genet Med. 2015;17(5):405–24.
14. McKenna A, Hanna M, Banks E, Sivachenko A, Cibulskis K, Kernytsky A, Garimella K, Altshuler D, Gabriel S, Daly M, et al. The genome analysis toolkit: a MapReduce framework for analyzing next-generation DNA sequencing data. Genome Res. 2010;20(9):1297–303.
15. Koboldt DC, Zhang Q, Larson DE, Shen D, McLellan MD, Lin L, Miller CA, Mardis ER, Ding L, Wilson RK. VarScan 2: somatic mutation and copy number alteration discovery in cancer by exome sequencing. Genome Res. 2012;22(3):568–76.

16. Cibulskis K, Lawrence MS, Carter SL, Sivachenko A, Jaffe D, Sougnez C, Gabriel S, Meyerson M, Lander ES, Getz G. Sensitive detection of somatic point mutations in impure and heterogeneous cancer samples. Nat Biotechnol. 2013;31(3):213–9.
17. Lai Z, Markovets A, Ahdesmaki M, Chapman B, Hofmann O, McEwen R, Johnson J, Dougherty B, Barrett JC, Dry JR. VarDict: a novel and versatile variant caller for next-generation sequencing in cancer research. Nucleic Acids Res. 2016;44(11):e108.
18. Sherry ST, Ward MH, Kholodov M, Baker J, Phan L, Smigielski EM, Sirotkin K. dbSNP: the NCBI database of genetic variation. Nucleic Acids Res. 2001;29(1):308–11.
19. Genomes Project C, Abecasis GR, Altshuler D, Auton A, Brooks LD, Durbin RM, Gibbs RA, Hurles ME, GA MV. A map of human genome variation from population-scale sequencing. Nature. 2010;467(7319):1061–73.
20. Futreal PA, Coin L, Marshall M, Down T, Hubbard T, Wooster R, Rahman N, Stratton MR. A census of human cancer genes. Nat Rev Cancer. 2004;4(3):177–83.
21. Liu X, Jian X, Boerwinkle E. dbNSFP v2.0: a database of human non-synonymous SNVs and their functional predictions and annotations. Hum Mutat. 2013;34(9):E2393–402.
22. Landrum MJ, Lee JM, Benson M, Brown G, Chao C, Chitipiralla S, Gu B, Hart J, Hoffman D, Hoover J, et al. ClinVar: public archive of interpretations of clinically relevant variants. Nucleic Acids Res. 2016;44(D1):D862–8.
23. Pruitt KD, Brown GR, Hiatt SM, Thibaud-Nissen F, Astashyn A, Ermolaeva O, Farrell CM, Hart J, Landrum MJ, McGarvey KM, et al. RefSeq: an update on mammalian reference sequences. Nucleic Acids Res. 2014; 42(Database issue):D756–63.
24. The UniProt C. UniProt: the universal protein knowledgebase. Nucleic Acids Res. 2017;45(D1):D158–69.
25. Finn RD, Coggill P, Eberhardt RY, Eddy SR, Mistry J, Mitchell AL, Potter SC, Punta M, Qureshi M, Sangrador-Vegas A, et al. The Pfam protein families database: towards a more sustainable future. Nucleic Acids Res. 2016;44(D1): D279–85.
26. Camon E, Magrane M, Barrell D, Lee V, Dimmer E, Maslen J, Binns D, Harte N, Lopez R, Apweiler R. The gene ontology annotation (GOA) database: sharing knowledge in Uniprot with gene ontology. Nucleic Acids Res. 2004; 32(Database issue):D262–6.
27. Kanehisa M, Goto S, Furumichi M, Tanabe M, Hirakawa M. KEGG for representation and analysis of molecular networks involving diseases and drugs. Nucleic Acids Res. 2010;38(Database issue):D355–60.
28. Law V, Knox C, Djoumbou Y, Jewison T, Guo AC, Liu Y, Maciejewski A, Arndt D, Wilson M, Neveu V, et al. DrugBank 4.0: shedding new light on drug metabolism. Nucleic Acids Res. 2014;42(Database issue):D1091–7.
29. Wagner AH, Coffman AC, Ainscough BJ, Spies NC, Skidmore ZL, Campbell KM, Krysiak K, Pan D, McMichael JF, Eldred JM, et al. DGIdb 2.0: mining clinically relevant drug-gene interactions. Nucleic Acids Res. 2016;44(D1): D1036–44.
30. Stenson PD, Mort M, Ball EV, Shaw K, Phillips A, Cooper DN. The human gene mutation database: building a comprehensive mutation repository for clinical and molecular genetics, diagnostic testing and personalized genomic medicine. Hum Genet. 2014;133(1):1–9.
31. Ginestet C. ggplot2: elegant graphics for data analysis. J R Stat Soc a Stat. 2011;174:245.
32. Wei T, Simko V: Corrplot:visualization of a correlation matrix. 2017. https://github.com/taiyun/corrplot. Accessed 20 Nov 2017.
33. Ramos AH, Lichtenstein L, Gupta M, Lawrence MS, Pugh TJ, Saksena G, Meyerson M, Getz G. Oncotator: cancer variant annotation tool. Hum Mutat. 2015;36(4):E2423–9.
34. Kusnoor SV, Koonce TY, Levy MA, Lovly CM, Naylor HM, Anderson IA, Micheel CM, Chen SC, Ye F, Giuse NB. My cancer genome: evaluating an educational model to introduce patients and caregivers to precision medicine information. AMIA Jt Summits Transl Sci Proc. 2016;2016:112–21.
35. Ng PC, Henikoff S. SIFT: predicting amino acid changes that affect protein function. Nucleic Acids Res. 2003;31(13):3812–4.
36. Adzhubei IA, Schmidt S, Peshkin L, Ramensky VE, Gerasimova A, Bork P, Kondrashov AS, Sunyaev SR. A method and server for predicting damaging missense mutations. Nat Methods. 2010;7(4):248–9.
37. Chun S, Fay JC. Identification of deleterious mutations within three human genomes. Genome Res. 2009;19(9):1553–61.
38. Schwarz JM, Rodelsperger C, Schuelke M, Seelow D. MutationTaster evaluates disease-causing potential of sequence alterations. Nat Methods. 2010;7(8):575–6.
39. Reva B, Antipin Y, Sander C. Predicting the functional impact of protein mutations: application to cancer genomics. Nucleic Acids Res. 2011;39(17):e118.
40. Shihab HA, Gough J, Cooper DN, Stenson PD, Barker GL, Edwards KJ, Day IN, Gaunt TR. Predicting the functional, molecular, and phenotypic consequences of amino acid substitutions using hidden Markov models. Hum Mutat. 2013;34(1):57–65.
41. Davydov EV, Goode DL, Sirota M, Cooper GM, Sidow A, Batzoglou S. Identifying a high fraction of the human genome to be under selective constraint using GERP++. PLoS Comput Biol. 2010;6(12):e1001025.
42. Pollard KS, Hubisz MJ, Rosenbloom KR, Siepel A. Detection of nonneutral substitution rates on mammalian phylogenies. Genome Res. 2010;20(1):110–21.
43. Xie YH: DT: a wrapper of the JavaScript library 'DataTables'. 2016. https://CRAN.R-project.org/package=DT. Accessed 20 Nov 2017.
44. Stransky N, Egloff AM, Tward AD, Kostic AD, Cibulskis K, Sivachenko A, Kryukov GV, Lawrence MS, Sougnez C, McKenna A, et al. The mutational landscape of head and neck squamous cell carcinoma. Science. 2011; 333(6046):1157–60.
45. Cerami E, Gao J, Dogrusoz U, Gross BE, Sumer SO, Aksoy BA, Jacobsen A, Byrne CJ, Heuer ML, Larsson E, et al. The cBio cancer genomics portal: an open platform for exploring multidimensional cancer genomics data. Cancer Discov. 2012;2(5):401–4.
46. Liberzon A. A description of the molecular signatures database (MSigDB) web site. Methods Mol Biol. 2014;1150:153–60.
47. Nishimura D: BioCarta. Biotech Software & Internet Report. Mary Ann Liebert, Inc; 2001;2:117–20.
48. Schaefer CF, Anthony K, Krupa S, Buchoff J, Day M, Hannay T, Buetow KH. PID: the pathway interaction database. Nucleic Acids Res. 2009;37(Database issue):D674–9.
49. Croft D, Mundo AF, Haw R, Milacic M, Weiser J, Wu G, Caudy M, Garapati P, Gillespie M, Kamdar MR, et al. The Reactome pathway knowledgebase. Nucleic Acids Res. 2014;42(Database issue):D472–7.
50. Skidmore ZL, Wagner AH, Lesurf R, Campbell KM, Kunisaki J, Griffith OL, Griffith M. GenVisR: genomic visualizations in R. Bioinformatics. 2016;32(19): 3012–4.
51. Kanehisa M, Sato Y, Kawashima M, Furumichi M, Tanabe M. KEGG as a reference resource for gene and protein annotation. Nucleic Acids Res. 2016;44(D1):D457–62.
52. Luo W, Brouwer C. Pathview: an R/bioconductor package for pathway-based data integration and visualization. Bioinformatics. 2013;29(14):1830–1.
53. Joly Y, Dove ES, Knoppers BM, Bobrow M, Chalmers D. Data sharing in the post-genomic world: the experience of the international cancer genome consortium (ICGC) data access compliance office (DACO). PLoS Comput Biol. 2012;8(7):e1002549.
54. Cancer Genome Atlas N. Comprehensive molecular characterization of human colon and rectal cancer. Nature. 2012;487(7407):330–7.
55. Sifrim A, Van Houdt JK, Tranchevent LC, Nowakowska B, Sakai R, Pavlopoulos GA, Devriendt K, Vermeesch JR, Moreau Y, Aerts J. Annotate-it: a Swiss-knife approach to annotation, analysis and interpretation of single nucleotide variation in human disease. Genome Med. 2012;4(9):73.
56. Ciriello G, Cerami E, Sander C, Schultz N. Mutual exclusivity analysis identifies oncogenic network modules. Genome Res. 2012;22(2):398–406.
57. Lawrence MS, Stojanov P, Polak P, Kryukov GV, Cibulskis K, Sivachenko A, Carter SL, Stewart C, Mermel CH, Roberts SA, et al. Mutational heterogeneity in cancer and the search for new cancer-associated genes. Nature. 2013; 499(7457):214–8.
58. Barretina J, Caponigro G, Stransky N, Venkatesan K, Margolin AA, Kim S, Wilson CJ, Lehar J, Kryukov GV, Sonkin D, et al. The cancer cell line encyclopedia enables predictive modelling of anticancer drug sensitivity. Nature. 2012;483(7391):603–7.
59. Dunn EJ, Moga PJ. Patient misidentification in laboratory medicine: a qualitative analysis of 227 root cause analysis reports in the veterans health administration. Arch Pathol Lab Med. 2010;134(2):244–55.
60. Trivedi UH, Cezard T, Bridgett S, Montazam A, Nichols J, Blaxter M, Gharbi K. Quality control of next-generation sequencing data without a reference. Front Genet. 2014;5:111.
61. Huang PJ, Lee CC, Tan BC, Yeh YM, Huang KY, Gan RC, Chen TW, Lee CY, Yang ST, Liao CS, et al. Vanno: a visualization-aided variant annotation tool. Hum Mutat. 2015;36(2):167–74.
62. Makarov V, O'Grady T, Cai G, Lihm J, Buxbaum JD, Yoon S. AnnTools: a comprehensive and versatile annotation toolkit for genomic variants. Bioinformatics. 2012;28(5):724–5.

63. Li MX, Gui HS, Kwan JS, Bao SY, Sham PC. A comprehensive framework for prioritizing variants in exome sequencing studies of Mendelian diseases. Nucleic Acids Res. 2012;40(7):e53.
64. Shetty AC, Athri P, Mondal K, Horner VL, Steinberg KM, Patel V, Caspary T, Cutler DJ, Zwick ME. SeqAnt: a web service to rapidly identify and annotate DNA sequence variations. BMC Bioinformatics. 2010;11:471.
65. Asmann YW, Middha S, Hossain A, Baheti S, Li Y, Chai HS, Sun Z, Duffy PH, Hadad AA, Nair A, et al. TREAT: a bioinformatics tool for variant annotations and visualizations in targeted and exome sequencing data. Bioinformatics. 2012;28(2):277–8.

Zeng et al. BMC Genomics 2018, 19(Suppl 2):84
DOI 10.1186/s12864-018-4459-6

BMC Genomics

RESEARCH

Open Access

Prediction of enhancer-promoter interactions via natural language processing

Wanwen Zeng[1,2], Mengmeng Wu[1,3] and Rui Jiang[1,2*]

From The Sixteenth Asia Pacific Bioinformatics Conference 2018
Yokohama, Japan. 15-17 January 2018

Abstract

Background: Precise identification of three-dimensional genome organization, especially enhancer-promoter interactions (EPIs), is important to deciphering gene regulation, cell differentiation and disease mechanisms. Currently, it is a challenging task to distinguish true interactions from other nearby non-interacting ones since the power of traditional experimental methods is limited due to low resolution or low throughput.

Results: We propose a novel computational framework EP2vec to assay three-dimensional genomic interactions. We first extract sequence embedding features, defined as fixed-length vector representations learned from variable-length sequences using an unsupervised deep learning method in natural language processing. Then, we train a classifier to predict EPIs using the learned representations in supervised way. Experimental results demonstrate that EP2vec obtains F1 scores ranging from 0.841~ 0.933 on different datasets, which outperforms existing methods. We prove the robustness of sequence embedding features by carrying out sensitivity analysis. Besides, we identify motifs that represent cell line-specific information through analysis of the learned sequence embedding features by adopting attention mechanism. Last, we show that even superior performance with F1 scores 0.889~ 0.940 can be achieved by combining sequence embedding features and experimental features.

Conclusions: EP2vec sheds light on feature extraction for DNA sequences of arbitrary lengths and provides a powerful approach for EPIs identification.

Keywords: Enhancer-promoter interactions, Three-dimensinal interactions, Natural language processing, Unsupervised learning

Background

One of the major discoveries in recent years is that non-coding DNAs are not “junk”. On the contrary, they fulfill a wide variety of crucial biological roles involving regulatory and signaling functions [1]. Enhancer is one of the most important noncoding elements that has a central role in controlling gene expression [2]. Recent studies have shown that noncoding single nucleotide polymorphisms (SNPs) that are associated with risk for numerous common diseases through genome-wide association studies (GWAS), frequently lie in cell line-specific enhancers [3, 4]. These GWAS SNPs are hard to interpret because we are unaware of how non-coding SNPs affect gene expression and disease transmission through the complicated regulatory relationship [5]. We can improve understanding of disease mechanisms if enhancers are accurately linked to the promoters/genes they regulate. For example, Guo et al. [6] identified mechanism of GWAS risk SNP rs7463708 in promoting prostate transformation. This SNP is located in the enhancer of long noncoding RNA (lncRNA) PCAT1 and significantly upregulates PCAT1 expression. PCAT1 interacts with the enhancers of prostate cancer genes GNMT and DHCR24,

* Correspondence: ruijiang@tsinghua.edu.cn
[1]MOE Key Laboratory of Bioinformatics; Bioinformatics Division and Center for Synthetic & Systems Biology, Beijing 100084, China
[2]Department of Automation, Tsinghua University, Beijing 100084, China
Full list of author information is available at the end of the article

and in turn promotes prostate tumorigenesis. Thus, the identification of true three-dimensional (3D) genome organization, especially EPIs across different cell lines constitutes important steps towards understanding of gene regulation, cell differentiation and disease mechanisms.

However, there are enormous technical challenges to obtain these 3D interactions in the entire genome. Chromosome conformation capture-based (3C) methods [7], including 4C [8] and 5C [9] have been developed to detect physical contacts in the 3D space but fail to capture whole genome interactions. Chromatin Interaction Analysis by Paired-End Tag Sequencing (ChIA-PET) [10] allows genome-wide measurements but is restricted to interactions mediated by a preselected protein of interest. The method of Hi-C [11] allows the genome-wide detections of interactions but its current resolution is not high enough (~ 10 kb) to capture individual EPIs. All these traditional experimental approaches for detecting 3D genome interactions remain time-consuming and noisy, motivating the development of computational approaches.

To bridge this growing gap between low-resolution experiments and high-resolution EPIs, some computational methods have been established, which mainly fall into two classes. One class is based on experimental features. For instance, IM-PET [12], RIPPLE [13], TargetFinder [14] and EpiTensor [15], aim to predict 3D genomic interactions in different cell lines by integrating numerous one-dimensional (1D) local chromatin states including genomic and epigenomic data. Among them, TargetFinder is the state-of-the-art computational method to identify true EPIs by collecting experimental data sets including histone modifications, TF binding, chromatin accessibility and gene expressions. The other class is based on sequence information only, which is represented by SPEID [16]. SPEID takes advantage of a convolutional Long Short-Term Memory (LSTM) network to learn the feature representation from input sequences automatically and can reliably predict EPIs.

Existing 3D genomic interaction prediction methods fail to exploit sequence information except SPEID. At the meantime, there are many inspiring methods for 1D chromatin states prediction [17, 18], including gkmSVM for enhancer prediction [19], DeepSEA for epigenomic state prediction [20] and DeepBind for DNA/RNA-binding proteins prediction [21], which extract sequence features and yield high performance. gkmSVM transforms variable-length sequences to fixed-length *k*-mer features to classify input DNA sequences. *k*-mer features are an unbiased, complete set of sequence features defined on arbitrary-length DNA sequences but lose the contextual information between adjacent *k*-mers. DeepSEA and DeepBind take advantage of powerful convolutional neural networks (CNN) but they require fixed-length sequences as input, which is also a limit for SPEID. Since DNA sequences are in variable length and contextual information is important for understanding the function of whole sequence, how to transform a variable-length sequence into a fixed-length vector representation conserving the context information remains challenging and crucial for improving sequence-based prediction methods.

It is well-known that learning a good representation of input data is an important task in machine learning. There is an analogous problem in natural language processing, which is to learn an embedding vector for a sentence, that is essentially, to train a model that is able to automatically transform a sentence to a vector and encodes its semantic meaning. Paragraph Vector [22] successfully solves the problem by mapping texts into a unified vector representation, and generates embedding representation which can be further used for different applications [23], such as machine translation [24], sentiment analysis [22], and information retrieval [25].

Inspired by the idea of sentence embedding, we present a novel 3D interactions prediction method, named EP2vec, in this paper. First, we utilize an unsupervised deep learning method, namely Paragraph Vector, to learn sequence embedding features. Concretely, we embed the enhancer sequences and promoter sequences into a vector space separately, and then every sequence can be represented as a vector, namely the sequence embedding features. Then, EP2vec uses the resulted features for subsequent classification of EPIs through supervised learning. Our experiments prove that we are able to accurately predict EPIs using only the sequence embedding features, which outperforms other existing computational methods. In addition, by combining both sequence embedding features and experimental features, we can further improve performance, which indicates sequence embedding features and experimental features are complementary to each other. Furthermore, by applying attention mechanism, we successfully interpret the meaning of sequence embedding features and find motifs that represent cell line information. The source code to implement EP2vec can be downloaded from https://github.com/wanwenzeng/ep2vec.

Methods

Datasets

The majority of our datasets were adapted from TargetFinder. Promoter and enhancer regions were identified using ENCODE Segway [26] and ChromHMM [27] annotations for K562, GM12878, HeLa-S3, and HUVEC cell lines, and using Roadmap [28] Epigenomics ChromHMM annotations for NHEK and IMR90 cell lines. Since EPIs could only happen between active enhancers and promoters, we used the full set of all enhancers and promoters as external resources to perform unsupervised feature extraction which would be elaborated in the next section. The total number of enhancers and the number of promoters for each cell line are reported in Table 1. The length distributions of enhancers and promoters in six cell lines are shown in Additional file 1: Figures S1 and S2.

To focus on distal interactions, enhancers closer than 10 kb to the nearest promoter were discarded. Using GENCODE [29] version 19 annotations and RNA-seq data from ENCODE, promoters were reserved if actively transcribed (mean FPKM > 0.3 [30] with irreproducible discovery rate < 0.1 [31]) in each cell line. Positive EPIs were annotated using high-resolution genome-wide Hi-C data [32]. These EPIs were assigned to one of five bins based on the distance between the enhancer and the promoter, such that each bin had the same number of interactions. Negative pairs were assigned to their corresponding distance bin and then subsampled within each bin, using one negative per positive. The number of positive or negative samples for each cell line is reported in Table 1.

In addition, we also collected a dataset from FANTOM5 project [4]. The FANTOM5 consortium extracted RNA transcripts from a multitude of different primary cells and tissues using the Cap Analysis of Gene Expression (CAGE) experiment. Because active enhancer regions were transcribed, they identified a distinct bidirectional CAGE pattern which could predict enhancer regions based on CAGE data not associated with promoters. The transcribed enhancer atlas held around 40,000 transcribed enhancers across the human body, which they called permissive enhancers. We collected the permissive enhancers and RefSeq promoters. Using statistical methods, FANTOM5 defined some enhancer-promoter interactions, which we considered as positive samples. Negative samples were generated as random pairs of enhancers and promoters based on the distance distribution of the positive samples.

Table 1 Details of each cell line dataset. The enhancers (or promoters) column indicates the number of all known active enhancers (or promoters) for each cell line, which are used for unsupervised feature learning for enhancer (or promoter) sequences

Dataset	enhancers	promoters	true EPIs	false EPIs
K562	82806	8196	1977	1975
IMR90	108996	5253	1254	1250
GM12878	100036	8453	2113	2110
HUVEC	65358	8180	1524	1520
HeLa-S3	103460	7794	1740	1740
NHEK	144302	5254	1291	1280
FANTOM	43011	49620	61542	61542

Workflow of EP2vec

The workflow of EP2vec contained two stages including the unsupervised feature extraction and supervised learning (Fig. 1). Sequences of active regulatory elements in a specific cell line have cell line-specific regulatory information. Hence, EP2vec could use unsupervised methods to extract useful information from the sequences set, which would benefit subsequent tasks such as EPIs prediction. EP2vec regarded DNA sequences as sentences with *k*-mers as words, and learned effective representations of these sequences based on the co-occurrence statistics of *k*-mers.

Stage 1 of EP2vec was unsupervised feature extraction which transforms enhancer sequences and promoter sequences in a cell line into sequence embedding features separately. Given a set of all known enhancers or promoters in a cell line, we first split all the sequences into *k*-mer words with stride $s = 1$ and assign a unique ID to each of them. Regarding the preprocessed sequences as sentences, we embedded each sentence to a vector by Paragraph Vector. Concretely, we used vectors of words in a context with the sentence vector to predict the next word in the context using softmax classifier. After training converges, we got embedding vectors for words and all sentences, where the vectors for sentences were exactly the sequence embedding features that we needed. Stage 2 is supervised learning for predicting EPIs. Given a pair of sequences, namely an enhancer sequence and a promoter sequence, we represented the two sequences using the pre-trained vectors and then concatenated them to obtain the feature representation. Lastly, we trained a Gradient Boosted Regression Trees classifier (GBRT) to predict whether this pair was a true EPI.

Feature extraction

In this section, we will illustrate how to apply Paragraph Vector to learn fixed-length feature representations from variable-length DNA sequences in Stage 1.

First, given a set of all known enhancers or promoters in a cell line, we assigned a unique ID for each sequence, and split it into *k*-mers. *k*-mers were split along a sequence using sliding window with stride *s*, meaning that two adjacent *k*-mers had a distance of *s* bps. Thus, in general, a sequence with *L* bps will be split into *floor* $\left[\frac{L-k}{s}\right]+1$ *k*-mers. For example, we could split "ATGCAACAC" into four 6-mers with stride $s = 1$ as "ATGCAA", "TGCAAC", "GCAACA" and "CAACAC" with ID "SEQEUNCE_i" (Fig. 1). From now on, we regarded the

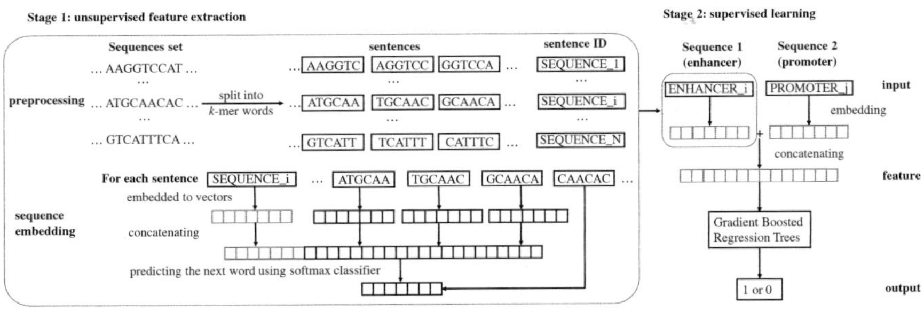

Fig. 1 The two-stage workflow of EP2vec. Stage 1 of EP2vec is unsupervised feature extraction which transforms enhancer sequences and promoter sequences in a cell line into sequence embedding features separately. Given a set of all known enhancers or promoters in a cell line, we first split all the sequences into k-mer words with stride $s = 1$ and assign a unique ID to each of them. Regarding the preprocessed sequences as sentences, we embed each sentence to a vector by Paragraph Vector. Concretely, we use vectors of words in a context with the sentence vector to predict the next word in the context using softmax classifier. After training converges, we get embedding vectors for words and all sentences, where the vectors for sentences are exactly the sequence embedding features that we need. Note that in sentence ID, SEQUENCE is a placeholder for ENHANCER or PROMOTER, and is the total number of enhancers or promoters in a cell line. Stage 2 is supervised learning for predicting EPIs. Given a pair of sequences, namely an enhancer sequence and a promoter sequence, we represent the two sequences using the pre-trained vectors and then concatenate them to obtain the feature representation. Lastly, we train a Gradient Boosted Gradient Trees classifier to predict whether this pair is a true EPI

split enhancers or promoters as sentences and the k-mers as words. Note that the vocabulary size of k-mers was 4^k.

Second, each sentence was mapped to a unique vector in a d-dimensional vector space, where d was the embedding dimension. Each word was also mapped to a unique vector in the same space. The basic training algorithms were greedy in nature, we followed general pratics to initialize all these vectors at random before training. For example, the sentence "SEQEUNCE_i" was mapped to a d-dimensional vector $x_i \in \mathbb{R}^d$, with each component initialized by a random value. Similarly, k-mers "ATGCAA", "TGCAAC", "GCAACA" and "CAACAC", which were indexed as $c_{i,\ 1}, c_{i,\ 2}, c_{i,\ 3}, c_{i,\ 4} \in [1, 4^k]$ in the k-mer vocabulary, were also mapped to four vectors $w^{c_{i,1}}, w^{c_{i,2}}, w^{c_{i,3}}, w^{c_{i,4}} \in \mathbb{R}^d$ with random initialization.

Third, we trained all these sentence vectors and word vectors by constructing the training loss function. In detail, we predicted the next word of a context in a sentence, by concatenating these vectors of words in this context and the sentence vector as predictive features. Since the vocabulary size was 4^k, the next word had 4^k possibilities. Generally speaking, the context has a fixed window length m and is sampled from the sentence in a sliding window fashion. For example, as shown in Fig. 1, we set the window length $m = 3$, and used the concatenated vectors of "ATGCAA", "TGCAAC", "GCAACA" and "SEQUENCE_i" to predicted the next word "CAACAC" by a 4096-way classification. Note that the sentence vector was shared across all contexts generated from this single sentence, while the word vector for one single k-mer was shared across all sentences.

More formally, given N sequences represented in N vectors $x_1, x_2, \ldots, x_N$. The i-th sequence contained T_i words represented in vectors $w^{c_{i,1}}, w^{c_{i,2}}, \ldots, w^{c_{i,T_i}}$, the objective of the model was to maximize the average log probability

$$\max \sum_{i=1}^{N} \frac{1}{T_i - m} \sum_{t=1}^{T_i - m} \log p\left(c_{i,t+m} | w^{c_{i,t}}, \cdots, , w^{c_{i,t+m-1}}, , x_i\right).$$

The prediction task was typically accomplished via a multiclass classifier, such as softmax classifier, which could be formulated as

$$p\left(c_{i,t+m} | w^{c_{i,t}}, \cdots, , w^{c_{i,t+m-1}}, , x_i\right) = \frac{e^{y_{c_{i,t+m}}}}{\sum_j e^{y_j}}.$$

Here, $j \in [1, 4^k]$ was an index to an output words, and y_j was the corresponding component of the un-normalized log-probability $y \in \mathbb{R}^{4^k}$ computed by

$$y = b + Uh,$$

where $U \in \mathbb{R}^{4^k \times (m+1)d}$ and $b \in \mathbb{R}^{4^k}$ are the softmax parameters, while $h = (w^{c_{i,t}}, \cdots, , w^{c_{i,t+m-1}}, x_i) \in \mathbb{R}^{(m+1)d}$ was the concatenation of the m word vectors and the sentence vector.

The N sentence vectors and 4^k word vectors were trained using stochastic gradient descent (SGD) together with the softmax parameters U and b, where the gradient was obtained via back propagation [33]. At every step of SGD, one could sample a fixed-length context from a random sentence, compute the error gradient and use the gradient to update the parameters in our model. In practice, hierarchical softmax [34–36] was preferred to softmax for fast training. In our study, the structure of the hierarchical softmax was a binary Huffman tree, where short codes were assigned to frequent words. This was a good speedup trick because common words were accessed quickly. This use of binary Huffman code for the hierarchical softmax was the same as Mikolov et al. [36].

After the training converges, words with similar meanings were expected to be mapped to adjacent positions in the vector space and the sentence vectors could be used as features for the sentence. In fact, the sentence vectors learned by the model were exactly the sequence embedding features which captured the sequence contextual information. Note that, we trained sequence embedding features for enhancers and promoters separately. We implemented these feature extraction based on the GENSIM packages [37].

Model training

In this section, we proceeded to interpret Stage 2 of EP2vec workflow (Fig. 1), namely supervised learning for EPIs classification. For each pair of an enhancer and a promoter, we first concatenated the sequence embedding features of the two sequences as the final features. Then based on this feature representation, we trained a GBRT classifier to predict the binary label, i.e., whether this pair was a true EPI. GBRT was a classifier which used decision trees as weak estimators and combines several weak estimators into ensemble as a single model, in a stage-wise fashion. The tree ensemble model was a set of classification and regression trees (CART). The prediction scores of each individual tree were summed up to get the final score.

GBRT performed gradient descent algorithm for the objective function for the binary classification of EPIs, and its performance mainly depended on three hyper-parameters: learning rate α, number of trees n, and tree-depth D. Smaller learning rates tended to result in better accuracy but require more iterations. Tree-depth D controlled the size of each decision tree. To yield the best performance, we figured out best hyper-parameter setting $\alpha = 1e-3$, $n = 4000$, $D = 25$, using grid search strategy. More details about training of GBRT could be found in the online codes.

Model evaluation

To examine the performance of EP2vec in predicting EPIs in specific cell line, we performed the stratified 10-fold cross-validation experiment in all datasets. We randomly partitioned training data into ten equal sized subsets and each subset contained roughly the same proportions of the two lines of class labels. One of the ten subsets was used for testing the model, and the remaining nine were used as training data. This validation process was repeated ten times, with each of the ten subsets used exactly once as test data.

We calculated F1 scores for each cross-validation, which considered both the precision p and the recall r of the test. Precision p is the number of correct positive results divided by the number of all positive results, and recall r is the number of correct positive results divided by the number of positive results that should have been returned. The F1 score could be interpreted as the harmonic mean of the precision and recall, as $F1 = 2rp/(r+p)$, which reaches its best value at 1 and worst at 0.

We compared the performance of EP2vec and several other baseline methods, including TargentFinder, gkmSVM, SPEID. We directly used the source codes their authors published online. TargetFinder definde three training sets. The first set included features for the enhancer and promoter only (E/P). The second set included features for an extended enhancer (using 3 kb of flanking sequence) and a non-extended promoter (EE/P). The last set included enhancers and promoters plus the window between them (E/P/W), which were up to thousands of base pairs. Since the performance of TargetFinder on the last set was consistently better than other two sets according to their publication, we only evaluated this method on the E/P/W set. For gkmSVM, we need to first transformed a pair of two sequences (enhancer and promoter) into a single sequence by concatenating them, and then used it as input for gkmSVM.

Attention mechanism

Not all words contribute equally to the representation of the sentence meaning. Hence, we introduced attention mechanism to find out such critical words that were most important to the meaning of the sentence. Considering the k-mer words as motifs, we essentially aimed to find motifs that contribute more to the vector representation of the enhancer/promoter. Take the i-th enhancer x_i as example, it contained T_i words $w^{c_{i,1}}, w^{c_{i,2}}, \ldots, w^{c_{i,T_i}}$. We measured the importance of each word by computing similarity between word vector $w^{c_{i,t}}$ and sentence vector x_i and got a normalized importance weight α_{it} through a softmax function, as

$$\alpha_{it} = \frac{\exp\left(x_i^T w^{c_{i,t}}\right)}{\sum_j \exp\left(x_i^T w^{c_{i,j}}\right)}.$$

Therefore, every word in the sentence had a weight representing its importance to the sentence. In order to validate that our model was able to select informative words in a sentence, we visualized the high-weight words. In detail, two sets of informative k-mers were obtained by picking out the most important words for enhancer and promoter respectively, in every sentence with positive label. Then we performed motif enrichment analysis using CentriMo [38] to compare these words against known motifs in the HOCOMOCO v9 dataset [39], and drew out top enriched motifs with sequence logo [40].

Results

Computational performance

To consolidate the importance of our work, we compare the performance of EP2vec against other three typical baseline methods, including TargetFinder, gkmSVM and SPEID. TargetFinder is based on experimental features obtained from biological sequencing experiments, and gkmSVM is based on k-mer features and SVM classifiers. SPEID is based on deep learning which uses LSTMs with sequence data to predict EPIs.

They all have their own advantages and disadvantages. (1) For TargetFinder, experimental features are rich of cell line-specific predictive information, but they are expensive and time-consuming to acquire. Besides, for some cell lines, the dimension of accessible experimental features is limited due to lack of biological experiments. (2) For gkmSVM, k-mer features are an unbiased, general, complete set of sequence features defined on arbitrary-length sequences. However, the k-mers can only capture local motif patterns because they only use the k-mer counts information without making full use of context information or co-occurrence information of k-mers. (3) For SPEID, LSTM is a powerful supervised deep learning technique which is able to capture long-range dependencies. Nonetheless, deep learning methods often have millions of parameters to learn in the training process which takes a long time, and special attention should be put on fine-tuning the network. Usually, it takes time to optimize the network structure for a specific dataset, but this optimal structure may be not applicable to other datasets due to overfitting problems.

Our paper proposes an innovative approach to represent a DNA sequence (or a pair of two DNA sequences) in a fixed-length vector, namely sequence embedding features, using the unsupervised method Paragraph Vector. The training of sequence embedding features utilizes the global statistics information of k-mers, and hence our features form a potentially better presentation for DNA sequences. Specifically, for EP2vec, we set $k = 6$, the stride $s = 1$, the context window size $m = 20$, and the embedding dimension $d = 100$. We report the F1 score statistics of the four methods in 10-fold cross-validation for each dataset in Table 2. In addition, we also calculate area under the Receiver Operating Characteristic curve (auROC) score (Additional file 1: Table S1) and area under the Precision Recall curve (auPRC) score (Additional file 1: Table S2).

The results in Table 2 show that EP2vec is slightly better than TargetFinder and significantly outperforms the other two sequence-based methods, namely gkmSVM and SPEID. For example, in the GM12878 cell line dataset, the average F1 scores of EP2vec, TargetFinder (on E/P/W), gkmSVM and SPEID are 0.867, 0.844, 0.779 and 0.809, respectively. On the whole, the F1 scores for six cell line datasets of the above four methods ranges from 0.867~ 0.933, 0.844~ 0.922, 0.731~ 0.822, 0.809~ 0.900, respectively. We are convinced that the sequence embedding features learned by EP2vec is comparable to experimental features and has superiority over the other two computational sequence features, because we are able to capture the global context information of DNA sequences.

Table 2 The mean values and the standard deviations of F1 scores for EP2vec and other three baseline methods in 10-fold cross-validation experiments. For FANTOM dataset, we do not evaluate TargetFinder due to lack of experimental features, and we do not evaluate SPEID since it is extremely time-consuming to run 10-fold cross validation of SPEID on so many samples

Dataset	EP2vec	TargetFinder	gkmSVM	SPEID
K562	0.882 (0.019)	0.881 (0.014)	0.821 (0.018)	0.846 (0.024)
IMR90	0.872 (0.020)	0.863 (0.017)	0.749 (0.026)	0.825 (0.032)
GM12878	0.867 (0.014)	0.844 (0.010)	0.779 (0.015)	0.809 (0.018)
HUVEC	0.875 (0.024)	0.878 (0.022)	0.731 (0.028)	0.809 (0.023)
HeLa-S3	0.920 (0.013)	0.913 (0.014)	0.822 (0.021)	0.888 (0.023)
NHEK	0.933 (0.015)	0.922 (0.018)	0.800 (0.024)	0.900 (0.019)
FANTOM	0.841(0.004)	/	0.803(0.017)	/

Sensitivity analysis

The goal of EP2vec is to capture global sequence information. In our approach, we must split sequences into words using a sliding window fashion to form sentences from which we could extract fixed-length embedding features. To evaluate the stability of EP2vec, we carry

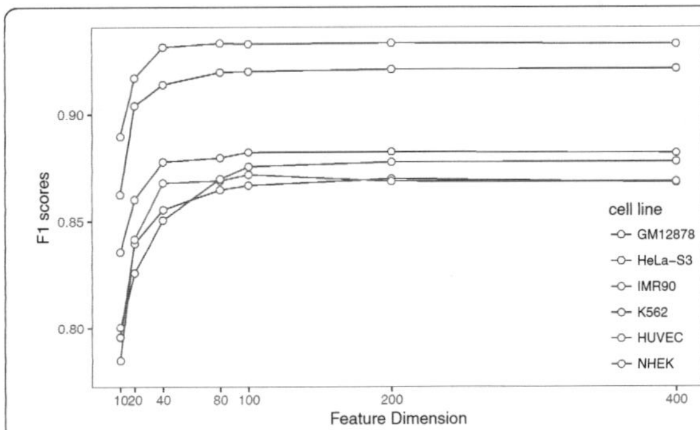

Fig. 2 The F1 scores of different embedding dimensions. As the embedding dimensions increase, the performance increses. And embedding dimension $d = 100$ is sufficient to obtain the near-optimal performances in all these datasets

out sensitivity analysis for hyper-parameters including k, the stride s and the embedding dimension d.

As shown in Fig. 2, we find that when the embedding dimension d decreases, our model degrades slightly. For example, the F1 score of EP2vec on HUVEC dataset is 0.875 when $d = 100$. Setting $d = 10$ and retaining the other hyper-parameters unchanged, we find the F1 score decreases to 0.800. In general, the performance improves with the increase of embedding feature dimension. We note that although the mean F1 scores are not similar across different cell lines, 100 is the common choice of embedding dimension to obtain the near-optimal performances for all datasets. Lower but acceptable performance requires embedding dimension of 40 in NHEK, IMR90 and K562 while 80 in the other cell lines.

Furthermore, we explore the performance of different settings of the model hyper-parameters including k, the stride s and the embedding dimension d. The sensitivity analysis of these hyper-parameters is shown in Additional file 1: Tables S3-S5. These results indicate that EP2vec is robust to all the three hyper-parameters and successful in capturing the information of whole sentences.

Visualizing motifs by attention mechanism

In order to interpret that our model is able to detect informative k-mers or motifs in a sequence, we visualize k-mers with high weights selected using the attention mechanism for K562 and HUVEC in Fig. 3. We consider the most informative k-mers as sequence motifs that determine sequence function. Consequently, we calculate the most informative k-mers in positive samples and present the top enriched known motifs in enhancers and promoters (Additional file 1: Tables S6 and S7).

For example, HUVECs are cells derived from the endothelium of veins from the umbilical cord and are reported to play an important role in hematopoiesis. Among the top five enriched motifs in HUVEC, MYB_f1, GFI1_f1, IKZF1_f1 and SOX15_f1 present some clues to HUVEC cell line-specific information. MYB_f1 will bind to MYB, which plays an essential role in the regulation of hematopoiesis. MYB may be aberrantly expressed or rearranged or undergo translocation in leukemias and lymphomas, and is considered to be an oncogene [41]. GFI1_f1 will bind to GFI1, which functions as a transcriptional repressor. This TF plays a role in diverse developmental contexts, including hematopoiesis and oncogenesis. It functions as part of a complex along with other cofactors to control histone modifications that lead to silencing of the target gene promoters [42]. IKZF1_f1 will bind to IKZF1, which belongs to the family of zinc-finger DNA-binding proteins associated with chromatin remodeling. Overexpression of some dominant-negative isoforms have been associated with B-cell malignancies, such as acute lymphoblastic leukemia [43]. SOX15_f1 will bind to SOX15, which is involved in the regulation of embryonic development and in the determination of the cell fate [44]. All of these top enriched motifs in HUVEC are experimentally proved to be related with hematopoiesis or other similar functions, which indicates that we successfully find informative motifs through applying attention mechanism in EP2vec.

As another example, K562 cells are of the erythroleukemia type, and the line is derived from a 53-year-old female chronic myelogenous leukemia patient in blast crisis. The top two enriched motifs in K562 is KLF6_si and TFE3_f1, which also give evidence to K562 specific information. KLF6_si will binding to KLF6. The TF is a transcriptional activator, and functions as a tumor suppressor. Multiple transcript variants encoding different isoforms have been found for this gene, some of which are implicated in carcinogenesis [45]. TFE3_f1 will bind to TFE3. This TF promotes the expression of genes downstream of transforming growth factor beta (TGF-beta) signaling. This gene may be involved in chromosomal translocations in renal cell carcinomas and other cancers, resulting in the production of usion proteins [46].

TF annotations for top five enriched motifs in all six cell lines are reported in Additional file 1: Tables S8 and S9. From these results, we conclude that sequence embedding features not only perform well but also are interpretable through motif enrichment analysis. Although deep learning is widely applied and always surpass conventional methods in various tasks, it is hard to interpret why deep models perform well. We make use of attention mechanism and try to find out why sequence embedding features

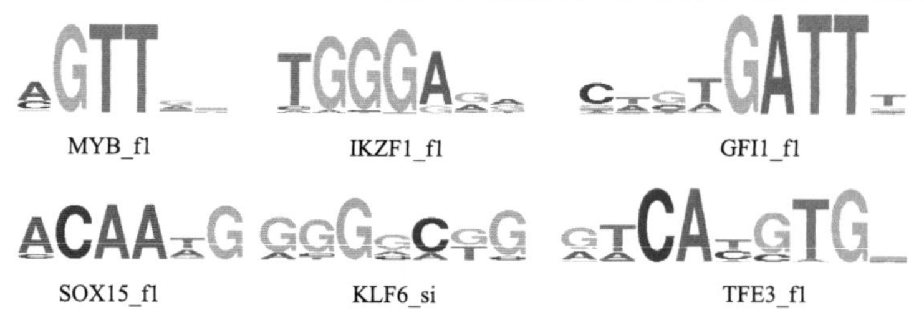

Fig. 3 The enriched motifs in HUVEC and K562. MYB_f1, IKZF1_f1, GFI1_f1 and SOX15_f1 are enriched in HUVEC. KLF6_si and TFE3_f1 are enriched in K562

outperform others methods. One reasonable explanation is that EP2vec can capture important motifs in a sequence that reveal sequence information.

Combination of two types of features

According to Table 2, we observe that our sequence embedding features outperform experimental features in TargetFinder and sequence features computed in gkmSVM and SPEID. Here, to further improve the prediction accuracy of our model, we attempt to combine our sequence embedding features and experimental features in TargetFinder.

Concretely, we concatenate the 200-dimensional sequence embedding features and the experimental features and then we use a GBRT with the same hyper-parameters as EP2vec to train a classifier for predicting EPIs. According to Fig. 4, we can see that sequence embedding features are better than experimental features in capturing useful sequence information, while the combination of both types of features generate even better performance. Consequently, sequence embedding features in enhancers and promoters and experimental features in windows between enhancers and promoters facilitate each other and combination of them performs better than all other feature sets. Finally, we conclude that sequence embedding features and experiment features can be complementary to each other and we could take advantage of existing experimental features and extracted sequence embedding features to predict true EPIs with high accuracy.

Discussion

Deep learning has successful applications in both computer vision and natural language processing (NLP). As is well known, Convolutional Neural Network (CNN) is a powerful deep learning model in computer vision area. Inspired by deep learning applied in image processing, DeepSEA and DeepBind first regard DNA sequences as binary images through one-hot encoding. They both preprocess the DNA sequences by transforming them into 4xL images (L is the length of a sequence), and then use a CNN to model DNA sequences. Many other deep learning approaches applied in sequence analysis recently all follow this idea and achieve excellent performance.

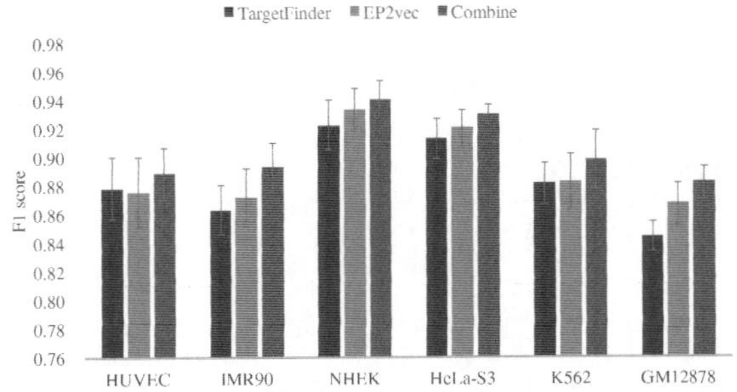

Fig. 4 The F1 scores of combined features and two single types of features in 10-fold cross-validation. The combination of both types of features generate even better performance, indicating sequence embedding features and experiment features can be complementary to each other

Our deep learning framework EP2vec is different from them. We solve this sequence analysis problem from a different perspective inspired from NLP. In fact, there are also many successful applications of deep learning in NLP area, such as word2vec which embeds words into a vector space. Paragraph Vector is based on word2vec, and it embeds whole sentences to vectors encoding their semantic meanings. We think it is more natural to treat a DNA sequence as a sentence other than an image, since the DNA sequence is only a one-dimensional data while images are often two-dimensional data. Hence, we regard a DNA sequence as a sentence which is comprised of *k*-mers (or words). We learn good representations of DNA sequences using Paragraph Vector as the results shown in Computational performance. In this unsupervised feature extraction stage, we apply deep learning to extract sequence embedding features which will be used in supervised classification. The superiority of our framework mainly lies in that we utilize the global statistics of *k*-mer relationships, and can learn a global representation of a DNA sequence.

Our method is innovative in using a different deep learning diagram from existing methods in the following several aspects:

First, we draw strength from recent advance in deep learning and successfully extract fixed-length embedding features for variable-length sequences. Our results suggest that it is possible to use only sequence embedding features instead of traditional genomic and epigenomic features to predict EPIs with competitive results, and that DNA sequences themselves provide enough information about what function they perform in different cell lines. Different from other computational features for DNA sequences, we learn the sequence embedding features on basis of the *k*-mer co-occurrence statistics using Paragraph Vector, and by learning an embedding vector directly for a sequence we can better represent the global sequence information.

Second, we carry out sensitivity analysis with regard to model hyper-parameters involved in the unsupervised feature learning stage. The result indicates that EP2vec is robust to its hyper-parameters and is effective in capturing the information of whole sequences. Even using only 10-dimensional sequence embedding features, EP2vec still yields satisfactory results.

Third, we explore important motifs that account for enhancers and promoter when mining the information in sequence embedding features. As we all know, deep learning often behaves like black box and people find it hard to explain what the extracted features mean. We illustrate the meaning of sequence embedding features by visualizing the motifs found by attention mechanism with sequence logo. These results indicate that sequence embedding features have underlying biological meanings which we need to pay more attention to.

Last but not the least, we train a hybrid model using both sequence embedding features and the experimental features, which generates better classification results than using a single type of features. We conclude that the two types of features are complementary to each other, and their combination is beneficial for prediction of EPIs.

Nevertheless, our approach can still be improved in the following several aspects. First, we treat every word equally without discrimination in the training. Nevertheless, using the attention mechanism, we pay more attention on important words in the visualizing process. Hence, we could adopt attention mechanism in the training process and gain better representation of the whole sequence. Second, in the unsupervised feature extraction stage of EP2vec workflow, we train sequence embedding features for enhancers and promoters separately, without using interaction information. In fact, we can inject the EPIs label information in this stage, so that we can encode not only the cell line specific information of enhancer and promoter sequences but also the paired information of enhancers and promoters in the feature representation. Third, we could combine sequence-based features and massive biological experiments data in the network training process for sequence embedding features. Although sequence features show good performance, they lose cell line specific information which is enriched in experimental features. We can fuse the cell line specific experimental features in training process and predict EPIs genome-wide.

Conclusions

In conclusion, EP2vec extracts sequence embedding features using unsupervised deep learning method and predicts EPIs accurately using GBRT classifier achieving state-of-the-art performance. Different from the previous sequence-based methods, EP2vec is innovative in extracting fixed-length embedding features for variable-length sequences and retaining the context information. Given the excellent performance of EP2vec, we will continue to improve our approach according to the above discussion. We expect EP2vec and the future revised version to play an important role in all kinds of sequence prediction tasks, such as identification of miRNA target sites and RNA-RNA interactions, and benefit further downstream analysis.

Additional file

Additional file 1: Supplementary Tables and Supplementary Figures. (DOCX 302 kb)

Abbreviations

1D: One-dimensional; 3C: Chromosome conformation capture-based; 3D: Three-dimensional; auPRC: area under the Precision Recall curve; auROC: area under the Receiver Operating Characteristic curve; CART: Classification and regression trees; ChIA-PET: Chromatin interaction analysis by paired-end tag sequencing; CNN: Convolutional neural networks; EPI: Enhancer-promoter interactions; GBRT: Gradient boosted regression trees; GWAS: Genome-wide association studies; lncRNA: long noncoding RNA; LSTM: Long short-term memory; NLP: Natural language processing; SGD: Stochastic gradient descent; SNP: Single nucleotide polymorphism

Acknowledgments

Rui Jiang is a RONG professor at the Institute for Data Science, Tsinghua University. We acknowledge the authors of TargetFinder, who provide us valuable data.

Funding

This research was partially supported by the National Natural Science Foundation of China (61573207, 61175002, 61721003). Publication costs were funded by National Natural Science Foundation of China (61573207, 61175002, 61721003).

Availability of data and materials

The source code coud be found in https://github.com/wanwenzeng/ep2vec. The data from TargetFinder could be found in https://github.com/shwhalen/targetfinder.

About this supplement

This article has been published as part of *BMC Genomics* Volume 19 Supplement 2, 2018: Selected articles from the 16th Asia Pacific Bioinformatics Conference (APBC 2018): genomics. The full contents of the supplement are available online at https://bmcgenomics.biomedcentral.com/articles/supplements/volume-19-supplement-2.

Authors' contributions

WWZ conducted all experiments. RJ designed the research. WWZ, MMW and RJ wrote this manuscript. All the authors read and approved the final manuscript.

Ethics approval and consent to participate

Not applicable.

Consent for publication

Not applicable.

Competing interests

The authors declare that they have no competing interests.

Publisher's Note

Springer Nature remains neutral with regard to jurisdictional claims in published maps and institutional affiliations.

Author details

[1]MOE Key Laboratory of Bioinformatics; Bioinformatics Division and Center for Synthetic & Systems Biology, Beijing 100084, China. [2]Department of Automation, Tsinghua University, Beijing 100084, China. [3]Department of Computer Science, Tsinghua University, Beijing 100084, China.

Published: 9 May 2018

References

1. Esteller M. Non-coding RNAs in human disease. Nat Rev Genet. 2011;12(12):861–74.

2. Shlyueva D, Stampfel G, Stark A. Transcriptional enhancers: from properties to genome-wide predictions. Nat Rev Genet. 2014;15(4):272–86.
3. Smemo S, Campos LC, Moskowitz IP, Krieger JE, Pereira AC, Nobrega MA. Regulatory variation in a TBX5 enhancer leads to isolated congenital heart disease. Hum Mol Genet. 2012;21(14):3255–63.
4. Andersson R, Gebhard C, Miguel-Escalada I, Hoof I, Bornholdt J, Boyd M, Chen Y, Zhao X, Schmidl C, Suzuki T, et al. An atlas of active enhancers across human cell types and tissues. Nature. 2014;507(7493):455–61.
5. Jiang R. Walking on multiple disease-gene networks to prioritize candidate genes. J Mol Cell Biol. 2015;7(3):214–30.
6. Guo H, Ahmed M, Zhang F, Yao CQ, Li S, Liang Y, Hua J, Soares F, Sun Y, Langstein J, et al. Modulation of long noncoding RNAs by risk SNPs underlying genetic predispositions to prostate cancer. Nat Genet. 2016; 48(10):1142–50.
7. Dekker J, Rippe K, Dekker M, Kleckner N. Capturing chromosome conformation. Science. 2002;295(5558):1306–11.
8. Simonis M, Klous P, Splinter E, Moshkin Y, Willemsen R, de Wit E, van Steensel B, de Laat W. Nuclear organization of active and inactive chromatin domains uncovered by chromosome conformation capture-on-chip (4C). Nat Genet. 2006;38(11):1348–54.
9. Dostie J, Richmond TA, Arnaout RA, Selzer RR, Lee WL, Honan TA, Rubio ED, Krumm A, Lamb J, Nusbaum C, et al. Chromosome conformation capture carbon copy (5C): a massively parallel solution for mapping interactions between genomic elements. Genome Res. 2006; 16(10):1299–309.
10. Fullwood MJ, Liu MH, Pan YF, Liu J, Xu H, Mohamed YB, Orlov YL, Velkov S, Ho A, Mei PH, et al. An oestrogen-receptor-alpha-bound human chromatin interactome. Nature. 2009;462(7269):58–64.
11. Lieberman-Aiden E, van Berkum NL, Williams L, Imakaev M, Ragoczy T, Telling A, Amit I, Lajoie BR, Sabo PJ, Dorschner MO, et al. Comprehensive mapping of long-range interactions reveals folding principles of the human genome. Science. 2009;326(5950):289–93.
12. He B, Chen C, Teng L, Tan K. Global view of enhancer-promoter interactome in human cells. Proc Natl Acad Sci U S A. 2014;111(21):E2191–9.
13. Roy S, Siahpirani AF, Chasman D, Knaack S, Ay F, Stewart R, Wilson M, Sridharan R. A predictive modeling approach for cell line-specific long-range regulatory interactions. Nucleic Acids Res. 2015;43(18):8694–712.
14. Whalen S, Truty RM, Pollard KS. Enhancer-promoter interactions are encoded by complex genomic signatures on looping chromatin. Nat Genet. 2016;48(5):488–96.
15. Zhu Y, Chen Z, Zhang K, Wang M, Medovoy D, Whitaker JW, Ding B, Li N, Zheng L, Wang W. Constructing 3D interaction maps from 1D epigenomes. Nat Commun. 2016;7:10812.
16. S Singh, Y Yang, B Poczos, J Ma. Predicting Enhancer-Promoter Interaction from Genomic Sequence with Deep Neural Networks. biorxiv 2016.
17. Duren Z, Chen X, Jiang R, Wang Y, Wong WH. Modeling gene regulation from paired expression and chromatin accessibility data. Proc Natl Acad Sci U S A. 2017;114(25):E4914–23.
18. Min X, Zeng W, Chen N, Chen T, Jiang R. Chromatin accessibility prediction via convolutional long short-term memory networks with k-mer embedding. Bioinformatics. 2017;33(14):i92–i101.
19. Ghandi M, Lee D, Mohammad-Noori M, Beer MA. Enhanced regulatory sequence prediction using gapped k-mer features. PLoS Comput Biol. 2014; 10(7):e1003711.
20. Zhou J, Troyanskaya OG. Predicting effects of noncoding variants with deep learning-based sequence model. Nat Methods. 2015;12(10):931–4.
21. Alipanahi B, Delong A, Weirauch MT, Frey BJ. Predicting the sequence specificities of DNA- and RNA-binding proteins by deep learning. Nat Biotechnol. 2015;33(8):831–8.
22. Le Q, Mikolov T. Distributed representations of sentences and documents. In: Proceedings of the 31st international conference on machine learning (ICML-14): 2014; 2014. p. 1188–96.
23. Gan M, Li W, Zeng W, Wang X, Jiang R. Mimvec: a deep learning approach for analyzing the human phenome. BMC Syst Biol. 2017;11(Suppl 4):76.
24. Sutskever I, Vinyals O, Le QV. Sequence to sequence learning with neural networks. In: Advances in neural information processing systems: 2014; 2014. p. 3104–12.
25. Huang P-S, He X, Gao J, Deng L, Acero A, Heck L: Learning deep structured semantic models for web search using clickthrough data. In: Proceedings of the 22nd ACM international conference on Conference on information & knowledge management: 2013. San Francisco: ACM; 2013: 2333-2338.
26. Hoffman MM, Buske OJ, Wang J, Weng Z, Bilmes JA, Noble WS. Unsupervised pattern discovery in human chromatin structure through genomic segmentation. Nat Methods. 2012;9(5):473–6.
27. Ernst J, Kellis M. ChromHMM: automating chromatin-state discovery and characterization. Nat Methods. 2012;9(3):215–6.
28. Bernstein BE, Stamatoyannopoulos JA, Costello JF, Ren B, Milosavljevic A, Meissner A, Kellis M, Marra MA, Beaudet AL, Ecker JR, et al. The NIH roadmap Epigenomics mapping consortium. Nat Biotechnol. 2010;28(10):1045–8.
29. Harrow J, Frankish A, Gonzalez JM, Tapanari E, Diekhans M, Kokocinski F, Aken BL, Barrell D, Zadissa A, Searle S, et al. GENCODE: the reference human genome annotation for the ENCODE project. Genome Res. 2012; 22(9):1760–74.
30. Ramskold D, Wang ET, Burge CB, Sandberg R. An abundance of ubiquitously expressed genes revealed by tissue transcriptome sequence data. PLoS Comput Biol. 2009;5(12):e1000598.
31. Li Q, Brown JB, Huang H, Bickel PJ. Measuring reproducibility of high-throughput experiments. Ann Appl Stat. 2011:1752–79.
32. Rao SS, Huntley MH, Durand NC, Stamenova EK, Bochkov ID, Robinson JT, Sanborn AL, Machol I, Omer AD, Lander ES, et al. A 3D map of the human genome at kilobase resolution reveals principles of chromatin looping. Cell. 2014;159(7):1665–80.
33. Rumelhart DE, Hinton GE, Williams RJ. Learning representations by back-propagating errors. Cogn Model. 1988;5(3):1.
34. Morin F, Bengio Y: Hierarchical Probabilistic Neural Network Language Model. In: Aistats: 2005. Citeseer; 2005: 246-252.
35. Mnih A, Hinton GE. A scalable hierarchical distributed language model. In: Advances in neural information processing systems: 2009, vol. 2009. p. 1081–8.
36. Mikolov T, Sutskever I, Chen K, Corrado GS, Dean J. Distributed representations of words and phrases and their compositionality. In: Advances in neural information processing systems: 2013; 2013. p. 3111–9.
37. Řehůřek R, Sojka P. Software framework for topic Modelling with large corpora. In: Proceedings of LREC 2010 workshop new challenges for NLP frameworks: 2010; 2010. p. 45–50.
38. Bailey TL, Machanick P. Inferring direct DNA binding from ChIP-seq. Nucleic Acids Res. 2012;40(17):e128.
39. Kulakovskiy IV, Vorontsov IE, Yevshin IS, Soboleva AV, Kasianov AS, Ashoor H, Ba-Alawi W, Bajic VB, Medvedeva YA, Kolpakov FA, et al. HOCOMOCO: expansion and enhancement of the collection of transcription factor binding sites models. Nucleic Acids Res. 2016;44(D1):D116–25.
40. Sebastian A, Contreras-Moreira B. footprintDB: a database of transcription factors with annotated cis elements and binding interfaces. Bioinformatics. 2014;30(2):258–65.
41. Ramsay RG, Gonda TJ. MYB function in normal and cancer cells. Nat Rev Cancer. 2008;8(7):523–34.
42. Hock H, Hamblen MJ, Rooke HM, Schindler JW, Saleque S, Fujiwara Y, Orkin SH. Gfi-1 restricts proliferation and preserves functional integrity of haematopoietic stem cells. Nature. 2004;431(7011):1002–7.
43. Virely C, Moulin S, Cobaleda C, Lasgi C, Alberdi A, Soulier J, Sigaux F, Chan S, Kastner P, Ghysdael J. Haploinsufficiency of the IKZF1 (IKAROS) tumor suppressor gene cooperates with BCR-ABL in a transgenic model of acute lymphoblastic leukemia. Leukemia. 2010;24(6):1200–4.
44. McLaughlin JN, Mazzoni MR, Cleator JH, Earls L, Perdigoto AL, Brooks JD, Muldowney JA 3rd, Vaughan DE, Hamm HE. Thrombin modulates the expression of a set of genes including thrombospondin-1 in human microvascular endothelial cells. J Biol Chem. 2005;280(23):22172–80.
45. DeKelver RC, Lewin B, Lam K, Komeno Y, Yan M, Rundle C, Lo MC, Zhang DE. Cooperation between RUNX1-ETO9a and novel transcriptional partner KLF6 in upregulation of Alox5 in acute myeloid leukemia. PLoS Genet. 2013; 9(10):e1003765.
46. Heimann P, El Housni H, Ogur G, Weterman MA, Petty EM, Vassart G. Fusion of a novel gene, RCC17, to the TFE3 gene in t(X;17)(p11.2;q25.3)-bearing papillary renal cell carcinomas. Cancer Res. 2001;61(10):4130–5.

Kim et al. BMC Genomics 2018, 19(Suppl 2):89
https://doi.org/10.1186/s12864-018-4460-0

BMC Genomics

RESEARCH Open Access

GRIM-Filter: Fast seed location filtering in DNA read mapping using processing-in-memory technologies

Jeremie S. Kim[1,6*], Damla Senol Cali[1], Hongyi Xin[2], Donghyuk Lee[3], Saugata Ghose[1], Mohammed Alser[4], Hasan Hassan[6], Oguz Ergin[5], Can Alkan[4*] and Onur Mutlu[6,1*]

From The Sixteenth Asia Pacific Bioinformatics Conference 2018
Yokohama, Japan. 15-17 January 2018

Abstract

Background: Seed location filtering is critical in DNA read mapping, a process where billions of DNA fragments (reads) sampled from a donor are mapped onto a reference genome to identify genomic variants of the donor. State-of-the-art read mappers 1) quickly generate possible mapping locations for seeds (i.e., smaller segments) within each read, 2) extract reference sequences at each of the mapping locations, and 3) check similarity between each read and its associated reference sequences with a computationally-expensive algorithm (i.e., sequence alignment) to determine the origin of the read. A seed location filter comes into play before alignment, discarding seed locations that alignment would deem a poor match. The ideal seed location filter would discard all poor match locations prior to alignment such that there is no wasted computation on unnecessary alignments.

Results: We propose a novel seed location filtering algorithm, GRIM-Filter, optimized to exploit 3D-stacked memory systems that integrate computation within a logic layer stacked under memory layers, to perform processing-in-memory (PIM). GRIM-Filter quickly filters seed locations by 1) introducing a new representation of coarse-grained segments of the reference genome, and 2) using massively-parallel in-memory operations to identify read presence within each coarse-grained segment. Our evaluations show that for a sequence alignment error tolerance of 0.05, GRIM-Filter 1) reduces the false negative rate of filtering by 5.59x–6.41x, and 2) provides an end-to-end read mapper speedup of 1.81x–3.65x, compared to a state-of-the-art read mapper employing the best previous seed location filtering algorithm.

Conclusion: GRIM-Filter exploits 3D-stacked memory, which enables the efficient use of processing-in-memory, to overcome the memory bandwidth bottleneck in seed location filtering. We show that GRIM-Filter significantly improves the performance of a state-of-the-art read mapper. GRIM-Filter is a universal seed location filter that can be applied to any read mapper. We hope that our results provide inspiration for new works to design other bioinformatics algorithms that take advantage of emerging technologies and new processing paradigms, such as processing-in-memory using 3D-stacked memory devices.

Keywords: High throughput sequencing, Genome sequencing, Seed location filtering, 3D-stacked DRAM, Processing-in-memory, Emerging memory technologies

*Correspondence: jeremiekim123@gmail.com;calkan@cs.bilkent.edu.tr; omutlu@gmail.com
[1]Department of Electrical and Computer Engineering, Carnegie Mellon University, Pittsburgh, PA, USA
[6]Department of Computer Science, ETH Zürich, Zürich, CH, Switzerland
Full list of author information is available at the end of the article

Background

Our understanding of human genomes today is affected by the ability of modern technology to quickly and accurately determine an individual's entire genome. The human genome is comprised of a sequence of approximately 3 billion bases that are grouped into deoxyribonucleic acids (*DNA*), but today's machines can identify DNA only in short sequences (i.e., *reads*). Determining a genome requires three stages: 1) cutting the genome into many short reads, 2) identifying the DNA sequence of each read, and 3) mapping each read against the reference genome in order to analyze the variations in the sequenced genome. In this paper, we focus on improving the third stage, often referred to as *read mapping*, which is a major computational bottleneck of a modern genome analysis pipeline. Read mapping is performed computationally by *read mappers* after each read has been identified.

Seed-and-extend mappers [1–6] are a class of read mappers that break down each read sequence into *seeds* (i.e., smaller segments) to find locations in the reference genome that closely match the read. Figure 1 illustrates the five steps used by a seed-and-extend mapper. First, the mapper obtains a read (❶ in the figure). Second, the mapper selects smaller DNA segments from the read to serve as seeds (❷). Third, the mapper indexes a data structure with each seed to obtain a list of possible locations within the reference genome that could result in a match (❸). Fourth, for each possible location in the list, the mapper obtains the corresponding DNA sequence from the reference genome (❹). Fifth, the mapper aligns the read sequence to the reference sequence (❺), using an expensive sequence alignment (i.e., verification) algorithm to determine the similarity between the read sequence and the reference sequence.

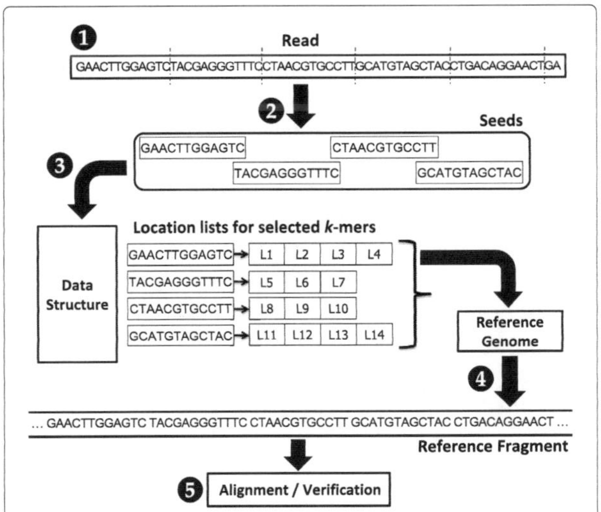

Fig. 1 Flowchart of a seed-and-extend mapper

To improve the performance of seed-and-extend mappers, we can utilize *seed location filters*, recently introduced by Xin et al. [7]. A seed location filter efficiently determines whether a candidate mapping location would result in an incorrect mapping *before* performing the computationally-expensive sequence alignment step for that location. As long as the filter can eliminate possible locations that would result in an incorrect mapping *faster* than the time it takes to perform the alignment, the entire read mapping process can be substantially accelerated [7–10]. As a result, several recent works have focused on optimizing the performance of seed location filters [7–12].

With the advent of seed location filters, the performance bottleneck of DNA read mapping has shifted from sequence alignment to seed location filtering [7–10]. Unfortunately, a seed location filter requires large amounts of memory bandwidth to process and characterize each of the candidate locations. Our goal is to reduce the time spent in filtering and thereby improve the speed of DNA read mapping. To this end, we present a new algorithm, *GRIM-Filter*, to efficiently filter locations with high parallelism. We design GRIM-Filter such that it is well-suited for implementation on 3D-stacked memory, exploiting the parallel and low-latency processing capability in the logic layer of the memory.

3D-stacked DRAM [13–22] is a new technology that integrates logic and memory in a three-dimensional stack of dies with a large internal data transfer bandwidth. This enables the bulk transfer of data from each memory layer to a logic layer that can perform simple parallel operations on the data.

Conventional computing requires the movement of data on the long, slow, and energy-hungry buses between the CPU processing cores and memory such that cores can operate on data. In contrast, processing-in-memory (PIM)-enabled devices such as 3D-stacked memory can perform simple arithmetic operations very close to where the data resides, with high bandwidth and low latency. With carefully designed algorithms for PIM, application performance can often be greatly improved (e.g., as shown in [19–21, 23]) because the relatively narrow and long-latency bus between the CPU cores and memory no longer impedes the speed of computation on the data.

Our goal is to develop a seed location filter that exploits the high memory bandwidth and processing-in-memory capabilities of 3D-stacked DRAM to improve the performance of DNA read mappers.

To our knowledge, this is the **first** seed location filtering algorithm that accelerates read mapping by overcoming the memory bottleneck with PIM using 3D-stacked memory technologies. GRIM-Filter can be used with any read mapper. However, in this work we demonstrate the effectiveness of GRIM-Filter with a hash table based mapper, mrFAST with FastHASH [7]. We improve the performance of hash table based read mappers while maintaining their high sensitivity and comprehensiveness (which were originally demonstrated in [3]).

Key mechanism. GRIM-Filter provides a quick method for determining whether a read will **not** match at a given location, thus allowing the read mapper to skip the expensive sequence alignment process for that location. GRIM-Filter works by counting the existence of small segments of a read in a genome region. If the count falls under a threshold, indicating that many small segments of a read are *not* present, GRIM-Filter discards the locations in that region before alignment. The existence of all small segments in a region are stored in a bitvector, which can be easily predetermined for each region of a reference genome. The bitvector for a reference genome region is retrieved when a read must be checked for a match in the region. We find that this regional approximation technique not only enables a high performance boost via high parallelism, but also improves filtering accuracy over the state-of-the-art. The filtering accuracy improvement comes from the finer granularity GRIM-Filter uses in counting the subsequences of a read in a region of a genome, compared to the state-of-the-art filter [7].

Key results. We evaluate GRIM-Filter qualitatively and quantitatively against the state-of-the-art seed location filter, *FastHASH* [7]. Our results show that GRIM-Filter provides a *5.59x–6.41x* smaller false negative rate (i.e., the proportion of locations that pass the filter, but that truly result in a poor match during sequence alignment) than the best previous filter with *zero* false positives (i.e., the number of locations that do not pass the filter, but that truly result in a good match during sequence alignment). GRIM-Filter provides an end-to-end performance improvement of *1.81x–3.65x* over a state-of-the-art DNA read mapper, *mrFAST* with *FastHASH*, for a set of real genomic reads, when we use a sequence alignment error tolerance of 0.05. We also note that as we increase the sequence alignment error tolerance, the performance improvement of our filter over the state-of-the-art increases. This makes GRIM-Filter more effective and relevant for future-generation error-prone sequencing technologies, such as nanopore sequencing [24, 25].

Motivation and aim

Mapping reads against a reference genome enables the analysis of the variations in the sequenced genome. As the throughput of read mapping increases, more large-scale genome analyses become possible. The ability to deeply characterize and analyze genomes at a large scale could change medicine from a reactive to a preventative and further personalized practice. In order to motivate our method for improving the performance of read mappers, we pinpoint the performance bottlenecks of modern-day mappers on which we focus our acceleration efforts. We find that across our data set (see "Methods" section), a state-of-the-art read mapper, mrFAST with FastHASH [7], on average, spends 15% of its execution time performing sequence alignment on locations that are found to be a match, and 59% of its execution time performing sequence alignment on locations that are discarded because they are not found to be a match (i.e., *false locations*).

Our *goal* is to implement a *seed location filter* that reduces the wasted computation time spent performing sequence alignment on such false locations. To this end, a seed location filter would quickly determine if a location will *not* match the read and, if so, it would avoid the sequence alignment altogether. The *ideal* seed location filter correctly finds all false locations without increasing the time required to execute read mapping. We find that such an ideal seed location filter would improve the *average* performance of mrFAST (with FastHASH) by *3.2x*. This speedup is primarily due to the reduced number of false location alignments. In contrast, most prior works [26–40] gain their speedups by implementing all or part of the read mapper in specialized hardware or GPUs, focusing mainly on the acceleration of the sequence alignment process, *not* the *avoidance* of sequence alignment. These works that accelerate sequence alignment provide orthogonal solutions, and could be implemented together with seed location filters, including GRIM-Filter, for additional performance improvement (see "Related work" section for more detail).

GRIM-Filter

We now describe our proposal for a new seed location filter, GRIM-Filter. At a high level, the key idea of GRIM-Filter is to store and utilize metadata on *short segments* of the genome, i.e., segments on the order of several hundred base pairs long, in order to quickly determine if a read will **not** result in a match at that genome segment.

Genome metadata representation

Figure 2 shows a reference genome with its associated metadata that is formatted for efficient operation by GRIM-Filter. The reference genome is divided into short contiguous segments, on the order of several hundreds of base pairs, which we refer to as *bins*. GRIM-Filter operates at the granularity of these bins, performing analyses on the metadata associated with each bin. This metadata is represented as a *bitvector* that stores whether or not a *token*, i.e., a short DNA sequence on the order of 5 base pairs, is present within the associated bin. We refer to each bit in the bitvector as an *existence bit*. To account for all

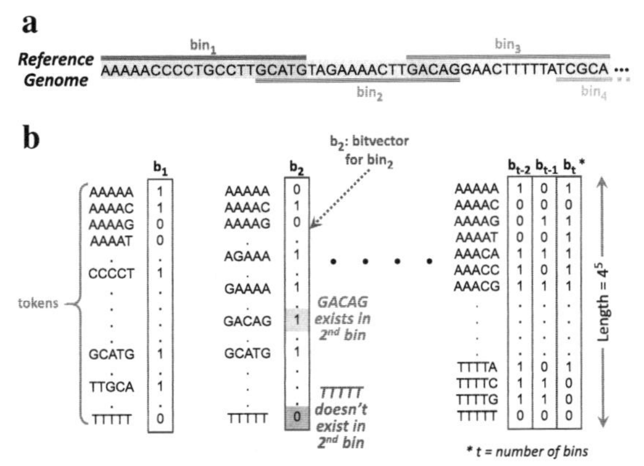

Fig. 2 GRIM-Filter has a 2D data structure where each bit at <row, column> indicates if a token (indexed by the row) is present in the corresponding bin (indicated by the column). **a** GRIM-Filter divides a genome into overlapping bins. **b** GRIM-Filter's metadata associated with a reference genome. Columns are indexed by the bin number of each location. Rows are indexed by the token value. In this figure, token size=5

possible tokens of length n, each bitvector must be 4^n bits in length, where each bit denotes the existence of a particular token instance. Figure 2 highlights the bits of two token instances of bin_2's bitvector: it shows that 1) the token GACAG (green) exists in bin_2, i.e., the existence bit associated with the token GACAG is set to 1 in the b_2 bitvector; and 2) the token TTTTT (red) is not present in bin_2, i.e., the existence bit associated with the token TTTTT is set to 0 in the b_2 bitvector.

Because these bitvectors are associated with the reference genome, the bitvectors need to be generated only once per reference, and they can be used to map any number of reads from other individuals of the same species. In order to generate the bitvectors, the genome must be sequentially scanned for every possible token of length n, where n is the selected token size. If bin_x contains the token, the bit in the b_x bitvector corresponding to the token must be set (1). If bin_x does not contain the token, then the same bit is left unset (i.e., 0). These bitvectors are saved and stored for later use when mapping reads to the same reference genome, i.e., they are part of the genome's metadata.

GRIM-Filter operation

Before sequence alignment, GRIM-Filter checks each bin to see if the bin contains a potential mapping location for the read, based on the list of potential locations provided by the read mapper. If the bin contains a location, GRIM-Filter then checks the bin to see if the location is likely to match the read sequence, by operating on the bitvector of the bin.

This relies on the *entire read* being contained within a given bin, and thus requires the bins to overlap with each other in the construction of the metadata (i.e., some base pairs are contained in multiple bins), as shown in Fig. 2a.

GRIM-Filter uses the described bitvectors to *quickly* determine whether a match within a given error tolerance is impossible. This is done before running the expensive sequence alignment algorithm, in order to reduce the number of unnecessary sequence alignment operations. For each location associated with a seed, GRIM-Filter 1) loads the bitvector of the bin containing the location; 2) operates on the bitvector (as we will describe shortly) to quickly determine if there will be no match (i.e., a poor match, given the error tolerance threshold); and 3) discards the location if it determines a poor match. If GRIM-Filter does **not** discard the location, the sequence at that location **must** be aligned with the read to determine the match similarity.

Using the circled steps in Fig. 3, we explain in detail how GRIM-Filter determines whether to discard a location z for a read sequence r. We use $bin\_num(z)$ to indicate the number of the bin that contains location z.

GRIM-Filter extracts every token contained within the read sequence r (❶ in the figure). Then, GRIM-Filter loads the bitvector for $bin_{bin\_num(z)}$ (❷). For each of the tokens contained in r, GRIM-Filter extracts the existence bit of that token from the bitvector (❸), to see whether the token exists somewhere within the bin. GRIM-Filter sums all of the extracted existence bits together (❹), which we refer to as the *accumulation sum* for location z (Sum_z). The accumulation sum represents the number of tokens from read sequence r that are present in $bin_{bin\_num(z)}$. A larger accumulation sum indicates that more tokens from r are present in the bin, and therefore the location is more likely to contain a match for r. Finally, GRIM-Filter compares Sum_z with a constant *accumulation sum threshold* value (❺), to determine whether location z is likely to match read sequence r. If Sum_z is greater than or equal to the threshold, then z is likely to match r, and the read

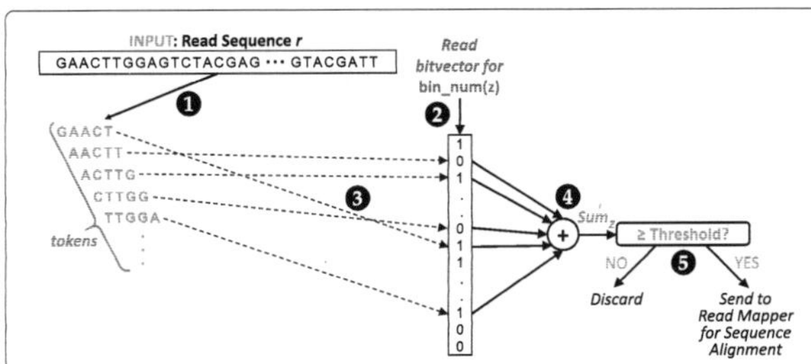

Fig. 3 Flow diagram for our seed location filtering algorithm. GRIM-Filter takes in a read sequence and sums the existence of its tokens within a bin to determine whether 1) the read sequence must be sequence aligned to the reference sequence in the bin or 2) it can be discarded without alignment. Note that token size = 5 in this example

mapper must perform sequence alignment on r to the reference sequence at location z. If Sum_z is less than the threshold, then z *will not* match r, and the read mapper skips sequence alignment for the location. We explain how we determine the accumulation sum threshold in "Determining the accumulation sum threshold" section.

Once GRIM-Filter finishes checking each location, it returns control to the read mapper, which performs sequence alignment on *only* those locations that pass the filter. This process is repeated for all seed locations, and it significantly reduces the number of alignment operations, ultimately reducing the end-to-end read mapping runtime (as we show in "Results" section). Our implementation of GRIM-Filter ensures a zero *false positive rate* (i.e., no locations that result in correct mappings for the read sequence are incorrectly rejected by the filter), as GRIM-Filter passes any seed location whose bin contains enough of the same tokens as the read sequence.

GRIM-Filter can also account for errors in the sequence, when some of the tokens do not match perfectly (see "Taking errors into account" section). Therefore, using GRIM-Filter to filter out seed locations does *not* affect the correctness of the read mapper.

Integration with a full read mapper

Figure 4 shows how we integrate GRIM-Filter with a read mapper to improve read mapping performance. Before the read mapper begins sequence alignment, it sends the read sequence, along with all potential seed locations found in the hash table for the sequence, to GRIM-Filter. Then, the *Filter Bitmask Generator* for GRIM-Filter performs the seed location filtering algorithm we describe in *GRIM-Filter Operation*, checking only the bins that include a potential seed location to see if the bin contains the same tokens as the read sequence (❶ in Fig. 4). For each location, we save the output of our threshold decision (the computation of which was shown in Fig. 3) as a bit within a *seed location filter bitmask*, where a 1 means that the location's accumulation sum was greater than or equal to the threshold, and a 0 means that the accumulation sum was less than the threshold. This bitmask is then passed to the *Seed Location Checker* (❷ in Fig. 4), which locates the reference segment corresponding to each seed location that passed the filter (❸) and sends the reference segment to the read mapper. The read mapper then performs sequence alignment on *only* the reference segments it receives from the seed location checker (❹), and outputs the correct mappings for the read sequence.

Determining the accumulation sum threshold

We now discuss in detail how to determine the threshold used to evaluate the accumulation sum (Sum_z). The threshold is used to determine whether or not a seed location should be sent to the read mapper for sequence alignment (shown as ❺ in Fig. 3). A greater value of Sum_z indicates that the seed location z is more likely to be a good match for the read sequence r. However, there are cases where Sum_z is high, but the read sequence results in a poor match with the seed location z. A simple example of this poor match is a read sequence that consists entirely of "A" base pairs, resulting in 100 AAAAA tokens, and a seed location that consists entirely of "G" base pairs except for a single AAAAA token. In this example, all 100 AAAAA tokens in the read sequence locate the one AAAAA token in the seed location, resulting in an accumulation sum of 100, even though the location contains only one AAAAA token. Because such cases occur, even though they may occur with low probability, GRIM-Filter cannot **guarantee** that a high accumulation sum for a seed location corresponds to a good match with a read sequence. On the other hand, GRIM-Filter can guarantee that a low accumulation sum (i.e., a sum that falls

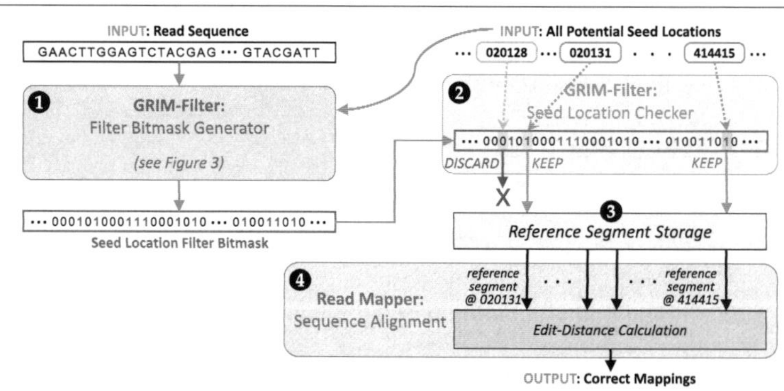

Fig. 4 GRIM-Filter integration with a read mapper. The *Filter Bitmask Generator* uses the bitvectors for each bin to determine whether any locations within the bin are potential matches with the read sequence, and saves potential match information into a *Seed Location Filter Bitmask*. The *Seed Location Checker* uses the bitmask to retrieve the corresponding reference segments for only those seed locations that match, which are then sent to the read mapper for sequence alignment

under the threshold) indicates that any reference sequence within the bin is a poor match with the read sequence.

This is because a lower sum means that fewer tokens from the read sequence are present in the bin, which translates directly to a greater number of errors in a potential match. For a low enough sum, we can guarantee that the potential read sequence alignment would have too many errors to be a good match.

Taking errors into account

If a read maps perfectly to a reference sequence in $bin_{bin\_num(z)}$, Sum_z would simply be the total number of tokens in a read, which is $read\_length - (n-1)$ for a token size of n. However, to account for insertions, deletions, and substitutions in the read sequence, sequence alignment has some error tolerance, where a read sequence and a reference sequence are considered a good match *even if* some differences exist. The accumulation sum threshold must account for this error tolerance, so we reduce the threshold below $read\_length - (n-1)$ to allow some tokens to include errors. Figure 5a shows the equation that we use to calculate the threshold while accounting for errors.

As shown in Fig. 5b, a token of size n in a bin overlaps with $n-1$ other tokens. We calculate the lowest Sum_z possible for a sequence alignment that includes only a single error (i.e., one insertion, deletion, or substitution) by studying these n tokens. If the error is an insertion, the insertion shifts at least one of the n tokens to the right, preserving the shifted token while changing the remaining tokens ($n-1$ in the worst case). If the error is a deletion or a substitution, the change in the worst case can affect all n tokens.

Figure 5b shows an example of how a substitution affects four different tokens, where $n = 4$. Therefore, for each error that we tolerate, we must assume the worst-case error (i.e., a deletion or a substitution), in which case up to n tokens will *not* match with the read sequence even when the location actually contains the read sequence.

The equation in Fig. 5a gives the accumulation sum threshold, accounting for the worst-case scenario for a sequence alignment error tolerance of e. This means that the maximum number of allowable errors is equal to the *ceiling* of the read size multiplied by the sequence alignment error tolerance. A sequence alignment error tolerance of $e = 0.05$ or less is widely used [2, 8, 41, 42]. For each allowable error, we assume that the worst-case number of tokens (equal to the token length n) are affected by the error. We also assume the worst case that each error affects a different set of tokens within the read, which results in the greatest possible number of tokens that may not match. We calculate this by multiplying the maximum number of allowable errors by n in the equation. Finally, we subtract the largest possible number of tokens that may *not* match from the total number of tokens in the read sequence, which is $read\_length - (n-1)$.

This leads to the threshold value that GRIM-Filter uses to determine the seed locations that the read mapper should perform sequence alignment on, as discussed in "GRIM-Filter operation" section and shown as ❺ in Fig. 3.

Candidacy for 3D-Stacked memory implementations

We identify **three** characteristics of the filter bitmask generator in GRIM-Filter that make it a strong candidate for implementation in 3D-stacked memory: 1) it requires only very simple operations (e.g., sums and comparisons); 2) it is highly parallelizable, since each bin can be operated on independently and in parallel; and 3) it is highly memory-bound, requiring a single memory access for approximately every three computational instructions (we determine this by profiling a software implementation of GRIM-Filter, i.e., GRIM-Software, which is described in "Methods" section). Next, we describe how we implement GRIM-Filter in 3D-stacked memory.

Mapping GRIM-Filter to 3D-stacked memory

In this section, we first describe the *3D-stacked DRAM* technology ("3D-stacked memory" section), which attempts to bridge the well-known disparity between

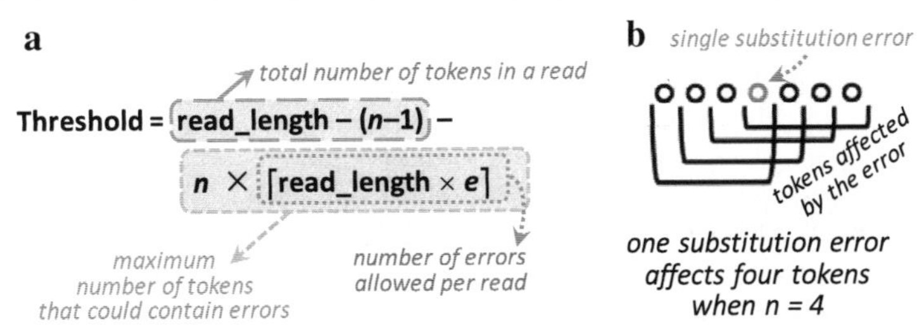

Fig. 5 a Equation to calculate the accumulation sum threshold for a read sequence, where n is the token length and e is the sequence alignment error tolerance. **b** Impact of a substitution error on four separate tokens, when $n = 4$. A single deletion or substitution error propagates to 4 consecutive tokens, while a single insertion error propagates to 3 consecutive tokens processor speed and memory bandwidth. Next, we describe how GRIM-Filter can be easily mapped to utilize this new memory technology *("Mapping GRIM-Filter to 3D-Stacked memory with PIM" section).* As the disparity between processor speed and memory bandwidth increases, memory becomes more of a bottleneck in the computing stack in terms of both performance and energy consumption [21, 43–46]. Along with 3D-stacked DRAM, which enables much higher bandwidth and lower latency compared to conventional DRAM, the disparity between processor and memory is alleviated by the re-emergence of the concept of *Processing-in-Memory* (PIM). PIM integrates processing units inside or near the main memory to 1) leverage high in/near-DRAM bandwidth, and low intra-DRAM latency; and 2) reduce energy consumption by reducing the amount of data transferred to and from the processor. In this section, we briefly explain the required background for these two technologies, which we leverage to implement GRIM-Filter in a highly-parallel manner.

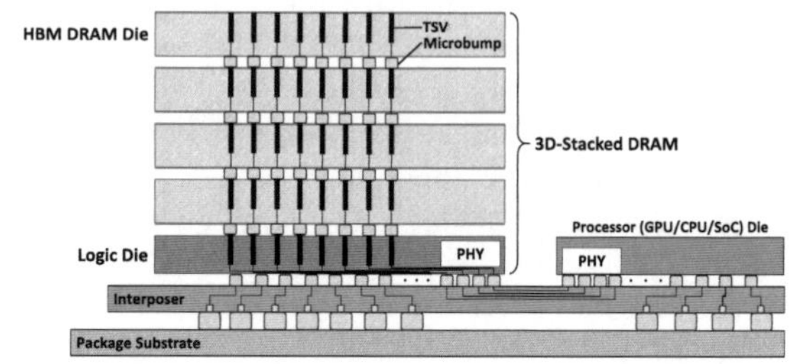

Fig. 6 3D-stacked DRAM example. High Bandwidth Memory consists of stacked memory layers (four layers in the picture) and a logic layer connected by high bandwidth through-silicon vias (TSVs) and microbumps [13, 14, 69]. The 3D-stacked memory is then connected to a processor die with an interposer layer that provides high-bandwidth between the logic layer and the processing units on the package substrate

3D-stacked memory

Main memory is implemented using the DRAM (dynamic random access memory) technology in today's systems [47–49]. Conventional DRAM chips are connected to the processors using long, slow, and energy-hungry PCB (printed circuit board) interconnects [47, 49–54]. The conventional DRAM chips do not incorporate logic to perform computation. For more detail on modern DRAM operation and architecture, we refer the reader to our previous works (e.g., [47, 50–52, 54–67]).

3D-stacked DRAM is a new DRAM technology that has a much higher internal bandwidth than conventional DRAM, thanks to the closer integration of logic and memory using the *through-silicon via* (TSV) interconnects, as seen in Fig. 6. TSVs are new, vertical interconnects that can pass through the silicon wafers of a 3D stack of dies [14, 22, 68]. A TSV has a much smaller feature size than a traditional PCB interconnect, which enables a 3D-stacked DRAM to integrate hundreds to thousands of these wired connections between stacked layers. Using this large number of wired connections, 3D-stacked DRAM can transfer bulk data simultaneously, enabling much higher bandwidth compared to conventional DRAM. Figure 6 shows a 3D-stacked DRAM (e.g., High Bandwidth Memory [13, 69]) based system that consists of four layers of DRAM dies and a logic die stacked together and connected using TSVs, a processor die, and a silicon interposer that connects the stacked DRAM and the processor. The vertical connections in the stacked DRAM are very wide and very short, which results in *high bandwidth* and *low power consumption*, respectively [14]. There are many different 3D-stacked DRAM architectures available today. High Bandwidth Memory (HBM) is already integrated into the AMD Radeon R9 Series graphics cards [15]. High Bandwidth Memory 2 (HBM2) is integrated in both the new AMD Radeon RX Vega^{64} Series graphics cards [70] and the new NVIDIA Tesla P100 GPU accelerators [71]. Hybrid Memory Cube (HMC) is developed by a number of different contributing companies [17, 18]. Like HBM, HMC also enables a logic layer underneath the DRAM layers that can perform computation [19–21]. HMC is already integrated in the SPARC64 XIfx chip [72]. Other new technologies that can enable processing-in-memory are also already prototyped in real chips, such as Micron's Automata Processor [73] and Tibco transactional application servers [74, 75].

Processing-in-memory (PIM). A key technique to improve performance (both bandwidth and latency) and reduce energy consumption in the memory system is to place computation units inside the memory system, where the data resides. Today, we see processing capabilities appearing inside and near DRAM memory (e.g., in the logic layer of 3D-stacked memory) [14, 19–21, 23, 53, 76–91]. This computation inside or near DRAM significantly reduces the need to transfer data to/from the processor over the memory bus.

PIM provides significant performance improvement and energy reduction compared to the conventional system architecture [19–21, 23, 76, 90, 92, 93], which must transfer *all data* to/from the processor since the processor is the only entity that performs all computational tasks.

3D-stacked DRAM with PIM. The combination of the two new technologies, 3D-stacked DRAM and PIM, enables very promising opportunities to build very high-performance and low-power systems. A promising design for 3D-stacked DRAM consists of multiple stacked memory layers and a tightly-integrated logic layer that controls the stacked memory, as shown in Fig. 6. As many prior works show [14, 19–22, 76, 80, 90, 92–95], the logic layer in 3D-stacked DRAM can be utilized not only for managing the stacked memory layers, but also for integrating application-specific accelerators or simple processing cores. Since the logic layer already exists and has enough space to integrate computation units, integrating application-specific accelerators in the logic layer requires modest design and implementation overhead, and little to no hardware overhead (see [20, 89] for various analyses). Importantly, the 3D-stacked DRAM architecture enables us to fully customize the logic layer for the acceleration of applications using processing-in-memory (i.e., processing in the logic layer) [20, 21, 76, 94].

Mapping GRIM-Filter to 3D-Stacked memory with PIM

We find that GRIM-Filter is a very good candidate to implement using processing-in-memory, as the filter is memory-intensive and performs simple computational operations (e.g., simple comparisons and additions). Figure 7 shows how we implement GRIM-Filter in a 3D-stacked memory. The *center block* shows each layer of an example 3D-stacked memory architecture, where multiple DRAM layers are stacked above a logic layer. The layers are connected together with several hundred TSVs, which enable a high data transfer bandwidth between the layers. Each DRAM layer is subdivided into multiple *banks* of memory. A bank in one DRAM layer is connected to banks in the other DRAM layers using the TSVs. These interconnected banks, along with a slice of the logic layer, are grouped together into a *vault*. Inside the 3D-stacked memory, we store the bitvector of each bin (see "GRIM-Filter" section) within a bank as follows: 1) each bit of the bitvector is placed in a different row in a consecutive manner (e.g., bit 0 is placed in row 0, bit 1 in row 1, and so on); and 2) all bits of the bitvector are placed in the same column, and the entire bitvector fits in the column (e.g., bitvector 0 is placed in column 0, bitvector 1 in column 1, and so on). We design and place customized logic to perform the GRIM-Filter operations within each logic layer slice, such that each vault can perform independent GRIM-Filter operations in parallel with every other vault. Next, we discuss how we organize the bitvectors within each bank. Afterwards, we discuss the customized logic required for GRIM-Filter and the associated hardware cost.

The *left block* in Fig. 7 shows the layout of bitvectors in a single bank. The bitvectors are written in column order (i.e., column-major order) to the banks, such that a DRAM access to a row fetches the existence bits of the same token across *many* bitvectors (e.g., bitvectors 0 to $t-1$ in the example in Fig. 7). When GRIM-Filter reads a row of data from a bank, the DRAM buffers the row within the bank's *row buffer* [50, 52, 96, 97], which resides in the same DRAM layer as the bank. This data is then copied into a *row data register* that sits in the logic layer, from which the GRIM-Filter logic can read the data.

This data organization allows each vault to compute the accumulation sum of multiple bins (e.g., bins 0 to $t-1$ in the example) simultaneously. Thus, GRIM-Filter can quickly and efficiently determine, across many bins, whether a seed location needs to be discarded or sequence aligned in any of these bins.

The *right block* in Fig. 7 shows the custom hardware logic implemented for GRIM-Filter in each vault's logic layer. We design a small *logic module* for GRIM-Filter, which consists of only an incrementer, accumulator, and comparator, and operates on the $bitvector_x$ of a single bin x. The incrementer adds 1 to the value in the accumulator, which stores the accumulated sum for bin x. In order to hold the final sum (i.e., Sum_z, shown as ❹ in Fig. 3), each accumulator must be at least $\lceil \log_2(read\_length) \rceil$ bits wide. Each comparator must be of the same width as the accumulator, as the comparator is used to check whether the accumulated sum exceeds the accumulated sum threshold. Because of the way we arrange the bitvectors in DRAM, a single read operation in a vault retrieves many (e.g., t) existence bits in parallel, from many (e.g., t) bitvectors, for the same token. These existence bits are copied from a DRAM bank's row buffer into a *row data register* within the logic layer slice of the vault. In order to

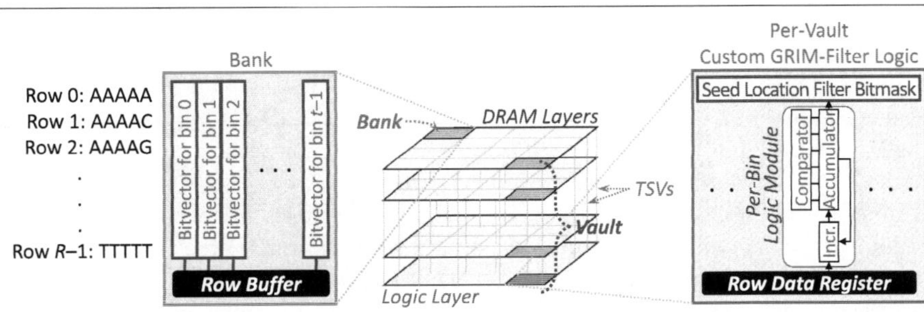

Fig. 7 Left block: GRIM-Filter bitvector layout within a DRAM bank. Center block: 3D-stacked DRAM with tightly integrated logic layer stacked underneath with TSVs for a high intra-DRAM data transfer bandwidth. Right block: Custom GRIM-Filter logic placed in the logic layer, for each vault

maximize throughput, we add a GRIM-Filter logic module *for each bin* to the logic layer slice. This allows GRIM-Filter to process all of these existence bits from multiple bitvectors in parallel.

Integration into the system and low-level operation
When GRIM-Filter starts in the CPU (spawned by a read mapper), it sends a read sequence r to the in-memory GRIM-Filter logic, along with a *range* of consecutive bins to check for a match. GRIM-Filter quickly checks the range of bins to determine whether or not to discard seed locations within those bins. In the logic layer, the GRIM-Filter Filter Bitmask Generator (see "Integration with a full read mapper" section) iterates through each token in read sequence r. For each token, GRIM-Filter reads the memory row in each vault that contains the existence bits for that token, for the bins being checked, into the row buffer inside the DRAM layer. Then, GRIM-Filter copies the row to the row data register in the logic layer. Each GRIM-Filter logic module is assigned to a single bin. The logic module examines the bin's existence bit in the row buffer, and the incrementer adds one to the value in the accumulator only if the existence bit is set. This process is repeated for all tokens in r. Once all of the tokens are processed, each logic module uses its comparator to check if the accumulator, which now holds the accumulated sum (Sum_z, shown as ❹ in Fig. 3) for its assigned bin, is greater than or equal to the accumulated sum threshold. If Sum_z is greater than or equal to the threshold, a *seed location filter bit* is set, indicating that the read sequence should be sequence aligned with the locations in the bin by the read mapper. To maintain the same amount of parallelism present in the bitvector operations, we place the seed location filter bits into a *seed location filter bitmask*, where each logic module writes to one bit in the bitmask once it performs the accumulator sum threshold comparison. The seed location filter bitmask is then written to the DRAM layer. Once the Seed Location Checker (see "Integration with a full read mapper" section) starts executing in the CPU, it reads the seed location filter bitmasks from DRAM, and performs sequence alignment for only those bits whose seed location filter bits are set to 1.

Hardware overhead. The hardware overhead of our GRIM-Filter implementation in 3D-stacked memory depends on the available bandwidth b between a memory layer and the logic layer. In HBM2 [16], this bandwidth is 4096 bits per cycle across all vaults (i.e., each clock cycle, 4096 bits from a memory layer can be copied to the row data registers in the logic layer). GRIM-Filter exploits all of this parallelism completely, as we can place b GRIM-Filter logic modules (4096 modules for HBM2) across all vaults within the logic layer. In total, for an HBM2 memory, and for a read mapper that processes reads consisting of 100 base pairs, GRIM-Filter requires 4096 incrementer lookup tables (LUTs), 4096 seven-bit counters (a seven-bit counter can hold the maximum accumulator sum for a 100-base-pair read sequence), 4096 comparators, and enough buffer space to hold the seed location filter bitmasks. With a larger bandwidth between the logic and memory layers, we would be able to compute the seed location filter bits for more bins in parallel, but this would also incur a larger hardware overhead in the logic layer.

While the read mapper performs sequence alignment on seed locations specified by one seed location filter bitmask, GRIM-Filter generates seed location filter bitmasks for a different set of seed locations. We find that a bitmask buffer size of 512 KB (stored in DRAM) provides enough capacity to ensure that GRIM-Filter and the read mapper never stall due to a lack of buffer space.

The overall memory footprint (i.e., the amount of storage space required) of the bitvectors for a reference genome is calculated by multiplying the number of bins by the size of a single bin. In "Sensitivity to GRIM-Filter Parameters" section, we show how we find a set of parameters that results in an effective filter with a low memory footprint (3.8 GB).

We conclude that GRIM-Filter requires a modest and simple logic layer, which gives it an advantage over other seed location filtering algorithms that could be implemented in the logic layer.

Results

We first profile the reference human genome in order to 1) determine a range of parameters that are reasonable to use for GRIM-Filter. We determine the points of diminishing returns for several parameter values. This data is presented in "Sensitivity to GRIM-Filter Parameters" section. Using this preliminary data, we reduce the required experiments to a reasonable range of parameters. Our implementation of GRIM-Filter enables the variation of runtime parameters (number of bins, token size, error tolerance, etc.) within the ranges of values that we determine from our experimentation for the best possible results. We then quantitatively evaluate GRIM-Filter's improvement in false negative rate and mapper runtime over the baseline mrFAST with FastHASH ("Full mapper results" section).

Sensitivity to GRIM-Filter parameters

In order to determine a range for the parameters for our experiments, we ran a series of analyses on the fundamental characteristics of the human reference genome. We perform these initial experiments to 1) determine effective parameters for GRIM-Filter and 2) compute its *memory footprint*. The memory footprint of GRIM-Filter depends directly on the number of bins that we divide the reference genome into, since each bin requires a bitvector to hold the token existence bits. Since the bitvector must contain a Boolean entry for each permutation of the token of size n, each bitvector must contain 4^n bits. The total memory footprint is then obtained by multiplying the bitvector size by the number of bins. In this section, we sweep the number of bins, token size, and error tolerance of GRIM-Filter while considering the memory footprint. To understand how each of the different parameters affect the performance of GRIM-Filter, we study a sweep on the parameters with a range of values that result in a memory footprint under 16 GB (which is the current capacity of HBM2 on state-of-the-art devices [71]).

Average read existence Figure 8 shows how varying a number of different parameters affects the *average read existence* across the bins. We define average read existence to be the ratio of bins with seed locations that pass the filter over all bins comprising the genome, for a representative set of reads. We would like this value to be as low as possible because it reflects the filter's ability to filter incorrect mappings. A lower average read existence means that fewer bins must be checked when mapping the representative set of reads. Across the three plots, we vary the token size from 4 to 6. Within each plot, we vary the number of bins to split the reference genome into, denoted by the different curves (with different colors and markers). The X-axis shows the error tolerance that is used, and the Y-axis shows average read existence. We plot the average and min/max across our 10 data sets (Table 1) as indicated, respectively, by the triangle and whiskers. We sweep the number of bins in multiples of 2^{16} because 2^{16} is an even multiple of the number of TSVs between the logic and memory layers in today's 3D-stacked memories (today's systems typically have 4096 TSVs). We want to use a multiple of 2^{16} so that we can utilize all TSVs each time we copy data from a row buffer in the memory layer to the corresponding row data register in the logic layer. This maximizes GRIM-Filter's internal memory bandwidth utilization within 3D-stacked memory.

We make *three* observations from the figure. First, looking across the three plots, we observe that increasing the *token size* from 4 to 5 provides a large (i.e., around 10x) reduction in average read existence, while increasing the token size from 5 to 6 provides a much smaller (i.e., around 2x) reduction in average read existence. The reduction in average read existence is due to the fact that, in a random pool of As, Cs, Ts, and Gs, the probability of observing a certain substring of size q is $\left(\frac{1}{4}\right)^q$. Because the distribution of base pairs across a reference genome and across a bin is *not* random, a larger token size does *not* always result in a large decrease, as seen when changing the token size from 5 to 6. We note that increasing the token size by one causes GRIM-Filter to use 4x the memory footprint. Second, we observe that in all three plots (i.e., for all token sizes), an increase in the number of bins results in a decrease in the average read existence. This is because the bin size decreases as the number of bins increases, and for smaller bins, we have a smaller sample size of the reference genome that any given substring could exist within. Third, we observe that for each plot, increasing the error tolerance results in an increase in the average read existence. This is due to the fact that if we allow more errors, fewer tokens of the entire read sequence must be present in a bin for a seed location from that bin to pass the filter. This increases the probability that a seed location of a random read passes the filter for a random bin. A poor sequence alignment at a location that passes the filter is categorized as a false negative. We conclude from this figure that using tokens of size 5 provides quite good filtering effectiveness (as measured by average read existence) without requiring as much memory footprint as using a token size of 6.

False negative rate. We choose our final bitvector size after sweeping the number of bins and the error tolerance (e). Figure 9 shows how varying these parameters affects the false negative rate of GRIM-Filter. The X-axis varies the number of bins, while the different lines represent different values of e.

We make two observations from this figure. First, we find that, with more bins (i.e., with a smaller bin size), the false negative rate (i.e., the fraction of locations that pass

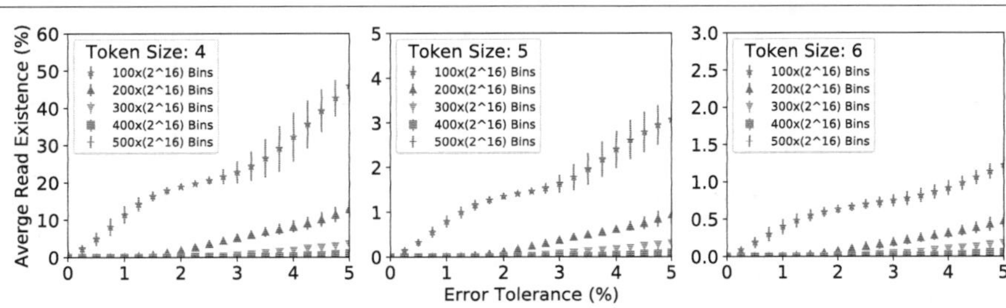

Fig. 8 Effect of varying token size, error tolerance, and bin count on average read existence. We use a representative set of reads to collect this data. A lower value of average read existence represents a more effective filter. Note that the scale of the Y-axis is different for the three different graphs

Table 1 Benchmark data, obtained from the 1000 Genomes Project [102]

	ERR240726_1	ERR240727_1	ERR240728_1	ERR240729_1	ERR240730_1
No. of reads	4031354	4082203	3894290	4013341	4082472
Read length	100	100	100	100	100
	ERR240726_2	ERR240727_2	ERR240728_2	ERR240729_2	ERR240730_2
No. of reads	4389429	4013341	4013341	4082472	4082472
Read length	100	100	100	100	100

the filter, but do not result in a mapping after alignment) decays exponentially. Above 300×2^{16} bins, we begin to see diminishing returns on the reduction in false negatives for all error tolerance values. Second, we observe that, as we increase the error tolerance, regardless of the other parameters, the false negative rate increases. We also find that the number of bins 1) minimally affects the runtime of GRIM-Filter (not plotted) and 2) linearly increases the memory footprint. Based on this study, we choose to use 450×2^{16} bins, which reflects a reasonable memory footprint (see below) with the other parameters.

Memory footprint. A larger number of bins results in more bitvectors, so we must keep this parameter at a reasonable value in order to retain a reasonable memory footprint for GRIM-Filter. Since we have chosen a token size of 5, GRIM-Filter requires t bitvectors with a length of $4^5 = 1024$, where t equals the number of bins we segment the reference genome into. We conclude that employing 450×2^{16} bins results in the best trade-off between memory footprint, filtering efficiency, and runtime. This set of parameters results in a total memory footprint of approximately 3.8 GB for storing the bitvectors of this mechanism, which is a very reasonable size for today's 3D-stacked memories [13, 15, 17, 18, 69–71]. We note that

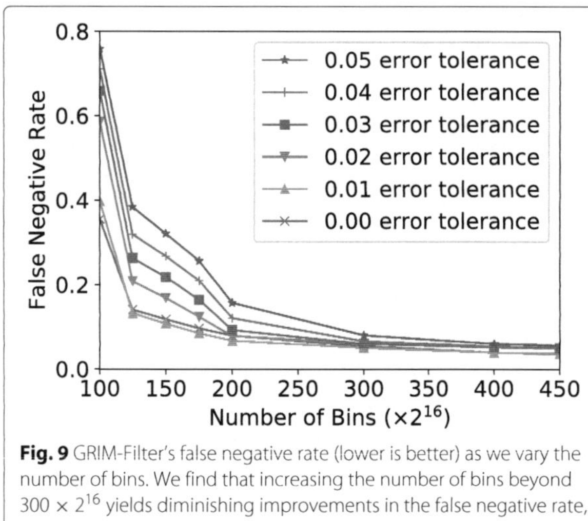

Fig. 9 GRIM-Filter's false negative rate (lower is better) as we vary the number of bins. We find that increasing the number of bins beyond 300×2^{16} yields diminishing improvements in the false negative rate, regardless of the error tolerance value

the time to generate the bitvectors is *not* included in our final runtime results, because these need to be generated only once per reference genome, either by the user or by the distributor. We find that, with a genome of length L, we can generate the bitvectors in $(9.03e-08) \times L$ seconds when we use 450×2^{16} bins (this is approximately 5 min for the human genome).

GRIM-Filter parallelization. GRIM-Filter operates on every bin independently and in parallel, using a separate logic module for each bin. Thus, GRIM-Filter's parallelism increases with each additional bin it operates on simultaneously. We refer to the set of consecutive bins that the GRIM-Filter logic modules are currently assigned to as the *bin window* (w). The internal bandwidth of HBM2 [16] enables copying 4096 bits from a memory layer to the logic layer every cycle, allowing GRIM-Filter to operate on as many as 4096 consecutive bins in parallel (i.e., it has a bin window of size $w = 4096$). GRIM-Filter must only check bin windows that contain at least one seed location (i.e., a span of 4096 consecutive bins with *zero* seed locations does not need to be checked). In contrast, if a consecutive set of 4096 bins contains many seed locations, GRIM-Filter can operate on every bin in parallel and quickly determine which seed locations within the 4096 bins can safely be discarded. In these cases, GRIM-Filter can most effectively utilize the parallelism available from the 4096 independent logic modules.

In order to understand GRIM-Filter's ability to parallelize operations on many bins, we analyze GRIM-Filter when using a bin window of size $w = 4096$, which takes advantage of the full memory bandwidth available in HBM2 memory. As we discuss in "Integration with a full read mapper" section, the read mapper generates a list of potential seed locations for a read sequence, and sends this list to GRIM-Filter when the filter starts. Several bins, which we call *empty bins*, do not contain *any* potential seed locations. When $w = 1$, there is only one logic module, and if the module is assigned to an empty bin, GRIM-Filter immediately moves on to the next bin without computing the accumulation sum. However, when $w = 4096$, some, but not all, of the logic modules may be assigned to empty bins. This happens because in order to simplify the hardware, GRIM-Filter operates all of the logic modules in lockstep (i.e., the filter fetches a *single row* from each bank of memory, which includes the existence bits *for a single token* across multiple rows, and all of the logic modules read and process the existence bits for the same token in the same cycle). Thus, a logic module assigned to an empty bin must wait for the other logic modules to finish before it can move onto another bin. As a result, GRIM-Filter with $w = 4096$ is not 4096x faster than GRIM-Filter with $w = 1$. To quantify the benefits of parallelization, we compare the performance of GRIM-Filter with these two bin window sizes using a representative set of reads. For 10% of the seeds, we find that GRIM-Filter with $w = 4096$ reduces the filtering time by 98.6%, compared to GRIM-Filter with $w = 1$. For the remaining seeds, we find that GRIM-Filter with $w = 4096$ reduces the filtering time by 10–20%. Thus, even though many of the logic modules are assigned to empty bins in a given cycle, GRIM-Filter reduces the filtering time by operating on many bins that contain potential seed locations in parallel.

Overlapping GRIM-Filter computation with sequence alignment in the CPU. In addition to operating on multiple bins in parallel, one benefit of implementing GRIM-Filter in 3D-stacked memory is that filtering operations can be parallelized with sequence alignment that happens on the CPU, since filtering no longer uses the CPU. Every cycle, for a bin window of size $w = 4096$, GRIM-Filter's Filter Bitmask Generator (❶ in Fig. 4) reads 4096 bits from memory, and updates the accumulation sums for the bins within the bin window that contain a potential seed location. Once the accumulation sums are computed and compared against the threshold, GRIM-Filter's Seed Location Checker (❷ in Fig. 4) can discard seed locations that map to bins whose accumulation sums do *not* meet the threshold (i.e., the seed locations that should not be sent to sequence alignment). The seed locations that are not discarded are sent to the read mapper for sequence alignment (❹ in Fig. 4), ending GRIM-Filter's work for the current bin window. While the read mapper aligns the sequences that passed through the filter from the completed bin window, GRIM-Filter's Filter Bitmask Generator moves onto another bin window, computing the seed location filter bits for a new set of bins. If GRIM-Filter can exploit enough parallelism, it can provide the CPU with enough bins to keep the sequence alignment step busy for at least as long as the time needed for the Filter Bitmask Generator to process the new bin window. This would allow the filtering latency to overlap *completely* with alignment, in effect hiding GRIM-Filter's latency. We find that a bin window of 4096 bins provides enough parallelism to *completely* hide the filtering latency while the read mapper running on the CPU performs sequence alignment.

Full mapper results

We use a popular seed-and-extend mapper, mrFAST [3], to retrieve all candidate mappings from the ten real data sets we evaluate (see "Methods" section). In our experiments, we use a token size of 5 and 450×2^{16} bins, as discussed in *Sensitivity to GRIM-Filter Parameters.* All remaining parameters specific to mrFAST are held at the default values across all of our evaluated read mappers.

False negative rate. Figure 10 shows the false negative rate of GRIM-Filter compared to the baseline FastHASH filter across the ten real data sets we evaluate. The six plots in the figure show false negative rates for error tolerance values (i.e., e) ranging from 0.00 to 0.05, in increments of 0.01 (An error tolerance of $e = 0.05$ is widely used in alignment during DNA read mapping [2, 8, 41, 42].). We make three observations from the figure. First, GRIM-Filter provides a much lower false negative rate than the baseline FastHASH filter for all data sets and for all error tolerance values. For an error tolerance of $e = 0.05$ (shown in the bottom graph), the false negative rate for GRIM-Filter is *5.97x* lower than for FastHASH filter, averaged across all 10 read data sets. Second, GRIM-Filter's false negative rate 1) increases as the error tolerance increases from $e = 0.00$ to $e = 0.02$, and then 2) decreases as the error tolerance increases further from $e = 0.03$ to $e = 0.05$. There are at least two conflicting reasons. First, as the error tolerance increases, the accumulation sum threshold decreases (as shown in Fig. 5) and thus GRIM-Filter discards *fewer* locations, which results in a *higher* false negative rate. Second, as the error tolerance increases, the number of acceptable (i.e., correct) mapping locations increases while the number of candidate locations remains the same, which results in a *lower* false negative rate. The interaction of these conflicting reasons results in the initial increase and the subsequent decrease in the false negative rates that we observe. Third, we observe that for higher error tolerance values, GRIM-Filter reduces the false negative rate compared to the FastHASH filter by a larger fraction. This shows that GRIM-Filter is much more effective at filtering mapping locations when we increase the error tolerance. We conclude that GRIM-Filter is very effective in reducing the false negative rate.

Execution time. Figure 11 compares the execution time of GRIM-3D to that of mrFAST with FastHASH across all ten different read data sets for the same error tolerance values used in Fig. 10. We make three observations. First, GRIM-3D improves performance for all of our data sets for all error tolerance values. For an error tolerance of $e = 0.05$, the average (maximum) performance improvement is 2.08x (3.65x) across all 10 data sets. Second, as the error tolerance increases, GRIM-3D's performance improvement also increases. This is because GRIM-Filter

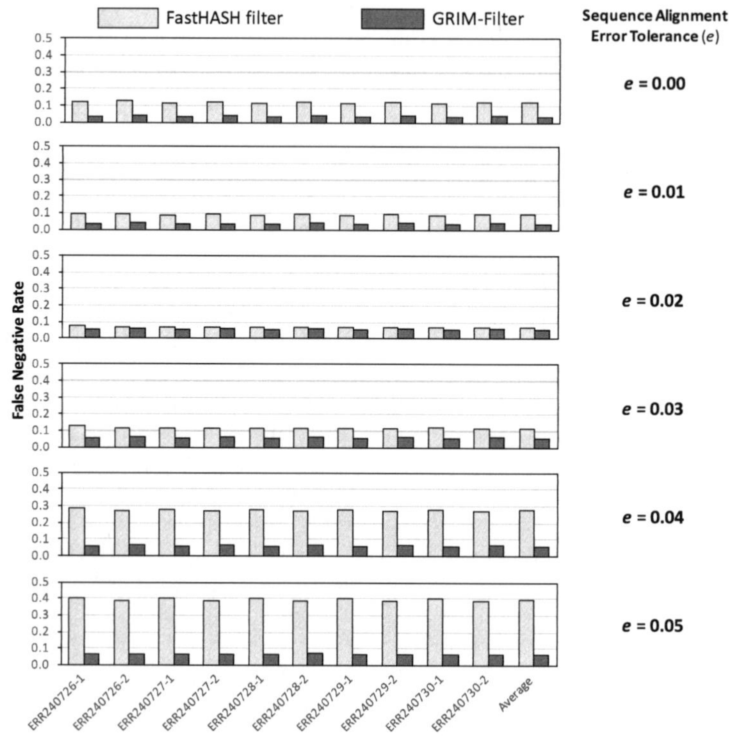

Fig. 10 False negative rates of GRIM-Filter and FastHASH filter across ten real data sets for six different error tolerance values

safely discards many more mapping locations than the FastHASH filter at higher error tolerance values (as we showed in Fig. 10). Thus, GRIM-Filter saves significantly more execution time than the FastHASH filter by ignoring many more unnecessary alignments. Third, based on an analysis of the execution time breakdown of GRIM-3D (not shown), we find that GRIM-3D's performance gains are mainly due to an 83.7% reduction in the average computation time spent on false negatives, compared to using the FastHASH filter for seed location filtering. We conclude that employing GRIM-Filter for seed location filtering in a state-of-the-art read mapper significantly improves the performance of the read mapper.

Related work

To our knowledge, this is the first paper to exploit 3D-stacked DRAM and its processing-in-memory capabilities to implement a new seed location filtering algorithm that mitigates the major bottleneck in read mapping, pre-alignment (i.e., seed location filtering). In this section, we briefly describe related works that aim to 1) accelerate pre-alignment algorithms, and 2) accelerate sequence alignment with hardware support.

Accelerating Pre-Alignment. A very recent prior work [9] implements a seed location filter in an FPGA, and shows significant speedup against prior filters. However, as shown in that work, the FPGA is still limited by the memory bandwidth bottleneck. GRIM-Filter can overcome this bottleneck on an FPGA as well.

Accelerating sequence alignment. Another very recent prior work [98] exploits the high memory bandwidth and the reconfigurable logic layer of 3D-stacked memory to implement an accelerator for sequence alignment (among other basic algorithms within the sequence analysis pipeline). Many prior works (e.g., [26–36]) use FPGAs to also accelerate sequence alignment. These works accelerate sequence alignment using customized FPGA implementations of different existing read mapping algorithms. For example, Arram et al. [28] accelerate the SOAP3 tool on an FPGA engine, achieving up to 134x speedup compared to BWA [99]. Houtgast et al. [32] present an FPGA-accelerated version of BWA-MEM that is 3x faster compared to its software implementation. Other works use GPUs [37–40] for the same purpose of accelerating sequence alignment. For example, Liu et al. [38] accelerate BWA and Bowtie by 7.5x and 20x, respectively. In contrast to GRIM-Filter, all of these accelerators focus on accelerating sequence alignment, whereas GRIM-Filter accelerates pre-alignment (i.e., seed location filtering). Hence, GRIM-Filter is orthogonal to these works, and can

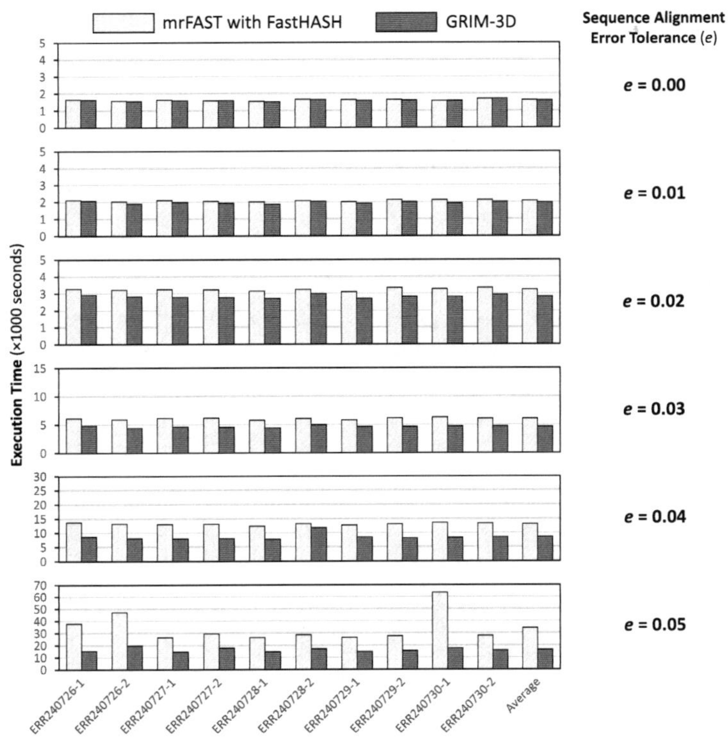

Fig. 11 Execution time of two mappers, GRIM-3D and mrFAST with FastHASH, across ten real data sets for six different error tolerance values Note that the scale of the Y-axis is different for the six different graphs

be combined with any of them for further performance improvement.

Discussion

We have shown that GRIM-Filter significantly reduces the execution time of read mappers by reducing the number of unnecessary sequence alignments and by taking advantage of processing-in-memory using 3D-stacked DRAM technology. We believe there are many other possible applications for employing 3D-stacked DRAM technology within the genome sequence analysis pipeline (as initially explored in [98]), and significant additional performance improvements can be obtained by combining future techniques with GRIM-Filter. Because GRIM-Filter is essentially a seed location filter to be employed before sequence alignment during read mapping, it can be used in any other read mapper along with any other acceleration mechanisms in the genome sequence analysis pipeline.

We identify three promising major future research directions. We believe it is promising to 1) explore the benefits of combining GRIM-Filter with other various read mappers in the field, 2) show the effects of mapping to varying sizes of reference genomes, and 3) examine how GRIM-Filter can scale to process a greater number of reads concurrently.

Conclusion

This paper introduces GRIM-Filter, a novel algorithm for seed location filtering, which is a critical performance bottleneck in genome read mapping. GRIM-Filter has three major novel aspects. First, it preprocesses the reference genome to collect metadata on large subsequences (i.e., *bins*) of the genome and stores information on whether small subsequences (i.e., *tokens*) are present in each bin. Second, GRIM-Filter efficiently operates on the metadata to quickly determine whether to discard a mapping location for a read sequence prior to an expensive sequence alignment, thereby reducing the number of unnecessary alignments and improving performance. Third, GRIM-Filter takes advantage of the logic layer within 3D-stacked memory, which enables the efficient use of processing-in-memory to overcome the memory bandwidth bottleneck in seed location filtering. We examine the trade-offs for various parameters in GRIM-Filter, and present a set of parameters that result in significant performance improvement over the state-of-the-art seed location filter, FastHASH. When running with a sequence alignment error tolerance of 0.05, we show that GRIM-Filter 1) filters seed locations with *5.59x–6.41x* lower false negative rates than FastHASH; and 2) improves the performance of the fastest read mapper, mrFAST with FastHASH, by *1.81x–3.65x*. GRIM-Filter is a universal seed location filter that can be applied to any read mapper.

We believe there is a very promising potential in designing DNA read mapping algorithms for new memory technologies (like 3D-stacked DRAM) and new processing paradigms (like processing-in-memory). We hope that the results from our paper provide inspiration for other works to design new sequence analysis and other bioinformatics algorithms that take advantage of new memory technologies and new processing paradigms, such as processing-in-memory using 3D-stacked DRAM.

Methods

Evaluated read mappers. We evaluate our proposal by incorporating GRIM-Filter into the state-of-the-art hash table based read mapper, mrFAST with FastHASH [7]. We choose this mapper for our evaluations as it provides high accuracy in the presence of relatively many errors, which is required to detect genomic variants within and across species [3, 7]. GRIM-Filter plugs in as an extension to mrFAST, using a simple series of calls to an application programming interface (API). However, we note that GRIM-Filter can be used with any other read mapper.

We evaluate two read mappers:

- *mrFAST with FastHASH* [7], which does *not* use GRIM-Filter;
- *GRIM-3D*, our 3D-stacked memory implementation of GRIM-Filter combined with mrFAST and the non-filtering portions of FastHASH.

Major evaluation metrics. We report 1) GRIM-Filter's *false negative rate* (i.e., the fraction of locations that pass through the filter but do not contain a match with the read sequence), and 2) the end-to-end *performance improvement* of the read mapper when using GRIM-Filter.

We measure the false negative rate of our filter (and the baseline filter used by the mapper) as the ratio of the number of locations that passed the filter but did not result in a mapping over all locations that passed the filter. Note that our implementation of GRIM-Filter ensures a zero *false positive rate* (i.e., it does *not* filter out any correct mappings for the read sequence), and, thus, GRIM-Filter does not affect the correctness of a read mapper.

Performance evaluation. We measure the performance improvement of GRIM-3D by comparing the execution time of our read mappers. We develop a methodology to *estimate* the performance of GRIM-3D, since real hardware systems that enable in-memory computation are unavailable to us at this point in time. To estimate GRIM-3D's execution time, we need to add up the time spent by three components (which we denote as t_x for component x):

- t_1: the time spent on read mapping,
- t_2: the time spent on coordinating which bins are examined by GRIM-Filter, and
- t_3: the time spent on applying the filter to each seed.

To obtain t_1 and t_2, we measure the performance of *GRIM-Software*, a software-only version of GRIM-Filter that does *not* take advantage of processing in 3D-stacked memory. We run GRIM-Software with mrFAST, and measure:

- *GRIM-Software-End-to-End-Time*, the end-to-end execution time for read mapping using GRIM-Software;
- *GRIM-Software-Filtering-Time*, the time spent only on applying the filter (i.e., the GRIM-Filter portions of the code shown in Fig. 4) using GRIM-Software.

The values of t_1 and t_2 are the same for GRIM-Software and GRIM-3D, and we can compute those by subtracting out the time spent on filtering from the end-to-end execution time: $t_1 + t_2 =$ GRIM-Software-End-to-End-Time − GRIM-Software-Filtering-Time. To estimate t_3, we use a validated simulator similar to Ramulator [47, 100], which provides us with the time spent by GRIM-3D on filtering using processing-in-memory. The simulator models the time spent by the in-memory logic to produce a seed location filter bitmask, and to store the bitmask into a buffer that is accessible by the read mapper.

Evaluation system. We evaluate the software versions of the read mappers (i.e., mrFAST with FastHASH and GRIM-Software) using an Intel(R) Core i7-2600 CPU running at 3.40 GHz [101], with 16 GB of DRAM for all experiments.

Data sets. We used ten real data sets from the 1000 Genomes Project [102]. We used the same data sets used by Xin et al. [7] for the original evaluation of mrFAST with FastHASH, in order to provide a fair comparison to our baseline. Table 1 lists the read length and size of each data set.

Code availability. The code for GRIM-Filter, GRIM-Software, and our simulator for 3D-stacked DRAM with processing-in-memory is freely available at https://github.com/CMU-SAFARI/GRIM.

Acknowledgements
An earlier version of this paper appears on arXiv.org [103]. An earlier version of this work was presented as a short talk at RECOMB-Seq [104]. We thank the anonymous reviewers for feedback.

Funding
This work was supported in part by a grant from the National Institutes of Health to O. Mutlu and C. Alkan (HG006004), the Semiconductor Research Corporation, and gifts from Google, Intel, Samsung, and VMware. Funding for the publication of this article was provided by a gift from Samsung.

Availability of data and materials
The datasets generated and/or analysed during the current study are available in the US National Library of Medicine, https://www.ncbi.nlm.nih.gov.
The code is available online at: https://github.com/CMU-SAFARI/GRIM.

About this supplement
This article has been published as part of *BMC Genomics* Volume 19 Supplement 2, 2018: Selected articles from the 16th Asia Pacific Bioinformatics Conference (APBC 2018): genomics. The full contents of the supplement are available online at https://bmcgenomics.biomedcentral.com/articles/supplements/volume-19-supplement-2.

Authors' contributions
JSK, DSC, and HX developed the GRIM-Filter algorithm. JSK and DSC implemented and applied GRIM-Filter to simulation experiments and real data. JSK, DSC, and DL developed and validated the processing-in-memory simulator. OM and CA conceived and planned the experiments, and supervised JSK, DSC, and HX for the algorithm development. All authors contributed to the writing of the manuscript. All authors read and approved the final manuscript.

Ethics approval and consent to participate
Not applicable.

Consent for publication
Not applicable.

Competing interests
The authors declare that they have no competing interests.

Publisher's Note

Springer Nature remains neutral with regard to jurisdictional claims in published maps and institutional affiliations.

Author details
[1]Department of Electrical and Computer Engineering, Carnegie Mellon University, Pittsburgh, PA, USA. [2]Department of Computer Science, Carnegie Mellon University, Pittsburgh, PA, USA. [3]NVIDIA Research, Austin, TX, USA. [4]Department of Computer Engineering, Bilkent University, Bilkent, Ankara, Turkey. [5]Department of Computer Engineering, TOBB University of Economics and Technology, Sogutozu, Ankara, Turkey. [6]Department of Computer Science, ETH Zürich, Zürich, CH, Switzerland.

Published: 9 May 2018

References

1. Hach F, Sarrafi I, Hormozdiari F, Alkan C, Eichler EE, Sahinalp SC. mrsFAST-Ultra: a compact, SNP-aware mapper for high performance sequencing applications. Nucleic Acids Res. 2014;42.W1:W494–W500.
2. Ahmadi A, Behm A, Honnalli N, Li C, Weng L, Xie X. Hobbes: Optimized Gram-Based Methods for Efficient Read Alignment. Nucleic Acids Res. 2012;40(6):e41–e41.
3. Alkan C, Kidd JM, Marques-Bonet T, Aksay G, Antonacci F, Hormozdiari F, Kitzman JO, Baker C, Malig M, Mutlu O, et al. Personalized Copy Number and Segmental Duplication Maps Using Next-Generation Sequencing. Nat Genet. 2009;41(10):1061–7.
4. Rumble SM, Lacroute P, Dalca AV, Fiume M, Sidow A, Brudno M. SHRiMP: accurate mapping of short color-space reads. PLoS Comput Biol. 2009;5.5:e1000386.
5. Hormozdiari F, Hach F, Sahinalp SC, Eichler EE, Alkan C. Sensitive and Fast Mapping of Di-Base Encoded Reads. Bioinformatics. 2011;27(14): 1915–21.
6. Weese D, Emde A-K, Rausch T, Döring A, Reinert K. RazerS—Fast Read Mapping with Sensitivity Control. Genome Res. 2009;19(9):1646–54.
7. Xin H, Lee D, Hormozdiari F, Yedkar S, Mutlu O, Alkan C. Accelerating Read Mapping with FastHASH. BMC Genomics. 2013;14(Suppl 1):S13.
8. Xin H, Greth J, Emmons J, Pekhimenko G, Kingsford C, Alkan C, Mutlu O. Shifted Hamming distance: a fast and accurate SIMD-friendly filter to accelerate alignment verification in read mapping. Bioinformatics. 2015;31.10:1553–1560.
9. Alser M, Hassan H, Xin H, Ergin O, Mutlu O, Alkan C. GateKeeper: a new hardware architecture for accelerating pre-alignment in DNA short read mapping. Bioinformatics. 2017;33.21:3355–3363.
10. Alser M, Mutlu O, Alkan C. MAGNET: Understanding and Improving the Accuracy of Genome Pre-Alignment Filtering. IPSI Trans Internet Res. 2017;13:33–42.
11. Hieu Tran N, Chen X. AMAS: optimizing the partition and filtration of adaptive seeds to speed up read mapping. IEEE/ACM Trans Comput Biol Bioinforma (TCBB). 2016;13.4:623–633. arXiv:1502.05041.
12. Xin H, Nahar S, Zhu R, Emmons J, Pekhimenko G, Kingsford C, Alkan C, Mutlu O. Optimal seed solver: optimizing seed selection in read mapping. Bioinformatics. 2015;32.11:1632–42.
13. Advanced Micro Devices, Inc. High Bandwidth Memory | Reinventing Memory Technology. http://www.amd.com/en-us/innovations/software-technologies/hbm. Accessed 26 Jan 2016.
14. Lee D, Ghose S, Pekhimenko G, Khan S, Mutlu O. Simultaneous multi-layer access: Improving 3D-stacked memory bandwidth at low cost. ACM Trans Archit Code Optim (TACO). 2016;12.4:63.
15. Advanced Micro Devices, Inc. AMD Radeon™ R9 Series Graphics Cards with High-Bandwidth Memory. http://www.amd.com/en-us/products/graphics/desktop/r9/. Accessed 26 Jan 2016.
16. O'Connor M. Highlights of the High-Bandwidth Memory (HBM) Standard. In: The Memory Forum. 2014.
17. Altera Corporation. Hybrid Memory Cube Controller IP Core User Guide. https://www.altera.com/en_US/pdfs/literature/ug/ug_hmcc.pdf. Accessed 26 Jan 2016.
18. Hybrid Memory Cube Consortium. Hybrid Memory Cube Member Tool Resources. http://hybridmemorycube.org/tool-resources.html. Accessed 26 Jan 2016.
19. Hsieh K, Ebrahimi E, Kim G, Chatterjee N, O'Connor M, Vijaykumar N, Mutlu O, Keckler SW. Transparent Offloading and Mapping (TOM): Enabling Programmer-Transparent Near-Data Processing in GPU Systems. In: International Symposium on Computer Architecture. IEEE Press; 2016. p. 204–16.
20. Hsieh K, Khan S, Vijaykumar N, Chang KK, Boroumand A, Ghose S, Mutlu O. Accelerating pointer chasing in 3D-stacked memory: Challenges, mechanisms, evaluation. In: Computer Design (ICCD), 2016 IEEE 34th International Conference on. IEEE; 2016. p. 25–32.
21. Ahn J, Hong S, Yoo S, Mutlu O, Choi K. A scalable processing-in-memory accelerator for parallel graph processing. In: Computer Architecture (ISCA), 2015 ACM/IEEE 42nd Annual International Symposium on. IEEE; 2015. p. 105–17.
22. Loh GH. 3D-Stacked Memory Architectures for Multi-Core Processors. Int Symp Comput Archit. 2008;36:453–64.
23. Seshadri V, Lee D, Mullins T, Hassan H, Boroumand A, Kim J, Kozuch MA, Mutlu O, Gibbons PB, Mowry TC. Ambit: In-memory accelerator for bulk bitwise operations using commodity DRAM technology. In: Proceedings of the 50th Annual IEEE/ACM International Symposium on Microarchitecture. ACM; 2017. p. 273–287.
24. David M, Dursi LJ, Yao D, Boutros PC, Simpson JT. Nanocall: An Open Source Basecaller for Oxford Nanopore Sequencing Data. Bioinformatics. 2016;33(1):49–55.
25. Senol D, Kim J, Ghose S, Alkan C, Mutlu O. Nanopore Sequencing Technology and Tools: Computational Analysis of the Current State, Bottlenecks and Future Directions. In: Pacific Symposium on Biocomputing Poster Session. 2017.
26. Aluru S, Jammula N. A Review of Hardware Acceleration for Computational Genomics. IEEE Des Test. 2014;31(1):19–30.
27. Arram J, Tsoi KH, Luk W, Jiang P. Hardware acceleration of genetic sequence alignment. In: International Symposium on Applied Reconfigurable Computing. Berlin, Heidelberg: Springer; 2013. p. 13–24.
28. Arram J, Tsoi KH, Luk W, Jiang P. Reconfigurable Acceleration of Short Read Mapping. In: International Symposium on Field-Programmable Custom Computing Machines; 2013. p. 210–217.

29. Ashley EA, Butte AJ, Wheeler MT, Chen R, Klein TE, Dewey FE, Dudley JT, Ormond KE, Pavlovic A, Morgan AA, et al. Clinical Assessment Incorporating a Personal Genome. The Lancet. 2010;375(9725):1525–35.
30. Chiang J, Studniberg M, Shaw J, Seto S, Truong K. Hardware accelerator for genomic sequence alignment. In: Engineering in Medicine and Biology Society, 2006. EMBS'06. 28th Annual International Conference of the IEEE. IEEE; 2006. p. 5787–9.
31. Hasan L, Al-Ars Z, Vassiliadis S. Hardware acceleration of sequence alignment algorithms-an overview. In: Design & Technology of Integrated Systems in Nanoscale Era, 2007. DTIS. International Conference on. IEEE; 2007. p. 92–7.
32. Houtgast EJ, Sima V-M, Bertels K, Al-Ars Z. An FPGA-based systolic array to accelerate the BWA-MEM genomic mapping algorithm. In: Embedded Computer Systems: Architectures, Modeling, and Simulation (SAMOS), 2015 International Conference on. IEEE; 2015. p. 221–7.
33. McMahon PL. Accelerating Genomic Sequence Alignment Using High Performance Reconfigurable Computers. PhD thesis., Univ. of California, Berkeley. 2008.
34. Olson CB, Kim M, Clauson C, Kogon B, Ebeling C, Hauck S, Ruzzo WL. Hardware acceleration of short read mapping. In: Field-Programmable Custom Computing Machines (FCCM), 2012 IEEE 20th Annual International Symposium on. IEEE; 2012. p. 161–8.
35. Papadopoulos A, Kirmitzoglou I, Promponas VJ, Theocharides T. FPGA-Based Hardware Acceleration for Local Complexity Analysis of Massive Genomic Data. VLSI J Integr. 2013;46(3):230–9.
36. Waidyasooriya HM, Hariyama M, Kameyama M. FPGA-Accelerator for DNA Sequence Alignment Based on an Efficient Data-Dependent Memory Access Scheme. In: Highly-Efficient Accelerators and Reconfigurable Technologies. 2014. p. 127–30.
37. Blom J, Jakobi T, Doppmeier D, Jaenicke S, Kalinowski J, Stoye J, Goesmann A. Exact and Complete Short-Read Alignment to Microbial Genomes Using Graphics Processing Unit Programming. Bioinformatics. 2011;27(10):1351–8.
38. Liu C-M, Wong T, Wu E, Luo R, Yiu S-M, Li Y, Wang B, Yu C, Chu X, Zhao K, et al. SOAP3: Ultra-Fast GPU-Based Parallel Alignment Tool for Short Reads. Bioinformatics. 2012;28(6):878–9.
39. Luo R, Wong T, Zhu J, Liu C-M, Zhu X, Wu E, Lee L-K, Lin H, Zhu W, Cheung DW, et al. SOAP3-dp: fast, accurate and sensitive GPU-based short read aligner. PLoS One. 2013;8.5:e65632.
40. Manavski SA, Valle G. CUDA Compatible GPU Cards as Efficient Hardware Accelerators for Smith-Waterman Sequence Alignment. BMC Bioinformatics. 2008;9(Suppl. 2):S10.
41. Cheng H, Jiang H, Yang J, Xu Y, Shang Y. BitMapper: An Efficient All-Mapper Based on Bit-Vector Computing. BMC Bioinformatics. 2015;16(1):192.
42. Hatem A, Bozdağ D, Toland AE, Çatalyürek ÜV. Benchmarking Short Sequence Mapping Tools. BMC Bioinformatics. 2013;14(1):184.
43. Mutlu O, Stark J, Wilkerson C, Patt YN. Runahead execution: An effective alternative to large instruction windows. IEEE Micro. 2003;23.6(6):20–5.
44. Mutlu O, Subramanian L. Research Problems and Opportunities in Memory Systems. Supercomputing Frontiers and Innovations. 2014;1(3): 19–55.
45. Mutlu O. Memory Scaling: A Systems Architecture Perspective. In: International Memory Workshop. 2013. p. 21–5.
46. Ipek E, Mutlu O, Martínez JF, Caruana R. Self-optimizing memory controllers: A reinforcement learning approach. In: Computer Architecture, 2008. ISCA'08. 35th International Symposium on. IEEE; 2008. p. 39–50.
47. Kim Y, Yang W, Mutlu O. Ramulator: A fast and extensible DRAM simulator. IEEE Comput Archit Lett. 2016;15.1:45–49.
48. Mutlu O. Main Memory Scaling: Challenges and Solution Directions. In: More than Moore Technologies for Next Generation Computer Design. Springer; 2015. p. 127–53.
49. Kim Y, Mutlu O. Memory Systems. In: Computing Handbook, Third Edition: Computer Science and Software Engineering. 2014.
50. Kim Y, Seshadri V, Lee v, Liu J, Mutlu O. A Case for Exploiting Subarray-Level Parallelism (SALP) in DRAM. In: International Symposium on Computer Architecture. IEEE; 2012. p. 368–79.
51. Liu J, Jaiyen B, Veras R, Mutlu O. RAIDR: Retention-Aware Intelligent DRAM Refresh. In: International Symposium on Computer Architecture. IEEE; 2012.
52. Lee D, Kim Y, Seshadri V, Liu J, Subramanian L, Mutlu O. Tiered-Latency DRAM: A Low Latency and Low Cost DRAM Architecture. In: International Symposium on High-Performance Computer Architecture. IEEE; 2013.
53. Seshadri V, Mutlu O. Simple Operations in Memory to Reduce Data Movement. In: Advances in Computers. 2017.
54. Lee D, Subramanian L, Ausavarungnirun R, Choi J, Mutlu O. Decoupled Direct Memory Access: Isolating CPU and IO Traffic by Leveraging a Dual-Data-Port DRAM. In: International Conference on Parallel Architectures and Compilation Techniques. IEEE; 2015. p. 174–187.
55. Hassan H, Pekhimenko G, Vijaykumar N, Seshadri V, Lee D, Ergin O, Mutlu O. ChargeCache: Reducing DRAM Latency by Exploiting Row Access Locality. In: International Symposium on High-Performance Computer Architecture. IEEE; 2016. p. 581–93.
56. Lee D, Khan S, Subramanian L, Ghose S, Ausavarungnirun R, Pekhimenko G, Seshadri V, Mutlu O. Design-Induced Latency Variation in Modern DRAM Chips: Characterization, Analysis, and Latency Reduction Mechanisms. Proc ACM Meas Anal Comput Syst. 2017;1(1):26.
57. Lee D, Kim Y, Pekhimenko G, Khan S, Seshadri V, Chang K, Mutlu O. Adaptive-Latency DRAM: Optimizing DRAM Timing for the Common-Case. In: International Symposium on High-Performance Computer Architecture. IEEE; 2015. p. 489–501.
58. Kim Y, Daly R, Kim J, Fallin C, Lee JH, Lee D, Wilkerson C, Lai K, Mutlu O. Flipping Bits in Memory without Accessing Them: An Experimental Study of DRAM Disturbance Errors. In: International Symposium on Computer Architecture. IEEE; 2014.
59. Chang KK, Yağlikçi AG, Ghose S, Agrawal A, Chatterjee N, Kashyap A, Lee D, O'Connor M, Hassan H, Mutlu O. Understanding reduced-voltage operation in modern dram devices: Experimental characterization, analysis, and mechanisms. Proc ACM Meas Anal Comput Syst. 2017;1.1:10. ACM.
60. Chang KK, Kashyap A, Hassan H, Ghose S, Hsieh K, Lee D, Li T, Pekhimenko G, Khan S, Mutlu O. Understanding Latency Variation in Modern DRAM Chips: Experimental Characterization, Analysis, and Optimization. In: SIGMETRICS. ACM; 2016. p. 323–36.
61. Liu J, Jaiyen B, Kim Y, Wilkerson C, Mutlu O. An Experimental Study of Data Retention Behavior in Modern DRAM Devices: Implications for Retention Time Profiling Mechanisms. In: International Symposium on Computer Architecture. IEEE; 2013.
62. Hassan H, Vijaykumar N, Khan S, Ghose S, Chang K, Pekhimenko G, Lee D, Ergin O, Mutlu O. SoftMC: A Flexible and Practical Open-Source Infrastructure for Enabling Experimental DRAM Studies. In: International Symposium on High-Performance Computer Architecture. IEEE; 2017. p. 241–52.
63. Patel M, Kim JS, Mutlu O. The Reach Profiler (REAPER): Enabling the Mitigation of DRAM Retention Failures via Profiling at Aggressive Conditions. In: International Symposium on Computer Architecture. IEEE; 2017. p. 255–68.
64. Chang KK-W, Lee D, Chishti Z, Alameldeen AR, Wilkerson C, Kim Y, Mutlu O. Improving DRAM Performance by Parallelizing Refreshes with Accesses. In: International Symposium on High-Performance Computer Architecture. IEEE; 2014. p. 356–367.
65. Chang KK. Understanding and Improving the Latency of DRAM-Based Memory Systems. PhD thesis, Carnegie Mellon Univ. 2017.
66. Kim Y. Architectural Techniques to Enhance DRAM Scaling. PhD thesis, Carnegie Mellon Univ. 2015.
67. Lee D. Reducing DRAM Energy at Low Cost by Exploiting Heterogeneity. PhD thesis, Carnegie Mellon Univ. 2016.
68. Kim DH, Athikulwongse K, Lim SK. A Study of Through-Silicon-Via Impact on the 3D Stacked IC Layout. In: International Conference on Computer-Aided Design. IEEE; 2009. p. 674–80.
69. JEDEC Solid State Technology Association. High Bandwidth Memory (HBM) DRAM. Standard JESD235. 2013.
70. Advanced Micro Devices, Inc. Radeon^{TM} RX Vega^{64}. https://gaming.radeon.com/en/product/vega/radeon-rx-vega-64/. Accessed 14 Oct 2017.
71. NVIDIA Corporation. Tesla P100 Data Center Accelerator. http://www.nvidia.com/object/tesla-p100.html. Accessed 14 Oct 2017.
72. Yoshida T. SPARC64^{TM} XIfx: Fujitsu's Next Generation Processor for HPC. In: Hot Chips 26 Symposium. IEEE; 2014. p. 1–31.
73. Dlugosch P, Brown D, Glendenning P, Leventhal M, Noyes H. An Efficient and Scalable Semiconductor Architecture for Parallel Automata Processing. Trans Parallel Distrib Syst. 2014;25(12):3088–98.

74. Tibco. In-Memory Computing. http://www.tibco.com/products/automation/in-memory-computing. Accessed 26 Jan 2016.
75. Micron. Micron Automata Processing. http://www.micronautomata.com/hardware. Accessed 26 Jan 2016.
76. Ahn J, Yoo S, Mutlu O, Choi K. PIM-Enabled Instructions: a Low-overhead, Locality-aware Processing-in-Memory Architecture. In: International Symposium on Computer Architecture. IEEE; 2015. p. 336–48.
77. Seshadri V, Hsieh K, Boroumand A, Lee D, Kozuch M, Mutlu O, Gibbons P, Mowry T. Fast bulk bitwise AND and OR in DRAM. IEEE Comput Archit Lett. 2015;14.2:127–131.
78. Seshadri V, Kim Y, Fallin C, Lee D, Ausavarungnirun R, Pekhimenko G, Luo Y, Mutlu O, Gibbons PB, Kozuch MA, et al. RowClone: Fast and Energy-Efficient In-DRAM Bulk Data Copy and Initialization. In: International Symposium on Microarchitecture. IEEE; 2013. p. 185–97.
79. Seshadri V, Mullins T, Boroumand A, Mutlu O, Gibbons PB, Kozuch MA, Mowry TC. Gather-Scatter DRAM: In-DRAM Address Translation to Improve the Spatial Locality of Non-Unit Strided Accesses. In: International Symposium on Microarchitecture. IEEE; 2015. p. 267–80.
80. Liu Z, Calciu I, Herlihy M, Mutlu O. Concurrent Data Structures for Near-Memory Computing. In: Symposium on Parallelism in Algorithms and Architectures. ACM; 2017. p. 235–45.
81. Pattnaik A, Tang X, Jog A, Kayiran O, Mishra AK, Kandemir MT, Mutlu O, Das CR. Scheduling Techniques for GPU Architectures with Processing-in-Memory Capabilities. In: International Conference on Parallel Architectures and Compilation Techniques. IEEE; 2016. p. 31–44.
82. Babarinsa OO, Idreos S. JAFAR: near-data processing for databases. In: Proceedings of the 2015 ACM SIGMOD International Conference on Management of Data. ACM; 2015. p. 2069–70.
83. Farmahini-Farahani A, Ahn JH, Morrow K, Kim NS. NDA: Near-DRAM acceleration architecture leveraging commodity DRAM devices and standard memory modules. In: High Performance Computer Architecture (HPCA), 2015 IEEE 21st International Symposium on. IEEE; 2015. p. 283–95.
84. Gao M, Ayers G, Kozyrakis C. Practical Near-Data Processing for In-Memory Analytics Frameworks. In: International Conference on Parallel Architectures and Compilation Techniques. IEEE; 2015. p. 113–24.
85. Gao M, Kozyrakis C. HRL: Efficient and Flexible Reconfigurable Logic for Near-Data Processing. In: International Symposium on High-Performance Computer Architecture. IEEE; 2016. p. 126–37.
86. Hassan SM, Yalamanchili S, Mukhopadhyay S. Near Data Processing: Impact and Optimization of 3D Memory System Architecture on the Uncore. In: International Symposium on Memory Systems. IEEE; 2015. p. 11–21.
87. Morad A, Yavits L, Ginosar R. GP-SIMD Processing-in-Memory. Trans Archit Code Optim. 2015;11(4):53.
88. Sura Z, Jacob A, Chen T, Rosenburg B, Sallenave O, Bertolli C, Antao S, Brunheroto J, Park Y, O'Brien K, et al. Data Access Optimization in a Processing-in-Memory System. In: International Conference on Computing Frontiers. ACM; 2015.
89. Zhang D, Jayasena N, Lyashevsky A, Greathouse JL, Xu L, Ignatowski M. TOP-PIM: Throughput-Oriented Programmable Processing in Memory. In: International Symposium on High-Performance Parallel and Distributed Computing. ACM; 2014. p. 85–98.
90. Boroumand A, Ghose S, Lucia B, Hsieh K, Malladi K, Zheng H, Mutlu O. LazyPIM: An Efficient Cache Coherence Mechanism for Processing-in-Memory. Comput Archit Lett. 2017;16(1):46–50.
91. Chang KK, Nair PJ, Lee D, Ghose S, Qureshi MK, Mutlu O. Low-Cost Inter-Linked Subarrays (LISA): Enabling Fast Inter-Subarray Data Movement in DRAM. In: International Symposium on High-Performance Computer Architecture. IEEE; 2016. p. 568–80.
92. Akin B, Franchetti F, Hoe JC. Data Reorganization in Memory Using 3D-Stacked DRAM. In: International Symposium on Computer Architecture. IEEE; 2015. p. 131–43.
93. Guo Q, Alachiotis N, Akin B, Sadi F, Xu G, Low TM, Pileggi L, Hoe JC, Franchetti F. 3D-stacked memory-side acceleration: Accelerator and system design. In: Workshop on Near-Data Processing (WoNDP)(Held in conjunction with MICRO-47); 2014.
94. Zhu Q, Akin B, Sumbul HE, Sadi F, Hoe JC, Pileggi L, Franchetti F. A 3D-stacked logic-in-memory accelerator for application-specific data intensive computing. In: 3D Systems Integration Conference (3DIC), 2013 IEEE International. IEEE; 2013. p. 1–7.
95. Zhu Q, Graf T, Sumbul HE, Pileggi L, Franchetti F. Accelerating sparse matrix-matrix multiplication with 3D-stacked logic-in-memory hardware. In: High Performance Extreme Computing Conference (HPEC), 2013 IEEE. IEEE; 2013. p. 1–6.
96. Mutlu O, Moscibroda T. Parallelism-Aware Batch Scheduling: Enhancing Both Performance and Fairness of Shared DRAM Systems. In: International Symposium on Computer Architecture. IEEE; 2008.
97. Mutlu O, Moscibroda T. Stall-Time Fair Memory Access Scheduling for Chip Multiprocessors. In: International Symposium on Microarchitecture. IEEE; 2007.
98. Liu P, Hemani A, Paul K, Weis C, Jung M, Wehn N. 3D-Stacked Many-Core Architecture for Biological Sequence Analysis Problems. Int J Parallel Prog. 2017;45(6):1420–60.
99. Li H, Durbin R. Fast and Accurate Long-Read Alignment with Burrows–Wheeler Transform. Bioinformatics. 2010;26(5):589–95.
100. SAFARI Research Group Ramulator: A DRAM Simulator Source Code. https://github.com/CMU-SAFARI/ramulator. Accessed 26 Jan 2016.
101. Intel Corporation. Intel Core i7-2600 Processor. https://ark.intel.com/products/52213. Accessed 14 Oct 2017.
102. 1000 Genomes Project Consortium. An Integrated Map of Genetic Variation from 1,092 Human Genomes. Nature. 2012;491(7422):56–65.
103. Kim JS, Senol D, Xin H, Lee D, Ghose S, Alser M, Hassan H, Ergin O, Alkan C, Mutlu O. GRIM-Filter: Fast Seed Filtering in Read Mapping Using Emerging Memory Technologies. 2017. arXiv:1708.04329.
104. Kim JS, Senol D, Xin H, Lee D, Alser M, Hassan H, Ergin O, Alkan C, Mutlu O. Genome Read In-Memory (GRIM) Filter: Fast Location Filtering in DNA Read Mapping with Emerging Memory Technologies. 2016. Presentation at RECOMB Satellite Workshop on Massively Parallel Sequencing.

Zhou *et al. BMC Genomics* 2018, **19**(Suppl 2):88
DOI 10.1186/s12864-018-4461-z

BMC Genomics

RESEARCH

Open Access

Computational analysis of the receptor binding specificity of novel influenza A/H7N9 viruses

Xinrui Zhou[1], Jie Zheng[1,2], Fransiskus Xaverius Ivan[1], Rui Yin[1], Shoba Ranganathan[3], Vincent T. K. Chow[4] and Chee-Keong Kwoh[1*]

From The Sixteenth Asia Pacific Bioinformatics Conference 2018
Yokohama, Japan. 15-17 January 2018

Abstract

Background: Influenza viruses are undergoing continuous and rapid evolution. The fatal influenza A/H7N9 has drawn attention since the first wave of infections in March 2013, and raised more grave concerns with its increased potential to spread among humans. Experimental studies have revealed several host and virulence markers, indicating differential host binding preferences which can help estimate the potential of causing a pandemic. Here we systematically investigate the sequence pattern and structural characteristics of novel influenza A/H7N9 using computational approaches.

Results: The sequence analysis highlighted mutations in protein functional domains of influenza viruses. Molecular docking and molecular dynamics simulation revealed that the hemagglutinin (HA) of A/Taiwan/1/2017(H7N9) strain enhanced the binding with both avian and human receptor analogs, compared with the previous A/Shanghai/02/2013(H7N9) strain. The Molecular Mechanics - Poisson Boltzmann Surface Area (MM-PBSA) calculation revealed the change of residue-ligand interaction energy and detected the residues with conspicuous binding preference.

Conclusion: The results are novel and specific to the emerging influenza A/Taiwan/1/2017(H7N9) strain compared with A/Shanghai/02/2013(H7N9). Its enhanced ability to bind human receptor analogs, which are abundant in the human upper respiratory tract, may be responsible for the recent outbreak. Residues showing binding preference were detected, which could facilitate monitoring the circulating influenza viruses.

Keywords: Influenza A/H7N9, Host specificity, Receptor binding, Molecular docking, Molecular dynamics simulation

Background

Influenza A viruses are undergoing continuous and rapid evolution, leading to the concerns about an outbreak or even a pandemic [1]. There have been four major pandemics in recent centuries caused by mutations and reassortment of influenza A viruses, which enabled these pathogens to break the barrier of cross-species transmission and evade the human immune system [2, 3]. As documented by the World Health Organization (WHO), the co-circulating strains of influenza viruses are mainly A/H1N1-pdm09, A/H3N2 and B viruses [4]. In addition, the novel and lethal A/H7N9 has attracted attention with a high case-fatality rate of 40% since the first wave of human infections in late March 2013. At least 489 deaths have been reported to WHO among 1307 laboratory-confirmed human infections as of 16 March 2017 [5]. The unusual increase of infected cases compared to earlier waves since September 2016 has caught the urgent attention of health authorities [6]. The fact that infections of H7N9 in poultry are subclinical makes it challenging to surveil its spreading among poultry and to measure the risk of human infection [7]. Most cases have been identified following their exposure to live birds or live poultry markets. Even with small clusters of

* Correspondence: ASCKKWOH@ntu.edu.sg
[1]School of Computer Science and Engineering, Nanyang Technological University, Singapore 639798, Singapore
Full list of author information is available at the end of the article

human infection cases being detected, the WHO declared low likelihood of human-to-human transmission of avian influenza A/H7N9, with the support of current epidemiological and virological evidence [5].

It is of profound concern that the deadly H7N9 may evolve to enhance its ability to spread among humans, although no sustained evidence has been discovered yet. To date, it is not clear how the mutations of influenza viruses lead to the shift of host specificity, the increase of pathogenicity, and the ability of airborne transmission [8], all of which are taken as requirements for avian influenza to cause pandemics. Great efforts have been made in annotating signatures for cross-species transmission and increased virulence, which can facilitate early detection of potential pandemic strains [9]. However, identifying such genetic markers is under debate, since the adaptation, characterized by positive and repeated selection of mutations, can hardly be distinguished conclusively from other ecological and evolutionary processes that drive the mutations [10]. To address this issue, parallel adaptive animal models are commonly applied to analyzing the process, which can identify repeated and probably gain-of-function mutations [11, 12]. Thus, the parallel adaptive animal models serve as a useful yet expensive tool to provide a basis for understanding host jump. Experiments have shown differential abilities of influenza viruses to bind host-specific sialic acid (SIA) -linked receptors influenced by HA protein [13–15]. However, the mechanisms of adaptive mutation leading to host-specificity shift still need to be elucidated.

With the development of high-throughput sequencing technology and the enriched protein structure databases, it is possible to computationally analyze the sequences and protein structures, and thereby optimize therapies [16]. Su et al. combined molecular docking and molecular dynamics simulation to analyze the conformational changes of drugs bound to neuraminidase (NA) of H7N9 under the mutation *R289K* [17]. Similarly, Pan et al. and Kannan et al. investigated the binding preferences of A/H1N1 HA protein for different host cell receptors, giving insight into the virulence enhanced by the HA mutation D222G [18, 19]. As to the emerging influenza A/H7N9 outbreak, one case declaring no exposure to any live bird, live poultry market or suspicious patient caught our attention (reported on 22 February 2017) [20]. The isolated strain of this patient is A/Taiwan/1/2017(H7N9), denoted as TW17. Here we intend to computationally and systematically investigate this new strain. We conducted both sequence and structure analyses, aiming to explore its potential to gain the ability to spread among humans. Our preliminary sequence analysis shows that the HA protein bears the most mutations among all viral proteins. Therefore, we pointed out the mutations in the HA protein and their corresponding functional domains. Besides, we analyzed the viral binding preference and the residue contributions, giving insight into how the mutations could change the protein structure, especially structures of the functional domains of the protein. First, we generated the HA structure of representative A/H7N9 strains by homology modelling. Then, molecular docking was applied to find the most favorable complexes with avian and human host receptor analogs respectively. Further, molecular dynamics simulation was conducted to reveal the change of residue-ligand interaction induced by the mutations of HA protein. The obtained results are novel and specific to the TW17 strain, providing deeper understanding about the impact of HA mutations and the mechanism of receptor recognition.

Results and discussion

Suspicious functional markers and binding regions of the TW17 strain

To select representative strains of influenza A/H7N9, we constructed phylogenetic trees for all segments of influenza A/H7N9. We observed that HA and NA of recent influenza A/H7N9 emerged from two major clusters as shown in Fig. 1. A bootstrap-based substitution model showed no conspicuous positive selection in each cluster. Moreover, Fig. 1 indicates that HA and NA of the current circulating strains probably originated from the A/Shanghai/2/2013(H7N9) strain (or SH13 in short). See Additional file 1 for phylogenetic trees of the other segments. Viral protein sequences of the target strain TW17 from the patient who denied any exposure to poultry and live market can be downloaded from the website of Global Initiative on Sharing All Influenza Data (GISAID) [21] by specifying the strain name A/Taiwan/1/2017(H7N9).

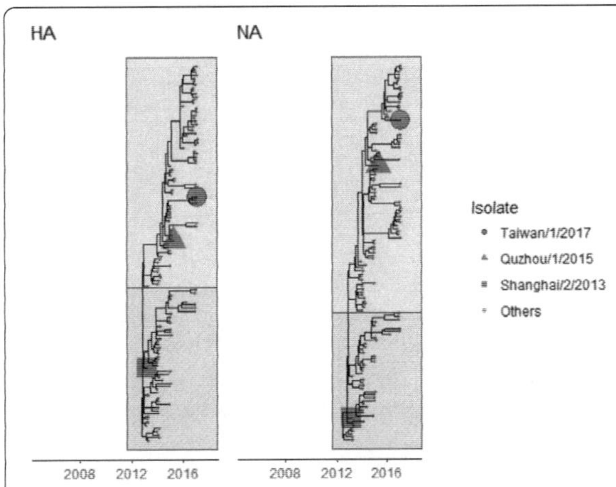

Fig. 1 Phylogenetic trees for influenza A/H7N9 HA and NA protein. The blue and green boxes show two major clusters from which HA and NA of recent influenza A/H7N9 emerged

We observed four insertions *RKRT* after site 337 and 15 mutations compared to the HA protein of the reference strain SH13. Phenotypically or epidemiologically interesting mutations of the other proteins are categorized by functional impact, summarized in Table 1 with annotations retrieved from FluSurver [22]. Since HA bears the most mutations related with protein functions, we focused on the impact of HA in TW17 strain. Most documented mutations of HA, 14 out of 15, are located at viral oligomerization interfaces, including A130P, S136N, I138T, L235Q and I335V, where the equivalent sites have also been reported as related to antigenic shifts or mild drug resistance [23]. A new potential N-glycosylation site with pattern *NGTR* at sites 136–139 was introduced by the mutations S136N and I138T, and the potential of *NNTY* at site 493 to be N-glycosylated is also likely to increase as predicted by NetNGlyc [24]. It has been reported that glycosylation of HA and NA is associated with host specificity, virulence and human immune response. Thus, glycosylation has been taken as an important way for the influenza viruses to evolve [25].

The results of FluSurver suggested possible significant change of viral binding with small ligands, especially host receptors (Table 1). Furthermore, we applied a carbon probe based approach implemented in SITEHOUND to identify putative ligand binding sites [26]. SITEHOUND calculated an affinity map for the carbon probe and then clustered the points with favorable interaction energies. The cluster with the highest total interaction energy is consistent with known equivalent receptor binding domain (RBD) in H3, including mutations A143V and L235Q which are associated with antibody recognition and host receptor binding. Hence, we used the equivalent RBD of H3 as tentative binding regions to dock host receptor analogs to HA protein, namely 130-loop (139–146), 190-helix (192–204), 220-loop (228–237) and some conserved residues (106, 152, 160–162).

Table 1 Mutations of the influenza TW17 compared to the SH13 strain

Impact	Protein	Mutations
Viral oligomerization interfaces or binding small ligands	HA	I56T, A130P, S136N, I138T, A143V, K182E, L235Q, M245I, A310T, I335V, G338A, E396A, E403K, S499R
	M2	E24D
	NA	M26I, M72I, Y166H, A210V, S242P, R289K, N322S
	PA	G66S
	[a]PB2	M570I, E627K
Host receptor binding	HA	L235Q, E396A
Host specificity shift	PB2	I292V, E627K
Glycosylation	HA	S136 N, I138T
Antibody recognition sites	HA	I56T, A130P, S136 N, I138T, A143V, L235Q, I335V, E396A, S499R
	NA	S242P
Drug binding	NA	S242P, R289K

[a]Best reference hit strain for PB2 is the influenza A/Duck/Guangdong/E1/2012(H10N8)

Molecular docking predicted the optimal conformations and indicated that the mutant TW17 strain enhanced binding with both avian receptor analogs *(LSTa) and* human receptor analogs *(LSTc)*

To obtain the optimal complexes of HA protein with host receptors, we conducted molecular docking. LSTa and LSTc are the avian and human receptor analogs respectively. Quick Vina 2 [27] was used to dock the LSTa and LSTc independently with the HA proteins. Besides the target strain TW17, receptor binding properties of two other strains, i.e. the HA of SH13 and A/Quzhou/1/2015(H7N9) (or QZ15 in short), were also investigated for comparison. The SH13 strain was detected as the best reference hit by FluSurver [22], and the HA sequence of QZ15 was the most similar to our target protein which was detected by BLAST as of 31 December 2016 [28]. Each docking experiment was conducted 500 times independently to obtain the best conformation and to analyze the group mean difference. Results of binding affinities for each group of experiments are presented in Fig. 2.

As observed, the HA protein acquired enhanced ability to bind both LSTa and LSTc as the viral strain evolved from SH13 to QZ15 and TW17. The observations were supported by Student's T-test, shown in Table 2, which was used to assess the significance of group mean difference of docking. Note that LSTc is abundant in human upper respiratory tract, while LSTa, the avian-like cell receptors, is abundant in human lower respiratory tract. The enhanced binding ability to LSTc suggests easier infection to human and a higher potential of airborne transmission, which may partially explain the current epidemic wave.

Furthermore, the best and worst conformations were superimposed as shown in Additional file 2. In each HA complex with LSTa, the optimal pose of LSTa has SIA close to the 220-loop, and the worst pose has SIA close to the 130-loop. It is the other way around for the HA complex with LSTc. The optimal pose of LSTc has SIA towards the 130-loop, while the worst pose has SIA towards the 220-loop. The superimposition demonstrates that the docking scores can differentiate reliable poses from non-reliable poses.

However, as tested in [29], docking score functions are good at searching for optimal conformations of ligands, but usually less accurate than atomic scale force fields for describing binding energy. Therefore, the docked complexes with the optimal binding affinity are used to conduct molecular dynamics simulations for further analysis (see Molecular dynamics simulation).

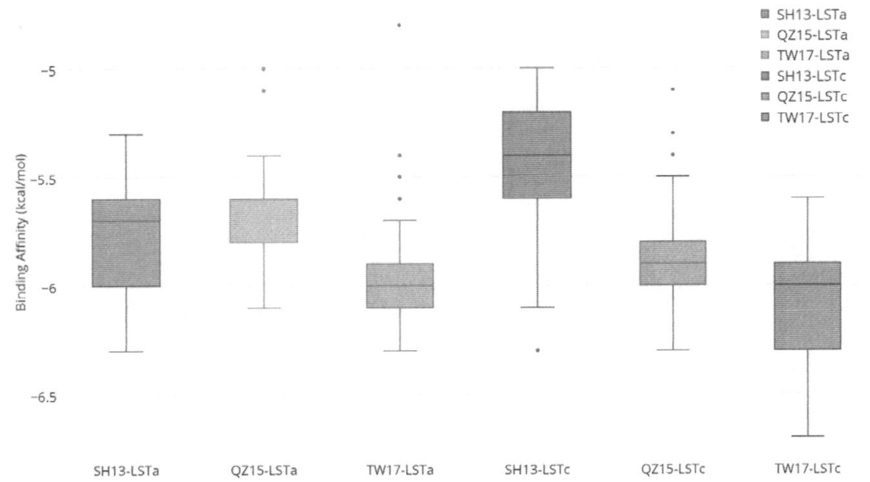

Fig. 2 Binding affinity of host receptor analogs with the H7N9 HA proteins. Binding affinity of 500 independent experiments docking LSTa and LSTc to the HA protein of SH13, QZ15 and TW17 strains respectively. SH13-LSTa stands for the docking of LSTa to the HA protein of SH13 strain and so forth

Molecular dynamics simulation reveals residues that contribute to enhanced binding of HA protein with host cell receptors

To observe the dynamic HA-receptor interaction and compare the change of interaction energy induced by substitutions on the protein, we conducted molecular dynamics simulation for the HA proteins to form complexes with receptor analogs. The result of root-mean-square deviation (RMSD) analysis of Cα atoms of HA-LSTa/LSTc complexes from the starting coordinates is shown in Fig. 3. The RMSD indicates that all the trajectories fluctuate within a small range. The TW17-LSTc complex has the largest fluctuation among the four trajectories, but still in a small range (around 0.45 nm).

The VdW energy, electrostatic energy and total interaction energy of SH13-LSTa, SH13-LSTc, TW17-LSTa and TW17-LSTc are visualized in Additional file 3. The comparison of their total energy is shown in Fig. 4. As observed, the total energy of TW17-LSTa fluctuates obviously during around 25 ns. Each trajectory fluctuated within a small range after 30 ns. Therefore, we chose the last 20 ns frames for further analysis. The average total binding energy values of SH13-LSTa, SH13-LSTc, TW17-LSTa and TW17-LSTc were calculated and listed in Table 3. The mutant TW17 HA obtained the largest binding energy with LSTc (- 664.779 kJ/mol) among the four complexes. In addition, the mutations enhanced binding of HA protein with both LSTa and LSTc. The binding with LSTc was increased by 69.67 kJ/mol, while the binding with LSTa was increased by 13.58 kJ/mol. Both the SH13 and TW17 strains have binding preferences for LSTc. The results may partially explain the outbreak in 2013 and the current epidemic wave in early 2017. To further test this hypothesis, we conducted another 50 ns of molecular dynamics simulation. The results are consistent. The mutant TW17 HA obtains the largest binding energy, and enhances the binding with two types of receptors, especially LSTc. Both SH13 and TW17 strains have binding preferences to LSTc. Additional file 4 shows the simulation results of each system and the average total binding energy.

To analyze the residue contribution to the enhanced binding, we visualize the receptor-ligand interactions in the optimally docked complexes in Fig. 5. For the SH13 strain, H192, L235 and S236 interacted with both LSTa and LSTc, the numbers have been converted to equivalent sites in the TW17 HA protein. For the TW17 strain, residues R139, T140, G142 and N164 interact with both LSTa and LSTc. Residue A143 in the HA of SH13

Table 2 T-test for docking experiments ($N = 500$)

Group 1	Group 2	[a]Mean Difference (kcal/mol)	99% Confidence Interval (kcal/mol)	*p*-value (one-tailed)
TW17-LSTa	SH13-LSTa	−0.205	(− 0.237, − 0.173)	< 0.0001
TW17-LSTc	SH13-LSTc	− 0.655	(− 0.696, − 0.614)	< 0.0001
SH13-LSTc	SH13-LSTa	− 0.345	(−0.307, − 0.383)	< 0.0001
TW17-LSTc	TW17-LSTa	−0.105	(− 0.141, − 0.069)	< 0.0001

[a]Mean difference = Mean (Group 1) − Mean (Group 2)

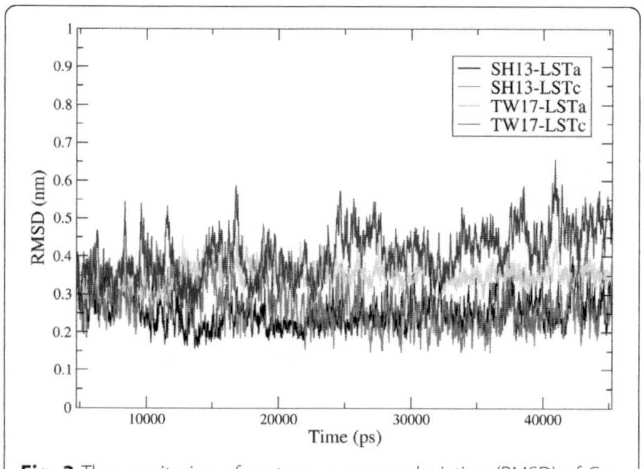

Fig. 3 The monitoring of root-mean-square deviation (RMSD) of Cα atoms from the starting coordinates

interacts with LSTa, but in the HA of TW17, V143 interacts with LSTc.

Furthermore, we decomposed the total interaction energy to observe the contribution of each residue. The convergence of energy contributions of the residues involved in interaction is shown in Additional file 5. To be consistent with the analysis of total energy, we also used the last 20 ns frames. The average contribution of each residue to the total binding energy was calculated. We focused on the residues that are involved in receptor-ligand interactions in the optimally docked complexes. Additional file 6 shows their average energy contribution in each complex. R139 enhances the binding of HA of TW17 strain with both LSTa and LSTc and it has binding preference for LSTa. Similarly, we quantified the binding preference of all residues by calculating the difference between energy contribution to binding LSTa and LSTc ($\Delta\Delta G = \Delta G_{HA\text{-}LSTa} - \Delta G_{HA\text{-}LSTc}$). For the HA of SH13 strain, the differences of energy contribution are all less than 10 kJ/mol. For instance, R139 had mild preference for LSTc. In contrast to the HA of SH13 strain, the mutant TW17 HA protein contains more residues showing apparent binding preference, including R139, V143, N164 (preferring LSTa) and K202 (preferring LSTc). The top 10 residues showing binding preference are visualized in Fig. 6. To investigate the impact of mutations on energy contribution, we also calculated the change of interaction energy before and after the mutations ($\Delta\Delta G = \Delta G_{SH13\text{-}receptor} - \Delta G_{TW17\text{-}receptor}$). The top 10 residues that affect the binding with receptors are presented in Fig. 7. R139 and K202 largely enhanced (with $\Delta\Delta G > 100$ kJ/mol) the binding the LSTa and LSTc respectively. Note that R139 is located at a new potential N-glycosylation sites introduced by the mutations S136 N and I138T, which may explain the increased binding with receptors. E199 and K202, which located at the 190-helix, have enhanced the binding with LSTc, but no mutation is observed in this region. The mutations nearby the 190-helix might be responsible for the change of residue contributions.

Conclusions

In this study, we analyzed the mutations of the current circulating influenza A/H7N9 virus strain, isolated from a patient claiming no exposure to any live bird, live poultry market or suspicious patient. We highlighted mutations bearing critical protein functions, namely host receptor binding, drug binding, antibody recognition and glycosylation. Furthermore, we focused on the HA binding with different host cell receptors by performing molecular docking and molecular dynamics simulation. The obtained results indicate that the mutant HA enhances its binding

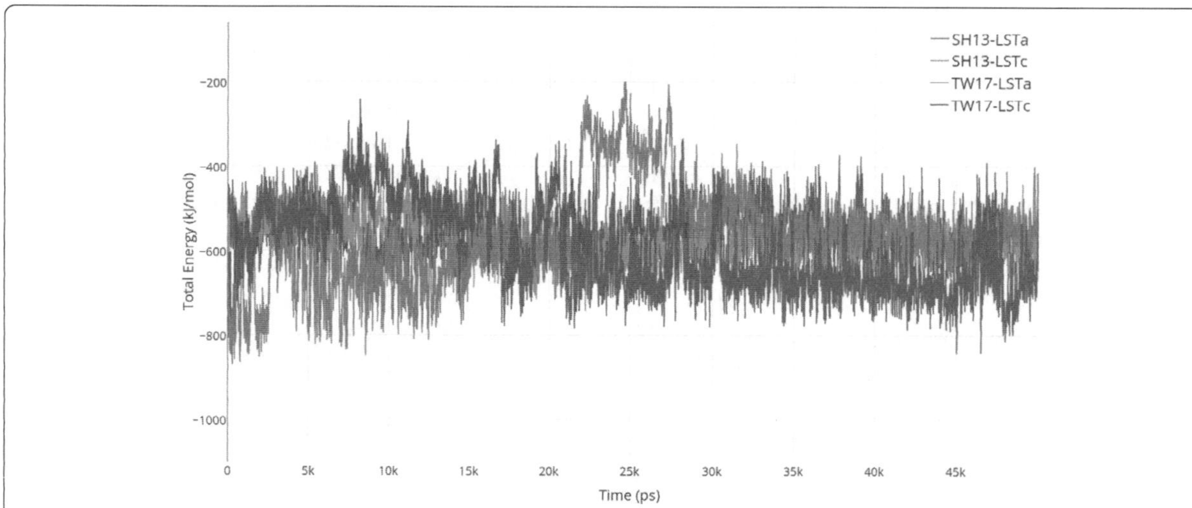

Fig. 4 Comparison of HA-SIA vacuum MM total energy. Visualize the fluctuation of total energy and its components for each complex (namely SH13-LSTa, SH13-LSTc, TW17-LSTa and TW17-LSTc) during the whole MD simulation process

Table 3 Average total binding energy (kJ/mol) of the HA-LSTa/LSTc complexes

	LSTa	LSTc	[a]ΔE_1
SH13	− 541.559	− 595.111	+ 53.553
TW17	− 555.135	- 664.779	+ 109.644
[b]ΔE_2	+ 13.576	+ 69.667	

[a]Binding preference of HA protein: $\Delta E_1 = \Delta E_{HA,\ LSTa} - \Delta E_{HA,\ LSTc}$
[b]Difference of HAs binding to receptors: $\Delta E_2 = \Delta E_{SH13,\ receptor} - \Delta E_{TW17,\ receptor}$

with both avian and human receptor analogs, especially human receptor analogs. Also, the MM-PBSA calculations of residue-ligand interaction revealed several critical residues showing binding preference, including residues R139, V143, N164 and K202. Similarly, major residues contributing to the change of interaction energy between HA and receptors were highlighted. We detected the mutations S136 N, I138T and mutations near the 190-helix as the most important substitutions in the HA protein. Although S136N, I138T do not cause direct change to the interaction energy, they introduce a new glycosylation pattern. Besides, their neighboring residue R139 enhances binding to both avian and human receptors.

The obtained results are novel and specific to the influenza A/Taiwan/1/2017(H7N9) strain, shedding light on the impacts of HA mutations and the mechanisms of receptor recognition. In addition, our pipeline of analysis should be applicable to analyzing the impacts of other mutations on the binding of proteins with small ligands.

Methods

Data

All protein sequences of influenza A/H7N9 were retrieved from Global Initiative on Sharing All Influenza Data (GISAID) EpiFlu Database [21] as of 8 April 2017. The virus strain in the patient claiming no exposure to any live bird or any live poultry market is influenza A/Taiwan/1/2017(H7N9) (Isolate ID: EPI_ISL_248778), denoted as TW17. HA of the strain from the first wave A/Shanghai/02/2013(H7N9) and the most similar HA sequence from A/Quzhou/1/2015(H7N9) were selected as references, denoted as SH13 and QZ15 respectively. The three representative strains were used to docking with host receptor analogs to show the changes of HA binding preference. The accession IDs are YP_009118475 and AKI82233 for HA proteins of SH13 and QZ15 respectively.

To conduct protein-ligand docking and molecular dynamics simulation, we need the structures of HA proteins and receptor analogs. Examining the Protein Data Bank (PDB), we found crystal structures of SH13 HA with LSTa (PDB ID: 4N5K) and LSTc (PDB ID:4N60) [30]. However, extra four residues were observed in the HA protein of the new TW17 strain, by comparing it with the sequences of available HA structures in PDB. Hence, we

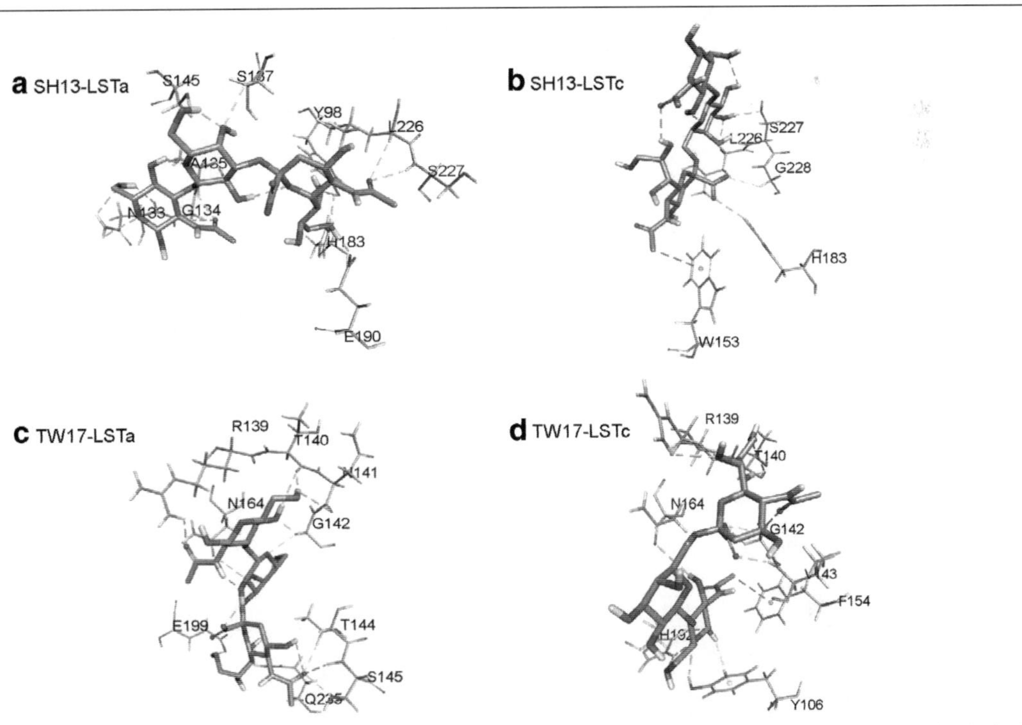

Fig. 5 The receptor-ligand interactions in the optimally docked complexes. **a** The optimally docked SH13-LSTa complex. **b** The optimally docked SH13-LSTc complex. **c** The optimally docked TW17-LSTa complex. **d** The optimally docked TW17-LSTc complex

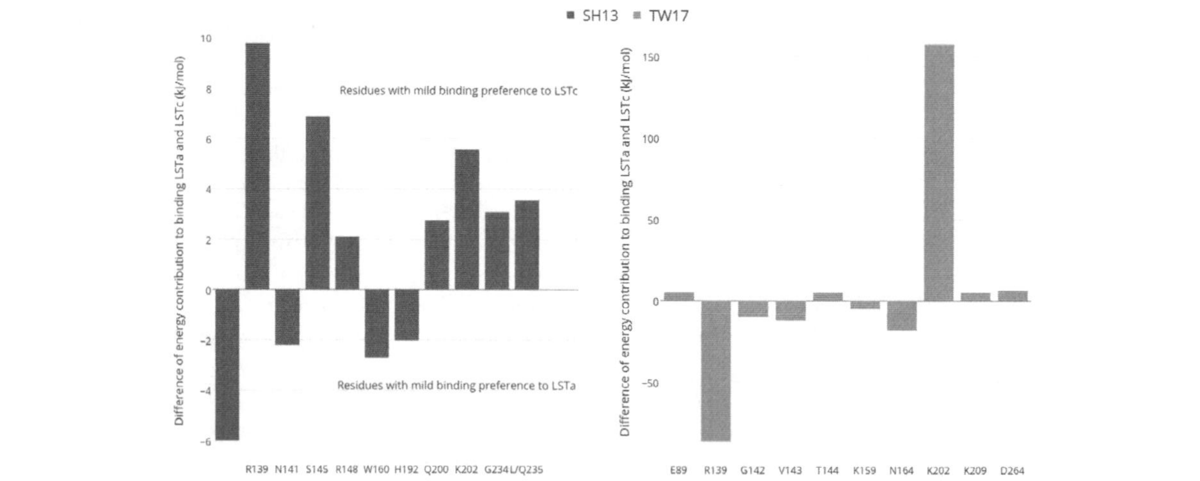

Fig. 6 The top 10 residues in HA that show binding preference for LSTa or LSTc

obtained the HA structures of the other representatives from homology modelling described in the next section. Avian and human receptor analogs were obtained from PDB:5E2Z [31] and 2YP3 [32] respectively.

Homology modelling

We used SWISS-MODEL for the homology modelling, constructing an atomic-resolution model of the target HA protein from its amino acid sequence [33, 34]. First, the primary sequence of the HA protein from the TW17 strain was searched with BLAST [35] and HHBlits [36] against the SWISS-MODEL template library [37], finding 1218 templates. Three templates with the highest quality were selected, two of which were found by HHBlits (PDBID: 3WHE.1.A and 4LN6.1.A) and one found by BLAST (PDBID: 4WSW.1.F). Models were built based on the target-template alignment using ProMod3 and measured with Global Model Quality Estimation (GMQE), a quality estimation combining properties from the target-template alignment while considering the QMEAN score of the obtained model [38, 39]. The model constructed from template 3WHE.1.A had the highest GMQE value of 0.72. Thus, it was selected for further analysis.

Phylogenetic tree construction

For phylogenetic tree analysis of PB2, PB1, PA, HA, NP, NA, M1 and NS1 genes of influenza H7N9 strains, 125 available isolates were collected (two genomes, whose isolate IDs are EPI ISL 148744 and EPI ISL 142188, were removed due to quality considerations). For each gene, the coding sequences (CDSs) were obtained from each

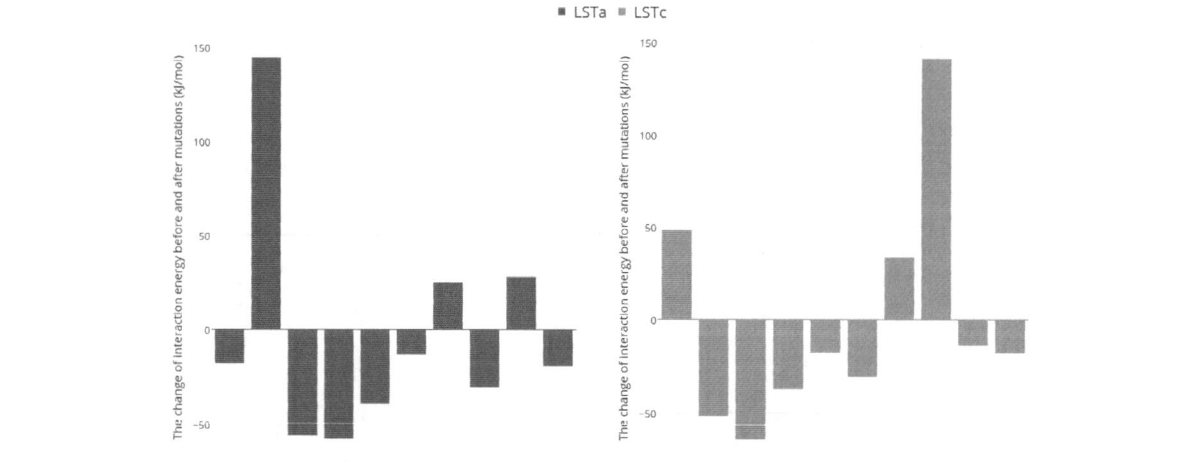

Fig. 7 The top 10 residues in HA that affect the HA binding with LSTa or LSTc

isolate using influenza virus sequence annotation tool provided by NCBI [40], and then the CDSs were aligned to codon positions using MUSCLE [41]. Then, the Bayesian Markov Chain Monte Carlo (MCMC) implemented in Beast v2.4.4 [42] was used to infer a time-tree from the alignment. The Beast software was run under the HKY substitution model, a Bayesian skyline coalescence model, and a strict molecular clock model to produce 2000 tree samples logged every 25,000 generations. Beast's outputs were assessed with TRACER v1.6 [43] and summarized as a maximum clade credibility (MCC) tree using TreeAnnotator v2.4.4 with 10% burn-in. The time-tree for each gene was visualized using ggtree package [44].

Protein-ligand docking

Molecular docking is commonly used to estimate the pose of conformation and roughly compare the binding affinity of a small number of complexes. With the structures of HA and host cell receptor analogs obtained, we applied protein-ligand docking using QuickVina 2, which optimized the local search of docked conformation candidates by a novel first-order-consistency-check heuristic [27]. All available rotatable bonds of receptor analogs, LSTa and LSTc, were activated to ensure flexibility. They were docked respectively to the receptor binding domain of the H7N9 HA protein (sites 106, 139–146, 152, 160–162, 192–204, 228–237). The top conformation with the optimal binding affinity among 500 independent docking experiments was selected to analyze the interactions between the host receptor analogs and the HA proteins.

Molecular dynamics simulation

The complexes of SH13 and TW17 HA proteins with LSTa and LSTc were used in molecular dynamics simulation. The complexes SH13-LSTa and SH13-LSTc were obtained from PDB (PDB ID: 4N5K and 4N60), while complexes TW17-LSTa and TW17-LSTc were from the optimal docking results.

GROMACS 5.1.2 [45] was used in the whole simulation process, energy minimization and equilibration of the system involved. The AMBER99SB-ILDN force field was used to describe the system [46]. All complexes were solvated by using the explicit TIP3P water model in a cubic box and counter ions were added to neutralize the system [47]. Steepest-descent energy minimization was applied to each complex. Position restrains for both NVT and NPT equilibration were conducted for 100 ps with modified Berendsen thermostat and Parrinello-Rahman pressure coupling [48, 49]. Temperature, pressure, density and total energy were all well equilibrated before running the production MD simulation for 50 ns.

Binding free energy and residue-ligand interaction energy calculation

Snapshots of MD simulations were recorded to analyze the binding free energy. We applied Molecular Mechanics - Poisson Boltzmann Surface Area (MM-PBSA) to estimate the binding free energy ΔG_{bind} under the assumption of no configurational rearrangement upon which the free energy changes. It is computed as:

$$\Delta G_{bind} = \Delta G_{bind,vac} + \Delta G_{bind,solv} \tag{1}$$

The binding free energy ΔG_{bind} comprises the binding free energy in the vacuum phase $\Delta G_{bind,\ vac}$ and the solvation free energy $\Delta G_{bind,\ solv}$, which is the difference of solvation free energy values of complex, receptor and ligand:

$$\Delta G_{bind,solv} = \Delta G_{solv,complex} - \left(\Delta G_{solv,receptor} + \Delta G_{solv,ligand}\right) \tag{2}$$

The solvation free energy for each component is composed of polar and non-polar energy derived from the PB equation and the SA method, computed as:

$$\Delta G_{solv} = \Delta G_{solv,polar} + \Delta G_{solv,nonpolar} \tag{3}$$

The binding free energy and decomposed interaction energy between residues of HA protein and receptor analogs were calculated using the g_mmpbsa package [50].

Additional files

Additional file 1: Phylogenetic trees of influenza A/H7N9. Time-scale phylogenetic trees of PB2, PB1, PA, HA, NP, NA, M1 and NS1 genes of influenza H7N9 strains. (PDF 117 kb)

Additional file 2: Superimpose the best and worst docked HA-ligand complexes. Visualize the structure of docked SH13-LSTa, SH13-LSTc, TW17-LSTa and TW17-LSTc complexes with the highest and the lowest binding affinities. (PDF 466 kb)

Additional file 3: Energy decomposition of HA-ligand complexes. VdW energy, electrostatic energy and total interaction energy of SH13-LSTa, SH13-LSTc, TW17-LSTa and TW17-LSTc. (PDF 702 kb)

Additional file 4: Two rounds of 50 ns molecular dynamics simulation for SH13-LSTa, SH13-LSTc, TW17-LSTa, TW17-LSTc. Average total binding energy (kJ/mol) of the HA-LSTa/LSTc complexes. (PDF 424 kb)

Additional file 5: Residues contribution during the whole simulation process. The convergence analysis of energy contribution for residues involved in HA-ligand interaction. (PDF 153 kb)

Additional file 6: Average energy contribution of residues that involved in receptor-ligand interactions in the optimally docked complexes. (PDF 24 kb)

Abbreviations

A/H1N1-pdm09: The 2009 pandemic strains of Influenza A/H1N1; GMQE: Global model quality estimation; HA: Hemagglutinin; LSTa: Avian receptor analogs with α(2,3) linked SIA; LSTc: Human receptor analogs with α(2,6) linked SIA; MM-PBSA: Molecular mechanics - poisson boltzmann surface area; NA: Neuraminidase; QZ15: Influenza A/Quzhou/1/2015(H7N9); RMSD: Root-mean square deviation; SH13: Influenza A/Shanghai/02/2013(H7N9); SIA: Sialic acid; TW17: Influenza A/Taiwan/1/2017(H7N9); WHO: World Health Organization

Acknowledgements
This work was supported by AcRF Tier 2 grant MOE2014-T2-2-023, Ministry of Education, Singapore. The authors also would like to thank the National Supercomputing Center (NSCC) Singapore for providing the computing resources.

Funding
The publication charges for this article were funded by AcRF Tier 2 grant MOE2014-T2-2-023, Ministry of Education, Singapore.

Availability of data and materials
All data and material are available in additional files.

About this supplement
This article has been published as part of *BMC Genomics* Volume 19 Supplement 2, 2018: Selected articles from the 16th Asia Pacific Bioinformatics Conference (APBC 2018): genomics. The full contents of the supplement are available online at https://bmcgenomics.biomedcentral.com/articles/supplements/volume-19-supplement-2.

Authors' contributions
XZ performed experiments, interpreted results, and wrote the manuscript. F.I. conducted phylogenetic tree analysis. JZ, FI, RY, SR, VC and CK revised the paper, provided overall supervision, direction and leadership to the research. All authors have read and approved the final manuscript.

Ethics approval and consent to participate
Not applicable.

Consent for publication
Not applicable.

Competing interests
The authors declare that they have no competing interests.

Publisher's Note

Springer Nature remains neutral with regard to jurisdictional claims in published maps and institutional affiliations.

Author details
[1]School of Computer Science and Engineering, Nanyang Technological University, Singapore 639798, Singapore. [2]Genome Institute of Singapore, A*STAR, Singapore 138672, Singapore. [3]Department of Chemistry and Biomolecular Sciences, Macquarie University, Sydney, NSW 2109, Australia. [4]Department of Microbiology and Immunology, Yong Loo Lin School of Medicine, National University of Singapore, Singapore 117545, Singapore.

Published: 9 May 2018

References

1. Tscherne DM, García-Sastre A. Virulence determinants of pandemic influenza viruses. J Clin Invest. 2011;121:6–13.
2. Smith GJD, et al. Dating the emergence of pandemic influenza viruses. Proc Natl Acad Sci. 2009;106(28):11709–12.
3. Viboud C, Simonsen L. Global mortality of 2009 pandemic influenza a H1N1. Lancet Infect Dis. 2012;12(9):651–3.
4. World Health Organization. Recommended composition of influenza virus vaccines for use in the 2016–2017 northern hemisphere influenza season. 2016; Available from: http://www.who.int/influenza/vaccines/virus/recommendations/en/ (accessed 17 Mar 2017).
5. World Health Organization. Monthly risk assessment summary. 2017; Available from: http://www.who.int/influenza/human_animal_interface/HAI_Risk_Assessment/en/ (accessed 23 Apr 2017).
6. Shen Y, Lu H. Global concern regarding the fifth case of human infection with avian influenza a (H7N9) virus in China. Bioscience Trends. 2017;11.1:120–21
7. Xiang N. Assessing change in avian influenza a (H7N9) virus infections during the fourth epidemic—China, September 2015–august 2016. MMWR Morb Mortal Wkly Rep. 2016;65
8. Taubenberger JK, Kash JC. Influenza virus evolution, host adaptation, and pandemic formation. Cell Host Microbe. 2010;7(6):440–51.
9. Tumpey TM, et al. A two-amino acid change in the hemagglutinin of the 1918 influenza virus abolishes transmission. Science. 2007;315(5812):655–9.
10. Pepin KM, et al. Identifying genetic markers of adaptation for surveillance of viral host jumps. Nat Rev Microbiol. 2010;8(11):802–13.
11. Ping J, et al. Genomic and protein structural maps of adaptive evolution of human influenza a virus to increased virulence in the mouse. PLoS One. 2011;6:e21740.
12. Narasaraju T, et al. Adaptation of human influenza H3N2 virus in a mouse pneumonitis model: insights into viral virulence, tissue tropism and host pathogenesis. Microbes Infect. 2009;11:2–11.
13. Ibricevic A, et al. Influenza virus receptor specificity and cell tropism in mouse and human airway epithelial cells. J Virol. 2006;80(15):7469–80.
14. Xiong X, McCauley JW, Steinhauer DA. Receptor binding properties of the influenza virus hemagglutinin as a determinant of host range. Current Topics Microbiol Immunol. 2014;385:63–91.
15. Edinger TO, Pohl MO, Stertz S. Entry of influenza a virus: host factors and antiviral targets. J Gen Virol. 2014;95(2):263–77.
16. Koday MT, et al. A computationally designed hemagglutinin stem-binding protein provides in vivo protection from influenza independent of a host immune response. PLoS Pathog. 2016;12(2):e1005409.
17. Su CT-T, et al. Structural analysis of the novel influenza a (H7N9) viral neuraminidase interactions with current approved neuraminidase inhibitors Oseltamivir, Zanamivir, and Peramivir in the presence of mutation R289K. BMC Bioinformatics. 2013;14(16):S7.
18. Pan D, et al. Molecular mechanism of the enhanced virulence of 2009 pandemic influenza a (H1N1) virus from D222G mutation in the hemagglutinin: a molecular modeling study. J Mol Modeling. 2012;18:4355–66.
19. Kannan S, Kolandaivel P. Computational studies of pandemic 1918 and 2009 H1N1 hemagglutinins bound to avian and human receptor analogs. J Biomol Struct Dyn. 2016;34(2):272–89.
20. World Health Organization. Human infection with avian influenza a(H7N9) virus - China (22-Feb-2017). 2017; Available from: http://www.who.int/csr/don/22-february-2017-ah7n9-china/en/ (accessed 27 Feb 2017).
21. Büch J, et al. Gisaid-a global initiative on sharing all influenza data. Influenza Other Respir Viruses. 2011;5:419.
22. A*STAR Bioinformatics Institute. *Influenza surveillance- prepared for the next wave*. Available from: http://flusurver.bii.a-star.edu.sg/ (accessed 17 Mar 2017).
23. Philpott M, et al. Hemagglutinin mutations related to attenuation and altered cell tropism of a virulent avian influenza a virus. J Virol. 1990;64(6):2941–7.
24. Gupta R, Brunak S. Prediction of glycosylation across the human proteome and the correlation to protein function. Pac Symp Biocomput. 2001;7:310–22.
25. Wu C-Y, et al. Influenza A surface glycosylation and vaccine design. Proc Natl Acad Sci. 2017;114(2):280–5.
26. Hernandez M, Ghersi D, Sanchez R. SITEHOUND-web: a server for ligand binding site identification in protein structures. Nucleic Acids Res. 2009;37(suppl 2):W413–6.
27. Alhossary A, et al. Fast, accurate, and reliable molecular docking with QuickVina 2. Bioinformatics. 2015;31(13):2214–6.
28. Johnson M, et al. NCBI BLAST: a better web interface. Nucleic Acids Res. 2008;36(suppl 2):W5–9.
29. Ramírez D, Caballero J. Is it reliable to use common molecular docking methods for comparing the binding affinities of enantiomer pairs for their protein target? Intl J Mol Sci. 2016;17(4):525.
30. Xu R, et al. Preferential recognition of avian-like receptors in human influenza a H7N9 viruses. Science. 2013;342(6163):1230–5.
31. Rose AS, et al. Web-based molecular graphics for large complexes. In: Proceedings of the 21st international conference on Web3D technology; 2016.
32. Lin YP, et al. Evolution of the receptor binding properties of the influenza a (H3N2) hemagglutinin. Proc Natl Acad Sci. 2012;109(52):21474–9.
33. Biasini M, et al. SWISS-MODEL: modelling protein tertiary and quaternary structure using evolutionary information. Nucleic Acids Res. 2014;42(W1):W252–W258.
34. Arnold K, et al. The SWISS-MODEL workspace: a web-based environment for protein structure homology modelling. Bioinformatics. 2006;22(2):195–201.
35. Altschul SF, et al. Gapped BLAST and PSI-BLAST: a new generation of protein database search programs. Nucleic Acids Res. 1997;25(17):3389–402.

36. Remmert M, et al. HHblits: lightning-fast iterative protein sequence searching by HMM-HMM alignment. Nat Methods. 2012;9(2):173–5.
37. Kiefer F, et al. The SWISS-MODEL repository and associated resources. Nucleic Acids Res. 2009;37(suppl 1):D387–92.
38. Benkert P, Biasini M, Schwede T. Toward the estimation of the absolute quality of individual protein structure models. Bioinformatics. 2010;27(3): 343–50.
39. Šali A, Blundell TL. Comparative protein modelling by satisfaction of spatial restraints. J Mol Biol. 1993;234(3):779–815.
40. Bao Y, et al. FLAN: a web server for influenza virus genome annotation. Nucleic Acids Res. 2007;35(suppl 2):W280–4.
41. Edgar RC. MUSCLE: multiple sequence alignment with high accuracy and high throughput. Nucleic Acids Res. 2004;32(5):1792–7.
42. Bouckaert R, et al. BEAST 2: a software platform for Bayesian evolutionary analysis. PLoS Comput Biol. 2014;10(4):e1003537.
43. Rambaut, A., et al. *Tracer v1.6*. Tracer (online 2015, may 29) 2014; Available from: http://beast.bio.ed.ac.uk (accessed 13 Apr 2017).
44. Yu G, et al. Ggtree: an r package for visualization and annotation of phylogenetic trees with their covariates and other associated data. Methods Ecol Evol. 2017;8(1):28–36.
45. Abraham MJ, et al. GROMACS: high performance molecular simulations through multi-level parallelism from laptops to supercomputers. SoftwareX. 2015;1:19–25.
46. Lindorff-Larsen K, et al. Improved side-chain torsion potentials for the amber ff99SB protein force field. Proteins: Structure, Function, Bioinformatics. 2010; 78(8):1950–8.
47. Jorgensen WL, et al. Comparison of simple potential functions for simulating liquid water. J Chem Phys. 1983;79(2):926–35.
48. Berendsen HJC, et al. Molecular dynamics with coupling to an external bath. J Chem Phys. 1984;81(8):3684–90.
49. Parrinello M, Rahman A. Polymorphic transitions in single crystals: a new molecular dynamics method. J Appl Phys. 1981;52(12):7182–90.
50. Kumari R, Kumar R, Lynn A. g_mmpbsa: a GROMACS tool for high-throughput MM-PBSA calculations. J Chem Inf Model. 2014;54(7):1951–62.

Zheng *et al. BMC Genomics* 2018, **19**(Suppl 2):100
https://doi.org/10.1186/s12864-018-4462-y

BMC Genomics

RESEARCH Open Access

Resolution effects in reconstructing ancestral genomes

Chunfang Zheng, Yuji Jeong, Madisyn Gabrielle Turcotte and David Sankoff*

From The Sixteenth Asia Pacific Bioinformatics Conference 2018
Yokohama, Japan.15-17 January 2018

Abstract

Background: The reconstruction of ancestral genomes must deal with the problem of resolution, necessarily involving a trade-off between trying to identify genomic details and being overwhelmed by noise at higher resolutions.

Results: We use the median reconstruction at the synteny block level, of the ancestral genome of the order Gentianales, based on coffee, *Rhazya stricta* and grape, to exemplify the effects of resolution (granularity) on comparative genomic analyses.

Conclusions: We show how decreased resolution blurs the differences between evolving genomes, with respect to rate, mutational process and other characteristics.

Keywords: Granularity, Gentianales, Gene order, Median, Synteny blocks, Rearrangements

Background

All comparative genomics, including the reconstruction of ancestral genomes must deal, implicitly or explicitly, with the problem of resolution, sometimes called "granularity", a topic that has preoccupied us since the mid-2000s (e.g., [1]), after the publication of the human and mouse genomes, and increasingly with the next few mammalian genomes [2].

Setting the level of resolution involves a trade-off between trying to decipher precise genomic details on one hand and being overwhelmed by noise at higher resolutions due to both mutational processes and methodological difficulties, on the other. Thus, despite some successes, notably with the ancient "boreoeutherian" mammalian genome [3], in most cases, a nucleotide by nucleotide reconstruction of even the coding region of an ancestral gene contains a large amount of uncertainty, while non-coding DNA reconstruction is much more refractory.

Starting few years ago, a number of algorithmic approaches to these questions have been proposed (e.g., [4–10], as reviewed by [11]).

*Correspondence: sankoff@uottawa.ca
Department of Mathematics and Statistics, University of Ottawa, 585 King Edward Avenue, Ottawa, K1N 6N5 Ontario, Canada

At the level of gene order, especially in plants, the processes of whole genome duplication and fractionation compound inherent inferential difficulties due to genome rearrangement. A telling aspect in the widely-used synteny block identifier SYNMAP [12, 13] is the default requirement that at least five pairs of orthologous genes in two genomes or five pairs of paralogous genes in a single genome, in close succession, are necessary before we have confidence that the two apparently homologous segments containing these genes are genuinely contemporaneous descendants of the same ancestral genomic fragment.

In this paper, we explore the difficulty in establishing synteny blocks for three genomes based on the pairwise output of SYNMAP, as a preliminary to a stepwise decrease in resolution in the calculation of the median estimate of the ancestor. (The median is a reconstructed genome that minimizes the sum of some distance measure from itself to three or more given genomes.)

There are a number of ways (e.g. [2, 4–11]) of establishing synteny blocks for more than two genomes and of changing the resolution of the analysis. Since our main goal in this paper is to detail the *consequences* of the loss of resolution, we do not make the assumptions or adopt the objective functions of any of these. Rather we carefully monitor the merger or split of overlapping pairwise synteny blocks, which may coincide on each genome without necessarily containing any genes in common. This procedure serves as preprocessing for the main exercise, which starts with the construction, at eight levels of resolution, of the median of three genomes, two in the order Gentianales: coffee (*Coffea canephora*, family Rubiaceae) [14] and *Rhazya stricta* (family Apocynaceae) [15], and one rosid outgroup: grape (*Vitis vinifera*, family Vitaceae, order Vitales) [16]. The interest in the two Gentianales species stems from our participation in the recent genome sequencing projects of both, and the fact that they are among the only sequenced asterid genomes to have escaped any whole genome duplication or triplication events more recent than the core eudicot triplication known as γ. This is true also of the conservative rosid, grape, which allows us to avoid reconstruction difficulties due to recently generated paralogous synteny blocks. The assemblies of the grape and coffee genomes are available in CoGe [12, 13] at the pseudo-molecule level, while the *Rhazya* scaffolds are of equal or better quality, but lack any ancillary data that could enable pseudo-molecule construction.

Each median construction is comprehensively analyzed in comparison with those of greater and lesser resolution, using DCJ (double-cut-and-join) distance [17], breakpoint distance, and a measure of interchromosomal mixing [1]. This analysis reveals the effects of resolution on the estimates of total median distance (i.e., the sum of the distances between the median and the three input genomes), the differential between the rates of gene-order evolution of the three plant genomes, the relative preponderance of translocation, reversals and other rearrangement mutations, and the breakpoint reuse rate. In addition, at full resolution, this work provides a detailed reconstruction of the ancestral Gentianales genome at the synteny block level.

Methods

The whole genome triplication γ, discovered by Jaillon et al. [16] while sequencing the grape genome, is broadly accepted to have occurred at the root of the core eudicots. This polyploidization produced a twenty-one chromosome genome from the seven chromosome ancestor. It is not too difficult to reconstruct the general structure of the immediate post-γ chromosomes since they differ from the nineteen-chromosome grape genome in only four or five large rearrangements. This baseline genome is the ultimate resource for deducing gene-order change in the core eudicots, serving as a close proxy for the ancestral hexaploid [18].

The main data for this paper consists of three-way synteny blocks between grape, *Rhazya* and coffee. As a first step, we used SynMap,

`https://www.genomevolution.org/CoGe/SynMap.pl`

(with default parameters) in the CoGe platform

`https://genomevolution.org/CoGe/`

to find pairwise synteny blocks between grape and *Rhazya* (806 blocks), between *Rhazya* and coffee (583 blocks), and between coffee and grape (722 blocks), drawing on the CDS sequences for these genomes also stored in CoGe.

In this construction, we used overall gene similarity and K_S scores to establish thresholds for distinguishing duplicate gene pairs due to speciation (orthologs) which were retained for our analysis, from the older, weaker and sparser gene pairs due to the gamma triplication ("out-paralogs"), which were discarded. We coloured (i.e., labelled) the blocks according to the 21 ancestral core eudicot chromosomes [18], and noted the strand (polarity) of the block in the two genomes.

The *Rhazya* genome sequence assembly [15] is made up of 980 scaffolds, 731 of which do not contain any genes. Among those that contain genes, many do not contain enough of them to establish synteny blocks with both grape and coffee. Others contained only one synteny block entirely contained in a grape-coffee block. None of these contain information helpful to describe the evolutionary divergence of *Rhazya* at the synteny block level from the other genomes. What is required are *Rhazya* scaffolds containing at least two successive synteny blocks that are not successive in either coffee or grape or both. Ultimately only 22 *Rhazya* scaffolds were large enough to satisfy this criterion. All of these contained at least 200 genes.

To construct the three-way blocks, we then sorted all the pairwise synteny blocks according to their position on the *Rhazya* genome. Because of the relatively small size of the *Rhazya* scaffolds, the coffee-grape blocks contained virtually no breakpoints that would further split any of the *Rhazya* blocks into two blocks. Some 70 triples constructed from the pairwise coffee-*Rhazya* and grape-*Rhazya* blocks were clearly coherent, usually missing at most a few genes at the start or end of each block, because of widespread fractionation. For almost twice as many, different parts of a *Rhazya* region were syntenic with two or more regions in coffee, in grape, or sometimes in both, resulting in two or more new contiguous syntenic blocks in *Rhazya* but non-contiguous blocks in one or the other or both of the other genomes. Parts of contiguous blocks in the other genomes homologous to non-contiguous parts in *Rhazya* would be determined in an analogous way.

Finally 204 three-way blocks were identified by this procedure, and formed the initial data set for our ancestral genome reconstruction. Note that these blocks are defined by pairwise orthologies between the genomes, and may involve very few orthologous triples, or none at all. Note as well that it is possible to use two-way blocks in addition to the three-way blocks in the type of analysis we undertook, and this is fairly reliable in identifying blocks in the ancestor, but it is not very useful in establishing their genomic position.

In this process, although conflicting paralogous identifications occurred at the level of individual gene pairs, these were not reflected in any confusion at the synteny block level. Within these blocks, the number of gene pairs and the average similarity levels between descendants of the same γ subgenome were clearly greater than between descendants of different subgenomes, and gene duplicates generated after γ had no effect. Such duplicates were either detected as tandem duplicates and removed automatically by SynMap or, isolated from syntenic context, did not show up at all in syntenic analysis.

We characterized this data set as 0-level resolution, because no data were excluded. In addition, we repeated the entire reconstruction at resolution levels $L = 10, 20, 30, 40, 50, 60$ and 70, at each level simply ignoring all blocks with less than L genes. Note that higher L corresponds to less resolution. In the figures presented here, the axis representing resolution is labelled "$\leftarrow$ resolution" to indicate that lower values of L indicate higher resolution.

Whenever removing smaller blocks from consideration revealed pairs of remaining blocks close together on all three genomes, i.e., with less than 250 genes (including those genes no longer considered to be in blocks) separating them on all three, we combined them. Because of the conservatism of the grape genome chromosomes, these combined blocks were virtually always of the same core eudicot "colour", but their gene orders, or polarity of the blocks, were not always consistent. We found that this "same-colour" tendency holds for a cutoff of 250 or less, but broke down if we merged blocks 300 genes or more apart. On the other hand, setting a cutoff less than 250 meant that large neighbouring blocks of the same colour sometimes failed to merge. Table 1 summarizes the the number of remaining blocks after small-block deletions and mergers, at each resolution, and the total number of genes in the remaining blocks.

Table 1 Loss of gene content from blocks retained as resolution decreases (as L increases)

Level L	Blocks Deleted	Blocks Remaining	After Mergers	Genes Not included	Genes Remaining
0	0	204	162	0	13862 (100%)
10	11	193	149	264	13599 (98%)
20	36	168	118	853	13009 (94%)
30	59	145	92	1550	12312 (89%)
40	75	129	76	2167	11695 (84%)
50	85	119	62	2689	11173 (81%)
60	91	113	56	3084	10778 (78%)
70	97	107	51	3546	10317 (74%)

To resolve the problem of conflicting orientations of merged blocks, the ensuing analysis was carried out repeatedly, each time changing the sign of one of the indeterminate blocks, keeping the change if it improved the objective function, until no further improvement was observed. Because neighbouring blocks in the original construction, where $L = 0$, could also be merged under the same "colour" criterion", the same procedure was also carried out at this level.

At each level of resolution, Xu's median solver for DCJ [19] was employed to infer the Gentianales ancestor. In producing an exhaustive list of medians, this was efficient enough to carry out several cycles of the polarity assignment described above.

For each median, we could then calculate the DCJ distance [17] to each of the three genomes, using a stepwise approach to reducing the cycle graph, prioritizing either translocations or reversals.

Table 2 shows the DCJ distance between the pairs of genomes, each represented as a series of blocks on chromosomes, at all resolution levels, as well as the same measure applied to the content of each block, and then summed. The latter is not directly pertinent to this paper, but is included for general interest. The table also displays the average positioning of the median with respect to the input genomes.

We also carried out the same calculations in the framework of the breakpoint distance. In the case of the breakpoint median, our method used a maximum weight matching algorithm [20].

Finally, we calculated the minimum number of translocations from the median to the three plant genomes according to a very conservative probabilistic model [1, 2].

Results

Figure 1 depicts the results of the median construction at three levels of resolution. The median is not unique (see "The set of medians" section below), but we used typical ones for these displays. The coloured regions reflect the seven core eudicot chromosomes before the γ triplication, while the three subgenomes after γ are distinguished by different border patterns around the same colour blocks.

Comparing the three genomic (DCJ) distances with evolutionary time (the Gentianales ancestor dating from 65 My ago, and the core eudicot represented by grape about 55 My older [21]) in the fully resolved analysis, reveals that coffee is much more conservative than *Rhazya*. As is well known, grape is extremely conservative at the level of syntenic blocks, and this is reflected by the DCJ distance from the median to grape, where the time span is about 165 My, including the time from the core eudicot ancestor to the Gentianales ancestor plus the time to the modern grape genome.

Table 2 Properties of rearrangement and median analyses. m = median, d = DCJ distance

Level	Distance between genomes		Average	Average	Average	Median
L	Block-internal	Whole blocks	$d(m,Rhazya)$	$d(m,\text{coffee})$	$d(m,\text{grape})$	Distance
0	68	322	74.5	38.6	58.9	172
10	72	293	68.5	34.0	54.5	157
20	88	237	52.5	34.1	42.4	129
30	92	186	38.7	31.1	34.2	104
40	92	148	31.0	21.0	28.0	80
50	100	116	27.6	14.3	22.1	64
60	100	105	21.5	16.9	19.6	58
70	96	97	18.9	1.0	19.6	40

The set of medians

Under the kind of distances used in comparative genomics, the median is generally not unique. For our eight levels of resolution, starting at level 0 and ending with level 70, the number of distinct medians was 543, 4296, 1710, 21016, 256, 324, 231 and 425.

Figure 2 shows that the evolutionary distance of each genome may vary slightly from one median to another. This variation is, however, not very large as can be seen from the low-variance distributions, and does not get worse with decreasing resolution. More important is that the distinction between the three genomes, very clear at level 0, is blurred and even reversed (for *Rhazya* and grape) at level 70.

What also changes more drastically are the proportions of translocations and inversions as the median scores decrease, also displayed in Fig. 2. At full resolution, *Rhazya* shows a large number of translocations, while coffee and, to a lesser extent, grape, show more reversals. As resolution decreases, these patterns are disrupted and even reversed. In other words, the structure of the DCJ distance in terms of translocations and reversals, is lost with decreasing resolution.

Distances

Figures 3 and 4 depict the decrease in the total of the DCJ and breakpoint distances between the three genomes and the median calculated from these genomes. In addition, Fig. 3 shows that the relative conservatism of coffee and grape compared to *Rhazya*, very clear at full resolution, is almost completely blurred by the time level 70 is reached.

In contrast, the breakpoint distance decreases more or less proportionately for the three genomes. This simply reflects the fact that the same blocks are deleted from the three genomes at each step,

Breakpoint reuse

In the context of genome rearrangement theory, a measure of the randomness of one genome relative to another is the statistic $r = 2d/b$, where d is the rearrangement distance and b is the number of breakpoints [22–24]. As resolution of the comparison of two genomes decreases, r increases from a minimum value of 1 to a maximum of 2. The quantity r is called "breakpoint reuse" because its increase is sometimes attributed (mistakenly in most cases) to a rearrangement history where almost all reversals and translocations, instead of creating two new breakpoints in the genome, introduce only one new breakpoint in a genome, while also "reusing" an existing breakpoint created by a previous translocation. A number of studies have shown that this interpretation is illusory (e.g. [25]), an artifact of decreasing resolution creating gene orders or block orders effectively random with respect to each other in the two genomes being compared.

Figure 5 suggests that reuse is not constant as resolution decreases, and Fig. 6 confirms the descent (or ascent) into randomness as resolution decreases.

Normalization

It might be thought that the decrease in distances seen in Fig. 5 could be attributable simply to the number of remaining synteny blocks as resolution decreases. The use of normalized distance has been suggested, and used successfully, in different contexts [26, 27], but in Fig. 7 we see that normalizing distances by the number of blocks involved explains only part of the decrease. The remaining decrease must be attributable to the increase of randomness of each genome with respect to the others as resolution decreases.

A non-constructive method

If a chromosome i in genome B consists of many alternating blocks corresponding to just two chromosomes j and k of genome A, it is likely that this i arose from a translocation or a fusion of j and k, followed by several reversals or transpositions only affecting the new chromosome. If on the other hand i contains many blocks, each corresponding to a different chromosome

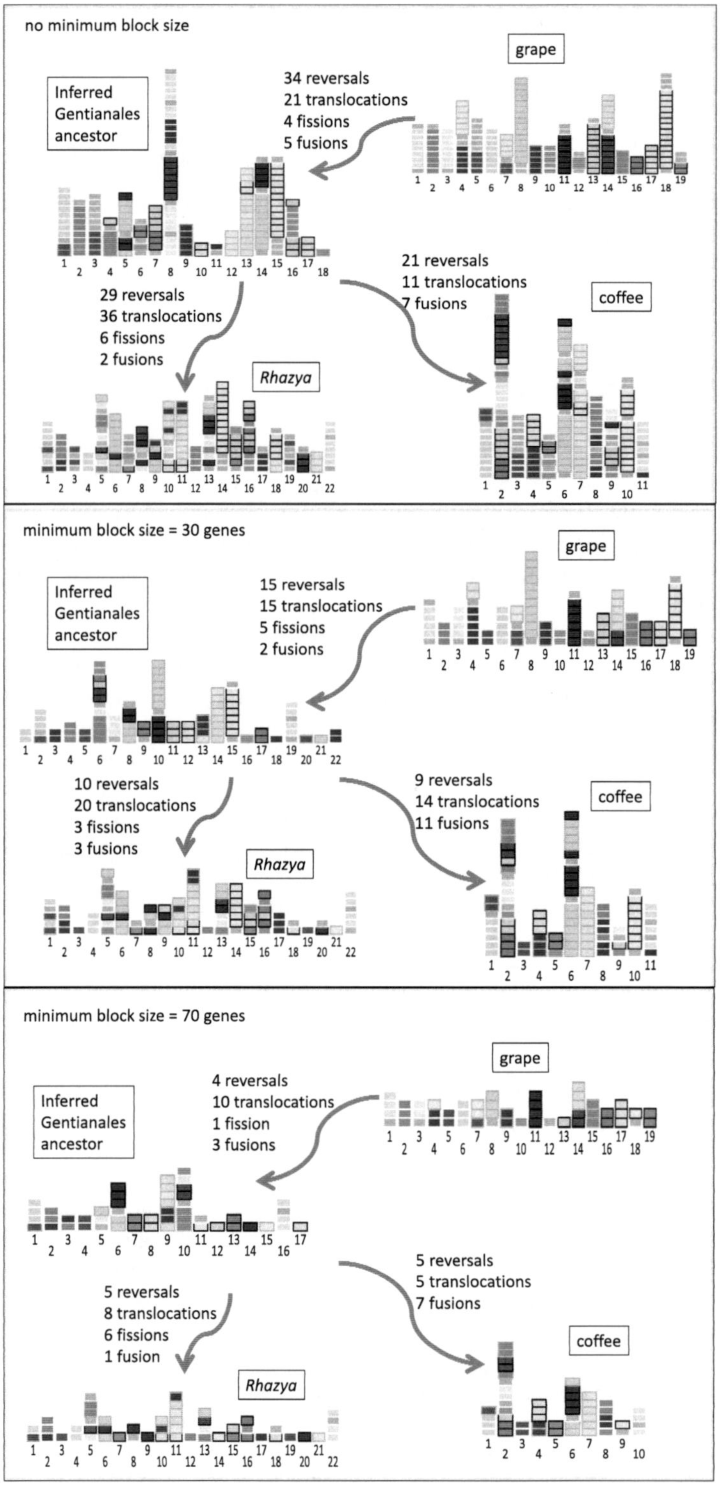

Fig. 1 Inference of Gentianales ancestor at 3 levels of resolution. Grape genome is proxy for the ancestral core eudicot genome. Colours reflect 7 pre-γ chromosomes; triple homeologous copies distinguished by black, white, and no borders. Grey blocks reflect regions of uncertain homeology. Uniform size blocks in each scheme not scaled by number of genes. For individual block identification, see http://albuquerque.bioinformatics.uottawa.ca/Softwares/resolution/index.html

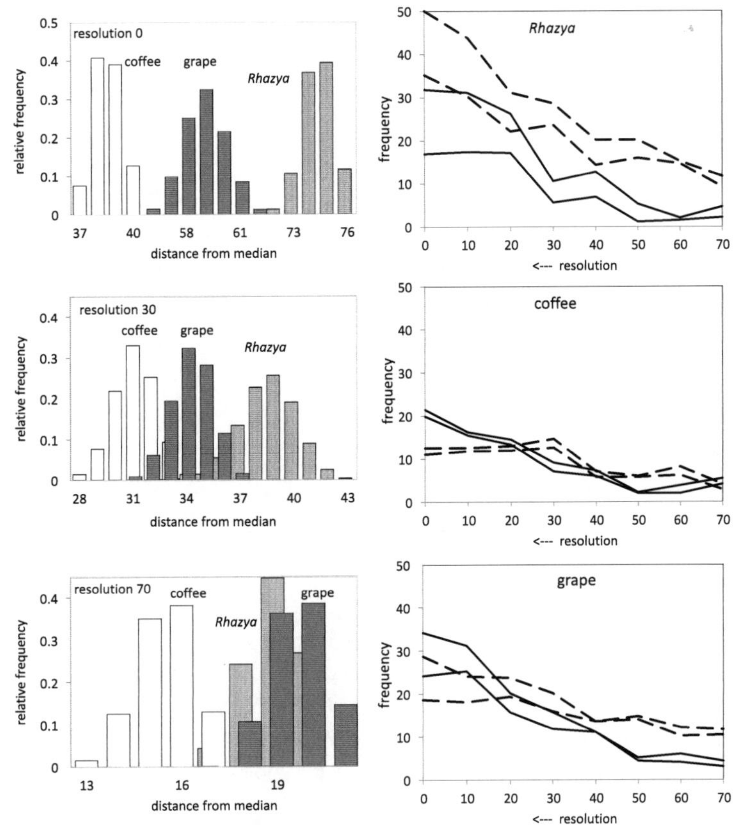

Fig. 2 Left: Effect of resolution on distance from median to *Rhazya*, coffee and grape. From top to bottom, with decreasing resolution, histograms for the three species overlap increasingly, blurring the gene order conservatism of coffee and grape versus the extensive rearrangement propensity of *Rhazya*. Right: Effect of resolution on numbers of translocations and inversions necessary to transform median to *Rhazya*, coffee and grape. Dotted line: translocations. Unbroken line: reversals. Upper and lower lines in each colour represent the maximum and minimum number found when prioritizing or delaying the corresponding operation versus the other operation. Fusions and fissions (not shown) are found at low, comparable, levels except for coffee which shows 5-10 fusions only, with no trend according to resolution

in genome A, then it is more likely that i results from many translocations or fusions of the corresponding chromosomes of A, followed by just a few inversions, or none.

Based on the idea, we developed an inference procedure for the number of rearrangements of various kinds separating genome B from genome A [1, 2]. See also [28, 29] for further work on this theory.

At the outset, assume the first translocation on the lineage from genome A to genome B involves chromosome i. The mild assumption of a uniform density of breakpoints across the genome implies that for any j the "partner" of i in the translocation will be chromosome j with probability $p_i(j) = \frac{p(j)}{1-p(i)}$, where $p(i)$ is the proportion of the genome covered by chromosome i.

If the translocation was with chromosome k, re-assign the labels i and k arbitrarily to the two new chromosomes. Thus the probability that the new chromosome labelled i contains no fragment of genome A chromosome j, where $j \neq i$, is $1 - p_i(j)$. For small $t^{(i)}$, after chromosome i has undergone $t^{(i)}$ translocations, the probability that it contains no fragment of the genome A chromosome j is approximately $(1 - p_i(j))^{t^{(i)}}$, neglecting second-order events, for example, the event that j previously translocated with one or more of the $t^{(i)}$ chromosomes that then translocated with i, and that a secondary transfer to i of material originally from j thereby occurred.

Then the probability that the genome B chromosome i contains at least one fragment from j is approximately $1-(1-p_i(j))^{t^{(i)}}$. Let $c^{(i)}$ be the number of genome A chromosomes with at least one fragment on i, i.e., the number of conserved syntenies on chromosome i, so that

$$c^{(i)} \leq c, \tag{1}$$

where c is the total number of chromosomes. Then

$$E\left(c^{(i)}\right) \approx 1 + \sum_{j \neq i} \left[1 - (1 - p_i(j))^{t^{(i)}}\right] \tag{2}$$

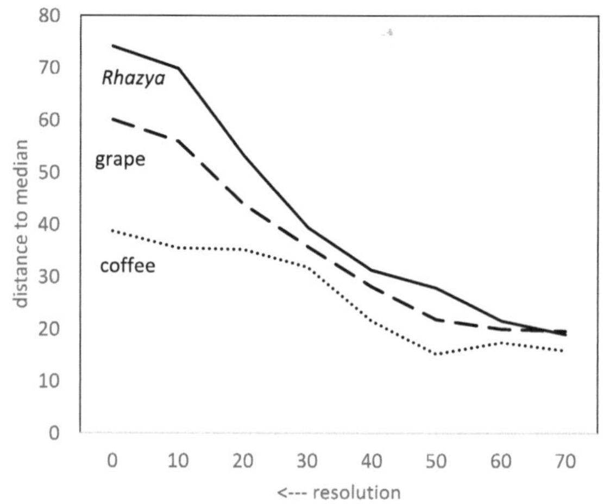

Fig. 3 Effect of resolution on DCJ distance from median to *Rhazya*, coffee and grape

so that

$$c - E\left(c^{(i)}\right) \approx \sum_{j \neq i} (1 - p_i(j))^{t^{(i)}} \tag{3}$$

where the leading 1 in (2) counts the fragment containing the left-hand endpoint of the genome A chromosome i itself. Suppose there have been a total of t translocations in the evolutionary history. Then, $\sum_i t^{(i)} = 2t$ and we can expect these to have been distributed among the chromosomes approximately as $t^{(i)} = 2tp(i)$ so that

$$c^2 - \sum_i E\left(c^{(i)}\right) \approx \sum_i \sum_{j \neq i} (1 - p_i(j))^{2tp(i)}. \tag{4}$$

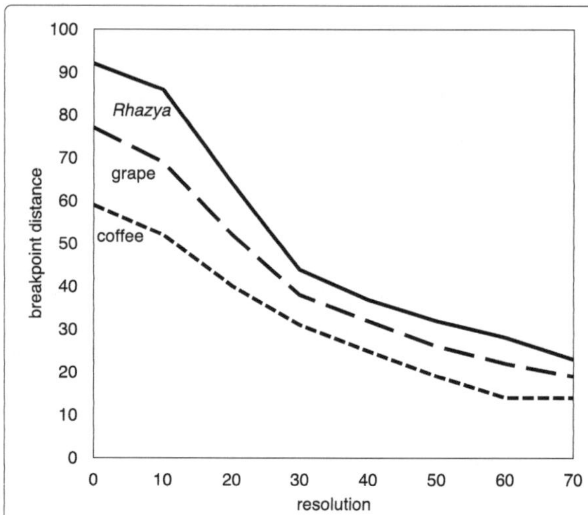

Fig. 4 Effect of resolution on breakpoint distance from median to *Rhazya*, coffee and grape

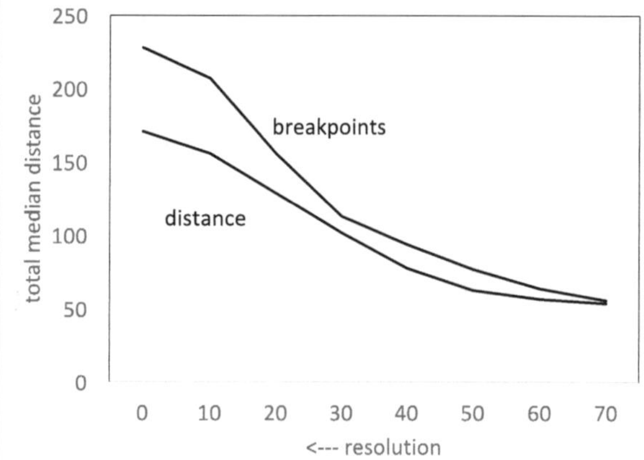

Fig. 5 Effect of resolution on total median scores

Substituting the $c^{(i)}$ for the $E\left(c^{(i)}\right)$ in Eq. (4) suggests solving

$$c^2 - \sum_i c^{(i)} = \sum_i \sum_{j \neq i} (1 - p_i(j))^{2\hat{t}p(i)} \tag{5}$$

to provide an estimator of t.

For the case where A and B have different numbers of chromosomes c and d, and where $q(j)$ is the proportion of B taken up by chromosome j, a better model is

$$cd - \sum_i c^{(i)} = \frac{d-1}{d} \sum_i \sum_j (1 - q(j))^{2p(i)\hat{t}}. \tag{6}$$

If, as with the plant genomes, chromosome lengths are not too variable within a genome, this reduces to

$$1 - \frac{\sum_i c^{(i)}}{cd} = \left(1 - \frac{1}{d}\right)^{1+2\hat{t}/c}. \tag{7}$$

The results of solving this are shown in Fig. 8. Given the high reuse rate for these comparisons, we omitted the

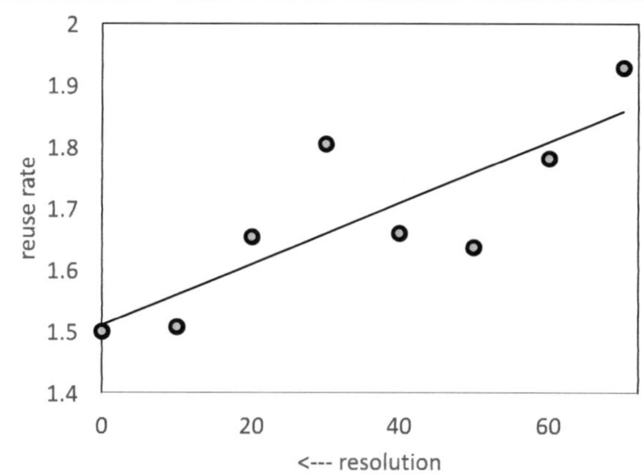

Fig. 6 Effect of resolution on breakpoint reuse. Least-squares fit line $R^2 = 0.68$

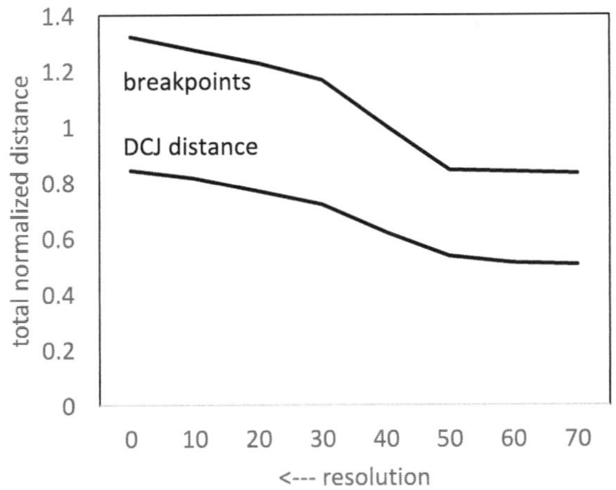

Fig. 7 Effect of resolution on total normalized median scores

factor 2 in solving Eq. (7). This only affects the scale of the y-axis in Fig. 8.

Although the absolute numbers are clearly underestimates, in comparison with Fig. 2, the relative tendencies are the same.

Discussion

This study only shows one side of the coin. The obverse would trace the effect of *increasing* the resolution, testing the consequences of increasing noise in the data. This could be done by changing the default criterion in CoGe for the number of genes required to establish a synteny block from 5 to 4 or even 3. Experience shows that this would allow an increasing proportion of spurious synteny blocks into the data, especially in plant genomes where co-linear paralogs from whole genome duplication and triplication abound.

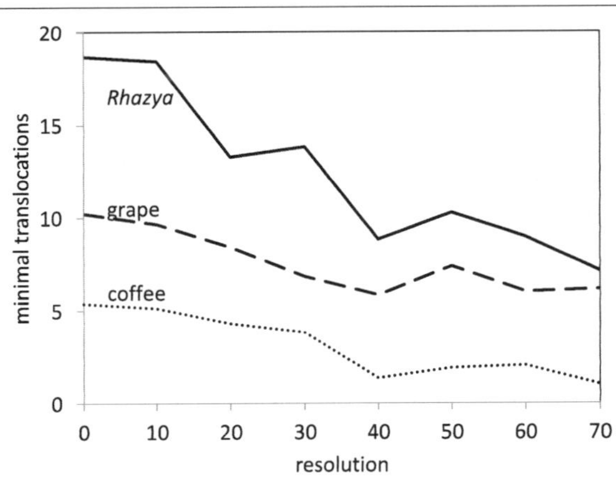

Fig. 8 Effect of resolution on (minimal) translocation number from median to *Rhazya*, coffee and grape

Conclusions

We have shown the multifaceted consequences of the loss of resolution in a typical comparative genomic study of three plant genomes. We reconstructed, at the synteny block level, the ancestral genome of the order Gentianales, based on coffee, *Rhazya stricta*, and outgroup grape. We showed how decreased resolution blurs the differences between the genomes, with respect to evolutionary rate and mutational process. We also showed how reuse rate increases, confirming the randomization of the genomes with respect to each other as resolution decreases.

Abbreviations

DCJ: Double cut and join; K_S: the number of synonymous substitutions per synonymous site

Acknowledgements

Research supported in part by grants to DS from the Natural Sciences and Engineering Research Council of Canada. DS holds the Canada Research Chair in Mathematical Genomics.

Funding

Funding for the publication of this article was provided by Natural Sciences and Engineering Research Council of Canada grant RGPIN-2016-05585.

Availability of data and materials

As indicated in text.

About this supplement

This article has been published as part of *BMC Genomics* Volume 19 Supplement 2, 2018: Selected articles from the 16th Asia Pacific Bioinformatics Conference (APBC 2018): genomics. The full contents of the supplement are available online at https://bmcgenomics.biomedcentral.com/articles/supplements/volume-19-supplement-2.

Authors' contributions

The study was planned by CZ and DS, who also wrote the paper. The breakpoint computations were carried out by MGT and the rearrangements under the probability model were computed by YJ. All authors have read and approved the paper.

Ethics approval and consent to participate

Not applicable.

Consent for publication

Not applicable.

Competing interests

The authors declare that they have no competing interests.

Publisher's Note

Springer Nature remains neutral with regard to jurisdictional claims in published maps and institutional affiliations.

Published: 9 May 2018

References

1. Mazowita M, Haque L, Sankoff D. Stability of rearrangement measures in the comparison of genome sequences. J Comput Biol. 2006;13(2):554–66.
2. Sankoff D, Mazowita M. Estimators of translocations and inversions in comparative maps. In: Lagergren J, editor. RECOMB Workshop on Comparative Genomics. Lecture notes in computer science. New York: Springer; 2004. p. 109–22.
3. Ma J, Zhang L, Suh BB, Raney BJ, Burhans RC, Kent WJ, Blanchette M, Haussler D, Miller W. Reconstructing contiguous regions of an ancestral genome. Genome Res. 2006;16(12):1557–65.

4. Rödelsperger C, Dieterich C. CYNTENATOR: progressive gene order alignment of 17 vertebrate genomes. PloS ONE. 2010;5(1):8861.
5. Pham SK, Pevzner PA. DRIMM-synteny: decomposing genomes into evolutionary conserved segments. Bioinformatics. 2010;26(20):2509–16.
6. Jean G, Nikolski M. SyDiG: uncovering synteny in distant genomes. Int J Bioinforma Res Appl. 2011;7(1):43–62.
7. Proost S, Fostier J, De Witte D, Dhoedt B, Demeester P, Van de Peer Y, Vandepoele K. i-ADHoRe 3.0 – fast and sensitive detection of genomic homology in extremely large data sets. Nucleic Acids Res. 2011;40(2):11.
8. Ghiurcuta CG, Moret BM. Evaluating synteny for improved comparative studies. Bioinformatics. 2014;30(12):9–18.
9. Grabherr MG, Russell P, Meyer M, Mauceli E, Alföldi J, Di Palma F, Lindblad-Toh K. Genome-wide synteny through highly sensitive sequence alignment: Satsuma. Bioinformatics. 2010;26(9):1145–51.
10. Minkin I, Patel A, Kolmogorov M, Vyahhi N, Pham S. Sibelia: A scalable and comprehensive synteny block generation tool for closely related microbial genomes. In: Darling AE, Stoye J, editors. Algorithms in Bioinformatics - 13th International Workshop, WABI 2013. Lecture notes in computer science. New York: Springer; 2013. p. 215–29.
11. Doerr D, Moret BME. Principled synteny using common intervals and hierarchical organization (PSyCHO) Presented at Bernard Moret Festschrift, UC Berkeley. 2017.
12. Lyons E, Freeling M. How to usefully compare homologous plant genes and chromosomes as DNA sequences. Plant J. 2008;53:661–73.
13. Lyons E, Pedersen B, Kane J, Freeling M. The value of nonmodel genomes and an example using SynMap within CoGe to dissect the hexaploidy that predates rosids. Trop Plant Biol. 2008;1:181–90.
14. Denoeud F, Carretero-Paulet L, Dereeper A, Droc G, Guyot R, Pietrella M, Zheng C, Alberti A, Anthony F, Aprea G, Aury JM, Bento P, Bernard M, Bocs S, Campa C, Cenci A, Combes MC, Crouzillat D, Da Silva C, Daddiego L, De Bellis F, Dussert S, Garsmeur O, Gayraud T, Guignon V, Jahn K, Jamilloux V, Joët T, Labadie K, Lan T, Leclercq J, Lepelley M, Leroy T, Li LT, Librado P, Lopez L, Muñoz A, Noel B, Pallavicini A, Perrotta G, Poncet V, Pot D, Priyono, Rigoreau M, Rouard M, Rozas J, Tranchant-Dubreuil C, VanBuren R, Zhang Q, Andrade AC, Argout X, Bertrand B, de Kochko A, Graziosi G, Henry RJ, Jayarama, Ming R, Nagai C, Rounsley S, Sankoff D, Giuliano G, Albert VA, Wincker P, Lashermes P. The coffee genome provides insight into the convergent evolution of caffeine biosynthesis. Science. 2014;345(6201):1181–84.
15. Sabir JS, Jansen RK, Arasappan D, Calderon V, Noutahi E, Zheng C, Park S, Sabir MJ, Baeshen MN, Hajrah NH, Khiyami MA, Baeshen NA, Obaid AY, Al-Malki AL, Sankoff D, El-Mabrouk N, Ruhlman TA. The nuclear genome of *Rhazya stricta* and the evolution of alkaloid diversity in a medically relevant clade of Apocynaceae. Scientific reports. 2016;6:33872.
16. Jaillon O, Aury JM, Noel B, Policriti A, Clepet C, Casagrande A, Choisne N, Aubourg S, Vitulo N, Jubin C, Vezzi A, Legeai F, Hugueney P, Dasilva C, Horner D, Mica E, Jublot D, Poulain J, Bruyére C, Billault A, Segurens B, Gouyvenoux M, Ugarte E, Cattonaro F, Anthouard V, Vico V, Del Fabbro C, Alaux M, Di Gaspero G, Dumas V, Felice N, Paillard S, Juman I, Moroldo M, Scalabrin S, Canaguier A, Le Clainche I, Malacrida G, Durand E, Pesole G, Laucou V, Chatelet P, Merdinoglu D, Delledonne M, Pezzotti M, Lecharny A, Scarpelli C, Artiguenave F, Pè ME, Valle G, Morgante M, Caboche M, Adam-Blondon AF, Weissenbach J, Quétier F, Wincker P. The grapevine genome sequence suggests ancestral hexaploidization in major angiosperm phyla. Nature. 2007;449(7161):463–7.
17. Yancopoulos S, Attie O, Friedberg R. Efficient sorting of genomic permutations by translocation, inversion and block interchange. Bioinformatics. 2005;21(16):3340–6.
18. Zheng C, Chen E, Albert VA, Lyons E, Sankoff D. Ancient eudicot hexaploidy meets ancestral eurosid gene order. BMC genomics. 2013;14(7):3.
19. Xu AW, Sankoff D. Decompositions of multiple breakpoint graphs and rapid exact solutions. In: Crandall KA, Lagergren J, editors. Algorithms in Bioinformatics, 8th International Workshop, WABI 2008. Lecture notes in computer science. New York: Springer; 2008. p. 25–37.
20. Tannier E, Zheng C, Sankoff D. Multichromosomal median and halving problems under different genomic distances. BMC Bioinformatics. 2009;10(1):120.
21. Kumar S, Stecher G, Suleski M, Hedges SB. Timetree: a resource for timelines, timetrees, and divergence times. Mol Biol Evol. 2017;34(7): 1812–9.
22. Bourque G, Pevzner PA, Tesler G. Reconstructing the genomic architecture of ancestral mammals: lessons from human, mouse, and rat genomes. Genome Res. 2004;14(4):507–16.
23. Sankoff D, Trinh P. Chromosomal breakpoint reuse in genome sequence rearrangement. J Comput Biol. 2005;12(6):812–21.
24. Sankoff D. The signal in the genomes. PLoS Comput Biol. 2006;2(4):35.
25. Attie O, Darling AE, Yancopoulos S. The rise and fall of breakpoint reuse depending on genome resolution. BMC Bioinformatics. 2011;12(S9):1.
26. Sankoff D, Bryant D, Deneault M, Lang BF, Burger G. Early eukaryote evolution based on mitochondrial gene order breakpoints. J Comput Biol. 2000;7(3–4):521–35.
27. Sankoff D, Deneault M, Bryant D, Lemieux C, Turmel M. Chloroplast gene order and the divergence of plants and algae, from the normalized number of induced breakpoints. In: Sankoff D, Nadeau JH, editors. Comparative Genomics. New York: Springer; 2000. p. 89–98.
28. Tittmann L. Calibration of a probabilistic model of DNA evolution. Master's thesis, Humboldt-Universität zu Berlin, Wirtschaftswissenschaftliche Fakultät. 2015.
29. Lucas J. Etude de l'évolution de l'ordre des gènes de vertébrés par simulation. PhD thesis, Université Pierre et Marie Curie-Paris VI. 2016.

Liu et al. BMC Genomics 2018, 19(Suppl 2):111
https://doi.org/10.1186/s12864-018-4457-8

BMC Genomics

RESEARCH Open Access

Genome-wide identification and comprehensive analysis of microRNAs and phased small interfering RNAs in watermelon

Li Liu[1†], Shuchao Ren[1,2†], Junqiang Guo[3], Qingyi Wang[3], Xiaotuo Zhang[4], Peiran Liao[1], Shipeng Li[1], Ramanjulu Sunkar[5] and Yun Zheng[2,3*]

From The Sixteenth Asia Pacific Bioinformatics Conference 2018
Yokohama, Japan. 15-17 January 2018

Abstract

Background: MicroRNAs (miRNAs) are a class of endogenous small non-coding RNAs involved in the post-transcriptional gene regulation and play a critical role in plant growth, development and stress responses. Watermelon (*Citrullus lanatus* L.) is one of the important agricultural crops worldwide. However, the watermelon miRNAs and phasiRNAs and their functions are not well explored.

Results: Here we carried out computational and experimental analysis of miRNAs and phased small interfering RNAs (phasiRNAs) in watermelon by analyzing 14 small RNA profiles from roots, leaves, androecium, petals, and fruits, and one published small RNA profile of mixed tissues. To identify the targets of miRNAs and phasiRNAs, we generated a degradome profile for watermelon leaf which is analyzed using the SeqTar algorithm. We identified 97 conserved pre-miRNAs, of which 58 have not been reported previously and 348 conserved mature miRNAs without precursors. We also found 9 novel pre-miRNAs encoding 18 mature miRNAs. One hundred and one 21 nucleotide (nt) PHAS loci, and two hundred and forty one 24 nt PHAS loci were also identified. We identified 127 conserved targets of the conserved miRNAs and TAS3-derived tasiRNAs by analyzing a degradome profile of watermelon leaf.

Conclusions: The presented results provide a comprehensive view of small regulatory RNAs and their targets in watermelon.

Keywords: *Citrullus lanatus* L., High-throughput sequencing, microRNA, PHAS, Phased small interfering RNA (phasiRNA), Bioinformatics

Background

Plant small RNAs (sRNAs), with 21 to 24 nucleotides (nt), play crucial roles in a variety of biological processes, including development, stress responses, defense and epigenetic modifications [1–3]. Based on their origin and biogenesis, small RNAs in plants can be divided into two main categories: microRNA (miRNA) and small interfering RNA (siRNA). The primary transcripts of MIR genes are transcribed by RNA polymerase II [4]. Such primary miRNA transcripts form typical hairpin-like structure that are cleaved twice by Dicer Like protein (DCL1) in the nucleus to excise a miRNA:miRNA* duplex [5]. Then, the duplex is exported to cytoplasm [6]. In the cytoplasm, miRNAs are loaded into an RNA-induced silencing complex (RISC) that normally contains an Argonaute (AGO) protein, and guide the RISC to cause site-specific cleavage or repression of the mRNA targets [7, 8]. The other strand of the duplex, i.e., miRNA*, is degraded [9]. Most plant miRNAs negatively regulate their target genes through homolog based mRNA cleavage or translation inhibition

*Correspondence: zhengyun5488@gmail.com
†Equal contributors
[2]Yunnan Key Laboratory of Primate Biomedical Research, Institute of Primate Translational Medicine, Kunming University of Science and Technology, Kunming 650500, China
[3]Faculty of Information Engineering and Automation, Kunming University of Science and Technology, Kunming 650500, China
Full list of author information is available at the end of the article

at post-transcriptional level [3, 6, 10], however some miRNAs, especially metazoan miRNAs, may activate their targets [11–13].

In addition to miRNAs, small interfering RNAs (siRNAs) also have important functions in plants. Especially, phasiRNA is a class of secondary siRNAs that are generated precisely in 21 or 24 nt phased pattern initiated at a specific position due to miRNA guided activity. A type of phasiRNAs are *trans*-acting siRNAs (tasiRNAs) because they repress their target transcripts from other loci of the genome at post-transcriptional level. The primary transcripts of tasiRNAs are non-coding and used to generate double strand RNAs (dsRNAs) by RDR6 (RNA-dependent RNA polymerase 6) [14]. The dsRNAs are then cleaved by DCL4 to form phased 21 nt segments [15–17] or by DCL5 to form 24 nt phased segments in rice [18, 19]. The precise phasing of tasiRNAs is guided by miRNAs [15] through either two [20] or one [18, 21–24] miRNA binding site on the primary tasiRNA transcripts. Four families of tasiRNA loci, named TAS1 to TAS4, have been identified in *Arabidopsis thaliana* [15, 17, 25]. Recent studies suggest that some coding genes, especially PPR [17, 24, 26], NB-LRR disease resistance proteins [24, 27–31], MYB transcription factors [24, 32, 33], also generate phasiRNAs, and their corresponding loci are called as PHAS loci [27]. These PHAS loci are also triggered by one or two miRNA binding sites [14, 27]. *TAS3* derived tasiARFs are the only phasiRNAs that have been validated to target ARF genes in *trans* [15, 20, 24]. The functions of most phasiRNAs are still largely unknown [14].

Watermelon is one of the 20 most important agricultural crops worldwide [34]. Although watermelon is economically important, only a few studies [34–38] have paid attention on the miRNAs in watermelon, and there are still no studies about phasiRNAs in watermelon.

To systematically investigate the miRNAs and phasiRNAs in watermelon, we sequenced 14 small RNA profiles from 5 tissues, i.e., roots, leaves, androecium, petals, and fruits. Comprehensive computational analyses of these small RNA profiles lead to identification of 97 pre-miRNAs of which 58 have not been reported, 101 PHAS loci encoding 21 nt phasiRNAs (including a new TAS3 locus), and 241 PHAS loci encoding 24 nt phasiRNAs. A degradome profile of watermelon leaf was generated for identifying targets of miRNAs and phasiRNAs. Totally, 127 conserved targets for conserved miRNAs and TAS3-derived tasiRNAs were identified in the analysis. These results significantly enhanced our understanding of small RNA guided gene regulations in watermelon.

Methods

Materials and small RNA sequencing profiles

The watermelon plants were grown in the greenhouse of Faculty of Life Science and Technology, Kunming University of Science and Technology (24° 51′ 0″ N, 102° 52′ 2″ E, altitude 1835 m), Yunnan, China. During the experiments, the daily average temperature was 25 °C, the daily maximum difference in temperature was 10 °C and humidity was 60-80%. Fourteen samples from five tissues, i.e., root, leaf, petal, androecium and pre-mature fruit, of watermelon plants without treatments were collected and frozen using liquid nitrogen immediately (Additional file 1: Supplementary Table S1). The 14 samples were stored at -80 °C until RNAs were extracted. Total RNAs were extracted from samples using the Trizol reagent according to the manufacturer's protocol. The integrities of the RNAs were checked using an ultraviolet spectrophotometer (Hoefer, MA, USA), based on the ratio of the optical density at 260 nm to that at 280 nm (OD260/280) and were also assessed by electrophoresis in a denaturing formaldehyde agarose gel, based on visual comparison of the 18S and 28S ribosomal RNAs. The RNA samples with OD260/280 between 1.8 and 2.0 were checked for the total quantities. Only samples with at least 20 μg were chosen for preparation of sRNA sequencing libraries. 20 μg total RNAs dissolved in 35 μl were delivered to the sequencing facility. The small RNAs of the samples were isolated from total RNAs and were sequenced using Illumina HiSeq 2000 sequencer. The obtained small RNA profiles had been deposited to NCBI under GEO the accession number, GSE102030. The qualities of the obtained sRNA profiles were evaluated with the FASTQC program (https://www.bioinformatics.babraham.ac.uk/projects/fastqc/) (in Additional file 2: Figure S1).

Identification of conserved and novel miRNAs

The computational pipelines for analyzing the obtained small RNA profiles were shown in supplementary materials (Additional file 2: Figure S2).

The mRNA and genome sequences of watermelon were downloaded from Cucumber Genome Database (http://cucurbitgenomics.org/). Watermelon (97103) genome and annotation (v6) were downloaded.

Our computational methods for analyzing the small RNA libraries was reported previously [39]. In brief, small RNAs were extracted after removing raw reads with more than 5 low scored nucleotides (< 25) and 3' adaptor sequences of the remaining reads (see Additional file 2: Figure S2d). Unique small RNAs were obtained after eliminating redundant sequences. The unique sRNAs were aligned to genome and mRNAs of watermelon to calculate the numbers of reads and unique sequences mapped to the genome and mRNAs. Then unique sequences mapped to known non-coding RNAs (rRNAs, tRNAs, snRNAs, snoRNAs) and repeats were removed from unique RNAs by aligning to databases Rfam (r11) [40], NONCODE (v3.0) [41], Silva [42], The TIGR Plant Repeat Databases

[43] and Repbase (r20) [44] using SOAP2 [45]. Remaining reads were mapped to the miRBase database (r21) [46] to calculate the frequencies of conserved miRNAs. The numbers of reads that were aligned to different types of molecules were counted and summarized (see Additional file 1: Table S2).

After this step, the remaining unique reads were aligned to genome to predict novel miRNAs. The surrounding sequences (down- and up-stream) were extended by 300 bp which were used to predict fold-back structure using RNAfold [47]. The structures with at least 18 paired nucleotides and a folding energy of smaller than -40 Kcal/mol were kept as putative precursors for further analysis. The small RNA reads were mapped to the obtained putative precursors. Finally, novel miRNA identification and annotation were strictly based on appearance of miRNA* and predictable fold back structures for the miRNA precursor sequences, as suggested by Meyers et al. [48].

As described previously [39], mature miRNA sequences from all plant species were downloaded from the miRBase (v21) [46] from which unique miRNA sequences were obtained (see Additional file 2: Figure S2a). Next, these unique miRNA sequences were used as queries against the watermelon genome using BLASTN to predict known miRNA homologs in watermelon that were not represented in small RNA libraries. Hits with no more than two mismatches were identified and the flanking regions (80nt, 130nt, 180nt down stream and upstream) to the mapped mature miRNAs were isolated and used to predict fold-back structures using the RNAfold. The predicted fold-back structures were examined for the presence of miRNA on the same arm of the hairpin as the known family members from other plants. These precursor sequences were further evaluated by MIRcheck and selected candidates that have ≤ 5 mismatches, ≤ 2 bulged or asymmetrically unpaired nucleotides, and ≤ 3 continuous mismatches within the mature miRNA.

We compared the identified conserved miRNAs with those reported previously [34–38]. If the sequences of a mature miRNA was the same as reported earlier, the conserved miRNA was named the same as reported previously. The remaining conserved miRNAs were named by using upper case MIR followed by the family name, and alphabetical letters in lower case if these have not been reported earlier.

Analysis of miRNA expression patterns in different tissues

The raw frequencies of mature miRNAs for samples of different tissues were compared with edgeR [49]. miRNAs with multiple-test corrected P-values of smaller than 0.05 were regarded as significantly deregulated miRNAs in the different tissues.

miRNA with abundances of at least 5 RPTM (Reads Per Ten Million sequencing tags) in at least two of the 14 samples and standard deviation of at least 1 were selected to perform hierarchical clustering. The normalized RPTM values plus one were log scaled to calculate correlation coefficient (CC) values between samples. Then the CC values were applied to the pheatmap function in the pheatmap library in R to perform hierarchical clustering. miRNAs with average abundances of at least 5 RPTM in the 14 samples were used to perform Principle Component Analysis (PCA). Log-scaled normalized RPTM values plus one were applied to the prcomp function in the psych library in R to perform PCA.

Identification of TAS3 loci in watermelon

As done previously [50], TAS3 derived tasiRNAs, siRNAs that target the ARF family genes, in Arabidopsis and rice were collected. Then, these tasiRNAs were aligned to the watermelon genome by allowing at most 2 mismatches. The neighboring sequences of the matched loci were cut out to include 250 nt from both 5' and 3' sides. Then we examined the typical miR390 complementary sites around the conserved tasiRNAs by predicting the miR390 sites on the cut-out sequences.

Identification of PHAS loci in watermelon

To predict PHAS loci and phasiRNAs, we used one more small RNA profile of mixed watermelon tissues in our previous work [35] in addition to 14 small RNA profiles generated in the current study. This sRNA profile was downloaded from the from the NCBI GEO database with the accession number GSM936361. This profile and the combined file of the 14 sRNA profiles generated in this study were used to predict PHAS loci and phasiRNAs, respectively.

The PHAS loci and phasiRNA were predicted as described previously [39, 51]. The unique sequences in the small RNA libraries were mapped to Repbase (r20) [44] and The TIGR Plant Repeat Databases [43] to remove sRNAs mapped to repeats, then the remaining sRNAs were mapped to the genome of watermelon using SOAP2 [45] (see Additional file 2: Figure S2b). A self-written program was used to scan the genome and cDNA sequences using a window of 210 nt or 240 nt (ten 21 nt or 24 nt) respectively. A two-nucleotide positive offset was used to calculate the positions of siRNAs on the anti-sense strand because the existence of two-nucleotide over-hang at the 3'-end of siRNA duplex [17, 24, 26, 27]. Then a P-value was calculated for each of the windows using a modified version of methods in [26],

$$P(X = k) = \sum_{X=k}^{m} \frac{\binom{20m}{n-k}\binom{m}{k}}{\binom{21m}{n}}, \quad (1)$$

where n was the number of unique 21 nt (or 24 nt) sRNAs mapped within a window, k was the number of phased unique 21 nt (or 24 nt) sRNAs within the window, and m was the number of phases. In this study, m was set to 10.

And a phase score was calculated for each position of the genome and cDNA sequences using the method in [52]. For a window started at a position with more than three phased unique sRNAs, i.e., when $k \geq 3$,

$$PhaseScore = \ln\left(1 + 10 \times \frac{\sum_{i=1}^{m} P_i}{1 + \sum_{i=1}^{m} U_i}\right)^{k-2}, \quad (2)$$

where P_i was the number of phased reads at the ith phase from the position, U_i was the number of non-phased reads at the ith phase from the position, and m was the number of phases in the window, and k was the number of unique phased siRNAs in the window. m was 10 in this study.

The window with a P-value less than 0.05 was extended 100 bp at both 5'- and 3'-ends, then the overlapped windows were merged. The P-values of the merged windows were used to calculate the false positive rates using the method in [53]. The merged windows with a maximal phase scores of larger than pre-determined threshold and multiple test corrected P-values of smaller than 0.05 were reported as PHAS loci. The predicted PHAS loci were named with its chromosome and a unique serial number for each chromosome. "P21" and "P24" were added at the beginning of the predicted PHAS loci encoding 21 and 24 nt phasiRNAs. The neighboring PHAS loci were predicted as PHAS clusters if the distances between individual PHAS loci were smaller than 2000 base pairs. The phased siRNAs of the predicted PHAS loci were reported as phasiRNAs. The phasiRNAs of a PHAS loci were named by adding siR and a serial number to the name of the PHAS loci.

Degradome sequencing

A leaf sample of a watermelon plant grown in greenhouse at Kunming University of Science and Technology, Kunming, Yunnan, China, was frozen in liquid nitrogen immediately after harvesting. The total RNA from the leaf of a watermelon plant was extracted using the Trizol reagent according to the manufacturer's protocol. The integrity of the RNA was checked with an ultraviolet spectrophotometry and 2100 BioAnalyzer. The degradome of polyadenylated transcripts was sequenced using Illumina HiSeq 4,000 sequencer. The degradome sequencing profile had been deposited into the NCBI GEO database under the accession number, GSE102030.

The quality of the obtained degradome profile was evaluated with the FASTQC program. The obtained degradome profile was processed with a self-developed pipeline (see Additional file 2: Figure S2c). Firstly, the raw degradome sequencing profile was filtered to remove low quality reads that have low scored nucleotides (< 20) in the first 20 nucleotides. Next, the first 20 nucleotides in remaining reads were used in later analysis. The unique sequences were obtained and the frequencies of the unique sequences were calculated. Then, similar to sRNA profiles, unique sequences were aligned to genome and mRNAs of watermelon, Rfam (r11) [40], NONCODE (v3.0) [41], Silva [42], The TIGR Plant Repeat Databases [43] and Repbase (r20) [44] using SOAP2 [45] to calculate the numbers of reads and unique sequences mapped to different categories of molecules as well as repeats. The SOAP2 alignment result of mapping degradome reads to watermelon mRNAs was used to predict miRNA and siRNA targets, as detailed below.

Identification of miRNA/siRNA targets in watermelon

The targets of miRNAs and tasiRNAs were predicted with the SeqTar algorithm [54]. For conserved miRNAs and tasiRNAs, the targets that had less than or equal to four mismatches were used for further analysis. For novel miRNAs, only targets with at least 1 valid read and less than 4 mismatches were used.

The targets of phasiRNAs were also predicted with the SeqTar algorithm. Only targets that have no mismatches or at least one valid degradome read and less than or equal to 3 mismatches were used for further analysis.

Identifying miRNA complementary sites on PHAS loci

The SeqTar algorithm was used to predict miRNA complementary sites on the original transcripts of PHAS loci. For conserved miRNAs, the targets that have at least one valid read, i.e., read started at the 9th to 11th positions of a miRNA binding site (as defined in [54]), or targets that have less than 4 mismatches were used for further analysis.

Results and discussion

High-throughput sequencing profiles of watermelon Small RNAs

We generated 14 small RNA libraries from the pooled RNA isolated from five different tissues (root, leaf, petal, androecium and premature fruit). Plants were grown under greenhouse conditions without special treatments. The small RNAs of these 14 samples were sequenced using Illumina HiSeq 2000 sequencer. Initially the 3' adapter from the raw reads was trimmed and small RNAs ranging between 18–30 nucleotides were extracted. Upon sequencing of these 14 small RNA libraries, we obtained approximately 20 million reads for each of the libraries. Totally, we obtained 280,698,505 total reads represented by 60,516,813 unique small RNAs (Additional file 1: Table S1). After examining the scores per nucleotides with FASTQC, the qualities of the obtained sRNA sequencing profiles were generally good with scores above 30 for the first 25 nt (see Additional file 2: Figure S1), which include most small RNAs with 21 to 24 nt. The small RNA sequences were aligned to mRNAs of watermelon, precursors of miRNAs in the miRBase (v21), other non-coding RNAs besides miRNAs, repeat elements and genome of watermelon with SOAP2 (see details in Figure S2 and in Materials and methods). Most reads could be aligned to the genome of watermelon (see Additional file 1: Table S2), suggesting high qualities of the obtained sRNA profiles. The total reads of 21 nt and 24 nt are over-represented while the unique reads of 23 nt and 24 nt are over-represented in all four libraries (Additional file 2: Figure S3). The 21 nt reads and unique sequences represents the only peak for those sequences mapped to the miRNA precursors (Additional file 2: Figure S3), consistent with the lengths of mature plant miRNAs.

Identification of conserved miRNAs and their expression patterns in different tissues

By aligning mature miRNAs from other species to watermelon genome (see details in Figure S2 and in Materials and methods), we identified 97 pre-miRNAs of which 58 have not been reported previously (Fig. 1a and Additional file 1: Table S3). We also found 348 mature miRNAs without precursor by aligning the sRNA profiles to the mature miRNAs in the miRBase (v21) (Additional file 1: Table S3). When compared to two model species, *Arabidopsis thaliana* and rice, we found 80 members that belong to 20 miRNA families that are highly conserved (Additional file 1: Table S4).

Because we have sequenced small RNAs from five different tissues of watermelon, it is feasible to assess the differences in miRNA expression patterns in root, leaf, petal, androecium and premature fruit. We performed Principle Component Analysis and Hierarchical Clustering based on the normalized frequencies of mature miRNAs (Additional file 1: Table S5). The samples from the same tissues were clustered together and samples from different tissues were clearly differentiated (Fig. 1b). We also found the strong correlation among the samples of the same tissues, meanwhile samples from different tissues have much lower correlation coefficient values (Fig. 1c). These results suggest that miRNAs in different tissues have important functions and perhaps involve in different physiological processes in different tissues.

We also compared the abundances of conserved miRNAs for samples from root, leaf, petal, and pre-mature fruit with edgeR [49] to identify deregulated miRNAs in different tissues. This analysis has identified 87 to 153 deregulated miRNAs between the compared tissues (Additional file 1: Table S6). For examples, MIR156i_5p has higher expression levels in leaves than in the roots, petals, and fruits; MIR319g has higher expression level in petals than in leaves; MIR167f-5p has higher expression level in petals than in roots (Additional file 1: Table S7).

We also identified 9 putative novel miRNAs in watermelon (Additional file 1: Table S8). Although miRNA and miRNA* were identified for these miRNAs, their expression levels are generally very low when being compared to conserved miRNAs, suggesting that these novel miRNAs may be non-functional or not functionally important.

Identifying miRNA targets in watermelon

To identify the miRNA targets, we generated a degradome profile with over 23 million reads, of which most could be aligned to watermelon genome (Additional file 1: Table S9). The nucleotide scores of the obtained degradome were of high quality since most nucleotides had scores above 35 (Additional file 2: Figure S4). We identified targets of conserved miRNA by analyzing the

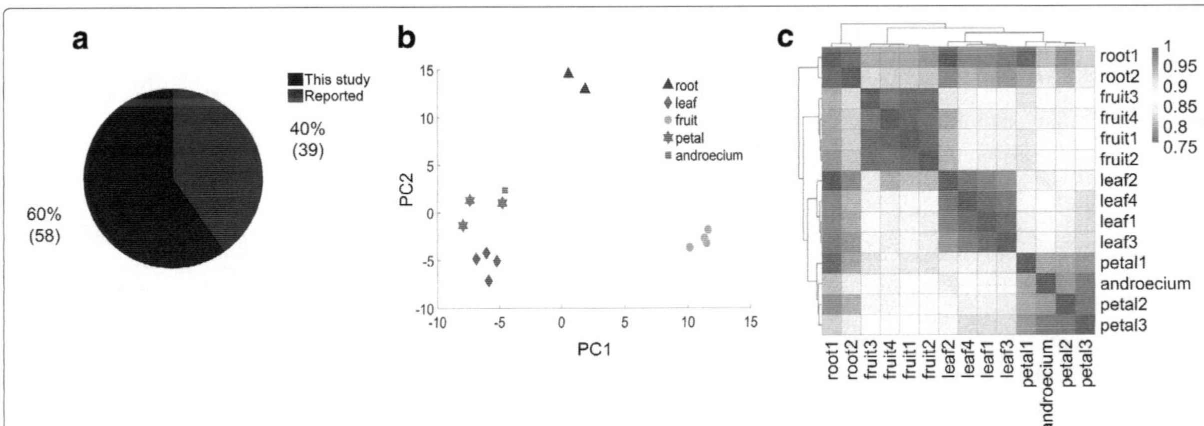

Fig. 1 The conserved miRNAs in watermelon and their expression patterns in different tissues. **a** The number of pre-miRNAs identified in this study and that reported in literature. **b** The PCA analysis of miRNA expression profiles in different tissues. **c** The hierarchical clustering of miRNA expression profiles in different tissues

obtained degradome profile of watermelon leaf using the SeqTar algorithm [54]. This analysis revealed 127 conserved targets for conserved miRNA families and TAS3 generated tasiARFs (Additional file 2: Tables S10 and S11).

Some of the conserved miRNA targets are shown in Fig. 2. The miR160 family targets 5 ARF genes (in Additional file 1: Table S10), of which two are shown in Fig. 2a and 2b. The miR166 family targets five HD-Zip transcription factors (in Additional file 1: Table S10), of

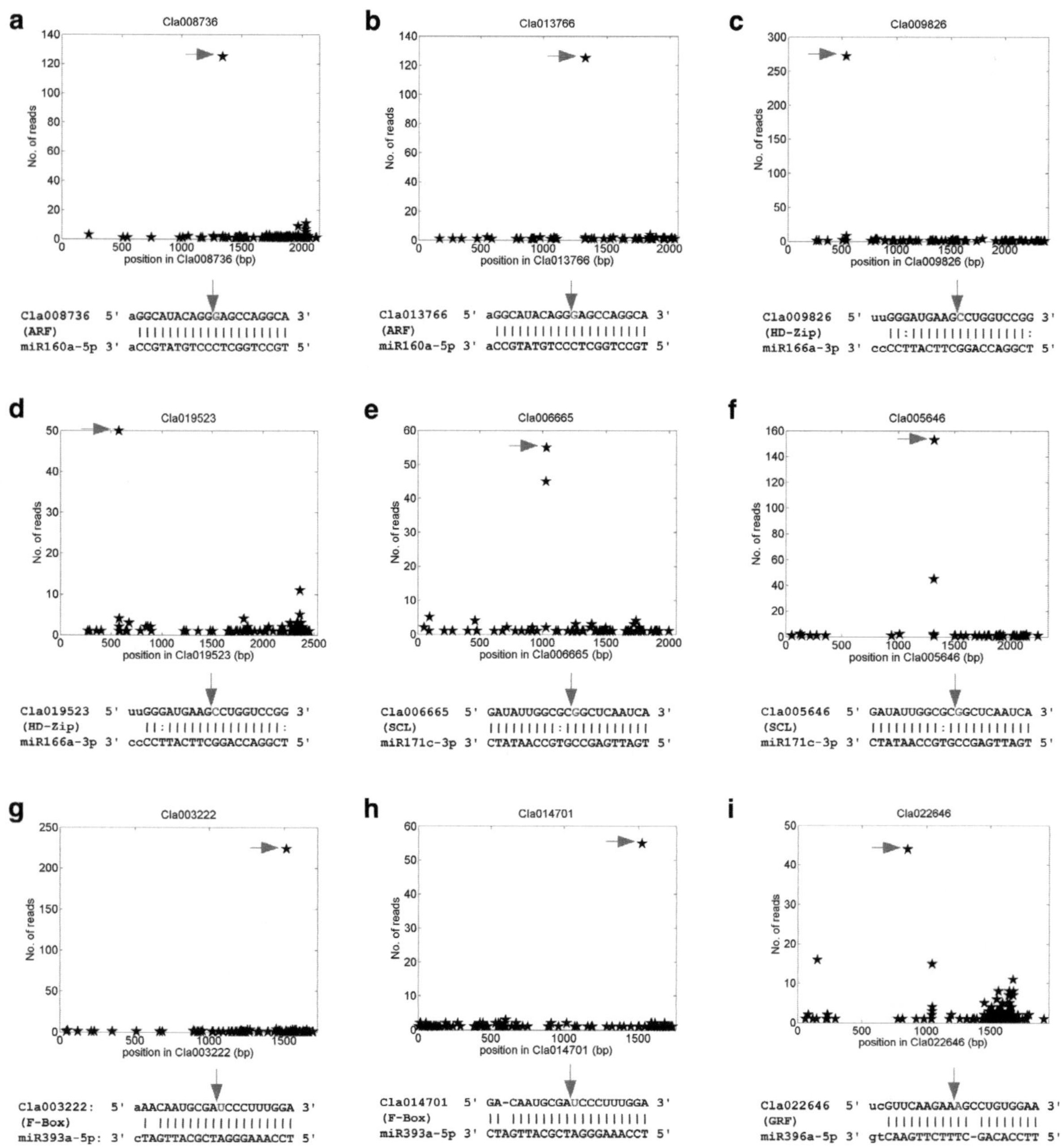

Fig. 2 Some of the identified conserved miRNA targets. The x-axis is the position on the transcript, and y-axis is the number of reads detected from a position. The arrows in the upper parts correspond to the positions pointed by the arrows of the same colors in the lower parts. **a** miR160a-5p:Cla008736, an ARF gene. **b** miR160a-5p:Cla013766, an ARF gene. **c** miR166a-3p:Cla009826, an HD-Zip gene. **d** miR166a-3p:Cla019523, an HD-Zip gene. **e** miR171c-3p:Cla006665, a GRAS/SCL transcription factor gene. **f** miR171c-3p:Cla005646, a GRAS/SCL transcription factor gene. **g** miR393a-5p:Cla003222, an F-box gene. **h** miR393a-5p:Cla014701, an F-box gene. **i** miR393a-5p:Cla003222, a GRF gene

which two are shown in Fig. 2c and 2d. The miR171 family targets eight SCL transcription factors (in Additional file 1: Table S10), of which two are shown in Fig. 2e and 2f. The miR393 family targets 8 F-box genes (in Additional file 1: Table S10), of which two are shown in Fig. 2g and 2h. The miR396 family targets 8 GRF transcriptional factors (in Additional file 1: Table S10), of which one is shown in Fig. 2i. Several other conserved miRNAs also cleaved their targets with significant number of valid degradome reads (see Additional file 2: Figure S5).

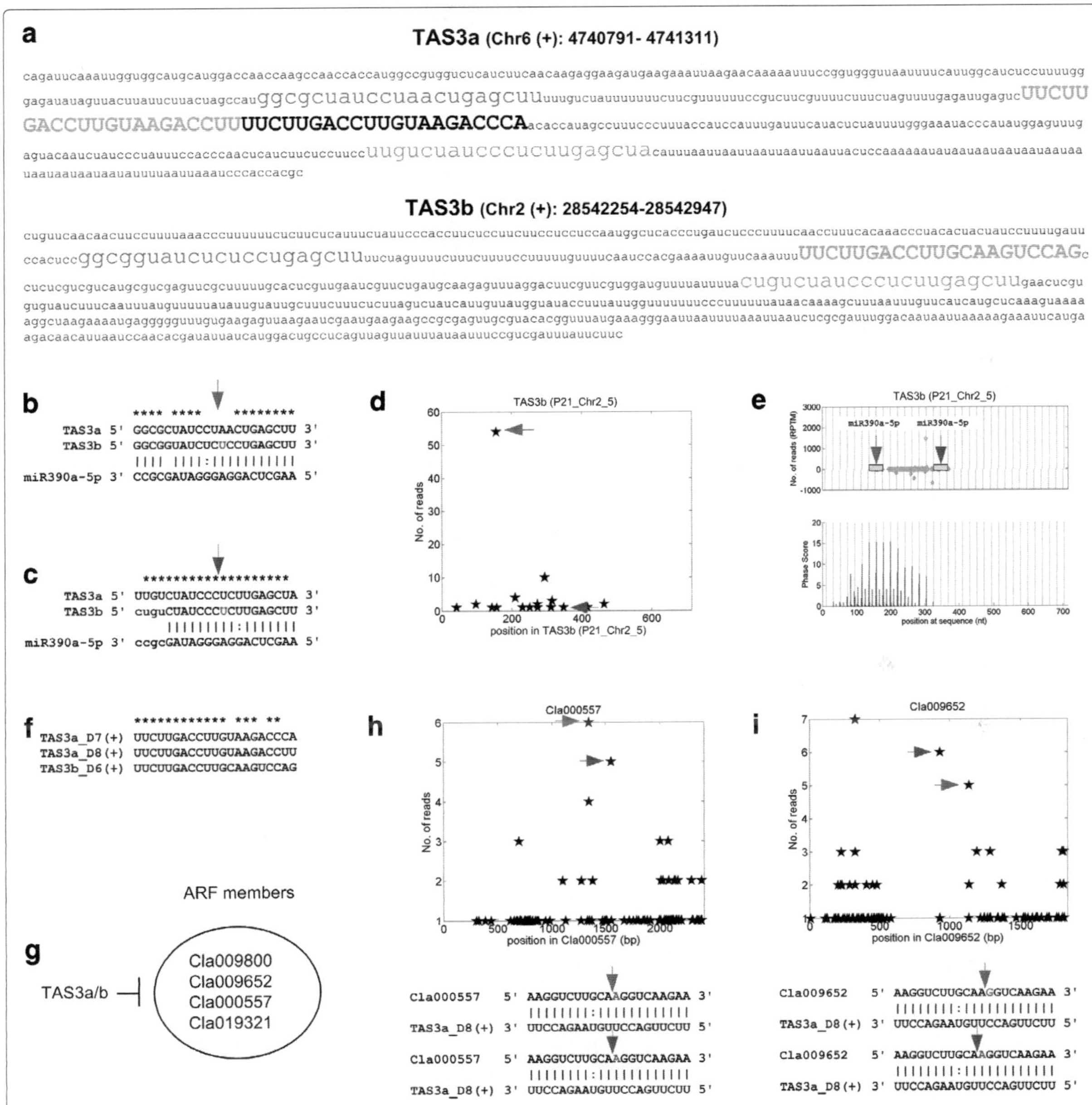

Fig. 3 The TAS3 loci identified in watermelon. **a** The sequence and genomic locus of TAS3a and TAS3b. The red and blue region are the 5′ and 3′ miR390 complementary sites. The yellow and black regions in upper cases are the tasiRNAs that target ARF genes. **b** The 5′ miR390 site on TAS3a and TAS3b. **c** The 3′ miR390 site on TAS3a and TAS3b. **d** The distribution of degradome reads on TAS3b transcript. The two positions pointed by red and blue arrows correspond to the 5′ and 3′ miR390 site shown in Part **b** and **c**, respectively. **e** The distribution of small RNA reads in the small RNA profile of mixed tissues (GSM936361) on TAS3b. **f** The tasiRNAs that target ARF genes in TAS3a and TAS3b, named by following the scheme in [15]. **g** Four ARF genes that are targeted by TAS3 derived tasiRNAs. **h** The T-plot of and complementary sites of TAS3a_D8(+):Cla000557. **i** The T-plot of and complementary sites of TAS3a_D8(+):Cla009652. In Part **h** and **i**, the x-axis is the position on the transcript, and y-axis is the number of reads detected from a position. The arrows in the upper parts correspond to the positions pointed by the arrows of the same colors in the lower parts

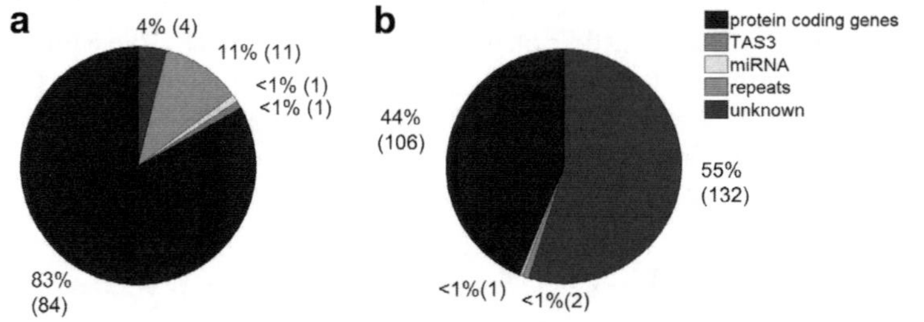

Fig. 4 The categories of predicted PHAS loci. **a** The categories of 21 nt PHAS loci. **b** The categories of 24 nt PHAS loci

In additional to those 127 conserved targets (in Additional file 1: Table S10), conserved miRNA may regulate non-conserved targets as well (Additional file 1: Table S12). Only 48 putative targets were identified for novel miRNAs and most of these targets have no or very limited valid reads at the identified miRNA complementary sites (Additional file 1: Table S13), again suggesting these novel miRNAs might be non-functional.

Identification of TAS3 loci in watermelon

We previously found one TAS3 in watermelon by analyzing ESTs (Expressed Sequencing Tags) of watermelon [35]. In this study, the previously identified TAS3 loci was found on chromosome 6 of watermelon, named as TAS3a (Fig. 3a). We found an additional TAS3 locus in this study, named as TAS3b, on chromosome 2 of watermelon in this study (Fig. 3b). Both of these two loci have typical miR390 sites beside the tasiRNA generation region (Fig. 3b to c). It is interesting that both the 5' and the 3' miR390 site on TAS3b could induce cleavage of TAS3b transcripts (Fig. 3d), and the 5' miR390 site is the major site that induces more cleavages on TAS3b transcripts. The 10 to 11 positions of 5' miR390 sites are complementary on TAS3b, but not matched on TAS3a (Fig. 3b), which might be the reason that the 5' miR390 site on TAS3b is cleavable, but non-cleavable on TAS3a. TAS3b shows a clear phased siRNA generation patterns in one of the sRNA profiles used (Fig. 3e). TAS3a and TAS3b encode two and one tasiRNAs, respectively (Fig. 3a and f). These three tasiRNAs target 4 ARF genes (Fig. 3g). Two ARF genes were cleaved by these TAS3-derived tasiRNAs at two different sites as revealed by the degradome analysis (Fig. 3h and i).

Typically, the 3' miR390 is the only or major site that could induce cleavage on TAS3 transcripts [20]. But Xia et al. [55] noticed that many spruce TAS3 genes include a cleavable 5' miR390 site. Thus, our results indicate that TAS3 loci of other species may also have cleavable 5' miR390 sites in addition to spruce.

PHAS loci in watermelon

The combined sRNA profiles of 14 samples in this study and one additional sRNA profile downloaded from NCBI GEO database were aligned to watermelon genome using SOAP2 [45], respectively. Then, we used a self-written program to predict PHAS loci and phasiRNAs from the

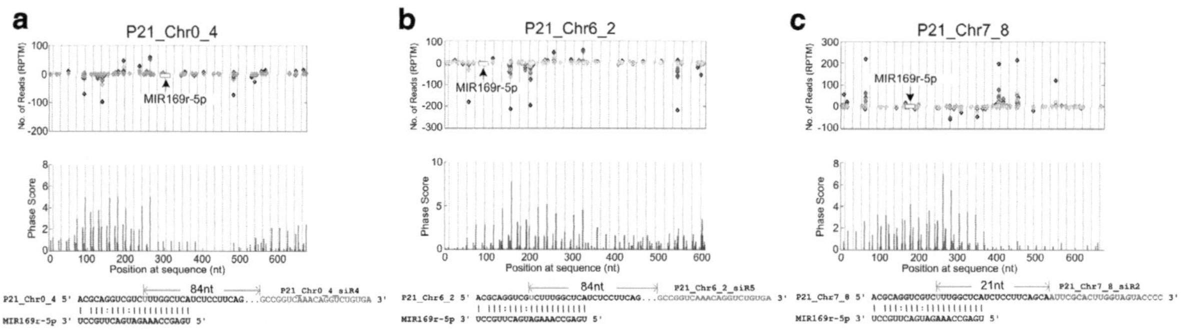

Fig. 5 The reads distribution and Phase Scores for three 21 nt PHAS loci potentially triggered by MIR169r-5p. The vertical gray lines with distances of 21 nt are the phased positions from the position with the highest phase scores of the PHAS loci. The yellow boxes in the read distribution panel represent the MIR169r-5p complementary sites. Sites pointed by miRNAs from above and under zero read line means miRNAs complement to the plus and minus strand of the predicted PHAS loci, respectively. The predicted MIR169r-5p complementary sites are shown below the phase score panel. The blue region after the complementary sites are one of the phasiRNAs generated from the PHAS loci. **a** P21_Chr0_4. **b** P21_Chr6_2. **c** P21_Chr7_8

alignment results of SOAP2 based on methods described earlier (see Materials and methods for details). We predicted one hundred and one 21nt PHAS loci by using a phase score threshold of 5 and a multiple-test corrected $P < 0.05$ (Additional file 1: Table S14). These PHAS loci produced at least four unique twenty-one nt phasiRNAs that were sequenced in the sRNA libraries used in study. We aligned the obtained PHAS sequences to NCBI Nucleotide Collection (nr/nt) database, the coding gene of watermelon and the TIGR Plant Repeat database to obtain putative annotation of the predicted PHAS loci (details are given in Additional file 1: Table S14). These PHAS loci were mainly originated from protein coding genes (Fig. 4a). The second largest category is repeats (as shown in Fig. 4a). One of the predicted 21 nt PHAS loci is a TAS3 locus.

Two hundred and forty one 24 nt PHAS were predicted by using a phase score threshold of 5 and a multiple-test corrected $P < 0.05$ (Additional file 1: Table S15). The 24 nt PHAS loci are different from the 21 nt PHAS loci in watermelon, with the largest category belonging to unknown category (Fig. 4b). The second largest category of 24 nt PHAS loci is protein coding genes (Fig. 4b).

As miRNAs are critical in generation of phasiRNAs, we identified miRNA triggers for these PHAS loci on both strands by using the degradome sequencing profile and the SeqTar algorithm. We found that the MIR169r-5p may trigger the generations of 21 phasiRNAs at three PHAS loci (Fig. 5). All of these three loci are unknown proteins (Additional file 1: Table S14). Except miR390 with 21 nt, most miRNA triggers of 21 nt PHAS loci are 22 nt, such as miR482/miR2118 [14, 19, 27, 56, 57], miR828

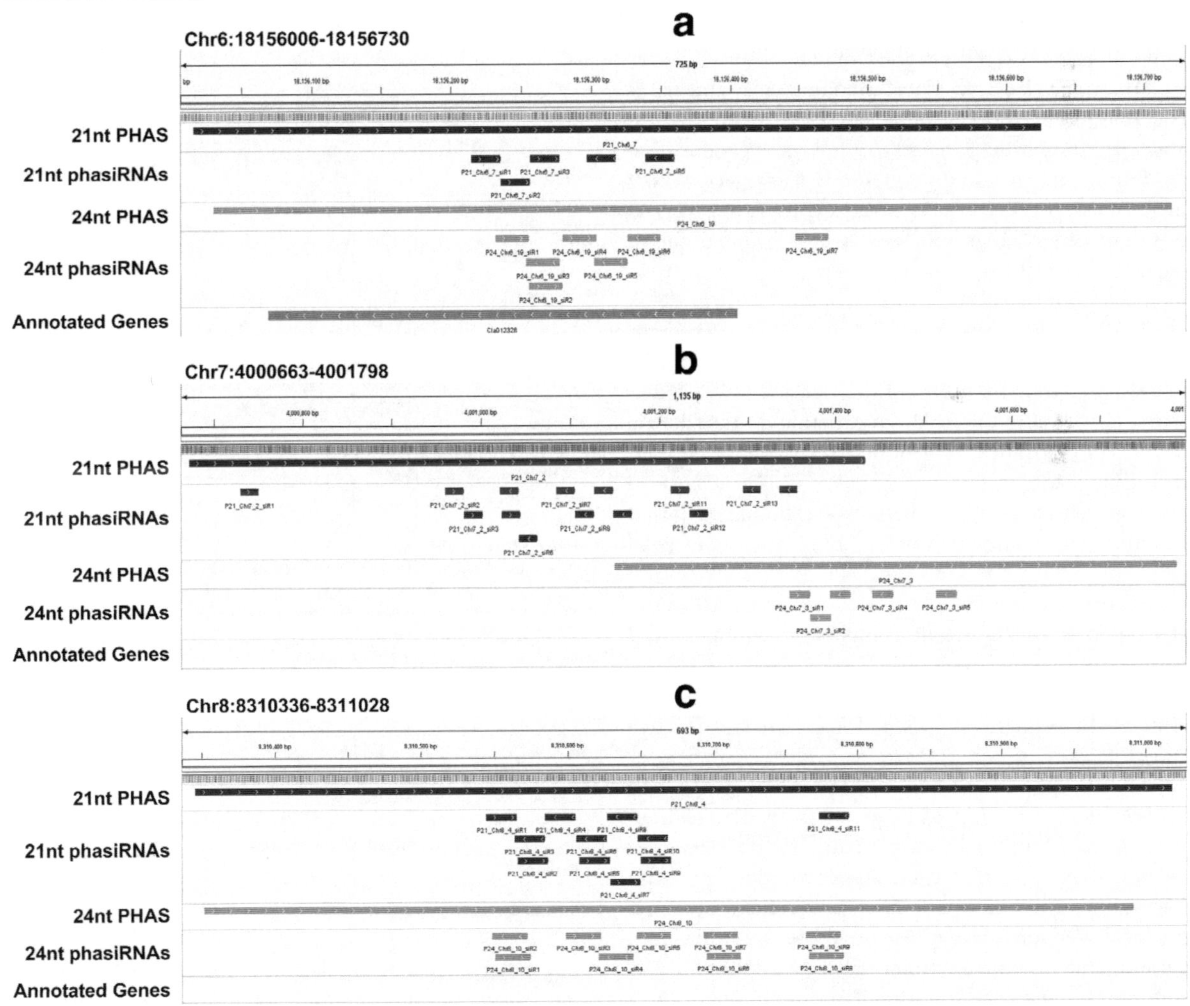

Fig. 6 Three loci that generate both 21 nt and 24 nt phasiRNAs in watermelon. The blue and orange lanes represent the 21 nt PHAS/phasiRNAs and 24 nt PHAS/phasiRNAs in watermelon, respectively. **a** P21_Chr6_7/P24_Chr6_19, from an unknown gene locus (Cla012328). **b** P21_Chr7_2/P24_Chr7_3 from an intergenic or un-annotated region. **c** P21_Chr8_4/P24_Chr8_10 from an intergenic or un-annotated region

[24, 25, 32, 33, 56], and miR7122, miR1509, miR173 [17, 24, 27]. Our results suggested that MIR169r-5p, with 21 nt, potentially triggers the generation of phasiRNAs.

In several previous studies, we found the some PHAS loci in other species may generate both 21 and 24 nt phasiRNAs [39, 56]. In this study, we found that three PHAS loci may generate both 21 and 24 nt phasiRNAs (Fig. 6). Only the first locus shown in Fig. 6a was annotated as an unknown gene, the other two loci in Figs. 6b and c were intergenic or un-annotated genes.

We did not find miR482 and miR2118 in watermelon, as in previous studies [34–38], suggesting that this miRNA family is non-conserved in watermelon. The miR482/miR2118 family is reported to trigger phasiRNAs by targeting NB-LRR genes in some plant species [24, 27–31, 51]. Overall, our results suggest that the miR482/miR2118 triggered phasiRNA generation pathway is not conserved in curcubits.

Putative phasiRNAs targets in watermelon

The 101 and 241 PHAS loci encoding 851 and 1227 twenty one and twenty four nt phasiRNAs, respectively (in Additional file 1: Table S16 and S17, respectively).

We found that in addition to some *cis*-targets, where phasiRNAs target their own transcripts, some phasiRNAs may target other genes in *trans* (see Additional file 1: Table S18 and S19 for 21 and 24 nt phasiRNAs, respectively). For examples (see Additional file 2: Figure S6), P21_Chr10_2_siR6 targets Cla021422, a BHLH transcription factor; P21_Chr1_12_siR7 target Cla005420, a photosystem II polypeptide; P21_Chr6_4_siR13 targets Cla006161, a Hydroxycinnamoyl transferase and P21_Chr5_3_siR3 targets Cla018830, a NADPH-cytochrome P450 reductase. These phasiRNAs induce significant cleavages on their corresponding target transcripts, suggesting that they might have *trans*-targets.

Conclusion

Watermelon is an important plants for economical and agricultural reason. In depth miRNA and PhasiRNA studies have been lacking for this important species. We systematically sequenced 14 small RNA profiles for five different tissues of watermelon, and identified 97 pre-miRNAs, 101 PHAS loci encoding 21 nt phasiRNAs, and 241 PHAS loci encoding 24 nt phasiRNAs. The expression patterns of miRNAs vary in different tissues and show to be very similar for samples of the same tissues. Samples from the same tissues could be corrected clustered based on the expression profiles of miRNAs, suggesting the important functions of miRNAs in different tissues. We also identified one additional TAS3 locus. One hundred and twenty seven targets of conserved miRNAs and TAS3 derived tasiRNAs were identified by analyzing a degradome profile. These results significantly improved our knowledge of sRNA guided gene regulations in watermelon.

Additional files

Additional file 1: This is an MS Excel file. This file includes 19 supplementary tables. (XLSX 3645 kb)

Additional file 2: This is a pdf file. This file includes 6 supplementary figures. (PDF 2335 kb)

Abbreviations

AGO: Argonaute; ARF: Auxin response factor; BHLH: Basic helix-loop-helix; BLAST: Basic local alignment search tool; BLASTN: Nucleotide BLAST; CC: Correlation coefficient; cDNA: Complementary DNA; DCL1: Dicer like-1; dsRNAs: Double strand RNAs; ESTs: Expressed sequencing tags; FDR: False discovery rate; GRF: Growth regulatory factors; HD-Zip: Homeodomain leucine zipper family of transcription factors; logFC: log2FoldChange; miRNAs: microRNAs; MYB: Myeloblastosis family of transcription factors; NADPH: Nicotinamide adenine dinucleotide phosphate; NB-LRR: Nucleoside-binding site leucine rich repeat genes; PCA: Principle component analysis; phasiRNAs: Phased small interfering RNAs; PPR: Pentatricopeptide repeats; RDR6: RNA-dependent RNA polymerase 6; RISC: RNA-induced silencing complex; RNA-seq: RNA sequencing; RPTM: Reads Per Ten Million sequencing tags; SCL: Scarecrow-like transcription factor; SeqTar: SEQuencing-based sRNA TARget prediction siRNAs: Small interfering RNAs; SOAP: Short oligonuclotide analysis program; sRNA: small RNA; tasiRNAs: *Trans*-acting small interfering RNAs

Acknowledgements

Not applicable.

Funding

Funding for the publication of this article was provided by a grant (No. 31460295) of National Natural Science Foundation of China (http://www.nsfc.gov.cn/). The research was supported in part by two grants (No. 31460295 and 31760314) of National Natural Science Foundation of China (http://www.nsfc.gov.cn/) and a grant (No. SKLGE-1511) of the Open Research Funds of the State Key Laboratory of Genetic Engineering, Fudan University, China, to YZ.

Availability of data and materials

The 14 obtained small RNA sequencing profiles were deposited into the NCBI GEO database and are accessible with series ID GSE102030.

About this supplement

This article has been published as part of *BMC Genomics* Volume 19 Supplement 2, 2018: Selected articles from the 16th Asia Pacific Bioinformatics Conference (APBC 2018): genomics. The full contents of the supplement are available online at https://bmcgenomics.biomedcentral.com/articles/supplements/volume-19-supplement-2.

Authors' contributions

YZ conceived and designed the research. LL and SR collected the samples and prepared RNA samples for sequencing. LL, YZ, RS, SR, JG, QW, XZ, PL, and SL analyzed the data. LL and YZ wrote the manuscript. All authors have read and approved the final manuscript.

Ethics approval and consent to participate

Not applicable.

Consent for publication

Not applicable.

Competing interests

The authors declare that they have no competing interests.

Publisher's Note

Author details
[1]Faculty of Life Science and Technology, Kunming University of Science and Technology, Kunming 650500, China. [2]Yunnan Key Laboratory of Primate Biomedical Research, Institute of Primate Translational Medicine, Kunming University of Science and Technology, Kunming 650500, China. [3]Faculty of Information Engineering and Automation, Kunming University of Science and Technology, Kunming 650500, China. [4]School of Life Sciences, Fudan University, Shanghai 200433, China. [5]Department of Biochemistry and Molecular Biology, Oklahoma State University, Stillwater 74078, OK, USA.

Published: 9 May 2018

References

1. Voinnet O. Origin, biogenesis, and activity of plant MicroRNAs. Cell. 2009;136(4):669–87.
2. Chen X. Small RNAs in development–insights from plants. Curr Opin Genet Dev. 2012;22(4):361–7.
3. Sunkar R, Li Y-F, Jagadeeswaran G. Functions of microRNAs in plant stress responses. Trends Plant Sci. 2012;17(4):196–203.
4. Chen X. Small RNAs and their roles in plant development. Annu Rev Cell Dev. 2009;25:21–44.
5. Park W, Li J, Song R, Messing J, Chen X. CARPEL FACTORY, a Dicer homolog, and HEN1, a novel protein, act in microRNA metabolism in *Arabidopsis thaliana*. Curr Biol. 2002;12(17):1484–95.
6. Bartel DP. MicroRNAs: genomics, biogenesis, mechanism, and function. Cell. 2004;116(2):281–97.
7. Jones-Rhoades MW, Bartel DP, Bartel B. MicroRNAs and their regulatory roles in plants. Annu Rev Plant Biol. 2006;57:19–53.
8. Sunkar R, Zhu J-K. MicroRNAs and short-interfering RNAs in plants. J Integr Plant Biol. 2007;49(6):817–26.
9. Pasquinelli AE. MicroRNAs and their targets: recognition, regulation and an emerging reciprocal relationship. Nat Rev Genet. 2012;13(4):271–82.
10. Axtell MJ. Classification and comparison of small RNAs from plants. Annu Rev Plant Biol. 2013;64:137–59.
11. Vasudevan S, Tong Y, Steitz JA. Switching from repression to activation: microRNAs can up-regulate translation. Science. 2007;318(5858):1931–4.
12. Vasudevan S. Posttranscriptional upregulation by microRNAs. Wiley Interdiscip Rev RNA. 2012;3(3):311–30.
13. Xiao M, Li J, Li W, Wang Y, Wu F, Xi Y, Zhang L, Ding C, Luo H, Li Y, et al. MicroRNAs activate gene transcription epigenetically as an enhancer trigger. RNA Biol. 2017;14(10):1326–34.
14. Fei Q, Xia R, Meyers BC. Phased, secondary, small interfering RNAs in posttranscriptional regulatory networks. Plant Cell. 2013;25(7):2400–15.
15. Allen E, Xie Z, Gustafson AM, Carrington JC. microRNA-directed phasing during trans-acting siRNA biogenesis in plants. Cell. 2005;121(2):207–21.
16. Xie Z, Allen E, Wilken A, Carrington JC. DICER-LIKE 4 functions in trans-acting small interfering RNA biogenesis and vegetative phase change in *Arabidopsis thaliana*. Proc Natl Acad Sci USA. 2005;102(36):12984–9.
17. Howell MD, Fahlgren N, Chapman EJ, Cumbie JS, Sullivan CM, Givan SA, Kasschau KD, Carrington JC. Genome-Wide Analysis of the RNA-DEPENDENT RNA POLYMERASE6/DICER-LIKE4 Pathway in *Arabidopsis* Reveals Dependency on miRNA- and tasiRNA-Directed Targeting. Plant Cell. 2007;19(3):926–42.
18. Johnson C, Kasprzewska A, Tennessen K, Fernandes J, Nan G-L, Walbot V, Sundaresan V, Vance V, Bowman LH. Clusters and superclusters of phased small RNAs in the developing inflorescence of rice. Genome Res. 2009;19(8):1429–40.
19. Song X, Li P, Zhai J, Zhou M, Ma L, Liu B, Jeong DH, Nakano M, Cao S, Liu C, et al. Roles of DCL4 and DCL3b in rice phased small RNA biogenesis. Plant J. 2012;69(3):462–74.
20. Axtell MJ, Jan C, Rajagopalan R, Bartel DP. A two-hit trigger for siRNA biogenesis in plants. Cell. 2006;127(3):565–77.
21. Chen HM, Chen L-T, Patel K, Li YH, Baulcombe DC, Wu SH. 22-Nucleotide RNAs trigger secondary siRNA biogenesis in plants. Proc Natl Acad Sci. 2010;107(34):15269–74.
22. Cuperus JT, Carbonell A, Fahlgren N, Garcia-Ruiz H, Burke RT, Takeda A, Sullivan CM, Gilbert SD, Montgomery TA, Carrington JC. Unique functionality of 22-nt miRNAs in triggering RDR6-dependent siRNA biogenesis from target transcripts in *Arabidopsis*. Nat Struct Mol Biol. 2010;17(8):997–1003.
23. Manavella PA, Koenig D, Weigel D. Plant secondary siRNA production determined by microRNA-duplex structure. Proc Natl Acad Sci. 2012;109(7):2461–6.
24. Xia R, Meyers BC, Liu Z, Beers EP, Ye S, Liu Z. MicroRNA superfamilies descended from miR390 and their roles in secondary small interfering RNA biogenesis in eudicots. Plant Cell. 2013;25(5):1555–72.
25. Rajagopalan R, Vaucheret H, Trejo J, Bartel DP. A diverse and evolutionarily fluid set of microRNAs in *Arabidopsis thaliana*. Genes Dev. 2006;20(24):3407–25.
26. Chen HM, Li Y-H, Wu SH. Bioinformatic prediction and experimental validation of a microRNA-directed tandem trans-acting siRNA cascade in *Arabidopsis*. PNAS. 2007;104(9):3318–23.
27. Zhai J, Jeong D-H, De Paoli E, Park S, Rosen BD, Li Y, González AJ, Yan Z, Kitto SL, Grusak MA, et al. MicroRNAs as master regulators of the plant *NB-LRR* defense gene family via the production of phased, *trans*-acting siRNAs. Genes Dev. 2011;25(23):2540–53.
28. Shivaprasad PV, Chen H-M, Patel K, Bond DM, Santos BA, Baulcombe DC. A microRNA superfamily regulates nucleotide binding site–leucine-rich repeats and other mRNAs. Plant Cell. 2012;24(3):859–74.
29. Li F, Pignatta D, Bendix C, Brunkard JO, Cohn MM, Tung J, Sun H, Kumar P, Baker B. MicroRNA regulation of plant innate immune receptors. Proc Natl Acad Sci. 2012;109(5):1790–5.
30. Källman T, Chen J, Gyllenstrand N, Lagercrantz U. A significant fraction of 21-nucleotide small RNA originates from phased degradation of resistance genes in several perennial species. Plant Physiol. 2013;162(2): 741–54.
31. Zhu QH, Fan L, Liu Y, Xu H, Llewellyn D, Wilson I. miR482 regulation of NBS-LRR defense genes during fungal pathogen infection in cotton. PloS ONE. 2013;8(12):e84390.
32. Xia R, Zhu H, An Y-q, Beers EP, Liu Z. Apple miRNAs and tasiRNAs with novel regulatory networks. Genome Biol. 2012;13(6):R47.
33. Zhu H, Xia R, Zhao B, An YQ, Dardick CD, Callahan AM, Liu Z. Unique expression, processing regulation, and regulatory network of peach (*Prunus persica*) miRNAs. BMC Plant Biol. 2012;12(1):149.
34. Liu N, Yang J, Guo S, Xu Y, Zhang M. Genome-wide identification and comparative analysis of conserved and novel microRNAs in grafted watermelon by high-throughput sequencing. PloS one. 2013;8(2):e57359.
35. Jagadeeswaran G, Nimmakayala P, Zheng Y, Gowdu K, Reddy UK, Sunkar R. Characterization of the small RNA component of leaves and fruits from four different cucurbit species. BMC Genomics. 2012;13(1):329.
36. Hu J, Sun L, Zhu Z, Zheng Y, Xiong W, Ding Y. Characterization of conserved microRNAs from five different cucurbit species using computational and experimental analysis. Biochimie. 2014;102:137–44.
37. Li H, Dong Y, Chang J, He J, Chen H, Liu Q, Wei C, Ma J, Zhang Y, Yang J, Zhang X. High-throughput microRNA and mRNA sequencing reveals that microRNAs may be involved in melatonin-mediated cold tolerance in *Citrullus lanatus* L. Front Plant Sci. 2016;7:1231.
38. Sun Y, Niu X, Fan M. Genome-wide identification of cucumber green mottle mosaic virus-responsive microRNAs in watermelon. Arch Virol. 2017;162:2591–2602.
39. Zheng Y, Li T, Xu Z, Wai CM, Chen K, Zhang X, Wang S, Ji B, Ming R, Sunkar R. Identification of microRNAs, phasiRNAs and their targets in pineapple. Trop Plant Biol. 2016;9(3):176–86.
40. Burge SW, Daub J, Eberhardt R, Tate J, Barquist L, Nawrocki EP, Eddy SR, Gardner PP, Bateman A. Rfam 11.0: 10 years of RNA families. Nucleic Acids Res. 2013;41(D1):226–32.
41. Bu D, Yu K, Sun S, Xie C, Skogerbø G, Miao R, Xiao H, Liao Q, Luo H, Zhao G, et al. NONCODE v3. 0: integrative annotation of long noncoding RNAs. Nucleic Acids Res. 2012;40(D1):210–5.
42. Quast C, Pruesse E, Yilmaz P, Gerken J, Schweer T, Yarza P, Peplies J, Glöckner FO. The SILVA ribosomal RNA gene database project: improved data processing and web-based tools. Nucleic Acids Res. 2013;41(D1): 590–6.
43. Ouyang S, Buell CR. The TIGR Plant Repeat Databases: a collective resource for the identification of repetitive sequences in plants. Nucleic Acids Res. 2004;32(suppl_1):360–3.
44. Bao W, Kojima KK, Kohany O. Repbase update, a database of repetitive elements in eukaryotic genomes. Mobile DNA. 2015;6(1):1.
45. Li R, Yu C, Li Y, Lam TW, Yiu S-M, Kristiansen K, Wang J. SOAP2: an improved ultrafast tool for short read alignment. Bioinformatics. 2009;25(15):1966–7.

46. Kozomara A, Griffiths-Jones S. miRBase: annotating high confidence microRNAs using deep sequencing data. Nucleic Acids Res. 2014;42(D1): 68–73.
47. Hofacker IL. Vienna RNA secondary structure server. Nucleic Acids Res. 2003;31(13):3429–31.
48. Meyers BC, Axtell MJ, Bartel B, Bartel DP, Baulcombe D, Bowman JL, Cao X, Carrington JC, Chen X, Green PJ, et al. Criteria for annotation of plant MicroRNAs. Plant Cell. 2008;20(12):3186–90.
49. Robinson MD, McCarthy DJ, Smyth GK. edgeR: a Bioconductor package for differential expression analysis of digital gene expression data. Bioinformatics. 2010;26(1):139–40.
50. Jagadeeswaran G, Zheng Y, Li Y-F, Shukla LI, Matts J, Hoyt P, Macmil SL, Wiley GB, Roe BA, Zhang W, Sunkar R. Cloning and characterization of small RNAs from Medicago truncatula reveals four novel legume-specific microRNA families. New Phytol. 2009;184(1):85–98.
51. Zheng Y, Chen K, Xu Z, Liao P, Zhang X, Liu L, Wei K, Liu D, Li YF, Sunkar R, et al. Small RNA profiles from panax notoginseng roots differing in sizes reveal correlation between mir156 abundances and root biomass levels. Sci Rep. 2017;7:9418.
52. De Paoli E, Dorantes-Acosta A, Zhai J, Accerbi M, Jeong DH, Park S, Meyers BC, Jorgensen RA, Green PJ. Distinct extremely abundant siRNAs associated with cosuppression in petunia. RNA. 2009;15(11):1965–70.
53. Benjamini Y, Hochberg Y. Controlling the false discovery rate: a practical and powerful approach to multiple testing. J R Stat Soc Ser B Methodol. 1995;57(1):289–300.
54. Zheng Y, Li Y-F, Sunkar R, Zhang W. SeqTar: an effective method for identifying microRNA guided cleavage sites from degradome of polyadenylated transcripts in plants. Nucleic Acids Res. 2012;40(4):28–28.
55. Xia R, Xu J, Arikit S, Meyers BC. Extensive families of miRNAs and PHAS loci in norway spruce demonstrate the origins of complex phasiRNA networks in seed plants. Mol Biol Evol. 2015;32(11):2905–18.
56. Zheng Y, Wang S, Sunkar R. Genome-wide discovery and analysis of phased small interfering RNAs in chinese sacred lotus. PLoS ONE. 2014;9(12):e113790.
57. Zhai J, Zhang H, Arikit S, Huang K, Nan G-L, Walbot V, Meyers BC. Spatiotemporally dynamic, cell-type–dependent premeiotic and meiotic phasiRNAs in maize anthers. Proc Natl Acad Sci. 2015;112(10):3146–51.

Chen et al. BMC Genomics 2018, 19(Suppl 2):112
DOI 10.1186/s12864-018-4458-7

BMC Genomics

RESEARCH Open Access

Comprehensive analysis of coding-lncRNA gene co-expression network uncovers conserved functional lncRNAs in zebrafish

Wen Chen[1,2†], Xuan Zhang[1†], Jing Li[1†], Shulan Huang[2], Shuanglin Xiang[2], Xiang Hu[2*] and Changning Liu[1*]

From The Sixteenth Asia Pacific Bioinformatics Conference 2018
Yokohama, Japan. 15-17 January 2018

Abstract

Background: Zebrafish is a full-developed model system for studying development processes and human disease. Recent studies of deep sequencing had discovered a large number of long non-coding RNAs (lncRNAs) in zebrafish. However, only few of them had been functionally characterized. Therefore, how to take advantage of the mature zebrafish system to deeply investigate the lncRNAs' function and conservation is really intriguing.

Results: We systematically collected and analyzed a series of zebrafish RNA-seq data, then combined them with resources from known database and literatures. As a result, we obtained by far the most complete dataset of zebrafish lncRNAs, containing 13,604 lncRNA genes (21,128 transcripts) in total. Based on that, a co-expression network upon zebrafish coding and lncRNA genes was constructed and analyzed, and used to predict the Gene Ontology (GO) and the KEGG annotation of lncRNA. Meanwhile, we made a conservation analysis on zebrafish lncRNA, identifying 1828 conserved zebrafish lncRNA genes (1890 transcripts) that have their putative mammalian orthologs. We also found that zebrafish lncRNAs play important roles in regulation of the development and function of nervous system; these conserved lncRNAs present a significant sequential and functional conservation, with their mammalian counterparts.

Conclusions: By integrative data analysis and construction of coding-lncRNA gene co-expression network, we gained the most comprehensive dataset of zebrafish lncRNAs up to present, as well as their systematic annotations and comprehensive analyses on function and conservation. Our study provides a reliable zebrafish-based platform to deeply explore lncRNA function and mechanism, as well as the lncRNA commonality between zebrafish and human.

Keywords: LncRNA, Zebrafish, Co-expression network, Gene ontology, KEGG, Conservation

Background

LncRNAs are termed as a heterogeneous class of transcripts with the length over 200 bp and without the potential of protein-coding [1–3]. By now, there are ten thousands of lncRNAs discovered across human, mouse, nematode, zebrafish etc. [4–6]. Meanwhile, an increasing number of studies demonstrated that lncRNAs have special functions in diverse biological processes and can extensively regulate gene expression and chromatin status, and thus influencing cellular homostasis [7–9]. As a result, these enormous amounts of lncRNAs can arrange to a complicated and fine-controlled regulation network for gene expression, mutation and dysregulation of them are likely linked to diverse human diseases and abnormal development [10–12].

Zebrafish as a powerful model for studying vertebrate biology has a relatively long history, on account of its clear development pattern and genetic background [13]. Large-scale genetic screens have identified hundreds of mutant phenotypes, many of which could imitate human

* Correspondence: huxiang@hunnu.edu.cn; liuchangning@xtbg.ac.cn
†Equal contributors
[2]State Key Laboratory of Developmental Biology of Freshwater Fish, School of Life Sciences, Hunan Normal University, Changsha, China
[1]Key Laboratory of Tropical Plant Resources and Sustainable Use, Xishuangbanna Tropical Botanical Garden, Chinese Academy of Sciences, Kunming, China

disease status, providing a simple and accessible system for investigating the corresponding human diseases [14–16]. At the early stage of this kind of research, the focal spot is primarily on protein-coding gene. The well-characterized cases of using zebrafish as a model for studying protein roles in human disease are coming from the hematopoietic diseases, like ALAS2 in a microcytic, hypochromic anemia, UROD in porphyria, and etc. [17]. The semblable studies are also illuminated in Cardiovascular-, Kidney- and other organ- disorders [18, 19]. However, the similar studies of using the zebrafish as a model to probe lncRNAs' function are still at infancy, because of the insufficient annotation and the lack of systematic survey.

To date, coupled with the progress of next generation sequencing in model organisms, it is feasible of large-scale discovering and annotating and integratively computational analyzing of zebrafish lncRNAs. By using chromatin marks, poly(A)-site mapping and RNA-Seq data, Ulitsky et.al had identified 550 distinct lncRNAs in zebrafish, with 29 lncRNAs having detectable sequence similarity with their putative mammalian orthologs and some having conserved genomic locations [20]. By a time-series of RNA-seq experiments, Andrea Pauli et al. had defined 1133 noncoding multi-exonic transcripts expressed at eight stages during early zebrafish embryogenesis, and found that zebrafish lncRNAs share many of the characteristics of their mammalian counterparts [21].

In this work, to better utilize the mature zebrafish system to uncover the commonality of lncRNAs' function and conservation between zebrafish and human, and further provide a research platform of zebrafish model organisms to explore the function of lncRNAs in development processes and human disease, we presented and analyzed the most complete collection and annotation of zebrafish lncRNAs, by comprehensively integrating various data resources from public RNA-seq datasets, databases and literatures. By establishing a co-expression network upon lncRNA- and coding- genes, we set out to optimize the present gene annotation of zebrafish lncRNAs, to predict their potential functions, and to hunt their putative mammalian orthologs, so as to illuminate their functional and mechanism conservation between species.

Results

Comprehensive collection and integrative data analysis of zebrafish lncRNAs

To acquire the zebrafish lncRNAs as comprehensive as possible, we made a pipeline for data collection, integration and annotation (Fig. 1a). First, we collected all possible lncRNAs of zebrafish from publicly available databases and literatures, and further integrated them with RNA-seq datasets covering 499 runs in 56 studies from NCBI SRA database (Additional file 1). Using the in-house "RNA-seq data analysis pipeline", we obtained our final dataset with more zebrafish lncRNAs, containing in total 13,604 lncRNA genes (21,128 transcripts) (Additional file 2). The major data sources of our lncRNA set are from RNA-seq data analysis and NONCODE [22], followed by NCBI, Ensembl and zflncRNApedia orderly [23–25] (Fig. 1b). Venn diagram shows that, although there is an obvious overlapping between different sources, a lot of unique lncRNAs emerged in RNA-seq datasets and NONCODE (Fig. 1c).

According to the genomic location, these zebrafish lncRNA genes are cataloged as intergenic, sense overlapping, antisense and intronic, in the percentage of 43.9%, 27.0%, 26.8% and 2.3% respectively (Fig. 2a). Similar to coding-gene, zebrafish lncRNA genes have a relatively uniform distribution on chromosome, but less density. There are about 100 lncRNA genes per 10 M sequence, while coding-genes are about 200 genes per 10 M sequence (Fig. 2b). As for the number of isoform for each gene, lncRNA genes are also in line with coding-genes: (1) about 76.1% of lncRNA genes are single-isoform genes, which occupy 66.5% in coding-genes; (2) lncRNA genes with isoform≤3 account to 94.2%, corresponding to 94.1% of coding-genes; (3) lncRNA genes with isoform≥4 only have 5.8% percentage, consistent with 5.9% of coding-genes (Fig. 2c). However, the exon number distribution of zebrafish lncRNAs is far from that of coding-gene, in that the amount of transcripts with multi-exons is significantly less (Fig. 2d). About 40% of zebrafish lncRNA transcripts have two exons, but for coding-genes, the percentage is only 6.7%; in reverse, transcripts with exon number ≥ 5 are increasing in coding-genes (75%), compared with lncRNAs (17.5%).

Construction of zebrafish coding-lncRNA gene co-expression network

To establish the coding-lncRNA gene co-expression network of zebrafish, we used all collected RNA-seq datasets to quantify the expression levels of both coding- and lncRNA-genes in different tissues and conditions (Additional file 3). The genes with high-expressional variance (top 75 percentile) were selected for identification of co-expressed gene-pairs using Spearman's correlation coefficient. The *P*-value was estimated by Fisher transformation, and the sets of *P*-value for each gene were adjusted by the Bonferroni method. Only gene-pairs with an adjusted *P*-value of 0.01 or less are retained in subsequent analyses.

By comparison of the correlation coefficients of gene expression between coding-genes, between lncRNA and coding-gene, and between lncRNAs, respectively (Fig. 3a), we found that the expression correlation between inter-lncRNAs is obviously weaker than inter-coding-genes (Kolmogorov-Smirnov test, *P*-value = 1.45e-232). Intriguingly, the expression correlation between lncRNA and

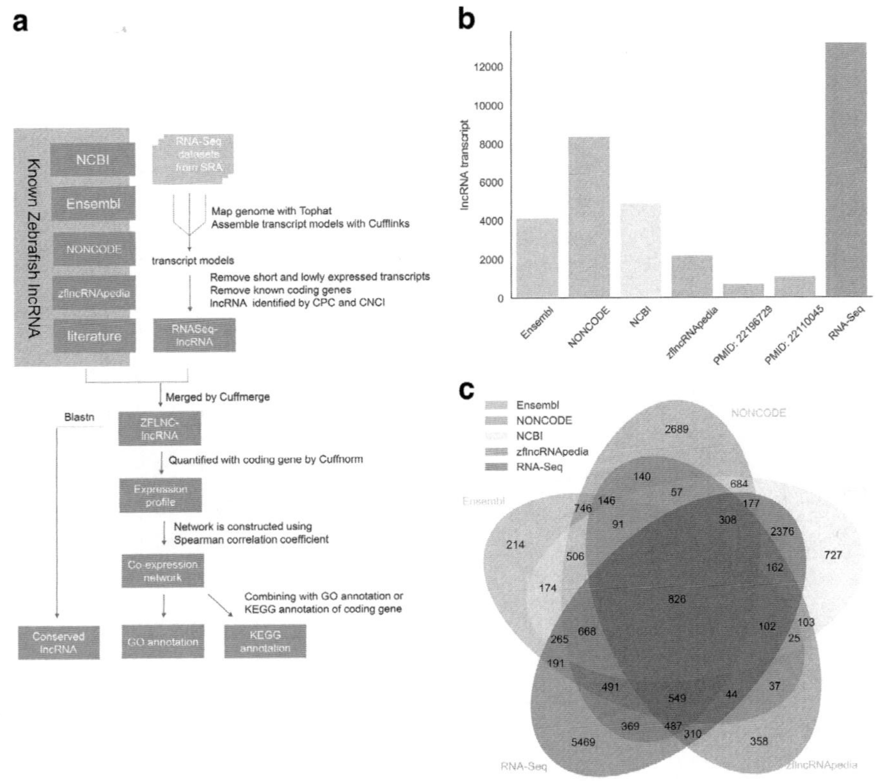

Fig. 1 Integration of all sources of zebrafish lncRNA. **a** Data sources and analysis pipeline. **b** The number of lncRNA transcripts from each source. **c** Venn diagram between different sources

coding-gene is visibly enhanced, as compared with inter-lncRNAs (Kolmogorov-Smirnov test, *P*-value = 3.49e-33), which is likely due to the functional synergia of lncRNAs with their protein partners.

We further evaluated the networks using different cutoffs of Spearman's correlation coefficient by several network parameters (Fig. 3b). With the ascending of cutoff value, the network size, that is consisting of both coding- and lncRNA- genes, naturally decreased. Yet, the average clustering coefficient and the transitivity scores, two parameters that reflect small-world characters of gene network, still held steady or slightly declined. Moreover, coding-genes having 10 or more known coding neighbors were used as a test set to evaluate the functional prediction performance for the network with different cutoffs (Fig. 3c). We found that, for each cutoff, its prediction performance is higher than randomly selected node; furthermore, the prediction sensitivity increases as the cutoff rises, since more GO annotations of coding-genes can be precisely predicted by their neighbor nodes. This indicated that the higher the cutoff, the more similar the annotated functions of connected gene-pairs in network.

According to the size and quality of these networks, we selected the network that has a Spearman's correlation coefficient cutoff of 0.5 for follow-up analysis. This coding-lncRNA gene co-expression network shows a scale-free distribution of connectivities (Fig. 3d). There were 9015 non-coding-genes and 24,425 coding-genes that were linked by 25,097,918 edges. Nearly 17,800,940 edges (70.93%) were connected among coding-genes, and 6,253,398 edges (24.92%) were connected among coding- and noncoding- genes, whereas another 1,043,580 edges (4.16%) were linked between pairs of noncoding- genes.

Functional annotation of zebrafish lncRNAs according to coding-lncRNA gene co-expression network

In our co-expression network established between coding- and lncRNA- gene, the connected gene-pairs tend to have more similar annotated functions, thereby we could use it to further annotate zebrafish lncRNAs' function, including GO and KEGG pathway. Of the 24,425 coding-genes in this network, 14,650 (59.98%) and 7022 (28.75%) were annotated with at least one GO Biological Process (BP) or KEGG term. Consequently, using functional enrichment analysis by hypergeometric distribution test (p-value < 0.05), we annotated each zebrafish lncRNA according to its immediate neighbor coding-genes that have known functional annotation. In this way, we achieved a set of GO (7345 genes) and

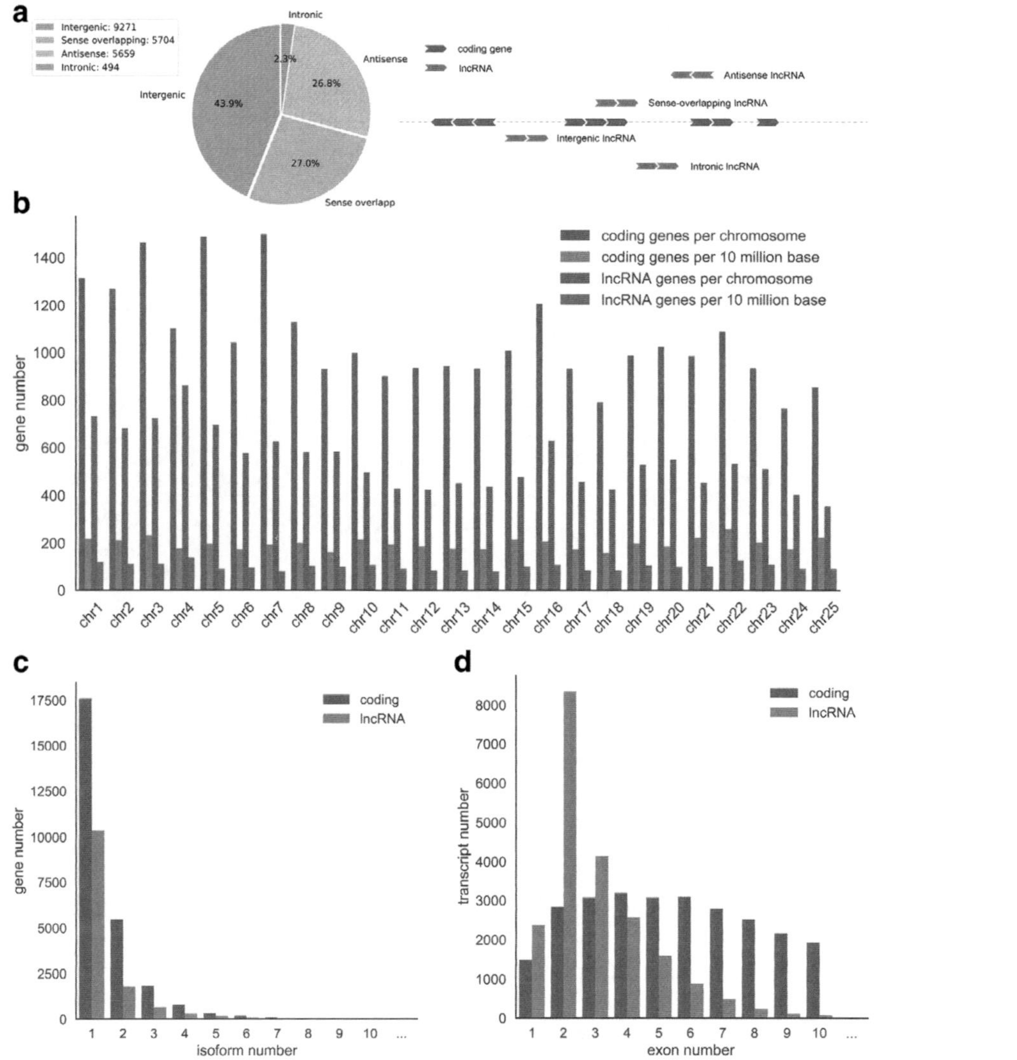

Fig. 2 Features of Zebrafish lncRNA. **a** Distribution of lncRNA subtypes. **b** Zebrafish lncRNA distribution in chromosomes. **c** Distribution of zebrafish lncRNA isoform number. **d** Distribution of zebrafish lncRNA exon number

KEGG (7055 genes) annotation of zebrafish lncRNAs, as listed in Additional file 4.

In terms of GO annotation, the top two enriched GO BP terms of zebrafish lncRNAs are associated with chromosome-related processes, as 'transposition' (*P*-value = 1.33e-11) and 'chromosome segregation' (*P*-value = 1.32e-10) (Fig. 4a). This finding is consistent with previous researches of mammalian lncRNAs, suggesting that lncRNAs can act in the nucleus to regulate the activity of chromosome. For instance, XIST lays over the X chromosome in cis and balances X-linked gene expression [26]; AIR regulates genomic imprinting of a cluster of autosomal genes in cis [27]; else, NORAD controls chromosome segregation and protects cell from becoming aneuploid [28].

In addition, compared with coding-genes, we found that the top ten GO BP terms of zebrafish lncRNAs more tend to be enriched in cell-signaling-related or nervous-system-related processes, such as 'cell-cell signaling' (*P*-value = 4.33e-10), 'cell junction organization' (*P*-value = 2.1e-10),'homeostatic process' (*P*-value = 3.56e-10) and 'neurological system process' (*P*-value = 4.5e-10). Interestingly, this finding is intensely echoed by the results of KEGG annotation, in that the top ten KEGG pathways of zebrafish lncRNAs are subtotally correlated to cell signaling and nervous system (Fig. 4b), such as 'Glutamatergic synapse (*P*-value = 5.9e-320), 'Morphine addiction' (*P*-value = 5.9e-320), 'Neuroactive ligand-receptor interaction' (*P*-value = 5.02e-186), and 'Retrograde endocannabinoid signaling' (*P*-value = 6.2e-182). Since previous studies have

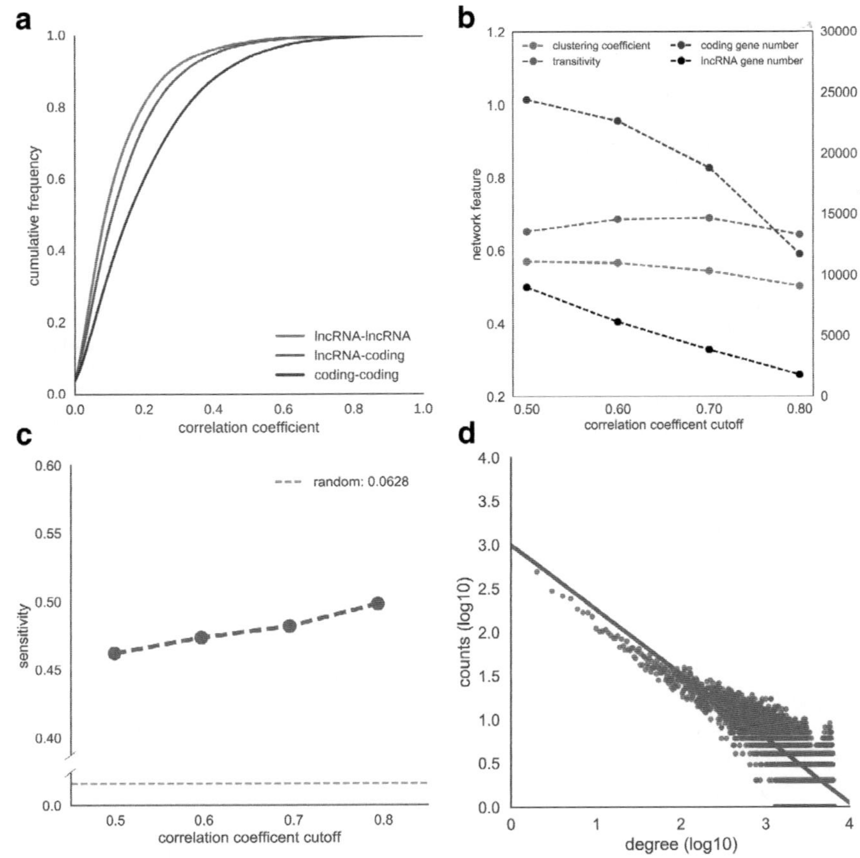

Fig. 3 Features of zebrafish coding-lncRNA gene co-expression network. **a** Cumulative distribution of gene expression Spearman's correlation coefficient. **b** Network statistics by different correlation coefficient cutoffs. **c** Evaluation of function prediction performance of the network with different cutoffs. **d** Network degree distribution (correlation coefficient cutoff = 0.5)

demonstrated that mammalian lncRNAs incline to be expressed specifically in the central nervous system and may play central roles in regulation of the development and function of nervous system [29, 30], we reasoned that zebrafish would be very compatible as a research platform for exploring lncRNA function in human nervous system.

Conservation analysis of zebrafish lncRNAs

For seeking out the putative mammalian orthologs of zebrafish lncRNAs, we used BLASTN to directly compare lncRNAs in zebrafish with that in human or mouse by turns. Those lncRNAs with best hits in bidirectional comparison were considered as lncRNA orthologs (E-value<=10−5). In total, we identified 1828 conserved zebrafish lncRNA genes (1890 transcripts). Therein, 1258 zebrafish lncRNA transcripts have their corresponding orthologs in human; while 1099 zebrafish lncRNA transcripts have mouse orthologs; moreover, 467 zebrafish lncRNA transcripts have the overlapped orthologs in both human and mouse (Additional file 5).

When compared with all lncRNAs, the top ten GO BP terms for conserved zebrafish lncRNA are analogously associated with cell-signaling-related and chromosome-related processes, such as 'cell-cell signaling' (*P*-value = 1.49e-10) and 'transposition' (*P*-value = 1.84e-05). There are a few of new GO BP terms added, such as 'macromolecular complex subunit organization' (*P*-value = 2.73e-10) and 'cellular component assembly' (*P*-value = 3.14e-10) (Fig. 4c). Correspondingly, in the top ten KEGG pathways, several novel roles, like 'Spliceosome' (*P*-value = 7.91e-29), 'RNA transport' (*P*-value = 6.05e-19) and 'Ribosome biogenesis in eukaryotes' (*P*-value = 2.7e-18), are present (Fig. 4d). This suggested that the conserved zebrafish lncRNAs are likely implicated in some of key biological processes in nucleus.

We also evaluated the full-transcript conservation levels of those conserved zebrafish lncRNAs using 8-way PhastCons scores (Fig. 5a). It was found that the PhastCons scores of conserved zebrafish lncRNA are close to that of zebrafish coding-gene, both of which are significantly greater than the entirety of all zebrafish lncRNAs (Kolmogorov-Smirnov test, *P*-value = 6.35e-155). Then, we examined the tissue

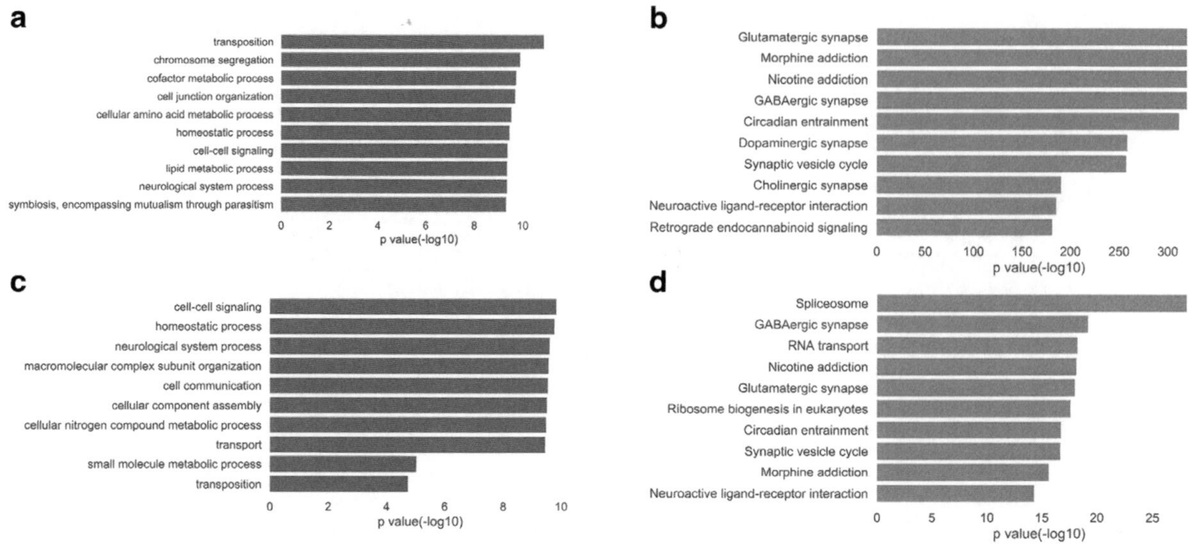

Fig. 4 Functional annotation of zebrafish lncRNAs. **a** LncRNA GO BP enrichment slim (top 10). **b** LncRNA KEGG pathway enrichment (top10). **c** Conserved lncRNA GO BP enrichment slim (top 10). **d** Conserved lncRNA KEGG pathway enrichment (top10)

expression specificity against diverse tissues and different conditions, amongst zebrafish lncRNA, conserved lncRNA and coding-gene. Unlike the unconserved ones, conserved lncRNAs are more resemblance of mRNA as displaying a less tissue-specific expression pattern (Kolmogorov-Smirnov test, *P*-value = 1.58e-12), thereby may express and execute functions in more extensive tissues (Fig. 5b).

Furthermore, we investigated the multi-tissue expression profiles of conserved zebrafish lncRNAs, as compared with their mammalian counterparts (Fig. 5c). The conserved zebrafish lncRNAs, along with their mammalian orthologs, have a more strengthened correlation in tissue-expression profiles at heart, liver, muscle, brain and blood tissues, in comparison of random lncRNA-pairs between zebrafish and human (Kolmogorov-Smirnov test, *P*-value = 1.44e-07). We have also evaluated the number of upstream transcription factor (TF) families shared by both conserved zebrafish lncRNA and its human counterpart (Fig. 5d). It was found that the conserved lncRNAs have a significant conformity with their mammalian counterparts in upstream regulatory pattern (Kolmogorov-Smirnov test, *P*-value = 1.95e-3). Our results indicated that, as compared with their mammalian counterparts, the conservative property of conserved zebrafish lncRNAs reflects not only on gene sequence, probably also on the aspects of tissue expression and transcription regulation pattern.

ZFLNCG05544 is a candidate lncRNA gene related to human neuron diseases

Based on our own results on GO and KEGG annotation, multi-tissue expression profile and coding-lncRNA gene co-expression network, we made an attempt to explore the potential function of zebrafish lncRNA. Previous studies showed that a considerable amount of mammalian lncRNAs are specifically expressed in central nervous system, and thus may play important roles in nervous system development and maintenance [29, 30]. Likewise, we discovered that there are 2423 zebrafish lncRNAs related to neuron, according to their GO and/or KEGG annotation. After confining the tissue type and the expression level of zebrafish lncRNAs (Brain, average FPKM> = 10), this number narrows down to 467. Further, we examined the co-expression state between the latter and known human-neuron-disease-related coding-genes in zebrafish (data collected from ZFIN, http://zfin.org) [13]. Consequently, one lncRNA gene, ZFLNCG05544, attracted our attention.

ZFLNCG05544 locates in chromosome 9, containing two transcripts ZFLNCT08573 and ZFLNCT08574 with the length of 288 nt and 402 nt respectively (Fig. 6a). ZFLNCG05544 is highly expressed in Brain (average FPKM = 27.11, look the details in Additional file 3). The top ten GO annotations of ZFLNCG05544 tend to be enriched in neuron-related processes, such as 'neuropeptide signaling pathway' (*P*-value = 2.5e-11), 'neurotransmitter transport' (*P*-value = 2.61e-11), 'modulation of synaptic transmission' (*P*-value = 3.59e-11), and 'synaptic signaling' (*P*-value = 5.12e-11) (Fig. 6c). Equally, the top ten KEGG annotations of ZFLNCG05544 are also linked to neuron, such as 'Neuroactive ligand-receptor interaction' (*P*-value = 7.3e-42), 'Nicotine addiction' (*P*-value = 4.9e-31), 'Glutamatergic synapse' (*P*-value = 2.0e-30), 'Circadian

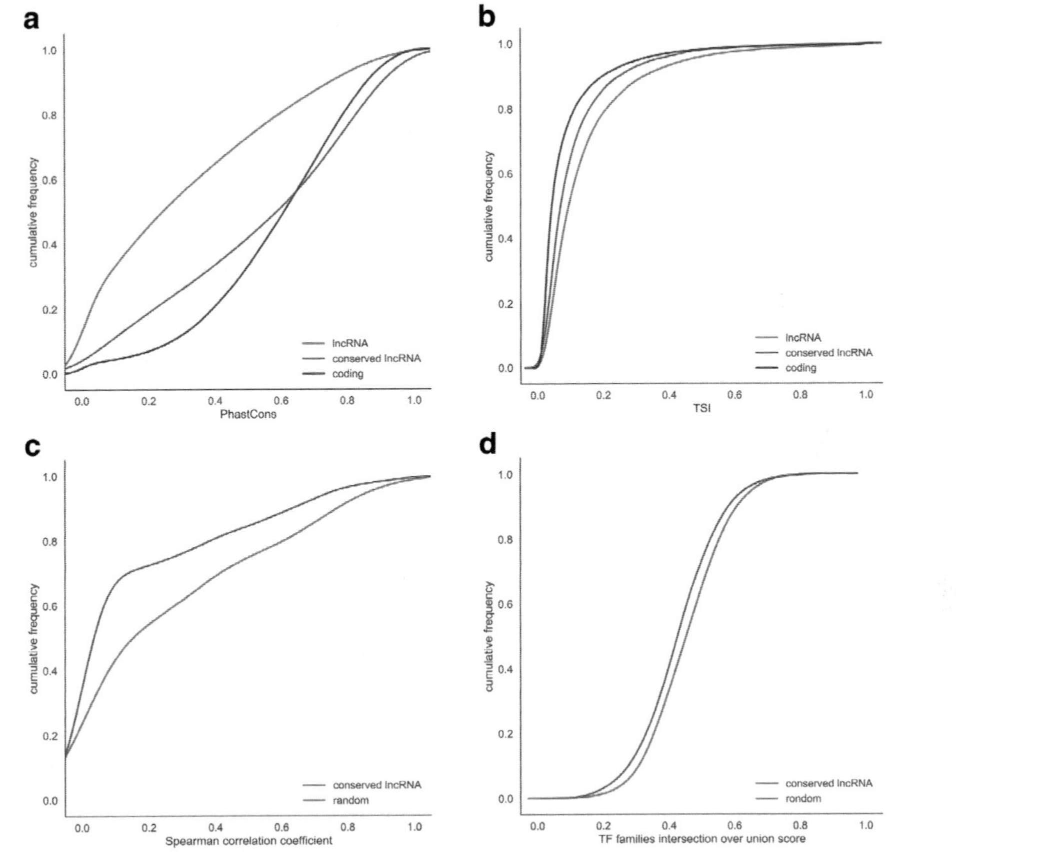

Fig. 5 Conservation analysis of zebrafish lncRNAs. **a** Cumulative distribution of conservation levels computed using PhastCons applied to the 8-way whole-genome. **b** Cumulative distribution of TSI (tissue specificity index). **c** Cumulative distribution of Spearman's correlation coefficient of gene expression. **d** Cumulative distribution of TF families' intersection over union score

entrainment' (*P*-value = 7.4e-23), and 'Retrograde endocannabinoid signaling' (*P*-value = 2.5e-19) (Fig. 6d).

As shown by the co-expression sub-network centered around ZFLNCG05544, there is a concurrent expression of ZFLNCG05544 with many neuron-related coding-genes, especially those relevant to known human neuron diseases, like ace, appa, appb, ctsf, dnajc5aa, dnajc5ab, kif1b and tfg (Fig. 6b). For example, the amyloid precursor protein (APP) has been a focus of intense investigation because of its association with Alzheimer's disease [31]. Previous studies have shown that the zebrafish APP homologue, appb, is required for motor neuron guidance and synapse formation, and is essential for the formation of Mauthner cell in the hindbrain during development [32, 33]. Another example is kinesin family member 1B (kif1b), which has been associated with susceptibility to multiple sclerosis. Lyons et al. have found that kif1b is essential for mRNA localization in oligodendrocytes and development of myelinated axons in zebrafish [34]. ZFLNCG05544 is intensely linked to these neruron-related coding-genes, as indicated by our coding-lncRNA gene co-expression network, implying its important roles in nervous system development and maintenance.

A very recent study gave a powerful support to our prediction of ZFLNCG05544 functions. Sarangdhar et al. demonstrated that zebrafish lncRNA durga modulates dendrite density through regulation of expression of kalrna, which is a coding-gene playing a key role in axonal development, nerve growth and synaptic remodeling [35]. Therein, lncRNA durga locates in the first exon of kalrna in an antisense orientation, and belongs to the catalogue of antisense lncRNA. By direct genomic mapping and BLAST alignment, we found that ZFLNCG05544 has the same location and strand direction in genome, when compared with lncRNA durga, but with a slight discrepancy in transcript length. Thus, they are probably different transcripts for the same gene. (Fig.6a). In view of the kalrna's association with many pathological conditions like schizophrenia and autism-spectrum disorders [36, 37], we thought that ZFLNCG05544 could be a proper candidate lncRNA gene related to multifarious human neuron diseases.

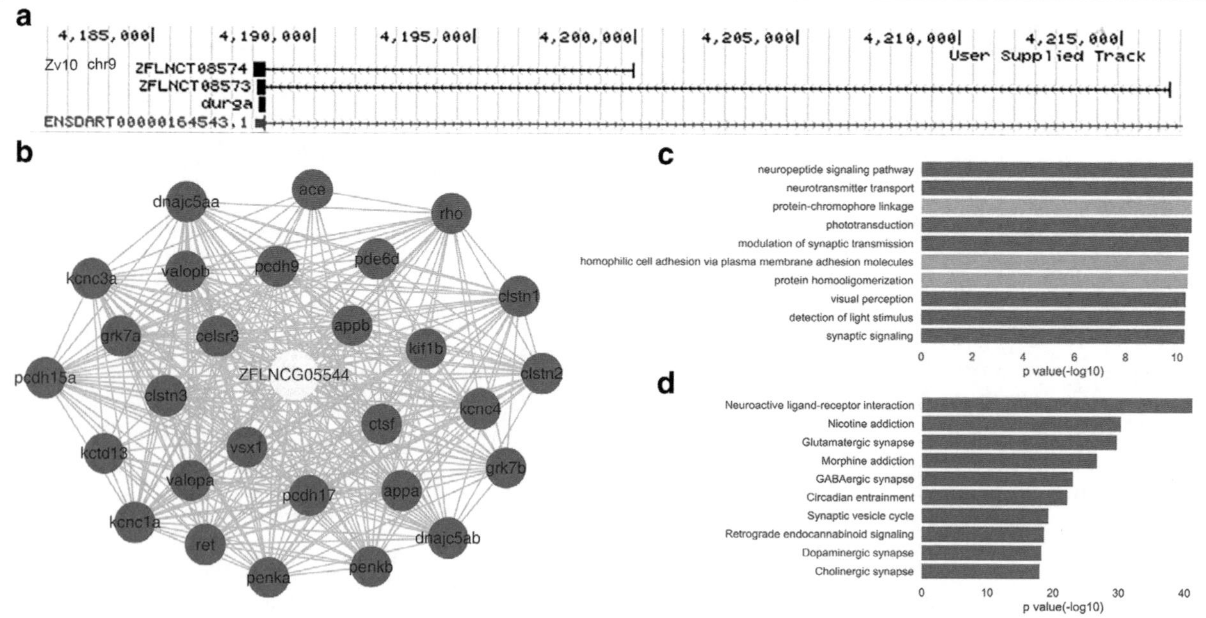

Fig. 6 ZFLNCG05544 is a candidate lncRNA gene related to human neuron diseases. **a** The two transcripts of ZFLNCG05544(ZFLNCT08573, ZFLNCT08573) and durga in UCSC genome browser. **b** ZFLNCG05544 co-expression subnetwork. **c** GO annotation of ZFLNCG05544 (top10). **d** KEGG annotation of ZFLNCG05544 (top10)

ZFLNCG08251 is a human MALAT1 homolog in zebrafish

Herein, we seek out 1890 conserved zebrafish lncRNAs, which have the corresponding mammalian orthologs. By further comparing them with those known functional human lncRNAs (101 entries in total) collected in lncRNAdb [38], twelve of conserved zebrafish lncRNAs in our own analysis hit their well-characterized human orthologs, including 7SK, 7SL, BC200, CCAT1, JPX, KRASP1, MALAT1, MASCRNA, MEGAMIND, OTX2OS1, PTENP1, RAB4B-EGLN2. Amongst them, MALAT1 can parallel a lncRNA ZFLNCG08251 in zebrafish chromosome 14, a transcript of 7540 nt length that is proximal to MALAT1 (8545 nt) (Additional file 2). This result is consistent with the discovery of Ulitsky et.al [20], in that a lncRNA transcript, in the same place of ZFLNCG08251, had been annotated to relevance of human and mouse's MATLAT1, yet fail to further function identification.

MALAT1 is a large, infrequently spliced non-coding RNA, which was closely related to various pathological processes, such as cancer, diabetes complications, and innate immunity [39–42]. Excitingly, the top ten GO annotations of ZFLNCG08251 are enriched in the exactly same categories of cancer, metabolism and immune (Fig. 7b), such as 'regulation of cell proliferation' (*P*-value = 1.7e-04), 'regulation of insulin-like growth factor receptor signaling pathway' (P-value = 3.8e-04), and 'immune system process' (P-value = 5.6e-05). Moreover, its corresponding top ten KEGG pathways are also encapsulated within cancer, metabolism, immune annotation (Fig. 7c), such as 'Central carbon metabolism in cancer' (*P*-value = 8.7e-04), 'Insulin resistance' (*P*-value = 4.6e-03), and 'Intestinal immune network for IgA production' (*P*-value = 6. 4e-04).

In addition, the co-expression sub-network of ZFLNCG08251 shows that it is likely to actively involve in the pathological processes of cancer, diabetes complications, and innate immunity etc., as its human counterpart (Fig. 7a). Amongst the neighbor genes co-expressed with ZFLNCG08251, we detected extensively three types of genes, which are correlated to cancerous signaling pathway (e.g. fas, fes, egfra, map3k5, junba, tgfb1a, tnfsf12, tnfrsf1a, tnfrsf18, jak3, inhbaa), immunity (e.g. il15, irf1b, cd40, tbk1, mavs, socs2, socs3a, stat1a, nod2), and glycometabolism (e.g. insulin-like growth factor-binding protein – igfbp3, igfbp5a, igfbp6b, insulin-like growth factor – igf1, insulin receptor substrate – irs2). All these results hint that ZFLNCG08251, along with its human homolog MALAT1, not only has the sequence conservation, but also the conservative function.

Discussion

Zebrafish is recognized as one of the most important vertebrate model organisms, based on the fact that the development and function of zebrafish organs are strikingly similar to human. Combined with the ease of creating mutant or transgenic individual, zebrafish has been served as an important model system for studying human-disease-related protein-coding genes [14–16]. Nevertheless, the comparable study on zebrafish lncRNAs is still scanty, resulting in

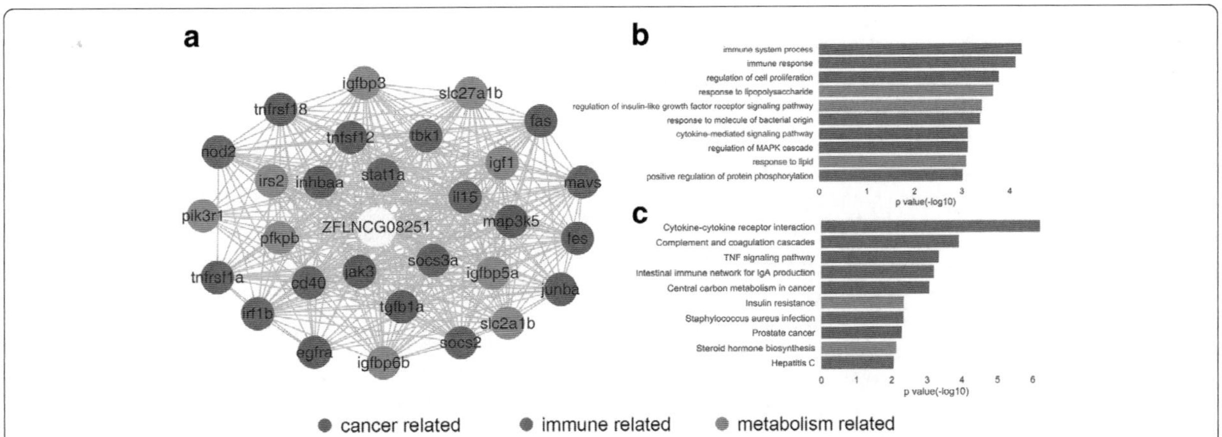

Fig. 7 ZFLNCG08251 is a human MALAT1 homolog in zebrafish. **a** ZFLNCG08251 co-expression subnetwork. **b**. GO annotation of ZFLNCG08251 (top10). **c** KEGG annotation of ZFLNCG08251 (top10)

part from the deficiency of systematic collection and characterization of lncRNAs, as well as the incomplete quantification of their expression profiles. On the other hand, the lack of systematic annotations of zebrafish lncRNAs' function and conservation also made a big obstacle to the follow-up study.

In this work, we provided by far the most comprehensive dataset of zebrafish lncRNAs, which is consist of multiple information of zebrafish lncRNAs, like expression profile, co-expression network, function and conservation annotation etc. Our functional analysis uncovered that, similar to mammalian lncRNA, zebrafish lncRNAs also tend to play roles in regulation of the development and function of nervous system; additionally, conservation analysis indicated that the conserved zebrafish lncRNAs have both sequential and functional conservative with their mammalian counterparts. Therefore, we inferred that zebrafish would be an applicable platform for exploring lncRNA function in development processes and human diseases.

Unlike the protein-coding gene, it is very difficult to directly annotate lncRNA's functions by sequence analysis, due to the lack of information about lncRNA functional and/or structural domain. The existing studies had revealed that mammalian lncRNAs often execute functions with their protein partners, this fact is verified by our analysis of coding-lncRNA gene co-expression in zebrafish. On account of this, the annotation for lncRNA often utilizes the relevant co-expressed coding-genes in network as to enrich lncRNA's function [43, 44]. Compared with human and mouse, the high-throughput sequencing data of zebrafish lncRNA transcriptome are still in less scale. We believed that, in the future, accumulating data of zebrafish transcriptome would further strength our power of lncRNA function prediction based on co-expression network.

Zebrafish shares a high genetic similarity to human, and lots of known human-disease-related coding-genes have functional homologs in zebrafish. However, the conservative of zebrafish lncRNAs is worthy to further gain insight [45]. Through direct BLASTN search, we found 1890 conserved zebrafish lncRNAs, accounting for about 8.9% of total zebrafish lncRNAs (1890/21,128), this result is consistent with the discovery on conservation analysis by Ulitsky et.al (5.3%, 29/550) [20]. In view of the low conservation of lncRNA sequence, syntenic conservation analysis will facilitate to search out the homologs of less conserved lncRNAs, other than the traditional direct BLASTN search [46]. Besides the increased sensitivity of conservation analysis, how to ensure the analysis' specificity is still an open question.

Conclusions

In summary, by integrative data analysis and construction of coding-lncRNA gene co-expression network, we gained by far the most comprehensive dataset of zebrafish lncRNAs, as well as their systematic annotations and comprehensive analyses on function and conservation. Our study provides some important insights into the functional roles of zebrafish lncRNAs and their possible application for studying human diseases. We anticipate that this study will provide a reliable zebrafish-based platform to deeply explore lncRNA function and mechanism, and make a roadmap to investigate the commonality of lncRNAs' function and conservation between zebrafish and human.

Methods

Data sources

RNA-Seq data are downloaded from NCBI SRA database; known zebrafish lncRNAs are collected from NCBI [24], Ensembl [25], NONCODEv4 [22], zflncRNApedia [23] and literatures [20, 21] (See Additional file 1 for details).

RNA-seq data analysis pipeline

SRA format files were dumped to FASTQ format files by SRA-Toolkit. Low quality reads were trimmed by Trimmomatic (Version 0.32) [47]. RNA-Seq reads were mapped to zebrafish genome (Zv9) using Tophat2 (Version 2.0.13) [48], then transcriptome was assembled by Cufflinks (Version 2.2.1) [49]. Multiple-exon transcripts were considered as being expressed if they had an FPKM greater than 0.1. For single-exon transcript, more rigorously, FPKM should be greater than 5 and transcript length greater than 2000. Those foregone coding-genes or transcripts with size less than 200 nt were filtered out. Then, lncRNA candidates were identified by CPC (Version 0.9-r2) [50] and CNCI (Version 2) [51]. At last, all zebrafish lncRNAs stemming from RNA-seq datasets and other publicly available sources were integrated together using the Cuffmerge program in the Cufflinks suite.

Coding-lncRNA gene co-expression network construction

The expression profile of zebrafish lncRNAs and coding-genes were quantified by Cuffnorm program in the Cufflinks suite and then scaled by upper-quartile normalization (Eq. 1). We then calculated the Spearman's correlation coefficient and its corresponding *P*-value (Eq. 2) between the expression profiles of each gene-pair using the in-house Perl script. Only gene-pair with an adjusted P-value of 0.01 or less and with a Spearman's correlation coefficient no less than 0.5 is regarded as co-expression in our coding-lncRNA gene co-expression network.

$$\rho_i = \frac{\frac{1}{n}\sum_i^n k_i}{k_i} \tag{1}$$

For each sample i, k_i is its upper quartile of all gene expression values different from 0, n is the number of all samples, ρ_i is the scaling factor for sample i. The upper quartile normalization for all gene expression values in sample i is obtained by multiplying them to the scaling factor ρ_i.

$$\begin{cases} \mathrm{Rs} = \dfrac{\sum_i (x_i - \bar{x})(y_i - \bar{y})}{\sqrt{\sum_i (x_i - \bar{x})^2 \sum_i (y_i - \bar{y})^2}} \\ \mathrm{F(Rs)} = \dfrac{1}{2}\ln\dfrac{1+\mathrm{Rs}}{1-\mathrm{Rs}} \\ \mathrm{Z} = \sqrt{\dfrac{\mathrm{n}-3}{1.06}}\mathrm{F(Rs)} \end{cases} \tag{2}$$

Where x or y represents the vector of the ranked expression value of each gene, Rs is the Spearman's correlation coefficient between x and y, x_i or y_i stands for the rank of each expression value, $\bar{x}$ or $\bar{y}$, is the mean value of these ranks. F (Rs) is the Fisher transformation of Rs, and n is the sample size i.e. the vector length. The corresponding *P*-value of each Rs is calculated from Z, which is a z-score for Rs that approximately follows a standard normal distribution under the null hypothesis of statistical independence [52, 53].

Network statistics

The clustering coefficient score C for a graph is the average of local clustering coefficient values C_i of each node (Eq. 3). The transitivity T of a graph is based on the relative number of triangles in the graph, compared to total number of connected triples of nodes (Eq. 4).

$$\begin{cases} \mathrm{C_i} = \dfrac{\textit{number of triangles connected to node i}}{\textit{number of triples centered around node i}} \\ C = \dfrac{1}{\mathrm{n}}\sum_{\mathrm{i}=1}^{\mathrm{n}} \mathrm{C_i} \end{cases} \tag{3}$$

Where If node i has two neighbor nodes and these two nodes are also connected, there is a triangle connected to node i, a triple centered around node i is a set of two edges connected to node i, n is the number of node in the network. By definition, $0 \leq C_i \leq 1$ and $0 \leq C \leq 1$.

$$T = \frac{3 \times \textit{number of triangles in the network}}{\textit{number of connected triples of nodes in the network}} \tag{4}$$

The factor of "3" in number accounts for the fact that each triangle contributes to three different connected triples in the graph, one centered at each node of the triangle. With this definition, $0 \leq T \leq 1$, and $T = 1$ if the network contains all possible edges.

LncRNA functional annotation

The GO annotation of zebrafish coding-gene was downloaded from Gene Ontology Consortium (only biological process annotations were considered). While, GO annotation of zebrafish lncRNA was predicted using the goatools (version 0.6.4) [54], which determines the GO annotation of one gene in our network according to the GO annotations of its immediate neighbor genes (*P*-value < 0.05).

The KEGG annotation of zebrafish coding-gene was obtained from KEGG Automatic Annotation Server using zebrafish coding-gene sequence [55]. While, KEGG annotation of zebrafish lncRNA was predicted using the in-house Python script. The KEGG annotation of one gene in our network was determined by the enrichment of KEGG annotations according to its immediate neighborhood (*p*-value < 0.05), when using hypergeometric distribution (Eq. 5).

$$P = 1-\sum_{i=0}^{k-1} \frac{\binom{M}{i}\binom{N-M}{n-i}}{\binom{N}{n}} \tag{5}$$

In this equation, N is the total number of genes in the network, M is the total number of genes having one certain KEGG annotation, n is the number of a gene's immediate neighbors and k is the number of neighbor genes having one certain KEGG annotation.

Analysis of conserved lncRNAs

Zebrafish lncRNAs' putative mammalian orthologs were inferred from bidirectional best hits in direct comparison of zebrafish lncRNA and human lncRNA or mouse lncRNA with BLASTN using a relatively non-stringent E-value threshold 10^{-5}. Zebrafish lncRNAs' conservation levels were evaluated using 8-way PhastCons score [56].

Using expression profiles of all SRA runs, the tissue specificity of zebrafish lncRNAs' gene expression was calculated by TSI (Eq. 6).

$$\mathrm{TSI} = \frac{\max(x_i)}{\sum_i x_i} \tag{6}$$

Where TSI is the tissue specificity index, x_i is the expression value.

To obtain the correlation of tissue expression among conserved lncRNAs, Spearman's correlation coefficient scores were calculated for each conserved zebrafish lncRNA and its human counterpart. The tissue-expression profile data of zebrafish come from five organs, including heart, liver, muscle, brain and blood, as reported by Kaushik et.al. [57]; while, that of human from Human body map.

To estimate the conformity of upstream (from − 500 to + 100) regulatory patterns of conserved lncRNAs, we evaluated the number of upstream transcription factor (TF) families shared by both conserved zebrafish lncRNA and its human counterpart using intersection over union score (Eq. 7). TF-binding site prediction was performed using the HOMER tool [58].

$$S = \frac{\cap(TF_Z, TF_H)}{\cup(TF_Z, TF_H)} \tag{7}$$

Where S is the intersection over union score of upstream TF families shared by both conserved zebrafish lncRNA (TF_Z) and its human counterpart (TF_H).

Additional files

Additional file 1: Data source. (XLSX 33 kb)

Additional file 2: lncRNA dataset. (ZIP 10705 kb)

Additional file 3: lncRNA-coding expression profiles. Replicate samples are merged. (ZIP 13305 kb)

Additional file 4: lncRNA GO and KEGG annotation. (ZIP 17127 kb)

Additional file 5: Conserved lncRNA. (XLSX 84 kb)

Abbreviations

BP: Biological Process; FPKM: Fragments Per Kilobase Million; GO: Gene Ontology; KEGG: Kyoto Encyclopedia of Genes and Genomes; LncRNA: Long non-coding RNA; TF: Transcription Factor; TSI: Tissue Specificity Index

Acknowledgements

RNA-Seq data analysis supported by HPC Platform, The Public Technology Service Center of Xishuangbanna Tropical Botanical Garden (XTBG), CAS, China.

Funding

This work was supported by the National Natural Science Foundation of China (No. 31471220, 91440113), Start-up Fund from Xishuangbanna Tropical Botanical Garden, 'Top Talents Program in Science and Technology' from Yunnan Province, the Opening Fund of State Key Laboratory of Developmental Biology of Freshwater Fish (2017KF003), the Scientific Research Fund of Hunan Provincial Education Department (14B114), and the Cooperative Innovation Center of Engineering and New Products for Developmental Biology of Hunan Province (20134486). Publication of this article was sponsored by the National Natural Science Foundation of China (No. 91440113).

Availability of data and materials

All data generated or analyzed during this study are included in this published article.

About this supplement

This article has been published as part of *BMC Genomics* Volume 19 Supplement 2, 2018: Selected articles from the 16th Asia Pacific Bioinformatics Conference (APBC 2018): genomics. The full contents of the supplement are available online at https://bmcgenomics.biomedcentral.com/articles/supplements/volume-19-supplement-2

Authors' contributions

CL and XH conceived, designed, and supervised this study. WC, XZ and JL performed the most data analysis and contributed equally to this work. SH and SX participate in the discussion and provide valuable advice and practical contributions. WC, JL and CL wrote the first draft of the article, all authors reviewed, edited and approved the manuscript.

Ethics approval and consent to participate

No animals or other organisms were used in this study.

Consent for publication

Not applicable.

Competing interests

The authors declare that they have no competing interests.

Published: 9 May 2018

References

1. Wang KC, Chang HY. Molecular mechanisms of long noncoding RNAs. Mol Cell. 2011;43(6):904–14.
2. Iyer MK, Niknafs YS, Malik R, Singhal U, Sahu A, Hosono Y, Barrette TR, Prensner JR, Evans JR, Zhao S, et al. The landscape of long noncoding RNAs in the human transcriptome. Nat Genet. 2015;47(3):199–208.
3. Ulitsky I, Bartel DP. lincRNAs: genomics, evolution, and mechanisms. Cell. 2013;154(1):26–46.
4. Bu D, Yu K, Sun S, Xie C, Skogerbo G, Miao R, Xiao H, Liao Q, Luo H, Zhao G, et al. NONCODE v3.0: integrative annotation of long noncoding RNAs. Nucleic Acids Res. 2012;40(Database issue):D210–5.

5. Zhao Y, Li H, Fang S, Kang Y, Wu W, Hao Y, Li Z, Bu D, Sun N, Zhang MQ, et al. NONCODE 2016: an informative and valuable data source of long non-coding RNAs. Nucleic Acids Res. 2016;44(D1):D203–8.
6. Liu C, Bai B, Skogerbo G, Cai L, Deng W, Zhang Y, Bu D, Zhao Y, Chen R. NONCODE: an integrated knowledge database of non-coding RNAs. Nucleic Acids Res. 2005;33(Database issue):D112–5.
7. Guttman M, Donaghey J, Carey BW, Garber M, Grenier JK, Munson G, Young G, Lucas AB, Ach R, Bruhn L, et al. lincRNAs act in the circuitry controlling pluripotency and differentiation. Nature. 2011;477(7364):295–300.
8. Cabili MN, Trapnell C, Goff L, Koziol M, Tazon-Vega B, Regev A, Rinn JL. Integrative annotation of human large intergenic noncoding RNAs reveals global properties and specific subclasses. Genes Dev. 2011;25(18):1915–27.
9. Gupta RA, Shah N, Wang KC, Kim J, Horlings HM, Wong DJ, Tsai MC, Hung T, Argani P, Rinn JL, et al. Long non-coding RNA HOTAIR reprograms chromatin state to promote cancer metastasis. Nature. 2010;464(7291):1071–6.
10. Li J, Xuan Z, Liu C. Long non-coding RNAs and complex human diseases. Int J Mol Sci. 2013;14(9):18790–808.
11. Wapinski O, Chang HY. Long noncoding RNAs and human disease. Trends Cell Biol. 2011;21(6):354–61.
12. Bhan A, Mandal SS. Long noncoding RNAs: emerging stars in gene regulation, epigenetics and human disease. ChemMedChem. 2014;9(9): 1932–56.
13. Bradford Y, Conlin T, Dunn N, Fashena D, Frazer K, Howe DG, Knight J, Mani P, Martin R, Moxon SA, et al. ZFIN: enhancements and updates to the Zebrafish model organism database. Nucleic Acids Res. 2011;39(Database issue):D822–9.
14. Zon LI. Zebrafish: a new model for human disease. Genome Res. 1999;9(2):99–100.
15. Dooley K, Zon LI. Zebrafish: a model system for the study of human disease. Curr Opin Genet Dev. 2000;10(3):252–6.
16. Lieschke GJ, Currie PD. Animal models of human disease: zebrafish swim into view. Nat Rev Genet. 2007;8(5):353–67.
17. Chiabrando D, Mercurio S, Tolosano E. Heme and erythropoieis: more than a structural role. Haematologica. 2014;99(6):973–83.
18. Bournele D, Beis D. Zebrafish models of cardiovascular disease. Heart Fail Rev. 2016;21(6):803–13.
19. Cirio MC, de Caestecker MP, Hukriede NA. Zebrafish models of kidney damage and repair. Curr Pathobiol Rep. 2015;3(2):163–70.
20. Ulitsky I, Shkumatava A, Jan CH, Sive H, Bartel DP. Conserved function of lincRNAs in vertebrate embryonic development despite rapid sequence evolution. Cell. 2011;147(7):1537–50.
21. Pauli A, Valen E, Lin MF, Garber M, Vastenhouw NL, Levin JZ, Fan L, Sandelin A, Rinn JL, Regev A, et al. Systematic identification of long noncoding RNAs expressed during zebrafish embryogenesis. Genome Res. 2012;22(3):577–91.
22. Xie C, Yuan J, Li H, Li M, Zhao G, Bu D, Zhu W, Wu W, Chen R, Zhao Y. NONCODEv4: exploring the world of long non-coding RNA genes. Nucleic Acids Res. 2014;42(Database issue):D98–103.
23. Dhiman H, Kapoor S, Sivadas A, Sivasubbu S, Scaria V. zflncRNApedia: a comprehensive online resource for Zebrafish long non-coding RNAs. PLoS One. 2015;10(6):e0129997.
24. Pruitt KD, Tatusova T, Brown GR, Maglott DR. NCBI reference sequences (RefSeq): current status, new features and genome annotation policy. Nucleic Acids Res. 2012;40(Database issue):D130–5.
25. Yates A, Akanni W, Amode MR, Barrell D, Billis K, Carvalho-Silva D, Cummins C, Clapham P, Fitzgerald S, Gil L, et al. Ensembl 2016. Nucleic Acids Res. 2016;44(D1):D710–6.
26. Plath K, Mlynarczyk-Evans S, Nusinow DA, Panning B. Xist RNA and the mechanism of X chromosome inactivation. Annu Rev Genet. 2002; 36:233–78.
27. Sleutels F, Zwart R, Barlow DP. The non-coding air RNA is required for silencing autosomal imprinted genes. Nature. 2002;415(6873):810–3.
28. Lee S, Kopp F, Chang TC, Sataluri A, Chen B, Sivakumar S, Yu H, Xie Y, Mendell JT. Noncoding RNA NORAD regulates genomic stability by sequestering PUMILIO proteins. Cell. 2016;164(1-2):69–80.
29. Mercer TR, Dinger ME, Sunkin SM, Mehler MF, Mattick JS. Specific expression of long noncoding RNAs in the mouse brain. Proc Natl Acad Sci U S A. 2008;105(2):716–21.
30. Chodroff RA, Goodstadt L, Sirey TM, Oliver PL, Davies KE, Green ED, Molnar Z, Ponting CP. Long noncoding RNA genes: conservation of sequence and brain expression among diverse amniotes. Genome Biol. 2010;11(7):R72.
31. O'Brien RJ, Wong PC. Amyloid precursor protein processing and Alzheimer's disease. Annu Rev Neurosci. 2011;34:185–204.
32. Abramsson A, Kettunen P, Banote RK, Lott E, Li M, Arner A, Zetterberg H. The zebrafish amyloid precursor protein-b is required for motor neuron guidance and synapse formation. Dev Biol. 2013;381(2):377–88.
33. Banote RK, Edling M, Eliassen F, Kettunen P, Zetterberg H, Abramsson A. Beta-Amyloid precursor protein-b is essential for Mauthner cell development in the zebrafish in a notch-dependent manner. Dev Biol. 2016; 413(1):26–38.
34. Lyons DA, Naylor SG, Scholze A, Talbot WS. Kif1b is essential for mRNA localization in oligodendrocytes and development of myelinated axons. Nat Genet. 2009;41(7):854–8.
35. Sarangdhar MA, Chaubey D, Bhatt A, Km M, Kumar M, Ranjan S, Pillai B. A novel long non-coding RNA, durga modulates dendrite density and expression of kalirin in Zebrafish. Front Mol Neurosci. 2017;10:95.
36. Deo AJ, Cahill ME, Li S, Goldszer I, Henteleff R, Vanleeuwen JE, Rafalovich I, Gao R, Stachowski EK, Sampson AR, et al. Increased expression of Kalirin-9 in the auditory cortex of schizophrenia subjects: its role in dendritic pathology. Neurobiol Dis. 2012;45(2):796–803.
37. Grubisha MJ, Lin CW, Tseng GC, Penzes P, Sibille E, Sweet RA. Age-dependent increase in Kalirin-9 and Kalirin-12 transcripts in human orbitofrontal cortex. Eur J Neurosci. 2016;44(7):2483–92.
38. Quek XC, Thomson DW, Maag JL, Bartonicek N, Signal B, Clark MB, Gloss BS, Dinger ME. lncRNAdb v2.0: expanding the reference database for functional long noncoding RNAs. Nucleic Acids Res. 2015;43(Database issue):D168–73.
39. Wu Y, Huang C, Meng X, Li J. Long noncoding RNA MALAT1: insights into its biogenesis and implications in human disease. Curr Pharm Des. 2015; 21(34):5017–28.
40. Gutschner T, Hammerle M, Eissmann M, Hsu J, Kim Y, Hung G, Revenko A, Arun G, Stentrup M, Gross M, et al. The noncoding RNA MALAT1 is a critical regulator of the metastasis phenotype of lung cancer cells. Cancer Res. 2013;73(3):1180–9.
41. Gast M, Schroen B, Voigt A, Haas J, Kuehl U, Lassner D, Skurk C, Escher F, Wang X, Kratzer A, et al. Long noncoding RNA MALAT1-derived mascRNA is involved in cardiovascular innate immunity. J Mol Cell Biol. 2016;8(2):178–81.
42. Liu JY, Yao J, Li XM, Song YC, Wang XQ, Li YJ, Yan B, Jiang Q. Pathogenic role of lncRNA-MALAT1 in endothelial cell dysfunction in diabetes mellitus. Cell Death Dis. 2014;5:e1506.
43. Liao Q, Xiao H, Bu D, Xie C, Miao R, Luo H, Zhao G, Yu K, Zhao H, Skogerbo G, et al. ncFANs: a web server for functional annotation of long non-coding RNAs. Nucleic Acids Res. 2011;39(Web Server issue):W118–24.
44. Liao Q, Liu C, Yuan X, Kang S, Miao R, Xiao H, Zhao G, Luo H, Bu D, Zhao H, et al. Large-scale prediction of long non-coding RNA functions in a coding-non-coding gene co-expression network. Nucleic Acids Res. 2011;39(9):3864–78.
45. Hezroni H, Koppstein D, Schwartz MG, Avrutin A, Bartel DP, Ulitsky I. Principles of long noncoding RNA evolution derived from direct comparison of transcriptomes in 17 species. Cell Rep. 2015;11(7):1110–22.
46. Necsulea A, Soumillon M, Warnefors M, Liechti A, Daish T, Zeller U, Baker JC, Grutzner F, Kaessmann H. The evolution of lncRNA repertoires and expression patterns in tetrapods. Nature. 2014;505(7485):635–40.
47. Bolger AM, Lohse M, Usadel B. Trimmomatic: a flexible trimmer for Illumina sequence data. Bioinformatics. 2014;30(15):2114–20.
48. Kim D, Pertea G, Trapnell C, Pimentel H, Kelley R, Salzberg SL. TopHat2: accurate alignment of transcriptomes in the presence of insertions, deletions and gene fusions. Genome Biol. 2013;14(4):R36.
49. Trapnell C, Williams BA, Pertea G, Mortazavi A, Kwan G, van Baren MJ, Salzberg SL, Wold BJ, Pachter L. Transcript assembly and quantification by RNA-Seq reveals unannotated transcripts and isoform switching during cell differentiation. Nat Biotechnol. 2010;28(5):511–5.
50. Kong L, Zhang Y, Ye ZQ, Liu XQ, Zhao SQ, Wei L, Gao G. CPC: assess the protein-coding potential of transcripts using sequence features and support vector machine. Nucleic Acids Res. 2007;35(Web Server issue):W345–9.
51. Sun L, Luo H, Bu D, Zhao G, Yu K, Zhang C, Liu Y, Chen R, Zhao Y. Utilizing sequence intrinsic composition to classify protein-coding and long non-coding transcripts. Nucleic Acids Res. 2013;41(17):e166.
52. Choi SC. Tests of equality of dependent correlation coefficients. Biometrika. 1977;64(3):645–7.
53. Fieller EC, Hartley HO, Pearson ES. Tests for rank correlation coefficients. I Biometrika. 1957;44(3-4):470–81.
54. Haibao T, Debra K, Brent P, Patrick F, Kenta S, Fidel R, Jeff Y, Chris M. GOATOOLS: tools for gene ontology. 2015. https://doi.org/10.5281/zenodo.31628.

55. Moriya Y, Itoh M, Okuda S, Yoshizawa AC, Kanehisa M. KAAS: an automatic genome annotation and pathway reconstruction server. Nucleic Acids Res. 2007;35(Web Server issue):W182–5.
56. Siepel A, Bejerano G, Pedersen JS, Hinrichs AS, Hou M, Rosenbloom K, Clawson H, Spieth J, Hillier LW, Richards S, et al. Evolutionarily conserved elements in vertebrate, insect, worm, and yeast genomes. Genome Res. 2005;15(8):1034–50.
57. Kaushik K, Leonard VE, Kv S, Lalwani MK, Jalali S, Patowary A, Joshi A, Scaria V, Sivasubbu S. Dynamic expression of long non-coding RNAs (lncRNAs) in adult zebrafish. PLoS One. 2013;8(12):e83616.
58. Heinz S, Benner C, Spann N, Bertolino E, Lin YC, Laslo P, Cheng JX, Murre C, Singh H, Glass CK. Simple combinations of lineage-determining transcription factors prime cis-regulatory elements required for macrophage and B cell identities. Mol Cell. 2010;38(4):576–89.

Liu et al. BMC Genomics 2018, 19(Suppl 2):171
DOI 10.1186/s12864-018-4456-9

BMC Genomics

RESEARCH

Open Access

Ouroboros resembling competitive endogenous loop (ORCEL) in circular RNAs revealed through transcriptome sequencing dataset analysis

Yu-Chen Liu[1,2,3], Hsiao-Chin Hong[2], Chi-Dung Yang[2,4], Wei-Hsiang Lee[2,5,6], Hsin-Tzu Huang[5] and Hsien-Da Huang[2,5*]

From The Sixteenth Asia Pacific Bioinformatics Conference 2018
Yokohama, Japan. 15-17 January 2018

Abstract

Background: Emerging evidence indicates that Circular RNAs (circRNAs) exert post-transcriptional regulation of gene expression. A subclass of circRNA was found enriched with miRNA target sites. This evidence suggests that this kind of circRNA functions as natural miRNA sponge. Noticing the potential impacts of circular RNA research, we were motivated to identify novel circRNAs as well as putative circRNA-miRNA interactions through retroactive sourced transcriptome sequencing samples.

Results: Through the analysis in 465 RNA-seq runs and 22 reports published in recent years, putatively circRNA sponged miRNA that had been experimentally verified targeting circRNA host gene were found. From this observation, supporting evidence of the competitive endogenous relationship of circRNAs and miRNAs targeting circRNA host genes can be observed. Given the self-regulation and self-induction nature of these circRNAs, this kind of hypothetical phenomenon was hereby called Ouroboros Resembling Competitive Endogenous Loop (ORCEL) in circular RNAs.

Conclusions: The fact that miRNA sponge circRNA originated from region miRNA target sites enriched regions, while genes encoded from these regions are conserved to be miRNA targets rationalize the existence of ORCEL.

Keywords: CircRNA, CircNet, CERNA, miRNA sponge, ORCEL

Background

More than 50 years have passed since H. Harris deduced that most nuclear RNAs are likely to be non-protein-coding in 1959 [1]. Existence of functional noncoding RNAs has become convention knowledge. Thanks to the dramatically expanded scope of transcriptomics research with high throughput sequencing technology developed in recent years [2], it is possible now to more accurately investigate the expression of non-coding RNA. Differential expressed long non coding RNAs have been found and reported in almost a weekly basis [3]. Among this trend of digging into retroactive sourced experimental data, circular RNAs emerge on stump eventually.

Circular RNAs (circRNAs) represent a type of regulatory noncoding RNA whose head 3′ and tail 5′ ends covalently bond together to result in a circular form. The circular form was verified with electron microscope in 1979 [4]. In 2012, Salzman et al. [5, 6] developed an algorithm to detect scrambled exons in RNA-Seq datasets, and reported that circular RNA isoforms are actually predominant in many human gene isoforms. Later, an improved version of the algorithm with exon splicing site AU/AC searching was applied to find the fact that

* Correspondence: bryan@mail.nctu.edu.tw
[2]Institute of Bioinformatics and Systems Biology, National Chiao Tung University, HsinChu, Taiwan
[5]Department of Biological Science and Technology, Institute of Bioinformatics, National Chiao Tung University, Hsin-Chu, Taiwan
Full list of author information is available at the end of the article

circRNAs serving as natural microRNA "sponges", which enriched with miRNA targeting site and serving competitive endogenous RNAs (ceRNAs) [7, 8]. A circRNA named CDR1as [9] expressed in human and mouse brain was shown to negatively regulate miR-7 in a post-transcriptional manner [7, 9, 10]; this mechanism appears to be evolutionarily conserved [7]. With the regulation potential of miRNAs, circRNAs became widely interested in the research field [11]. Circular RNA can be enriched within the sample through treating samples with RNase R before conducting RNA-Seq [12, 13].

In the following years, extended identification of circRNAs in mouse [14], fly [15] and other animals [16] suggests that circRNA ubiquity is evolutionally-conserved. The reported circRNAs are not results of singular case. These experiment results tend to be reproducible. Further evidence indicates that human circRNA expression exhibits tissue specificity, and now tens of thousands of circRNAs have been found and reported across human tissues [14, 17–22].

Aware of the significance of circRNAs research, in 2015, we constructed a database called CircNet [23] to not only collect the published data, but also extend the scope of reported circRNAs and provide resources to aid in field.

To acquire an enriched collection of human circular RNAs, previously reported and newly identified human circRNAs were collected in this study. Reported human back-spliced junction sites were also collected from 22 recent studies [6, 7, 13, 20, 22, 24–39]. In addition, 465 transcriptome sequencing data sets were collected from NCBI Sequence Read Archive [23, 40], including datasets used in recent publications [14, 18–22, 38].

Circular RNA back-spliced junction sites were identified within these samples. To acquire the expression patterns of circRNAs within these samples, a pipeline was developed to annotate the sequence of the circRNAs. With the putative sequences of circRNAs, miRNA target prediction was conducted on these sequences. Transcripts abundance within the samples was conducted through the transcript deconvolution algorithm [41].

Through analyzing the collect circRNA, combined with the annotation and miRNA sponge prediction, we found that in 728 human circRNA host genes, the generated circRNAs were found to have high affinity to sponge miRNAs which were reported to target these genes. The calculated *P*-value of the circRNA and miRNA pairs are lower than 0.005. From this observation, existence of competitive endogenous loop of circRNAs and their host gene can be observed. Given the self-regulation and self-induction nature of these circRNAs, it was hereby named Ouroboros Resembling Competitive Endogenous Loop (ORCEL) in circular RNAs.

Results

Through the expression profiling of circRNA, combined with the previously reported back spliced junction sites, we found 2747 circRNAs originated from 2693 back spliced junction sites that meet the defined threshold. From the analysis, we found that in 728 human circRNA host genes, including 909 circRNA transcripts, the generated circRNAs were found to have high affinity to sponge miRNAs. Meanwhile these miRNAs were reported to target the circRNA originated genes.

Annotation of circRNA sequence

From our analysis, we found 2693 back splice junction sites that meet the defined threshold on human genome. The combined amount of peer review reports and RNA-Seq samples in which these junction sites were found or reported ranges from 10 to 126. The miRNA sponge CDR1as [9] was found in over 100 of our collected samples. A detail list of these circRNAs, as well as their corresponding SRPBM and FPKM values within the samples can be found in Additional file 1. The data was also updated into our public database CircNet [23] in late 2016.

Self-regulation and self-induction nature of ORCEL

With the observation of most circRNAs originated from circularization of coding gene exons, it was deducted that circRNA biogenesis competes with pre-mRNA splicing [14]. As an isoform originated from part of gene locus, expression of circRNAs correlates with their originated genes. While on the other hand, the post-transcriptional roles of miRNAs to most of the genes has become a conventional knowledge [42]. Result of our analysis suggest the existence of competitive endogenous loop of circRNAs and their host genes. In the sequences of 909 circRNA transcripts from 728 human genes, we found enriched miRNA targeting sites. Echo to the recent evidences suggesting that one circRNA can sponge multiple different miRNAs [34], we found 1112 miRNAs that can be sponged by these circRNAs. Meanwhile these miRNAs were also found to target the circRNA originated genes. Hence a looping regulative relationship among circRNA, circRNA originated gene and miRNA targeting the gene was found. Given the self-regulation and self-induction nature of these circRNAs, the phenomenon was hereby named Ouroboros Resembling Competitive Endogenous Loop (ORCEL) in circular RNAs. The term was inspired by Friedrich August Kekulé and his famous discovery of benzene ring. A complete list of ORCEL is available in Additional file 2.

Enrichment analysis of the ORCEL genes

An enrichment analysis of the 728 ORCEL genes was conducted through DAVID [43]. In the result of this analysis we found that many ORCEL genes participate in important KEGG [44] pathways such as Ubiquitin mediated proteolysis, Pathways in cancer, Focal adhesion and Progesterone-mediated oocyte maturation, as summarized in Table 1. In the 32 genes participated in the hsa05200: Pathways in cancer, 115 miRNAs and 45 circRNA transcripts were found participated in the ORCEL, as illustrated in the network of Fig. 1. The network was generated through Cytoscape [45].

From the network illustrated in Fig. 1, it can be noticed that among the 32 genes enriched in the pathways in cancer, only the ORCEL of HIA1 doesn't involve miRNAs targeting multiple other genes involved in the pathway. A complete table of the enrichment can be found in Additional file 3. In the network illustrated in Fig. 2, for the 19 genes participated in the hsa04120: Ubiquitin mediated proteolysis, 52 miRNAs and 22 circRNA transcripts were found participated in the ORCEL. It is also worth mentioned that 61 ORCEL genes were enriched in the GO term GO:0046907~intracellular transport and 121 ORCEL genes were enriched in GO:0031974~membrane-enclosed lumen. The ORCEL phenomenon can potential be correlated to intracellular transport.

Discussion

In this study, putative roles of circRNAs serving as endogenous miRNA sponges were investigated through analysis of transcriptome sequencing data sets and published results. With the head-tail junction structure, circular RNAs are more stable than other kind of long noncoding RNAs. Hence circRNAs are hypothetically easier to acuminate in cells. The longer half-lives of circRNAs allow the presence of miRNA target sites increase within the cells. Through statistical analysis of the abundance of the miRNA target seeds on circRNA sequences, putative circRNA serving as miRNA sponges were identified. Recent years experiment results suggest certain threshold of miRNA target sites needs to be reached for the ceRNAs to have physiological effects [46, 47], henceforth the analysis was focused on the abundance of miRNA target sites. The application of FPKM of the putative circRNA sequences, and SRPBM of the back spliced junction sites as thresholds should increase the prudence of the circRNA sequence prediction in this study. With the further compliance with results of recent year studies, results of our analysis suggest that only a certain subset of expressed circRNAs potentially serve as nature miRNA sponges. Among the miRNAs predicted to be sponged by these identified circRNAs, many had been experimentally verified to target the circRNA source genes. From these observations we hypothesize that genes targeted by miRNAs tend to be conserved with enriched miRNA targeting site in the coding region. CircRNAs coded from these regions can henceforth sponge the miRNAs when overexpressed. This phenomenon was hereby named Ouroboros Resembling Competitive Endogenous Loop (ORCEL) in circular RNAs. The term was inspired by Friedrich August Kekulé and his famous discovery of benzene ring [48]. Given the observation of this phenomena in genes involved in cancer pathways, validation of this hypothetical phenomena shall significantly impact the research prospects in medical science. ORCEL can potentially serving as a kind of control mechanism to resist miRNAs overdose. The fact that miRNA sponge circRNA originated from region miRNA target sites enriched regions, while genes encoded from these regions are conserved to be miRNA targets rationalize the existence of ORCEL.

Table 1 KEGG pathway enrichment of ORCEL genes

KEGG Pathway	No. of Genes	*P*-value
hsa04120:Ubiquitin mediated proteolysis	19	2.14E-05
hsa05200:Pathways in cancer	32	2.59E-05
hsa04510:Focal adhesion	23	4.83E-05
hsa04914:Progesterone-mediated oocyte maturation	13	2.97E-04

The ORCEL genes enriched in KEGG pathways with significant *P* values

Conclusions

Through the bioinformatics analysis it was found that for certain subset of circRNAs, putatively sponged miRNA had been experimentally verified targeting circRNA host gene. From this observation, the existence of competitive endogenous loop of circRNAs and their host gene can be observed. Given the self-regulation and self-induction nature of these circRNAs, this kind of phenomenon was hereby called Ouroboros Resembling Competitive Endogenous Loop (ORCEL) in circular RNAs.

Methods

The data analysis process of this research is summarized in Fig. 3. First, to identify circRNA, transcriptome sequencing data sets were obtained from the NCBI Sequence Read Archive (SRA). The back-spliced junction sites in each RNA-seq sample were identified using a circRNA discovery pipeline adapting the scripts provided on circBase [7, 49], which was referred as find_circ [50]. Detected back-spliced junction sites, along with the collected junction sites from previous reports, were further compared with the hg19 human genome annotation from RefSeq to annotate circRNA isoform sequence. The annotated sequence were then applied in the prediction of circRNA-miRNA interactions. Occurrence of miRNA target seeds in circRNA isoforms were examined and normalized by isoform length. The significance of interactions was evaluated by referring to the background distribution of miRNA seeds in all transcripts and only circRNA-miRNA

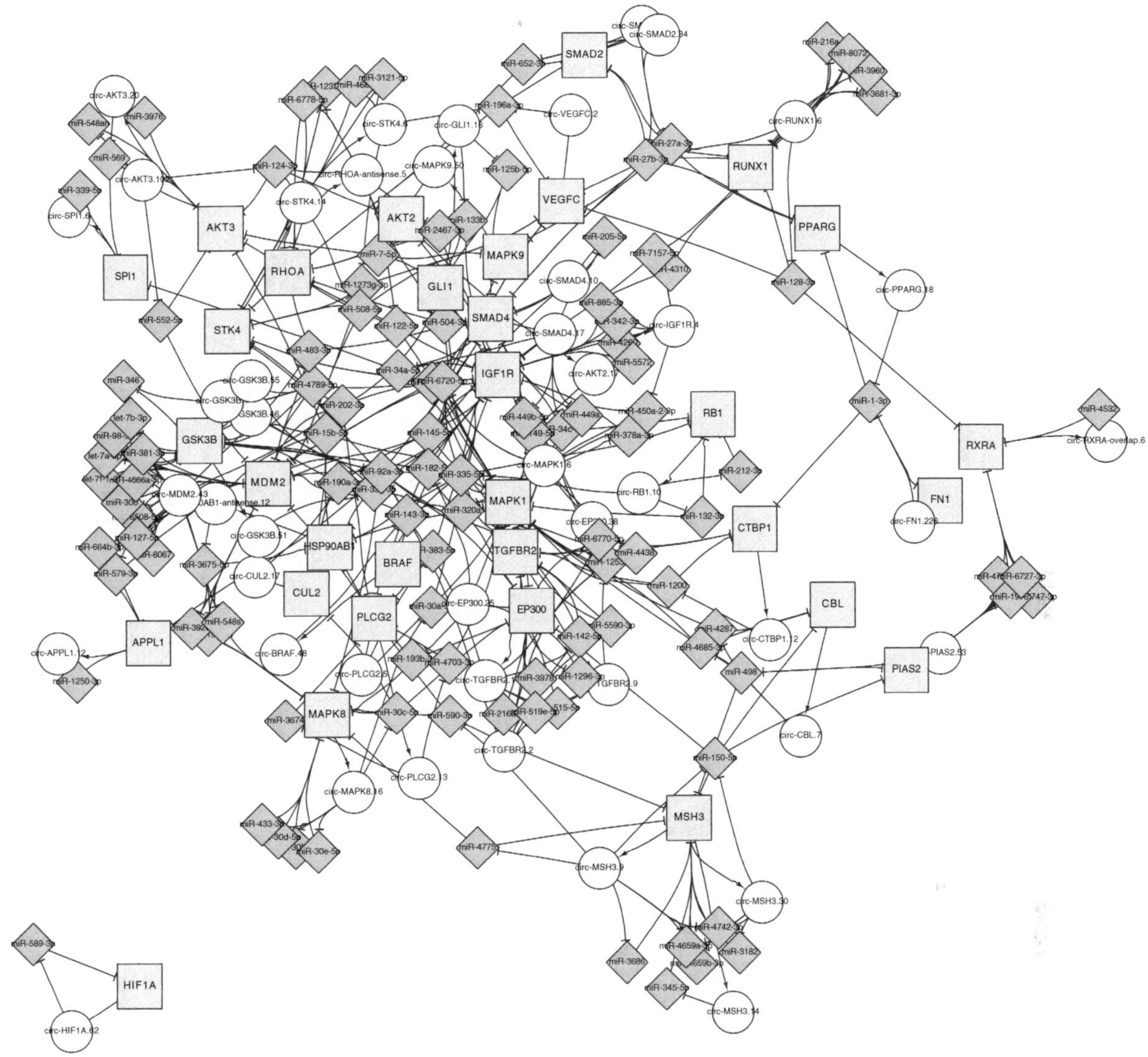

Fig. 1 Network of 32 ORCEL genes enriched in pathways in cancer. The genes are illustrated as green rectangle nodes. The circRNAs are illustrated as yellow circles. The miRNAs involved are illustrated as pink diamond nodes

interactions with *P*-values < 0.005 were collected. Expression profiling of the circRNAs within each of the samples collected from SRA was conducted in two different approaches: normalized counts of reads spanning the back spliced junction sites SRPBM [13] and normalized counts of reads aligned on the annotated sequences of circRNAs in units of FPKM. Only the circRNAs with estimated expression level over the threshold and found in multiple researches or samples were analyzed in this study.

Detection of the back spliced junction sites

Reported human back-spliced junction sites were collected from 22 recent studies [6, 7, 13, 20, 22, 24–39]. In addition, 465 transcriptome sequencing data sets were collected from NCBI Sequence Read Archive [23, 40]. The back-spliced junction sites in each RNA-seq sample were identified using a circRNA discovery pipeline referred as find_circ [7, 49, 50]. We apply the criteria defined in the pipeline hence the detected junction sites met same standards as those in the previous reports, as described in the Memczak et al. 2013 study.

Annotation of circRNA full sequence

The method was described in our previous study [23]. The method was further applied on the updated data from reports between year 2014 and 2016. To acquire the full length nucleotide sequence from RNA-

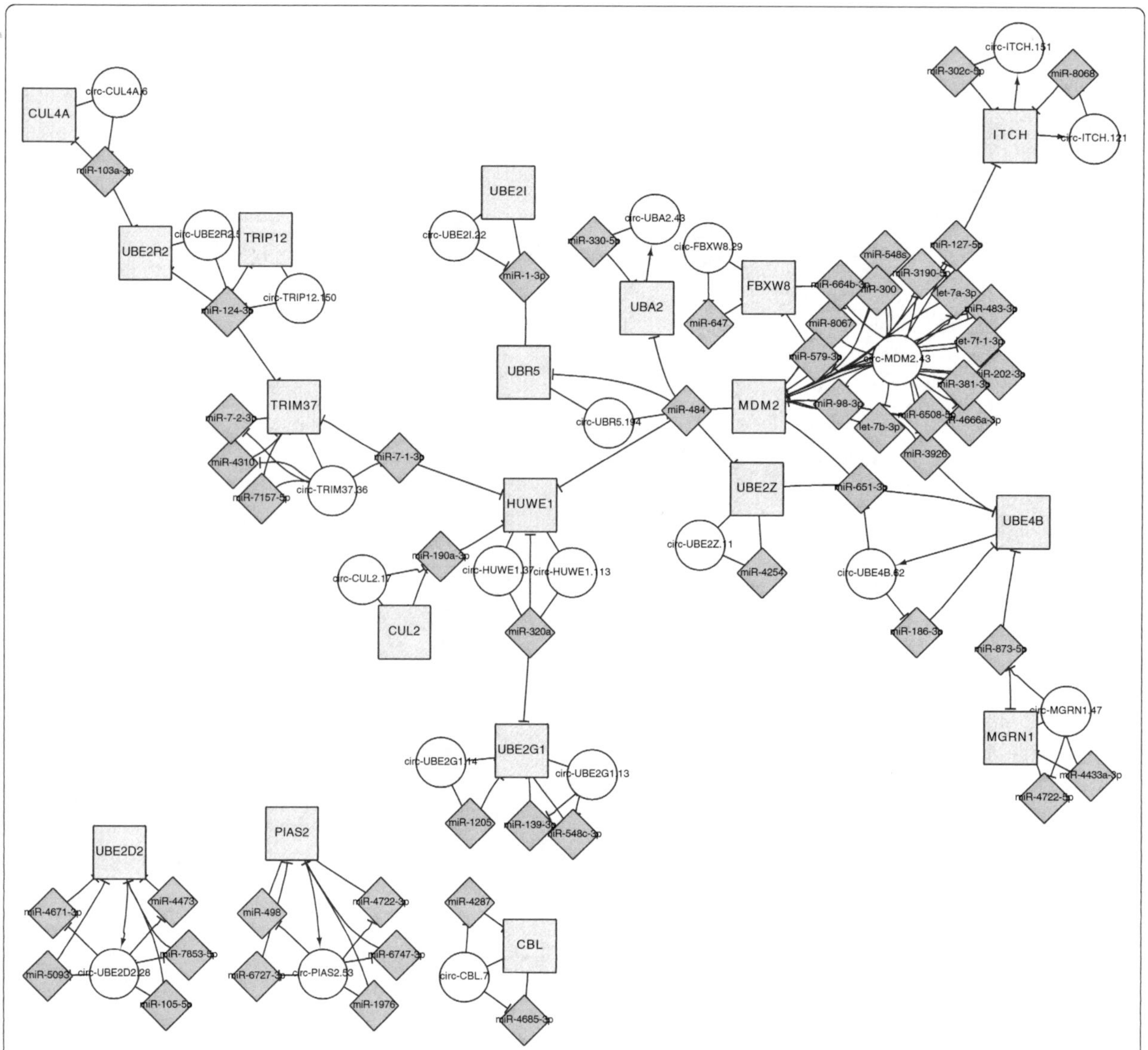

Fig. 2 Network of 19 ORCEL genes enriched in ubiquitin mediated proteolysis. The genes are illustrated as green rectangle nodes. The circRNAs are illustrated as yellow circles. The miRNAs involved are illustrated as pink diamond nodes

seq reads, back-spliced junction sites were compared with the hg19 human genome annotation as obtained from UCSC genome browser and RefSeq [51, 52]. Given the results of recent research, multiple circRNA isoforms might originate from the same back-spliced junction site [6, 53]. Hence we annotated multiple circRNA isoforms for one back-spliced junction site. The annotation was conducted following the guideline:

(1) For the back spliced junction sites locate on exact "head" and "tail" locus of exons from same transcript from RefSeq [51, 52], all the flanking exons of the transcript were considered as part of the same circRNA.

$$SRPBM = \frac{Reads\ count \times 10^9}{Read\ lenth \times Mapped\ reads} \quad (2)$$

(2) For those back spliced junction sites flanked multiple isoforms from RefSeq [51, 52], existence of multiple isoform of circRNA was assumed.

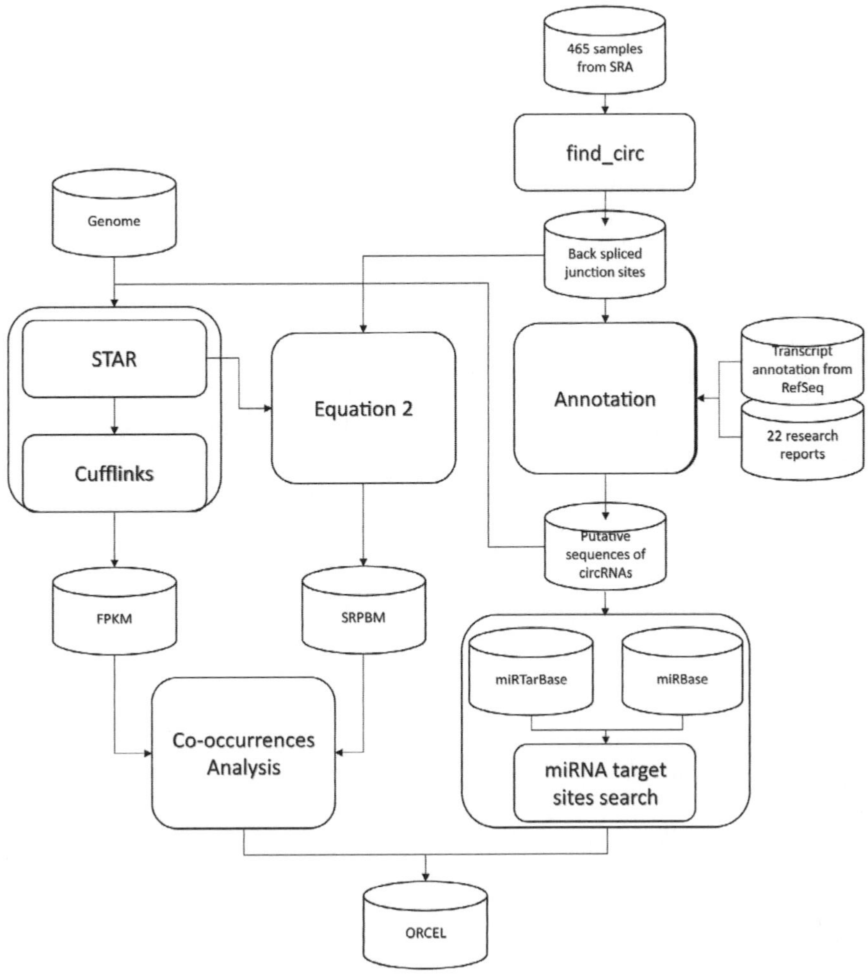

Fig. 3 Overview of the data analysis process in this research. The general view of the process of ORCEL discovery is illustrated in this figure

(3) For those circRNAs associated with these back-spliced junction sites having small misalignments to exon locations, flanked exons and a small portion of intron sequence locating in the head or tail locations were considered as parts of the isoforms.
(4) For the junction sites that were found to be located in intergenic positions while others, despite overlapping with certain genes, localized to their antisense strands, the entire flanked sequence was considered as the sequence of the circRNA.

The resulted sequence annotation was took for expression analysis and miRNA target search. The annotation along with the gene transcripts was recorded in gtf format for the expression profiling.

Identification of potential miRNA sponges

Developed from results of our previous study [23, 54], to find the potential miRNA and circRNA interactions, we conducted a statistical analysis on the amount of miRNA binding sites on the annotated circRNAs sequences.

The miRNA target sequences deemed typical: 6mer, 7mer-A1, 7mer-m8 and 8mer sequences [42] were extracted from miRBase [55]. Perfect complementarity sites were found on the annotated circRNA as well as gene transcripts sequences through iterative searching. To normalize the number of occurrences of these sites by the length of the transcripts, the following formula was used:

$$\text{Frequency of Nmer} = \frac{Number\ of\ target\ seeds \times 1000}{N \times Length\ of\ CircRNA} \tag{1}$$

Where the 'N' is the length of the seed. $N = 6$ for 6mer, $N = 7$ for 7mers and 8 for 8mer. With this formula, four frequency numbers can be acquired from each pair of circRNA and miRNA. To distinguish circRNA from linear isoforms, frequency values were also calculated for gene transcripts mRNA. We calculated all the frequency value of the circRNAs as well as the linear isoforms pairing to miRNAs, and then converted the Z-score of the normal distribution into one tail *P*-value through survival function.

The circRNA-miRNA pair with *P*-value < 0.005 was considered high regulatory potential between the circRNA and miRNA. The miRNAs and experimentally verified gene targets were collected from miRTarBase [54].

Expression profiling of circRNAs

As previously described, the abundance of the circRNA within the collected samples were estimated through the transcript deconvolution algorithm of the Cufflinks pipeline [41]. To further increase the prudence of circRNA detection within the transcriptome, the normalized counts of sequence reads spanning the back spliced junctions were considered.

The normalized count of reads on back spliced junctions

To normalize the amount of the normalized sequence reads spanning the junction sites, a concept of spliced reads per billion mapping (SRPBM) was applied [13]. Amount of reads mapped onto hg19 human genome was acquired through the tool STAR [56]. The equation applied to calculate SRPBM is as illustrated in Eq. 2. The junction sites with the value of SRPBM larger than 1.0 were selected.

Transcript deconvolution of circRNAs

RNA-seq aligner STAR [56] was applied to realign the sequence reads from the 465 RNA-seq samples on human genome. With the forth-mentioned gtf file containing annotated exon locus of circRNAs and mRNAs, and the bam files generated from STAR, we estimated the abundance within the sample of the annotated sequence through Cufflinks [41]. The resulted transcripts with FPKM over 1.0 were selected.

Co-occurrence analysis of circRNA

From the result of recent year comparison study [50], we deducted that inconsistency between different circRNA detection tools and false discovery of highly abundant circRNA could occur in the result of our analysis. Hence in addition to the two values of estimated abundance of circRNA, we further applied the following conditions:

The amount of previous peer review reports in which the back spliced junction sites were reported.

The amount of samples among the 465 collected samples in which the back spliced junction sites were found meeting the criteria defined in find_circ [7, 49, 50].

Only the circRNAs with the combined amount of these two values over 10 were considered in the analysis of this report.

Additional files

Additional file 1: A detail list of these circRNAs, as well as their corresponding SRPBM and FPKM values within the samples. (XLSX 528 kb)

Additional file 2: A complete list of ORCEL (Ouroboros Resembling Competitive Endogenous Loop) reported in this article. (XLSX 59 kb)

Additional file 3: A complete table of the enrichment of ORCEL genes. (XLSX 90 kb)

Abbreviations

CeRNA: Competing endogenous RNAs; circRNA: Circular RNA; FPKM: Fragments Per Kilobase of transcript per Million mapped reads; ORCEL: Ouroboros resembling competitive endogenous loop; SRPBM: Spliced reads per billion mapping

Acknowledgements

The authors would like to thank the Ministry of Science and Technology, the National Chiao Tung University and Ministry of Education, Taiwan, and University System of Taiwan (VGHUST), for financially supporting this research. This work is supported by multiple programs from the Ministry of Science and Technology, Taiwan [MOST103-2628-B-009-001-MY3, MOST105-2627-M-009-007-, MOST105-2319-B-400-002-, MOST104-2911-I-009-509 and MOST 106-2633-B-009-001].

Funding

Publication charges for this work, is supported by multiple programs from the Ministry of Science and Technology, Taiwan [MOST103–2628-B-009-001-MY3, MOST105–2627-M-009-007-, MOST105–2319-B-400-002-, MOST104–2911-I-009-509 and MOST 106–2633-B-009-001]. This paper (work) is particularly supported by "Aiming for the Top University Program" of the National Chiao Tung University and Ministry of Education, Taiwan, R.O.C.

Availability of data and materials

All the analyzed data mentioned in this report is available in the additional files.

About this supplement

This article has been published as part of BMC Genomics Volume 19 Supplement 2, 2018: Selected articles from the 16th Asia Pacific Bioinformatics Conference (APBC 2018): genomics. The full contents of the supplement are available online at https://bmcgenomics.biomedcentral.com/articles/supplements/volume-19-supplement-2.

Authors' contributions

YCL and HDH designed the study. CDY, WHL, HTH and HDH reviewed and selected the referenced samples. YCL and CDY analyzed and interpreted the data. YCL and HDH wrote the manuscript. All authors read and approved the final manuscript.

Ethics approval and consent to participate

Not applicable.

Consent for publication

Not applicable.

Competing interests

The authors declare that they have no competing interests.

Publisher's Note

Springer Nature remains neutral with regard to jurisdictional claims in published maps and institutional affiliations.

Author details
[1]Institute of Engineering in Medicine, University of California, La Jolla, San Diego, CA, USA. [2]Institute of Bioinformatics and Systems Biology, National Chiao Tung University, HsinChu, Taiwan. [3]Institute of Pharmacology, National Yang-Ming University, Taipei, Taiwan. [4]Institute of Population Health Sciences, National Health Research Institutes, Miaoli, Taiwan. [5]Department of Biological Science and Technology, Institute of Bioinformatics, National Chiao Tung University, Hsin-Chu, Taiwan. [6]Department of Medical Research, Chung Shan Medical University Hospital, Taichung, Taiwan.

Published: 9 May 2018

References

1. Harris H. Turnover of nuclear and cytoplasmic ribonucleic acid in two types of animal cell, with some further observations on the nucleolus. Biochem J. 1959;73(2):362.
2. Wang Z, Gerstein M, Snyder M. RNA-Seq: a revolutionary tool for transcriptomics. Nat Rev Genet. 2009;10(1):57–63.
3. Kung JT, Colognori D, Lee JT. Long noncoding RNAs: past, present, and future. Genetics. 2013;193(3):651–69.
4. Hsu M-T, Coca-Prados M: Electron microscopic evidence for the circular form of RNA in the cytoplasm of eukaryotic cells. 1979.
5. Salzman J, Gawad C, Wang PL, Lacayo N, Brown PO. Circular RNAs are the predominant transcript isoform from hundreds of human genes in diverse cell types. PLoS One. 2012;7(2):e30733.
6. Salzman J, Chen RE, Olsen MN, Wang PL, Brown PO. Cell-type specific features of circular RNA expression. PLoS Genet. 2013;9(9):e1003777.
7. Memczak S, Jens M, Elefsinioti A, Torti F, Krueger J, Rybak A, Maier L, Mackowiak SD, Gregersen LH, Munschauer M, et al. Circular RNAs are a large class of animal RNAs with regulatory potency. Nature. 2013;495(7441):333–8.
8. Tay Y, Rinn J, Pandolfi PP. The multilayered complexity of ceRNA crosstalk and competition. Nature. 2014;505(7483):344–52.
9. Hansen TB, Wiklund ED, Bramsen JB, Villadsen SB, Statham AL, Clark SJ, Kjems J. miRNA-dependent gene silencing involving Ago2-mediated cleavage of a circular antisense RNA. EMBO J. 2011;30(21):4414–22.
10. Hansen TB, Jensen TI, Clausen BH, Bramsen JB, Finsen B, Damgaard CK, Kjems J. Natural RNA circles function as efficient microRNA sponges. Nature. 2013;495(7441):384–8.
11. Wilusz JE, Sharp PA. A circuitous route to noncoding RNA. Science. 2013;340(6131):440–1.
12. Suzuki H, Zuo Y, Wang J, Zhang MQ, Malhotra A, Mayeda A. Characterization of RNase R-digested cellular RNA source that consists of lariat and circular RNAs from pre-mRNA splicing. Nucleic Acids Res. 2006;34(8):e63.
13. Jeck WR, Sorrentino JA, Wang K, Slevin MK, Burd CE, Liu J, Marzluff WF, Sharpless NE. Circular RNAs are abundant, conserved, and associated with ALU repeats. RNA. 2013;19(2):141–57.
14. Ashwal-Fluss R, Meyer M, Pamudurti NR, Ivanov A, Bartok O, Hanan M, Evantal N, Memczak S, Rajewsky N, Kadener S. circRNA biogenesis competes with pre-mRNA splicing. Mol Cell. 2014;56(1):55–66.
15. Westholm JO, Miura P, Olson S, Shenker S, Joseph B, Sanfilippo P, Celniker SE, Graveley BR, Lai EC. Genome-wide analysis of drosophila circular RNAs reveals their structural and sequence properties and age-dependent neural accumulation. Cell Rep. 2014;9(5):1966–80.
16. Wang PL, Bao Y, Yee M-C, Barrett SP, Hogan GJ, Olsen MN, Dinneny JR, Brown PO, Salzman J: Circular RNA is expressed across the eukaryotic tree of life. PLoS One 2014, 9(3):e90859.
17. Gao Y, Wang J, Zhao F. CIRI: an efficient and unbiased algorithm for de novo circular RNA identification. Genome Biol. 2015;16(1):4.
18. Tian M, Chen R, Li T, Xiao B. Reduced expression of circRNA hsa_circ_0003159 in gastric cancer and its clinical significance. J Clin Lab Anal. 2017.
19. Bachmayr-Heyda A, Reiner AT, Auer K, Sukhbaatar N, Aust S, Bachleitner-Hofmann T, Mesteri I, Grunt TW, Zeillinger R, Pils D. Correlation of circular RNA abundance with proliferation–exemplified with colorectal and ovarian cancer, idiopathic lung fibrosis, and normal human tissues. Sci Rep-Uk. 2015;5:8057.
20. Zhang X-O, Wang H-B, Zhang Y, Lu X, Chen L-L, Yang L. Complementary sequence-mediated exon circularization. Cell. 2014;159(1):134–47.
21. Guo JU, Agarwal V, Guo H, Bartel DP. Expanded identification and characterization of mammalian circular RNAs. Genome Biol. 2014;15(409):10.1186.
22. Caiment F, Gaj S, Claessen S, Kleinjans J. High-throughput data integration of RNA–miRNA–circRNA reveals novel insights into mechanisms of benzo [a] pyrene-induced carcinogenicity. Nucleic Acids Res. 2015;43(5):2525–34.
23. Liu Y-C, Li J-R, Sun C-H, Andrews E, Chao R-F, Lin F-M, Weng S-L, Hsu S-D, Huang C-C, Cheng C. CircNet: a database of circular RNAs derived from transcriptome sequencing data. Nucleic Acids Res. 2016;44(D1):D209–15.
24. Boeckel J-N, Jaé N, Heumüller AW, Chen W, Boon RA, Stellos K, Zeiher AM, John D, Uchida S, Dimmeler S. Identification and characterization of hypoxia-regulated endothelial circular RNA. Circ Res. 2015;117(10):884–90.
25. Bahn JH, Zhang Q, Li F, Chan T-M, Lin X, Kim Y, Wong DT, Xiao X. The landscape of microRNA, Piwi-interacting RNA, and circular RNA in human saliva. Clin Chem. 2015;61(1):221–30.
26. Bachmayr-Heyda A, Reiner AT, Auer K, Sukhbaatar N, Aust S, Bachleitner-Hofmann T, Mesteri I, Grunt TW, Zeillinger R, Pils D. Correlation of circular RNA abundance with proliferation-exemplified with colorectal and ovarian cancer, idiopathic lung fibrosis, and normal human tissues. Scientific reports. 2015;5:8057.
27. Conn SJ, Pillman KA, Toubia J, Conn VM, Salmanidis M, Phillips CA, Roslan S, Schreiber AW, Gregory PA, Goodall GJ. The RNA binding protein quaking regulates formation of circRNAs. Cell. 2015;160(6):1125–34.
28. Memczak S, Papavasileiou P, Peters O, Rajewsky N. Identification and characterization of circular RNAs as a new class of putative biomarkers in human blood. PLoS One. 2015;10(10):e0141214.
29. Alhasan AA, Izuogu OG, Al-Balool HH, Steyn JS, Evans A, Colzani M, Ghevaert C, Mountford JC, Marenah L, Elliott DJ. Circular RNA enrichment in platelets is a signature of transcriptome degradation. Blood. 2016;127(9):e1–e11.
30. Cheng J, Metge F, Dieterich C. Specific identification and quantification of circular RNAs from sequencing data. Bioinformatics. 2016;32(7):1094–6.
31. Zhang XO, Dong R, Zhang Y, Zhang JL, Luo Z, Zhang J, Chen LL, Yang L: Diverse alternative back-splicing and alternative splicing landscape of circular RNAs. Genome Res. 2016;26(9):1277–87.
32. Song X, Zhang N, Han P, Moon BS, Lai RK, Wang K, Lu W. Circular RNA profile in gliomas revealed by identification tool UROBORUS. Nucleic Acids Res. 2016;44(9):e87.
33. Dang Y, Yan L, Hu B, Fan X, Ren Y, Li R, Lian Y, Yan J, Li Q, Zhang Y. Tracing the expression of circular RNAs in human pre-implantation embryos. Genome Biol. 2016;17(1):1.
34. Zheng Q, Bao C, Guo W, Li S, Chen J, Chen B, Luo Y, Lyu D, Li Y, Shi G. Circular RNA profiling reveals an abundant circHIPK3 that regulates cell growth by sponging multiple miRNAs. Nature communications. 2016;7.
35. Zhang Y, Zhang X-O, Chen T, Xiang J-F, Yin Q-F, Xing Y-H, Zhu S, Yang L, Chen L-L. Circular intronic long noncoding RNAs. Mol Cell. 2013;51(6):792–806.
36. Rybak-Wolf A, Stottmeister C, Glažar P, Jens M, Pino N, Giusti S, Hanan M, Behm M, Bartok O, Ashwal-Fluss R. Circular RNAs in the mammalian brain are highly abundant, conserved, and dynamically expressed. Mol Cell. 2015;58(5):870–85.
37. Guo JU, Agarwal V, Guo H, Bartel DP. Expanded identification and characterization of mammalian circular RNAs. Genome Biol. 2014;15(7):1.
38. Gao Y, Wang J, Zhao F. CIRI: an efficient and unbiased algorithm for de novo circular RNA identification. Genome Biol. 2015;16(1):1.
39. Kelly S, Greenman C, Cook PR, Papantonis A. Exon skipping is correlated with exon circularization. J Mol Biol. 2015;427(15):2414–7.
40. Leinonen R, Sugawara H, Shumway M, International Nucleotide Sequence Database C. The sequence read archive. Nucleic Acids Res. 2011; 39(Database issue):D19–21.
41. Trapnell C, Roberts A, Goff L, Pertea G, Kim D, Kelley DR, Pimentel H, Salzberg SL, Rinn JL, Pachter L. Differential gene and transcript expression analysis of RNA-seq experiments with TopHat and cufflinks. Nat Protoc. 2012;7(3):562–78.
42. Bartel DP. MicroRNAs: target recognition and regulatory functions. Cell. 2009;136(2):215–33.
43. Dennis G, Sherman BT, Hosack DA, Yang J, Gao W, Lane HC, Lempicki RA. DAVID: database for annotation, visualization, and integrated discovery. Genome Biol. 2003;4(9):R60.
44. Kanehisa M, Goto S. KEGG: kyoto encyclopedia of genes and genomes. Nucleic Acids Res. 2000;28(1):27–30.
45. Shannon P, Markiel A, Ozier O, Baliga NS, Wang JT, Ramage D, Amin N, Schwikowski B, Ideker T. Cytoscape: a software environment for integrated models of biomolecular interaction networks. Genome Res. 2003;13(11):2498–504.
46. Denzler R, Agarwal V, Stefano J, Bartel DP, Stoffel M. Assessing the ceRNA hypothesis with quantitative measurements of miRNA and target abundance. Mol Cell. 2014;54(5):766–76.

47. Thomson DW, Dinger ME. Endogenous microRNA sponges: evidence and controversy. Nat Rev Genet. 2016;17(5):272–83.
48. Brock WH. August Kekulé (1829–1896): theoretical chemist. Endeavour. 1996; 20(3):121–5.
49. Glažar P, Papavasileiou P, Rajewsky N. circBase: a database for circular RNAs. RNA. 2014;20(11):1666–70.
50. Hansen TB, Venø MT, Damgaard CK, Kjems J. Comparison of circular RNA prediction tools. Nucleic Acids Res. 2016;44(6):e58.
51. Karolchik D, Baertsch R, Diekhans M, Furey TS, Hinrichs A, Lu Y, Roskin KM, Schwartz M, Sugnet CW, Thomas DJ. The UCSC genome browser database. Nucleic Acids Res. 2003;31(1):51–4.
52. Pruitt KD, Tatusova T, Maglott DR. NCBI reference sequences (RefSeq): a curated non-redundant sequence database of genomes, transcripts and proteins. Nucleic Acids Res. 2007;35(suppl 1):D61–5.
53. You X, Vlatkovic I, Babic A, Will T, Epstein I, Tushev G, Akbalik G, Wang M, Glock C, Quedenau C. Neural circular RNAs are derived from synaptic genes and regulated by development and plasticity. Nat Neurosci. 2015;18(4):603–10.
54. Chou C-H, Chang N-W, Shrestha S, Hsu S-D, Lin Y-L, Lee W-H, Yang C-D, Hong H-C, Wei T-Y, Tu S-J. miRTarBase 2016: updates to the experimentally validated miRNA-target interactions database. Nucleic Acids Res. 2016; 44(D1):D239–47.
55. Griffiths-Jones S, Grocock RJ, Van Dongen S, Bateman A, Enright AJ. miRBase: microRNA sequences, targets and gene nomenclature. Nucleic Acids Res. 2006;34(suppl 1):D140–4.
56. Dobin A, Davis CA, Schlesinger F, Drenkow J, Zaleski C, Jha S, Batut P, Chaisson M, Gingeras TR. STAR: ultrafast universal RNA-seq aligner. Bioinformatics. 2013;29(1):15–21.

Noutahi and El-Mabrouk *BMC Genomics* 2018, **19**(Suppl 2):102
https://doi.org/10.1186/s12864-018-4455-x

BMC Genomics

RESEARCH

Open Access

GATC: a genetic algorithm for gene tree construction under the Duplication-Transfer-Loss model of evolution

Emmanuel Noutahi* and Nadia El-Mabrouk

From The Sixteenth Asia Pacific Bioinformatics Conference 2018
Yokohama, Japan. 15-17 January 2018

Abstract

Background: Several methods have been developed for the accurate reconstruction of gene trees. Some of them use reconciliation with a species tree to correct, *a posteriori*, errors in gene trees inferred from multiple sequence alignments. Unfortunately the best fit to sequence information can be lost during this process.

Results: We describe GATC, a new algorithm for reconstructing a binary gene tree with branch length. GATC returns optimal solutions according to a measure combining both tree likelihood (according to sequence evolution) and a reconciliation score under the Duplication-Transfer-Loss (DTL) model. It can either be used to construct a gene tree from scratch or to correct trees infered by existing reconstruction method, making it highly flexible to various input data types. The method is based on a genetic algorithm acting on a population of trees at each step. It substantially increases the efficiency of the phylogeny space exploration, reducing the risk of falling into local minima, at a reasonable computational time. We have applied GATC to a dataset of simulated cyanobacterial phylogenies, as well as to an empirical dataset of three reference gene families, and showed that it is able to improve gene tree reconstructions compared with current state-of-the-art algorithms.

Conclusion: The proposed algorithm is able to accurately reconstruct gene trees and is highly suitable for the construction of reference trees. Our results also highlight the efficiency of multi-objective optimization algorithms for the gene tree reconstruction problem. GATC is available on Github at: https://github.com/UdeM-LBIT/GATC.

Keywords: Gene tree, Genetic algorithm, Phylogeny, Reconciliation, Gene duplication, Horizontal gene transfer

Background

Most biological discoveries can only be made in the light of evolution. In particular, functional annotation of genes is usually deduced from the orthology and paralogy relation between genes, which is inferred from the comparison of a gene tree with a species tree. Therefore, phylogenetic tree reconstruction is an important component of most bioinformatic pipelines. In this paper, we focus on the reconstruction of gene trees.

Standard phylogenetic tools are based on maximum likelihood (ML) or bayesian methods reconstructing a tree from gene sequences (e.g. PhyML [1], RAxML [2], MrBayes [3], PhyloBayes [4]). However, for a variety of reasons due, not only to technical limitations but also to the data itself (sequences too close to each other or conversely too divergent), sequence-only methods often do not allow to confidently discriminate one tree from another.

To address this limitation, more recent gene tree reconstruction methods, designated here as *integrative methods*, include information from the species tree. The idea is to consider, in addition to a maximum likelihood value measuring the fitness of a tree to a sequence alignment (according to a model of sequence evolution), a measure reflecting the evolution of a whole gene family

*Correspondence: fmr.noutahi@umontreal.ca
Département d'Informatique et de Recherche Opérationnelle, Université de Montréal, Montréal, Canada

through gene gain and loss. A standard measure of fitness between a gene tree and a species tree is computed in terms of a "reconciliation" score. In a probabilistic framework, the reconciliation score corresponds to the probability density of the gene tree given some parameters (rates of events and branch lengths) under a birth-death and gain model of evolution. For the Most Parsimonious Reconciliation model (MPR), this score corresponds to the optimal number of gene gain and loss events, weighted by their costs, explaining the incongruence between a gene tree and a species tree.

Most integrative methods for gene tree reconstruction assume a simplified model of gene family evolution where gene gain events are reduced to gene duplication (e.g. TreeBeST [5], TreeFix [6], ProfileNJ [7], NOTUNG [8], SPIMAP [9], Giga [10]). In fact, the MPR problem for the Duplication-Loss (hereafter denoted DL) model of gene family evolution is linear-time solvable [11]. By introducing horizontal gene transfer (HGT) events, the Duplication-Transfer-Loss (DTL) model becomes NP-hard in general if time consistency is required for inferred events (unless the species tree is fully dated) [12–14]. However the MPR problem for the DTL model, with an undated species tree, can still be computed in polynomial time if the time consistency requirement is relaxed [15–17]. Due to this reasonable time-complexity, some recent phylogenetic softwares have extended the gene family evolution model to transfers (MowgliNNI [18], ecceTERA [19], TreeFix-DTL [20]). Continuous effort is also made for developing fast probabilistic frameworks capturing HGT events (see [21] for a review of these models).

Integrative methods report gene trees with better accuracy compared with sequence-only methods [18, 20, 22, 23], but they still leave space for improvement, both on tree quality and on computation time. In fact, most of them rely on a two-steps approach, first computing a tree with the best fit to the sequences, and then exploring a tree space surrounding the initial tree to select one minimizing the considered reconciliation distance. From an accuracy point of view, this two step methodology does not guarantee that the output tree optimizes both the likelihood given the sequence alignment, and the reconciliation measure, as the best fit to the sequences may be lost at the second step. In addition, by considering a single tree at a time, the risk of ignoring a large part of the tree space and falling into a local minimum is high. Other integrative methods (see for example PhylDog [24] and PrIME-DLTRS [23]) compute the joint likelihood associated with a substitution model and DTL event rates, given a fixed, dated and utrametric species tree and a gene family alignment. They use tree exploration heuristics similar to those found in sequence-only programs for phylogenetic tree reconstruction to explore the solution space, often in a bayesian-MCMC framework. These methods come at a high computational cost, especially when HGT events are considered, and they are still subject to the risk of being stuck in a local optimum.

In this paper, we present *GATC* (*G*enetic *A*lgorithm for gene *T*ree *C*onstruction), a new software for gene tree reconstruction under the DTL model that can take as input completely unresolved, partially unresolved or fully resolved trees, and outputs a tree minimizing a measure combining both tree likelihood (according to sequence evolution) and a reconciliation score. In other words, it can either be used as a two-step correction method, when input trees are the output of other phylogenetic methods, or as a one-step method resolving a full polytomy (star tree) in a way optimizing fit to both the species tree and the sequences.

With GATC, we explore a new methodological framework based on a Genetic Algorithm (GA), a global search metaheuristic that mimics biological evolution [25]. The ability of GAs to find near-optimal solutions quickly, even for complex models and data makes them suitable for the problem of phylogenetic inference. In fact, the GA methodology has been previously applied to the phylogenetic inference problem, starting with Matsuda in 1996 [26] using a maximum likelihood criterion, Lewis [27] who introduced a subtree swap crossover operator, and other more recent algorithms (e.g. self-adaptive GA [28], Ga-mt [29], METAPIGA [30], GARLI [31]). However, all these algorithms are solely based on sequence information and, as discussed above, are often error-prone in the case of gene tree reconstruction. To our best knowledge, GAs have never been applied to species tree-aware gene tree reconstruction, although the technique is suitable to the resolution of Multi-Objective Optimization Problems (MOOP).

To measure the performance of GATC, we compared it to current state-of-the-art softwares on a dataset of simulated cyanobacterial phylogenies. Our results show that GATC is more accurate than existing methods, suggesting that it substantially increases the efficiency of the phylogeny space exploration. We also evaluated GATC's ability to infer accurate homology relationships between genes on a standardized, manually curated, dataset of real trees. The predicted relationships were mostly in agreement with the ones inferred from a reference tree, highlighting the efficiency of the framework.

Methods

Notation on trees

All considered trees are rooted unless explicitely stated. A tree is *binary* if all its internal nodes have exactly two children, and *non-binary* otherwise. Unless stated differently, all trees are considered binary.

We denote by $V(T)$ the nodeset, by $E(T)$ the edgeset, by $\mathcal{L}(T)$ the leafset and by $r(T)$ the root of a tree T. An edge e of $E(T)$ is written as a pair (x, y) of two adjacent nodes where e is an outgoing edge of x. For $e = (x, y)$, x is the parent $p(y)$ of y, while y is a child of x. A node x is an *ancestor* of y, which is denoted $x <_T y$, if it is on the path from y to the root (excluding y). In this case, y is called a *descendant* of x. Similarly, an edge $e' = (x', y')$ is an ancestor of an edge $e = (x, y)$ if it is on the path from y to the root. Given a node x, $T[x]$ is the subtree of T rooted at x and $\mathcal{L}(x)$ the leafset of $T[x]$. Two subtrees $T[x]$ and $T[y]$ are *separated* in T if $x \neq y$, $x \not<_T y$ and $x \not>_T y$. In this case, $\mathcal{L}(x) \cap \mathcal{L}(y) = \emptyset$, if the leaves of T are uniquely labeled by the elements of $\mathcal{L}(T)$.

A *species tree* is a tree S with leaves uniquely labeled and $\mathcal{L}(S)$ being a set of species. Likewise, a *gene tree* is a tree G with leaves uniquely labeled and $\mathcal{L}(G)$ corresponding to a set of genes where each gene g belong to a genome $s(g)$. We denote by $\overline{G}$ the tree obtained from G by replacing each leaf label g_i by its genome $s(g_i)$. Notice that the mapping $s : \mathcal{L}(G) \rightarrow \mathcal{L}(S)$ does not have to be injective nor surjective. In particular, $\overline{G}$ may have several equally labeled leaves.

A *reconciliation* of G (or similarly $\overline{G}$) with S (see Fig. 1) is an extension of s from $V(G)$ to $V(S)$ with additional labels on each internal node x of G, describing the type of evolutionary event that has led to $G[x]$ (duplication, speciation or transfer). G can be expanded to include lost genes.

Finally, we refer to the process of removing a leaf l and its associated edge $(p(l),\ l)$ from a tree T as the *deletion of l from T*.

Vocabulary of Genetic Algorithms

A Genetic Algorithm (GA) is an algorithmic framework mimicking biological evolution. The vocabulary of a GA is filled with biological metaphors. It begins with a *population* of *individuals* whose *chromosomes or genomes* encode specific solutions to the problem of interest. Performance of theses individuals in solving the problem is measured by their *fitness score*. To avoid confusion, throughout this paper the word "chromosome" will be used solely to designate the data structure of a genetic algorithm, and the word "genome" will be used in its biological meaning to designate the macromolecules containing the genes under study.

At each step, starting from an initial population, a new population is generated using three operators: *selection*, *crossover* and *mutation* [25], which are defined according to the nature of the problem and the encoding of the solution. During selection, the fitness score is used to select individuals for reproduction. Selected candidates are combined using the crossover operator to create new individuals that are then modified by the mutation operator in order to introduce diversity and avoid local optima. With *elitism*, the less fit individuals of the newly obtained population are replaced by the best fit of the previous *generation*, in order to conserve the best solutions found so far. The process described above is repeated through multiple

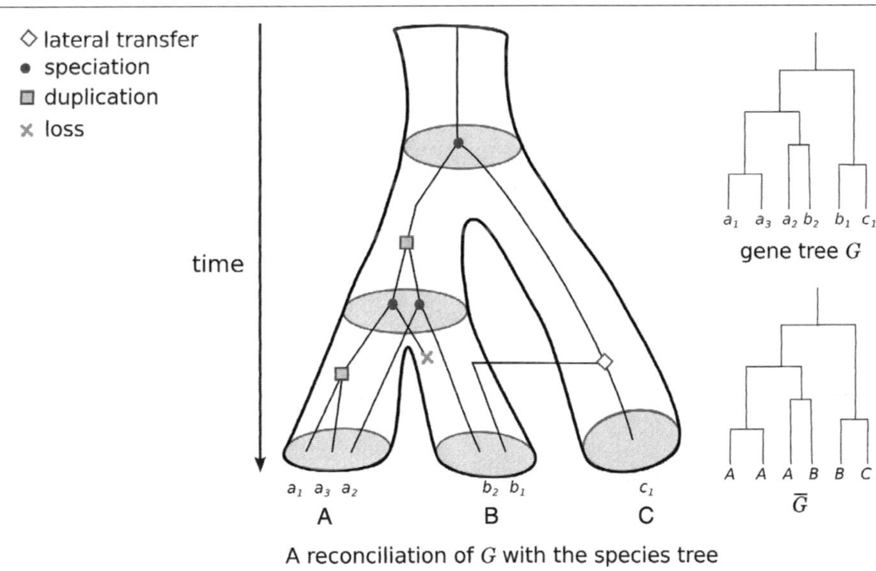

Fig. 1 A reconciliation between a gene tree G and a species tree. The reconciliation represents a history of the gene family evolution through speciation, gene duplication, gene loss and HGT, in a way that is consistent with the species tree. Both G and $\overline{G}$ are shown on the figure

generations, until an optimal solution (chromosome of the best individual) is obtained or a stop criterion is reached.

This natural selection process generally leads to the improvement of the average population fitness over generations. While GAs often converge to an optimal or near optimal solution, their performance mainly depends on the mechanism for balancing two potentially contradictory objectives: keeping the best solutions found so far, and at the same time efficiently exploring the search space for promising solutions.

The GATC algorithm description

In the rest of the manuscript, we will loosely refer to the tree likelihood given a multiple sequence alignment as *sequence likelihood*.

Given a sequence alignment D and a species tree S, our objective is to find the gene tree G or a set $\mathcal{G}$ of gene trees, with branch length, that are (near) optimal for both the sequence likelihood and the reconciliation score. To solve this problem, our fitness function should reflect both objectives. We will present different ways for computing the fitness score, either by a linear combination of the two scores, or by trying to reach a *pareto optimality*. We start by presenting the general framework of the GA.

Solution encoding

A chromosome σ is defined as (G, θ) where G is a rooted binary gene tree and θ is the set of hyperparameters underlying the evolutionary model. Namely, $\theta = (\lambda, \delta, \tau, e, l, m)$, representing respectively the duplication rate, the loss rate, the transfer rate, the substitution rates across the gene tree edges, branch lengths and the substitution model. Some of these parameters might be kept fixed during the evolutionary process. For example, the substitution model m is usually fixed for all generations, whereas duplication, loss and transfer rates can vary when a probabilistic model is used to compute the reconciliation score. When parsimony is preferred, they correspond to fixed event cost. Branch lengths and edge substitution rates are usually optimized during sequence likelihood computation.

For the probabilistic model of reconciliation, the initial values of the hyperparameters λ, δ, τ and e are randomly drawn from a uniform distribution, unless explicitely provided. The default substitution model used for nucleotide sequences is the GTR model [32] with a gamma distribution to account for rate variation [33], whereas for proteins, the JTT model [34] with gamma-distributed rates is used.

Gene trees in the initial population

When starting trees are available (from any other tree construction method, integrative or not), they can be used as the population of the first generation in our GA. Otherwise the trees of the initial population are generated, either randomly or according to a predefined procedure. GATC implementation allows generating the initial trees from a star tree, using PolytomySolver [35] which outputs the most parsimonious trees for the DL-reconciliation score (but not necessary optimal for the DTL-reconciliation score or the sequence likelihood), or using bootstrapped trees obtained with RAxML [2]. These two methods should be prefered to the initialization with random trees, which can affect the algorithm's convergence.

Notice that two trees G_1, G_2 such that $\overline{G_1} = \overline{G_2}$ have the same reconciliation score, and thus if G_1 is a solution of PolytomySolver, minimizing the DL-reconciliation score, then G_2 is also a solution. Therefore, in this case, to increase the initial population of the GA, additional trees can be obtained by permutation of the genes at the leaves of G_1 in a way respecting the mapping function s.

Computing the sequence likelihood and reconciliation score

To evaluate the fitness of each chromosome σ_i , $1 < i < n$, in a population of size n, we first compute a vector $\vec{z}_i$ of two components, called the *raw score vector*, containing the sequence likelihood and the reconciliation score. Note that when the objective is to optimize only sequence likelihood, the second component corresponding to the reconciliation score is set to zero.

The sequence likelihood scores $p(D|G, l, m)$ can be computed using the Felsenstein algorithm [36] and the further computationnal enhancement described by Stamatakis et al. (2004) [37]. In fact, GATC use subroutines from RAxML to compute the sequence likelihood, thus benefiting from both its high computational speed and its large set of substitution models.

As for the reconciliation score, it can be computed under either the probabilistic or MPR model. For the MPR scoring model, we implemented the Bansal algorithm [15] which computes the DTL reconciliation cost between a binary gene tree G and a binary species tree S in time $O(|G||S|)$. Notice that, as explicit transfer pathways are not specified, a DTL scenario is not necessarily possible as it may violate temporal constraints [14]. In fact, a donating and a receiving species must have coexisted at the time of the transfer. Moreover, in contrast to duplications and losses, HGT are inter-dependent, and can induce contradictory temporal constraints on ancestral species. However, as the reconciliation problem for DTL using undated species tree with the constraint of respecting temporal constraints is NP-hard, the Bansal algorithm remains a good alternative for computing a reasonable DTL reconciliation score. In absence of HGTs, we compute the DL reconciliation score using a linear-time algorithm [11], to speed up calculations.

For the probabilistic scoring model, we have implemented the DTL model first described by Tofigh [38, 39] and used by PrIME-DLRS [23]. It is based on a birth-death model of evolution including rates for gene duplication, transfer and loss that requires discretization of a dated species tree and numerical resolution of ordinary differential equations. We refer the readers to the Supplementary Material of [23] for a thorough description of how the probability density of the reconciliation is computed.

Rather than minimizing the reconciliation score and maximizing likelihood, it is easier to simultaneously minimize both measures. For this reason, we take the negative log value when likelihood is used for any of the two scores. Therefore, it has to be understood that the best adapted individuals will be those with the lowest fitness.

Computing the fitness score

Given a raw score vector $\vec{z}_i$ for a chromosome σ_i, a weight vector $\vec{w}$ and a scaling function ϕ, we define the fitness score f_i of σ_i as $f_i = \vec{w} \cdot \phi(\vec{z}_i)$. In other words, f_i corresponds to the weighted sum of the two components of the raw score vector, scaled by a function ϕ. The simplest definition of ϕ is the identity function $\phi(\vec{z}) = \vec{z}$. An alternative is to standardize each score to a zero-minimum resulting in the following formulation : $\phi\left(z_i^k\right) = z_i^k - \min_i(z_i)^k$ for $1 \leq k \leq 2$ and $1 \leq i \leq n$. However, for this latter scaling function, fitness is not comparable between individuals of different generations.

Using the method described above for computing f_i transforms our problem into a single objective minimization problem and is suitable when both components of z_i are log likelihood values, since it is related to the joint weighted probability density for sequences data and reconciliation to the species tree.

When the reconciliation score is computed using parsimony, combining the two scores this way might not be optimal. Instead, we compute a set of *pareto optimal solutions* for this multi-objective optimization problem (MOOP). Several evolutionary based techniques have been developed for MOOP [40]. Here we will use a technique similar to the widely known NSGA (Non-dominated Sorting Genetic Algorithm) [41].

A raw score vector $\vec{z}_i = \left(z_i^1, z_i^2\right)$ is said to *dominate* another vector $\vec{z}_j = \left(z_j^1, z_j^2\right)$, denoted as $\vec{z}_i \prec \vec{z}_j$, iff $\vec{z}_i \neq \vec{z}_j$ and $z_i^1 < z_j^1$,$z_i^2 < z_j^2$. We are interested in finding the set of non-dominated solutions called *pareto set* (PS) and denoted as :

$$PS = \{\sigma_i \mid \nexists\, \sigma_j, \vec{z}_j \prec \vec{z}_i\}$$

At the end of the GA's evolutionary process, the pareto set represents the set of pareto optimal solutions. In contrast with classical genetic algorithms, computing the pareto set requires to consider simultaneously a parent population P_i and its offspring P_i', as optimal solutions from P_i can be lost if we use P_i' as the population P_{i+1} of the next generation.

Algorithm 1 illustrates the way fitness is computed for all individuals of a generation. It proceeds in a *wave* fashion, selecting the non-dominated individuals from the population $P^* = P_i \cup P_i'$, assigning them a shared fitness score, and then removing them from P^*. This process is repeated while increasing the fitness score for the individuals in the new waves, until the expected population size per generation is met or there is no non-dominated individuals anymore. In the latter case, the fitness of the remaining individuals is computed as the sum of their *dominance rank* (number of individuals that dominates an individual) and the fitness of the last wave. This process ensures that individuals belonging to the same wave have the same fitness and as such the same probability to reproduce. The n individuals with the best fitness constitutes P_{i+1}. Selection, crossover and mutation operators can be applied to P_{i+1} resulting in offspring P_{i+1}'.

Algorithm 1 Compute next generation population P_{k+1} from P_k

procedure COMPUTENEXTPOP(P_k)
  Compute P_k', the offspring population of P_k
  $P_k^* \leftarrow P_k \cup P_k'$
  Evaluate z_i for all $\sigma_i \in P_k^*$
  Compute the dominance rank d_i for each $\sigma_i \in P_k^*$
  $w \leftarrow 1$
  while $\exists\, \sigma_i \in P_k^* \mid d_i = 0$ **do**
    $Wave_w \leftarrow \{\sigma_i \mid d_i = 0\}$
    Set a shared fitness for all $\sigma_i \in Wave_w$ as w
    $P_k^* \leftarrow P_k^* \setminus Wave_w$
    Compute the dominance rank d_i for each $\sigma_i \in P_k^*$
    $w \leftarrow w + 1$
  end while
  for $\sigma_i \in P_k^*$ **do**
    set the fitness of σ_i as $w + d_i$
  end for
  $P_{k+1} \leftarrow \bigcup_w Wave_w \cup P_k^*$
  return the first $|P_k|$ of P_{k+1} according to fitness
end procedure

Selection

GATC implements multiple classical selection methods. Individuals can either be selected for crossover using the tournament selector [42] or using the roulette wheel selector which chooses individuals with probability inversely proportionnal to their fitness values (recall that the best indivuals have the smallest fitness value). Alternatively, the random uniform selector can be used, which gives equal reproduction probability to all individuals regardless of their fitness. Selected indivuals are used in the crossover operator to produce the individuals of the next generation.

Crossover

In the crossover operators implemented in GATC, two offsprings are created from two parent chromosomes. Each offspring inherits its hyperparameter θ from one of its parents, while its gene tree is obtained from the combination of the two parental trees.

Given trees G_i and G_j respectively from parent σ_i and σ_j, the first offspring is obtained with the subtree swap crossover operator [27], achieved by the following actions:

1 Select a subtree $G_i[x]$ from G_i (the root is excluded)
2 Delete all leaves from G_j that are also in $\mathcal{L}(x)$;
3 Regraft $G_i[x]$ to a random edge of G_j to obtain the offspring tree G'_j.

The second offspring tree G'_i, is obtained in a similar way by selecting a subtree from G_j and regrafting it in G_i. The crossover operator is illustrated on Fig. 2a.

In the special case where the objective is to only optimize the sequence likelihood, under the hypothesis that the reconciliation score is already optimal, this crossover operator is not applicable as it does not preserve the reconciliation score. Instead, the offspring trees are created by exchanging two subtrees $G_1[x]$ and $G_2[y]$ such that $\overline{G_1[x]}$ and $\overline{G_2[y]}$ are isomorphic with respect to the labels at their leaves (see Fig. 2b).

Mutation

For a chromosome $\sigma_i = (G_i, \theta_i)$, a mutation is performed either on the tree G_i or on the rates λ, δ, τ, e unless their values are fixed. Mutations on the rate parameters consist in drawing a new value from their distribution. On the other hand, a mutation operates on G_i by applying a topological modification. GATC uses SPR (Subtree Pruning and Regrafting) and re-rooting operations (see Fig. 3a-b) to generate a new tree topology. As with the crossover operator, when only sequence likelihood has to be optimized, reconciliation score should be preserved after mutations. For this purpose, mutation are performed by permuting the genes assignment to the leaves of G_i in such a way that only genes belonging to the same species are allowed to switch places (see Fig. 3c).

Stop criteria

We proposed several criteria to stop the GA evolution. The simplest ones are to terminate when a maximum number of generations or a time limit are met, or when all individuals converge to a single fitness value. Aside from these criteria, we propose another simple termination criterion called *population-AU criterion* that is based on the use of a reference ML tree. Under this criterion, evolution is stopped when all the individuals in the current population are statistically equivalent to the known ML tree, according to the AU test [43]. This stop criterion allows for a good performance when the objective is restricted to the optimization of sequence likelihood.

Results and discussion

To measure the efficiency of GATC in reconstructing accurate gene trees, we compared its performance, on a simulated dataset, to four different gene tree reconstruction methods: ecceTERA [19], TreeFix-DTL [20], ProfileNJ [7], MowgliNNI [18] and RAxML [2]. In contrast to RAxML which is a sequence-only method, the former four methods use both sequences and species tree information. We also used GATC to reconstruct the gene trees of three gene families for which reference trees have been proposed [44]. We will entirely focus on evaluating

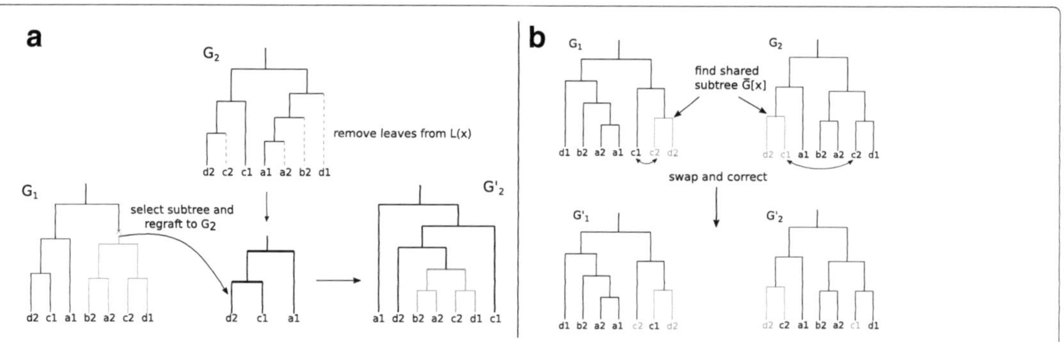

Fig. 2 Crossover operator. **a** Subtree swap. A subtree $G_1[x]$ (in red) is pruned from G_1 then regrafted to a random branch of G_2 after deleting from G_2 its leaves that also appear in $G_1[x]$ (shown in dotted lines). To obtain the second child, a similar operation is performed from G_2 to G_1 **b** Subtree swap preserving reconciliation. Two subtrees $G_1[x]$ and $G_2[y]$, respectively from G_1 and G_2, such that $\overline{G_1[x]} = \overline{G_2[y]}$ are swapped and the remaining leaves are corrected to conserve the same leafset as the parent

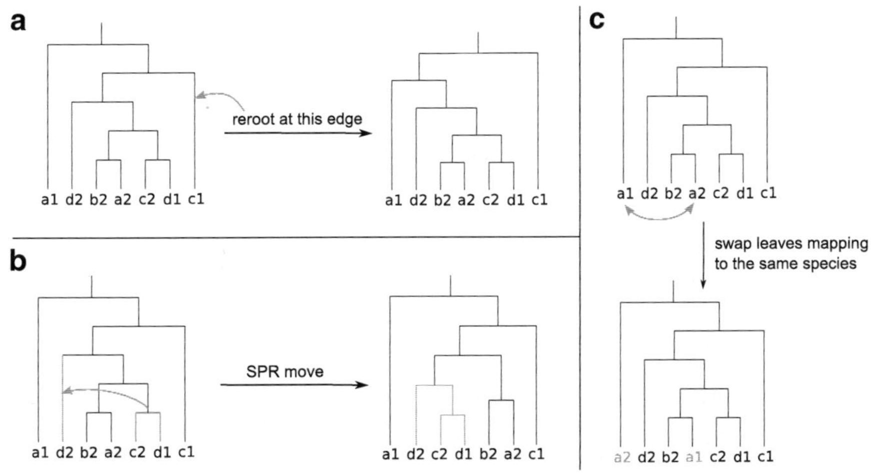

Fig. 3 Mutation operator. **a** Re-rooting. The tree is rerooted at a random edge. **b** SPR move. A subtree is pruned from the tree and regrafted to another edge. **c** Mutation preserving reconciliation cost. Two leaves l_1 and l_2 such that $s(l_1) = s(l_2)$ are swapped. This mutation only alter the sequence likelihood

GATC's performance under the MPR model, as it is our main contribution and also because DTL-reconciliation scores can be computed significantly faster under this model.

Evaluation on a simulated Cyanobacteria histories dataset

We used the simulated cynobacteria dataset of Szöllosi et al. (2013) [22] publicly available at http://datadryad.org/resource/doi:10.5061/dryad.pv6df. This dataset consists of 1099 gene families from 39 cyanobacteria species along with a well-resolved dated species tree. To construct the dataset, the gene families were retrieved from HOGENOM [45] and multiple alignments were performed on these families with Muscle [46]. For each alignment, an MCMC sample of at least 3000 trees was obtained with PhyloBayes [47] and used to reconstruct an amalgamated tree with ALE [22]. These trees were used to simulate new multiple alignments of artifical sequences under the LG model with a gamma distribution. We refer to [22] for a more detailed description on the construction of the dataset.

From each of the 1099 simulated artificial sequence alignments, we reconstructed an inital tree using RAxML (LG + Gamma, 100 bootstraps). The RAxML trees (with bootstrap values) were used as input for all programs being compared against GATC.

For all programs, we used fixed DTL rates ($\lambda = 2, \tau = 3, \delta = 1$) except for ProfileNJ which supports only a DL model of reconciliation and for which we took $\tau = \infty$. We ran TreeFix-DTL with default parameters and LG + Gamma as model of evolution. As it requires rooted trees, the input RAxML trees were rooted using the midpoint rooting method [48]. MowgliNNI, ecceTERA and ProfileNJ were run with a threshold of 0.7 for the contraction of for weak edges. Note that ProfileNJ and ecceTERA can output several solutions from which users can later select a tree according to some other measure (sequence likelihood for example). In our comparison, we only consider the first solution returned by both methods, as this selection process is not part of the methods, and also removes their running time advantage. We ran GATC with the following parameters : a maximum of 50 generations, a time limit of 90 minutes per gene family, LG + Gamma as the model of evolution and parsimony for DTL-reconciliation. We used the tournament selector as the selection operator and set the crossover and mutation rates to 0.8 and 0.5 respectively (see Additional file 1: Section 1 and Figure S1, for a discussion on the effect of crossover and mutation rates on accuracy). To construct the initial population of the GA, we used PolytomySolver's resolutions of RAxML trees after contraction of edges with support less than 0.7. In order to keep the GA population size fixed at 30, we randomly removed or duplicated chromosomes from the initial population until its size became 30. We also used the population-AU as additional stopping criterion with the RAxML tree being the known best ML tree and a significance level $\alpha = 0.05$. When there were more than one tree in the pareto optimal set, the tree with the lowest DTL-reconciliation score was returned as GATC final solution.

We measured the accuracy, defined as the normalized Robinson-Foulds distance between each reconstructed tree and the true tree. As shown in Fig. 4, trees reconstructed with species tree-aware algorithms were more accurate than RAxML's trees. This result was expected, since it has been shown several times that integration

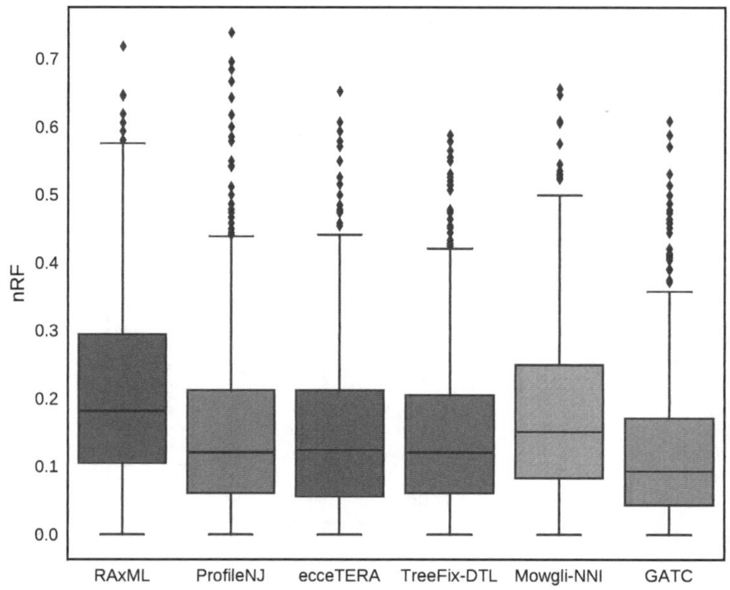

Fig. 4 Accuracy of RAxML, ProfileNJ, ecceTERA, TreeFix-DTL, Mowgli and GATC on a dataset of simulated Cyanobacteria histories: we measure the normalized Robinson-Foulds distance of the reconstructed trees to the true gene trees for all 1099 gene families. GATC achieves the best accuracy on the simulated dataset, followed by TreeFix-DTL

of species tree information usually improves gene trees reconstruction. GATC, in particular, achieves a better accuracy than other methods, due to its improved tree space search efficiency. The algorithm also appears to be robust, in some extent, to errors in the species tree topology (see Additional file 1: Section 2 and Figure S2). However, it should be noted that in order to obtain accurate results, there is a need to allocate a considerable time for the evolution of the GA. As such, the algorithm is much slower, in comparison to ProfileNJ and ecceTERA which can output solutions in a few seconds. To our surprise, ProfileNJ was almost as accurate as the second best method (TreeFix-DTL), although it only supports a DL model of reconciliation and HGT were present in the dataset. It is possible that most edges with weak support were not involved in HGT events, which can explain the observed performance of ProfileNJ.

Evaluation on an empirical dataset

In an attempt to establish a benchmark for comparing orthology prediction methods, Boeckmann et al. (2011) [44] proposed manually curated "gold standard" gene trees for three well-conserved gene families : the Popeye-domain containing family (POP), the NOX 'ancestral-type' subfamily of NADPH oxidases (NOX) and the V-type ATPase beta subunit (VATP).

These gene families have been re-analyzed here to assess the performance of GATC on an empirical dataset. The reference species tree used was obtained from SwissTree [49]. Protein sequences from genomes not found in the species tree were removed and the remaining sequences aligned with Muscle [46]. GATC was used to reconstruct the corresponding trees for each gene family with initial population of trees obtained from bootstrap replicates. We used the same parameters as above except for the DTL events cost, which was changed to: ($\lambda = 1$, $\tau = \infty$, $\delta = 1$). Here, we prohibit HGT events since they are not expected in the dataset. We also set the maximum number of generations to 300 and the maximum time of evolution to 3h per gene family. For comparison, the average time needed by RAxML to obtain the best ML trees is 2.4h.

In order to measure the accuracy of GATC, we investigated how close the reference trees were to the set of pareto optimal trees. Figure 5 shows the distribution of individuals' scores, over generations, during the GA evolution for each gene family. We were able to retrieve the reference tree for the NOX and VATP gene families, whereas the reference tree for the POP family was located close to a cluster of pareto optimal trees. From the same figure, it can also be seen that even though the ML and MPR trees theoretically belong to the pareto optimal set of the complete tree space, they are often located far from the desired optimal result.

Since the reference gene tree was not obtained for the POP family, we report the precision and recall of orthologous and parologous genes inference from the solutions returned by GATC, compared to the proposed reference

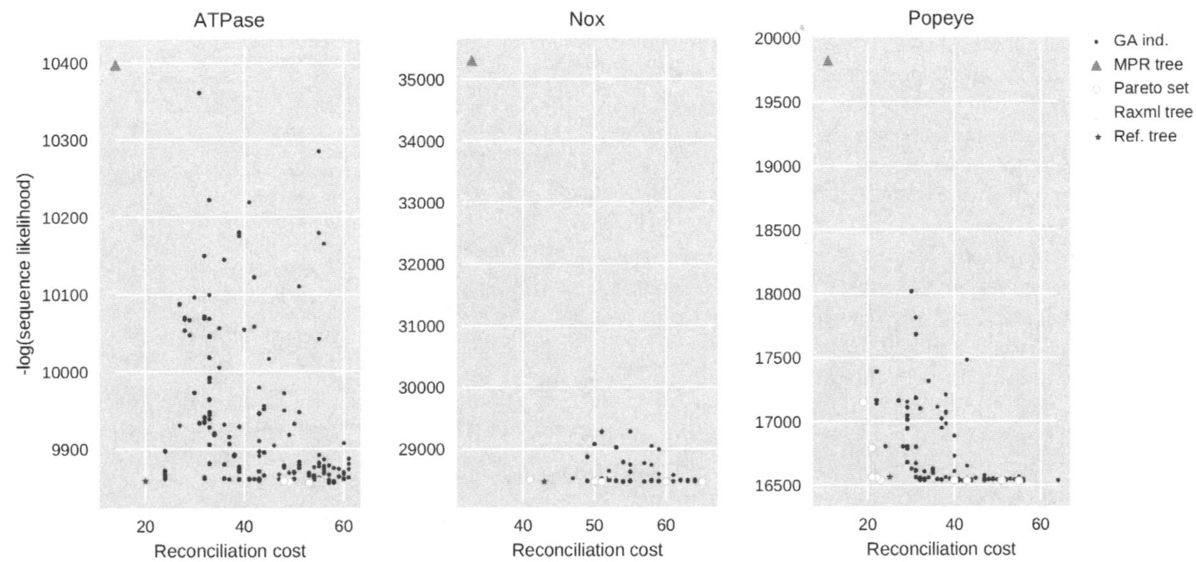

Fig. 5 Distribution of individuals' raw scores during evolution on three "gold standard" gene families. The scores of the ML tree obtained with RAxML, the MPR tree for the DL score, and the reference gene tree of [44] are also shown. Note that for fair comparison, the RAxML tree reconciliation score correspond to the best rooting score, whereas the MPR tree sequence likelihood correspond to the tree with the minimum negative log likelihood in the set of equivalent MPR trees. For the sake of visibility, we increased the size of each data point. The "best tree" is expected to be located in the lower left corner. For the ATPase and Nox families, the reference tree was present in the set of pareto optimal trees returned by GATC. For the Popeye gene families, the reference tree was located in the proximity of a cluster of pareto optimal solutions

tree (Table 1). Note that GATC only outputs ten trees from the 30 individuals of the final population resulting in four unique trees (see Additional file 1: Figures S4-S7). We computed precision and recall for the two types of gene relationships as follows:

$$Precision = \frac{TP}{TP + FP}, \quad Recall = \frac{TP}{TP + FN}$$

where TP corresponds to the number of shared pairs of orthologs/paralogs with the reference tree, FP corresponds to the number of predicted pairs of orthologs/paralogs not present in the reference tree, and FN to the number of missed orthologs/paralogs. As shown on Table 1, the precision and recall for the inferred gene relationships were high for all four solutions. Difference between GATC's solutions and the reference POP gene family tree (Additional file 1: Figure S3) can mostly be explained by the fact that duplication nodes were often placed lower in the solutions, resulting in fewer number of losses and consequently lower reconciliation scores.

It is hard to argue whether the proposed reference tree represents the true evolutionary history of the gene family over our pareto optimal solutions. In fact, from Fig. 5, it can be seen that some pareto optimal solutions were better than the reference POP gene tree for both scores, suggesting that they could be of higher quality. As the true evolutionary histories of gene families are hardly known, relying on high-quality phylogenetic gene trees for biological analyses is preferable.

In summary, our results on the empirical dataset demonstrate how a GA framework can be used for the inference of gene trees with high accuracy.

Conclusion

Algorithms for constructing gene trees from multiple sequence alignments are widely used. However when a

Table 1 Comparison between the reference tree of the Popeye family and the pareto optimal trees returned by GATC

	NormRF distance	Orthologs		Paralogs	
		Prec.	Rec.	Prec.	Rec.
Tree 1	0.260	0.763	0.942	0.971	0.871
Tree 2	0.260	0.765	0.941	0.971	0.873
Tree 3	0.087	0.902	0.983	0.992	0.894
Tree 4	0.109	0.829	0.866	0.940	0.922

reliable species tree is available, it is preferable to use species tree-aware methods which are often more accurate. In this work, we have presented a GA framework for the reconstruction of gene trees using both sequences and species tree information. From the comparison with existing methods, we have shown that this framework, implemented in a software called GATC, outputs more accurate gene trees.

As the true evolutionary history of a gene family does not always correspond to the most parsimonious one, GATC assumes instead that the true gene tree can most likely be found in the pareto optimal set of the search space. Therefore, given enough time, the algorithm will converge to a set of candidates containing that tree. Although this hypothesis was supported by our results on the empirical dataset, it does not necessary hold for all gene families. For example, since our reconciliation model does not consider Incomplete Lineage Sorting (ILS), the efficiency of GATC is expected to decrease in presence of ILS. Indeed, signals of deep coalescence leading to incongruence between species and gene tree would be explained by DTL events, possibly resulting in incorrect trees. Moreover, another problem still persists when there are several trees in the final pareto set, as alternative criteria for discriminating between these equivalent candidates are required. In its current implementation, GATC outputs solutions sorted by either the sequence likelihood or the reconciliation score.

Despite the good results we obtained by using GATC, one fundamental aspect that should be adressed in order to improve efficiency is the required evolution time. Indeed, running time cannot be accurately estimated especially when the starting trees have poor quality. When ML or bayesian trees have been inferred beforehand, it may be appropriate to set the maximum evolution time to the time required to find the best ML tree. As the underlying idea behind GAs allows for easy parallelism, running time can be dramatically reduced. Balance between scalability to large datasets and search efficiency would likely be achieved by carefully selecting the different genetic operators and the stopping criteria. Finally, to avoid being trapped in local optima, multiple replicate searches, using different settings (such as DTL, crossover and mutations rates, population size and initialization) can be performed in parallel with exchange of information through a migration operator.

Additional file

Additional file 1: Contains supplementary information on the effect of operator rates (**Figure S1**) and errors in the species tree (**Figure S2**) on reconstruction accuracy. It also contains the original reference tree of the Poyeye family (**Figure S3**) and the four alternative trees obtained by GATC (**Figure S4-S7**). (PDF 315 kb)

Abbreviations

AU: Approximately unbiased test; DL: Duplication-loss; DTL: Duplication-Transfer-Loss; FN: False negative; FP: False positive; GA: Genetic algorithm; GATC: Genetic algorithm for gene tree construction; HGT: Horizontal gene transfer; ILS: Incomplete lineage sorting; JTT: Jones, Taylor and Thornton substitution matrix; LG: Le and Gascuel substitution matrix; ML: Maximum likelihood; MOOP: Multi-objective optimization problem; MPR: Most parsimonious reconciliation; NSGA: Non-dominated sorting genetic algorithm; SPR: Subtree pruning and regrafting; TP: True positive

Acknowledgments

Computations were made on the supercomputer Briarée from Université de Montréal managed by Calcul Québec and Compute Canada. The authors also thank the three anonymous reviewers for their insightful comments and constructive suggestions that helped improve the manuscript.

Funding

Publication charges for this work was funded by The Natural Sciences and Engineering Research Council of Canada (NSERC).

Availability of data and materials

GATC is available for download at https://github.com/UdeM-LBIT/GATC. The simulated cynobacteria dataset analysed during the current study is publicly available at http://datadryad.org/resource/doi:10.5061/dryad.pv6df. The empirical dataset of reference gene trees is available at https://swisstree.vital-it.ch/gold_standard. Full outputs of all programs are available from the corresponding author on reasonable request.

About this supplement

This article has been published as part of *BMC Genomics* Volume 19 Supplement 2, 2018: Selected articles from the 16th Asia Pacific Bioinformatics Conference (APBC 2018): genomics. The full contents of the supplement are available online at https://bmcgenomics.biomedcentral.com/articles/supplements/volume-19-supplement-2.

Authors' contributions

EN and NEM designed the research and wrote the manuscript. EN implemented GATC, performed the experiments and analysed the results. All authors read and approved the final manuscript.

Ethics approval and consent to participate

Not applicable.

Consent for publication

Not applicable.

Competing interests

The authors declare that they have no competing interests.

Published: 9 May 2018

References

1. Guindon S, Gascuel O. A simple, fast and accurate algorithm to estimate large phylogenies by maximum likelihood. Syst Biol. 2003;52:696–704.
2. Stamatakis A. RAxML-VI-HPC: Maximum likelihood-based phylogenetic analysis with thousands of taxa and mixed models. Bioinformatics. 2006;22:2688–90.
3. Ronquist F, Huelsenbeck JP. MrBayes3: Bayesian phylogenetic inference under mixed models. Bioinformatics. 2003;19:1572–4.
4. Lartillot N, Philippe H. A bayesian mixture model for across-site heterogeneities in the amino-acid replacement process. Mol Biol Evol. 2004;21(6):1095–109. https://doi.org/10.1093/molbev/msh112.
5. Schreiber F, Patricio M, Muffato M, Pignatelli M, Bateman A. Treefam v9: a new website, more species and orthology-on-the-fly. Nucleic Acids Res. 2013. https://doi.org/10.1093/nar/gkt1055.
6. Wu YC, Rasmussen MD, Bansal MS, Kellis M. TreeFix: Statistically informed gene tree error correction using species trees. Syst Biol. 2013;62(1):110–20.

7. Noutahi E, Semeria M, Lafond M, Seguin J, Boussau B, Guéguen L, El-Mabrouk N, Tannier E. Efficient gene tree correction guided by genome evolution. PLoS One. 2016;11(8):e0159559. https://doi.org/10.1371/journal.pone.0159559. eCollection 2016.
8. Chen K, Durand D, Farach-Colton M. Notung: Dating gene duplications using gene family trees. J Comput Biol. 2000;7:429–47.
9. Rasmussen MD, Kellis M. A bayesian approach for fast and accurate gene tree reconstruction. Mol Biol Evol. 2011;28(1):273–90.
10. Thomas PD. GIGA: a simple, efficient algorithm for gene tree inference in the genomic age. BMC Bioinformatics. 2010;11:312.
11. Zhang L. On a mirkin-muchnik-smith conjecture for comparing molecular phylogenies. J Comput Biol. 1997;4(2):177–87.
12. Hallett MT, Lagergren J. Efficient algorithms for lateral gene transfer problems. In: Proceedings of the fifth annual international conference on computational biology (RECOMB 2001). New York: ACM; 2001. p. 149–56.
13. Ovadia Y, Fielder D, Conow C, Libeskind-Hadas R. The cophylogeny reconstruction problem is NP-complete. J Comput Biol. 2011;18(1):59–65. https://doi.org/10.1089/cmb.2009.0240.
14. Doyon JP, Scornavacca C, Gorbunov KY, Szöllősi GJ, Ranwez V, Berry V. An efficient algorithm for gene/species trees parsimonious reconciliation with losses, duplications and transfers. In: Proceedings of the 14th International Conference on Research in Computational Molecular Biology (RECOMB). Berlin/Heidelberg: LNCS, Springer; 2010. p. 93–108.
15. Bansal MS, Eric JA, Kellis M. Efficient algorithms for the reconciliation problem with gene duplication, horizontal transfer and loss. Bioinformatics. 2012;28(12):283–91. https://doi.org/10.1093/bioinformatics/bts225.
16. Tofigh A, Hallett M, Lagergren J. Simultaneous identification of duplications and lateral gene transfers. IEEE/ACM Trans Comput Biol Bioinform. 2011;8(2):517–35. https://doi.org/10.1109/TCBB.2010.14.
17. Chen ZZ, Deng F, Wang L. IEEE/ACM Trans Comput Biol Bioinform (TCBB). 2012;9(5):1515–28.
18. Nguyen TH, Ranwez V, Pointet S, Chifolleau AMA, Doyon JP, Berry V. Reconciliation and local gene tree rearrangement can be of mutual profit. Algorithms Mol Biol. 2013;8(12). https://doi.org/10.1186/1748-7188-8-12.
19. Jacox E, Chauve C, Szöllősi GJ, Ponty Y, Scornavacca C. eccetera: comprehensive gene tree-species tree reconciliation using parsimony. Bioinformatics. 2016;32(13):2056–8. https://doi.org/10.1093/bioinformatics/btw105.
20. Bansal MS, Wu Y, Alm EJ, Kellis M. Improved gene tree error-correction in the presence of horizontal gene transfer. Bioinformatics. 2015;31(8):1211–8. https://doi.org/10.1093/bioinformatics/btu806.
21. Szöllősi GJ, Tannier E, Daubin V, Boussau B. The inference of gene trees with species trees. Syst Biol. 2014;64(1):42–62.
22. Szöllősi GJ, Rosikiewicz W, Boussau B, Tannier E, Daubin V. Efficient exploration of the space of reconciled gene trees. Syst Biol. 2013;62(6):901–12. https://doi.org/10.1093/sysbio/syt054.
23. Sjöstrand J, Tofigh A, Daubin V, Arvestad L, Sennblad B, Lagergren J. A bayesian method for analyzing lateral gene transfer. Syst Biol. 2014;63(3):409–20.
24. Boussau B, Szöllősi GJ, Duret L, Gouy M, Tannier E, Daubin V. Genome-scale coestimation of species and gene trees. Genome Res. 2013;23:323–30.
25. Goldberg DE. Genetic Algorithms in Search, Optimization and Machine Learning, 1st edn. Inc. Boston: Addison-Wesley Longman Publishing Co., Inc.; 1989.
26. Matsuda H. Protein phylogenetic inference using maximum likelihood with a genetic algorithm. In: Pacific Symposium on Biocomputing. London: World Scientific. 1996. p. 512–23.
27. Lewis PO. Genetic algorithm for maximum-likelihood phylogeny inference using nucleotide sequence data. Mol Biol Evol. 1998;15(3):277–83.
28. Skourikhine A. Phylogenetic tree reconstruction using self-adaptive genetic algorithm. In: Proceedings of the 1st IEEE International Symposium on Bioinformatics and Biomedical Engineering (BIBE '00). Washington, DC: IEEE Computer Society; 2000. p. 129–34. https://doi.org/10.1109/BIBE.2000.889599.
29. Katoh K, Kuma K, Miyata T. Genetic algorithm-based maximum-likelihood analysis for molecular phylogeny. J Mol Evol. 2001;53:477–84.
30. Lemmon AR, Milinkovitch MC. The metapopulation genetic algorithm: An efficient solution for the problem of large phylogeny estimation. Proc Nat Acad Sci USA. 2002;99(16):10516–21.
31. Zwickl DJ. Genetic algorithm approaches for the phylogenetic analysis of large biological sequence datasets under the maximum likelihood criterion. PhD thesis, The University of Texas at Austin. 2006.
32. Lanave C, Preparata G, Sacone C, Serio G. A new method for calculating evolutionary substitution rates. J Mol Evol. 1984;20(1):86–93.
33. Yang Z. Maximum likelihood phylogenetic estimation from dna sequences with variable rates over sites: approximate methods. J Mol Evol. 1994;39(3):306–14.
34. Jones DT, Taylor WR, Thornton JM. The rapid generation of mutation data matrices from protein sequences. Bioinformatics. 1992;8(3):275–82.
35. Lafond M, Noutahi E, El-Mabrouk N. Efficient non-binary gene tree resolution with weighted reconciliation cost. In: Grossi R, Lewenstein M, editors. 27th Annual Symposium on Combinatorial Pattern Matching (CPM 2016), Leibniz International Proceedings in Informatics (LIPIcs), vol. 54. Dagstuhl: Schloss Dagstuhl-Leibniz-Zentrum fuer Informatik; 2016. p. 14:1–14:12.
36. Felsenstein J. Evolutionary trees from dna sequences: a maximum likelihood approach. J Mol Evol. 1981;17(6):368–76.
37. Stamatakis A, Ludwig T, Meier H. Raxml-iii: a fast program for maximum likelihood-based inference of large phylogenetic trees. Bioinformatics. 2004;21(4):456–63.
38. Tofigh A. Using trees to capture reticulate evolution: lateral gene transfers and cancer progression. PhD thesis, KTH. 2009.
39. Sjöstrand J. Reconciling gene family evolution and species evolution. PhD thesis, Numerical Analysis and Computer Science (NADA), Stockholm University. 2013.
40. Coello CAC, Lamont GB, Veldhuizen DAV. Evolutionary Algorithms for Solving Multi-Objective Problems (Genetic and Evolutionary Computation). Secaucus: Springer; 2006.
41. Srinivas N, Deb K. Muiltiobjective optimization using nondominated sorting in genetic algorithms. Evol Comput. 1994;2(3):221–48.
42. Miller BL, Goldberg DE, et al. Genetic algorithms, tournament selection, and the effects of noise. Complex Syst. 1995;9(3):193–212.
43. Shimodaira H. An approximately unbiased test of phylogenetic tree selection. Syst Biol. 2002;51(3):492–508.
44. Boeckmann B, Robinson-Rechavi M, Xenarios I, Dessimoz C. Conceptual framework and pilot study to benchmark phylogenomic databases based on reference gene trees. Brief Bioinform. 2011;12(5):423–35.
45. Penel S, Arigon AM, Dufayard JF, Sertier AS, Daubin V, Duret L, Gouy M, Perrière G. Databases of homologous gene families for comparative genomics. BMC Bioinformatics. 2009;10(6):3.
46. Edgar RC. Muscle: multiple sequence alignment with high accuracy and high throughput. Nucleic Acids Res. 2004;32(5):1792–7.
47. Lartillot N, Lepage T, Blanquart S. Phylobayes 3: a bayesian software package for phylogenetic reconstruction and molecular dating. Bioinformatics. 2009;25(17):2286–8.
48. Swofford DL, Olsen GJ, Waddell PJ, Hillis DM. Phylogenetic inference. In: Hillis DM, Moritz C, Mable BK, editors. Molecular systematics. 2nd edn. Sunderland: Sinauer Associates, Inc.; 1996. p. 407–514.
49. Boeckmann B, Marcet-Houben M, Rees JA, Forslund K, Huerta-Cepas J, Muffato M, Yilmaz P, Xenarios I, Bork P, Lewis SE, et al. Quest for orthologs entails quest for tree of life: in search of the gene stream. Genome Biol Evol. 2015;7(7):1988–99.

Chow *et al. BMC Genomics* 2018, **19**(Suppl 2):85
DOI 10.1186/s12864-018-4469-4

BMC Genomics

RESEARCH Open Access

Delineation of condition specific *Cis*- and *Trans*-acting elements in plant promoters under various Endo- and exogenous stimuli

Chi-Nga Chow[1], Yi-Fan Chiang-Hsieh[1], Chia-Hung Chien[1], Han-Qin Zheng[1], Tzong-Yi Lee[2], Nai-Yun Wu[1], Kuan-Chieh Tseng[1], Ping-Fu Hou[1] and Wen-Chi Chang[1*]

From The Sixteenth Asia Pacific Bioinformatics Conference 2018
Yokohama, Japan. 15-17 January 2018

Abstract

Background: Transcription factors (TFs) play essential roles during plant development and response to environmental stresses. However, the relationships among transcription factors, *cis*-acting elements and target gene expression under endo- and exogenous stimuli have not been systematically characterized.

Results: Here, we developed a series of bioinformatics analysis methods to infer transcriptional regulation by using numerous gene expression data from abiotic stresses and hormones treatments. After filtering the expression profiles of TF-encoding genes, 291 condition specific transcription factors (CsTFs) were obtained. Differentially expressed genes were then classified into various co-expressed gene groups based on each CsTFs. In the case studies of heat stress and ABA treatment, several known and novel *cis*-acting elements were identified following our bioinformatics approach. Significantly, a palindromic sequence of heat-responsive elements is recognized, and also obtained from a 3D protein structure of heat-shock protein-DNA complex. Notably, overrepresented 3- and 4-mer motifs in an enriched 8-mer motif could be a core *cis*-element for a CsTF. In addition, the results suggest DNA binding preferences of the same CsTFs are different according to various conditions.

Conclusions: The overall results illustrate this study may be useful in identifying condition specific *cis*- and *trans*-regulatory elements and facilitate our understanding of the relationships among TFs, *cis*-acting elements and target gene expression.

Keywords: Abiotic stress, *Cis*-acting elements, Co-expression, Hormone, Microarray, Transcription factors

Background

A series of gene regulation is critical for plant to adapt the environmental changes [1–3]. The regulation of spatial and temporal transcription via transcription factors (TFs) derives plant survival with short- and long- term impacts on plant physiology and development [4, 5]. Thus, identification of TF binding sites on promoter sequences of their target genes is essential to characterize TF function [6].

* Correspondence: sarah321@mail.ncku.edu.tw
[1]College of Biosciences and Biotechnology, Institute of Tropical Plant Sciences, National Cheng Kung University, Tainan 701, Taiwan
Full list of author information is available at the end of the article

In the past few years, several computational and experimental methods have been used to identify the relationship between TFs and *cis*-acting elements. Some useful resources have been also developed to construct transcriptional regulatory networks, such as JASPAR and AGRIS [7, 8]. However, the DNA binding information of most TFs is still very limited. For example, among 1717 *Arabidopsis* TFs from PlantTFDB, only 64 TFs and three TF complexes with their target genes have been characterized. Furthermore, the corresponding TFs of several *cis*-acting elements collected in recent databases are not available [7–12]. Recently, protein-binding microarrays are applied to study the DNA binding sequences of 63 and 313 TFs from *Arabidopsis* in two previous studies, respectively [13, 14]. Other methods such as chromatin immunoprecipitation combining with sequencing (ChIP-seq) and microarray (ChIP-chip) are also broadly employed in TF binding sites discovery under a given condition. For example, ABA-elicited transcriptional regulation for 21 ABA-related TFs were constructed by using ChIP-seq and RNA sequencing [15]. Moreover, DNA affinity purification sequencing (DAP-seq) was introduced to investigate the *cis*-acting elements and DNA modifications pattern for 529 (30%) *Arabidopsis* TFs [16].

In addition, some studies indicate that co-expressed members shared similar biological functions, which suggest a potential framework of transcriptional regulation [17, 18]. This concept has been widely used to investigate the functional elements involved in regulating transcriptional activity [19–21]. For instance, several functional *cis*-acting elements have been identified as key regulatory components related to the stress or pathogen-responsive pathways on the basis of gene expression clusters [22].

Although these researches illustrate the possibility of genome–wide analysis, current understanding of the relationships among TFs, *cis*-acting elements and target gene expression is still limited. In this work, we have developed a new bioinformatics approach for identification of condition specific *cis*- and *trans*- regulatory elements by using microarray expression data and genomic promoter sequences. Based on 344 *Arabidopsis* microarray samples, 291 condition specific TFs (CsTFs) were defined. Furthermore, the potential *cis*-acting elements of each CsTF were examined to reveal possible regulatory map of the CsTF. The results of the enriched 8-mer motif analysis from co-expressed clusters exhibit high consistency with the known condition responsive *cis*-acting elements.

Methods

Microarray data collection and processing

The *Arabidopsis thaliana* microarray data were obtained from the AtGenExpress and the NASCArrays [23, 24]. A total of 344 microarray samples including nine abiotic stresses (cold, osmotic, salt, drought, genotoxic, UV-B, wounding, heat, and oxidative) extracted from shoot and root and eight hormones treatments (IAA, cytokinin, gibberellin, brassinolide, ABA, methyl jasmonate, GA-3 and ACC) were used. The platform for all samples was Affymetrix ATH1 microarray chip (GPL198). For all conditions, two replicate samples were analysed. Array intensities were background adjusted and quantile normalized by using the justRMA function in the *affy* package of Bioconductor in R statistical language [25, 26]. The probe set annotation table downloaded from TAIR database (v10) was used to identify corresponding genes of each probe set ids [27]. Any ambiguous probe set ids which were associated with more than two genomic loci were discarded. A total of 20,922 genes were applied for further analysis.

Identification of condition specific transcription factors

A list of *Arabidopsis* TF-coding genes and their regulatory information (i.e. experimental binding matrices and annotation) were retrieved from PlantTFDB and PlantPAN 2.0 [9, 28]. Among 1717 TFs from PlantTFDB, only 1367 TFs could be identified in GPL198 platfrom. To recognize CsTFs for each condition, differentially expressed TFs (DETFs) were selected by using Students *t*-test between control and treatment with the confidence interval 0.99 and $\log_2$ fold change large than 1. The *p*-value of Students *t*-test statistic method was performed by using t.test () function in R. Furthermore, the z-score of fold change were used as a measurement to choose CsTFs from DETFs. The formula is as follows:

$$z_{condition\ a} = \frac{x-\mu}{\sigma}$$

where *condition a* denotes the corresponding condition what a TF is defined as DETF, whereas x is the fold change values in the treatment relative to control for *condition a*. μ and σ are mean and standard deviation of the fold changes for all conditions in the dataset of *condition a*, respectively. Please see Additional file 1: Table S1 for the treatment, control, and dataset of each condition. Totally, 291 DETFs were deified as CsTFs with z-score larger than 2.

Co-expressed gene groups of CsTFs

To identify the co-expression gene groups, differentially expressed genes (DEGs) ($P < 0.01$; $|\log_2$ (fold change)$| \geq |\log_2$ (1.5)$|$) were selected from the same induced condition of a CsTF. Then, the expressions data of all samples from the corresponding dataset were used to assess co-expression between a CsTF and a DEG (in Additional file 1: Table S1). The co-expressed genes of each CsTF were selected from the DEGs based on Pearson correlation coefficients (PCC ≥ 0.8). The PCC was calculated by using the cor () function in R.

Construction of genomic promoter dataset for *Arabidopsis* genes

A promoter was defined as the 1000 bp upstream of the transcription start site. All promoter sequences as well as the transcription start sites of 41,671 *Arabidopsis* transcripts (33,602 genes) were obtained from the TAIR [27]. To refine promoter sequences, we eliminated 25 promoters that contain uncertain bases (i.e. N, S, K, M, R, and W). Among 33,577 genes with annotated promoters, 5884 (18%) genes have multiple transcripts with the same TSSs or the TSSs closed to each other. Since the probe annotation of GPL198 is based on detectable genes rather than transcripts, promoters from different transcripts should be merged to avoid multiple calculation of one motif. We merged their promoter sequences according to their locations in the genome (Additional file 1: Figure S1). Finally, 33,715 promoter sequences for 33,577 genes were constructed as genomic promoter dataset for *Arabidopsis*.

Enrichment analysis of *cis*-acting elements in condition-specific promoters

The *cis*-acting elements of a CsTF were identified from promoter regions of its co-expressed gene group. To construct background calibration required for motif enrichment analysis, all possible 4–8-mer motifs (87,296 motifs) were scanned in the genomic promoter dataset of *Arabidopsis*. Since many motifs can be found in one promoter many times, both the presence/absence and frequency of motifs were taken into consideration. We thus evaluate motif enrichment using the following two criteria:

"Presence enrichment" assesses whether a motif significantly arise in a co-expressed gene group by comparing to whole genome background. As such, for each motif (*motif a*) assigned to a given CsTF its probability (*p*-value) was calculated by using a hypergeometric distribution, based on the following formula:

$$p_{motif\ a} = \sum_{i=x}^{M} \frac{\binom{M}{i}\binom{N-M}{n-i}}{\binom{N}{n}}$$

where x is the number of co-expressed genes whose promoters contain *motif a*, n is the number of co-expressed genes, N is the total number of genes in the background population, and M is the number of genes whose promoters contain *motif a* in the background set.

"Number enrichment" is used to estimate the preferred frequency of one motif in the promoters of the co-expressed gene group related to a CsTF in comparison with the background promoter sets. Different to presence enrichment, number enrichment for *motif a* was computed using the above formula of hypergeometric distribution where x is the number of *motif a* resides in the promoters of co-expressed genes, n is the number of respective bases for *motif a* in the promoters of co-expressed genes, N is the total number of respective bases for *motif a* in the background population, and M is the number of *motif a* in the background promoter sequences. An example of calculating "presence enrichment" and "number enrichment" is illustrated in Additional file 1: Figure S2.

Identification of all possible 4–8-mer DNA motifs in the promoter regions was performed with Bowtie [29]. In both cases, the *P*-value of a motif lower than 0.001 is defined as significantly enriched. The dhyper () and phyper () functions in R were used to obtain the hypergeometric *P*-values.

Motif alignments and sequence logos of each CsTF

To clarify the DNA binding sequences of a given CsTF, position specific scoring matrices were utilized to describe the frequency of each base at a certain position. In case study of heat stress and ABA treatment, the critical *cis*-acting elements for heat stress and ABA responsiveness have a highly overrepresented 3-mer and 4-mer in the enriched 8-mer motifs, respectively (in Additional file 2: Table S2-S3). Therefore, we designed three steps to discover DNA binding sequences of a given CsTF under all conditions:

Step 1: collection of enriched 8-mer motifs with top 10 3-mer and 4-mer motifs

Eight-mer motifs with both presence enrichment and number enrichment were defined as enriched motifs. Subsequently, the occurrences of 3-mer and 4-mer in all enriched 8-mer motifs were calculated and ranked. The enriched 8-mer motifs containing the 3-mer and 4-mer with top10 frequencies were collected.

Step 2: assembly of enriched 8-mer motifs on their original promoters

All enriched 8-mer motifs collected above were mapped back to the promoter regions of the co-expressed genes. If the enriched 8-mer motifs overlap with others at least one base, they will be merged. Then, we collected the sequences from each overlapping region.

Step 3: generation of sequence logos

The position specific scoring matrices were created based on the alignments of the sequences from the step 2. Multiple sequence alignment of the overlapping motifs was performed to check conserved bases by using the ClustalW program (version 2.1) [30]. The sequence logos for consensus DNA-binding sites were then displayed using the WebLogo tool [31].

Top10 3-mer and 4-mer motifs and sequence logos of each CsTF on all other conditions can be retrieved from our online database (http://wcchang.itps.ncku.edu.tw/CsTFAnalysis/). The motif-motif similarity between

CsTFs and Plant Cistrome Database were estimated by using the Tomtom program [16, 32].

Expression and genomic sequence resources for rice

The microarray expressions data including 276 samples for six hormones treatments were downloaded from RiceXPro [33]. A list of rice TF-coding genes were retrieved from PlantTFDB [9]. Additional file 1: Table S4 illustrates the treatment, control, and dataset of each condition. Promoter sequence for rice were downloaded from RAP-DB [34]. All analysis was used the same analytical flow and cut-off values as above.

Results

The system flow of this research is shown in Fig. 1. Following the analysis processes, CsTFs selection, co-expression classification from DEGs, promoter element scanning and motif enrichment analysis were applied to infer CsTF regulations.

Condition specific transcription factors

291 CsTFs in responses to specific abiotic stresses and hormone treatments were identified based on gene expression data (Fig. 2). All CsTFs under 26 conditions are listed in Additional file 2: Table S5. Several CsTFs are consistent with known essential regulators in their corresponding conditions. For example, under heat stress in shoot, four CsTFs are heat-shock transcription factor (HSF) family proteins which have reported as key factors in heat responsiveness. All of these, At2g26150 (Heat stress transcription factor A-2, HSFA2), AT3G51910 (Heat stress transcription factor A-7a, HSFA7A), AT4G11660 (Heat stress transcription factor B-2b, HSFB2B) and AT5G62020 (Heat stress transcription factor B-2a, HSFB2A), are essential regulatory components to mediate the heat tolerance pathway during heat shock [35, 36]. Moreover, three cold-specific TFs, AT4G25470 (C-repeat/DRE binding factor 2, CBF2), AT4G25490 (C-repeat/DRE binding factor 1, CBF1) and AT4G25480 (C-

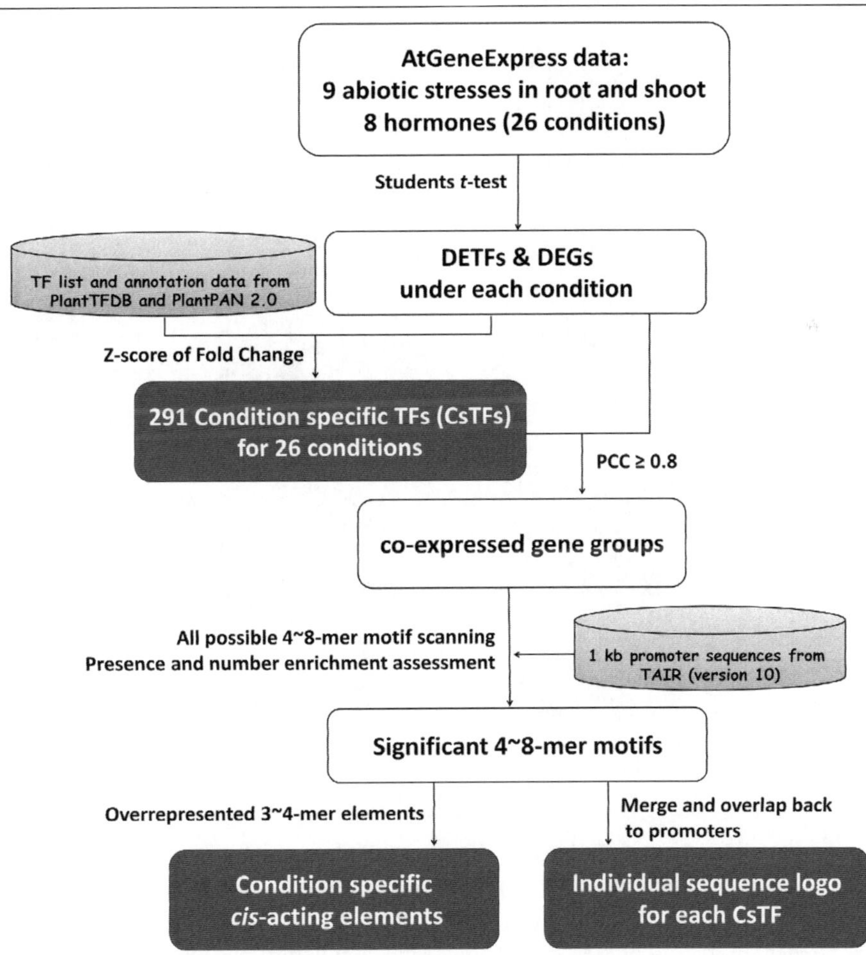

Fig. 1 A schematic of the analysis processes to identify significant *trans-* and *cis-* acting elements. TF: transcription factor; DETF: differentially expressed TF; DEG: differentially expressed genes; PCC: Pearson correlation coefficient

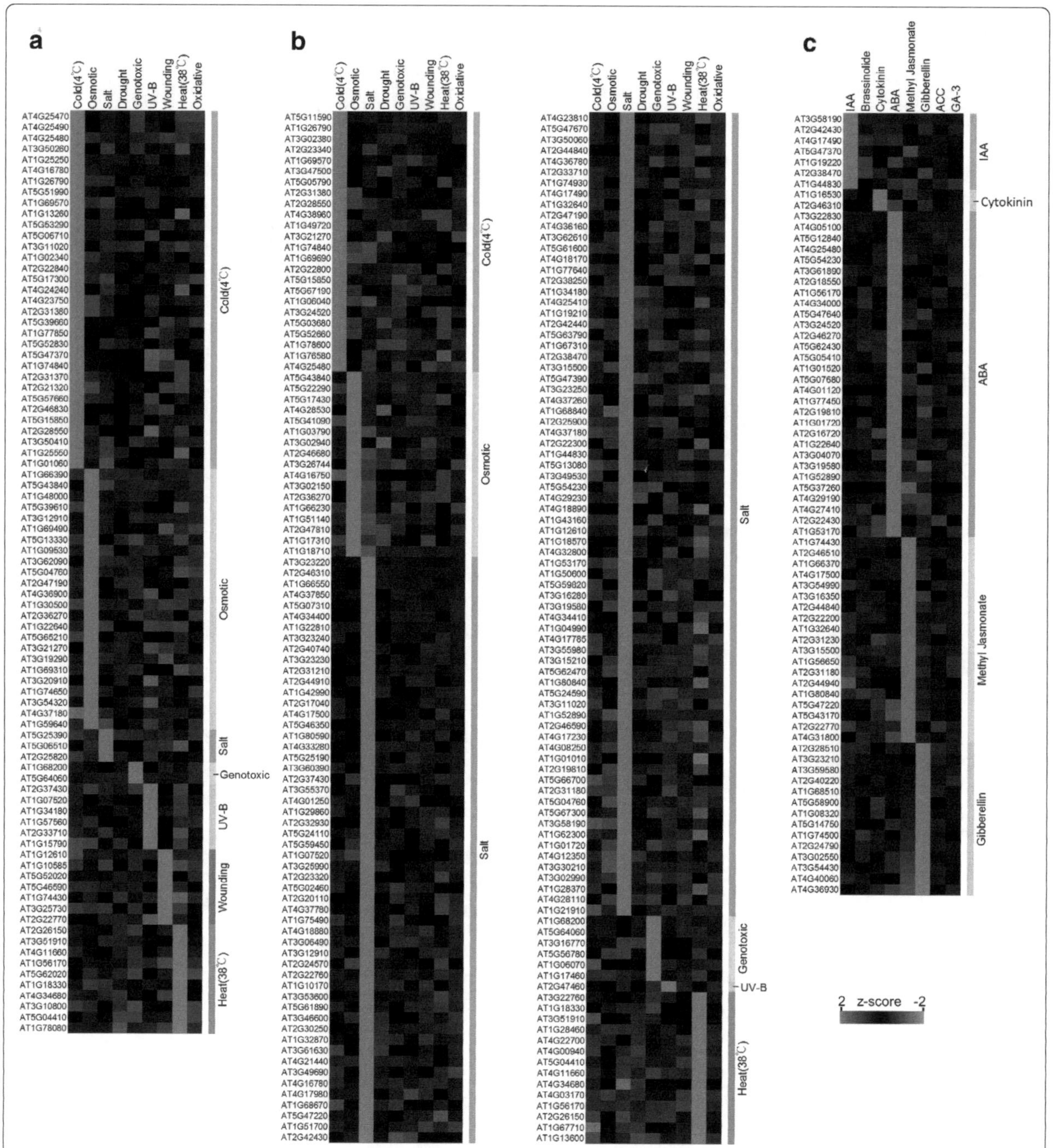

Fig. 2 Expression pattern of CsTFs. The z-score profile indicated that each CsTF is strongly induced in one condition. CsTFs are ordered according to their condition clusters (marked by different colors in the right of heat maps). **a** and **b** show expression patterns of CsTFs under abiotic stresses in shoot and root, respectively. **c** shows expression patterns under eight hormone treatments

repeat/DRE binding factor 3, CBF3) have been reported to regulate over 100 genes when plants are exposed to low temperature [36, 37]. For ABA responsiveness, AT4G34000 (ABRE binding protein, ABF3) is a major transcription factor in plant ABA signalling transduction [38]. However, over half of CsTFs have little information about their regulatory mechanisms and DNA binding sites (summarized in Additional file 1: Table S6-S8). These CsTFs could be novel candidates for further experimental validation under specific conditions.

Numerous osmotic-specific TFs shown higher expression levels in salt stress than other conditions, and revealed strong correlation between salt and osmotic stress. This phenomenon was also identified in salt-specific TFs (Fig. 2). Several known and novel CsTFs related to cross talk between abiotic stresses and hormones were investigated. For instance, AT1G74430 (MYB95) and AT2G22770 (NAI1) are expected to regulate methyl jasmonate- and wounding-induced genes expression. This is consistent with the previous research which displayed that methyl jasmonate mediates plant responses against mechanical wounding cause by insects and pathogens [39, 40]. Moreover, NAI1 is a well-known regulator which affects the formation of ER body after plants are wounded [41, 42]. In addition, the correlations between methyl jasmonate and salt stress in root shows that three members (AT2G44840 (ERF13), AT4G17500 (ERF1A), and AT5G47220 (ERF2)) of ethylene response factor (ERF) family might play important roles in stress and hormone cross-talking.

Motif preferences among co-expressed promoters (number and presence enrichment analysis)

The highest percentage of annotated TFs related to plant heat responsiveness was discovered in the heat-specific TF group, and more TF binding sites are known when comparing to other conditions (in Additional file 1: Table S6). Heat stress was thus selected as a case to examine the motif preferences of CsTFs. Table 1 summarizes the results for ten heat stress specific TFs (CsTFs under heat stress) belonging to six families. To investigate the transcriptional regulation mechanisms, 4–8-mer motif enrichments were evaluated on the promoters of co-expression genes for each CsTF. Due to the limited number of co-expressed genes, AT1G78080 (ethylene-responsive transcription factor RAP2.4) was discarded before further analysis.

The morphological and physiological alterations usually depends on the presence of stress-responsive *cis*-acting elements and their occurrence frequency on target promoters when plant adapt to environmental changes [43]. To set up a reasonable criterion that could distinguish actual *cis*-acting elements from promoters, we assessed the enrichment of the presence and number of 4–8-mer motifs among co-expressed genes promoters (see Additional file 1: Figure S3-S7). Interestingly, 8-mer motifs with presence enrichment also shown significant number enrichment (Fig. 3). The presence enriched motifs underwent number enrichment with markedly low *p*-values (under 10e-4). By contrast, the motifs lacking presence enrichment did not occupy high percentages of the co-expressed group compared with whole genome. In the case of four HSF proteins, even though more motifs fitting the heat shock element (HSE; GAANNTTC) were found in the second quadrant of scatter plots than the first quadrant, numerous motif in the second quadrant might be false positives (Fig. 3). These results indicate that if a motif is important for a CsTF, the significant presence of the motif seems to be required in the promoters of the co-expressed genes. The presence enrichment thus is a good index to evaluate the important of motifs for a CsTF. Based on above, such presence and number enrichments show the applicability to select motifs from promoter sequences.

Enriched motifs reveal characteristic of actual stress responsive regulations

A conserved repetitive palindromic motif (GAAnnTTC) is a well-known HSE in upstream promoters of heat-inducible genes. Previous studies also indicated that the

Table 1 Summary of 10 CsTFs under heat stress in shoot

TF	co.G+ DEG1.5[a]	A[b]	Name	Family	Name of TFBS	Sequence of TFBS
AT1G18330	66	–	MYB-related transcription factor EPR1	Myb/SANT; MYB-related	–	–
AT1G56170	67	–	Nuclear transcription factor Y subunit C-2	NF-YC	–	CCAAT
AT1G78080	0	yes	Ethylene-responsive transcription factor RAP2–4	AP2; ERF	GCC-box pathogenesis-related promoter element	–
AT2G26150	72	yes	Heat stress transcription factor A-2	HSF	heat shock elements (HSE)	5′-AGAAnnTTCT-3′
AT3G10800	30	–	–	bZIP	–	–
AT3G51910	43	yes	Heat stress transcription factor A-7a	HSF	heat shock elements (HSE)	5′-AGAAnnTTCT-3′
AT4G11660	80	–	Heat stress transcription factor B-2b	HSF	heat shock elements (HSE)	5′-AGAAnnTTCT-3′
AT4G34680	67	yes	GATA transcription factor 3	GATA	–	5′-GATA-3′;5′-GAT-3′
AT5G04410	55	–	NAC domain-containing protein 78	NAC; NAM	–	–
AT5G62020	66	–	Heat stress transcription factor B-2a	HSF	heat shock elements (HSE)	5′-AGAAnnTTCT-3′

[a]the number of co-expressed genes (intersection of co-expressed genes ($PCC \geq 0.8$) and DEGs under heat stress (p-value < 0.01; $|\log_2 (FC)| \geq 1.5$))
[b]activator

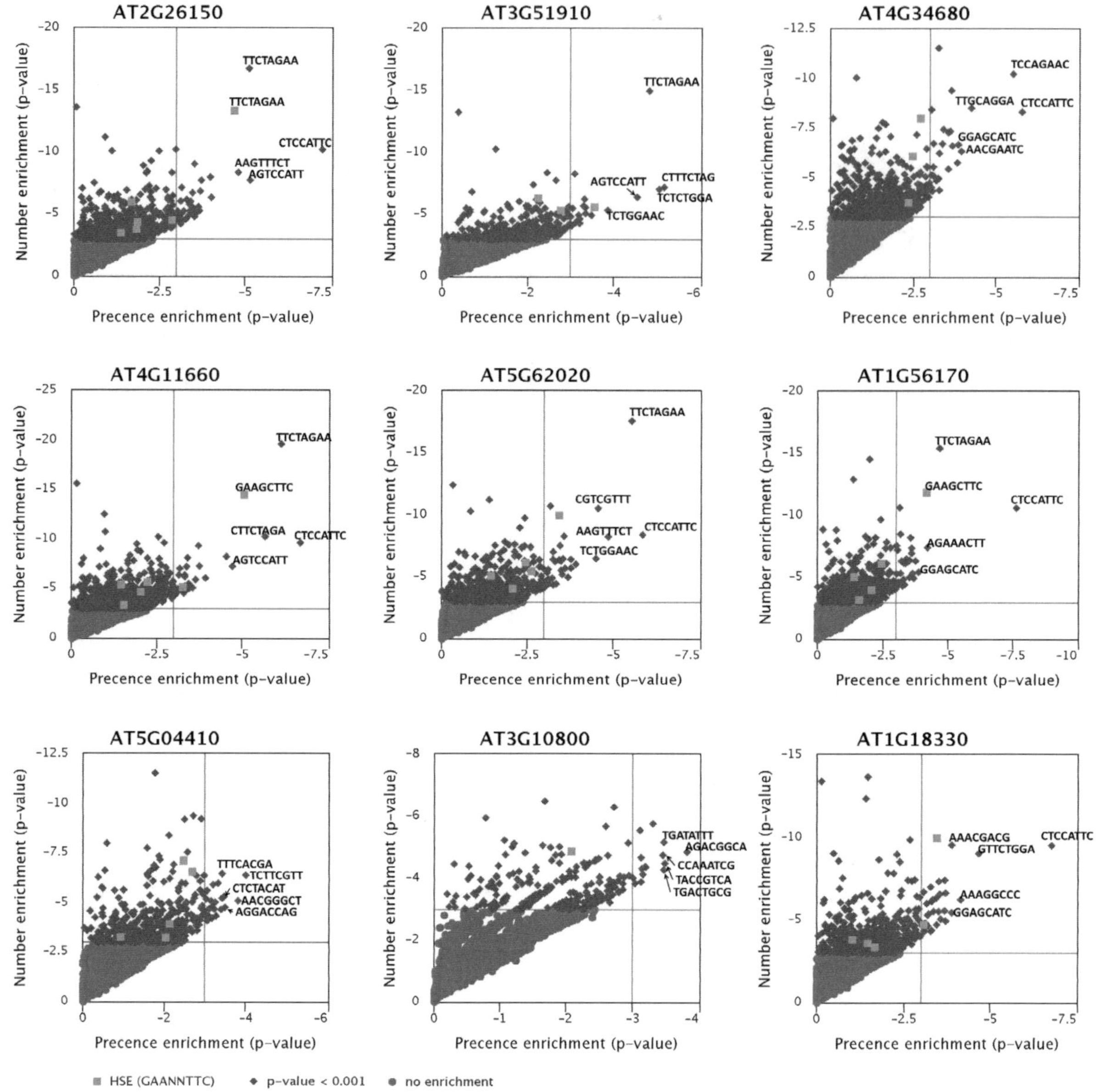

Fig. 3 Scatter plots show *p*-values of presence and number enrichment for all observed 8-mer motifs. *P*-values are performed under $\log_{10}$ scale. Vertical and horizontal red lines are the thresholds of enrichment, 10e-3. Based on these two lines, each scatter plot can separate in four quadrants. Eight-mer motifs locating in first quadrant are defined as enriched motifs, which are both presence enrichment and number enrichment. The motifs in second quadrants are number enrichment instead of presence enrichment. The 8-mer sequences marked on sequences indicate the top five most-enriched motifs

recognition of HSE is required for the regulation of HSFs [44, 45]. Therefore, we are interested in whether palindromic elements are significant in the enriched 8-mer motifs. Among the enriched 8-mer motifs of seven heat stress specific TFs, at least three types of palindromic motifs have been found (in Additional file 1: Table S9). Interestingly, most of these were (GAAnnTTC), a typical HSE, or (TTCnnGAA) which exchanged the order of two 3-mer core sequence of HSE. This palindromic sequence also indicates that the heat stress specific TFs might function as a dimer. The structure of the HSF-DNA complex from *Kluyveromyces lactis* (PDB Id: 3HTS) also illustrates the same issue (Additional file 1: Figure S8) [46].

The members from the same TF family which were classified based on similar DNA binding domains tend to display similar DNA binding specificity [13, 14, 47, 48]. The position specific scoring matrices for six families of heat stress specific TFs (bZIP, NAC; NAM, Myb/SANT; MYB-related, GATA, NF-YC, HSF) from PlantPAN 2.0 show that most of the matrices for the same TF family were similar to each other (in Additional file 1: Table S10) [28]. These matrices also illustrated that the conserved core sequences for half of heat stress responsive families may prefer to be 3-mer motifs (Additional file 1: Figure S9A). To examine the properties observed in experimental binding sites, we investigated the occurrences of 3-mer in all enriched 8-mer for 9 CsTFs. The top 10 significant entries revealed that a sequence (GAA) was common in four HSF proteins and was highly overrepresented (in Additional file 2: Table S2). In addition, the other two sequences (TTC) and (TCT) were also highly presented in the enriched 8-mer in HSF proteins. For other CsTFs (AT1G18330, AT5G04410, AT4G34680, and AT1G56170), the sequence (GAA) shown higher frequency than the other 3-mer motifs. This reveals that GAA motif is a vital *cis*-acting element for gene transcriptional regulation during heat stress responsiveness. Taken together, the palindromic feature and the heat-responsive overrepresented 3-mer demonstrate that our bioinformatics approach could successfully identify critical *cis*-acting elements.

Determination of DNA binding sequences for CsTFs

The interaction preferences among TFs and target genes are crucial for transcriptional regulation. To reveal the potential DNA binding sites for a CsTF, the position specific scoring matrices were generated according to the occurrence and overlapping of the enriched 8-mer motifs in the promoters. In co-expressed gene promoters, the overlapping motifs are enriched particularly within ~ 500 bp upstream of the transcriptional start site (Additional file 1: Figure S10). This position bias of the overlapping motifs was similar to the experimentally verified motifs in plants [49, 50]. The sequence logos for each CsTF are displayed in Fig. 4. Motif comparisons of eight matrices from five heat stress specific TFs show that three matrices corresponding to HSFB2B and HSFB2A are similar to those from the published database (Fig. 4) [16]. For the four HSF proteins, their sequences present the perfect type HSE (GAANNTTC). Among these, HSFA7A was slightly different from the others, with additional AC-rich flanking sequences at its 5'end of the sequence logo. Interestingly, a member of the GATA family, AT4G34680 (GATA transcription factor 3, GATA3) also preferred to recruit this novel motif in the promoters of its co-expressed genes.

A case study ABA-CsTF identification

To further demonstrate the applications of our approach to hormone treatment, the CsTFs for ABA treatment were taken as an example, due to more known motifs can be referred (in Additional file 1: Table S8). Following the analysis procedure, 30 CsTFs were identified as ABA specific regulators belonging to 16 families (in Additional file 2: Table S5). According to the confirmed binding matrixes of TF families, the significant 4-mer in the enriched 8-mer motif might be efficient sequences to recognize the essential *cis*-acting elements for a specific condition (Additional file 1: Figure S9). Notably, by assessing the overrepresented 4-mer in enriched 8-mer motifs, a sequence (ACGT), the core flanking of ABRE, is recognized in all ABA-CsTFs (in Additional file 2: Table S3). An "ACGT" is a top one overrepresented 4-mer sequence identified in 80% of ABA-CsTFs, and shows within top five 4-mer sequence in the other 20%. This suggests that the presence of the ACGT motif in gene promoters is necessary for ABA responsive regulation in *Arabidopsis*, which is consistent with current studies [51, 52].

To further validate the candidate *cis*-elements under ABA treatment, the developed methods were employed in *Oryza sativa*. The microarray expressions data from six hormones treatments (ABA, auxin, brassinosteroid, cytokinin, gibberellin, and jasmonic acid) (in Additional file 1: Table S4) were used. Following our analytical methods, 60 and 69 rice ABA-CsTFs were identified from shoot and root, respectively. Expectedly, a well-known ABRE core motif (ACGT) were investigated in top ten overrepresented 4-mer of ABA-CsTF in rice. Interestingly, the top ten overrepresented 4-mer were slightly different when comparing between rice and *Arabidopsis* (Additional file 1: Figure S11). Although "ACGT" was identified in the top ten overrepresented 4-mer for rice ABA-CsTFs, the higher percentage of "CACG" and "CCAC" were recognized and might be novel *cis*-elements for ABA responsiveness. The 3′ and 5′ flank sequences of ABRE core motif (ACGT) was different between *Arabidopsis* (AC**ACGT**GTC) and rice (CC**ACGT**), which suggests the specificity of species. Except for known ABRE, the other top ten overrepresented 4-mer also performed various preference between *Arabidopsis* and rice (Additional file 1: Figure S11). For example, "GTGG", "AATA", and "TACG" were especially for *Arabidopsis*, but not for rice. Together, the rice ABA-CsTFs and ABRE core motifs suggest that our approach can be widely applied to explore CsTFs and condition specific *cis*-elements in plant. The rice/*Arabidopsis* comparisons also demonstrate that the variance among species can also be retrieved in CsTF analysis.

With regard to the potential DNA binding sites for an individual CsTF, the sequence logos of *Arabidopsis* ABA-CsTFs were demonstrated in Fig. 5. Significantly, the sequence logos of AT4G34000 (ABA responsive

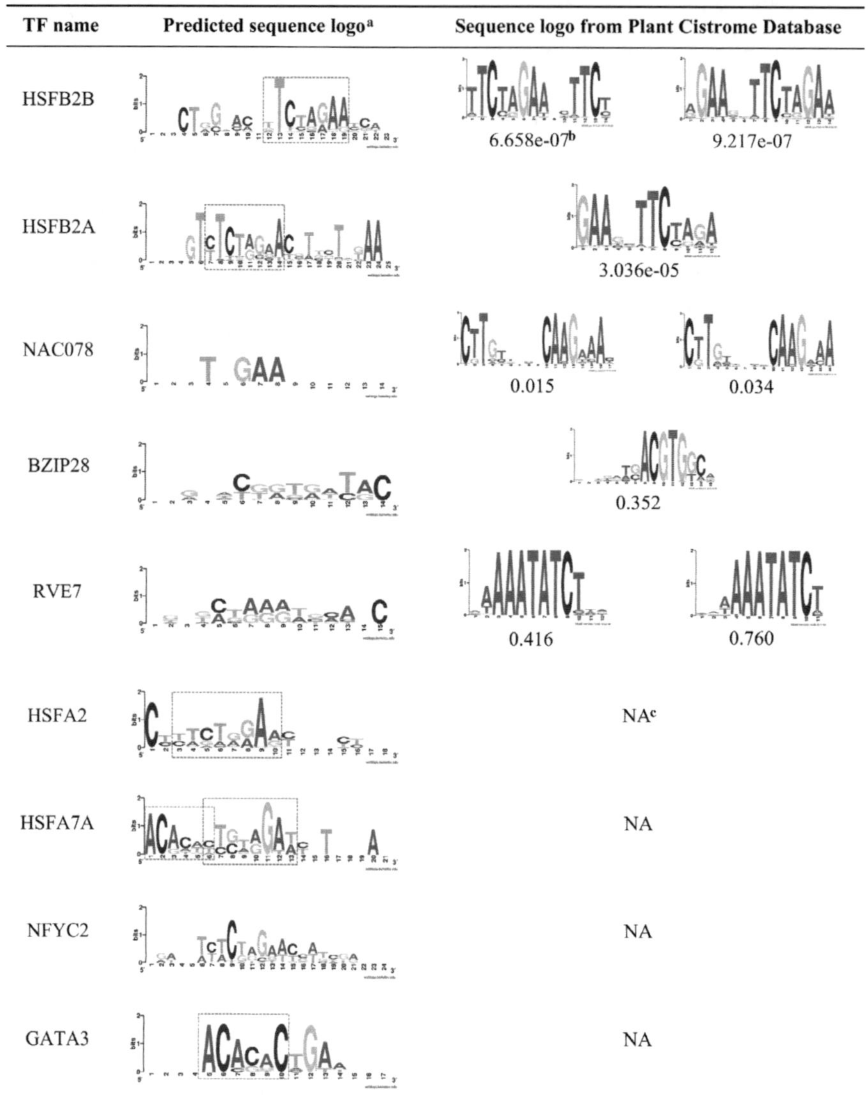

Fig. 4 Sequence logos of CsTFs under heat stress. [a]:HSEs (GAANNTTC) are marked in blue dotted rectangles. AC-rich flanking sequences are mark in orange dotted rectangles. [b]: motif-motif similarity (*P*-value) is calculated by Tomtom. [c]: No sequence logo could be obtained from Plant Cistrome Database

elements-binding factor 3, ABF3), which has been experimentally confirmed to regulate the ABA signalling pathway, was consistently matched to the preference of known binding sites (ACACGTGT). Except for AT3G19580 (zinc-finger protein 2, ZF2), other eight ABA-CsTFs, bind to G-box motifs, which have been verified by ChIP-seq (Fig. 5) [15]. In summary, the results of ABA and heat stress analysis illustrated that our approach is suitable to explore and uncover the *cis*-acting regulation for abiotic stresses and hormone treatments. Information on all other conditions can be retrieved from our online database (http://wcchang.itps.ncku.edu.tw/CsTFAnalysis/).

Discussion

In this study, we developed a series of computational methods to discover specific *trans-* and *cis-* regulatory elements under a particular condition. Among 291 CsTFs, several of these were confirmed with regard to their corresponding conditions (in Additional file 2: Table S5). Overlapping the CsTF list between root and shoot, we observed that 37% of the overlapping CsTFs act as salt-responsive regulators in root while they were induced due to different abiotic stresses in shoot. This may illustrates that the plant root needs more direct and specific regulations to resist salt stress than shoot. The other possibility is that shoot response to salt stress is

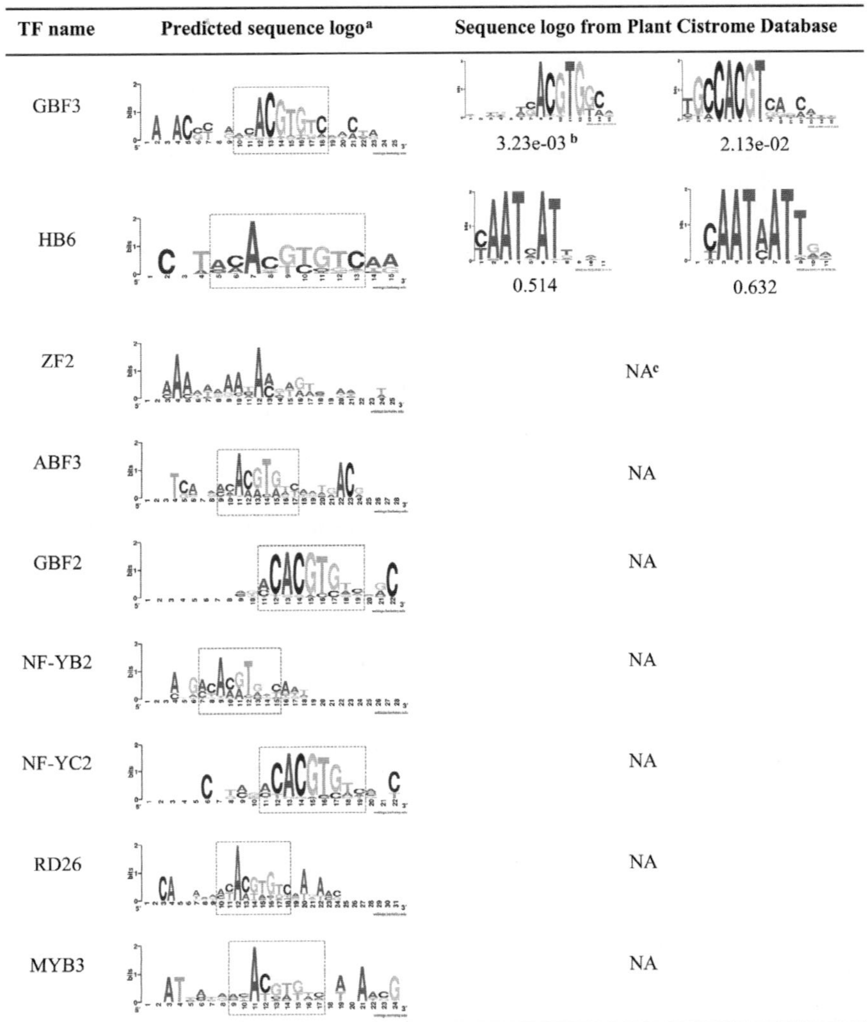

Fig. 5 Sequence logos of nine CsTFs under ABA treatment. [a]: G-box motifs (ACACGTGTC) are marked in dotted rectangles. [b]: motif-motif similarity (*P*-value) calculated by Tomtom. [c]: No sequence logo could be obtained from Plant Cistrome Database

indirect and affected by second messengers. More protein partners or co-factors might be required for TFs in shoot to co-regulate their target genes under salt stress. Additionally, the overlapping between abiotic stresses and hormone treatment shows that plant resistant abiotic stress is usually related to endogenous stimulus.

Although more advanced computational approaches and experiments have been developed to investigate stress-responsive TFs and their regulatory networks, the relationships between TFs and their binding sites remain unknown. The results of sequence logos demonstrate that our method can successfully generate potential binding sites for CsTFs. In the cases of heat stress specific TFs, GO enrichment analysis of the co-expressed genes reveals that the four HSF proteins show significant enrichment in major functions, such as response to heat and response to stress (in Additional file 1: Table S11). However, they play different roles in several subfunctions (in Additional file 1: Table S12). These differences thus demonstrate the complexity of heat stress regulation, since some of the HSF proteins generally cooperate with other TFs to activate distinct gene functions in plant cells [44, 53]. In the overlapping CsTFs list, with regard to responses to heat stress in shoot and root, we found that, in root, heat stress-TFs tend to bind promoter elements which contain three repetitive core sequences (GAA) of HSE compared to only two repetitions in shoot (data not shown). This finding suggests that the integration of condition-specific *cis*-elements under various conditions can provide further characteristics to distinguish TF binding sites in distinct plant organs (tissues).

To examine the reliability of the results, we compared the ABA specific DNA-binding sequences from our computational approaches to those from in vivo experiments. Among 10 ABA-CsTF which have been reported as ABA-related TFs in the previous study, the sequence logos of eight ABA-CsTF show the same preferences (A**CACGTG**TC) of ChIP-seq experiments [15]. They contain a core sequence of G-box motif (CACGTG) and two flanking sequence, A- and –TC at 5′ and 3′ sides, respectively. This finding shows our bioinformatics approach for genome-wide promoter analysis can be used to infer valuable *cis*-acting regulation in response to a stimulus with high consistent with the in vivo verified motifs. Though we demonstrated the application of our approach to abiotic stress and hormone treatments, it may be useful for studying the relationships among TFs and *cis*-acting elements using other gene expression data, such as RNA-sequencing or derived conditions, e.g. developmental stages and biotic stresses.

Conclusions

This study provides new bioinformatics approach combining microarray expression data and genomic promoter sequences for identification of condition specific *cis*- and *trans*- regulatory elements. Several known and novel *cis*-acting elements were identified for 291 CsTFs and 26 conditions. The results of heat stress and ABA treatment suggest that overrepresented 3- and 4-mer motifs in an enriched 8-mer motif could be a core *cis*-element for a CsTF. The overall results illustrate this study may be useful in identifying condition specific *cis*- and *trans*- regulatory elements and facilitate our understanding of the relationships among TFs, *cis*-acting elements and target gene expression.

Additional files

Additional file 1: Tables S1, S4, and S6 to S12. Figures S1 to S11. Supplementary Tables and Figures. (DOC 4939 kb)

Additional file 2: Tables S2, S3, and S5. Top 10 significant entries of overrepresented 3-mer and 4-mer sequences for heat stress specific TFs and ABA-CsTFs, respectively, and CsTFs lists in three condition datasets. (XLS 143 kb)

Abbreviations

ChIP-chip: Chromatin immunoprecipitation combining with microarray; ChIP-seq: Chromatin immunoprecipitation combining with sequencing; CsTF: Condition specific transcription factor; DAP-seq: DNA affinity purification sequencing; DEG: Differentially expressed gene; DFTF: Differentially expressed TF; ERF: Ethylene response factor; HSE: Heat shock element; HSF: Heat-shock transcription factor; PCC: Pearson correlation coefficients; TF: Transcription factor

Acknowledgements

The authors would like to thank the Ministry of Science and Technology (MOST 105-2311-B-006-004 -MY3 and MOST 104-2313-B-006-002) and Academia Sinica (Innovative Translational Agricultural Research Grant) of the Republic of China for financially supporting this research.

Funding

Funding for the publication of this article was provided by the Ministry of Science and Technology of the Republic of China grant.

Availability of data and materials

All data generated or analysed during this study are included in this published article and its supplementary information files. The datasets generated and analysed during the current study are available in the online database, http://wcchang.itps.ncku.edu.tw/CsTFAnalysis/.

About this supplement

This article has been published as part of *BMC Genomics* Volume 19 Supplement 2, 2018: Selected articles from the 16th Asia Pacific Bioinformatics Conference (APBC 2018): genomics. The full contents of the supplement are available online at https://bmcgenomics.biomedcentral.com/articles/supplements/volume-19-supplement-2.

Authors' contributions

WCC conceived and designed the experiments and revised the paper. CNC, YFCH, CHC, and PFH analyzed the data and performed the experiments. HQZ, NYW and KCT developed the database and webpage. CNC wrote the paper. All authors read and approved the final manuscript.

Ethics approval and consent to participate

Not applicable.

Consent for publication

Not applicable.

Competing interests

The authors declare that they have no competing interests.

Publisher's Note

Author details

[1]College of Biosciences and Biotechnology, Institute of Tropical Plant Sciences, National Cheng Kung University, Tainan 701, Taiwan. [2]School of Science and Engineering, The Chinese University of Hong Kong, Shenzhen, China.

Published: 9 May 2018

References

1. Berkowitz O, De Clercq I, Van Breusegem F, Whelan J. Interaction between hormonal and mitochondrial signalling during growth, development and in plant defence responses. Plant Cell Environ. 2016;39(5):1127–39.
2. Ha S, Vankova R, Yamaguchi-Shinozaki K, Shinozaki K, Tran LS. Cytokinins: metabolism and function in plant adaptation to environmental stresses. Trends Plant Sci. 2012;17(3):172–9.
3. Peleg Z, Blumwald E. Hormone balance and abiotic stress tolerance in crop plants. Curr Opin Plant Biol. 2011;14(3):290–5.
4. Riechmann JL, Heard J, Martin G, Reuber L, Jiang C, Keddie J, Adam L, Pineda O, Ratcliffe OJ, Samaha RR, et al. Arabidopsis transcription factors: genome-wide comparative analysis among eukaryotes. Science. 2000; 290(5499):2105–10.
5. Wellmer F, Riechmann JL. Gene network analysis in plant development by genomic technologies. Int J Dev Biol. 2005;49(5–6):745–59.
6. Spitz F, Furlong EE. Transcription factors: from enhancer binding to developmental control. Nat Rev Genet. 2012;13(9):613–26.
7. Mathelier A, Zhao X, Zhang AW, Parcy F, Worsley-Hunt R, Arenillas DJ, Buchman S, Chen CY, Chou A, Ienasescu H, et al. JASPAR 2014: an extensively expanded and updated open-access database of transcription factor binding profiles. Nucleic Acids Res. 2014;42(Database issue):D142–7.
8. Yilmaz A, Mejia-Guerra MK, Kurz K, Liang X, Welch L, Grotewold E. AGRIS: the Arabidopsis gene regulatory information server, an update. Nucleic Acids Res. 2011;39(Database issue):D1118–22.
9. Jin J, Zhang H, Kong L, Gao G, Luo J. PlantTFDB 3.0: a portal for the functional and evolutionary study of plant transcription factors. Nucleic Acids Res. 2014;42(Database issue):D1182–7.

10. Hieno A, Naznin HA, Hyakumachi M, Sakurai T, Tokizawa M, Koyama H, Sato N, Nishiyama T, Hasebe M, Zimmer AD, et al. Ppdb: plant promoter database version 3.0. Nucleic Acids Res. 2014;42(Database issue):D1188–92.
11. Higo K, Ugawa Y, Iwamoto M, Korenaga T. Plant cis-acting regulatory DNA elements (PLACE) database: 1999. Nucleic Acids Res. 1999;27(1): 297–300.
12. Matys V, Kel-Margoulis OV, Fricke E, Liebich I, Land S, Barre-Dirrie A, Reuter I, Chekmenev D, Krull M, Hornischer K, et al. TRANSFAC and its module TRANSCompel: transcriptional gene regulation in eukaryotes. Nucleic Acids Res. 2006;34(Database issue):D108–10.
13. Franco-Zorrilla JM, Lopez-Vidriero I, Carrasco JL, Godoy M, Vera P, Solano R. DNA-binding specificities of plant transcription factors and their potential to define target genes. Proc Natl Acad Sci U S A. 2014;111(6):2367–72.
14. Weirauch MT, Yang A, Albu M, Cote AG, Montenegro-Montero A, Drewe P, Najafabadi HS, Lambert SA, Mann I, Cook K, et al. Determination and inference of eukaryotic transcription factor sequence specificity. Cell. 2014; 158(6):1431–43.
15. Song L, Huang SC, Wise A, Castanon R, Nery JR, Chen H, Watanabe M, Thomas J, Bar-Joseph Z, Ecker JR. A transcription factor hierarchy defines an environmental stress response network. Science. 2016:**354**(6312).
16. O'Malley RC, Huang SC, Song L, Lewsey MG, Bartlett A, Nery JR, Galli M, Gallavotti A, Ecker JR. Cistrome and Epicistrome features shape the regulatory DNA landscape. Cell. 2016;166(6):1598.
17. Wong DC, Schlechter R, Vannozzi A, Holl J, Hmmam I, Bogs J, Tornielli GB, Castellarin SD, Matus JT. A systems-oriented analysis of the grapevine R2R3-MYB transcription factor family uncovers new insights into the regulation of stilbene accumulation. DNA Res. 2016;
18. Maruyama K, Todaka D, Mizoi J, Yoshida T, Kidokoro S, Matsukura S, Takasaki H, Sakurai T, Yamamoto YY, Yoshiwara K, et al. Identification of cis-acting promoter elements in cold- and dehydration-induced transcriptional pathways in Arabidopsis, rice, and soybean. DNA Res. 2012;19(1):37–49.
19. Zou C, Sun K, Mackaluso JD, Seddon AE, Jin R, Thomashow MF, Shiu SH. Cis-regulatory code of stress-responsive transcription in Arabidopsis Thaliana. Proc Natl Acad Sci U S A. 2011;108(36):14992–7.
20. Sarkar NK, Kim YK, Grover A. Coexpression network analysis associated with call of rice seedlings for encountering heat stress. Plant Mol Biol. 2014;84(1–2):125–43.
21. Serin EA, Nijveen H, Hilhorst HW, Ligterink W. Learning from co-expression networks: possibilities and challenges. Front Plant Sci. 2016;7:444.
22. Chen W, Provart NJ, Glazebrook J, Katagiri F, Chang HS, Eulgem T, Mauch F, Luan S, Zou G, Whitham SA, et al. Expression profile matrix of Arabidopsis transcription factor genes suggests their putative functions in response to environmental stresses. Plant Cell. 2002;14(3):559–74.
23. Kilian J, Whitehead D, Horak J, Wanke D, Weinl S, Batistic O, D'Angelo C, Bornberg-Bauer E, Kudla J, Harter K. The AtGenExpress global stress expression data set: protocols, evaluation and model data analysis of UV-B light, drought and cold stress responses. Plant J. 2007;50(2):347–63.
24. Craigon DJ, James N, Okyere J, Higgins J, Jotham J, May S. NASCArrays: a repository for microarray data generated by NASC's transcriptomics service. Nucleic Acids Res. 2004;32(Database issue):D575–7.
25. Gautier L, Cope L, Bolstad BM, Irizarry RA. Affy–analysis of Affymetrix GeneChip data at the probe level. Bioinformatics. 2004;20(3):307–15.
26. Irizarry RA, Hobbs B, Collin F, Beazer-Barclay YD, Antonellis KJ, Scherf U, Speed TP. Exploration, normalization, and summaries of high density oligonucleotide array probe level data. Biostatistics. 2003;4(2): 249–64.
27. Lamesch P, Berardini TZ, Li D, Swarbreck D, Wilks C, Sasidharan R, Muller R, Dreher K, Alexander DL, Garcia-Hernandez M, et al. The Arabidopsis information resource (TAIR): improved gene annotation and new tools. Nucleic Acids Res. 2012;40(Database issue):D1202–10.
28. Chow CN, Zheng HQ, Wu NY, Chien CH, Huang HD, Lee TY, Chiang-Hsieh YF, Hou PF, Yang TY, Chang WC. PlantPAN 2.0: an update of plant promoter analysis navigator for reconstructing transcriptional regulatory networks in plants. Nucleic Acids Res. 2016;44(D1):D1154–60.
29. Langmead B, Trapnell C, Pop M, Salzberg SL. Ultrafast and memory-efficient alignment of short DNA sequences to the human genome. Genome Biol. 2009;10(3):R25.
30. Goujon M, McWilliam H, Li W, Valentin F, Squizzato S, Paern J, Lopez R. A new bioinformatics analysis tools framework at EMBL-EBI. Nucleic Acids Res. 2010;38(Web Server issue):W695–9.
31. Crooks GE, Hon G, Chandonia JM, Brenner SE. WebLogo: a sequence logo generator. Genome Res. 2004;14(6):1188–90.
32. Gupta S, Stamatoyannopoulos JA, Bailey TL, Noble WS. Quantifying similarity between motifs. Genome Biol. 2007;8(2):R24.
33. Sato Y, Takehisa H, Kamatsuki K, Minami H, Namiki N, Ikawa H, Ohyanagi H, Sugimoto K, Antonio BA, Nagamura Y. RiceXPro version 3.0: expanding the informatics resource for rice transcriptome. Nucleic Acids Res. 2013; 41(Database issue):D1206–13.
34. Sakai H, Lee SS, Tanaka T, Numa H, Kim J, Kawahara Y, Wakimoto H, Yang CC, Iwamoto M, Abe T, et al. Rice annotation project database (RAP-DB): an integrative and interactive database for rice genomics. Plant Cell Physiol. 2013;54(2):e6.
35. Yu HD, Yang XF, Chen ST, Wang YT, Li JK, Shen Q, Liu XL, Guo FQ. Downregulation of chloroplast RPS1 negatively modulates nuclear heat-responsive expression of HsfA2 and its target genes in Arabidopsis. PLoS Genet. 2012;8(5):e1002669.
36. Ikeda M, Mitsuda N, Ohme-Takagi M. Arabidopsis HsfB1 and HsfB2b act as repressors of the expression of heat-inducible Hsfs but positively regulate the acquired thermotolerance. Plant Physiol. 2011;157(3):1243–54.
37. Gehan MA, Park S, Gilmour SJ, An C, Lee CM, Thomashow MF. Natural variation in the C-repeat binding factor cold response pathway correlates with local adaptation of Arabidopsis ecotypes. Plant J. 2015; 84(4):682–93.
38. Yoshida T, Fujita Y, Maruyama K, Mogami J, Todaka D, Shinozaki K, Yamaguchi-Shinozaki K. Four Arabidopsis AREB/ABF transcription factors function predominantly in gene expression downstream of SnRK2 kinases in abscisic acid signalling in response to osmotic stress. Plant Cell Environ. 2015;38(1):35–49.
39. Zhang T, Poudel AN, Jewell JB, Kitaoka N, Staswick P, Matsuura H, Koo AJ. Hormone crosstalk in wound stress response: wound-inducible amidohydrolases can simultaneously regulate jasmonate and auxin homeostasis in Arabidopsis Thaliana. J Exp Bot. 2016;67(7):2107–20.
40. Wasternack C, Hause B. Jasmonates: biosynthesis, perception, signal transduction and action in plant stress response, growth and development. An update to the 2007 review in annals of botany. Ann Bot. 2013;111(6):1021–58.
41. Hara-Nishimura I, Matsushima R. A wound-inducible organelle derived from endoplasmic reticulum: a plant strategy against environmental stresses? Curr Opin Plant Biol. 2003;6(6):583–8.
42. Yamada K, Nagano AJ, Nishina M, Hara-Nishimura I, Nishimura M. Identification of two novel endoplasmic reticulum body-specific integral membrane proteins. Plant Physiol. 2013;161(1):108–20.
43. Hayami N, Sakai Y, Kimura M, Saito T, Tokizawa M, Iuchi S, Kurihara Y, Matsui M, Nomoto M, Tada Y, et al. The responses of Arabidopsis early light-induced Protein2 to ultraviolet B, high light, and cold stress are regulated by a transcriptional regulatory unit composed of two elements. Plant Physiol. 2015;169(1):840–55.
44. Scharf KD, Berberich T, Ebersberger I, Nover L. The plant heat stress transcription factor (Hsf) family: structure, function and evolution. Biochim Biophys Acta. 2012;1819(2):104–19.
45. Sugio A, Dreos R, Aparicio F, Maule AJ. The cytosolic protein response as a subcomponent of the wider heat shock response in Arabidopsis. Plant Cell. 2009;21(2):642–54.
46. Littlefield O, Nelson HC. A new use for the 'wing' of the 'winged' helix-turn-helix motif in the HSF-DNA cocrystal. Nat Struct Biol. 1999; 6(5):464–70.
47. Yamasaki K, Kigawa T, Seki M, Shinozaki K, Yokoyama S. DNA-binding domains of plant-specific transcription factors: structure, function, and evolution. Trends Plant Sci. 2013;18(5):267–76.
48. Lindemose S, Jensen MK, Van de Velde J, O'Shea C, Heyndrickx KS, Workman CT, Vandepoele K, Skriver K, De Masi F. A DNA-binding-site landscape and regulatory network analysis for NAC transcription factors in Arabidopsis Thaliana. Nucleic Acids Res. 2014;42(12):7681–93.
49. Korkuc P, Schippers JH, Walther D. Characterization and identification of cis-regulatory elements in Arabidopsis based on single-nucleotide polymorphism information. Plant Physiol. 2014;164(1):181–200.
50. Yu CP, Lin JJ, Li WH. Positional distribution of transcription factor binding sites in Arabidopsis Thaliana. Sci Rep. 2016;6:25164.
51. Zhang W, Ruan J, Ho TH, You Y, Yu T, Quatrano RS. Cis-regulatory element based targeted gene finding: genome-wide identification of abscisic acid- and abiotic stress-responsive genes in Arabidopsis Thaliana. Bioinformatics. 2005;21(14):3074–81.

52. Hattori T, Totsuka M, Hobo T, Kagaya Y, Yamamoto-Toyoda A. Experimentally determined sequence requirement of ACGT-containing abscisic acid response element. Plant Cell Physiol. 2002;43(1):136–40.
53. Soares-Cavalcanti NM, Belarmino LC, Kido EA, Pandolfi V, Marcelino-Guimaraes FC, Rodrigues FA, Pereira GA, Benko-Iseppon AM. Overall picture of expressed heat shock factors in Glycine max, Lotus Japonicus and Medicago Truncatula. Genet Mol Biol. 2012;35(1 (suppl)):247–59.

Valenzuela *et al. BMC Genomics* 2018, **19**(Suppl 2):87
https://doi.org/10.1186/s12864-018-4465-8

BMC Genomics

RESEARCH

Open Access

Towards pan-genome read alignment to improve variation calling

Daniel Valenzuela[1], Tuukka Norri[1], Niko Välimäki[2], Esa Pitkänen[3] and Veli Mäkinen[1*]

From The Sixteenth Asia Pacific Bioinformatics Conference 2018
Yokohama, Japan. 15-17 January 2018

Abstract

Background: Typical human genome differs from the reference genome at 4-5 million sites. This diversity is increasingly catalogued in repositories such as ExAC/gnomAD, consisting of >15,000 whole-genomes and >126,000 exome sequences from different individuals. Despite this enormous diversity, resequencing data workflows are still based on a single human reference genome. Identification and genotyping of genetic variants is typically carried out on short-read data aligned to a single reference, disregarding the underlying variation.

Results: We propose a new unified framework for variant calling with short-read data utilizing a representation of human genetic variation – a pan-genomic reference. We provide a modular pipeline that can be seamlessly incorporated into existing sequencing data analysis workflows. Our tool is open source and available online: https://gitlab.com/dvalenzu/PanVC.

Conclusions: Our experiments show that by replacing a standard human reference with a pan-genomic one we achieve an improvement in single-nucleotide variant calling accuracy and in short indel calling accuracy over the widely adopted Genome Analysis Toolkit (GATK) in difficult genomic regions.

Keywords: Pan-genome reference, Variation calling, Read alignment

Background

Accurate identification and genotyping of genetic variation, or variation calling, in high-throughput resequencing data is a crucial phase in modern genetics studies. Read aligners [1–3] have been successful at aligning short reads to a reference genome (e.g. GRCh37). Among the many analyses downstream of read alignment, here we focus on variation calling. Variation calling is the process of characterizing one individual's genome by finding how it differs from the other individuals of the same species. The standard approach is to obtain a set of reads from the donor and to align them against a single reference genome. The most recent human reference genome, GRCh38, improves on the previous reference version GRCh37 in many respects, including mitochondrial and centromeric sequence quality. Despite containing alternative haplotypes for certain loci, GRCh38 is still largely a haploid consensus reference sequence. Thus, it has been meant to be supplemented by the various databases capturing human genetic variation. Following the alignment of short reads to the reference, multiple tools may be utilized to call variants with respect to the genome (e.g., [4–6]).

However, our current knowledge about the human genome is pan-genomic [7]: after the first human genome was sequenced, the cost of sequencing has decreased dramatically, and today many projects are curating huge genomic databases. These efforts include the 1000 Human Genomes Project [8], UK10K [9], and the Exome Aggregation Consortium and the genome Aggregation Database (ExAC/gnoMAD) [10], the latter consisting of 126,216 exome sequenced and 15,136 whole-genome sequenced individuals. These efforts have already had a significant impact on population and disease genetics. For instance,

*Correspondence: veli.makinen@helsinki.fi
[1]Department of Computer Science, Helsinki Institute for Information Technology HIIT, University of Helsinki, P.O. Box 68 (Gustaf Hällströmin katu 2b), 00014 Helsinki, Finland
Full list of author information is available at the end of the article

the pathogenicity of many suspected predisposition variants has been questioned after the discovery of the variants to be relatively frequent in the human population [10]. Supplementing this burgeoning data are the sequencing efforts focusing on phenotypes, for example cancer [11].

In order to align reads to the pan-genome we use pan-genomic indexing [12–20]. That is, instead of having one reference sequence, an entire collection of sequences is indexed, allowing the reads to be mapped against any genome of the reference set or even to some recombination of them.

There is no consensus about how to represent a pan-genome [7]. Previous efforts can roughly be categorized into three classes: one can consider (i) a graph representing a reference and variations from it, (ii) a set of reference sequences, or (iii) a modified reference sequence.

An example of class (i) approach to pan-genomic indexing is to represent the pan-genome as a graph that recognizes all possible variation combinations (population automaton), and then use an extension of the Burrows-Wheeler Transform to support efficient read alignment [16]. Experiments on variation-rich regions of human genome show that the read alignment accuracy is greatly improved over the standard approach [16]. An important caveat of this approach is the indexing phase: the size of the index is exponential in the worst case. Thus, typically it is necessary to drop some variants to achieve a good expected case behavior [16]. Alternatively, one can enumerate all close-by variant combinations and index the resulting variant contexts (i.e. short subpaths in population automaton) in addition to the reference [12, 14, 17, 18]. Yet, in these approaches, the context length needs to be short to avoid exponential blowup.

Class (ii) approaches consider the pan-genome as a set of individual genomic sequences [13, 15, 21]. The Burrows-Wheeler Transform of those sequences is of linear size and the shared content among individuals translates into highly compressed indexes. Lately, there have been proposals to use Lempel-Ziv indexing to obtain an extremely well compressed index that support efficient read alignment [15, 21, 22].

Class (iii) approaches aim at modifying the reference or encoding variants into the reference to improve read alignment accuracy [14, 20].

The scalability of indexed approaches building on the simple class (ii) model of a set of sequences makes them attractive choice as a basis of variation calling. Unfortunately, unlike with class (i) and class (iii) approaches, the literature on them has primary concentrated on the time and space efficiency aspects, neglecting the final goal of enhancing variation calling. This article aims to fill this gap: We propose a model that relies on the class (ii), and we show that by adding little structure to it we can design a flexible pipeline for variation calling that can be seamlessly incorporated into sequencing data analysis workflows.

We represent the pan-genome reference as a multiple sequence alignment and we index the underlying set of sequences in order to align the reads to the pan-genome. After aligning all the reads to the pan-genome we perform a read pileup on the multiple sequence alignment of reference genomes. The multiple sequence alignment representation of the pan-genome lets us extract a linear ad hoc reference easily (see "Methods" section). Such a linear ad hoc reference represents a possible recombination of the genomic sequences present in the pan-genome that is closer to the donor than a generic reference sequence. The ad hoc reference is then fed to any standard read alignment and variation detection workflow. Finally, we need to normalize our variants: after the previous step, the variants are expressed using the ad hoc reference instead of the standard one. The normalization step projects the variants back to the standard reference. Our overall scheme to call variants is illustrated in Fig. 1.

Results

PanVC, our method for variant calling aligns the reads against multiple reference genomes (represented as a multiple sequence alignment) using by default CHIC aligner, a read aligner that specializes in repetitive collections [23]. Using those alignments, it generates an ad hoc reference which is given to GATK workflow instead of the standard reference (See Fig. 1 and "Methods" section). In our experiments, this approach is labeled MSA_{chic}. As an alternative, we implemented a PanVC version that does not rely on CHIC Aligner, but instead, uses BWA to align against each sequence in the reference. This approach is labelled MSA_{base}

Additionally, we also compare against the pan-genome reference graph approach [16], which we modified also to output an ad hoc reference (see "Methods" section), so that one can apply the same GATK workflow also for that. This approach is labelled GRAPH.

Finally, as a baseline, we considered GATK workflow [4] that aligns the reads against a reference genome using BWA and analyses the resulting read pileup. This baseline approach is labelled GATK.

Experimental setup

Our experimental setup consists of a hidden donor genome, out of which a set of sequencing reads is given as input to the variation calling prediction workflows. Our framework PanVC, and also the graph-based approach will use reference set of 20, 50 and 186 genomes. GATK baseline method is limited to use only one reference.

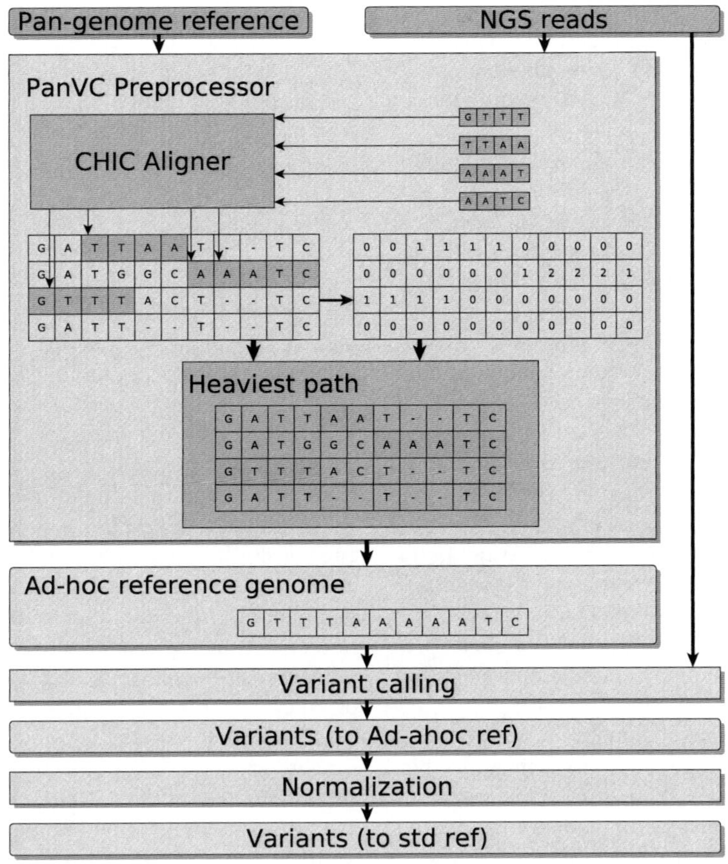

Fig. 1 Schematic view of our PanVC workflow for variation calling, including a conceptual example. The pan-genomic reference comprises the sequences GATTATTC, GATGGCAAATC, GTTTACTTC and GATTTTC, represented as a multiple sequence alignment. The set of reads from the donor individual is GTTT, TTAA, AAAT and AATC. CHIC aligner is used to find the best alignment of each read. In the example, all the alignments are exact matches starting in the first base of the third sequence, the third base of the first sequence, the seventh base of the second sequence, and on the eight base of the second sequence. After all the reads are aligned, the score matrix is computed by incrementing the values of each position where a read aligns. With those values, the heaviest path algorithm extracts a recombination that takes those bases with the highest scores. This is the ad hoc genome which is then used as a reference for variant calling using GATK. Finally the variants are normalized so that they are using the standard reference instead of the ad hoc reference

Our experiments focus on variation calling on complex regions with larger indels and/or densely located simpler variants, where significant improvements are still possible. The reason for that is that graph-based pan-genome indexing has been already thoroughly evaluated [16] for mapping accuracy on human genome data. From those results one can infer that on areas with isolated short indels and SNVs, a regular single-reference based indexing approach with a highly engineered alignment algorithm might be already sufficient.

Therefore, we based our experimental setup on the analysis of highly-polymorphic regions of the human genome [24, 25] that was created in a previous study [16]. This test setup consists of variation-rich regions from 93 genotyped Finnish individuals (1000 genomes project, phase 1 data). The 93 diploid genomes gave us a multiple alignment of 186 strains plus the GRCh37 consensus reference.

We chose variation-rich regions that had 10 SNVs within 200 bases or less. The total length of these regions was 2.2 MB. To produce the ground-truth data for our experimental setup, we generated 221559 100 bp single-end reads from each of the Finnish individuals giving an average coverage of 10*x*.

Evaluation

All evaluated methods output variation calling results that are projected with respect to the standard reference genome. Our hidden donor genome can also be represented as a set of variants with respect to the standard reference genome. This means that we can calculate the standard prediction success measures such as precision and recall. For this, we chose to define the prediction events per base, rather than per variant, to tolerate better invariances of variant locations as have been found to be critical in a recent study [26] (See "Methods" section, "Experimental set-up").

In addition to precision and recall, we also compute the unit cost edit distance of the true donor and the predicted donor. This is defined as the minimum amount of single base substitutions, insertions, or deletions required to convert the predicted donor into the true donor. Here the sequence content of the true donor is constructed by applying its set of variants to the standard reference and the sequence content of the predicted donor is constructed by applying the predicted variants to the standard reference.

There are good incentives to use this evaluation measure to complement precision and recall: first, it gives a single number reflecting how close the predicted sequence is to the ground truth. Second, the projection from the ad hoc reference to the standard reference may loose information. Third, repeat- and error-aware direct comparison of indel variant predictions is non-trivial and only handled properly on deletions [26].

As our experiments are on human data, where genomes are diploids, the heterozygous variants may overlap, which causes some changes to the evaluation measures above. That is, when applying the variants to the reference, we omit variants that overlap already processed ones, and the result is thus a single sequence consisting of all compatible variants. We follow this approach also when computing the precision and recall measures to make the "per base" prediction events well-defined. The results are illustrated in Tables 1 and 2. Row GATK of Table 1 stands for the GATK workflow. Rows MSA + GATK of Table 1 stand for the multiple sequence alignment -based pan-genome indexing scheme specified in the "Methods" section. Row Graph + GATK of Table 1 is using the graph-based indexing of [16] modified to make it compatible with our workflow. The results are averages over all the donors.

Table 1 Edit distance from the predicted donor sequence to the true donor. The average distance between the true donors and the reference is 95193,9

	Pan-genome reference size			
	1	20	50	100
GATK	74695.9	-	-	-
MSA_{base} + GATK	-	2885.5	1956.9	1204.7
MSA_{chic} + GATK	-	1349.3	1117.4	1099.3
Graph +GATK	-	3230.4	3336.8	2706.9

Table 2 Precision and recall of our method MSA_{chic} compared to GATK

Measure	GATK	20	50	100
SNV precision	0.992161	0.998585	0.998863	0.998773
SNV recall	0.904897	0.997098	0.998695	0.999072
Indel precision	0.364853	0.996514	0.99731	0.997778
Indel recall	0.0624981	0.982659	0.985723	0.985958

Discussion

Our results indicate that using pan-genome indexing improves variation calling significantly on highly-polymorphic regions of the human genome: the edit distance between the predicted donor and the true donor is much smaller already when 10 references are used in place of one, and it keeps decreasing when more references are used. When the evaluation metric is precision and recall, the same behavior is observed. In particular, indel calls are improved significantly after the use of pan-genome indexing. Our results reconfirm previous findings about the graph-based approach to pan-genome indexing for specific problems [12, 18]. The approach of tailoring the reference has recently been reported to be beneficial even without using any pan-genomic information; an iterative process to augment a reference and realign has been studied in [19].

A unique feature of our proposal is its genericity. For example, our approach works both on graph representations and on multiple alignment representations of a pan-genome. Earlier studies on pan-genome indexing have mostly focused on read alignments, which are then normalized to the reference to achieve compatibility with the existing variant calling workflows. Instead, here we proposed to globally analyse all read alignments and to produce an ad hoc reference that can be used in place of the standard reference. We keep the projection between the ad hoc reference and the standard reference, so that the variation calling results can always be normalized to the standard reference afterwards.

In addition to variation calling, our methods could be extended to other applications such as to support haplotype analysis in a similar way to a previous study [18]. Namely, one can modify the heaviest path algorithms to produce two predictions. One way to do this is to remove the coverages along the path of the first ad hoc reference and run the heaviest path algorithm again to produce a second ad hoc reference. We leave as future work to make our method fully scalable. We have tested it on multiple alignments of size 1000 times a human chromosome, and with such enormous data sets our analysis pipeline takes weeks to run on a high-performance computer with 1.5 TB of main memory. The current version of our software already contains several engineering solutions to optimize the space usage of intermediate result files and exploit parallelism for maximum speed. Together with our collaborators we are also working on a fully distributed version of the pan-genome analysis pipeline. However, already in its current shape, our software is fully functional in restricted settings, such as calling variants in difficult regions of moderate size. Such feature can be incorporated in a full genome analysis workflow, that processes easy regions using more standard techniques.

Conclusions

Prior work has focused on graph representations of pan-genomes, usually for specific regions [18]. We show that a multiple sequence alignment can be used as a practical alternative, to keep the structure of a pan-genomic reference.

Our experiments show that by replacing a standard human reference with a pan-genomic one we achieve an improvement in single-nucleotide variant calling accuracy and in short indel calling accuracy over the widely adopted Genome Analysis Toolkit (GATK) in difficult genomic regions.

Methods

In the following we provide a detailed description of each component of our workflow (Fig. 1). Our scheme is designed to be modular, and to be used in combination with any variation calling workflow.

The first part of our workflow is the generation of the ad hoc reference. This is done by the preprocessor, using as an input the raw reads of the donor as an input and the pan-genome reference.

The second part is to actually call the variants. We don't provide any details on how to do it because we resort to a variant calling workflow, using our ad hoc reference instead of the standard one. In our experiments, we resort to GATK [4].

Finally, we need to normalize our variants. After the previous step the variants are expressed using the ad hoc reference instead of the standard. The normalization step uses metadata generated from the preprocessor to project the variants back to the standard reference.

Pan-genome preprocessor

The main role of the pan-genome preprocessor is to extract an ad hoc reference sequence from the pan-genome using the reads from the donor as an input.

Pan-genome representation

Following the literature reviewed in the Background section, the existing pan-genome indexing approaches for read alignment could be classified as follows. Some approaches consider the input as a set of sequences, some build a graph or an automata that models the population, and others consider the specific case of a reference sequence plus a set of variations. However, the boundaries between these categories are loose, as a set of sequences could be interpreted as a multiple sequence alignment, which in turn could be turned into a graph. Our scheme can work with different pan-genome representations and indexes provided that it is possible to model recombinations. The multiple sequence alignment and graph representations are versatile enough, but just a collection of sequences is not.

We consider our input pan-genome as a multiple sequence alignment and we store all the positions with a gap. In this way we decouple the problem of book keeping the structure of the pan-genome (in our case, as a multiple sequence alignment) and the problem of indexing the set of underlying sequences.

To transform one representation into the other and to be able to map coordinates we store bitmaps to indicate the positions where the gaps occur. Consider our running example of a multiple alignment

```
GATTAAT--TC
GATGGCAAATC
GTTTACT--TC
GATT--T--TC
```

We may encode the positions of the gaps by four bitvectors:

```
11111110011
11111111111
11111110011
11110010011
```

Let these bitvectors be B_1, B_2, B_3, and B_4. We extract the four sequences omitting the gaps, and preprocess the bitvectors for constant time `rank` and `select` queries [27–29]: $\texttt{rank}_1(B_k, i) = j$ tells the number of 1s in $B_k[1..i]$ and $\texttt{select}_1(B_k, j) = i$ tells the position of the j-th 1 in B_k. Then, for $B_k[i] = 1$, $\texttt{rank}_1(B_k, i) = j$ maps a character in column i of row k in the multiple sequence alignment to its position j in the k-th sequence, and $\texttt{select}_1(B_k, j) = i$ does the reverse mapping, i.e. the one we need to map a occurrence position of a read to add the sum in the coverage matrix.

These bitvectors with rank and select support take $n + o(n)$ bits of space for a multiple alignment of total size n [27–29]. Moreover, since the bitvectors have long runs of 1s (and possibly 0s), they can be compressed efficiently while still supporting fast rank and select queries [30, 31].

Pan-genome indexing and read alignment

Now, the problem of indexing the pan-genome is reduced to index a set of sequences.

To demonstrate our overall scheme, we first use a naive approach to index the pan-genome as a baseline: we index each of the underlying sequences individually using BWA [1]. This approach does not offer a scalable pan-genome indexing solution, but it provides a good baseline for the accuracy that one can expect from a true pan-genome indexing solution to provide. In our experiments, this approach is labeled MSA_{base}.

For a scalable solution that can manage large and highly repetitive set of references we resort to CHIC aligner [23], which combines Lempel-Ziv compression to remove the redundancy with a Burrows-Wheeler index to align the reads. In our experiments, this approach is labeled MSA_{chic}.

Heaviest path extraction

After aligning all the reads to the multiple sequence alignment, we extract a recombined (virtual) genome favoring the positions where most reads were aligned. To do so we propose a generic approach to extract such a heaviest path on a multiple sequence alignment. We define a score matrix S that has the same dimensions as the multiple sequence alignment representation of the pan-genome. All the values of the score matrix are initially set to 0.

We use CHIC aligner to find the best alignment for each donor's read. Then we process the output as follows. For each alignment of length m that starts at position j in the genome i of the pan-genome, we increment the scores in $S[i][j], S[i][j+1] \ldots S[i][j+m-1]$ (adjusting the indexes using the bit-vector representations considered in the previous subsection). When all the reads have been processed we have recorded in S that the areas with highest scores are those where more reads were aligned. An example of this is shown in Fig. 1.

Then we construct the ad hoc reference as follows: we traverse the score matrix column wise, and for each column we look for the element with the highest score. Then, we take the nucleotide that is in the same position in the multiple sequence alignment and append it to the ad hoc reference. This procedure can be interpreted as a heaviest path in a graph: each cell (i,j) of the matrix represents a node, and for each node (i,j) there are N outgoing edges to nodes $(i+1,k)$, $k \in \{1,\ldots,N\}$. We add an extra node A with N outgoing edges to the nodes $(1,k)$, and another node B with N ingoing edges from nodes (L,k). Then the ad hoc reference is the sequence spelled by the heaviest path from A to B. The underlying idea of this procedure is to model structural recombinations among the indexed sequences.

A valid concern is that the resulting path might contain too many alternations between sequences in order to maximize the weight.

To address this issue there is a simple dynamic programming solution to extract the heaviest path, constrained to have a limited number of jumps between sequences: Consider a table $V[1 \ldots L][1 \ldots N][0 \ldots Z]$ initially set to 0. The values $V[i,j,k]$ correspond to the weight of the heaviest path up to character i, choosing the last character from sequence j, that has made exactly k changes of sequences so far. The recursion for the general case ($k > 0, i > 1$) is as follows: $V[i,j,k] = S[i,j] + max\{V[i-1,j,k], max_{j' \neq j} V[i-1,j',k-1]\}$, and the base case for $k = 0$, $i > 1$ is: $V[i,j,0] = S[i,j] + V[i-1,j]$, and for $k = 0$, $i = 1$: $V[1,k,0] = S_{1,j}$.

Once the table is fully computed, the weight of the heaviest path with at most k^* changes is given by $max_j\{V[L,j,k^*]\}$. To reconstruct the path we need to traceback the solution.

However, in our experiments we noticed that the unconstrained version that just selects a maximum weight path without additional constraints performs better than the constrained version, and so we use the former by default in our pipeline.

It is worth noting that as opposed to a graph representation of the pan-genome where the possible recombinations are limited to be those pre-existing in the pan-genome, our multiple sequence alignment representation can also generate novel recombinations by switching sequences in the middle of a pre-existing variant. This happens in our example in Fig. 1, where the ad hoc reference could not be predicted using the graph representation of the same pan-genome shown in Fig. 2.

Variant calling

Variant calling can be in itself a complex workflow, and it might be tailored for specific type of variants (SNVs, Structural Variants), etc. We aim for a modular and flexible workflow, so any workflow can be plugged in it. The only difference is that we will feed it the ad hoc reference instead of the standard one.

In our experiments, we used GATK [4] version 3.3, following the Best Practices: first we aligned the reads to the reference using BWA, and next we used Picard to sort the reads and remove duplicates. Then we performed indel realignment using GATK RealignerTargetCreator and IndelRealigner, and finally we called variants using GATK HaplotypeCaller using parameters `genotyping mode = DISCOVERY`, `standemit conf = 10` and `standcall conf = 30`.

Normalizer

Finally we need to normalize our set of variants. To do so we apply the variants to the ad hoc reference, so that we obtain an alignment between the ad hoc reference and the predicted sequence. The metadata generated in the preprocessor stage – while extracting the heaviest path – includes an alignment between the standard reference and the ad hoc reference. Using those, we can run a linear-time algorithm to obtain an alignment between the

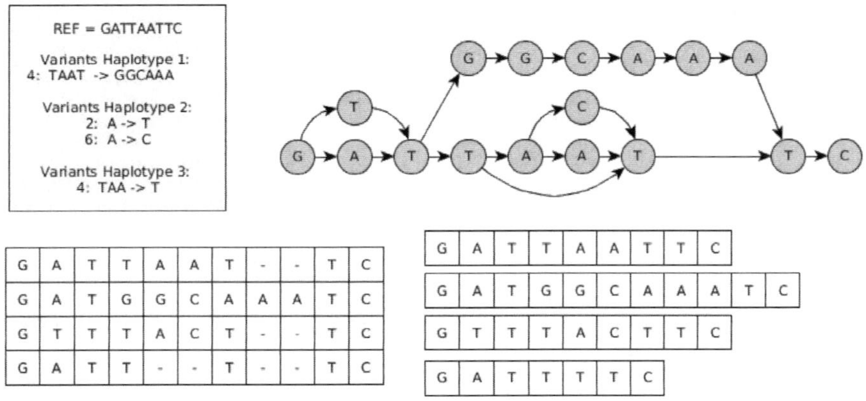

Fig. 2 Four different representations of a pan-genome that corresponds to the same set of individuals. Top left: a reference sequence plus a set of variants to specify the other individuals. Top right: a (directed acyclic) graph representation. Bottom left: a multiple sequence alignment representation, Bottom right: a set of sequences representations

standard reference and the predicted sequence. From this alignment, we can generate a vcf file that expresses the predicted sequence as a set of variants from the standard reference.

Experimental set-up

Evaluation metric

We separate the single nucleotide variant (SNV) calls from indel calls as the results differ clearly for these two subclasses. A true positive (TP) SNV call is a SNV in the true donor and in the predicted donor. A false positive (FP) SNV call is not a SNV in the true donor but is a SNV in the predicted donor. A false negative (FN) SNV call is a SNV in the true donor but is not a SNV in the predicted donor. A true positive (TP) indel call is either an inserted base in the true donor with an identical inserted base in the predicted donor, or a deleted base in both the true and predicted donor. A false positive (FP) indel call is neither inserted nor deleted base in the true donor but is either inserted or deleted base in the predicted donor. A false negative (FN) indel call is an inserted or deleted base in the true donor but is neither inserted nor deleted base in the predicted donor. We report precision=TP/(TP+FP) and recall=TP/(TP+FN).

Modification to graph representation of pan-genome

In our approach we have used a multiple sequence alignment to represent the pan-genomic reference, but it is relatively easy to to use a graph representation [16] instead. A graph representation of a pan-genome usually use a vertex-labeled directed acyclic graph (labeled DAG), and reads are aligned to the paths of this labeled DAG. After all the reads have been aligned to the pan-genome, instead of our score matrix, we can store for each vertex the number of read alignments spanning it. Then the heaviest path can be easily computed using dynamic programming in a topological ordering of the graph: the weight of the heaviest path $h(v)$ to a vertex v is $\max_{v' \in N^-(v)} h(v') + w(v)$, where $w(v)$ is the weight of a vertex and $N^-(v)$ is the set of vertices connected with a in-coming arc to v.

The difference to the multiple alignment heaviest path is that the number of recombinations cannot be limited when using the graph representation.

Another part that is different is the normalizer module to map the variants predicted from the ad hoc reference to the standard reference. For this, the original proposal in [16] already records the path spelling the standard reference, so while extracting the heaviest path one can detect the intersection to the standard reference path and store the corresponding projection as an alignment. Thus, one can use the same evaluation metrics as in the case of multiple sequence alignment -based variation calling.

Data availability

The datasets generated during and/or analysed during the current study are available from the corresponding author on reasonable request; most of the data and scripts to replicate the experiments, as well as a pre-built pan-genome index for the 1000 Human Genomes project data, are available online: https://www.cs.helsinki.fi/gsa/panVC

Code availability

Our tools are open source and available online: https://gitlab.com/dvalenzu/PanVC

Abbreviations

DAG: Directed acyclic graph; FN: False negative; FP: False positive; GATK: Genome analysis toolkit; MSA: Multiple sequence alignment; SNV: Single nucleotide variant; TN: True negative; TP: True positive

Acknowledgements
This work has been supported by the Academy of Finland (grants 284598 and 287665).

Funding
Funding for the publication of this article was provided by grant 284598.

Availability of data and materials
The datasets used and/or analysed during the current study are available from the corresponding author on reasonable request.

About this supplement
This article has been published as part of *BMC Genomics* Volume 19 Supplement 2, 2018: Selected articles from the 16th Asia Pacific Bioinformatics Conference (APBC 2018): genomics. The full contents of the supplement are available online at https://bmcgenomics.biomedcentral.com/articles/supplements/volume-19-supplement-2.

Authors' contributions
The original idea is by VM. The idea was further developed by all the authors. Most of the implementation is due to DV. TN also contributed to the implementation. All of the authors have read and approve the final manuscript.

Ethics approval and consent to participate
Not applicable.

Consent for publication
Not applicable.

Competing interests
The authors declare that they have no competing interests.

Author details
[1]Department of Computer Science, Helsinki Institute for Information Technology HIIT, University of Helsinki, P.O. Box 68 (Gustaf Hällströmin katu 2b), 00014 Helsinki, Finland. [2]Department of Medical and Clinical Genetics, Genome-Scale Biology Program, University of Helsinki, Helsinki, Finland. [3]European Molecular Biology Laboratory, Genome Biology Unit, Heidelberg, Germany.

Published: 9 May 2018

References

1. Li H, Durbin R. Fast and accurate short read alignment with burrows–wheeler transform. Bioinformatics. 2009;25(14):1754–60.
2. Langmead B, Trapnell C, Pop M, Salzberg SL, et al. Ultrafast and memory-efficient alignment of short dna sequences to the human genome. Genome Biol. 2009;10(3):25.
3. Li R, Li Y, Kristiansen K, Wang J. Soap: short oligonucleotide alignment program. Bioinformatics. 2008;24(5):713–4.
4. Auwera GA, Carneiro MO, Hartl C, Poplin R, del Angel G, Levy-Moonshine A, Jordan T, Shakir K, Roazen D, Thibault J, et al. From fastq data to high-confidence variant calls: the genome analysis toolkit best practices pipeline. Curr Protocol Bioinform. 2013;43:11.10.1–33.
5. Li H. A statistical framework for snp calling, mutation discovery, association mapping and population genetical parameter estimation from sequencing data. Bioinformatics. 2011;27(21):2987–93.
6. Garrison E, et al. FreeBayes . 2016. https://github.com/ekg/freebayes.
7. Consortium CP-G, et al. Computational pan-genomics: status, promises and challenges. Brief Bioinform. 2008;19(1):118–135.
8. Consortium TGP. A global reference for human genetic variation. Nature. 2015;526(7571):68–74.
9. Consortium TU. The UK10K project identifies rare variants in health and disease. Nature. 2015;526(7571):82–90.
10. Consortium EA. Analysis of protein-coding genetic variation in 60,706 humans. Nature. 2016;536(7616):285–91.
11. International Cancer Genome Consortium, et al. International network of cancer genome projects. Nature. 2010;464(7291):993–8.
12. Schneeberger K, Hagmann J, Ossowski S, Warthmann N, Gesing S, Kohlbacher O, Weigel D. Simultaneous alignment of short reads against multiple genomes. Genome Biol. 2009;10:98.
13. Mäkinen V, Navarro G, Sirén J, Välimäki N. Storage and retrieval of highly repetitive sequence collections. J Computat Biol. 2010;17(3):281–308.
14. Huang L, Popic V, Batzoglou S. Short read alignment with populations of genomes. Bioinformatics. 2013;29(13):361–70.
15. Ferrada H, Gagie T, Hirvola T, Puglisi SJ. Hybrid indexes for repetitive datasets. Philosophical Trans R Soc A. 2014;372.
16. Sirén J, Välimäki N, Mäkinen V. Indexing graphs for path queries with applications in genome research. IEEE/ACM Trans Comput Biol Bioinform. 2014;11(2):375–88.
17. Danek A, Deorowicz S, Grabowski S. Indexing large genome collections on a pc. PLoS ONE. 2014;9(10):e109384.
18. Dilthey A, Cox C, Iqbal Z, Nelson MR, McVean G. Improved genome inference in the mhc using a population reference graph. Nat Genet. 2015;47:682–8.
19. Schröder J, Girirajan S, Papenfuss AT, Medvedev P. Improving the power of structural variation detection by augmenting the reference. PLOS ONE. 2015;10(8):1–10.
20. Maciuca S, del Ojo Elias C, McVean G, Iqbal Z. A natural encoding of genetic variation in a burrows-wheeler transform to enable mapping and genome inference. In: Algorithms in Bioinformatics - 16th International Workshop, WABI 2016, Aarhus, Denmark, August 22-24, 2016. Proceedings, Lecture Notes in Computer Science, vol. 9838. Switzerland: Springer; 2016. p. 222–33.
21. Deorowicz S, Danek A, Grabowski S. Genome compression: a novel approach for large collections. Bioinformatics. 2013;29(20):2572–8.
22. Valenzuela D. CHICO: A compressed hybrid index for repetitive collections. In: Proc. 15th International Symposium on Experimental Algorithms (SEA), LNCS. Switzerland: Springer; 2016. p. 326–38.
23. Valenzuela D, Mäkinen V. CHIC: a short read aligner for pan-genomic references. bioRxiv. 2017. Cold Spring Harbor Laboratory. https://doi.org/10.1101/178129. https://www.biorxiv.org/content/early/2017/08/18/178129.full.pdf.
24. Horton R, et al. Variation analysis and gene annotation of eight MHC haplotypes: The MHC haplotype project. Immunogenetics. 2007;60(1):1–18.
25. Khurana E, et al. Integrative annotation of variants from 1092 humans: Application to cancer genomics. Science. 2013;342(6154):1235587.
26. Wittler R, Marschall T, Schönhuth A, Mäkinen V. Repeat- and error-aware comparison of deletions. Bioinformatics. 2015;31(18):2947–54.
27. Jacobson G. Space-efficient static trees and graphs. In: Proc. FOCS. Washington, DC: IEEE Computer Society; 1989. p. 549–54.
28. Clark D. Comxpact pat trees. PhD thesis, University of Waterloo, Canada. 1996.
29. Munro I. Tables. In: Proc. FSTTCS. LNCS v. 1180. Berlin: Springer; 1996. p. 37–42.
30. Raman R, Raman V, Rao S. Succinct indexable dictionaries with applications to encoding k-ary trees and multisets. In: Proc. SODA. Philadelphia: SIAM; 2002. p. 233–42.
31. Navarro G, Mäkinen V. Compressed full-text indexes. ACM Comput Surv. 2007;39(1):2.

Anselmetti *et al. BMC Genomics* 2018, **19**(Suppl 2):96
https://doi.org/10.1186/s12864-018-4466-7

BMC Genomics

RESEARCH Open Access

Phylogenetic signal from rearrangements in 18 Anopheles species by joint scaffolding extant and ancestral genomes

Yoann Anselmetti[1,2], Wandrille Duchemin[2,3], Eric Tannier[2,3], Cedric Chauve[4] and Sèverine Bérard[1*]

From The Sixteenth Asia Pacific Bioinformatics Conference 2018
Yokohama, Japan. 15-17 January 2018

Abstract

Background: Genomes rearrangements carry valuable information for phylogenetic inference or the elucidation of molecular mechanisms of adaptation. However, the detection of genome rearrangements is often hampered by current deficiencies in data and methods: Genomes obtained from short sequence reads have generally very fragmented assemblies, and comparing multiple gene orders generally leads to computationally intractable algorithmic questions.

Results: We present a computational method, ADSEQ, which, by combining ancestral gene order reconstruction, comparative scaffolding and *de novo* scaffolding methods, overcomes these two caveats. ADSEQ provides simultaneously improved assemblies and ancestral genomes, with statistical supports on all local features. Compared to previous comparative methods, it runs in polynomial time, it samples solutions in a probabilistic space, and it can handle a significantly larger gene complement from the considered extant genomes, with complex histories including gene duplications and losses. We use ADSEQ to provide improved assemblies and a genome history made of duplications, losses, gene translocations, rearrangements, of 18 complete *Anopheles* genomes, including several important malaria vectors. We also provide additional support for a differentiated mode of evolution of the sex chromosome and of the autosomes in these mosquito genomes.

Conclusions: We demonstrate the method's ability to improve extant assemblies accurately through a procedure simulating realistic assembly fragmentation. We study a debated issue regarding the phylogeny of the *Gambiae complex* group of *Anopheles* genomes in the light of the evolution of chromosomal rearrangements, suggesting that the phylogenetic signal they carry can differ from the phylogenetic signal carried by gene sequences, more prone to introgression.

Keywords: Scaffolding, Comparative genomics, Mosquito genomics

Background

The promises of using genes as evolutionary markers for phylogeny, introduced half a century ago by Zuckerkandl and Pauling [1], have been unfulfilled so far [2, 3]. At every scale of evolution, gene histories can differ from organisms tree-like phylogeny due to non-tree like evolutionary mechanisms such as incomplete lineage sorting, horizontal gene transfer, hybridization, symbiosis, among others. This happens for example in the *Gambiae complex*, composed of several African *Anopheles* mosquito species, whose phylogeny is important to shed light on the origin of malaria transmission to humans [4], but is difficult to trace because of apparent extensive gene introgression within this complex [5, 6]. Chromosomal rearrangements have been recognized, for an even longer time [7], as valuable phylogenetic markers, due to several reasons, including their lower occurrence rate

*Correspondence: Severine.Berard@umontpellier.fr
[1]ISEM, Université de Montpellier, CNRS, IRD, EPHE, Montpellier, France
Full list of author information is available at the end of the article

compared to sequence evolution. They have proved to be of great interest for understanding mosquito evolution for example [8, 9], due to the fact introgression of whole chromosomes is much less frequent than introgression of genes [10]. Moreover, in terms of functional and ecological implications, chromosomal rearrangements have also been shown to be involved in important adaptation processes [11–13].

However chromosome evolution is still challenging to study, especially from short reads sequence data, and current methods have severe limitations that we outline now. First, some methods are limited to consider only chromosomal regions which are highly similar and whose differences are detected by genetic mapping, polytene chromosome banding, *in situ* hybridization or targeted sequencing [8]. Among methods which can handle whole genome sequence data [14–19], some consider only a small number of markers (often genes) with simple evolutionary histories (typically duplication-free histories and one-to-one orthologous gene families), and most of them require fully assembled extant genomes, aside of the recently published method DESCHRAMBLER [20]. As a consequence existing methods are hardly applicable to currently available genomic data, characterized by very short sequencing reads that can not resolve genomic repeats [21, 22], leading to highly fragmented genome assemblies, often in the form of hundreds or thousands of contigs or scaffolds, where evolutionary breakpoints can not be distinguished from assembly artifacts.

Various types of data can help to improve the contiguity of genome assemblies obtained from short reads. For example Third-Generation Sequencing (TGS) technologies generate long, albeit noisy, sequencing reads that can resolve ambiguities due to repeats [23]; alternatively, chromosome conformation capture technologies [24, 25] or genome maps [26–28] have also been used successfully for scaffolding large genomes. However in the absence of long range sequence data or genome maps, the most widely used approach to scaffold contigs is the comparative approach, using one or several related genomes to guide the scaffolding. The principle of comparative scaffolding is to align contigs of a fragmented genome assembly onto one, or a set of, assembled reference genome(s) and to deduce contig adjacencies from the contiguity of the corresponding alignments along the reference(s). Most comparative scaffolding methods rely on a single reference genome, assumed to be closely related enough that contiguity along the reference can confidently imply contiguity in the newly assembled genome [29–35]. Such methods have mostly been used to assemble pathogen genomes using closely related assembled references, but have also been shown to be applicable in wider contexts, such as scaffolding an antelope genome using a cow genome as reference [34]. There exists few methods that can handle several reference genomes at once, that can be distinguished between methods that do require that the phylogenetic relation between the considered genomes are provided [36–39] and methods that do not need such information [40]. Moreover, only two methods make use of sequencing data that might not appear in the contigs to be scaffolded but can provide a valuable scaffolding signal that complements the comparative signal [35, 37]. All such methods are also limited to handle contigs containing repeats and discard repeated contigs.

An important feature of most of these methods is that they assume a hypothesis of genome rearrangement parsimony or near-parsimony to transfer contiguity information from the reference(s) to the genome of interest, this hypothesis being either explicit [30, 31, 33, 38] or implicit [36, 39]. This points at the fact that the comparative approach is a kind of conundrum: to scaffold genomes, comparative methods rely, at least implicitly, on a framework to compare genomes and detect conserved synteny and chromosomal rearrangements, while whole genome evolution methods do not fare well when provided with fragmented genome assemblies.

We introduce a new computational method, ADSEQ, that addresses the issues raised above, regarding both genome evolution by rearrangements and comparative genome scaffolding; we apply it to simultaneously study the chromosome evolution and improve the scaffolding of 18 *Anopheles* genomes, 16 of them recently sequenced by Neafsey et al. [4], including several important malaria vector species. The method ADSEQ does not need a fully assembled reference genome, as is required by most comparative scaffolding methods, but takes as input a set, that can be arbitrarily large, of fragmented genome assemblies, together with a species phylogeny. It also takes advantage of sequencing data such as paired-end reads, for species for which it is available. From this input, ADSEQ computes ancestral genome segments, as well as extant scaffolding adjacencies. Additionally, ADSEQ allows the user to infer evolutionary scenarios in terms of gene duplication, gene loss, gene displacement and genome rearrangement along each branch of the species phylogeny. A Gibbs-Boltzmann probabilistic framework based on the cost of adjacencies gain and breaks in evolutionary scenarios provides a statistical support on all ancestral and extant inferred adjacencies, with sequencing data used to define a prior on extant gene adjacencies. To handle genes whose history involves duplication and loss, ADSEQ relies on the use of reconciled gene trees, in terms of gene duplication and losses, which allows to use a much larger gene set than existing comparative scaffolding methods relying on one-to-one orthologous genes or gene families with simple duplication/loss histories. We present, together with the ADSEQ method, a validation procedure for the extant genome scaffolding aspect of ADSEQ relying on a realistic framework to generate artificially fragmented genome assemblies.

Using ADSEQ, we provide an analysis of whole genome evolution in a large set of *Anopheles* species with an unprecedented precision, being able to quantify duplications, losses, gene displacements between chromosomes, and chromosomal rearrangements. We work at a much finer scale than cytogenetics methods [41–47], rely on a larger gene complement than traditional genome rearrangement methods based on rothologous genes, and we provide a refined evolutionary analysis compared to [4] due to the improvement of extant genome scaffolding. In particular we find that gene displacements between chromosomes are much more frequent for genes belonging to families with duplication/loss histories, that previous studies handling only one-to-one orthologous genes had mostly outlooked. We also use our method to compare two alternative *Anopheles* phylogenies. We find that *Anopheles* genomes are compartmentalized between autosomes and sex chromosome according to duplications and chromosomal rearrangements, just as they were found to be according to gene sequences. We provide an alternative hypothesis to the conclusions of Fontaine et al. [5] about introgression of the major part of the genome. Indeed our measures of rearrangements and duplications are in favour of the phylogeny supported by most genes.

Methods

We first describe the main methodological contribution of our work, the ADSEQ tool, followed by its application to the specific *Anopheles* genomes data set we analyze in detail. We begin by introducing some simple but important terminology. A gene, extant or ancestral, is seen as a directed DNA segment with two *extremities*. Genes are parts of larger segments containing one or several genes, which are contig, super-contigs, scaffolds or even chromosomes for well assembled genomes. Two genes that are contiguous along such a segment define an *adjacency* between one extremity of each gene, that we call a *gene adjacency*. Thus *segments of genes*, either extant or ancestral, are encoded by linearly organized sets of gene adjacencies. For extant genomes, such segments, corresponding to contigs, scaffolds or chromosomes, are observed, while for ancestral genomes, segments are reconstructed and are to be considered as hypothetical as they are not directly supported by sequence data. Ancestral segments have previously been termed Contiguous Ancestral Regions (CARs) [19, 48, 49]. However, to stress the similarity between scaffold in extant genomes and CAR in ancestral genome, we use the generic terminology "segment" for both throughout this paper.

Assembly recovery through detection of Coevolution with sequencing data (ADSEQ)

ADSEQ builds upon a family of methods aimed at reconstructing the evolutionary history of gene adjacencies introduced with the DECO algorithm [50] and extended along several lines in [51, 52]. It is implemented within the package DECOSTAR [53]. The aim of ADSEQ is to jointly scaffold extant genomes and reconstruct ancestral gene orders, through the joint analysis of phylogenomics and sequencing data.

ADSEQ takes as minimal input a species tree and a set of extant genomes data:gene adjacencies, homologous gene families and their associated reconciled gene trees (see for example [54] for background on reconciled gene trees). Reconciled gene trees implicitly yield the gene content of ancestral genomes. A key feature of ADSEQ is that the extant gene adjacencies can originate from genomes in various state of assembly, from fully assembled genomes —where gene adjacencies encode the gene order along the chromosomes—, to ambiguously assembled genomes represented as scaffolding graphs, through fragmented genomes assembled into contigs or scaffolds. Each extant gene adjacency is assigned a *prior* score in $[0, 1]$, expected to represent the confidence that the adjacency actually occurs in the genome of interest. This prior can be obtained from sequencing data or genome maps for example; so adjacencies between genes in fully assembled genomes will have a high prior score, while a potential gene adjacency in a poorly assembled genome and that is not supported by a strong signal in terms of sequencing data will likely be assigned a low prior score.

ADSEQ processes independently all pairs of gene families for which at least one extant adjacency is observed between genes from these families. A *solution* of ADSEQ on such an instance is a set of extant and ancestral gene adjacencies between extant and ancestral genes of the two considered families, that are consistent with the given reconciled gene trees. Taking the obtained solutions over all pairs of gene families defining ADSEQ instances, ancestral adjacencies link ancestral genes extremities into ancestral genome segments, while extant adjacencies improve the scaffolding of the provided extant genomes and reduce their fragmentation. This, together with the reconciled gene trees, in turn provides an important input material for studying whole-genome evolution through mechanisms such as gene duplication, loss and transfer, introgression, or genome rearrangement. Figure 1 provides a high-level overview of the ADSEQ algorithm.

An important feature of ADSEQ is that, for each considered instance, it does not compute a single solution, but samples solutions from the, often large, associated search space. In order to sample solutions, ADSEQ relies on a *score*, function of the number of evolutionary events and the prior score of adjacencies. Any solution indeed yields

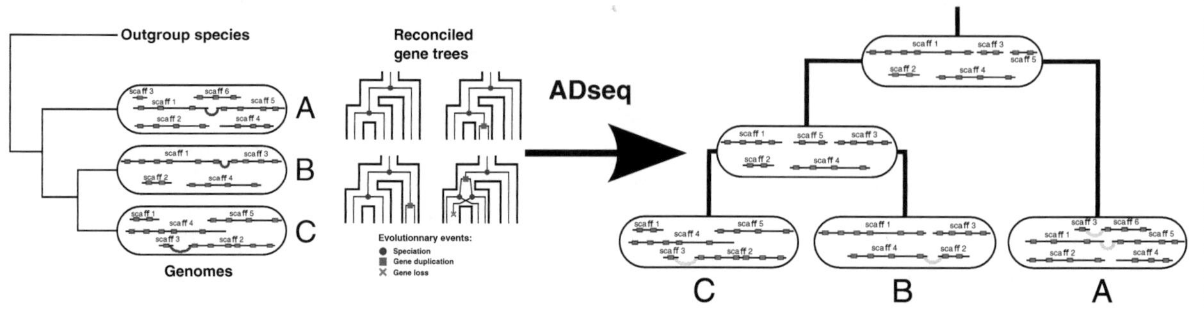

Fig. 1 Input and output of ADSEQ. (Left) Input data: (1) a species tree with extant genomes (A, B and C) containing observed adjacencies (black link) and scaffolding gene adjacencies with a *prior* score (blue link); each grey box represents a gene. (2) reconciled gene trees representing evolutionary histories of gene families annotated by evolutionary events. (Right) Typical output: gene order across ancestral and extant genomes including new extant gene adjacencies with a *posterior* score (green link) between genes located at fragments extremities in the initial genome assemblies

a number of gains and breakages of gene adjacencies, that model genome rearrangement events consistent with the provided reconciled gene trees (see a description of the propagation rules that allow to infer gains/breakages from reconciled gene trees in the DECO algorithm [50]). The *score* of a solution S is defined as $n(S) = g(S) + b(S) + c(S)$ where $g(S)$ is the number of gene adjacency gains scaled according to a user-defined unit cost of a gain; $b(S)$ is the number of gene adjacency breakages again scaled according to a user-defined unit cost; $c(S)$ is the cost of including or discarding extant adjacencies, based on their prior score: for an adjacency of *prior* score p, including the adjacency in a solution costs $-kT_0 \log(p)$ while discarding it from the solution costs $-kT_0 \log(1-p)$, where kT_0 is a pseudo-temperature that we discuss in details in Additional file 1.

A polynomial time and space Dynamic Programming (DP) algorithm samples solutions for a given instance with a probability proportional to their score. More precisely, ADSEQ can sample solutions under a Gibbs-Boltzmann probability distribution defined as follows: the Gibbs-Boltzmann score of a solution S is equal to $exp^{-n(S)/kT}$, where kT is a user-defined pseudo-temperature, and this score defines implicitly a probability distribution over the set of all solutions. Tuning the pseudo-temperature kT provides a control over the probability to sample parsimonious solutions: a low pseudo-temperature tends to increase the probability to sample most parsimonious solutions while a large pseudo-temperature skews the Gibbs-Boltzmann distribution toward the uniform distribution over the set of all solutions (we refer to [51] for more details on the Gibbs-Boltzmann framework applied to the DECO algorithm).

In the present work, the *prior* scores of extant adjacencies are either 1.0 for adjacencies that are observed in a contig or scaffold, or a scaffolding score obtained from sequencing data using the scaffolding software BESST. Pairs of genes located at the extremities of contigs and for which sequencing data do not provide any evidence for a scaffolding adjacency receive a small prior score as described in [52] (see also Additional file 1). The *posterior* scores are defined as the frequency out of a sample of 100 solutions with temperatures $kT = kT_0 = 0.1$ that skews the Gibbs-Boltzmann distribution toward optimal solutions.

Linearization of extant and ancestral components

After ADSEQ is applied to all considered pairs of gene families, the obtained result is a set of ancestral and extant gene adjacencies, each adjacency being assigned a *posterior* score. False positives – i.e. pairs of genes predicted inaccurately to be contiguous in an extant or ancestral genome–, can be due to errors in the data (for example errors in gene families or reconciled gene trees) or to errors in the inference process (for example the parsimony assumption might be wrong for some gene adjacencies). As a result, it is possible that a given gene (or gene extremity) belongs to more than two (more than one) adjacencies, which is incompatible with the expected linear structure of chromosomes.

To address this issue, we process the set of ADSEQ ancestral and extant adjacencies in such a way that they define linear ancestral and extant segments. To do so, we apply, independently for each species, a method used both in ancestral genome reconstruction [55] and scaffolding algorithms [56]. It consists in extracting, for each species, a Maximum-Weight Matching (MWM) in the graph whose vertices are gene extremities and edges are gene adjacencies, weighted by their posterior score. This MWM can still include circular segments, that are linearized by removing the least-weight edge of each such circular segment. Moreover, prior to this linearization step, we discard adjacencies whose posterior score is below a user-defined threshold, that was set to 0.1

after simulations aimed at measuring the accuracy of the ADSEQ algorithm (see Fig. 2). This linearization step is done independently for each species, ancestral or extant.

Application to the *Anopheles* data set

We now describe how we generated the data for the 18 *Anopheles* genomes.

Species trees

The main species phylogeny we considered was taken from [5]. We call it the "X phylogeny" as it is based on the X chromosome genes. It is the species tree used by default unless another is specified. We also considered an alternative "Whole Genome (WG) phylogeny", that was introduced in [5] as the most frequently observed among trees built using sequences from the autosomes (Fig. 3).

Genomic data

Most genomic data we used were produced by the *Anopheles 16 Genomes project* described in [4] and retrieved from VectorBase (https://www.vectorbase.org): genome assemblies (contigs, scaffolds, or chromosome arms), gene annotations, and gene sequences (CDS). For 16 out of the 18 considered *Anopheles* species, we retrieved from the NCBI Sequence Read Archive (SRA) paired-end Illumina libraries with an insert size of 180bp ('fragment' libraries) and mate-pair libraries with an insert size of 1.5kbp ('jump' libraries), both obtained from a single female individual. Additional low-coverage long-range sequencing libraries ('Fosill' libraries) were obtained from pools of individuals. Details are available in Additional file 17: Table S1.

Gene families and trees

We retrieved homologous gene families from orthologous gene groups of 21 *Culicidae* species recovered from the OrthoDBmoz2 database (http://cegg.unige.ch/orthodbmoz2) generated using OrthoDB [57]. We generated a multiple sequence alignment for each family, used RAXML [58] to compute draft gene trees with bootstrap supports, and then corrected these draft gene trees using PROFILENJ [59]. PROFILENJ contracts branches with low bootstrap support and, using the species tree, resolves the polytomies in a way that minimizes the number of duplications and losses in a reconciliation. This resulted in 14,940 gene trees containing 183,680 genes. We refer to Additional file 1, Additional file 2: Figure S1, Additional file 3: Figure S2, Additional file 4: Figure S3, and Additional file 5: Figure S4, and Additional file 18: Table S2 for a description of our preprocessing pipeline (including preprocessing of these gene families), a comparison of the newly inferred gene trees with the original ones computed by the *Anopheles* consortium and for statistics on the inferred gene duplications and losses in these gene trees. As a species phylogeny is required to reconcile gene trees, we repeated the process described above for both the X phylogeny and the WG phylogeny.

Extant adjacencies and prior scores from sequencing data

Sequencing reads from all libraries were filtered to discard low quality reads and trimmed to 75bp length using TRIMMOMATIC [60], then mapped onto the contigs or scaffolds of the considered species using BOWTIE2 [61]; for reads with multiple mappings, all of them were conserved. The scaffolding software BESST [62, 63] was then used to detect potential scaffolding adjacencies between pairs of contigs containing at least one annotated gene. Scaffolding adjacencies that were not supported by at least four pairs of reads were discarded. For remaining scaffolding adjacencies, we assigned a score defined as the arithmetic mean of the two scores computed by BESST, the link variation score and the link dispersity score. Detailed statistics on the scaffolding adjacencies so obtained are available in Additional file 6: Figure S5 and Additional file 7: Figure S6.

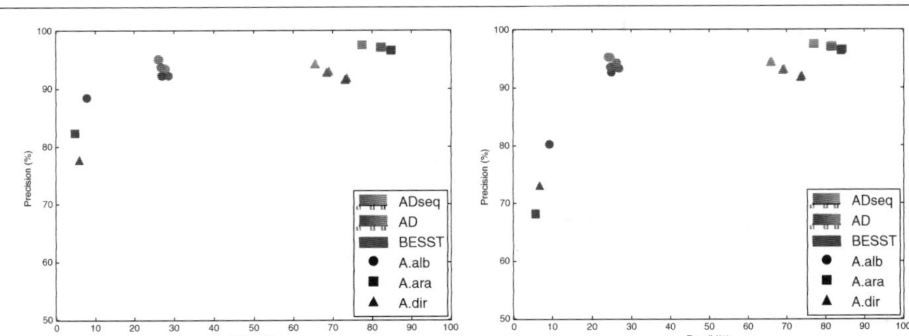

Fig. 2 Precision and recall statistics for scaffolding adjacencies on three artificially fragmented genomes (A.alb: *A. albimanus*, A.ara: *A. arabiensis* and A.dir: *A. dirus*). **Left graph:** results with 50% of reads. **Right graph:** results with all reads. The different methods results are plotted with the precision on the Y axis and the recall on the X axis. For ADSEQ and AD, results for three different adjacency support thresholds (0.1, 0.5 and 0.8) before genome linearization are plotted and represented with a color gradient. Note: A True Positive (TP) adjacency requires the proper orientation of both genes

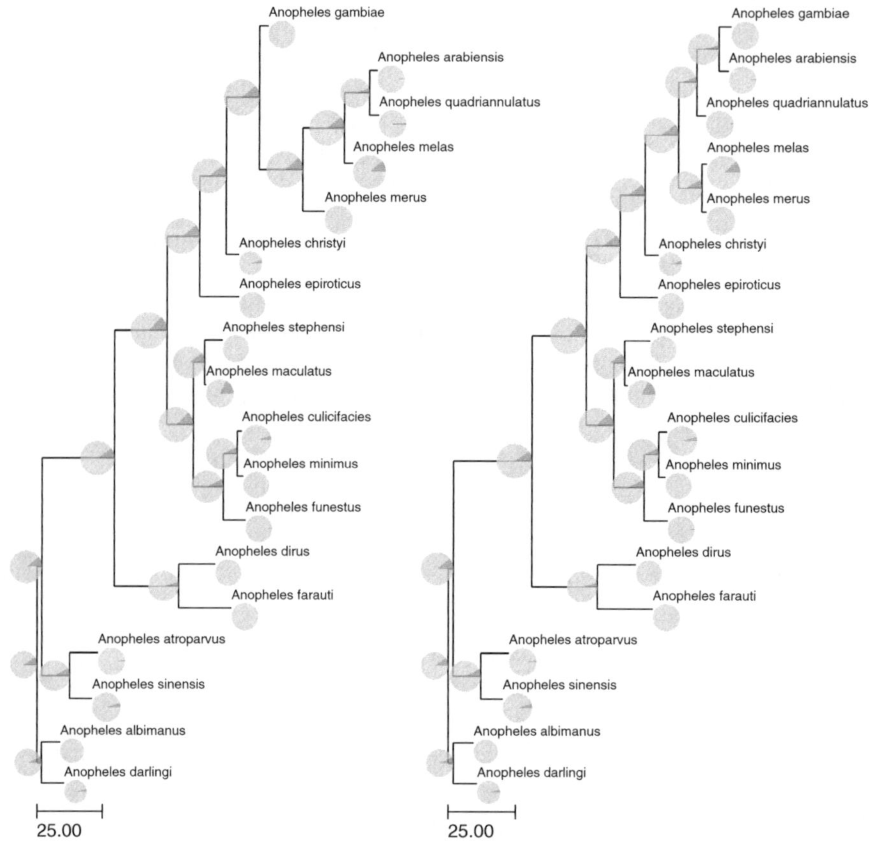

Fig. 3 A. species trees (left: X phylogeny, right: WG phylogeny) with rearrangements per adjacency as branch lengths ($\times 10^{-3}$). The pie chart for a given species represents the adjacency degree of the genes of this species: orange represents genes with no adjacency, light blue genes with adjacency degree of 1 and green genes with adjacency degree of 2. Moreover, the diameter of each chart is proportional to the number of genes in the corresponding species

Genome fragmentation simulations for measuring the accuracy of ADSEQ for extant scaffolding

We developed a validation protocol of ADSEQ to measure its ability to propose reliable extant scaffolding adjacencies (see Additional file 8: Figure S7 for an illustration of the protocol and Additional file 19: Table S3 for assembly statistics at different steps of the protocol). The key element is to provide to ADSEQ a genome whose assembly is more fragmented than the reference assembly, in order to verify that ADSEQ can retrieve the lost adjacencies. Moreover our simulation framework aims at generating a realistic fragmentation, as relying on a random fragmentation, as used in other validation protocols [16, 52], generates data that are in general easy to scaffold using comparative methods.

To avoid this pitfall we simulated a fragmented assembly by re-assembling the considered genomes using KMERGENIE [64] and MINIA [65]. MINIA was chosen due to its stringency in handling repeats, that leads to more conservative and fragmented assembly compared to other contig assemblers. We applied this protocol to a randomly chosen either all or half of the raw sequence reads, independently three times, with the species *A. albimanus*, *A. arabiensis* and *A. dirus*, whose positions in the species tree allow to consider various evolutionary contexts.

We ran ADSEQ as described above on these new assemblies and compared its results (scaffolding adjacencies) with the reference assemblies. We call a True Positive (TP) adjacency an adjacency inferred by ADSEQ and present in the initial genome assembly. A False Positive (FP) adjacency corresponds to an adjacency inferred by ADSEQ and not present in the reference genome assembly. A FP can however be a true adjacency not found by the reference assembly (e.g., connecting two scaffolds), so we call Certain False Positive (CFP) a FP adjacency which extremities are not scaffold or contig extremities in initial genome assembly. Finally a False Negative (FN) is a pair of gene extremities that are contiguous in the initial assembly but are not inferred as an adjacency by ADSEQ. From these values we compute the usual *precision* and *recall* statistics, but using CFP for the false positives count.

Gene order evolution analysis

Assignment of chromosome segments

To compare the evolution of *Anopheles* chromosomes, especially the apparent differences between the X chromosome and the autosomes described in [4], it is necessary to assign extant and ancestral chromosome segments to either the X chromosome or the autosomes. As *A. gambiae* is the only fully assembled genome in our data set, this is also the only genome for which such information is readily available; in all other species the genomes are assembled into scaffolds with no indication of whether this scaffold belongs to the X chromosome or an autosome, unless additional data is available, such as genome maps. We assigned extant and ancestral genes and segments to the X chromosome or autosomes using the following probabilistic method. For each gene g (ancestral or extant), a set of *An. gambiae* orthologs is defined as all *An. gambiae* genes from the same gene family than g whose last common ancestor with g in the reconciled gene tree is a speciation node. Note that this set might be empty, and that this definition includes the case where g is an ancestor of a *An. gambiae* gene. The probability of g being on the X chromosome is then defined as the frequency of orthologs located onto *An. gambiae* X chromosome, or, if no ortholog is present on this X chromosome, by a background probability, defined as the global frequency of *An. gambiae* genes on the X chromosome. Then each segment is given a probability to be located on the X chromosome as the mean of probabilities for all genes it contains. Each gene then inherits the probability of being on the X chromosome from the segment it belongs to.

Very recently an assignment of *An. albimanus* genes to chromosomes was published together with a new assembly [66] that we could use to verify that our assignment method is accurate: out of 8840 genes assigned to a chromosome in the novel assembly, we correctly predicted the autosomal/X placement of 8837 genes (comparing the assignment of higher probability and the assignment in the new assembly).

Gene movements (translocations)

For every couple of genes for which one is a direct descendant of the other, we inferred a gene movement (between the X chromosome and an autosome) if the probability of the ancestral gene being on the X chromosome is ≤ 0.2 while the probability of the descendant gene being on the X chromosome is ≥ 0.8, or conversely.

Detecting chromosomal rearrangements

For every branch of the species tree, genes with exactly one exemplar in their family both in the ancestor and descendant species were selected. Then conserved adjacencies were computed, which are adjacencies present between ancestral genes and descendant homologous genes. In order to discard gene displacements, which are counted elsewhere, we filtered also genes which are not involved in any common adjacency. A *rearrangement* (gain or breakage of a gene adjacency) is counted every time two gene extremities are contiguous on a segment (with respect to the reduced selected set of genes) of the ancestor, but not on the descendant, or conversely. When they are not contiguous, in order to detect a certain genome rearrangement, we require also that they are not both the extremities of their segments, to avoid counting as a rearrangement a potentially undetected scaffolding adjacency. Rearrangements were not directly counted as gains and breakages output from ADSEQ because this count can be blurred by adjacencies gained or broken by gene duplications and gene losses. As a consequence, gene duplications and losses are not counted as rearrangements, that are limited to synteny breakages due to balanced rearrangements (inversions, transpositions, large translocations), that do not change the gene content.

We stress that this method to detect genome rearrangements is conservative and underestimates rearrangement counts, as it does not detect the rearrangements hidden by the assembly fragmentation of the considered genomes. Moreover, this underestimation can be biased by the degree of fragmentation of the compared species, so two numbers of rearrangements are not necessarily comparable in biological terms, even for species closely located in the species phylogeny. However, given the same genomes in the input, the numbers of rearrangements for two different phylogenies are comparable, as are the number of rearrangements in sex chromosomes and autosomes.

Results

The results are organized in three parts. First, we describe the results of the simulation-based evaluation of the accuracy of ADSEQ to recover extant scaffolding adjacencies. Then we describe extant and ancestral genomes obtained with ADSEQ, and analyze important aspects of their evolutionary history. Finally we use ADSEQ to evaluate different species trees and re-examine the conclusions of [5] regarding species evolution.

Validation of the ADSEQ algorithm for extant scaffolding

We compared the scaffolding performance of ADSEQ with two other methods, on the same data set of realistically fragmented genomes (see Methods). One is using sequence information only, BESST [63], while the other one is using the comparative approach only, on the same phylogenomic data, AD (ADSEQ, where the possibility of using sequence information is turned off). Figure 2 shows the precision statistic in function of the recall statistic for the two data sets (sampling 50% of the reads or using them all). For ADSEQ and AD, different values of the threshold for filtering adjacencies prior to linearization (see Methods) were tested.

ADSEQ outperforms BESST in precision and recall independently of the threshold. This shows that in the majority of cases, adding phylogenomic information improves the recall without affecting the precision compared to a method purely based on sequence data. Additional results, where the gene orientation is not considered to determine an inferred adjacency as TP, are provided in Additional file 9: Figure S8 et Additional file 10: Figure S9. These results show that BESST has an equivalent precision and recall statistics compared to results of Fig. 2 where gene orientation is considered. This comparison shows that for most of inferred adjacencies the three methods inferred the right orientation for both genes involved in adjacency. In summary the combined approach using comparative signal and sequence data (ADSEQ) is giving significantly better results than a method based on sequence alone (BESST).

We now compare ADSEQ and AD, for a better illustration of the Fig. 2 to compare precision and recall statistics between ADSEQ and AD see subfigures zooming in Additional file 11: Figure S10. Recall statistic is slightly better with ADSEQ than with AD for all considered threshold values. For most of the condition the precision statistic is slightly better ADSEQ than AD (except in some conditions for *A. albimanus* and *A. dirus*, see Additional file 11: Figure S10). For *A. albimanus*, with all reads considered, ADSEQ outperforms AD for recall statistic for a threshold fixed to 0.1 and 0.5. So from a quantitative point of view, adding sequence data seems to have a smaller impact on the recall and precision statistics compared to using synteny evolution. Note however that the combination of both supports for extant scaffolding adjacencies (sequence data and synteny evolution) is an important by-product of ADSEQ. A phylogenetic method alone is more difficult to trust in the absence of sequence data. So even if the general statistics are comparable, the additional support brought by the sequence data is an important feature. Moreover, additional results in Additional file 12: Figure S11 and Additional file 13: Figure S12 strongly support that a joint combination of phylogenetic and sequence signals (ADSEQ) overpasses an *a posteriori* combination of phylogenetic signal and sequence data (AD + BESST) for scaffolding improvement. These results show indeed that combining AD + BESST slightly overpasses ADSEQ in term of recall statistic (stronger TP adjacencies) but at the expense of a strong decrease of the precision.

Improved scaffolding of *Anopheles* extant and ancestral genomes

Properties of the improved assemblies for *Anopheles* extant genomes and of the reconstructed *Anopheles* ancestral genomes segments are summarized in Fig. 4. We describe three runs of ADSEQ: one without proposing extant scaffolding adjacencies (which amounts to use the DECLONE algorithm [51] to reconstruct ancestral genomes without improving extant genomes) and two with ADSEQ using the X and WG species phylogenies. The first observation that can be made is that the ability to create extant scaffolding adjacencies has a very significant impact on the ability to reconstruct ancestral segments,

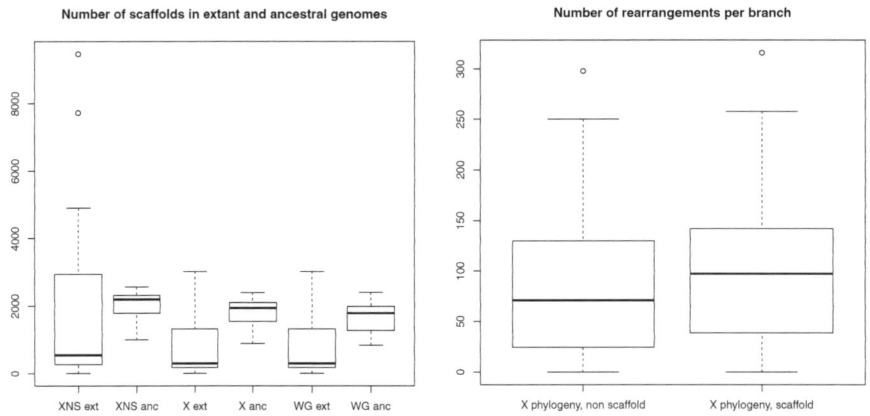

Fig. 4 Left: Number of segments in extant and ancestral genomes, according to three runs of ADSEQ in three different conditions. In the first run, we turn off the scaffolding mode on the X phylogeny, that is, it only constructs ancestral segments. The first column "XNS ext" thus describes the initial assembly, and "XNS anc" the assembly of ancestral genomes when reconstructed without extant scaffolding. In the second and third runs, the scaffolding mode was turned on, and run with the X phylogeny ("X ext" and "X anc") and the WG phylogeny ("WG ext" and "WG anc"). **Right:** Number of rearrangements over all branches of the X phylogeny, with and without the scaffolding mode

that define ancestral genomes at a similar level of fragmentation than the improved extant genomes. This effect is important toward refined genome evolution analysis that rely on the ancestral segments as input material, especially to detect chromosomal rearrangements.

Additional file 20: Table S4 and Additional file 14: Figure S13, Additional file 15: Figure S14, and Additional file 16: Figure S15, provide more detailed illustration and statistics on the improved scaffolding. We observe that from 36,634 initial extant segments (contigs, supercontigs and scaffolds after the various filterings steps described in Methods and SI text), we scaffold the extant genomes into 13,525 segments, with an average number of 94 genes per segment up from 37 before running ADSEQ. Very similar results are obtained for all genomes independently of the chosen phylogeny, confirming the overall picture described in Fig. 4-Left.

On the right side of Fig. 4, we can observe that we retrieve a significantly (Wilcoxon paired test, p-value $< 10^{-4}$) higher number of rearrangements by the joint scaffolding technique than by just constructing ancestral genomes without scaffolding extant genomes. So the joint scaffolding of extant and ancestral genomes is beneficial to both. In particular scaffolding extant genomes while reconstructing ancestral genomes gives access to more information regarding the evolutionary history.

An interesting feature of ADSEQ is the possibility that the linearization step does delete an observed gene adjacency in an extant genome. This is unlikely as observed adjacencies have the highest score in the linearization procedure, but it can happen if it is in conflict with other adjacencies with high posterior probability. It happened only once in our dataset, for an adjacency between two *A. culcifacies* genes. The two genes were predicted in the reverse order, or equivalently in the reverse orientation, because all identified homologs were arranged similarly. This can be explained either by two inversions, one encompassing each gene, or by an assembly or annotation error. This shows that our approach can also detect questionable adjacencies in the given extant assemblies.

Evolution and phylogeny

ADSEQ is not a phylogenetic method *per se*, as it requires a given species phylogeny and does not include an extension to search an optimal phylogeny according to some evolutionary criterion. However as a method which infers ancestral gene orders and evolutionary events, and is computationally efficient (all steps that require a species phylogeny, including the correction of gene trees with ProfileNJ, the reconciliation of the gene trees with the species trees, and the joint scaffolding / ancestral genome reconstruction takes five hours on a laptop), it can be used to compare a few selected competing phylogenies. To this aim we compared several measures obtained by the same methods using the two phylogenies, WG and X. They are the two main examined topologies in [5], where X is preferred as depicting the species history.

Duplications

Our pipeline using PROFILENJ to correct gene trees allows to record gene duplications. We counted a total of 6461 duplications for the X phylogeny, against 6159 duplications for the WG phylogeny (see Table 1). This means that for many gene families, a duplication was identified in the X phylogeny and not in the WG phylogeny. For these families, a well supported branch (100% bootstrap with RaxML) was compatible with the WG phylogeny but not with the WG phylogeny, indicating that well supported branches are more often compatible with the WG phylogeny. This supports the result of [5] that most genes follow the WG phylogeny. The fact that this is observed on the autosomes and not on the X phylogeny also supports that the genomes evolve with two compartments.

Scaffolding and ancestral genome reconstruction

On the left side of Fig. 4, we can observe a first difference between the results obtained with the X and the WG phylogenies: the extant scaffolding is slightly better (in terms of fragmentation level of the extant genomes) with the X phylogeny (mean segment number is 835 genes for the X phylogeny, versus 840 for the WG phylogeny), while the ancestral scaffolding is better with the WG phylogeny (mean segment number is 1860 for the X phylogeny, versus 1756 for the WG phylogeny). The better extant scaffolding with the X phylogeny can be attributed to the basal position of the genome with best assembly (*A. gambiae*) in the *Gambiae complex*. Indeed in ADSEQ the sister species can be assembled according to *A. gambiae*, but outgroup species cannot, so the assembly is necessarily better if a fully assembled genome has more sisters species and less outgroups as it is the case with *A. gambiae* in the X phylogeny. Interestingly ancestral genomes are better scaffolded with the WG phylogeny, even with this sister-branch artifact that concerns extant genomes. This better ancestral genome reconstruction obtained with the WG phylogeny could be considered as a first signal contradicting the hypothesis that the X phylogeny is the true species phylogeny, although it does not allow to draw definitive conclusions.

Table 1 Numbers of inferred rearrangements and duplications in the X chromosome and in the autosomes, according to the phylogeny (X or WG) used as a parameter of ADSEQ

Event	X phylogeny		WG phylogeny	
	X chr.	Autosomes	X chr.	Autosomes
Duplications	604	5857	606	5553
Rearrangements	415	2949	416	2760

Conflict

With both phylogenies we also measured the level of syntenic conflicts, defined as the sum of the posterior scores of the adjacencies discarded during the linearization phase (data shown in SI text). We observe a higher level of syntenic conflicts in the X phylogeny (7665) than in the WG phylogeny (6319). According to simulations (described in SI text), the level of conflict is higher with a wrong phylogeny, even if it is not with the same order of magnitude than what we observe on our data. This could be seen as a second element contradicting the sequence-based hypothesis that the X phylogeny is the true species phylogeny, although the high level of conflict observed with both phylogenies here again does not allow to draw reliable conclusions.

Gene movements (translocations)

As it is believed there have been no large-scale rearrangement between the X chromosome and the autosomes in the *Anopheles* history [4, 67], we could assign most extant and ancestral segments (at least almost all that contain more than one gene) either to the X chromosome or to an autosome, with high accuracy (see Methods). Then we identified which genes moved from the X chromosome to an autosome, or conversely, by screening all couples of direct ancestor/descendant genes, one being in a segment assigned to the X chromosome and the other to an autosome. We found 429 genes having moved from the X chromosome to an autosome, and 469 from an autosome to the X chromosome, which confirms the trend found by Neafsey et al. [4] (59 over 132 gene movements originated from the X chromosome), although we draw our conclusion from experiments using now many more genes than in [4].

Genome rearrangements

We now turn to detecting genome rearrangements, as defined in the Methods section. In particular, we stress that we look for breaks and gains of gener adjacencies due to genome rearrangements such as inversions, transpositions and translocations, excluding duplications and losses, as well as adjacency breaks and gains due to duplications and losses. Moreover, due to the fragmented nature of many ancestral genomes, we expect to underestimate the true number of synteny breaks and gains.

We detect 3364 gains or breakages of adjacencies (860 in the *Gambiae complex*) using the X phylogeny, and 3176 using the WG phylogeny (590 in the *Gambiae complex*). The difference is illustrated in Fig. 3. Between the two competing phylogenies, one can observe a 30% decrease in the number of rearrangements within the *Gambiae complex* with the WG phylogeny compared to the X phylogeny. These gains/breaks of adjacencies can be combined along each branch to detect inversions, defined by pairs of breaks in the ancestor and pairs of gains in the descendant involving the same four gene extremities. This lead to the identification of 242 inversions in the X phylogeny (including 16 inversions in the *Gambiae complex*, including 4 on the single lineage to *A. gambiae*) and 240 inversions with the WG phylogeny, with only 4 in the *Gambiae complex*, two of them on the branch leading to *A. gambiae*.

Comparison of sex chromosomes and autosomes evolution

Sex chromosome and autosomes have different evolutionary modes, according to duplications and rearrangements. Table 1 summarizes the number of inferred events of gene duplication and genome rearrangement in the sex chromosomes and autosomes, depending on the chosen phylogeny.

A striking observation is the different behavior of the X chromosome and of the autosomes regarding duplications and genome rearrangements. We do not count loss events to compare phylogenies because the absence of genes can be due to the fragmented assembly and not necessarily to actual gene losses during evolution. This compartmentalization was observed by [5] for genes and attributed to introgression in the autosomes; it was also noticed in [4] that the genome rearrangement rate was much higher in the X chromosome than in autosomes. We observe here a similar trend. We computed genome rearrangement rates by normalizing the number of observed gains and breaks of gene adjacencies by the number of gene adjacencies in the whole set of extant and ancestral genomes; with the X phylogeny we could observe that the X chromosome has a rearrangement rate equal to 1.46 times the rate observed in the autosomes, a figure that is similar (1.57) using the WG phylogeny. The observed higher reate of rearrangement in the X chromosome is in fact likely higher, as the relative fragmentation of the chromosome X is higher compared to the autosomes in most species both extant and ancestral (data not shown).

Moreover, we can observe interesting differences between the X and WG phylogenies. Constantly less events are found on the X chromosome with the X phylogeny, while less events are found on the autosomes with the WG phylogeny. It seems indeed that not only do genes follow different histories because of introgression [5], but also entire chromosomes do. However, the observed compartmentalization alone does not allow us to specify which part of the genome has followed the species diversification. As the *Gambiae complex* is estimated to be 2.2 million years old, it is reasonable to use parsimony arguments concerning rearrangements (see argument in the next paragraph). If we do so, we find less rearrangements in total in the WG phylogeny: even normalized by the number of adjacencies (because an increase in the number of rearrangements might be the effect of a higher number of adjacencies): 9.15 10^{-3} for the WG phylogeny versus 9.68 10^{-3} for the X phylogeny. This means that rearrangements do not yield the same phylogenetic signal than the one suggested in [5]), which puzzles the evolutionary scenario in the *Gambiae complex* [68].

Assessing the relevance of rearrangements parsimony

The fact that parsimony can give a good account on the phylogeny can be questioned. Indeed, rearrangements in *Anopheles* are not uniformly distributed [69], they can show some degree of convergence, and rearrangements can show inter-species polymorphism. To test whether in the *Gambiae complex* we are in the domain of validity of parsimony, we compared the gene order of *A. gambiae* with *A. albimanus*, which, following the recent improved assembly of *A. albimanus*, are the two genomes which have their genes assigned to chromosomes. We selected all genes with an assignment to a chromosome, and applied the EM2 distance estimator [70]. It is based on a non uniform model of genome rearrangements which has proved to give the most reliable results on mammalian genomes, whose evolution spans a similar amount of time than the *Anopheles* genomes. We found an estimation of 1313 inversions with the statistical estimator, while the parsimony solution was 1300 (data not shown). So the parsimony result is within in the 1% interval of the statistical method, far from saturation. As *A. gambiae* and *A. albimanus* are separated by approximately 79 million years of evolution, we may suppose that, in the 2 or 3 million years that have shaped the *Gambiae complex*, rearrangements were not numerous enough to contradict parsimony.

Discussion

An important contribution of our work is the unification of two domains of research, namely genome scaffolding, and the evolution of gene order and ancestral genome reconstruction. They are usually separated despite the similarity of their objectives (reconstructing ancestral gene order is akin to a scaffolding procedure if ancestral genes are considered as contigs). Our work improves on previous works (especially [37, 39, 71]) in several aspects. In particular, we integrate elements coming from more traditional phylogenetic methods, such as gene trees and reconciliations, in order to be able to handle a large gene complement that includes gene families with complex evolutionary histories. Another important aspect of our work is the validation procedure of the scaffolding method. We propose a novel simulation procedure which takes real sequence data but lowers the coverage and rely on a conservative contigs assembler to obtain a realistic fragmentation. Using this validation method, we show that combining both sequence data and comparison with related genomes in a phylogenetic context produce better scaffolds, at least in the context of *Anopheles* genomes.

The benefit of the joint approach that considers in the same framework scaffolding extant genomes and reconstructing ancestral genomes is evident from both the improved extant genomes assemblies, where we reduce the fragmentation from roughly 36,000 segments to below 14,000 segments, and the detection of genome rearrangements, where we observe again a much better resolution of ancestral genomes. This allows us to detect a statistically significantly larger number of genome rearrangements that can not be confused with assembly artifacts. To the best of our knowledge, ADSEQ is currently the only method that can process such a data set with many genomes, most of them provided with fragmented assemblies, while using a large complement of gene families without being limited by the nature of the evolution of these families in terms of duplications and losses, and using also sequencing data. Regarding extant genomes scaffolding, the quality of our results depends of a set of factors, such as the quality of the initial extant assemblies and the position in the species phylogeny; we do not gain much for well assembled species such as *A. albimanus* which is almost an outgroup, while we refine very well the assembly of the genomes of species such as *A. minimus* or *A. dirus*. It is important to note that while we rely on sequencing data in the present work, other sources of data such as genome maps for example could be used to define a prior score for scaffolding assemblies. In terms of genome rearrangements, we likely underestimates their actual number due to the fragmentation of the reconstructed ancestral genomes. This is a consequence the very conservative approach we follow that detects only rearrangements for which there is a clear support. It remains to see if more realistic models of genome rearrangements that do not rely on reconstructed ancestral gene orders would be able to cope, in terms of computational complexity and of robustness of the detected rearrangements, with both the large number of species considered here and the level of fragmentation of the extant genomes assemblies. Nevertheless, the results we obtain support strongly the observation of [4] that the X chromosome evolves by genome rearrangements at a much higher rate than the autosomes.

Conclusions

Genes sequences and nucleotide substitutions are questionable markers for phylogeny because of introgression, due to hybridization or horizontal transfer, as it has been several times demonstrated recently, and in particular on *Anopheles* species. Genome rearrangements can help decipher phylogenetic relationships. However, they are difficult to infer from genomes with fragmented assemblies, a common feature for genomes obtained from short sequencing reads. We introduce a computational and statistical method for the inference of genome rearrangement events in the presence of fragmented assemblies, proposing improved scaffolding based on sequencing data and an evolutionary model of gain/breakage of genes synteny along a given species phylogeny. We provide an evolutionary study of rearrangements, providing alternative hypotheses on the introgression process in the *Gambiae complex*.

Finally, our work opens the way to several research avenues. Generally, our general approach that relies on the joint analysis of sequencing data and the comparative approach to improve the quality of extant genome data could be extended to correct other types of errors that assembly breakpoints. To cite a specific example, it could be extended to account for the well known problem of unassembled genes [72], that create apparent gene loss and rearrangement breakpoints. Other avenues could include the development of metrics to compare alternative species phylogenies or the introduction in the evolutionary model of introgression events.

Additional files

Additional file 1: Supplementary text. (PDF 181 kb)

Additional file 2: Figure S1. Extant and ancestral genome gene content (left) and ancestral gene degree (right). Left: Number of genes of extant species (left), ancestral species using the reconciled VectorBase gene trees (middle), and ancestral species using the reconciled PROFILENJ gene trees (right). Right: Gene degree distribution of ancestral genes after applying ADSEQ with the RAW gene trees (blue graph) and the PROFILENJ gene trees (red graph), compared to the expected gene degree distribution for theoretical perfectly assembled genomes (black graph). The degree of a gene is defined as the sum of the ADSEQ *posterior* scores of adjacencies involving this gene. Here the value at coordinate x is the sums of all degrees in the interval $[x, x+1]$. (PDF 92 kb)

Additional file 3: Figure S2. The pipeline is split into two parts: The first part (Blue one) processes genome content data to obtain extant genome adjacencies. **Step 1** detects genes that are included in other genes. In **Step 2**, gene families containing these genes are filtered out to avoid ambiguity in defining observed extant gene adjacencies. In **Step 3**, gene trees are inferred from the gene sequences, one gene tree per gene family (see Additional file 5: Figure S4 for more information on gene trees inference pipeline). Finally, in **Step 4** genes contained in gene families for which gene trees have not been inferred are discarded from the analysis (41 gene families containing representing 1,039 genes). The second part of the pipeline (Green one) processes sequencing data to obtain scaffolding adjacencies that will be used to improve extant genome assembly with the ADSEQ algorithm. **Step A** trims reads with TRIMMOMATIC to remove low qualities reads and remaining adapters. In **Step B**, trimmed reads are mapped onto their respective genome with BOWTIE2 considering all multiple mappings. In **Step C**, pairs of contigs for which paired-end reads suggest a possible contiguity along their chromosome are linked with the scaffolding software BESST, and the resulting potential scaffolding adjacencies are scored according to the BESST model. Then, in **Step 5** scaffolding gene adjacencies are determined from contigs adjacencies obtained from sequencing data processing part with genes present in gene trees. This results in scaffolds with observed scored scaffolding adjacencies that are used as input of DECOSTAR (**Step 6**). See Additional file 18: Table S2 for a description per species on dataset used for DECOSTAR. Pipeline to produce input data for ADSEQ on 18 *Anopheles* dataset. The pipeline takes as input a species tree for the 18 *Anopheles* species, the whole set of gene families and gene trees for these species and genomic data (contigs, scaffolds and chromosomes). The goal of the pipeline is to produce input data for the ADSEQ algorithm to reconstruct ancestral genome structure and evolution and to improve extant genome scaffolding. (PDF 103 kb)

Additional file 4: Figure S3. Distribution of gene families number (red bars) and number of genes (blue bars) per families containing x species. **Left graph:** distribution of the 17,780 raw input gene trees corresponding to 212,800 genes. **Middle graph:** distribution of the 14,981 gene families, containing 184,719 genes, after discarding families containing included genes (after step 2 of Additional file 3: Figure S2). **Right graph:** distribution of the 14,940 gene trees, composed of 183,680 genes, after gene trees inference pipeline (after steps 3 and 4 of Additional file 5: Figure S4). (PDF 83 kb)

Additional file 5: Figure S4. Pipeline to improve gene trees inference from homologous gene family. CDS sequences of genes in gene trees have been obtained from VectorBase database and homologous gene families deduced from the 14,981 gene trees resulting of step 2 of Additional file 5: Figure S4. **Step A** consists to multiple align homologous genes with MUSCLE with parameter "-maxiters 2" if a gene sequence have a size upper than 32,000 bp. In **Step B**, GBLOCKS was applied on alignments to select high confidence alignment sites. At this step, 41 gene families have been discarded due to sequences that were not present in a selected blocks. For **Step C**, RAxML have been used to infer maximum likelihood gene trees with the GTR-GAMMA model and 100 bootstrap iterations. Finally in **Step D**, the maximum likelihood gene trees are processed with PROFILENJ to potentially changing branches with bootstrap support lower than 100% in a DL reconciliation model (min(Duplication,Loss)) with the species tree [73]. (PDF 35 kb)

Additional file 6: Figure S5. Distributions of scaffolding adjacencies scores computed by BESST for scaffolding adjacencies supported by at least 3 paired reads. **Left graph:** adjacency scores distribution between all contigs or scaffolds, over 405,939 scaffolding adjacencies. **Right graph:** adjacency scores distribution for contigs and scaffolds with gene corresponding to the 68,876 scaffolding gene adjacencies considered by DECOSTAR. Blue bars represent the link variation score, red bars the link dispersity score and purple bars the mean of the two link scores. For more information on the link scores see SI text and [62]. (PDF 37 kb)

Additional file 7: Figure S6. Distributions of scaffolding adjacencies link scores computed by BESST for scaffolding adjacencies supported by at least 3 paired reads, for each of the 18 *Anopheles* species. **Upper graphs:** distribution of scores all 405,939 potential scaffolding adjacencies. **Lower graphs:** distribution of scores for all 68,876 scaffolding gene adjacencies used as input by DECOSTAR. **Left graphs:** distribution of link dispersity scores. **Middle graphs:** distribution of link variation scores. **Right graphs:** distribution of the mean of link variation and dispersity scores. Each color corresponds to one species and the number between parenthesis in the legend indicates the number of scaffolding adjacencies inferred by BESST for each species. (PDF 64 kb)

Additional file 8: Figure S7. The ADSEQ validation protocol. (PDF 61 kb)

Additional file 9: Figure S8. Precision and recall statistics for scaffolding adjacencies on three artificially fragmented genomes (A.alb: *Anopheles albimanus*, A.ara: *Anopheles arabiensis* and A.dir: *Anopheles dirus*), when gene orientations are not accounted for. **Left graph:** results with 50% of reads. **Right graph:** results with all reads. The different methods results are plotted with the precision on the Y axis and the recall on the X axis. For ADSEQ and AD, results for three different adjacency support threshold (0.1, 0.5 and 0.8) before genome linearization are plotted and represented with a color gradient. These results show similar results to Fig. 2 showing that for most of the predicted adjacencies the three methods infer the correct gene orientation. (PDF 106 kb)

Additional file 10: Figure S9. Subfigures zooming of Additional file 9: Figure S8 to compare precision and recall statistics between ADSEQ and AD. **Upper graphs:** zoom of results with 50% of reads. **Lower graphs:** results with all reads. (PDF 154 kb)

Additional file 11: Figure S10. Subfigures zooming of Additional file 3: Figure S2 to compare precision and recall statistics between ADSEQ and AD. **Upper graphs:** zoom of results with 50% of reads. **Lower graphs:** results with all reads. (PDF 136 kb)

Additional file 12: Figure S11. Venn diagrams showing adjacencies shared by the three scaffolding methods ADSEQ, AD and BESST with a sample of 50% of the reads. **Upper Venn diagrams:** results for *Anopheles albimanus*. **Middle Venn diagrams:** results for *Anopheles arabiensis*. **Lower Venn diagrams:** results for *Anopheles dirus*. **Left diagrams:** False Negative (FN) adjacencies, corresponding to adjacencies created by the fragmentation process and that have not been recovered. **Center diagrams:** results for True Positive (TP) adjacencies. Here, an adjacency is considered TP if the pair of genes is adjacent in the reference assembly and the orientation of genes involved in the adjacency is properly recovered. **Right diagrams:** results for Certain False Positive (CFP) adjacencies. An adjacency is determined as CFP when the pair of gene does not belong to the reference assemblies and one of the two genes is not located at a contig extremity in reference genome, or if the recovered orientation of genes is incorrect. If we consider method individually, these results show that ADSEQ outperforms AD and BESST with the lowest number of FN adjacencies, the largest number of TP adjacencies and the lowest number of CFP adjacencies (except for *An. albimanus* where AD has the lowest number of CFP (224 vs. 231 for ADSEQ). However, if we combine *a posteriori* AD and BESST, this performs better than ADSEQ in terms of recall (higher number of TP adjacencies) but at the expense of a strong decreases of precision (much higher number of CFP adjacencies). (PDF 20 kb)

Additional file 13: Figure S12. Similar to Additional file 12: Figure S11 with all reads included. (PDF 20 kb)

Additional file 14: Figure S13. Scatter plot exhibiting scaffolding improvement of the 18 *Anopheles* genomes by ADSEQ with X species tree phylogeny. Right plot is a zoom of a small part of the left graph. Each color corresponds to one species. For each species, upper part of vertical line corresponds to number of segments in initial genome assembly and lower part the number of segments after scaffolding improvement by ADSEQ. Circle diameter is proportional to the % of scaffolding improvement of the genome where scale is displayed in lower right part of the graphs. The X axis represent the sum of *a posteriori* scores of discarded adjacencies during linearization step representing the degree of syntenic conflicts in adjacencies prediction of ADSEQ (see paragraph "Conflict" in section "Results"). (PDF 56 kb)

Additional file 15: Figure S14. Similar to Additional file 14: Figure S13 with WG species tree phylogeny. (PDF 56 kb)

Additional file 16: Figure S15. Similar to Additional file 14: Figure S13 with RAW gene trees instead of PROFILENJ gene trees. (PDF 48 kb)

Additional file 17: Table S1. Summary of genome assemblies and sequencing data information. 16 on the 18 *Anopheles* species have been sequenced in [4] and data are available on the SRA database of the NCBI (see column 4 and 6 for BioProject and SRA ID). FASTQ files of paired sequencing data have been obtained with SRA-TOOLKIT. After mapping of paired reads on reference genome assemblies (column 2), median insert size of libraries have been determined with package "CollectInsertSizeMetrics" of PICARD TOOLS (v1.61) (column 7). Column "Library name" give information on the sequencing strategies employed in [4]. Where 'fragment' library corresponds to a Paired-End library with an expected insert size of 180bp and FR orientation (→ ←). 'jump' library corresponds to a Mate-Pair library with an insert size of 1.5kbp and RF orientation (←→). And 'fosill' corresponds to a library generated from a pool of hundred mosquitoes to improve the scaffolding with an expected insert size around 38kbp and FR orientation (→ ←). Column 3 gives the ID of gene set used in this study. (PDF 38 kb)

Additional file 18: Table S2. Assembly statistics on the 18 *Anopheles* genomes. Statistics before before processing are displayed in columns 2-7 and after the pipeline to produce input data for the DECOSTAR algorithm in columns 8-11 (see Additional file 3: Figure S2 for illustration of the data preprocessing step). For initial dataset assembly statistics, columns 2 and 3 present contigs number and N50 statistic in bp for all contigs in genome assemblies. In columns 4-7, only contigs with at least one gene are considered. Column 4 corresponds to contigs number with gene in reference assemblies. Columns 5 & 6 represent N50 statistics respectively in bp and in gene number. Column 7 represent the number of gene in reference genome assemblies. For genome assemblies used as input of DECOSTAR, all contigs contains at least one gene. Column 8 gives the number of contigs after step 4 of Additional file 3: Figure S2. Columns 9 & 10 represent N50 statistics respectively in bp and in gene number. And column 11 represents the number of gene in genomes taken as input of DECOSTAR. The input dataset of DECOSTAR is composed of 14,940 gene trees (see Additional file 4: Figure S3 and Additional file 5: Figure S4 for more information on gene trees) and 68,876 gene adjacencies with sequence support (scaffolding gene adjacencies) (see Additional file 6: Figure S5 and Additional file 7: Figure S6 for more information on scaffolding adjacencies). (PDF 33 kb)

Additional file 19: Table S3. Assembly statistics at various stages of the gene annotation step of validation protocol of ADSEQ (step 4 of Additional file 8: Figure S7). For each species and annotation step, table gives different assembly statistics (column2): the number of contigs in the assembly, the size of the assembly in bp and in gene number, the N50 statistics of the assembly in bp and in gene number (if available) and gene trees number corresponding to gene present in the assembly (for the two last columns). Column 3 (Initial assembly) corresponds to the assembly statistics of reference genomes. Columns 4-8 of upper and lower table corresponds to MINIA assembly statistics at different filtering step respectively with 50% reads sampling and without reads sampling. Column "initial" corresponds to assembly in output of MINIA algorithm assembly. MINIA contigs are then mapped on reference genome to annotate gene of reference assembly on MINIA contigs. Assembly statistics after filter1 corresponds to MINIA contigs that have been mapped on reference assembly with an identity and a coverage >= 90%. Filter2 consists to keep only contig with an unique optimal alignment (to avoid uncertainty in gene annotation). Column 7 corresponds to MINIA assembly statistics after merging of MINIA contigs overlapping a same gene (simulating RNA-seq scaffolding). Then last column corresponds to statistics after filter3 that consists to discard gene families of genes that have not been mapped on MINIA contigs. (PDF 51 kb)

Additional file 20: Table S4. Scaffolding statistics on the 18 *Anopheles* genomes before and after ADSEQ (with the X (upper table) and WG (lower table) species phylogenies). The columns 2-5 correspond to assemblies statistics before running the ADSEQ algorithm. Column 2 corresponds to the number of contigs in reference assemblies. The N50 statistic corresponding to the contig size where 50% of the total assembly length is comprised in contigs with size superior or equal to this value. This metric is computed with size considerd both in bp (in column 3) and in gene number (in column 4). Column 5 gives the number of genes in genome assemblies give as input to ADSEQ. Columns 6-9 and 10-13 represent scaffolding statistics of ADSEQ respectively for X chromosome species tree topology and Whole-Genome topology. Columns 6 & 10 represent scaffolds number after ADSEQ. Columns 7 & 11, and 8 & 12 represent N50 statistics respectively for size in bp and size in gene number. Columns 9 & 13 represent new adjacencies inferred by ADSEQ (#scaff adj) represent the number of new adjacencies that are scaffolding adjacencies (i.e. adjacencies with sequence signal proposed by BESST and inferred by ADSEQ). (PDF 56 kb)

Abbreviations

CARs: Contiguous ancestral regions; CDS: Coding DNA sequence; CFP: Certain false positive; DP: Dynamic programming; FN: False negative; FP: False positive; MWM: Maximum-weight matching; SRA: Sequence read archive; TGS: Third-generation sequencing; TP: True positive; WG: Whole genome

Acknowledgements

Data computation and analyses benefited from the Compute Canada Western Canada Research Grid (Westgrid) and Montpellier Bioinformatics Biodiversity (MBB) platform services.

Funding

This work is funded by the Agence Nationale pour la Recherche, Ancestrome project ANR-10-BINF-01-01, the MITACS Globalink program and the Natural Sciences and Engineering Research Council of Canada (NSERC), Discovery Grant RGPIN-249834 and a resource allocation from Compute Canada. Funding for the publication of this article was provided by the Agence Nationale pour la Recherche, Ancestrome project ANR-10-BINF-01-01.

Availability of data and materials

Data and scripts used in this study are available on the GitHub repository: https://github.com/YoannAnselmetti/ADseq-Anopheles-APBC2018.

About this supplement
This article has been published as part of *BMC Genomics* Volume 19 Supplement 2, 2018: Selected articles from the 16th Asia Pacific Bioinformatics Conference (APBC 2018): genomics. The full contents of the supplement are available online at https://bmcgenomics.biomedcentral.com/articles/supplements/volume-19-supplement-2.

Authors' contributions
ET, CC and SB designed the study. YA and WD implemented the method. YA applied the method to data. YA, ET, CC and SB analyzed the results and wrote the manuscript. All authors read and approved the manuscript.

Ethics approval and consent to participate
Not applicable.

Consent for publication
Not applicable.

Competing interests
The authors declare that they have no competing interests.

Publisher's Note

Author details
[1]ISEM, Université de Montpellier, CNRS, IRD, EPHE, Montpellier, France. [2]Univ Lyon, Université Lyon 1, CNRS, Laboratoire de Biométrie et Biologie Evolutive UMR5558, 43 Boulevard du 11 novembre 1918, 69622 Villeurbanne cedex, France. [3]INRIA Grenoble - Rhône-Alpes, 655 Avenue de l'Europe, 38330 Montbonnot-Saint-Martin, France. [4]Department of Mathematics, Simon Fraser University, 8888 University Drive, V5A1S6 Burnaby, BC, Canada.

Published: 9 May 2018

References

1. Zuckerkandl E, Pauling L. Molecules as documents of evolutionary history. J Theor Biol. 1965;8(2):357–66.
2. Springer MS, Gatesy J. The gene tree delusion. Mol Phylogenet Evol. 2016;94:1–33.
3. Doolittle WF, Brunet TDP. What Is the Tree of Life? PLOS Genet. 2016;12(4):e1005912.
4. Neafsey DE, Waterhouse RM, Abai MR, Aganezov SS, Alekseyev MA, Allen JE, et al. Mosquito genomics. Highly evolvable malaria vectors: the genomes of 16 Anopheles mosquitoes. Science. 2015;347(6217):1258522.
5. Fontaine MC, Pease JB, Steele A, Waterhouse RM, Neafsey DE, Sharakhov IV, et al. Mosquito genomics. Extensive introgression in a malaria vector species complex revealed by phylogenomics. Science. 2015;347(6217):1258524.
6. Wen D, Yu Y, Hahn MW, Nakhleh L. Reticulate evolutionary history and extensive introgression in mosquito species revealed by phylogenetic network analysis. Mol Ecol. 2016;25:2361–72.
7. Dobzhansky T, Sturtevant AH. Inversions in the Chromosomes of Drosophila Pseudoobscura. Genetics. 1938;23(1):28–64.
8. Kamali M, Xia A, Tu Z, Sharakhov IV. A new chromosomal phylogeny supports the repeated origin of vectorial capacity in malaria mosquitoes of the Anopheles gambiae complex. PLoS Pathog. 2012;8:e1002960.
9. Sharakhov IV. Chromosome phylogenies of malaria mosquitoes. Tsitologiia. 2013;55:238–40.
10. della Torre A, Merzagora L, Powell JR, Coluzzi M. Selective introgression of paracentric inversions between two sibling species of the Anopheles gambiae complex. Genetics. 1997;146:239–44.
11. Ayala D, Ullastres A, González J. Adaptation through chromosomal inversions in Anopheles. Front Genet. 2014;5:129.
12. Ayala D, Acevedo P, Pombi M, Dia I, Boccolini D, Costantini C, et al. Chromosome inversions and ecological plasticity in the main African malaria mosquitoes. Evolution. 2017;71(3):686–701.
13. Love RR, Steele AM, Coulibaly MB, Traore SF, Emrich SJ, Fontaine MC, et al. Chromosomal inversions and ecotypic differentiation in *Anopheles gambiae*: the perspective from whole-genome sequencing. Mol Ecol. 2016;25(23):5889–906.
14. Tang J, Moret BME. Scaling up accurate phylogenetic reconstruction from gene-order data. Bioinformatics. 2003;19(Suppl 1):i305–i312.
15. Larget B, Simon D, Kadane J. Bayesian phylogenetic inference from animal mitochondrial genome arrangements. J R Stat Soc Ser B. 2002;64(4):681–693.
16. Avdeyev P, Jiang S, Aganezov S, Hu F, Alekseyev MA. Reconstruction of Ancestral Genomes in Presence of Gene Gain and Loss. J Comput Biol. 2016;23(3):150–64.
17. Hu F, Lin Y, Tang J. MLGO: phylogeny reconstruction and ancestral inference from gene-order data. BMC Bioinformatics. 2014;15:354.
18. Zheng C, Sankoff D. On the PATHGROUPS approach to rapid small phylogeny. BMC Bioinformatics. 2011;12(S-1):S4.
19. Bradley R, Jones ET, Rajaraman A, Chauve C. ANGES: reconstructing ANcestral GEnomeS maps. Bioinformatics. 2012;28(18):2388–90.
20. Kim J, Farré M, Auvil L, Capitanu B, Larkin DM, Ma J, et al. Reconstruction and evolutionary history of eutherian chromosomes. Proc Natl Acad Sci. 2017;114(27):E5379–E5388.
21. Nagarajan N, Pop M. Sequence assembly demystified. Nat Rev Genet. 2013;14(3):157–67.
22. Treangen TJ, Salzberg SL. Repetitive DNA and next-generation sequencing: computational challenges and solutions. Nat Rev Genet. 2011;13(1):36–46.
23. Berlin K, Koren S, Chin CS, Drake JP, Landolin JM, Phillippy AM. Assembling large genomes with single-molecule sequencing and locality-sensitive hashing. Nat Biotechnol. 2015;33(6):623–30.
24. Dudchenko O, Batra SS, Omer AD, Nyquist SK, Hoeger M, Durand NC, et al. De novo assembly of the *Aedes aegypti* genome using Hi-C yields chromosome-length scaffolds. Science. 2017. Epub ahead of print. https://doi.org/10.1126/science.aal3327.
25. Putnam NH, O'Connell BL, Stites JC, Rice BJ, Blanchette M, Calef R, et al. Chromosome-scale shotgun assembly using an in vitro method for long-range linkage. Genome Res. 2016;26(3):342–50.
26. Lewin HA, Larkin DM, Pontius J, O'Brien SJ. Every genome sequence needs a good map. Genome Res. 2009;19(11):1925–8.
27. Tang H, Zhang X, Miao C, Zhang J, Ming R, Schnable JC, et al. ALLMAPS: robust scaffold ordering based on multiple maps. Genome Biol. 2015;16(1):3+.
28. Fierst JL. Using linkage maps to correct and scaffold de novo genome assemblies: methods, challenges, and computational tools. Front Genet. 2015;6:220.
29. Assefa SA, Keane TM, Otto TD, Newbold C, Berriman M. ABACAS: algorithm-based automatic contiguation of assembled sequences. Bioinformatics. 2009;25(15):1968–9.
30. Rissman AI, Mau B, Biehl BS, Darling AE, Glasner JD, Perna NT. Reordering contigs of draft genomes using the Mauve Aligner. Bioinformatics. 2009;25(16):2071–3.
31. Dias Z, Dias U, Setubal JC. SIS: a program to generate draft genome sequence scaffolds for prokaryotes. BMC Bioinformatics. 2012;13:96.
32. Bao E, Jiang T, Girke T. AlignGraph: algorithm for secondary de novo genome assembly guided by closely related references. Bioinformatics. 2014;30(12):319–28.
33. Lu CL, Chen K, Huang S, Chiu H. CAR: contig assembly of prokaryotic draft genomes using rearrangements. BMC Bioinformatics. 2014;15:381.
34. Tamazian G, Dobrynin P, Krasheninnikova K, Komissarov A, Koepfli KP, O'Brien SJ. Chromosomer: a reference-based genome arrangement tool for producing draft chromosome sequences. GigaScience. 2016;5(1):38.
35. Shaik S, Kumar N, Lankapalli AK, Tiwari SK, Baddam R, Ahmed N. Contig-Layout-Authenticator (CLA): A Combinatorial Approach to Ordering and Scaffolding of Bacterial Contigs for Comparative Genomics and Molecular Epidemiology. PLoS ONE. 2016;11(6):1–19.
36. Husemann P, Stoye J. Phylogenetic comparative assembly. Algoritm Mol Biol. 2010;5:3.
37. Kim J, Larkin DM, Asan CQ, Zhang Y, Ge RL, et al. Reference-assisted chromosome assembly. Proc Natl Acad Sci. 2013;110(5):1785–90.
38. Kolmogorov M, Raney BJ, Paten B, Pham SK. Ragout - a reference-assisted assembly tool for bacterial genomes. Bioinformatics. 2014;30(12):302–9.
39. Aganezov S, Alekseyev MA. In: Bourgeois A, Skums P, Wan X, Zelikovsky A, editors. Multigenome scaffold co-assembly based on the analysis of gene orders and genomic repeats. vol. 9683. Cham: Springer International Publishing; 2016. p. 237–249.
40. Bosi E, Donati B, Galardini M, Brunetti S, Sagot M, Lió P, et al. MeDuSa: a multi-draft based scaffolder. Bioinformatics. 2015;31(15):2443–51.

41. Sharakhov IV, Serazin AC, Grushko OG, Dana A, Lobo N, Hillenmeyer ME, et al. Inversions and gene order shuffling in Anopheles gambiae and A. funestus. Science. 2002;298:182–5.
42. Wang-Sattler R, Blandin S, Ning Y, Blass C, Dolo G, Touré YT, et al. Mosaic genome architecture of the Anopheles gambiae species complex. PLoS ONE. 2007;2(11):e1249.
43. Coluzzi M, Sabatini A, della Torre A, Di Deco MA, Petrarca V. A polytene chromosome analysis of the Anopheles gambiae species complex. Science. 2002;298:1415–8.
44. Caccone A, Min GS, Powell JR. Multiple origins of cytologically identical chromosome inversions in the Anopheles gambiae complex. Genetics. 1998;150:807–14.
45. Gentile G, Slotman M, Ketmaier V, Powell JR, Caccone A. Attempts to molecularly distinguish cryptic taxa in Anopheles gambiae s.s. Insect Mol Biol. 2001;10:25–32.
46. Mathiopoulos KD, della Torre A, Santolamazza F, Predazzi V, Petrarca V, Coluzzi M. Are chromosomal inversions induced by transposable elements? A paradigm from the malaria mosquito Anopheles gambiae. Parassitologia. 1999;41:119–23.
47. Appawu MA, Baffoe-Wilmot A, Afari EA, Nkrumah FK, Petrarca V. Species composition and inversion polymorphism of the Anopheles gambiae complex in some sites of Ghana, west Africa. Acta Trop. 1994;56:15–23.
48. Ma J, Zhang L, Suh BB, Raney BJ, Burhans RC, Kent WJ, et al. Reconstructing contiguous regions of an ancestral genome. Genome Res. 2006;16(12):1557–65.
49. Chauve C, Tannier E. A methodological framework for the reconstruction of contiguous regions of ancestral genomes and its application to mammalian genomes. PLoS Comput Biol. 2008;4(11):e1000234.
50. Bérard S, Gallien C, Boussau B, Szöllosi GJ, Daubin V, Tannier E. Evolution of gene neighborhoods within reconciled phylogenies. Bioinformatics. 2012;28(18):i382–i388.
51. Chauve C, Ponty Y, Zanetti JPP. Evolution of genes neighborhood within reconciled phylogenies: an ensemble approach. BMC Bioinformatics. 2015;16(Suppl 19):S6.
52. Anselmetti Y, Berry V, Chauve C, Chateau A, Tannier E, Bérard S. Ancestral gene synteny reconstruction improves extant species scaffolding. BMC Genomics. 2015;16(Suppl 10):S11.
53. Duchemin W, Anselmetti Y, Patterson M, Ponty Y, Berard S, Chauve C, et al. DeCoSTAR: Reconstructing the ancestral organization of genes or genomes using reconciled phylogenies. Genome Biol Evol. 2017;9(5): 1312–1319.
54. Jacox E, Chauve C, Szöllosi GJ, Ponty Y, Scornavacca C. ecceTERA: comprehensive gene tree-species tree reconciliation using parsimony. Bioinformatics. 2016;32(13):2056–2058.
55. Manuch J, Patterson M, Wittler R, Chauve C, Tannier E. Linearization of ancestral multichromosomal genomes. BMC Bioinformatics. 2012;13(Suppl 19):S11.
56. Mandric I, Zelikovsky A. ScaffMatch: scaffolding algorithm based on maximum weight matching. Bioinformatics. 2015;31(16):2632–8.
57. Kriventseva EV, Tegenfeldt F, Petty TJ, Waterhouse RM, Pozdnyakov IA, Ioannidis P, et al. OrthoDB v8 : update of the hierarchical catalog of orthologs and the underlying free software. Nucleic Acids Res. 2015;43(Database issue):250–6.
58. Stamatakis A. RAxML-VI-HPC: Maximum likelihood-based phylogenetic analysis with thousands of taxa and mixed models. Bioinformatics. 2006;22:2688–90.
59. Noutahi E, Semeria M, Lafond M, Seguin J, Boussau B, Guéguen L, et al. Efficient gene tree correction guided by genome evolution. PLoS ONE. 2016;11(8):e0159559.
60. Bolger AM, Lohse M, Usadel B. Trimmomatic: a flexible trimmer for Illumina sequence data. Bioinformatics. 30;2014:2114.
61. Langmead B, Salzberg SL. Fast gapped-read alignment with Bowtie 2. Nat Methods. 2012;9(4):357–9.
62. Sahlin K, Vezzi F, Nystedt B, Lundeberg J, Arvestad L. BESST - Efficient scaffolding of large fragmented assemblies. BMC Bioinformatics. 2014;15(1):281.
63. Sahlin K, Chikhi R, Arvestad L, Science C. Genome scaffolding with PE-contaminated mate-pair libraries. bioRxiv preprint. 2015;1–13.
64. Chikhi R, Medvedev P. Informed and automated k-mer size selection for genome assembly. Bioinformatics. 2014;30(1):31–7.
65. Chikhi R, Rizk G. Space-efficient and exact de Bruijn graph representation based on a Bloom filter. Algoritm Mol Biol. 2013;8(1):22+.
66. Artemov GN, Peery AN, Jiang X, Tu Z, Stegniy VN, Sharakhova MV, et al. The Physical Genome Mapping of Anopheles albimanus Corrected Scaffold Misassemblies and Identified Interarm Rearrangements in Genus Anopheles. G3 Genes Genomes Genet. 2017;7(1):155–64.
67. Sharakhov IV, Artemov GN, Sharakhova MV. Chromosome evolution in malaria mosquitoes inferred from physically mapped genome assemblies. J Bioinforma Comput Biol. 2016;14(2):1630003.
68. Clark AG, Messer PW. Conundrum of jumbled mosquito genomes. Science. 2015;6217(347):27–8.
69. Pombi M, Caputo B, Simard F, Di Deco MA, Coluzzi M, della Torre A, et al. Chromosomal plasticity and evolutionary potential in the malaria vector Anopheles gambiae sensu stricto: insights from three decades of rare paracentric inversions. BMC Evol Biol. 2008;8:309.
70. Biller P, Guéguen L, Knibbe C, Tannier E. Breaking good: accounting for the diversity of fragile regions for estimating rearrangement distances. Genome Biol Evol. 2016;8:1427–39.
71. Lin Y, Nurk S, Pevzner PA. What is the difference between the breakpoint graph and the de Bruijn graph? BMC Genomics. 2014;15(Suppl 6):S6.
72. Denton JF, Lugo-Martinez J, Tucker AE, Schrider DR, Warren WC, Hahn MW. Extensive Error in the Number of Genes Inferred from Draft Genome Assemblies. PLoS Comput Biol. 2014;10(12):e1003998+.
73. Goodman M, Czelusniak J, Moore GW, Romero-Herrera AE, Matsuda G. Fitting the gene lineage into its species lineage, a parsimony strategy illustrated dy cladograms constructed from globin sequences. Syst Zool. 1979;28(2):132–63.

Choi *et al. BMC Genomics* 2018, **19**(Suppl 2):101
DOI 10.1186/s12864-018-4467-6

BMC Genomics

RESEARCH Open Access

Identification of usual interstitial pneumonia pattern using RNA-Seq and machine learning: challenges and solutions

Yoonha Choi[1], Tiffany Ting Liu[1], Daniel G. Pankratz[1], Thomas V. Colby[2], Neil M. Barth[1], David A. Lynch[3], P. Sean Walsh[1], Ganesh Raghu[4], Giulia C. Kennedy[1] and Jing Huang[1*]

From The Sixteenth Asia Pacific Bioinformatics Conference 2018
Yokohama, Japan. 15-17 January 2018

Abstract

Background: We developed a classifier using RNA sequencing data that identifies the usual interstitial pneumonia (UIP) pattern for the diagnosis of idiopathic pulmonary fibrosis. We addressed significant challenges, including limited sample size, biological and technical sample heterogeneity, and reagent and assay batch effects.

Results: We identified inter- and intra-patient heterogeneity, particularly within the non-UIP group. The models classified UIP on transbronchial biopsy samples with a receiver-operating characteristic area under the curve of ~ 0.9 in cross-validation. Using in silico mixed samples in training, we prospectively defined a decision boundary to optimize specificity at ≥85%. The penalized logistic regression model showed greater reproducibility across technical replicates and was chosen as the final model. The final model showed sensitivity of 70% and specificity of 88% in the test set.

Conclusions: We demonstrated that the suggested methodologies appropriately addressed challenges of the sample size, disease heterogeneity and technical batch effects and developed a highly accurate and robust classifier leveraging RNA sequencing for the classification of UIP.

Background

Interstitial lung disease (ILD) consists of a group of diseases affecting the pulmonary interstitium with similar clinical presentation; idiopathic pulmonary fibrosis (IPF) is the most common ILD with the worst prognosis. By definition, the cause of IPF is unknown; this can make an accurate, confident and timely diagnosis challenging. An accurate diagnosis for IPF requires multidisciplinary evaluation of clinical, radiologic and histopathologic features [1, 2], with patients frequently subject to an uncertain and lengthy diagnostic process. In particular, determining the presence of usual interstitial pneumonia (UIP), a hallmark characteristic of IPF, often requires histopathology via invasive surgery that may not be an option for sick or elderly patients. Furthermore, the quality of the histopathology reading is highly variable across clinics, and access to expert pathology review may be limited [3]. Thus, a consistent, accurate, non-invasive diagnostic tool to distinguish UIP from non-UIP without the need for surgery is critical to reduce patient suffering and enable physicians to reach earlier, confident clinical diagnoses and better treatment decisions.

To build this new diagnostic tool, we used exome-enriched RNA sequencing data from transbronchial biopsy samples (TBBs) collected via bronchoscopy, a less invasive procedure compared to surgery. Several studies have revealed that genomic information in transcriptomic data is indicative of phenotypic variation in cancer and other chronic disease [4, 5]. Complex traits are driven by large numbers of genes distributed across the genome, including those with no apparent relevance to disease [6]. More importantly, the feasibility of recapitulating the UIP pattern

* Correspondence: jing@veracyte.com
[1]Veracyte, Inc, 6000 Shoreline Court, Suite 300, South San Francisco, CA 94080, USA
Full list of author information is available at the end of the article

using transcriptomic data has been established [7]. Here we describe the algorithmic challenges we encountered and the analytical solutions we developed to address them.

Machine learning methods have been applied extensively to solve biomedical problems, and have deepened our understanding of diseases such as breast cancer [8] and glioblastoma [9] by allowing researchers to construct biological pathways, identify biologically relevant phenotypes and better predict disease risk. However, recent advances in machine learning are often designed for large (e.g. at least thousands of samples) data sets such as medical imaging data and sequential data [10, 11]. Clinical studies often have limited sample sizes due to the challenges in accruing patients to clinical trials. The issue is compounded in our study since many patients are too medically fragile to undergo a surgical biopsy. Further, among the ones collected, a substantial proportion yielded non-diagnostic results, rendering them unsuitable for supervised learning.

Inter- and intra-patient heterogeneity adds significant complexity to classification. The non-UIP category is not one disease, but a collection of heterogeneous diseases with a wide range of phenotypes, all of which are encountered in the clinic. This, coupled with the small sample size, resulted in limited numbers of samples in each non-UIP disease category. Another unique feature of this study is heterogeneity within a patient: histopathology features are not uniform across the entire lung. Not surprisingly, genomic signatures also vary depending on the location of the biopsy sample [12]. To better understand heterogeneity, multiple samples (up to 5) per patient were collected and sequenced separately for patients in the training set. This multiple sampling represents both a challenge and an opportunity, which are described in detail in later sections.

Thus far, we have described challenges specific to machine learning. However, since the classification models we developed serve as the foundation for a diagnostic test to be used clinically, there are two additional requirements we need to address. First, for cost-effectiveness, only one sequencing run per patient is commercially viable and the independent test set needs to reflect this reality. This necessitated analytically bridging individual samples used in the training set to pooled samples in the test set. Secondly, we need to ensure that the final locked classifier not only performs well on the independent test set, but will maintain this performance on all incoming future samples. Therefore, developing a classifier that is highly robust to foreseeable batch effects in the future becomes critically important.

In the following sections, we illustrate some of the challenges with quantitative analysis, describe practical solutions to overcome those challenges, show evidence of improvement, and discuss limitations of these approaches as well as directions of our future work (Fig. 1a).

Methods

Study design

Patients under medical evaluation for ILD that were 18 years of age or older and were undergoing a planned, clinically indicated lung biopsy procedure to obtain a histopathology diagnosis were eligible for enrollment in a multi-center sample collection study (BRonchial sAmple collection for a noVel gEnomic test; BRAVE) [7]. Patients for whom a bronchoscopy procedure was not indicated, not recommended or difficult were not eligible for participation in the BRAVE study. Patients were grouped based on the type of biopsy being performed for pathology: BRAVE-1 patients underwent surgical lung biopsy (SLB), BRAVE-2 patients underwent TBB, and BRAVE-3 patients underwent transbronchial cryobiopsy. The study was approved by institutional review boards, either centrally (Western IRB) or at each institution, and all patients provided informed consent, prior to their participation.

During study accrual, 201 BRAVE patients were prospectively divided into a group of 113 considered for use in training (enrolled December 2012 to July 2015) and 88 used in validation (enrolled August 2014 through May 2016). The training group ultimately yielded 90 patients with usable RNA sequence data and reference labels derived from pathology (described below) that were used to train and cross-validate the models. The independent test group yielded 49 patients that met prospectively defined inclusion criteria related to sample handling, sample adequacy, and the determination of reference truth labels. All clinical information related to the test set, including reference labels and associated pathology, were blinded to the algorithm development team until after the classifier parameters were finalized and locked. The test set was prospectively scored by an independent third party who was not involved in algorithm development, sequencing data generation or reference label generation.

Pathology reviews and label assignment

Histopathology diagnoses were determined centrally by a consensus of three expert pathologists using biopsies and slides collected specifically for pathology, following processes described previously [7, 12]. The central pathology diagnoses were determined separately for each lung lobe sampled for pathology. A reference label was then determined for each patient from the aggregated lobe-level diagnoses according to the following rules: If any lobe was diagnosed as any UIP subtype, e.g. Classic UIP (all features of UIP present), Difficult UIP (not all features of classic UIP well represented), Favor UIP (fibrosing interstitial process with UIP leading the differential), or combinations of these, then 'UIP' was assigned as the reference label for that patient. If any lung lobe

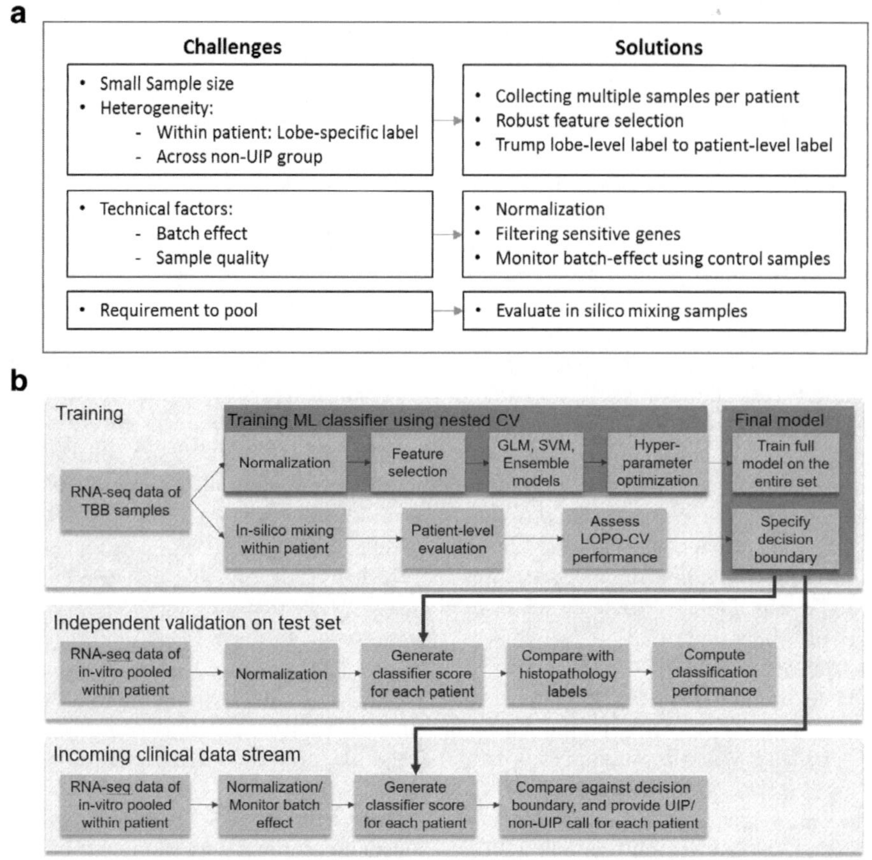

Fig. 1 An overview of challenges and solutions in our machine learning application, and our analysis pipeline. **a** Challenges and solutions in machine learning application. **b** Analysis pipeline in the development and evaluation of a molecular genomic classifier to predict usual interstitial pneumonia (UIP) pattern in interstitial lung disease (ILD) patients

was diagnosed with a 'non-UIP' pathology condition [7] and the other lobe was non-diagnostic or diagnosed with unclassifiable fibrosis (e.g. chronic interstitial fibrosis, not otherwise classified (CIF, NOC)), then 'non-UIP' was assigned as the patient level reference label. When all lobes were diagnosed with unclassifiable fibrosis or were non-diagnostic, then no reference label could be assigned and the patient was excluded. This patient-level reference label process was identical between training and test sets.

Molecular testing, sequencing pipeline, and data QC

Up to five TBB samples were sampled from each patient and typically, two upper lobe and three lower lobe samples were collected. TBB samples for molecular testing were placed into a nucleic acid preservative (RNAprotect, QIAGEN, Valencia, CA) and stored at 4 °C for up to 18 days, prior to and during shipment to the development laboratory, followed by frozen storage. Total RNA was extracted (AllPrep DNA/RNA Micro Kit, QIAGEN), quantitated (QuantiFluor RNA System, Promega, Madison, WI), pooled by patient where appropriate, and 15 ng input into the TruSeq RNA Access Library Prep procedure (Illumina, San Diego, CA), which enriches for the coding transcriptome using multiple rounds of amplification and hybridization to probes specific to exonic sequences. Libraries which met in-process yield criteria were sequenced on NextSeq 500 instruments (2 × 75 bp paired-end reads) using the High Output kit (Illumina, San Diego, CA). Raw sequencing (FASTQ) files were aligned to the Human Reference assembly 37 (Genome Reference Consortium) using the STAR RNA-seq aligner software [13]. Raw read counts for 63,677 Ensembl-annotated gene-level features were summarized using HTSeq [14]. Data quality metrics were generated using RNA-SeQC [15]. Library sequence data which met minimum criteria for total reads, mapped unique reads, mean per-base coverage, base duplication rate, the percentage of bases aligned to coding regions, the base mismatch rate, and uniformity of coverage within genes were accepted for use in downstream analysis.

Normalization

Sequence data were filtered to exclude any features that were not targeted for enrichment by the library assay, resulting in 26,268 genes. For the training set, expression count data for 26,268 Ensembl genes were normalized by *sizefactor* estimated with the median-of-ratio method and transformed by parametric variance-stabilizing transformation (VST), asymptotically equal to the logarithm to base 2 (DESeq2 package [16]). To mimic the processing of future clinical patients, the vector of geometric means and the parameters of VST from the training set were frozen and separately reapplied to the independent test set for the normalization.

For algorithm training and development, RNA sequence data was generated separately for each of 354 individual TBB samples from 90 patients. Eight additional TBB samples ('sentinels') were replicated in each of eight processing runs, from total RNA through to sequence data, to monitor potential batch effects. For independent validation, total RNA extracted from a minimum of three and a maximum of five TBBs per patient were mixed by equal mass within each patient prior to library preparation and sequencing. Each patient in the training set thus contributes up to five sequence samples to training, whereas each patient in the test set is represented by a single sequenced sample, analogous to the planned testing of clinical samples.

Differential expression analysis

We first explored whether differentially expressed genes found using a standard pipeline [17] could be applied directly to classify UIP from non-UIP samples. Differentially expressed genes were identified using DESeq2, a Bioconductor R package [16]. Raw gene-level expression counts of 26,268 genes in the training set were used to perform the differential analysis. A cutoff of p-value < 0.05 after multiple-testing adjustment and fold change > 2 was used to select differentially expressed genes. Within the training set, we performed pairwise differential analyses between all non-UIP and UIP samples, and between UIP samples and each non-UIP pathology subtype with more than 10 samples available, including bronchiolitis ($N = 10$), hypersensitivity pneumonitis (HP) ($N = 13$), nonspecific interstitial pneumonia (NSIP) ($N = 12$), organizing pneumonia (OP) ($N = 23$), respiratory bronchiolitis (RB) ($N = 16$), and sarcoidosis ($N = 11$). Principal component analysis (PCA) plots of all the training samples were generated using differentially expressed genes identified above.

Gene expression correlation heatmap

The correlation r^2 values of samples for six representative patients were computed using their VST gene expression, and a heatmap of the correlation matrix with patient order preserved was plotted to visualize intra- and inter-patient heterogeneity in gene expression. The 6 patients were selected to represent the full spectrum of within patient heterogeneity including two non-UIP and two UIP patients with the same or similar pathology subtypes between upper and lower lobes, as well as one UIP and one non-UIP patient each having different pathology subtypes, one being UIP and the other being non-UIP, in upper versus lower lobes. The heatmap was generated using the gplots R package.

Classifier development

We summarize the development and evaluation of a classifier in Fig. 1b. Our specific goal is to build a robust binary classifier on TBB samples to provide accurate and reproducible UIP/non-UIP predictions. We designed a high specificity test (specificity > 85%) to ensure a high positive predictive value. Thus, when the test predicts UIP, that result is associated with high confidence.

Feature filtering for classifier development

First, features that are not biologically meaningful or less informative were removed due to low expression level without variation among samples. We filtered genes annotated in Ensembl as pseudogenes, ribosomal RNAs, individual exons in T-cell receptor or Immunoglobulin genes, and excluded low expressed genes with raw counts expression level < 5 for the entire training set or expressed with count > 0 for less than 5% of samples in the training set.

We also excluded genes with highly variable expression in identical samples processed in multiple batches, as this would suggest sensitivity to technical, rather than biological factors. To identify such genes, we fitted a linear mixed effect model on the sentinel TBB samples that were processed across multiple assay plates. We fit this model for each gene separately

$$g_{ij} = \mu + \beta sample_{ij} + batch_i + e_{ij} \tag{1}$$

where g_{ij} is the gene expression of sample j and batch i, μ is the average gene expression for the entire set, $sample_{ij}$ is a fixed effect of biologically different samples, and $batch_i$ is the batch-specific random effect. The total variation was used to identify highly variable genes; the top 5% of genes by this measure were excluded (See Additional file 1: Figure S1). Thus, 17,601 Ensembl genes remained as candidates for the downstream analysis.

In silico mixing within patient

The classifiers were trained and optimized on individual TBB samples to maximize sampling diversity and the information content available during the feature selection and weighting process. However, for patients in the test set and in commercial setting, ideally multiple TBB samples would be pooled at the post-extraction stage, as RNA, and processed as pooled RNA in a single reaction through library prep, sequencing and classification [7]. To evaluate whether a classifier developed on individual samples could maintain high performance on pooled samples, we developed a method to simulate pooled samples "in silico" from individual sample data. First, raw read counts were normalized by *sizefactor* computed using geometric means across genes within the entire training set. The normalized count C_{ij} for sample $i = 1, \ldots, n$ and gene $j = 1, \ldots, m$ is computed by

$$C_{ij} = K_{ij}/S_j$$

where $s_j = \text{median}_i \frac{K_{ij}}{\left(\prod_{v=1}^{m} K_{iv}\right)^{1/m}}$ and K_{ij} is the raw count for sample i and gene j. Then, for each training patient $p = 1, \ldots, P$, in silico mixed count K_{ij}^p is defined by

$$K_{ij}^p = \frac{1}{n_p} \sum_{i \in I(p)} C_{ij}$$

where $I(p)$ is the index set of individual sample i that belongs to patient p. The frozen variance stabilizing transformation (VST) in the training set was applied to K_{ij}^p.

Training classifiers

As the test is intended to recognize and call a reference label defined by pathology [1], we defined the reference label to be the response variable in classifier training [4], and the exome-enriched, filtered and normalized RNA sequence data as the predictive features. We evaluated multiple classification models, including random forest, support vector machine (SVM), gradient boosting, neural network and penalized logistic regression [18]. Each classifier was evaluated based on 5-fold cross-validation and leave-one-patient-out cross-validation (LOPO CV) [19]. Ensemble models were also examined by combining individual machine learning methods via weighted average of scores of individual models [20].

To minimize overfitting during training and evaluation, we stratified each cross-validation fold such that all data from a single patient was either included or held out from a given fold. Hyper-parameter tuning was performed within each cross-validation split in a nested-cross validation manner [21]. We chose random search and one standard error rule [19] for selection of best parameters from inner CV to further minimize potential overfitting. Ultimately, we repeated the hyper-parameter tuning on the full training set to define the parameters in the final, locked classifier.

Best practices for a fully independent validation require that all classifier parameters, including the test decision boundary, are prospectively defined. This therefore must be done using only the training set data. Since the test set would be composed of TBBs pooled at the patient level, an in silico mixing model was used to simulate the distribution of patient-level pooled scores within the training set. We simulated within-patient mixtures 100 times at each LOPO CV-fold, with gene-level technical variability added to the VST gene expression. The gene-level technical variability was estimated using the mixed effect model (Eq. (1)) on the TBB samples replicated across multiple processing batches. The final decision boundary was chosen to optimize specificity (> 0.85) without severely compromising sensitivity (≥ 0.65). Performance was estimated using patient-level LOPO CV scores from replicated in silico mixing simulation. To be conservative for specificity, we used a criterion for averaged specificity of greater than 90% to choose a final decision boundary. For decision boundaries with similar estimated performance in simulation the decision boundary with highest specificity was chosen (See Additional file 1: Figure S2).

Evaluation of batch effects and monitoring scheme for future samples

To ensure the extensibility of classification performance to a future, unseen clinical patient population, it is crucial to ensure there are no severe technical factors, referred to as batch effects, that may cause significant shifts, rotations, compressions, or expansions of score distributions over time. To quantify batch effects in existing data and to evaluate the robustness of the candidate classifiers to observable batch effects, we scored nine different TBB pools, triplicated within each batch and processed across three different processing batches, and used a linear mixed effect model to evaluate variability of scores for each classifier. The model that is more robust against batch effects, as indicated by low score variability in the linear mixed model, was chosen as the final model for independent validation.

To monitor batch effects, we processed UIP and non-UIP control samples in each new processing batch. To capture a potential batch effect, we compared the scores of these replicated control samples and monitored whether estimated score variability remains smaller than the pre-specified threshold, σ_{sv}, determined in training using the in silico patient-level LOPO CV scores (see Additional file 1 for details regarding how acceptable variability levels were determined).

Independent validation

The final candidate classifier was prospectively validated on a blinded, independent test set of pooled TBB samples from 49 patients. Classification scores on the test set were derived using the locked algorithm and compared against the pre-set decision boundary to give the binary prediction of UIP vs. non-UIP calls: classification scores above the decision boundary were called UIP, and those equal to or below the decision boundary were called non-UIP. The continuous classification scores were compared against the histopathology labels to construct the receiver-operator characteristic curve (ROC) and calculate the area under the curve (AUC). The binary classification predictions were compared against the histopathology reference labels to calculate sensitivity and specificity.

Results

Distribution of ILD patients

Table 1 summarizes the distribution of ILD patients within UIP and non-UIP groups. Among collected patients, the prevalence of patients with a UIP pattern is higher in the training set (59%) than in the test set (47%), however this difference is not statistically significant (p-value 0.27). Three patients in the training set and one patient in the test set have potential within patient disease heterogeneity: one lobe was labeled as one of non-UIP subtypes (nonspecific interstitial pneumonia, pulmonary hypertension, or favor hypersensitivity pneumonitis), while the other lobe was labeled as a UIP pattern, driving the final patient-level label to a designation of UIP.

The non-UIP group includes a diversity of heterogeneous subtypes commonly encountered in clinical practice. Due to the small sample size, several subtypes have only one or two patients. Three subtypes—amyloid or light chain deposition, exogenous lipid pneumonia, and organizing alveolar hemorrhage—are present in the test set, which do not exist in the training set.

Intra-patient heterogeneity

Heterogeneity in samples from the same patient was observed in both histopathologic diagnosis and gene expression. Three such patients with pathology diagnoses spanning UIP and non-UIP groups posed a computational challenge for patient-level diagnostic classification. The correlation matrix of samples from six patients also revealed prominent intra- and inter-patient variability in expression profiles (See Additional file 1: Figure S3). We found two non-UIP patients with the same labels across different lobes and similar gene expression pattern (patients 1 and 2 in Additional file 1: Figure S3), two UIP patients with the same or similar labels and highly

Table 1 The distribution of patients and samples.

Representative histopathology types	Training set		Test set
	# samples	# patients	# patients
UIP Total	212 (60%)	53 (59%)	23 (47%)
Usual Interstitial pneumonia (UIP)	136	34	11
Difficult UIP	40	11	7
Favor UIP	22	5	4
UIP (lower lobe) + Nonspecific interstitial pneumonia (NSIP) (upper lobe)	5	1	
Difficult UIP (lower lobe) + NSIP (upper lobe)	4	1	
UIP (lower lobe) + Pulmonary hypertension (upper lobe)	5	1	
Favor HP (lower lobe) + Difficult UIP (upper lobe)			1
Non-UIP Total	142 (40%)	37 (41%)	26 (53%)
Respiratory bronchiolitis (RB); Smoking-related interstitial fibrosis	26	7	7
Hypersensitivity pneumonitis; Favor HP	19	4	4
Sarcoidosis	17	5	4
NSIP; Cellular NSIP; Favor NSIP	18	5	3
Diffuse alveolar damage; DAD with hemosiderosis	2	1	2
Amyloid or light chain deposition			1
Bronchiolitis	12	3	1
Eosinophilic pneumonia (EP)	5	1	1
Exogenous lipid pneumonia			1
Organizing alveolar hemorrhage			1
Organizing pneumonia (OP)	29	7	1
Pneumocystis pneumonia (PP)	4	1	
Emphysema	10	3	
Total	354	90	49

correlated expression profiles (patients 5 and 6 in Additional file 1: Figure S3), as well as one UIP and one non-UIP patient with dissimilar labels and heterogeneous expression (patients 3 and 4 in Additional file 1: Figure S3), providing a representative visualization of the full spectrum of heterogeneity we observe within and across patients.

DE analysis between UIP and non-UIP

We first investigated whether differentially expressed genes found by DESeq2 between UIP and non-UIP were predictive of the two diagnostic classes. We identified 151 differentially expressed genes between UIP and non-UIP with statistical significance (adjusted $p < 0.05$, fold change > 2): 55 were up-regulated and 96 were down-regulated genes in UIP (Fig. 2a and Table 2). However, using these differentially expressed genes alone was insufficient to separate the two classes perfectly, as shown by the PCA plot (Fig. 2b). In contrast, PCA spanned by the 190 classifier genes selected for the final classifier could separate the two classes much better (Fig. 2c).

Heterogeneity in non-UIP patients

We observed heterogeneity in gene expression of non-UIP samples, consisting of more than a dozen different histopathology pattern diagnoses. We identified genes that were significantly different (adjusted $p < 0.05$, fold change > 2) between UIP samples and each non-UIP pathology subtype with a sample size greater than 10 (Table 2). The higher the number of differentially expressed genes, the more dissimilar the non-UIP pathology subtype is from UIP. A comparison of the list of differential genes in each non-UIP subtype with that from all non-UIP samples showed that the number of overlapping genes was highly dependent on the number of differential genes identified in the individual non-UIP subtype, indicating that some non-UIP pathologies may have more dominant effects on the overall differential genes found between all non-UIP and UIP samples (Table 2). Moreover, there are few overlapping differential genes among those identified in individual non-UIP pathology patterns. For example, 172 genes were common between 1174 differential genes in Sarcoidosis and 701 in RB, and 6 common genes were found among differential genes from sarcoidosis, RB and NSIP. There are no common genes among differential genes from bronchiolitis, NSIP and HP. This suggests distinct molecular expression patterns within different non-UIP pathology subtypes.

PCA plots using the differentially expressed genes between selected non-UIP subtypes and UIP samples showed that the specific non-UIP pathology subtypes tend to be well-separated from UIP samples for subtypes such as RB and HP (See Additional file 1: Figures S4A, C), while other non-UIP samples may be interspersed with UIP samples (See Additional file 1: Figures S4B, E). This demonstrated that differential genes derived from one non-UIP subtype may not be generalizable to other non-UIP subtypes.

Comparison between in silico mixing and in vitro pooling within patient

In silico mixed samples within each patient were developed to model in vitro pools using the training set samples. To ensure in silico mixed and in vitro pooled samples were reasonably matched, the pooled samples of 11 patients were sequenced and compared with in silico mixed samples (data not shown). The average r-squared value based on expression level of 26,268 genes for the pairs of in silico mixed and in vitro pooled samples was 0.99 (standard deviation

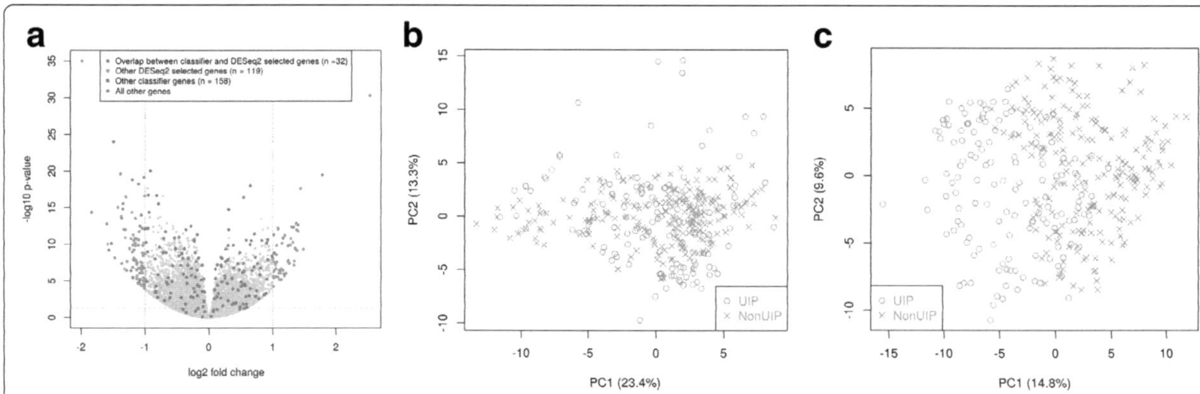

Fig. 2 Gene selection using DESeq2 and our classifier. **a** Volcano plot showing 151 genes selected by DESeq2 (adjusted *p*-value < 0.05 and fold change > 2) and 190 predictive genes in our classifier, with 32 common ones (pink) between the two sets of genes. **b** PCA plot of all TBB samples using only DESeq2 selected genes showing that these genes are not sufficient to separate UIP samples (orange circle) from non-UIP samples (blue cross). **c** PCA plot of all TBB samples using classifier genes illustrating that TBB samples can be classified into UIP (orange circle) and non-UIP (blue cross) samples using these genes

Table 2 The number of significantly expressed genes (multiple-testing adjusted p-value < 0.05, fold change > 2) between each non-UIP subtype and UIP samples (*N* = 212). The number of differentially expressed genes overlapping with those between UIP and non-UIP samples was summarized in the third columns

	Up-regulated genes	Down-regulated genes	Total number of differentially expressed genes	# genes overlapping with those from all non-UIP samples
All non-UIP (*N* = 147)	55	96	151	151 (100%)
Bronchiolitis (*N* = 10)	41	34	75	6 (8%)
Hypersensitivity pneumonitis (HP) (*N* = 13)	32	53	85	14 (16%)
Nonspecific interstitial pneumonia (NSIP) (*N* = 12)	37	49	86	13 (15%)
Organizing pneumonia (OP) (*N* = 23)	1	15	16	31 (52%)
Respiratory bronchiolitis (RB) (*N* = 16)	549	152	701	64 (9%)
Sarcoidosis (*N* = 11)	448	726	1174	93 (8%)

(SD) = 0.003), which indicated that the simulated expression level of in silico mixed samples was well-matched with that of in vitro pooled samples, considering the average r-squared values were 0.98 (SD = 0.008) for technical replicates and 0.94 (SD = 0.04) for biological replicates.

The classification scores of in silico and in vitro mixed samples by two candidate classifiers, the ensemble and penalized logistic regression models (described below) were also compared in a scatterplot (Fig. 3). The number of replicates for each in vitro pooled sample ranges from 3 to 5, so the mean score of the multiple replicates was used. The classification scores of in silico mixed samples were highly correlated with those of in vitro pooled samples with a Pearson's correlation of 0.99 for both classifiers (Fig. 3). The points fell right around the line of X = Y with no obvious shift or rotation.

Cross-validation performance on the training set

We evaluated multiple methods of feature selection and machine learning algorithms on our training set of 354 TBB samples from 90 patients. As an initial attempt, individual and ensemble models were evaluated separately based on 5-fold CV and cross-validated AUC (cvAUC), as estimated using the mean of the empirical AUC of each fold. Overall, the linear models such as the penalized regression model (cvAUC = 0.89) outperformed non-linear tree-based models such as random forest (cvAUC = 0.83) and gradient boosting (cvAUC = 0.84). The cvAUC of a neural network classifier was under 0.8. The best performance was achieved by (1) the ensemble model of SVMs with linear and radial kernels, and (2) penalized logistic regression; both of which had cvAUC = 0.89. However, with the heterogeneity among diseases and the small sample size, CV performance on all models was found to vary significantly depending on the split.

In LOPO CV, we evaluated the patient-level performance by using 100 replicates of in silico mixed samples for each patient within LOPO CV folds. The computed classification scores of individual samples and averaged

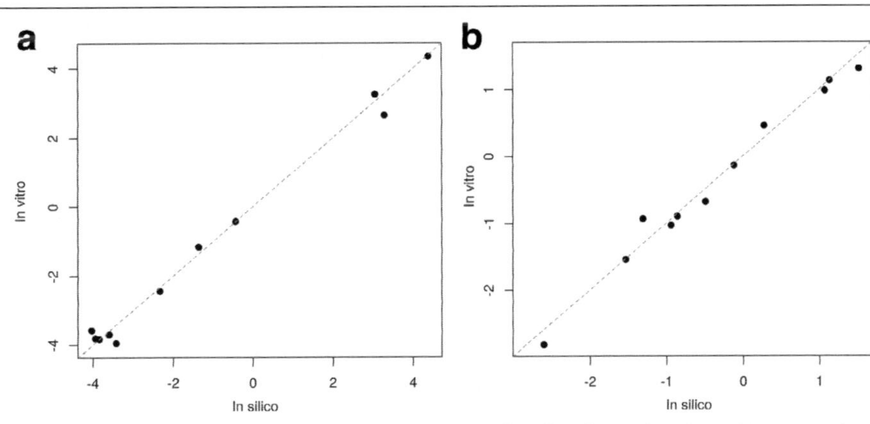

Fig. 3 Comparison between in silico and in vitro mixing within patient. Scatterplot of in silico and in vitro mixing comparison scored by (**a**) an ensemble classifier with an R-squared value of 0.99, and (**b**) a penalized logistic regression classifier with an R-squared value of 0.98

scores of in silico mixed samples are shown in Fig. 4. Overall, the patient-level performance was slightly higher compared to the sample-level performance. Based on combined scores across LOPO CV folds, the ensemble model and the penalized logistic regression model achieved the best performance, with an AUC of 0.9 [0.87–0.93] and 0.87 [0.83–0.91] at sample-level and 0.93 [0.88–0.98] and 0.91 [0.85–0.97] at in silico mixing patient-level, respectively (Fig. 5a and Table 3).

Robustness of classifiers

The estimated score variability was 0.46 and 0.22 for the ensemble model and the penalized logistic regression model, respectively (Table 4). Both were less than 0.9 and 0.48, the pre-specified thresholds of acceptable score variability (See Additional file 1: Figures S5 and S6). Considering that the score range of the ensemble classifier is wider than that of the penalized logistic regression classifier, the proportion of the variability to the range of 5% and 95% quantiles of scores was compared. Overall, the penalized logistic regression classifier had less variability in scores than the ensemble model. This implied that the penalized logistic regression model was more robust to technical (reagent/laboratory) effects and thus offered more consistent scores for technical replicates (Table 4). With high cross-validation performance and robustness, the penalized logistic regression model was chosen as our final candidate model for the independent validation.

Independent validation performance

Using the locked penalized logistic classifier with a pre-specified decision boundary at 0.87 (See Additional file 1: Figure S2), the validation performance was evaluated based on the independent test set of samples that were prospectively pooled and processed. The final classifier achieved specificity of 0.88 [0.70–0.98] and sensitivity of 0.70 [0.47–0.87] with an AUC of 0.87 [0.76–0.98] (Fig. 5b and Table 3). The point estimate of the validation performance was lower than in silico patient-level training CV performance, but *p*-values of 0.6, 0.7, and 1 for AUC, sensitivity, and specificity, respectively, indicate negligible differences.

Discussion

In this study, we demonstrate that accurate and robust classification is achievable even when critical challenges exist. By leveraging appropriate statistical methodologies, machine learning approaches, and RNA sequencing technology, we provide a meaningful diagnostic test to improve the care of patients with interstitial lung diseases.

Machine learning, particularly deep learning, has experienced revolutionary progress in the last few years. Empowered with these recently developed and highly sophisticated tools, classification performance is dramatically improved in many applications [10]. However, most of these tools require readily available and high-confidence labels as well as large sample sizes: the magnitude of the performance improvement is directly and positively related to the number of samples with high-quality labels [22, 23]. In this project, like many other clinical studies based on patient samples, the sample size is limited: we have a total of only 90 patients in the training set (Table 1). Additionally, the non-UIP group is a collection of many pathology subtypes, each with its own distinct biology, several of which have only one or two patients in the training set [24] (Table 1). Not surprisingly, these various pathology subtypes are different at the molecular and genomic level. We attempted to use the training samples to identify common features across non-UIP pathology with respect to differentiating them from the UIP group, but none emerged (Table 2 and See Additional file 1: Figure S4). Furthermore, three pathology subtypes (Amyloid or light chain deposition, Exogenous lipid pneumonia, and Organizing alveolar hemorrhage) present in the test set were not encountered in the training set (Table 1). We also observed a change in UIP proportions between training (59%) and testing (47%). The last two factors may help explain the slightly lower performance in the test set compared to the cross-validation performance of the training set. With this work showing the feasibility of successfully separating UIP vs. non-UIP pathology using genomic information on minimally-invasive samples, it may pave the path and make it easier to collect required diagnostic samples from these patients soon. If that's the case, one direction of our future work will potentially be about collecting sufficient samples from each distinct non-UIP pathology subtypes to better represent the heterogeneity of that group and further enhance the classification performance. Unfortunately, as of now, the sample size is very limited and the recent advances in machine learning that leverage large sample sizes are not applicable in this situation. For this reason, our focus has been on more traditional linear models or tree-based models. It may also explain why, among our candidates, the linear models outperform the non-linear tree-based models, since we have too small a sample size in individual non-UIP pathology subtypes to power any interaction which the tree-based model attempts to capture.

To directly address the small training size, up to 5 distinct TBB samples within the same patient were run from RNA extraction to sequencing and were included to successfully expand the 90-patient set to encompass 354 samples (Table 1). This, in concept, is similar to the concept of data augmentation [25], but instead of simulating or extrapolating the augmented data, we generated sequencing data from real experiments on multiple TBB samples from the same patient. The goal is to provide

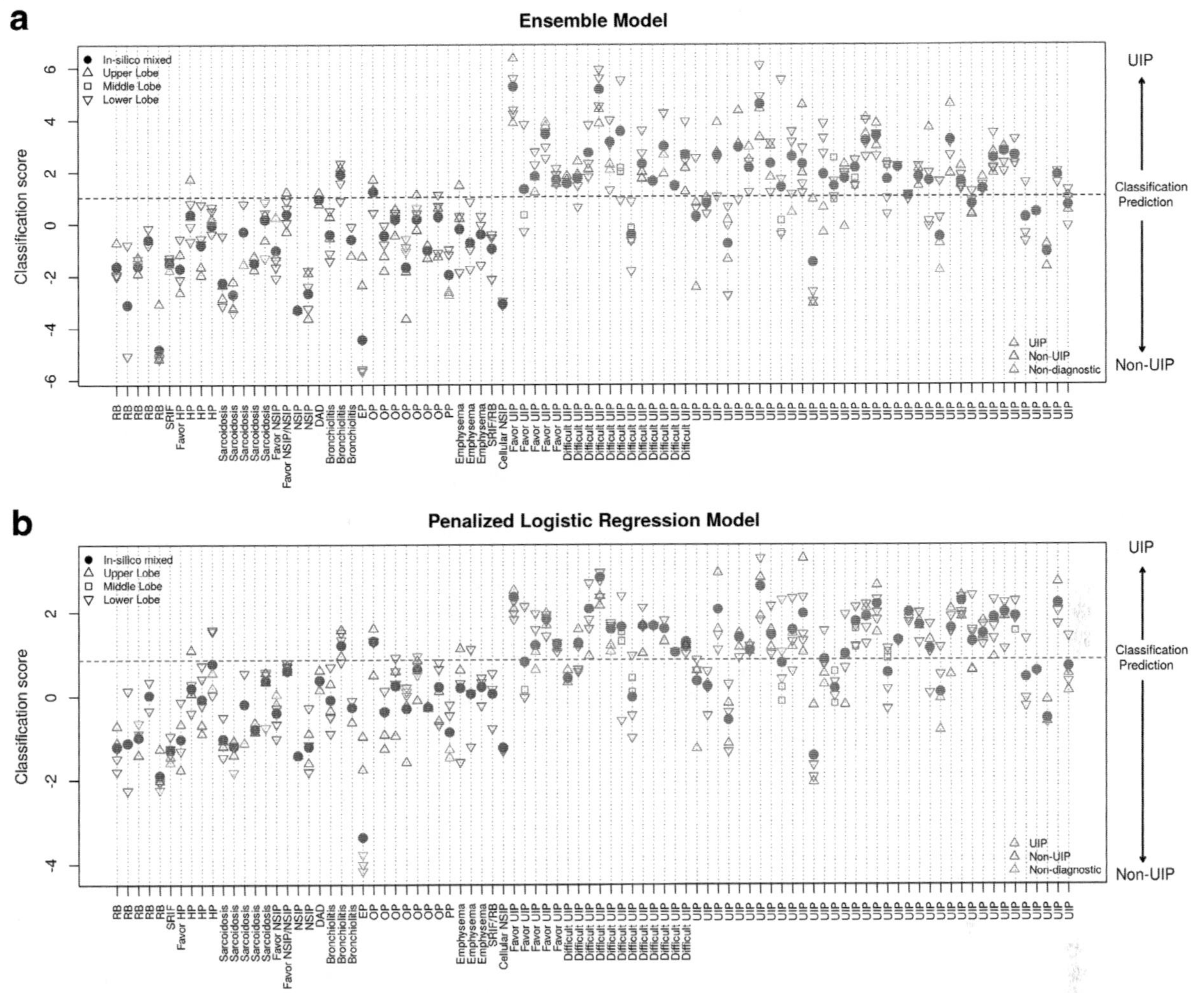

Fig. 4 Cross-validated classification scores for the two candidate models. Classification scores of (**a**) Ensemble Model and (**b**) Penalized Logistic Regression Model from leave one patient out (LOPO) cross validation. Red, blue and gray indicate samples with histopathology UIP, non-UIP, and non-diagnostic respectively. Circle, up-pointing triangle, square and down-pointing triangle indicate in silico mixed sample, upper, middle and lower lobe samples respectively. Each vertical line represents one distinct patient; all the points on the same line are the multiple samples from the same patient. The X-axis is the patient-specific pathology label where non-UIP patients are on the left and UIP patients are on the right. The Y-axis is the cross-validated classification score. The purple horizontal dash line is the prospectively defined classifier decision boundary. Scores above the decision boundary is predicted as UIP and scores below or equal to the decision boundary is predicted as non-UIP

additional information to enhance classification performance. We took special precautions to use a single patient as the smallest unit when defining the cross-validation fold and evaluating performance. This prevents patients with more samples from having higher weight, or samples from the same patient straddling on both sides of model building and model evaluation, causing over-fitting. We also applied nested cross-validation as well as the one SD (standard deviation) rule for model selection and parameter optimization, to factor-in the high variability on performance due to small sample size, and to aggressively trim down the model complexity to guard against overfitting.

While running multiple TBB samples per patient in the training set helps with the sample size limitation, it creates a new challenge. The commercial setting is economically viable only if we can limit the test to one sequencing run per patient. To achieve that, RNA material from multiple TBB samples within one patient needs to be pooled together before sequencing. However, whether a classifier trained on individual TBB samples is applicable to pooled TBB samples becomes a critical question to answer before performing the validation experiment. To address this, we designed a series of in-silico mixing simulations to mimic patient-level in-vitro pools of the test set. This approach is also the

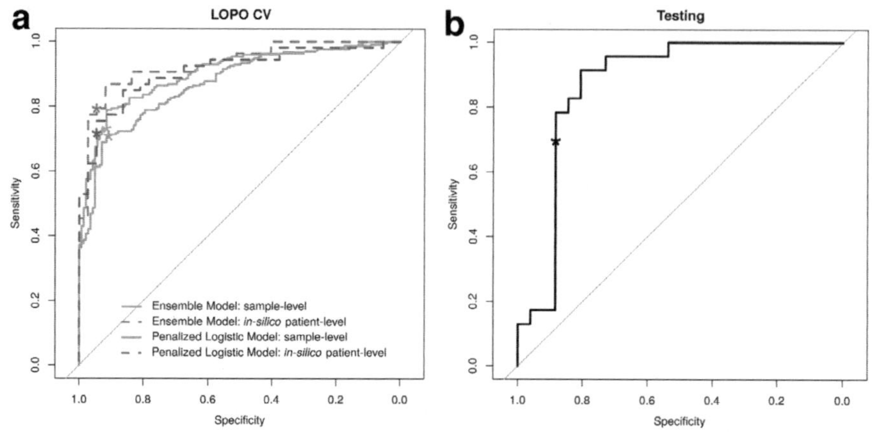

Fig. 5 Evaluation of classifier performance. Receiver operating characteristics (ROC) curves of (**a**) classification performance from leave-one-patient-out (LOPO) cross-validation and (**b**) validation on independent test set. The asterisk on each ROC curve corresponds to the prospectively defined decision boundary of each proposed model

fundamental building block for defining the prospective decision boundary of the classifier as well as the optimal number of TBBs required to achieve the best classification performance (data show in supplementary Figure E1 in [7]). The simulated in-silico data agrees well with the experimental in-vitro data (Fig. 3) giving us confidence in using this approach to extrapolate expected performance to pooled samples and proceed with the validation experiments with the pooled setting. This in-silico approach works well in this study since samples pooled together are the same type (TBB) and from the same patient, thus have similar characteristics such as the rate of duplicated reads or the total number of reads. However, we found that it may be tricky to extend the proposed in-silico mixing model to mix samples of different characteristics or qualities, for example UIP vs non-UIP samples or TBB mixed with different type of samples such as blood. In those cases, samples with substantially higher total number of reads tend to dominate the expressions of combined samples violating the basic assumptions of the mixed model proposed here. More sophisticated methodology is required to accurately model such complex mixtures.

A successful validation that meets the required clinical performance (Fig. 5 and Table 3) is only the first step towards a useful commercial product aiming to improve patient care. Equally important is the importance of providing consistent and reliable performance for the future patient stream. This requires us to proactively anticipate and address any potential batch effects of sequencing data from incoming patients that may cause systematic changes in classification scores. We tackle this important issue starting from the upstream feature selection step (See Additional file 1: Figure S1) where genes that are highly sensitive to batch effects were removed from downstream analysis. Furthermore, additional experimental data were generated for 9 distinct TBB pools in three different batches that were distinct from those used to process samples for training. We leverage this experiment to directly evaluate the robustness of each candidate model against unseen batches and help select the final model.

Conclusion

Limited sample size and high heterogeneity within the non-UIP class are two major classification challenges we

Table 3 The performance of the two models on the training set and our final selected model on the test set, evaluated using area under the curve (AUC), specificity and, sensitivity

	LOPO CV				Independent Testing
	Ensemble Model		Penalized Logistic Model		Penalized Logistic Model
	Sample-level	In silico mixing	Sample-level	In silico mixing	In vitro mixing
AUC	0.90 [0.87–0.93]	0.93 [0.88–0.98]	0.87 [0.83–0.91]	0.91 [0.85–0.97]	0.87 [0.76–0.98]
Specificity	0.92 [0.86–0.96]	0.95 [0.82–0.99]	0.91 [0.85–0.95]	0.95 [0.82–0.99]	0.88 [0.70–0.98]
Sensitivity	0.73 [0.67–0.79]	0.79 [0.66–0.89]	0.71 [0.64–0.77]	0.72 [0.58–0.83]	0.70 [0.47–0.87]

Table 4 Estimation of variability of scores from the two classifiers using linear mixed effect models. The percentage (%) is the ratio of estimated variability to the range between 5% and 95% quantiles in classification scores

Classifier	Between-run	Intra-run (Residual)	Inter-run (Total)
Ensemble	0.28 (4.0%)	0.37 (5.3%)	0.46 (6.5%)
Penalized logistic regression	0.10 (2.6%)	0.19 (4.9%)	0.22 (5.6%)

faced in this project and which commonly exist in clinical studies. In addition, a successful commercial product needs to perform economically and consistently for all future incoming samples, which requires the underlying classification model to be applicable to pooled samples and highly robust against assay variability. We demonstrated that it is feasible to achieve highly accurate and robust classification despite these difficulties and we described practical solutions to each of these challenges. The methodologies have proven to be successful in this study and could be applicable to other clinical scenarios facing similar difficulties.

Additional file

Additional file 1: Supplementary document including a description of score variability simulation and supplementary Figures S1 to S6. (PDF 3639 kb)

Abbreviations

AUC: Area under the curve; CV: Cross-validation; DAD: Diffuse alveolar damage; EP: Eosinophilic pneumonia; HP: Hypersensitivity pneumonitis; ILD: Interstitial lung disease; IPF: Idiopathic pulmonary fibrosis; LOPO: Leave-one-patient-out; NSIP: Nonspecific interstitial pneumonia; OP: Organizing pneumonia; PCA: Principal component analysis; PP: Pneumocystis pneumonia; RB: Respiratory bronchiolitis; ROC: Receiver operating characteristic; SD: Standard deviation; SLB: Surgical lung biopsy; SVM: Support vector machine; TBB: Transbronchial biopsy; UIP: Usual interstitial pneumonia; VST: Variance-stabilizing transformation

Acknowledgements

We'd like to thank Jilei Yang for his effort in classifier algorithm development, Grazyna Fedorowicz for performing experiments, Hannah Neville for coordinating the clinical sample collection of BRAVE study, and Srinivas Jakkula for performing the independent validation of the final classifier.

Funding

Funding for the publication of this article was provided by Veracyte Inc.

Availability of data and materials

All data generated and supportive for the conclusion in this study are included in this published article.

About this supplement

This article has been published as part of *BMC Genomics* Volume 19 Supplement 2, 2018: Selected articles from the 16th Asia Pacific Bioinformatics Conference (APBC 2018): genomics. The full contents of the supplement are available online at https://bmcgenomics.biomedcentral.com/articles/supplements/volume-19-supplement-2 .

Authors' contributions

YC, DGP, PSW, NMB, GCK and JH designed, performed, analyzed and interpreted the study. NMB coordinated the clinical sample collection. TVC, DAL and GR performed clinical patient review and provided key study feedback. YC, TTL and JH implemented the analysis, prepared figures and drafted the manuscript. YC, TTL, DGP, PSW, GCK and JH edited and revised the manuscript. All authors have read and approved the final manuscript.

Ethics approval and consent to participate

Ethics approval was obtained prior to the initiation of the studies described in this report. All lung tissue specimen including TBB were collected with central (Western IRB, #20130158, #20141155, and #20130705) and institutional review board approval and written informed consent from all participants.

Consent for publication

Not applicable

Competing interests

YC, TTL, DGP, NMB, PSW, GCK and JH are employed by Veracyte Inc.

Publisher's Note

Author details

[1]Veracyte, Inc, 6000 Shoreline Court, Suite 300, South San Francisco, CA 94080, USA. [2]Department of Laboratory Medicine and Pathology, Mayo Clinic, Scottsdale, AZ, USA. [3]Department of Radiology, National Jewish Health, Denver, CO, USA. [4]Department of Medicine and Laboratory Medicine, University of Washington Medical Center, Seattle, WA, USA.

Published: 9 May 2018

References

1. Flaherty KR, King TE Jr, Raghu G, Lynch JP 3rd, Colby TV, Travis WD, Gross BH, Kazerooni EA, Toews GB, Long Q, et al. Idiopathic interstitial pneumonia: what is the effect of a multidisciplinary approach to diagnosis? Am J Respir Crit Care Med. 2004;170:904–10.
2. Travis WD, Costabel U, Hansell DM, King TE Jr, Lynch DA, Nicholson AG, Ryerson CJ, Ryu JH, Selman M, Wells AU, et al. An official American Thoracic Society/European Respiratory Society statement: update of the international multidisciplinary classification of the idiopathic interstitial pneumonias. Am J Respir Crit Care Med. 2013;188:733–48.
3. Flaherty KR, Andrei AC, King TE Jr, Raghu G, Colby TV, Wells A, Bassily N, Brown K, du Bois R, Flint A, et al. Idiopathic interstitial pneumonia: do community and academic physicians agree on diagnosis? Am J Respir Crit Care Med. 2007;175:1054–60.
4. Tuch BB, Laborde RR, Xu X, Gu J, Chung CB, Monighetti CK, Stanley SJ, Olsen KD, Kasperbauer JL, Moore EJ, et al. Tumor transcriptome sequencing reveals allelic expression imbalances associated with copy number alterations. PLoS One. 2010;5:e9317.
5. Twine NA, Janitz K, Wilkins MR, Janitz M. Whole transcriptome sequencing reveals gene expression and splicing differences in brain regions affected by Alzheimer's disease. PLoS One. 2011;6:e16266.
6. Boyle EA, Li YI, Pritchard JK. An expanded view of complex traits: from polygenic to omnigenic. Cell. 2017;169:1177–86.
7. Pankratz DG, Choi Y, Imtiaz U, Fedorowicz GM, Anderson JD, Colby TV, Myers JL, Lynch DA, Brown KK, Flaherty KR, et al. Usual interstitial pneumonia can be detected in transbronchial biopsies using machine learning. Ann Am Thorac Soc. 2017;14:1646.
8. Sorlie T, Tibshirani R, Parker J, Hastie T, Marron JS, Nobel A, Deng S, Johnsen H, Pesich R, Geisler S, et al. Repeated observation of breast tumor subtypes in independent gene expression data sets. Proc Natl Acad Sci U S A. 2003; 100:8418–23.
9. Brennan CW, Verhaak RG, McKenna A, Campos B, Noushmehr H, Salama SR, Zheng S, Chakravarty D, Sanborn JZ, Berman SH, et al. The somatic genomic landscape of glioblastoma. Cell. 2013;155:462–77.
10. LeCun Y, Bengio Y, Hinton G. Deep learning. Nature. 2015;521:436–44.
11. Rajpurkar P, Hannun AY, Haghpanahi M, Bourn C, Ng AY: Cardiologist-level arrhythmia detection with convolutional neural networks. arXiv preprint arXiv:170701836 2017.
12. Kim SY, Diggans J, Pankratz D, Huang J, Pagan M, Sindy N, Tom E, Anderson J, Choi Y, Lynch DA, et al. Classification of usual interstitial pneumonia in

patients with interstitial lung disease: assessment of a machine learning approach using high-dimensional transcriptional data. Lancet Respir Med. 2015;3:473–82.
13. Dobin A, Davis CA, Schlesinger F, Drenkow J, Zaleski C, Jha S, Batut P, Chaisson M, Gingeras TR. STAR: ultrafast universal RNA-seq aligner. Bioinformatics. 2013;29:15–21.
14. Anders S, Pyl PT, Huber W. HTSeq–a python framework to work with high-throughput sequencing data. Bioinformatics. 2015;31:166–9.
15. DS DL, Levin JZ, Sivachenko A, Fennell T, Nazaire MD, Williams C, Reich M, Winckler W, Getz G. RNA-SeQC: RNA-seq metrics for quality control and process optimization. Bioinformatics. 2012;28:1530–2.
16. Love MI, Huber W, Anders S. Moderated estimation of fold change and dispersion for RNA-seq data with DESeq2. Genome Biol. 2014;15:550.
17. Anders S, DJ MC, Chen Y, Okoniewski M, Smyth GK, Huber W, Robinson MD. Count-based differential expression analysis of RNA sequencing data using R and bioconductor. Nat Protoc. 2013;8:1765–86.
18. Dobson AJ, Barnett A. An introduction to generalized linear models. 2nd ed. New York: Chapman & Hall; 2002.
19. Friedman J, Hastie T, Tibshirani R. The elements of statistical learning. New York: Springer series in statistics; 2001.
20. Rokach L. Ensemble-based classifiers. Artif Intell Rev. 2010;33:1–39. %@ 0269-2821.
21. Krstajic D, Buturovic LJ, Leahy DE, Thomas S. Cross-validation pitfalls when selecting and assessing regression and classification models. J Cheminform. 2014;6:10.
22. Gu B, Hu F, Liu H. Modelling classification performance for large data sets. Adv Web Age Inf Manag. 2001:317–28.
23. Sun C, Shrivastava A, Singh S, Gupta A: Revisiting unreasonable effectiveness of data in deep learning era. arXiv preprint arXiv:170702968 2017.
24. Libbrecht MW, Noble WS. Machine learning applications in genetics and genomics. Nat Rev Genet. 2015;16:321–32.
25. Wong SC, Gatt A, Stamatescu V, McDonnell MD. Understanding data augmentation for classification: when to warp? In: IEEE; 2016. p. 1–6. %@ 150902896X.

Chou *et al. BMC Genomics* 2018, **19**(Suppl 2):103
DOI 10.1186/s12864-018-4463-x

BMC Genomics

RESEARCH

Open Access

The aquatic animals' transcriptome resource for comparative functional analysis

Chih-Hung Chou[1,2†], Hsi-Yuan Huang[1,2†], Wei-Chih Huang[1,2†], Sheng-Da Hsu[1], Chung-Der Hsiao[3], Chia-Yu Liu[1], Yu-Hung Chen[1], Yu-Chen Liu[1,2], Wei-Yun Huang[2], Meng-Lin Lee[2], Yi-Chang Chen[4] and Hsien-Da Huang[1,2*]

From The Sixteenth Asia Pacific Bioinformatics Conference 2018
Yokohama, Japan. 15-17 January 2018

Abstract

Background: Aquatic animals have great economic and ecological importance. Among them, non-model organisms have been studied regarding eco-toxicity, stress biology, and environmental adaptation. Due to recent advances in next-generation sequencing techniques, large amounts of RNA-seq data for aquatic animals are publicly available. However, currently there is no comprehensive resource exist for the analysis, unification, and integration of these datasets. This study utilizes computational approaches to build a new resource of transcriptomic maps for aquatic animals. This aquatic animal transcriptome map database dbATM provides de novo assembly of transcriptome, gene annotation and comparative analysis of more than twenty aquatic organisms without draft genome.

Results: To improve the assembly quality, three computational tools (Trinity, Oases and SOAPdenovo-Trans) were employed to enhance individual transcriptome assembly, and CAP3 and CD-HIT-EST software were then used to merge these three assembled transcriptomes. In addition, functional annotation analysis provides valuable clues to gene characteristics, including full-length transcript coding regions, conserved domains, gene ontology and KEGG pathways. Furthermore, all aquatic animal genes are essential for comparative genomics tasks such as constructing homologous gene groups and blast databases and phylogenetic analysis.

Conclusion: In conclusion, we establish a resource for non model organism aquatic animals, which is great economic and ecological importance and provide transcriptomic information including functional annotation and comparative transcriptome analysis. The database is now publically accessible through the URL http://dbATM.mbc.nctu.edu.tw/.

Background

Aquatic animals have significant economic benefits for humans and also play key roles in the development of medical applications [1, 2]. Aquaculture transforms natural aquatic resources such as fish, shrimp, and mussels into socially-valued commodities [3]. Chemicals derived from aquatic organisms are used to develop pharmaceutical compounds with important clinical applications [2]. In most of the aquaculture industries, prophylactic antibiotics have been used to prevent bacterial and virus infections [4].

To investigate the genomic resource of aquatic animals, next generation sequencing (NGS) approach has been adapted to uncover new genes and biological mechanisms. To effectively increase the our knowledge on aquatic animal gene discovery at mRNA level, RNA-seq has been applied to several important aquaculture species such as, *Plecoglossus altivelis* [5], *Cyprinus carpio* [6], *Penaeus monodon* [7], *Hyriopsis cumingii* [8], *Mytilus galloprovincialis* [9], and *Anguilla anguilla* [10]. To gain insights into the immunogenetics and immune response system, RNA-seq has been successfully adapting to

* Correspondence: bryan@mail.nctu.edu.tw
†Equal contributors
[1]Institute of Bioinformatics and Systems Biology, National Chiao Tung University, Hsinchu 300, Taiwan
[2]Department of Biological Science and Technology, National Chiao Tung University, Hsinchu 300, Taiwan
Full list of author information is available at the end of the article

study the host defense gene activities against bacteria and viruses [11]. *Clupea pallasii, Fundulus grandis, Pandalus latirostri,* etc. [12–15], have recently come under significant additional anthropogenic and environmental pressure, and SNPs and a set of genes associated with immune function in these species have been identified as providing high differentiation between two regions, and this insight can be applied in developing ecological monitoring systems. Evolutionary research on species including *Sinocyclocheilus angustiporus* (a comparative investigation between surface and cave species) revealed reduced expression of a series of visual photo-transduction and retinal disease-related genes in cave-dwelling species [16]. *Astyanax mexicanus* was also used to clarify heritable genetic changes governing adaptation to cave environments [17]. Beyond NGS transcriptomic data analysis, the MitoFish database focuses on fish mitochondrial genome annotation to explore a new resource for resolving fish phylogenies and identifying new fish species [18].

NGS technology has emerged as a powerful tool to illustrate the blueprints of novel species. Whole genome sequencing (WGS) and RNA sequencing (RNA-seq) can be used to investigate the individual differences, evolution and gene function [19, 20]. However, whole genome sequencing is more costly in terms of expense and computing resources than RNA-seq, and cannot obtain comprehensive information for transcriptomes directly by gene prediction. A better strategy is using RNA-seq to elucidate the molecular basis of biological functions [21]. Given the economical importance of these aquatic animals, a comprehensive resource of transcriptomes should be establsihed despite the lack of draft genomes of these non model organisms.

RNA-seq technology pave the way for transcriptome data investigation. With the techniques, transcripts can not only be detected in accordance with the reference genome but also can be used for de novo the assembly of RNA-seq reads to discover genes and profile their expression in organisms without a reference genome [22–24]. However, even if these data are deposited in the NCBI Sequence Read Archive (SRA) [25] which provides high-throughput unassembled and un-annotated raw NGS reads for more than 1000 Terabase pairs, public access is not provided to allow for searches of the well-annotated data. To address this problem, we have created a queryable database from the annotated transcriptomic database, called dbATM (http://dbATM.mbc.nctu.edu.tw). A similar approach was reported for ASGARD [26], which collects the annotated transcriptomes from three model arthropod species and NHPRTR [21], focusing on non-human primate NGS transcriptomic sequencing and analysis. dbATM is an easily-accessible database which integrates functional analysis applications.

Aquatic animals have high economic value and also play important roles in the monitoring of environmental pollution in ecosystems. Aquatic animals' related studies have seen exponentially increased in recent years (Additional file 1: Figure S1). Many studies have used NGS approaches to investigate genomic diversity among aquatic animals at genomic and transcriptomic levels. However, these studies largely focus on single species, and no databases exist so far to facilitate homologous gene mining among multiple aquatic species. In this study, we aim to establish an integrated aquatic animal transcriptomic database to facilitate studies in the fields of genomics, evolution, and phylogeny. We initially performed whole organism RNA-seq on four aquatic animals and collected RNA-seq data for eighteen species from the public domain. Reference genome free RNA-seq data of aquatic animals download from the three common NGS reads database, NCBI Sequence Read Archive (SRA), European Nucleotide Archive (ENA), and DDBJ Sequence Read Archive (DRA). Then, a computational pipeline was developed to annotate these NGS transcriptomes. Finally, all data were summarized and provided in the dbATM database. Figure 1 illustrates how the database can be used for mining transcriptomics, functions such Gene Ontology and KEGG pathways and comparative analysis of aquatic animals. The dbATM collects transcriptomes for twenty-two aquatic animals' transcriptomes and provides an invaluable tool for homologous evolution and evo-devo studies.

Results

Overview of dbATM content

Additional file 1: Table S1 displays all aquatic animal RNA-seq data sources and original data types. Following systematic analysis (Fig. 2), the individual species information and comparative analysis can be accessed from dbATM. Table 1 shows the species annotation of the transcripts, unigenes, proteins, and homologous genes. To facilitate evolutionary studies and comparative analysis across the different aquatic animal species, a taxonomic tree was constructed according to biological classification (Fig. 3), for shellfish, shrimp, and fish from invertebrate to vertebrate.

Web interface

The dbATM provides various query interfaces and graphical visualization pages to access to aquatic transcriptome data (Fig. 4). The summary table shows species information including scientific name (or common name), followed by total sequencing read size (in bp), sequencing platform, type of read sequences, number of reads and a link to NCBI SRA and NCBI Taxonomy and Wikipedia. Several browser functions were designed to allow users to obtain transcriptomic information. Users can browse by species, range of gene expression in FPKM, gene length, Gene Symbol, and Gene Ontology [27] based

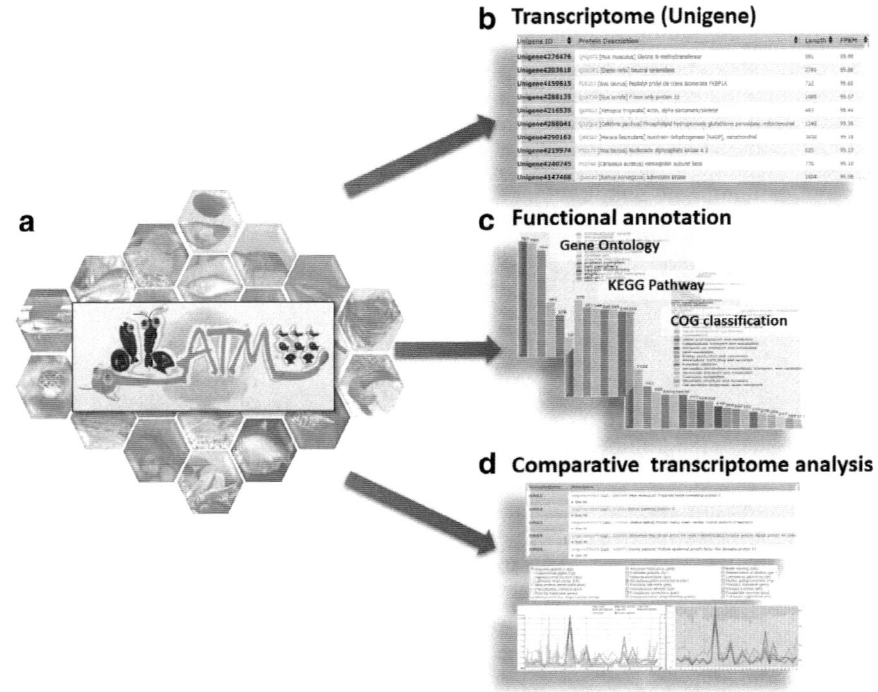

Fig. 1 Aquatic animal transcriptomes maps database. **a** The database collects RNA-seq data for more than 20 different aquatic animals. The database consists of three parts: **b** Detailed information for individual genes such as gene name and description, length, expression (FPKM), and sequence. **c** Functional annotation of all gene groups in individual species. **d** Evolutionary studies of all species transcriptomic data by constructing a comparative analysis system for homologous genes

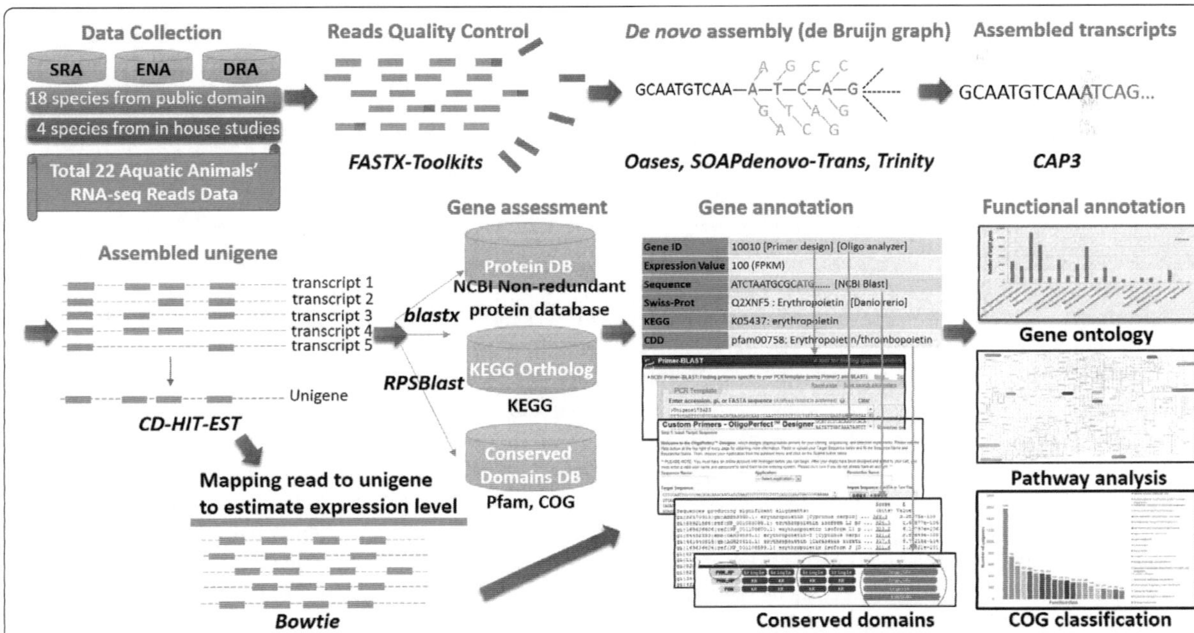

Fig. 2 dbATM database system and analysis. First, publicly available RNA-seq data were collected and combined with additional data produced for this study. Next, the NGS raw data were trimmed and filtered to remove low quality reads. Third, NGS reads were assembled based on the de Bruijn graphs and clustered into transcripts or unigenes. Finally, assembled reads were represented to genes and gene functions by BLAST against external database

Table 1 Gene annotation statistics in dbATM

Species	No. of transcripts	No. of proteins	No. of UniGene	No. of homologous genes
Anguilla japonica	99,696	38,815	13,600	9893
Astyanax mexicanus	59,723	23,085	10,443	7838
Clupea harengus	28,360	15,975	7713	5773
Crassostrea gigas	47,472	33,491	1149	377
Fundulus grandis	32,405	17,620	7388	4707
Gasterosteus aculeatus	77,510	43,170	14,035	6394
Ictalurus punctatus	57,435	28,850	13,400	10,452
Lateolabrax japonicus	27,075	15,130	6502	4227
Microphysogo biobrevirostris	50,827	31,923	14,839	11,295
Mytilus galloprovincialis	10,738	5944	815	141
Neocaridina denticulate	59,513	23,033	6593	1406
Pandalus latirostris	39,062	11,894	4127	1065
Penaeus monodon	36,979	14,740	4796	1162
Planorbarius corneus	61,364	20,951	3802	1004
Plecoglossus altivelis	43,524	22,836	9115	6544
Poecilia formosa	47,579	22,728	8271	3836
Poecilia mexicana	39,598	27,485	10,322	6423
Protopterus annectens	26,187	14,498	6050	2463
Pundamilia nyererei	19,205	13,762	5287	2749
Sinocyclocheilus angustiporus	51,104	41,391	11,505	8168
Sinocyclocheilus anophthalmus	81,704	48,526	13,583	10,452
Tetraodon nigroviridis	53,394	32,410	11,781	4238
Total	1,023,379	533,127	185,166	19,363[a]

[a]The total number of homologous gene groups show here is the real homologous gene groups in the database. Not the summation from individual species

on homologous evidence, KEGG orthology pathway [28], COG [29] and Pfam [30]. An introduction page presents assembled transcriptomic information for each species, including statistics for raw sequence data, assembly quality and annotated genes (Fig. 4a, c). The embedded NCBI BLAST page allows users to input sequences or upload a text file containing sequences in FASTA format to search all genes sequences in the dbATM based on nucleotide or protein similarity (Fig. 4b). dbATM also allows users to browse homologous genes in homologous groups (Fig. 4d), and provides an interface for comparing KEGG pathway mapping results with the average FPKM of genes for a given pathway in different species, allowing users to select a maximum of 10 species for expressional profiling comparison.

Investigation of aquatic animals' homologous genes

Most current transcriptomic studies focus on single species, and these "standalone" transcriptomic databases do not allow for the exploration of homologous gene identities and their relative expression levels. To overcome this constraint, we provide a novel function to comparatively analyze potential homologous genes among 22 aquatic animal species. In total, 19,369 homologous genes were identified by OrthoMCL [31] and their gene name. Using homologous search function in dbATM, we can discover that several homologous genes (like *yars, cars, nop14, acsbg2, trip12, rab3gap2,* and *herc2*) are distributed from invertebrates (mollusca and arthropoda) to vertebrates (ostariophysi, euteleostei, actinopterygii, and chordate). The results shown that the essential genes such as the tRNA synthetase *yars* and *cars*, are evolutionary conserved. We also found there 21 homologous genes shared across at least 17 species: *yars, cars, nop14, acsbg2, trip12, rab3gap2, herc2, ddost, vwa8, cdk11b, ascc3, aplp2, loc563777, nup98, nup188, sdad1, nup205, dlat, acadm, rtn4ip1, kansl3*. The number of homologous genes in each clade were shown in Table 2. The homologous gene lists in each clade were shown in Additional file1: Table S2. Information on homologous genes across different animal phyla will provide good material for future evo-devo studies.

Investigation of gene expression profiles

To profile the gene expression level among different aquatic animals, we compare the gene expression value (displayed by FPKM) across different KEGG pathway

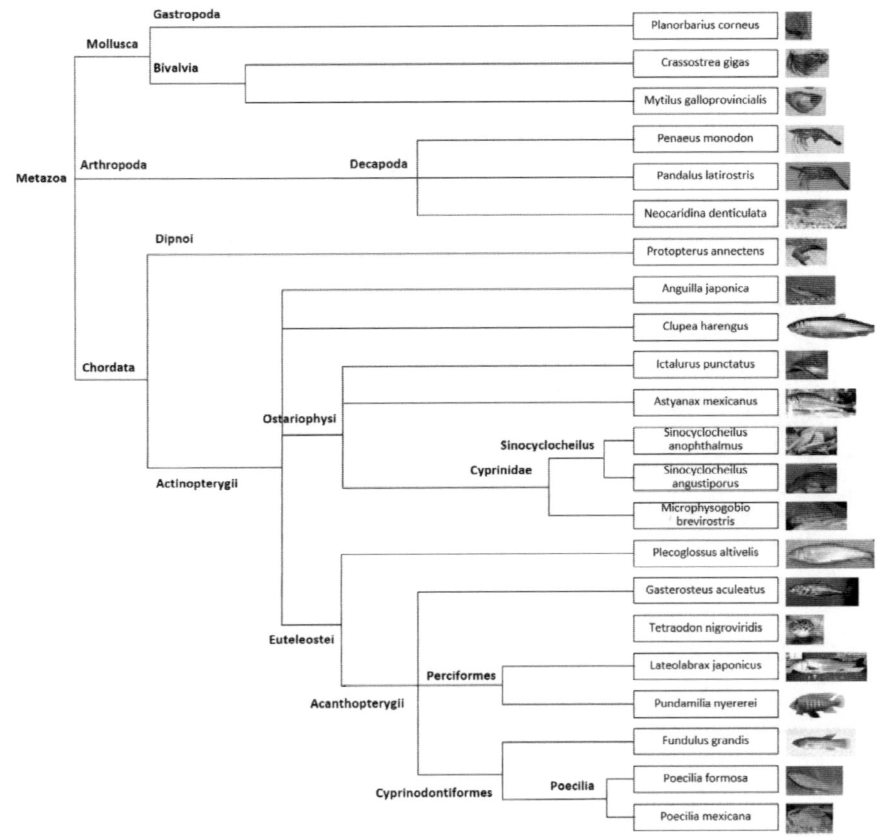

Fig. 3 Species in dbATM and their biological classification. We categorize organisms into sub-groups and represent their rough evolutionary distances for further evolutionary and phylogenetic investigation

categories. Results show the gene expression profile of a particular KEGG pathway stands out in some species. Distinct gene expressional profiling provides a good entry point to compare the gene expression level across different species or different tissue/organs among the same species. For example, the condition of the three species *Sinocyclocheilus angustiporus*, *Sinocyclocheilus anophthalmus*, and *Tetraodon nigroviridis* are from the brain tissue, and show similar expression profile trends (Fig. 5a). However, tissue for another four different condition species *Clupea harengus*, *Gasterosteus aculeatus*, *Protopterus annectens*, *Tetraodon nigroviridis* are whole organism, muscle, liver, and brain, respectively. The expression profiles panel of the four conditions (Fig. 5b) demonstrate quite shock and distinct.

Discussion

We present a new systematic approach for de novo RNA-seq dataset analysis and annotation including improvement of de novo assembly quality and construct a new resource for the transcriptomic map aquatic animals for evolution and phylogenetic study. The system flow is illustrated in Fig. 2 and further detail is provided in the Materials and Methods section. First, we optimized the de novo assembly pipeline of RNA-seq data [32] by combining Oases, SOAPdenovo-Trans, and Trinity assemblers. To obtain complete transcriptomes from various assemblers [33, 34], we used the CD-HIT-EST cluster tool to merge Oases, SOAPdenovo-Trans, and Trinity assembling results. This combinational approach significantly reduce the transcript number and keeping a representative sequence for each unigene. Second, the gene annotation by BLASTX shows that some transcripts are mapped to the same gene or gene family. By following NCBI UniGene approach, we clustered the same gene to a group to reduce redundancy. Third, to facilitate the use of transcriptomes data for evolution and phylogenetic investigations, unigenes derived from all collected species were used to construct a homologous gene group. Besides, we provided the nucleotide (unigene) and protein sequence of all the 22 species on "Download Page". The sequence data may apply the machine learning approach that could extract new features to develop gene

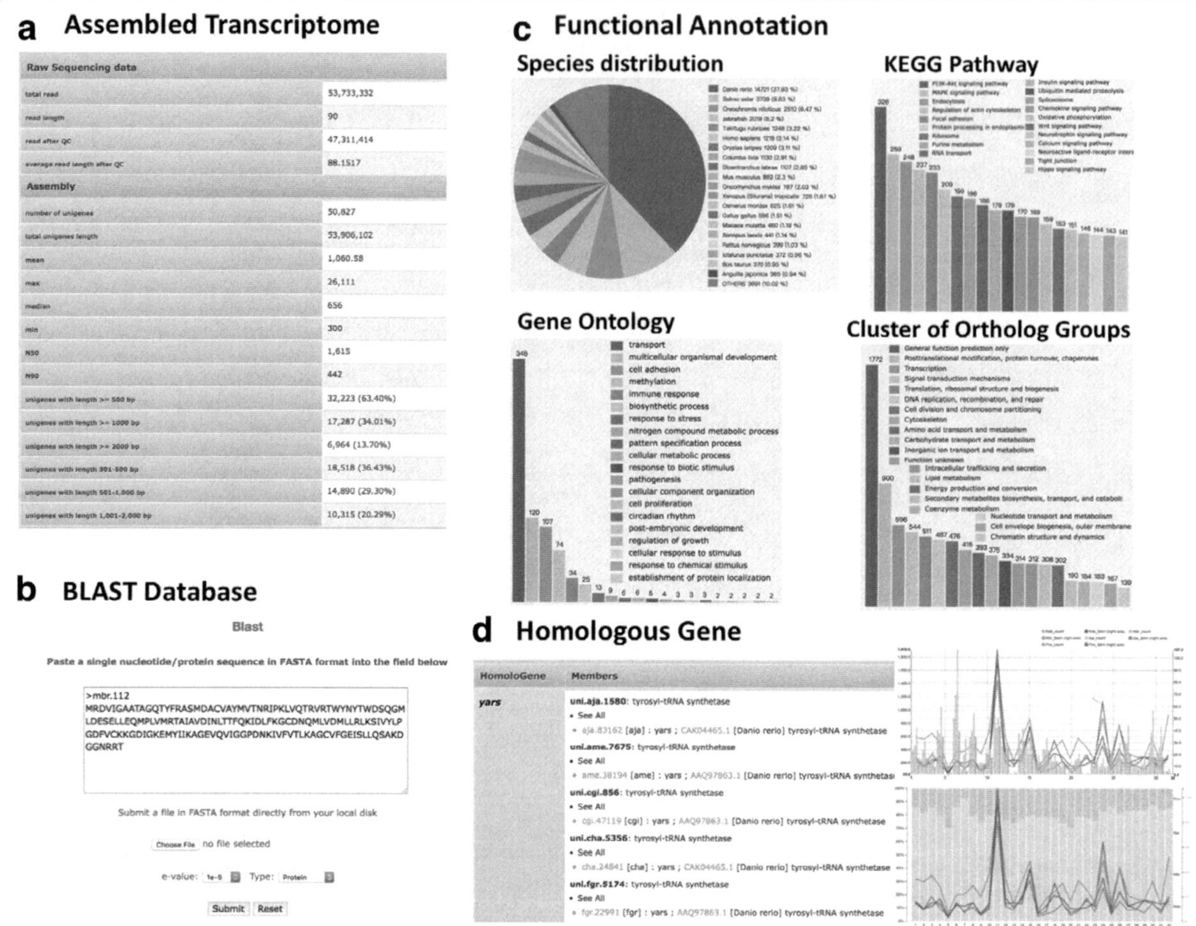

Fig. 4 dbATM web interface. The dbATM provides various query interfaces and graphical visualizations to facilitate access to aquatic transcriptomic data. **a** Summary table of the reads, assembly quality, and annotation statistics. **b** A web interface to facilitate unigene mining by entering either sequence or keyword. **c** The functional annotation interface including annotated species distribution, KEGG pathway, Gene Ontology, and cluster of ortholog groups. **d** The homologous gene search and browsing interface for comparative analysis

prediction tools for aquatic animals. Finally, we develop a comparative analysis system by incorporating KEGG pathway analysis and gene expression profiles. This system assesses species and conditional diversity using the gene expression profile in specific biological pathways. dbATM not only improves de novo assembly quality of RNA-seq data but also constructs a database of the homologous genes of aquatic animals to allow for comparative study.

Conclusions

dbATM is the first database to contain comprehensive transcriptomes annotations for more than 20 aquatic

Table 2 Homologous genes statistics in each clade

	Mollusca[a]	*Arthropoda*[b]	*Ostariophysi*[c]	*Euteleostei*[d]	*Actinopterygii*[e]	*Chordate*[f]
Numbers	27	582	3027	85	19	8

[a]*Mollusca* clade: *Planorbarius corneus, Crassostrea gigas, Mytilus galloprovincialis*
[b]*Arthropoda* clade: *Penaeus monodon, Pandalus latirostris, Neocaridina denticulate*
[c]*Ostariophysi* clade: *Ictalurus punctatus, Astyanax mexicanus, Sinocyclocheilus angustiporus, Sinocyclocheilus anophthalmus, Microphysogo biobrevirostris*
[d]*Euteleostei* clade: *Plecoglossus altivelis, Gasterosteus aculeatus, Tetraodon nigroviridis, Lateolabrax japonicas, Pundamilia nyererei, Fundulus grandis, Poecilia Formosa, Poecilia mexicana*
[e]*Actinopterygii* clade: *Ictalurus punctatus, Astyanax mexicanus, Sinocyclocheilus angustiporus, Sinocyclocheilus anophthalmus, Microphysogo biobrevirostris, Plecoglossus altivelis, Gasterosteus aculeatus, Tetraodon nigroviridis, Lateolabrax japonicas, Pundamilia nyererei, Fundulus grandis, Poecilia Formosa, Poecilia Mexicana. Anguilla japonica, Clupea harengus*
[f]*Chordate* clade: *Ictalurus punctatus, Astyanax mexicanus, Sinocyclocheilus angustiporus, Sinocyclocheilus anophthalmus, Microphysogo biobrevirostris, Plecoglossus altivelis, Gasterosteus aculeatus, Tetraodon nigroviridis, Lateolabrax japonicas, Pundamilia nyererei, Fundulus grandis, Poecilia formosa, Poecilia mexicana. Anguilla japonica, Clupea harengus, Protopterus annectens*

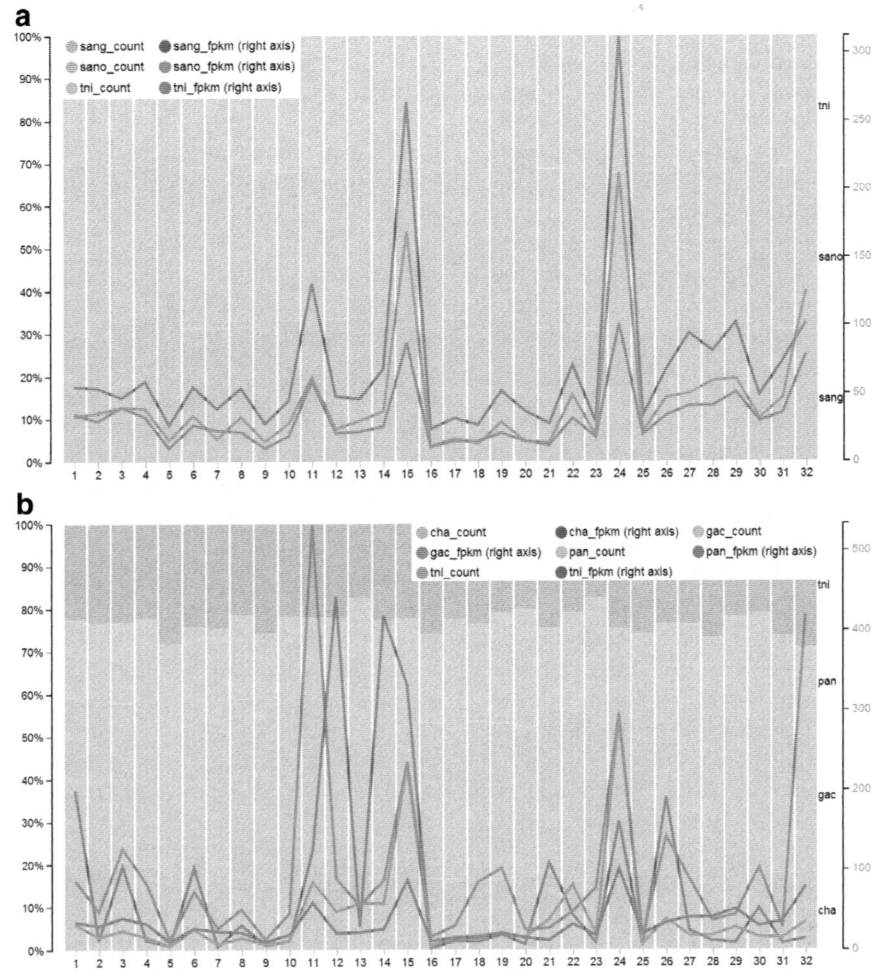

Fig. 5 Gene expression profiles panel. The dbATM provides a function for comparative analysis of different tissues, conditions or species. **a** All three species samples are taken from the brain, presenting similar expression profile trends. **b** All four species are taken from different tissue, showing distinct expression profiles. Abbreviation code for species: *cha, Clupea harengus;* gac, *Gasterosteus aculeatus;* pan, *Protopterus annectens;* sang, *Sinocyclocheilus angustiporus;* sano, *Sinocyclocheilus anophthalmus; tni, Tetraodon nigroviridis*

animals. The user-friendly web interface allows public access to valuable information for transcriptome assembly of each organism. The newly constructed homologous genes database and comparative analysis system also give aquatic researchers insights into the fields of eco-toxicity, animal physiology, comparative genomics and phylogenetic. dbATM also serves as an important repository for the aquatic animal transcriptomes by analyzing RNA-seq data. The assembly gene information provided could be a potentially valuable resource for designing expression microarrays to detect gene expression profiles across many conditions, thus facilitating studies in ecology and molecular biology. Given recent advances in NGS technologies, more non-model aquatic animal RNA-seq experiments will be deposited in the public domain.

Methods

System overview

First, RNA-seq data for 18 species of aquatic animals that were not well-annotated were collected from public resources such as SRA, ENA and DRA (the data from Illumina paired-end NGS sequencing platform were collected), with an additional four datasets produced for this study (Fig. 3 and Additional file 1: Table S1). Second, the quality of the raw reads from NGS was validated to trim and filter low quality reads. Third, the trimmed reads were then assembled through the designed combinational approach. Finally, genes functionally annotated by BLAST were compared against external databases (Fig. 3). All of the annotation results were soundly provided in the dbATM database.

RNA-seq dataset collection

We collected eighteen RNA-seq datasets from the NCBI Sequence Read Archive (SRA), European Nucleotide Archive (ENA), and DDBJ Sequence Read Archive (DRA). RNA-seq data deposited to these datasets were generated from Illumina paired-end NGS sequencing platform. All RNA-seq data collected in dbATM were reference genome-free (including scaffolds or contig status) in the NCBI taxonomy database. We also performed RNA-seq on four additional aquatic species. Additional file 1: Table S1 present detailed information for the 22 aquatic animal species profiled in our dbATM database.

Cleaning NGS read data

The NGS raw data for the RNA-seq in SRA format were converted to FASTQ format using the SRA-Toolkit v2.2.2 [35] and the FASTQ format reads were cleaned to increase read quality by FASTX-Toolkit v0.0.13 (http://hannonlab.cshl.edu/fastx_toolkit/). First, reads with adapters were removed. Second, the sequencing reads were scanned for quality at the reads tails. If the Phred quality scores were below 20 at the nucleotide end, the nucleotide sequences were removed. Third, if a read's average Phred quality score was below 20, the read was discarded. Fourth, reads were discarded if their length was less than 70% nucleotides after tail trimming and reads filtering. Finally, for paired-end datasets, if one paired read had been removed in a previous step, the other read was also removed to synchronize the read pairs.

De novo assembly

To ensure more complete assembly, we used three de novo assembly tools: Oases (v0.2.08, requiring Velvet v1.2.09) [36, 37], SOAPdenovo-Trans (release 1.02) [38], and Trinity (release 2013–02-25) [24] that based on the de Bruijn graphs. Previous reports suggest that coverage in transcriptomes assembly is not uniform. Higher *k*-mer lengths could obtain good quality assemblies for highly expressed transcripts. Lower *k*-mer lengths are used for poorly expressed transcripts [39]. This has been experimentally verified in model organisms [40]. We used multiple *k*-mer assemblies in Oases and SOAP-denovoTrans (for all species, $k = 21$, 29, 37, 45, 53, and 61, except for read lengths in *Fundulus grandis* smaller than 60, where $k = 21$, 27, 33, 39, 45, and 51). The insertion length of paired-end RNA-seq data was estimated using the observed-insert-length.pl program (included in the Oases package). Trinity was applied with default parameters. After the first step assembly in three assembler, we selected high read coverage and relatively long length as the representative transcript and conducted further assembly using CAP3 Version Date: 2007–10-15 [41] with an overlap length cutoff of 200 and overlap percent identity cutoff of 99 (-o 200 -p 99) [32]. To obtain comprehensive transcriptomes from various assemblers [33, 34], we used the CD-HIT-EST v4.6 [42] cluster tool with a sequence identity cutoff of 90% to merge results from Oases, SOAPdenovo-Trans, and Trinity.

Abundance estimation

To determine the expression level in each species, we calculated the relative abundance of previously assembled results by RSEM (v1.2.3) [43] requiring bowtie (v1.0.0) [44] to accurately quantify transcript expression with or without reference genomes. The relative abundances of transcripts were measured in unit of normalized reads count aligned on de novo assembly, FPKM (Fragments Per Kilobase of transcript per Million mapped reads) [45]. After de novo assembly and expression calculation, we select FPKM > = 1 and gene length > = 300 as a cutoff for further analysis such as functional annotation and comparative analysis.

Functional annotation

BLASTX (v2.2.28+) [46] with E-value cutoff of 1.0e-5 were used to homology search the NCBI non-redundant protein database (nr, v2012–11-19) for the assembled genes. The best blast hit protein was annotated to the representative gene and protein name. The gene annotation by BLASTX indicates that some transcripts were mapped to the same gene or gene family. To reduce redundancy, we used the NCBI UniGene approach which clusters the same gene to a group. The results for the BLASTX annotation genes were mapped to the Gene Ontology database. We also extracted the open reading frames (ORFs) from all genes according to the best BLASTX match hit results. To identify genes involved in KEGG pathway, we used the KEGG Automatic Annotation Server, KAAS [47]. Finally, we identified the putative protein domains on translated protein sequences extracted from genes by aligning to CDD (conserved domain and protein classification) [48] using RPS-BLAST.

Comparative analysis

Based on a comparative analysis, different species were grouped into three main categories. First, we used OrthoMCL [31] with an E-value cutoff of 1.0e-10 [49] to perform a similarity search orthologous by computational approach, and then defined the orthologous groups using the representative gene name or description, and called each group a homologous gene group, indicating that the different species shared a common ancestor. Second, we compared the abundant gene expressions among the different species using the functional annotation KEGG results and the D3: Data-Driven Documents [50] to provide enhanced graphical representations. Considerable amounts of annotation information was obtained from the KEGG pathway category [28] with average of gene expressions (FPKM) in different species. In Fig. 5, the bar is the read count aligned to each KEGG pathway category, and the line is the average FPKM of the genes mapped to this pathway. In the KEGG pathway category, we excluded Human Diseases, Drug Development, and Global Maps from the metabolic category. The graph visualization of comparative analysis represents the gene expression profile in different tissues, conditions or species using KEGG pathway. For example, the Fig. 5a illustrates the samples from the brain, so the expression profile trends (gene count and average FPKM) are similar. In contrast, the Fig. 5b displays the samples from different tissue, so the expression profile trends are distinct. Third, a web interface of BLAST against dbATM to allow the user using blastp/ blastn to search protein/nucleotide data by pasting the query sequence or by submitting a file in FASTA format.

Additional file

Additional file 1: Figure S1. The cumulative publications of related articles of aquatic animal in PubMed. **Table S1.** RNA-seq datasets from twenty-two aquatic animals were analyzed in dbATM. **Table S2.** Statistics of homologous genes in each clade and their gene lists. (PDF 280 kb)

Abbreviations

COG: Clusters of Orthologous Groups; DRA: DDBJ Sequence Read Archive; ENA: European Nucleotide Archive; FPKM: Fragments per kilobase of transcript per million mapped fragments; KEGG: Kyoto Encyclopedia of Genes and Genomes; NGS: Next-generation sequence; RNA-seq: RNA sequencing; SRA: Sequence Read Archive

Acknowledgements

Not applicable.

Funding

This work is supported by the Ministry of Science and Technology, Taiwan (MOST) [MOST 106-2627-M-009-002, MOST 106-2319-B-400-001, MOST 106-2633-B-009-001]. This work is particularly supported by "Aiming for the Top University Program" of the National Chiao Tung University and Ministry of Education, Taiwan [MOHW106-TDU-B-212-144005 and MMH-HB-10602]. Funding for the publication of this article was provided by MOST grant [MOST-106-2627-M-009-002].

Availability of data and materials

The RNA sequencing data sets generated in this study have been deposited in the NBCI SRA database under accession SRR1185326 for *Anguilla japonica*, SRR1185341 for *Microphysogobio brevirostris*, SRR118532 for *Neocaridina denticulate*, and SRR1185333 for *Planorbarius corneus*. Others accession codes of RNA-seq data download from public domain database were listed in Additional file 1: Table S1. The database is now publically accessible through the URL http://dbATM.mbc.nctu.edu.tw/.

About this supplement

This article has been published as part of *BMC Genomics* Volume 19 Supplement 2, 2018: Selected articles from the 16th Asia Pacific Bioinformatics Conference (APBC 2018): genomics. The full contents of the supplement are available online at https://bmcgenomics.biomedcentral.com/articles/supplements/volume-19-supplement-2.

Authors' contributions

CHC, CDH and HDH conceived and designed the experiments. CHC, HYH, WCH, SDH, CYL, YHC, YCL, WYH analyzed the datasets. CHC, HYH, SDH, CYL, YHC establish the database. CHC, HYH, SDH, CYL, YHC, YCL, WYH, WCH, MLL, YCC, CDH and HDH wrote or revised the paper. All authors have read and approved the final manuscript.

Ethics approval and consent to participate

Not applicable.

Consent for publication

Not applicable.

Competing interests

All authors declare that they have no competing interests.

Publisher's Note

Springer Nature remains neutral with regard to jurisdictional claims in published maps and institutional affiliations.

Author details

[1]Institute of Bioinformatics and Systems Biology, National Chiao Tung University, Hsinchu 300, Taiwan. [2]Department of Biological Science and Technology, National Chiao Tung University, Hsinchu 300, Taiwan. [3]Department of Bioscience Technology, Chung Yuan Christian University, Chungli 320, Taiwan. [4]Institute of Molecular Medicine and Bioengineering, National Chiao Tung University, Hsinchu 300, Taiwan.

Published: 9 May 2018

References

1. Balmford A, Bruner A, Cooper P, Costanza R, Farber S, Green RE, Jenkins M, Jefferiss P, Jessamy V, Madden J, et al. Economic reasons for conserving wild nature. Science. 2002;297(5583):950–3.
2. Schwartsmann G. Marine organisms and other novel natural sources of new cancer drugs. Ann Oncol. 2000;11(Suppl 3):235–43.
3. Naylor RL, Goldburg RJ, Primavera JH, Kautsky N, Beveridge MC, Clay J, Folke C, Lubchenco J, Mooney H, Troell M. Effect of aquaculture on world fish supplies. Nature. 2000;405(6790):1017–24.
4. Cabello FC. Heavy use of prophylactic antibiotics in aquaculture: a growing problem for human and animal health and for the environment. Environ Microbiol. 2006;8(7):1137–44.
5. Lu XJ, Hang XY, Yin L, He YQ, Chen J, Shi YH, Li CH. Sequencing of the first ayu (Plecoglossus Altivelis) macrophage transcriptome and microarray development for investigation the effect of LECT2 on macrophages. Fish Shellfish Immunol. 2013;34(2):497–504.
6. Ji P, Liu G, Xu J, Wang X, Li J, Zhao Z, Zhang X, Zhang Y, Xu P, Sun X. Characterization of common carp transcriptome: sequencing, de novo assembly, annotation and comparative genomics. PLoS One. 2012;7(4):e35152.
7. Christie AE, Roncalli V, Lona PB, McCoole MD, King BL, Bucklin A, Hartline DK, Lenz PH. In silico characterization of the insect diapause-associated protein couch potato (CPO) in Calanus Finmarchicus (Crustacea: Copepoda). Comp Biochem Physiol Part D Genomics Proteomics. 2013;8(1):45–57.
8. Bai Z, Zheng H, Lin J, Wang G, Li J. Comparative analysis of the Transcriptome in tissues secreting purple and white nacre in the pearl mussel Hyriopsis cumingii. PLoS One. 2013;8(1):e53617.
9. An HS, Lee JW. Development of microsatellite markers for the Korean mussel, Mytilus Coruscus (Mytilidae) using next-generation sequencing. Int J Mol Sci. 2012;13(8):10583–93.
10. Minegishi Y, Henkel CV, Dirks RP, van den Thillart GE. Genomics in eels–towards aquaculture and biology. Marine Biotechnol. 2012;14(5):583–90.
11. Hu H, Bandyopadhyay PK, Olivera BM, Yandell M. Elucidation of the molecular envenomation strategy of the cone snail Conus geographus through transcriptome sequencing of its venom duct. BMC Genomics. 2012;13:284.
12. Roberts SB, Hauser L, Seeb LW, Seeb JE. Development of genomic resources for Pacific herring through targeted transcriptome pyrosequencing. PLoS One. 2012;7(2):e30908.
13. Garcia TI, Shen Y, Crawford D, Oleksiak MF, Whitehead A, Walter RB. RNA-Seq reveals complex genetic response to deepwater horizon oil release in Fundulus Grandis. BMC Genomics. 2012;13:474.
14. Kawahara-Miki R, Wada K, Azuma N, Chiba S. Expression profiling without genome sequence information in a non-model species, Pandalid shrimp (Pandalus Latirostris), by next-generation sequencing. PLoS One. 2011;6(10):e26043.
15. Kelley JL, Passow CN, Plath M, Arias Rodriguez L, Yee MC, Tobler M. Genomic resources for a model in adaptation and speciation research: characterization of the Poecilia Mexicana transcriptome. BMC Genomics. 2012;13:652.

16. Meng F, Braasch I, Phillips JB, Lin X, Titus T, Zhang C, Postlethwait JH. Evolution of the eye transcriptome under constant darkness in Sinocyclocheilus cavefish. Mol Biol Evol. 2013;30(7):1527–43.
17. Gross JB, Furterer A, Carlson BM, Stahl BA. An integrated Transcriptome-wide analysis of cave and surface dwelling Astyanax Mexicanus. PLoS One. 2013;8(2):e55659.
18. Iwasaki W, Fukunaga T, Isagozawa R, Yamada K, Maeda Y, Satoh TP, Sado T, Mabuchi K, Takeshima H, Miya M, et al. MitoFish and MitoAnnotator: a mitochondrial genome database of fish with an accurate and automatic annotation pipeline. Mol Biol Evol. 2013;30(11):2531–40.
19. Werner T. Next generation sequencing in functional genomics. Brief Bioinform. 2010;11(5):499–511.
20. Martin JA, Wang Z. Next-generation transcriptome assembly. Nat Rev Genet. 2011;12(10):671–82.
21. Pipes L, Li S, Bozinoski M, Palermo R, Peng X, Blood P, Kelly S, Weiss JM, Thierry-Mieg J, Thierry-Mieg D, et al. The non-human primate reference transcriptome resource (NHPRTR) for comparative functional genomics. Nucleic Acids Res. 2013;41(Database issue):D906–14.
22. Wang Z, Gerstein M, Snyder M. RNA-Seq: a revolutionary tool for transcriptomics. Nat Rev Genet. 2009;10(1):57–63.
23. Mortazavi A, Williams BA, McCue K, Schaeffer L, Wold B. Mapping and quantifying mammalian transcriptomes by RNA-Seq. Nat Methods. 2008;5(7):621–8.
24. Grabherr MG, Haas BJ, Yassour M, Levin JZ, Thompson DA, Amit I, Adiconis X, Fan L, Raychowdhury R, Zeng Q, et al. Full-length transcriptome assembly from RNA-Seq data without a reference genome. Nat Biotechnol. 2011;29(7):644–52.
25. Coordinators NR. Database resources of the National Center for biotechnology information. Nucleic Acids Res. 2013;41(Database issue):D8–D20.
26. Zeng V, Extavour CG. ASGARD: an open-access database of annotated transcriptomes for emerging model arthropod species. Database. 2012;2012:bas048.
27. Ashburner M, Ball CA, Blake JA, Botstein D, Butler H, Cherry JM, Davis AP, Dolinski K, Dwight SS, Eppig JT, et al. Gene ontology: tool for the unification of biology. The gene ontology consortium. Nat Genet. 2000;25(1):25–9.
28. Kanehisa M, Goto S, Kawashima S, Okuno Y, Hattori M. The KEGG resource for deciphering the genome. Nucleic Acids Res. 2004;32(Database issue):D277–80.
29. Tatusov RL, Koonin EV, Lipman DJ. A genomic perspective on protein families. Science. 1997;278(5338):631–7.
30. Punta M, Coggill PC, Eberhardt RY, Mistry J, Tate J, Boursnell C, Pang N, Forslund K, Ceric G, Clements J, et al. The Pfam protein families database. Nucleic Acids Res. 2012;40(Database issue):D290–301.
31. Li L, Stoeckert CJ Jr, Roos DS. OrthoMCL: identification of ortholog groups for eukaryotic genomes. Genome Res. 2003;13(9):2178–89.
32. Yang Y, Smith SA. Optimizing de novo assembly of short-read RNA-seq data for phylogenomics. BMC Genomics. 2013;14:328.
33. Garg R, Patel RK, Jhanwar S, Priya P, Bhattacharjee A, Yadav G, Bhatia S, Chattopadhyay D, Tyagi AK, Jain M. Gene discovery and tissue-specific transcriptome analysis in chickpea with massively parallel pyrosequencing and web resource development. Plant Physiol. 2011;156(4):1661–78.
34. Ghangal R, Chaudhary S, Jain M, Purty RS, Chand Sharma P. Optimization of de novo short read assembly of seabuckthorn (Hippophae Rhamnoides L.) transcriptome. PLoS One. 2013;8(8):e72516.
35. Kodama Y, Shumway M, Leinonen R, International Nucleotide Sequence Database C. The sequence read archive: explosive growth of sequencing data. Nucleic Acids Res. 2012;40(Database issue):D54–6.
36. Schulz MH, Zerbino DR, Vingron M, Birney E. Oases: robust de novo RNA-seq assembly across the dynamic range of expression levels. Bioinformatics. 2012;28(8):1086–92.
37. Zerbino DR, Birney E. Velvet: algorithms for de novo short read assembly using de Bruijn graphs. Genome Res. 2008;18(5):821–9.
38. Luo R, Liu B, Xie Y, Li Z, Huang W, Yuan J, He G, Chen Y, Pan Q, Liu Y, et al. SOAPdenovo2: an empirically improved memory-efficient short-read de novo assembler. GigaScience. 2012;1(1):18.
39. Surget-Groba Y, Montoya-Burgos JI. Optimization of de novo transcriptome assembly from next-generation sequencing data. Genome Res. 2010;20(10):1432–40.
40. Gibbons JG, Janson EM, Hittinger CT, Johnston M, Abbot P, Rokas A. Benchmarking next-generation transcriptome sequencing for functional and evolutionary genomics. Mol Biol Evol. 2009;26(12):2731–44.
41. Huang X, Madan A. CAP3: a DNA sequence assembly program. Genome Res. 1999;9(9):868–77.
42. Li W, Godzik A. Cd-hit: a fast program for clustering and comparing large sets of protein or nucleotide sequences. Bioinformatics. 2006;22(13):1658–9.
43. Li B, Dewey CN. RSEM: accurate transcript quantification from RNA-Seq data with or without a reference genome. BMC bioinformatics. 2011;12:323.
44. Langmead B, Trapnell C, Pop M, Salzberg SL. Ultrafast and memory-efficient alignment of short DNA sequences to the human genome. Genome Biol. 2009;10(3):R25.
45. Trapnell C, Williams BA, Pertea G, Mortazavi A, Kwan G, van Baren MJ, Salzberg SL, Wold BJ, Pachter L. Transcript assembly and quantification by RNA-Seq reveals unannotated transcripts and isoform switching during cell differentiation. Nat Biotechnol. 2010;28(5):511–5.
46. Camacho C, Coulouris G, Avagyan V, Ma N, Papadopoulos J, Bealer K, Madden TL. BLAST+: architecture and applications. BMC bioinformatics. 2009;10:421.
47. Moriya Y, Itoh M, Okuda S, Yoshizawa AC, Kanehisa M. KAAS: an automatic genome annotation and pathway reconstruction server. Nucleic Acids Res. 2007;35(Web Server issue):W182–5.
48. Marchler-Bauer A, Zheng C, Chitsaz F, Derbyshire MK, Geer LY, Geer RC, Gonzales NR, Gwadz M, Hurwitz DI, Lanczycki CJ, et al. CDD: conserved domains and protein three-dimensional structure. Nucleic Acids Res. 2013;41(Database issue):D348–52.
49. Matthews LR, Vaglio P, Reboul J, Ge H, Davis BP, Garrels J, Vincent S, Vidal M. Identification of potential interaction networks using sequence-based searches for conserved protein-protein interactions or "interologs". Genome Res. 2001;11(12):2120–6.
50. Bostock M, Ogievetsky V, Heer J. D(3): Data-Driven Documents. IEEE Trans Vis Comput Graph. 2011;17(12):2301–9.

He *et al. BMC Genomics* 2018, **19**(Suppl 2):110
https://doi.org/10.1186/s12864-018-4464-9

BMC Genomics

RESEARCH Open Access

Efficient algorithms for polyploid haplotype phasing

Dan He[1*], Subrata Saha[2], Richard Finkers[3] and Laxmi Parida[2]

From The Sixteenth Asia Pacific Bioinformatics Conference 2018
Yokohama, Japan. 15-17 January 2018

Abstract

Background: Inference of haplotypes, or the sequence of alleles along the same chromosomes, is a fundamental problem in genetics and is a key component for many analyses including admixture mapping, identifying regions of identity by descent and imputation. Haplotype phasing based on sequencing reads has attracted lots of attentions. Diploid haplotype phasing where the two haplotypes are complimentary have been studied extensively. In this work, we focused on Polyploid haplotype phasing where we aim to phase more than two haplotypes at the same time from sequencing data. The problem is much more complicated as the search space becomes much larger and the haplotypes do not need to be complimentary any more.

Results: We proposed two algorithms, (1) Poly-Harsh, a Gibbs Sampling based algorithm which alternatively samples haplotypes and the read assignments to minimize the mismatches between the reads and the phased haplotypes, (2) An efficient algorithm to concatenate haplotype blocks into contiguous haplotypes.

Conclusions: Our experiments showed that our method is able to improve the quality of the phased haplotypes over the state-of-the-art methods. To our knowledge, our algorithm for haplotype blocks concatenation is the first algorithm that leverages the shared information across multiple individuals to construct contiguous haplotypes. Our experiments showed that it is both efficient and effective.

Background

Haplotype, or the sequence of alleles residing on the same chromosome, is the fundamental unit of genetic variation. Inference of haplotypes plays an important role in many analyses, including identifying regions of IBD (Identity-by-descent) [1–3], admixture mapping [4], imputation of uncollected genetic variation [5, 6]. Molecular methods [7] are expensive and not amenable to high throughput technologies for obtaining haplotypes. Therefore most studies rely on genotype information and infer haplotypes from genotypes, referred to as haplotype inference or haplotype phasing.

Next-generation sequencing (NGS) technologies have been applied to haplotype phasing as each sequencing read originates from a single chromosome and alleles spanned by that read are on the same haplotype. Phasing diploid haplotypes, especially human haplotypes, has been studied extensively. The human genome is diploid and the two copies of each chromosome are mostly homozygous, namely the alleles at the same positions are mostly identical. However, there are some variations between the pairs and if there are different alleles at the same positions between the pair of chromosomes, they are referred to as heterozygous alleles. For diploid haplotype phasing, only heterozygous alleles are considered and thus the two haplotypes are complimentary to each other.

For diploid haplotype phasing, since many reads overlap with each other, most methods infer haplotypes by partitioning the reads into two sets corresponding to chromosomal origin in such a way that the number of conflicts between the reads and the predicted haplotypes is minimized (such objective function is called Minimum Error Correction, or MEC). Many methods have been proposed for the diploid haplotype phasing problem: HASH

*Correspondence: hedanus@gmail.com
[1]College of Computer Science and Software, Shenzhen University, Shenzhen 518060, China
Full list of author information is available at the end of the article

[8] and HAPCUT [9] are based on graph structure. He et al. [10] proposed a dynamic programming method as well as a Max-SAT formulation. Deng et al. [11] combined this dynamic programming method with a heuristic approach. Mousavi et al. [12] suggested a HapSat model by converting the haplotype determination problem to a Max-2-SAT problem. HapAssembly converts the haplotype assembly problem to an integer linear programming problem for optimization [13]. Harsh [14] applied a Gibbs Sampling algorithm to phase diploid haplotypes. WhatsHap [15] proposed a fixed parameter tractable approach with coverage as the parameter and can optimize weighted MEC in runtime linear in the number of SNPs.

Recently polyploid haplotype phasing, where more than two haplotypes are phased at the same time, has attracted lots of attention. Polyploid haplotypes mainly come from plant genomes as the plants usually contain more than two haplotypes. Examples of such organisms include potato (which is tetraploid) and wheat (hexaploid). Compared with diploid haplotype phasing, polyploid haplotype phasing is much more challenging as the search space of possible haplotypes increase quadratically with the number of haplotypes. For k-haplotypes each with n SNPs, the search space is $O\left(2^{(k-1)n}\right)$. Therefore it is very challenging to find the optimal haplotypes.

Various polyploid haplotype phasing methods have been developed. HapCompass [16] relied on a graphical approach to develop a scheme which resolves conflicts arising from incorrect haplotype phasing. HapTree [17] investigated the polyploid setup using a branch-and-bound scheme. SDhaP [18] formulates the haplotype assembly problem as a semi-definite program and exploits its special structure, the low rank of the underlying solution, to solve it rapidly and with high accuracy. H-PoP and H-PoPG [19] conducted Polyploid Balanced Optimal Partition (PBOP).

In this work, we proposed a Gibbs Sampling based algorithm Poly-Harsh. The Gibbs Sampling algorithm considers the polyploid haplotype phasing problem as sampling k haplotypes from a huge search space ($O\left(2^{(k-1)n}\right)$, where k is the ploidy, n is the number of SNPs). The k haplotypes minimizes certain objective function (In this work, we take MEC (Minimum Error Correction) as objective function). Poly-Harsh samples the conditional probability of read assignment with fixed haplotypes and haplotype values with fixed read assignment alternatively. We derived a formula for the conditional probability which depends not only on the correct assignment of a read to a haplotype, but also depends on correctly not assigning a read to a haplotype. Our experiments on simulated data showed that the alternative sampling converges fast (usually in less than 100 iterations) and our method achieves a better performance than the state-of-the-art methods.

Haplotype phasing algorithms usually produce blocks of haplotypes due to low coverage or sequencing error. How the blocks should be concatenated is a very challenging problem and has never been resolved. When a single individual is phased, we in general do not have enough information to concatenate the haplotype blocks and the blocks can be only concatenated randomly. However, when a set of individuals inherited from the same founders are phased together, we could leverage the haplotype blocks from all the individuals to better concatenate them. Thus we also proposed an efficient algorithm to concatenate haplotype blocks of a set of individuals from their sequencing data simultaneously. The algorithm consists of three steps (1) candidates generation (2) frequent candidates detection (3) true candidates detection. We showed that our algorithm achieves a much better accuracy than a baseline method, the random concatenation.

Methods

Gibbs-sampling

Our method is based on Gibbs sampling and we first introduce the general idea of Gibbs sampling below. Consider the following distribution typically used to perform optimization in graphical models:

$$P(X) = \frac{1}{Z} exp\left(\mu \sum_{i=1} \sum_{j=1} \phi_{ij}(x_i, x_j)\right) \quad (1)$$

where $X = (x_1, x_2, \ldots, x_d)$ is a d-dimensional vector and Z is a normalization factor. The function ϕ specifies the edge potential for two variables with an edge between them. We would like to collect samples of X based on this distribution $P(X)$.

Gibbs sampler is a special case of Monte Carlo Markov Chain (MCMC) method [20], which is guaranteed to converge to the equilibrium distribution after sufficient burn-in iterations. In each iteration, it randomly samples one variable x_i based on the conditional probability $P(x_i|x_{[-i]})$ where all other variables $x_{[-i]} = (x_1, x_2, \ldots, x_{i-1}, x_{i+1}, \ldots, x_d)$ are fixed. Formally, this conditional probability can be written based on bayesian rule:

$$P(x_i = t|x_{[-i]}) = \frac{P(x_i = t, x_{[-i]})}{\sum_{t'} P(x_i = t', x_{[-i]})} \quad (2)$$

Readers can refer to [21] for a more detailed description of Monte Carlo Markov Chain.

Polyploid haplotype phasing

The inputs for the polyploid haplotype phasing problem is the ploidy k (the number of haplotypes to be phased), the set of aligned sequencing reads X (We assume the raw reads have been aligned to a reference sequence and thus the SNPs spanned by the reads are identified already), a sequencing error rate ϵ. The VCF (Variant Call Format) file containing the SNP positions and dosages could be optional. The dosage information gives the number of reference alleles and alternative alleles for a given SNP position and therefore can be used to reduce the phasing search space and to improve the phasing accuracy. Notice for some programs such as HapCompass, the dosage information is mandatory.

The output of the phasing algorithms is the k phased haplotypes. There are a few popular metrics to evaluate the performance of the phasing algorithms, such as MEC (minimum error correction) [22], minimum fragment removal, MSR (minimum single nucleotide polymorphism (SNP) removal) [23], and two recent models MFC (maximum fragments cut) [24] and BOP (balanced optimal partition) [25]. In this work, we focused on MEC as our metric.

MEC

We focus on minimizing MEC between the phased haplotypes and the input read matrix, which is calculated as the total number of mismatches between the reads and their assigned haplotypes. The following formula is the MEC for polyploid haplotypes:

$$MEC(X,H) = \sum_{j=1}^{m}\sum_{k=1}^{n} r_{jk} \times D(x_j, h_k) \tag{3}$$

where X is a set of m sequencing reads, H is the set of n haplotypes, x_j is the j-th read, h_k is the k-th haplotype, $D(x_j, h_k)$ is the number of mismatches between x_j and h_k, r_{jk} is 1 if the j-the read is assigned to the k-the haplotype and 0 vice versa. A read is assigned to a haplotype which minimizes its number of mismatches. Notice mismatches only occur at SNP positions as all other positions are homozygous. Thus the polyploid haplotype phasing problem is to phase k haplotypes H given the set of reads X and the SNP positions so that the objective function $MEC(X,H)$ is minimized. It is known [8] that minimizing MEC is NP-hard even for diploid haplotype phasing when the length of the reads is greater than one.

Recently various methods have been proposed to the polyploid haplotype phasing problem. HapCompass [16] builds a compass graph from the sequencing reads, which is an undirected weighted graph. In the compass graph, the vertices are the SNPs and an edge between a pair of SNPs indicates that at least one read spans the two SNPs. There is an integer weight associated with each edge. It is shown that a compass graph has a unique phasing if it has no conflicting cycles, which is a simple cycle that contains either an odd number of negative edges or at least one 0-weight edge or both. Haplotype phasings correspond to spanning trees in the graph. The phasing problem is converted to a minimum weighted edge removal optimization on the graph and an algorithm based on cycle basis local optimizations for resolving conflicting cycles is proposed.

HapTree [17] aims to trim down the search space for all the possible haplotypes to a much smaller set of more likely solutions. It takes an inductive approach, generating a collection of likely phasing solutions for the first two SNPs in the genome, and then extending those to phasing solutions of the first three SNPs, and those to the first four SNPs, and so on. When extending any particular solution, HapTree chooses (based on computing likelihoods) how the alleles of the newly added SNP may be assigned to chromosomes; it includes only those assignments that are sufficiently likely. Upon including all SNPs to be phased, HapTree randomly chooses a solution of maximum likelihood from amongst the solutions it has found. It is shown [18] that the trimming process might be time consuming for some cases.

SDHaP [18] formulates the haplotype assembly problem as a semi-definite program and exploits its special structure, namely the low rank of the underlying solution, to solve the problem rapidly and with high accuracy. A graph is defined where the nodes are the reads, the edge between two nodes indicate that the two corresponding reads overlap by at least one SNP. A weight is associated with each edge and the weight is computed as the following:

$$W_{ij} = \frac{k_{sim} - k_{dissim}}{k_{sim} + k_{dissim}} \tag{4}$$

where k_{sim} denotes the number of overlapping positions where the reads have an identical alleles and k_{dissim} is the number of positions where they are different. Then giving the graph and the ploidy, SDHaP aims to find $k-1$ cuts such that the sum of intra-partition edge weights is maximized and inter-partition edge weights is minimized and the problem is solved via correlation clustering. SDhaP formulates the problem as a semi-definite program (SDP), and employs a low-rank Lagrangian scheme followed by randomized projections and a greedy refinement of the k-ploid haplotypes to solve the SDP.

H-PoP and H-PoPG [19] try to partition the DNA reads sequenced from a k-ploid organism into k groups such that the reads of the same group share the same alleles on as many SNP loci as possible and the reads from different groups are different on as many loci as possible. Heuristic strategies are proposed by limiting the number of intermediate solutions at each iteration of a dynamic programming algorithm. Notice H-PoP assumes no VCF file, which contains the dosage information of the variants. is provided while H-PoPG accepts VCF file as an input.

The polyploid haplotype phasing method we proposed in this work is an extension of Harsh [14], which applies the Gibbs Sampling algorithm to phase diploid haplotypes (namely two haplotypes). For diploid scenario, the two haplotypes are complimentary. Therefore indeed only one haplotype needs to be phased. For polyploid scenario, the haplotypes are more than two and they can share common segments and are not necessarily complimentary. Thus the problem is much more challenging. Harsh can not be applied to polyploid haplotype phasing in that the conditional probability estimation for the Gibbs Sampling process needs to be completely re-invented.

Poly-Harsh

Haplotype blocks

Notice that there are cases where haplotype components are disconnected, i.e., we need to identify haplotype blocks that are not connected by any reads. There are two possible reasons for disconnected haplotype components, or blocks: the adjacent SNPs might be far from each other, namely their distance is longer than the length of the reads and thus they will not be spanned by any read; the sequencing coverage is low and thus not all SNPs are covered. To identify the haplotype blocks, we can create a graph where the nodes are SNPs and an edge between two SNPs indicates that the two SNPs are connected by some reads. Then we identify the connected components of the graph, which are the SNPs contained in each haplotype block. There would not be any read spanning two blocks and every read only covers the SNPs from a single block. We next phase each block independently, using only the reads covering the SNPs for that specific block.

Gibbs sampling

In this work, we developed a Gibbs Sampling based method Poly-Harsh for polyploid haplotype phasing. Our algorithm consists of two major steps: fix the haplotypes, compute read assignments; then fix the read assignments, compute the haplotypes. Here when we compare a haplotype to the reference haplotype, if the allele is the same as the reference, the genotype value is 0. If the allele is alternative, the genotype value is 1. In cases where the allele is neither the reference nor the alternative, we simply assign the genotype value as 1. For each SNP position i, we define a genotype value vector $h_i = [g_{1,i}, g_{2,i}, \ldots, g_{k,i}]$, where $g_{j,i}$ is the genotype value of the j-th haplotype at the i-th SNP. For a given ploidy k (for illustration purpose and without losing generality, assuming $k = 4$), the genotype value vector could be one of 2^k binary vectors [1, 0, 0, 0], [1, 1, 0, 0], ..., [0, 1, 1, 1]. Notice we ignore vectors [0,0,0,0] and [1,1,1,1] as they indicate the allele is homozygous. We also define a read assignment vector r as a k-dimensional vector $r = [a_1, a_2, \ldots, a_k]$, where $a_i = 1$ if the read is assigned to the i-th haplotype and $a_i = 0$ if the read is not assigned to the i-th haplotype. For example, for $k = 4$, if the read is assigned to the first haplotype, $r = [1, 0, 0, 0]$. If the read is assigned to the second haplotype, $r = [0, 1, 0, 0]$ and so on so forth. Notice a read can not be assigned to more than one haplotype and on the other hand it has to be assigned to one haplotype. Given the genotype value vector h_i for the i-th SNP and a read assignment vector r_j for the j-th read x_j, we define a function as below:

$$\begin{aligned} \theta(h_i, r_j, x_j) &= ln(1-\epsilon)^t + ln(\epsilon)^{k-t} \\ t &= match(h_i, r_j \times x_{j,i}) \end{aligned} \quad (5)$$

where ϵ is the sequencing error rate, $x_{j,i}$ is the i-th value of the x_j, $match(A, B)$ is the vector-wise matches between two vectors A and B where the number of matches is increased by 1 if two vector elements are identical (either both 1 or both 0). For example, for $h_i = [1, 1, 0, 0]$, $r_j = [1, 0, 0, 0]$, $x_{j,i} = 1$, we have $r_j \times x_{j,i} = [1, 0, 0, 0]$ and thus $t = match(h_i, r_j \times x_{j,i}) = 3$. For $h_i = [1, 1, 0, 0]$, $r_j = [1, 0, 0, 0]$, $x_{j,i} = 0$, we have $r_j \times x_{j,i} = [0, 0, 0, 0]$ and thus $t = match(h_i, r_j \times x_{j,i}) = 2$. The θ function essentially models the probability of the correct read assignment given the matches between the read and the haplotypes. Notice the function considers the mismatches due to sequencing error.

Given ploidy as k, the set of genotype value vectors $H = [h_1, h_2, \ldots, h_n]$ where $h_i = [g_{1,i}, g_{2,i}, \ldots, g_{k_i}]$ as we have defined and n is the number of SNPs, the set of read assignment vectors $R = [r_1, r_2, \ldots, r_n]$ where $r_i = [a_1, a_2, \ldots, a_k]$ is the assignment vector for the i-th read as we have defined, the sampling process proceeds as follows: We first randomly initiate H, then we fix H and compute the conditional probability $P(R|H)$. We sample the reads assignment R based on $P(R|H)$. Next we fix R and compute the conditional probability $P(H|R)$. We sample the genotype value vectors H based on $P(H|R)$. For ploidy k, we have 2^k haplotype values for a specific SNP. Assuming the genotype value vector is h and the set of reads $X = [x_1, x_2, \ldots, x_n]$, we could compute its probability as

$$P(h|R) = \frac{exp\left(\sum_{j=1}^{n} \theta(h, r_j, x_j)\right)}{exp\left(\sum_{i=1,j=1}^{i=2^k,j=n} \theta(h_i, r_j, x_j)\right)} \quad (6)$$

where the function θ is defined in Eq. 5, h_i is the i-th genotype value vector, r_j is the assignment vector for the j-th read x_j. Then we apply sampling to update the genotype values of the SNP. We do a similar Gibbs sampling step for read origin given fixed haplotypes. Again, we conduct sampling to update the read assignment vector on the fixed haplotypes H for a given read x as below:

$$P(r|H) = \frac{exp\left(\sum_{j=1}^{2^k} \theta(h_j, r, x)\right)}{exp\left(\sum_{i=1,j=1}^{i=k,j=2^k} \theta(h_j, r_i, x)\right)} \quad (7)$$

where the function θ is defined in Eq. 5, h_j is the j-th genotype value vector, r_i is the assignment vector when the read x is assigned to the i-th haplotype.

Given H and R, we can easily compute the MEC of the phasing. Notice the k haplotypes can be constructed from the genotype value vectors H, by concatenating all the genotype values from the same haplotype. For example, haplotype one can be constructed as $[g_{1,1}, g_{1,2}, \ldots, g_{1,n}]$. Therefore, the haplotype phasing problem becomes identifying the optimal H that minimizes MEC.

Algorithm 1 Algorithm Poly-Harsh

Require: ploidy k, set of aligned reads X, error rate ϵ
Ensure: k phased haplotypes

1: Randomly Initialize k haplotypes H
2: For fixed haplotype H, sample read origin R
3: For fixed read origin R, sample haplotype H
4: $mec \leftarrow MEC(H, R)$
5: Repeat steps 2 and 3 for sufficient rounds until equilibrium
6: Collect haplotypes and the corresponding MEC by repeating steps 2 and 3, and output the one with the minimum MEC.

Notice for diploid scenario and polyploid scenario, the computations for $P(r|H)$ and $P(h|R)$ are significantly different, as for polyploid scenario, r and h are k-dimensional vectors. The probabilities not only depends on the haplotype the read is assigned to, but also depends on the remaining haplotypes: if one haplotype has a mismatch at a SNP position to a read and the read is not assigned to the haplotype, it is considered as a correct operation to not assign the read to the haplotype.

We repeat the above two steps iteratively. For each iteration, we compute the MEC score of the reads. The process converges when the MEC does not improve or we have reached certain number of iterations. As Gibbs Sampling is sampling based and its performance is affected by the initial random simulation of H, it may fall into local optimum solutions. Therefore we re-run the program multiple times, each time started from a different random seed H. This helps the program to escape from local optimum. A pseudocode of the algorithm is shown in Algorithm 1. Notice the algorithm shows only one run of the procedure. If we would like to run the algorithm multiple times, we need to repeat steps 1 to 5 multiple times, each time with a different randomly initialized H.

Contiguous haplotype reconstruction

As discussed previously each sample consists of 4 haplotypes. The phasing algorithms may not always produce contiguous sequence of haplotypes for each sample. Instead it produces broken haplotypes due to low coverage depth and/or errors. We can think of the output as blocks of sequences where each block contains contiguous subsequences of 4 broken haplotypes. If the cardinality of blocks is k for a particular sample, the number of possible candidates will be 4^k. As there is no prior information available, it is computationally impossible to construct all the 4 haplotypes from 4^k candidates. Fortunately we have multiple samples and by extracting information from shared haplotypes among samples, we can detect true haplotypes with a very high level of confidence. In this article we propose a randomized algorithm for constructing all 4 haplotypes of each sample.

Here we briefly summarize the 3 fundamental steps of our algorithm. At the beginning for each sample it builds all the candidate haplotypes by concatenating the subsequences in each possible ways from the ordered list of blocks. In the second step the algorithm finds the set of candidate haplotypes which occurs at least twice across the entire set of samples. By utilizing the pruned set of candidate haplotypes, we detect all the 4 haplotypes of each sample. We now describe our algorithm next.

Generate candidates

At first we recursively construct all possible candidate haplotypes for each sample. Let the number of samples and length of each candidate be n and t, respectively. Without loss of generality, let each sample consists of k blocks of broken haplotypes. If each block contains 4 haplotypes, the number of possible candidates will be 4^k as stated above. We recursively construct candidates for each sample. The time complexity for constructing all possible candidate haplotypes for m samples is $O\left(4^k mt\right)$ which is exponential with respect to k. It is computationally very expensive task when k is large and the execution time can be very large. However in reality k is very small and we can safely assume that $1 \leq k \leq 8$.

Detect frequent candidates

Frequent candidates are those candidate haplotypes which occurs multiple times (i.e., at least twice) across m samples. This set of candidates contains highly accurate haplotypes because of their multiple occurrences. Exact algorithm is quadratic in the number of candidate haplotypes $\left(\text{i.e., } O\left(\left(4^k m\right)^2 t\right)\right)$. To reduce the runtime we are proposing a randomized algorithm. The expected runtime is sub-quadratic in the number of candidate haplotypes and at the same time the algorithm is also highly accurate in computing the set of frequent candidates. At first we illustrate how to find most similar pair of candidates. By naturally extending it, we will describe the process of finding the set of frequent candidates. We describe our randomized algorithm next.

Suppose we are given n vectors $\hat{b}_1, \hat{b}_2, \ldots, \hat{b}_n$ each of length t. The problem is to find the pair of vectors that are the most similar (i.e., the Hamming distance between them is the smallest). Note that, given two vectors, we can find the Hamming distance between them in $O(t)$ time. A straight forward algorithm to identify the most correlated pair of vectors takes $O\left(n^2t\right)$ time. This algorithm computes the Hamming distance between every pair of vectors. We can achieve a better run time using randomization. We say that the correlation between a pair of strings is p if the Hamming distance between them is $t(1-p)$. Let p_1 be the correlation between the most correlated pair of strings and p_2 be the correlation between the second most correlated pair of strings.

The idea of our algorithm is to iteratively collect pairs of strings that are candidates to be the most correlated. Once we collect enough pairs, we compute the distance between each pair in this collection and output the closest. In each iteration we pick q columns randomly. For any vector (or string), the values in these columns can be concatenated to get a q-bit integer. We hash the vectors based this integer value. Subsequently, we generate pairs as follows: Consider any bucket in the hash table. If there are m vectors in this bucket, then each pair of vectors in this bucket is added as a candidate to a list C. There are $O\left(n^{\frac{\log p_1}{\log p_2}}\log n\right)$ iterations in the algorithm. We can show that after $O\left(n^{\frac{\log p_1}{\log p_2}}\log n\right)$ iterations, C will have the most correlated pair of bulbs with a high probability (i.e., with a probability of $1-n^{-\Omega(1)}$). For details readers are refereed to [26].

We can naturally extend the above algorithm to get the frequent candidate haplotypes. In each iteration we compute similarity coefficients for all the the pairs in each hash bucket. The similarity coefficient is defined as $SC=\frac{t-d}{t}$ where t is the length of the haplotype and d is the Hamming distance between a pair of interest. The more the similarity coefficient, the more similar will be the pair. We retain that pair which has the similarity coefficient $\geq$ a threshold, V. The individual candidate haplotype belonging in each pair is then hashed into a hash map H. The keys and values of hash map H is the candidate strings and frequency of occurrences, respectively. Finally we sort H with respect to its values (i.e., the frequency of occurrences) and output candidates occurring at least twice across m samples. For each iteration the expected number of pairs generated is $O(n)$ as described in [26]. The expected time to build the hash map H will be also $O(n)$ for a particular iteration. As the expected time to resolve a collision (i.e., look up and update) for H is $O(1)$, the total time spent for all the iterations will be $O\left(n^{1+\frac{\log p_1}{\log p_2}}\right)$. Since we can sort the map by using any integer sorting algorithm, the expected runtime of this step is $O\left(n^{1+\epsilon}t\log n\right)$ where $n=4^km$ and $0<\epsilon<1$. We have used fixed number of iterations (i.e., 50) in the experimentations. Let the number of frequent candidates be s.

Detect true candidates

Frequent set of candidates contains all the haplotypes shared twice across the samples. Each sample has 4^k candidate haplotypes. We need to find 4 haplotypes for each sample. Let S' and S'' be the sets of 4^k candidate haplotypes of a particular sample and frequent candidates, respectively. Intuitively if we find the closest (i.e., most similar) pair of haplotypes ($c'\in S', c''\in S''$), c' will be one of the 4 haplotypes of a particular sample. Haplotypes c' and c'' may be identical but it is not guaranteed. c' may not be in the set of frequent candidates due to large number of errors. In this case the closest one (i.e., c'') from s gives us the best possible information to find c'. Next we illustrate this step algorithmically.

Each sample has 4^k candidates and the size of pruned set is s as described in 3. For each candidate $c'\in S'$ and $c''\in S''$, we compute pair-wise similarity coefficient. We then sort all the pairs (c', c'') with respect to SC in non-increasing order and then rank of frequent candidates in non-decreasing order. We traverse the sorted list of pairs and collect first 4 candidate haplotypes c' with the following restrictions: (1) No candidate haplotype c' will be chosen more than once; and (2) Each subsequence from each block will be used exactly once. If the size of each candidate is t, we need $O\left(4^kst\right)$ time to compute pair-wise similarity coefficients (since the number such pairs is 4^ks). Sorting the list of pairs with respect to similarity score and then by rank of frequent candidate haplotypes will take no more than $O\left(4^ks(k+\log s)\right)$ time. After sorting detecting first 4 haplotypes could take $O\left(4^kst\right)$ time. In total the time complexity to find 4 candidate haplotypes for m samples will be $O\left(4^kms(k+t+\log s)\right)$.

Results and discussion

Polyploid haplotype phasing

In this work, we focused on phasing relatively short regions such as genes, rather than the whole genome. Our simulation pipeline is as below: We randomly simulate one gene of length 1300bp which contains 30 biallelic SNPs. For tetraploid case ($k=4$) we simulate four haplotypes for each individual. Then we constructed a VCF for each individual according to their simulated haplotypes. Given the VCF and the haplotypes, we next use Mason [27] to simulate the paired-end reads. Mason is a read simulator software for Illumina, 454 and Sanger reads. Its features include coverage, read length, position specific error rates and base quality values. We feed Mason with the VCF with dosage information and set the error rate as 0.01 across all SNP positions and vary the coverage and the read length as shown in Table 1. Illumina paired-end reads are simulated.

In order to show that re-running Poly-Harsh multiple times in usual leads to better performance as the program could escape from the local optimum, we show in Fig. 1

Table 1 Different parameters for the simulated data

Parameters	Coverage	Read Length
Parameter Set 1	40	100
Parameter Set 2	80	100
Parameter Set 3	40	200
Parameter Set 4	80	200
Parameter Set 5	100	100

the minimum, the mean and the standard deviation of 8 rounds of running of Poly-Harsh with respect to each parameter settings. As we can see that for all parameter settings, with 8 rounds of running, the minimum MEC we obtained is in usual much better than the mean MEC, especially when the MEC is relatively large. This indicates that running the program multiple times could in general improve the performance. In our experiments, we see that in general the performance converges in 8 rounds of running. Also we can observe that larger coverage and longer reads lead to larger MEC and larger standard deviation.

Next we compare the phasing performance of Poly-Harsh with Hapcompass [16] and H-PoP/H-PoPG [19]. We again take the parameter settings specified in Table 1. For each parameter setting, we randomly simulate 10 data sets and we show the average performance. For Poly-Harsh, we conducted 8 rounds of running, each with 100 iterations. Notice that HapCompass does require the dosage information while both Poly-Harsh and H-PoPG (the version of H-PoP that takes VCF) take the dosage information as optional. Therefore, for a fair comparison, we first feed the dosage information to all three methods. We show the MEC of the three methods in Fig. 2. We can see that both Poly-Harsh and H-PoPG achieved better results compared with HapCompass. Poly-Harsh achieved the best performance for all experiments.

Next we feed only the aligned sequencing reads information to H-PoP and Poly-Harsh for comparison purpose. HapCompass is excluded in the experiments as it requires dosage information. We show that the MEC of both methods in Fig. 3. We can see again Poly-Harsh achieved better results compared with H-PoP. Also the MEC is in general larger when dosage is not provided, indicating that without dosage, the phasing becomes less accurate.

Finally, the execution time of Poly-Harsh depends on the number of rounds of running. For one round of running, on our data sets, the execution time for all three methods are comparable, all less than 1 second. The execution time of Poly-Harsh increases linearly with respect to the number of rounds of running.

Contiguous haplotype reconstruction

We created 15 simulated datasets each having 30 samples. As described above each sample contains 4 haplotypes. Each haplotype is a binary string of length 1,300 bps. Binary string is created by concatenating randomly chosen binary values (i.e., either 0 or 1) under uniform distribution. As the phased haplotypes usually contain errors, we introduce errors in the haplotypes by flipping bases according to given error rates. For example, if an error rate is 1%, the probability that a base in a haplotype will be flipped is 0.01. Each sample is randomly partitioned into a number of blocks. The number of such blocks will be 1

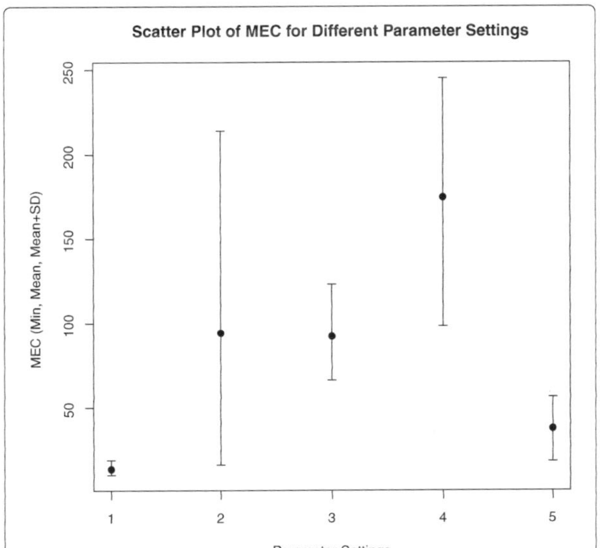

Fig. 1 The min, mean and mean+sd of the MEC for Poly-Harsh on 8 rounds of running with respect to different parameter settings

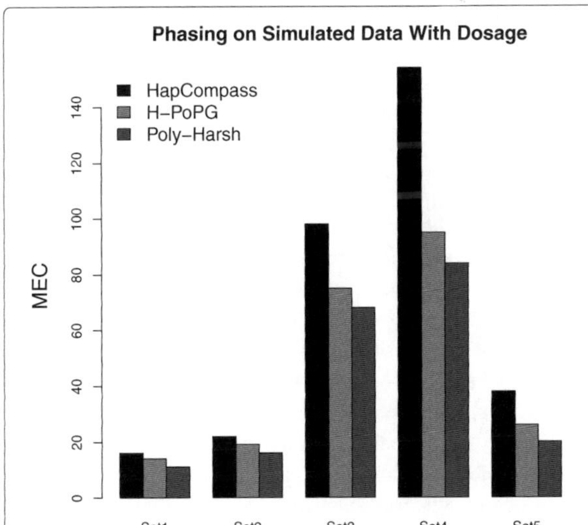

Fig. 2 The comparison of MEC for HapCompass, H-PoPG and Poly-Harsh on simulated data with parameter settings specified in Table 1. We feed all methods with the VCF file and dosage information

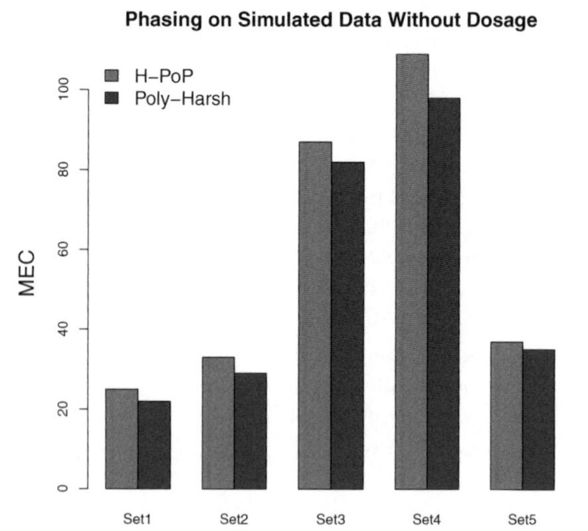

Fig. 3 The comparison of MEC for H-PoP and Poly-Harsh on simulated data with parameter settings specified in Table 1. No VCF file is provided

Table 2 Performance evaluations by varying *shared* and *error rates*. The length of each haplotype is 1,300 bp. Each dataset contain 30 samples. Each sample contains contiguous subsequence of 4 broken haplotypes

Dataset	% Shared	% Error rate	% Sensitivity	Time in seconds
D1	100	0	100.00	19.51
D2		1	100.00	22.02
D3		5	100.00	25.75
D4		10	96.67	35.13
D5		20	96.67	25.88
D6	80	0	96.67	25.53
D7		1	91.66	25.69
D8		5	91.66	26.20
D9		10	94.16	23.51
D10		20	90.00	12.33
D11	60	0	82.50	23.44
D12		1	76.67	19.45
D13		5	72.71	10.49
D14		10	77.50	9.91
D15		20	71.50	12.04

through 8. As described we are able to construct a haplotype correctly, if it occurs at least twice across the dataset of interest. To make the datasets more realistic, we introduce the concept of *Shared*. If a dataset is told to be X% shared, it means X haplotypes occurs $2\times$ across all the 30 samples.

We measure the effectiveness of our proposed algorithm using two different metrics. These metrics are defined below.

1 **Sensitivity:** The fraction of the haplotypes correctly constructed. Let the number of constructed and true haplotypes be P and T, respectively. Now, the Sensitivity can be written as $SN = \frac{P}{T}$
2 **Time:** Measured elapsed time using total number of CPU clock cycles consumed by each of the algorithm.

At first consider datasets D1-D5 (please, see Table 2). These datasets are generated in a way such that each haplotype must occur $2\times$ across the samples. By varying the error rate we measure the performance of our algorithm. For error rate $1-2$% our algorithm could able to correctly construct all the haplotypes from the set of broken subsequences. For the high error rate (such as, 10% or 20%), it correctly constructs 96.7% haplotypes.

Now consider datasets D6-D10. 80% haplotypes in each dataset occur twice across the samples. Thus the *sensitivity* should be at least 80%. But the correctness of our algorithm is above 90%. It is due to the fact that if we can construct 3 haplotypes of a sample correctly, the rest will be formed accurately. The same observation applies to datasets D11-D15. We visualize the results in Figs. 4 and 5.

As our method is the very first algorithm that tries to construct contiguous haplotypes from the phased haplotype blocks, we compared our method against a simple baseline method: concatenate the blocks randomly. The baseline method has a sensitivity of around 25% among all scenarios, as the random concatenation doesn't rely on any information from the dataset. Thus our method is much more accurate than the random concatenation algorithm.

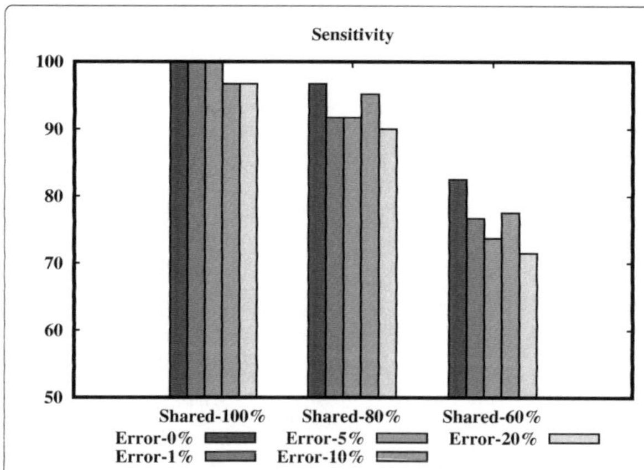

Fig. 4 Sensitivity of our algorithm

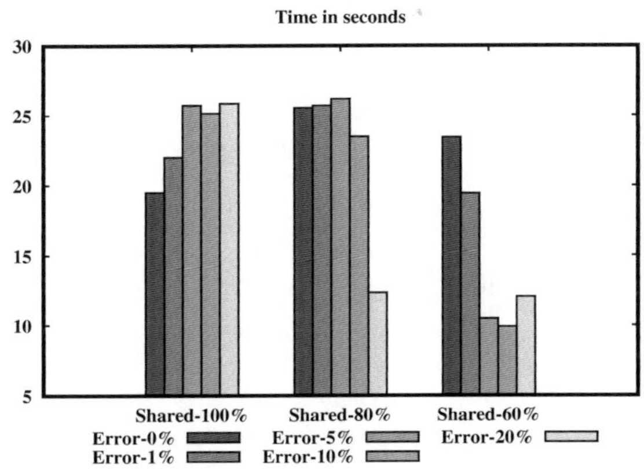

Fig. 5 Elapsed time of our algorithm

The experimental evaluations show that our algorithm is indeed effective and efficient in terms of both accuracy and runtime. For synthetic dataset, our algorithm achieves nearly 100% accuracy where the datasets have less errors and modest number of shared haplotypes. Here accuracy is defined as the fraction of haplotypes constructed correctly. In some datsets the median of accuracy is $70-80\%$. This is due to the fact that the haplotype construction may be affected if the dataset of interest has very low discriminative power. In this case, there is no sufficient information to distinguish true and false haplotypes accurately. It has cumulative effects also. Each sample has 4 haplotypes as stated above. Suppose we are not able to correctly construct the first haplotype. Then there is a high chance that the rest will be constructed incorrectly. This case arises when (1) dataset contains large number of errors and/or (2) there is little shared information across the samples. In case 2 it is impossible to construct all the haplotypes accurately. In order to construct haplotypes correctly, 3 out of 4 haplotypes of each sample must be occurred at least twice across the samples. In case 1 shared information may be lost because of noise (such as, missing values, phasing errors) in the dataset.

Conclusion

In this work we proposed a novel polyploid haplotype phasing algorithm that is applicable to any ploidy. The algorithm is based on Gibbs Sampling, where given the set of sequencing reads, we fix the haplotypes and the read assignments alternatively, then estimate the conditional probability of each other and sample their values based on their corresponding conditional probabilities. The Gibbs sampling method has been shown to work well on diploid haplotype phasing [14]. Our experiments illustrate that for tetraploid haplotypes, our method is able to improve the quality of the phased haplotypes (based on Minimum Error Correction) over the state-of-the-art methods. Due to low coverage or sequencing errors, the phased haplotypes usually contain isolated blocks. We proposed an algorithm to construct contiguous haplotypes from the phased haplotype blocks of a set of individuals whose haplotypes are inherited from the same set of founders. To our knowledge, this is the first algorithm that leverages the shared information across multiple individuals to construct contiguous haplotypes. Our experiments showed that our method is both efficient and effective.

Also in our future work, we will evaluate our methods on different metrics like switch error, hamming distance etc. We would also like to evaluate our method on different ploidies to investigate its performance regarding to the number of phased haplotypes.

Abbreviations

MCMC: Monte carlo markov chain; MEC: Minimum error correction; NGS: Next generation of sequencing; SNP: Single nucleotide polymorphism; VCF: Variant call format

Acknowledgements

The authors want to acknowledge Enrico Siragusa and Nina Haiminen from IBM T.J. Watson Research for their help on generating the simulated data.

Funding

Funding for the publication of this article was provided by IBM T.J. Watson Research

Availability of data and materials

The data is simulated and the simulation procedure is described in the manuscript.

About this supplement

This article has been published as part of *BMC Genomics* Volume 19 Supplement 2, 2018: Selected articles from the 16th Asia Pacific Bioinformatics Conference (APBC 2018): genomics. The full contents of the supplement are available online at https://bmcgenomics.biomedcentral.com/articles/supplements/volume-19-supplement-2.

Authors' contributions

DH designed and implemented the Poly-harsh algorithm for polyploid haplotype phasing. SS designed and implemented the algorithm to concatenate haplotype blocks into contiguous haplotypes. RF introduced the polyploid haplotype phasing problem and helped on the simulation. LP helped design both algorithms and suggested the integration of the two algorithms. DH, SS and LP made contributions to writing the manuscript, with additional input from all remaining authors. All authors read and approved the final manuscript.

Ethics approval and consent to participate

Not Applicable.

Consent for publication

Not Applicable.

Competing interests

The authors declare that they have no competing interests.

Author details

[1]College of Computer Science and Software, Shenzhen University, Shenzhen 518060, China. [2]IBM T.J. Watson Research Center, 1101 Kitchawan Rd, Yorktown Heights, NY 10598, USA. [3]Wageningen University & Research, 6708 PB, Wageningen, Netherlands.

Published: 9 May 2018

References

1. Gusev A, Lowe JK, Stoffel M, Daly MJ, Altshuler D, Breslow JL, Friedman JM, Pe'er I. Whole population, genome-wide mapping of hidden relatedness. Genome Res. 2009;19(2):318–26.
2. Browning SR, Browning BL. High-resolution detection of identity by descent in unrelated individuals. Am J Hum Genet. 2010;86(4):526–39. https://doi.org/10.1016/j.ajhg.2010.02.021.
3. Browning BL, Browning SR. A fast, powerful method for detecting identity by descent. Am J Hum Genet. 2011;88(2):173–82. https://doi.org/10.1016/j.ajhg.2011.01.010.
4. Patterson N, Hattangadi N, Lane B, Lohmueller KE, Hafler DA, Oksenberg JR, Hauser SL, Smith MW, OBrien SJ, Altshuler D, et al. Methods for high-density admixture mapping of disease genes. Am J Hum Genet. 2004;74(5):979–1000.
5. Marchini J, Howie B, Myers S, McVean G, Donnelly P. A new multipoint method for genome-wide association studies by imputation of genotypes. Nat Genet. 2007;39(7):906–13.
6. Kang HM, Sul JH, Service SK, Zaitlen NA, Kong S-y, Freimer NB, Sabatti C, Eskin E, et al. Variance component model to account for sample structure in genome-wide association studies. Nat Genet. 2010;42(4):348–54.
7. Patil N, Berno AJ, Hinds DA, Barrett WA, Doshi JM, Hacker CR, Kautzer CR, Lee DH, Marjoribanks C, McDonough DP, et al. Blocks of limited haplotype diversity revealed by high-resolution scanning of human chromosome 21. Science. 2001;294(5547):1719–23.
8. Bansal V, Halpern AL, Axelrod N, Bafna V. An MCMC algorithm for haplotype assembly from whole-genome sequence data. Genome Res. 2008;18(8):1336.
9. Bansal V, Bafna V. HapCUT: an efficient and accurate algorithm for the haplotype assembly problem. Bioinformatics. 2008;24(16):153.
10. He D, Choi A, Pipatsrisawat K, Darwiche A, Eskin E. Optimal algorithms for haplotype assembly from whole-genome sequence data. Bioinformatics. 2010;26(12):183–90.
11. Deng F, Cui W, Wang L. A highly accurate heuristic algorithm for the haplotype assembly problem. BMC Genomics. 2013;14(2):1.
12. Mousavi SR, Mirabolghasemi M, Bargesteh N, Talebi M. Effective haplotype assembly via maximum boolean satisfiability. Biochem Biophys Res Commun. 2011;404(2):593–8.
13. Chen ZZ, Deng F, Wang L. Exact algorithms for haplotype assembly from whole-genome sequence data. Bioinformatics. 2013;29(16):1938–45.
14. Yang WY, Hormozdiari F, Wang Z, He D, Pasaniuc B, Eskin E. Leveraging reads that span multiple single nucleotide polymorphisms for haplotype inference from sequencing data. Bioinformatics. 2013;29(18):2245–52.
15. Patterson M, Marschall T, Pisanti N, Van Iersel L, Stougie L, Klau GW, Schönhuth A. Whatshap: weighted haplotype assembly for future-generation sequencing reads. J Comput Biol. 2015;22(6):498–509.
16. Aguiar D, Istrail S. Hapcompass: a fast cycle basis algorithm for accurate haplotype assembly of sequence data. J Comput Biol. 2012;19(6):577–90.
17. Berger E, Yorukoglu D, Peng J, Berger B. Haptree: A novel bayesian framework for single individual polyplotyping using ngs data. PLoS Comput Biol. 2014;10(3):1003502.
18. Das S, Vikalo H. Sdhap: haplotype assembly for diploids and polyploids via semi-definite programming. BMC Genomics. 2015;16(1):1.
19. Xie M, Wu Q, Wang J, Jiang T. H-pop and h-popg: heuristic partitioning algorithms for single individual haplotyping of polyploids. Bioinformatics. 2016;32(24):3735–44.
20. Geman S, Geman D. Stochastic relaxation, gibbs distributions, and the bayesian restoration of images. IEEE Trans Pattern Anal Mach Intell. 1984;PAMI-6(6):721–41.
21. Liu JS. Monte Carlo Strategies in Scientific Computing. New York: Springer; 2008.
22. Lippert R, Schwartz R, Lancia G, Istrail S. Algorithmic strategies for the single nucleotide polymorphism haplotype assembly problem. Brief Bioinform. 2002;3(1):23.
23. Lancia G, Bafna V, Istrail S, Lippert R, Schwartz R. SNPs problems, complexity, and algorithms. In: Proceedings of the 9th Annual European Symposium on Algorithms. Lecture Notes in Computer Science. New York: Springer-Verlag; 2001. p. 182–93.
24. Duitama J, Huebsch T, McEwen G, Suk EK, Hoehe MR. Refhap: a reliable and fast algorithm for single individual haplotyping. In: Proceedings of the First ACM International Conference on Bioinformatics and Computational Biology. Niagara Falls: ACM; 2010. p. 160–9.
25. Xie M, Wang J, Jiang T. A fast and accurate algorithm for single individual haplotyping. BMC Syst Biol. 2012;6(Suppl 2):8.
26. Rajasekaran S, Saha S. Efficient algorithms for the two locus problem in genome-wide association study: Algorithms for the two locus problem. In: Proceedings of the 25th ACM International on Conference on Information and Knowledge Management. Seattle: ACM; 2016. p. 2305–10.
27. Holtgrewe M. Mason–a read simulator for second generation sequencing data. Technical report FU Berlin. Berlin: Freie University; 2010. http://publications.imp.fu-berlin.de/962/. Accessed 8 Oct 2011.

Taguchi *BMC Bioinformatics* 2018, **19**(Suppl 4):99
https://doi.org/10.1186/s12859-018-2068-7

BMC Bioinformatics

RESEARCH Open Access

Tensor decomposition-based and principal-component-analysis-based unsupervised feature extraction applied to the gene expression and methylation profiles in the brains of social insects with multiple castes

Y.-H. Taguchi

From The Sixteenth Asia Pacific Bioinformatics Conference
Yokohama, Japan. 15-17 January 2018

Abstract

Background: Even though coexistence of multiple phenotypes sharing the same genomic background is interesting, it remains incompletely understood. Epigenomic profiles may represent key factors, with unknown contributions to the development of multiple phenotypes, and social-insect castes are a good model for elucidation of the underlying mechanisms. Nonetheless, previous studies have failed to identify genes associated with aberrant gene expression and methylation profiles because of the lack of suitable methodology that can address this problem properly.

Methods: A recently proposed principal component analysis (PCA)-based and tensor decomposition (TD)-based unsupervised feature extraction (FE) can solve this problem because these two approaches can deal with gene expression and methylation profiles even when a small number of samples is available.

Results: PCA-based and TD-based unsupervised FE methods were applied to the analysis of gene expression and methylation profiles in the brains of two social insects, *Polistes canadensis* and *Dinoponera quadriceps*. Genes associated with differential expression and methylation between castes were identified, and analysis of enrichment of Gene Ontology terms confirmed reliability of the obtained sets of genes from the biological standpoint.

Conclusions: Biologically relevant genes, shown to be associated with significant differential gene expression and methylation between castes, were identified here for the first time. The identification of these genes may help understand the mechanisms underlying epigenetic control of development of multiple phenotypes under the same genomic conditions.

Keywords: Tensor decomposition, Principal component analysis, Feature extraction, Gene expression, Methylation, Social insect

Correspondence: tag@granular.com
Department of Physics, Chuo University, 1-13-27 Kasuga, Bunkyo-ku, 112-8551 Tokyo, Japan

Background

Organisms often exhibit different phenotypes despite a common genomic background. For example, juveniles and adults frequently have different body plans (e.g., tadpoles and frogs, caterpillars and butterflies, and megalopas and clubs). Nonetheless, juvenile and adult organisms have different sizes or must survive in distinct environments, and these conditions require different phenotypes. More striking examples are castes of social insects, such as ants and bees, which can form two distinct forms: queens and workers, both female [1]. They are usually closely related, but queens and workers have different sizes and lifespans. The mechanism that potentially allows social insects to form castes with distinct body plans is the epigenome [2, 3], which is flexible and can lead to the formation of different phenotypes without genomic alterations. Therefore, it is important to determine the correlation between an epigenome and phenotype by analyzing gene expression.

In actuality, there have been many discussions regarding how an epigenome can affect a phenotype [4]. Because a gene can affect the phenotype through the regulation of gene expression, it is natural to expect that the epigenome can also affect the phenotype through the regulation of gene expression [5]. The epigenome is even expected to be heritable and thus affect phenotypes through generations [6]. In this field, the relation between the epigenome and phenotype has been comprehensively investigated; through regulation of gene expression, epigenetic mechanisms have the potential to determine and alter cell phenotypes, and epigenetic mechanisms also mediate dosage compensation, chromosomal silencing, and imprinting [7]. In this regard, castes of social insects are expected to be affected by various epigenomic alterations.

Some pioneering studies in this field have been conducted [8, 9], but statistical analyses in these studies have not been satisfactory, for example,

1. Gene ontology (GO) enrichment analyses had an insufficiently small false discovery rate (FDR): < 0.5 [8];
2. identification of differentially expressed genes (DEGs) was performed at insufficiently large $q > 0.6$, corresponding to FDR < 0.4 [9];
3. highly methylated regions significantly different between phenotypes (castes) were not identified [8].

Even though these issues do not always reduce quality of the studies, addressing them should increase the confidence in the conclusions.

Inadequate statistical analyses may be due to disregarding the multivariate nature of variables. All the performed statistical analyses have been single–gene–based, meaning that group behaviors were considered only *after* identification of genes. Recently, I proposed a principal component analysis (PCA)-based unsupervised feature extraction (FE) as a method that can perform multivariate analysis *before* gene selection and applied it to various bioinformatic problems [10–31]. Therefore, applying PCA-based unsupervised FE to the analysis of datasets may yield more reliable results. PCA-based unsupervised FE was also extended to tensor decomposition (TD) to integrate multiway [32–34] and multiview [35–37] datasets. TD-based unsupervised FE applied to the integrated analysis of gene expression and methylation profiles may allow for identification of relations between gene expression and methylation, essential for identification of the mechanisms behind epigenetic regulation of phenotype development.

Methods

A flow chart showing the experimental design is presented in Fig. 1.

Gene expression and methylation profiles

All gene expression and methylation profiles [8] were retrieved from Gene Expression Omnibus (GEO). Gene expression profiles of *P. canadensis* and *D. quadriceps* are available as Supplementary Files in GEO ID GSE59525: GSE59525_RPKM_Pcan.txt.gz and SE59525_RPKM_Dqua.txt.gz. Methylation profiles of *P. canadensis* and *D. quadriceps* are in GSM14388XX_PcanYYY_CX.txt.gz and GSM14388XX_DquaYYY_CX.txt.gz, which are also available as Supplementary Files in GEO ID GSE59525 (XX and YYY are presented in Table 1). An additional gene expression profile of *P. canadensis* [9],

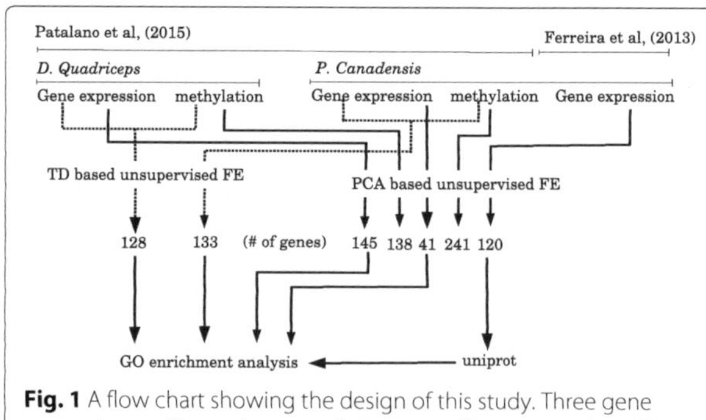

Fig. 1 A flow chart showing the design of this study. Three gene expression profiles and two methylation profiles were retrieved from two studies on two social insect species, *Polistes canadensis* and *Dinoponera quadriceps*. All of them were processed by PCA-based unsupervised FE, whereas two pairs of gene expression and methylation profiles were analyzed by TD-based unsupervised FE. Differential expression between castes was analyzed in seven obtained gene sets. Analysis of GO term enrichment was performed on three sets of genes derived from gene expression analysis and on two sets of genes generated by TD-based unsupervised FE. The full list of the selected genes is presented in Additional file 1

Table 1 The list of files containing methylation profiles retrieved from GEO

XX	YYY	XX	YYY	Description
P. canadensis		*D. quadriceps*		
50	_Synthetic	57	_Synthetic	Control
51	21Q	58	1AQ	Queen 1st replicate
52	43Q	59	2AQ	Queen 2nd replicate
53	75Q	60	3AQ	Queen 3rd replicate
54	26W	61	1CW	Worker 1st replicate
55	42W	62	3CW	Worker 2nd replicate
56	76W	63	3DW	Worker 3rd replicate

13059_2012_3057_MOESM9_ESM.CSV, was retrieved from the supplementary file presented in the study (Additional file 9). Gene expression values were used as-is, but methylation profile values were integrated so that they represented the relative methylation within genes. Assuming m_{s1} and m_{s2} are methylation and nonmethylation values respectively at locus s, then the relative methylation within the ith gene can be defined as

$$\frac{\sum_{s\in i} m_{s1}}{\sum_{s\in i}(m_{s1}+m_{s2})},$$

where $\sum_{s\in i}$ is taken over s bases within DNA sequences corresponding to the ith gene body.

PCA-based unsupervised FE

A flow chart showing PCA- and TD-based unsupervised FE is presented in Fig. 2.

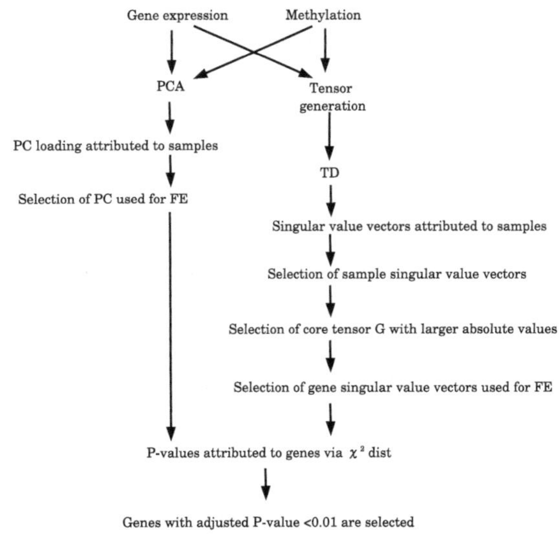

Fig. 2 A flow chart of PCA- and TD-based unsupervised FE. Gene expression and methylation profiles were examined by PCA or TD. For PCA, gene expression and methylation profiles were processed separately, whereas TD was applied after generating a tensor from them. For PCA, principal component (PC) loading attributed to samples was studied and selected for FE. Because PC loadings and PC scores attributed to genes show one-to-one correspondence, PC scores corresponding to the selected PC loadings were subjected to FE. For TD, one-sample singular value vectors used for FE were selected. Afterwards, during analysis of core tensors, G, gene singular value vectors associated with Gs with larger absolute values were selected. By means of the identified PC scores or gene singular value vectors, P-values were determined for genes, assuming a χ^2 distribution, and genes associated with adjusted P-values less than 0.01 were finally selected

PCA

Assume that $N \times M$ matrix X represents gene expression or methylation of the ith gene in the jth sample, $x_{ij} \in \mathbb{R}^{N\times M}$, which can be standardized as $\sum_i x_{ij} = 0, \sum_i x_{ij}^2 = N$. The kth PC score, $u_{ki} \in \mathbb{R}^{\min(N,M)\times N}$, attributed to the ith gene can be obtained as the ith component of the kth eigenvector u_k of XX^T, where X^T is the transposed matrix of X, so that

$$XX^T\boldsymbol{u}_k = \lambda_k \boldsymbol{u}_k,$$

where λ_k is the kth eigenvalue. The kth PC loading, $v_{kj} \in \mathbb{R}^{\min(N,M)\times M}$, attributed to the jth sample can be obtained as the jth component of the kth vector v_k, which is defined as

$$\boldsymbol{v}_k = X^T\boldsymbol{u}_k.$$

This is also the kth eigenvector of matrix X^TX because

$$X^TX\boldsymbol{v}_k = X^TXX^T\boldsymbol{u}_k = X^T\lambda_k\boldsymbol{u}_k = \lambda_k\boldsymbol{v}_k.$$

PCA-based unsupervised FE

To carry out FE by PCA, PC loadings, v_k, of interest that can be used for FE must be identified, and there are several approaches. Suppose that Ω represents a set of ks of the identified v_ks. Then, gene i, primarily contributing to the kth ($k \in \Omega$) PC score, u_k, should be identified, and this task can be accomplished by selecting the outliers within the space spanned by the PC scores:

$$span(\boldsymbol{u}_k : k \in \Omega).$$

$u_{ki}(k \in \Omega, 1 \le i \le N)$ was assumed to follow a multiple normal distribution, and P-values, P_is, were attributed to each gene i via the χ^2 distribution,

$$P_i = P_{\chi^2}\left[> \sum_{k\in\Omega}\left(\frac{u_{ki}}{\sigma_k}\right)^2\right], \quad (1)$$

where $P_{\chi^2}[> x]$ represents the cumulative probability of the χ^2 distribution that the argument is greater than x, whereas σ_k is standard deviation. Afterwards, genes *is*, associated with the Benjamini–Hochberg criterion–adjusted P_is [38] lower than the threshold value, e.g., 0.01, were selected as outliers.

TD-based unsupervised FE

When two sets of experimental factors are affecting each sample, e.g., tissues and diseases, gene expression and methylation profiles should be presented as a tensor $x_{ij\ell} \in \mathbb{R}^{N\times M\times L}$, where j and ℓ correspond to the tissue and disease, respectively.

Equivalence of PCA and singular value decomposition (SVD)

PCA can be extended to a tensor as follows. PCA is known to be equivalent to SVD,

$$U^T XV = \Lambda, \quad (2)$$

where U and V are $N \times M$ and $M \times M$ orthogonal matrices, respectively. Λ represents a $M \times M$ diagonal matrix. Assume that $U = (\boldsymbol{u}_1, \ldots, \boldsymbol{u}_M)$ and $V = (\boldsymbol{v}_1, \ldots, \boldsymbol{v}_M)$. The diagonal component of Λ can be written as λ_k. Then, using $\boldsymbol{v}_k = X^T\boldsymbol{u}_k$ and $XX^T\boldsymbol{u}_k = \lambda\boldsymbol{u}_k$,

$$U^T XV = U^T XX^T U = U^T \Lambda U = \Lambda.$$

Therefore, U and V composed of PC scores and loadings satisfy Eq. (2). Subsequently,

$$UU^T XVV^T = X = U\Lambda V^T,$$

or this relation can be written as

$$x_{ij} = \sum_k \lambda_k u_{ki} v_{kj}. \quad (3)$$

Extension to TD

Equation (3) can be easily generalized to TD [32] by extending the matrix to a tensor,

$$x_{ij\ell} = \sum_{k_1=1}^{N} \sum_{k_2=1}^{M} \sum_{k_3=1}^{L} G(k_1, k_2, k_3) u_{k_1 i} u_{k_2 j} u_{k_3 \ell},$$

where $u_{k_1 i} \in \mathbb{R}^{N\times N}$, $u_{k_2 j} \in \mathbb{R}^{M\times M}$, $u_{k_1 \ell} \in \mathbb{R}^{L\times L}$ and $U_{K_i} = (u_1, \ldots, u_{K_i})$ with $(K_1, K_2, K_3) = (N, M, L)$ were assumed to be orthogonal matrices. Hereafter, U_{K_i} and u_{k_i} are referred to as singular value matrix and singular value vector, respectively. Given that the core tensor, $G(k_1, k_2, k_3) \in \mathbb{R}^{N\times M\times L}$, is as large as x_{ijk}, this situation represents an overcomplete problem, i.e., there is no unique decomposition. In this study, higher-order SVD (HOSVD) [39], which is known to frequently give a global minimum [40], was employed to perform TD, assuming $\sum_i x_{ij\ell} = 0, \sum_i x_{ij\ell}^2 = N$, as in the PCA cases.

TD-based unsupervised FE

First, (k_2, k_3) of interest, which were attributed to samples, were selected, and, as for PCA, there are different approaches. Next, $G(k_1, k_2, k_3)$s associated with selected k_2, k_3 were ranked based on their absolute values, and by means of top-ranked k_1s with the set defined as Ω, the same procedure applied to Eq. (1) was repeated by replacing PC score u_{ki} with a gene singular value vector, $u_{k_1 i}$. Currently, we do not have any specific criterion specifying how many k_1s should be considered.

TD-based unsupervised FE for integrated analysis of multi-omics data

For two distinct omics datasets, such as a gene expression profile, x_{ij}, and a methylation profile, $x_{i\ell}$, TD-based unsupervised FE can be used for the integrated analysis by generating a tensor,

$$x_{ij\ell} = x_{ij} x_{i\ell},$$

to which TD-based unsupervised FE can be applied [35]. The subsequent procedure was the same as the standard one described in the previous subsection.

GO enrichment analysis

To perform GO enrichment analysis of the previously described dataset [8], the list of genes associated with GO terms was downloaded [41]: PCAN.v01.GO.tsv for *P. canadensis* and DQUA.v01.GO.tsv for *D. quadriceps*.

Because genes presented in the second study we used [9] were not fully annotated but contained protein sequence gene IDs, a list of gene IDs was uploaded to UniProt [42] and GO term associations were downloaded. GO enrichment analyses were performed on the retrieved list.

Fisher's exact test was selected for evaluation of the overlaps between the set of provided genes and genes associated with a specific GO term. The obtained P-values were adjusted in accordance with the Benjamini–Hochberg criterion [38]. GO terms associated with adjusted P-values less than 0.01 were selected.

Results

PCA-based unsupervised FE applied to the dataset provided by Patalano et al. [8]

PCA-based unsupervised FE was performed on the dataset presented in another study [8].

The methylation profile

PCA-based unsupervised FE was applied to the methylation profiles of *P. canadensis* and *D. quadriceps*. In Fig. 3(a) and (c), the first PC loadings, v_1, are presented, attributed to seven samples, comprising one control, three queen samples, and three worker samples of *P. canadensis* and *D. quadriceps* each. For both, v_1 mainly denotes the difference between the control and queen or worker samples. In Fig. 3(b) and (d), the histogram of the first PC score, u_1, attributed to genes is presented, and red areas represent the selected genes, with adjusted P-values computed using Eq. (1) lower than 0.01 (241 and 138 selected genes for *P. canadensis* and *D. quadriceps*, respectively).

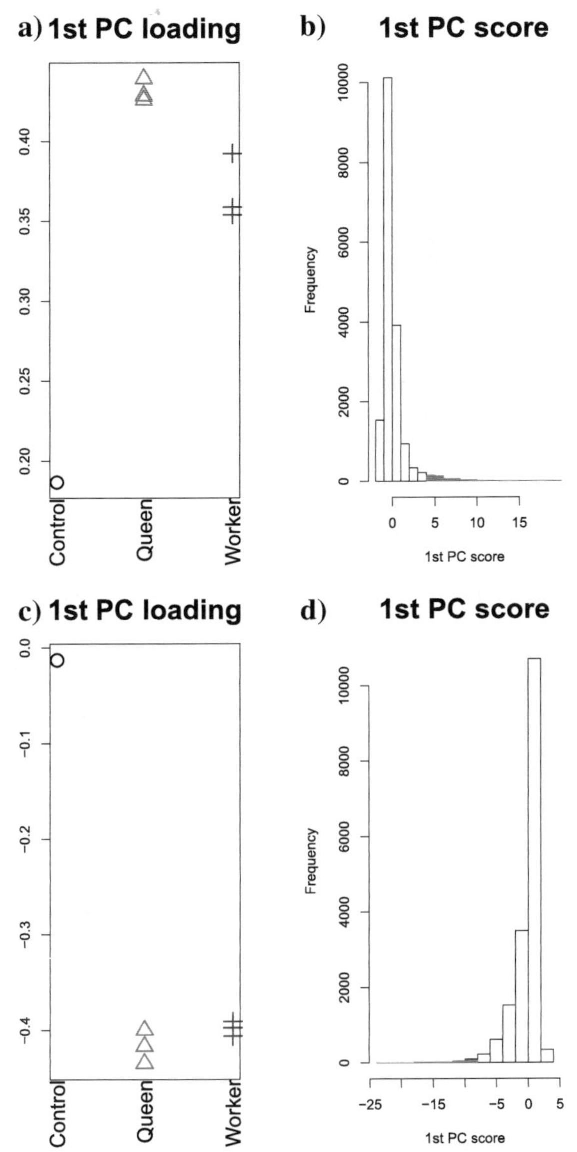

Fig. 3 The results of PCA performed on methylation profiles. PCA-based unsupervised FE was tested on methylation profiles. **a** and **b**: *P. canadensis*. **c** and **d**: *D. quadriceps*. **a** and **c**: A boxplot of PC loadings v_1, (**b**) and (**d**): A PC score histogram, u_1. Red areas in (**b**) and (**d**) represent selected genes

Because in Fig. 3(a) hypermethylated genes are presented, and the results in Fig. 3(b) suggest that only the genes with positive PC scores are selected, all the selected genes were found to be hypermethylated. Similarly, because in Fig. 3(c), hypomethylation is presented, and the results in Fig. 3(d) indicate that only the genes with negative PC scores are selected, all the selected genes were found to be associated with hypermethylation as well. This finding is in agreement with the results of the other study [8].

Patalano et al. [8] did not find genes associated with the emergence of distinct methylation patterns between queens and workers, but the results presented in Fig. 3(a) and (c) show minor differences between queens and workers, suggesting that the selected genes may have different methylation profiles between queens and workers overall. Three statistical tests were carried out to analyze the differences between queens and workers (Table 2) and demonstrated that PCA-based unsupervised FE could identify gene-associated methylation patterns between queens and workers, unlike the analyses in the other study [8]. Given that in Fig. 3(a) and (c), the upregulation and downregulation of methylation in queens are presented, whereas in Fig. 3(b) and (d), genes associated with positive and negative PC scores are shown, the selected genes were found to be associated with hypermethylation in queens. This finding suggests that hypermethylated genes are associated with relative hypermethylation in queens.

Gene expression

PCA-based unsupervised FE was performed on the gene expression profiles of *P. canadensis* and *D. quadriceps*. Because the gene expression profile of *P. canadensis* was log 2-ratio converted, it was scaled back to the original one as $2^{x_{ij}}$ before the application of PCA. In Fig. 4(a) and (c), v_3 and v_4 are presented, showing the most significant differences in gene expression levels between queens and workers of *P. canadensis* and *D. quadriceps*, respectively. In Fig. 4(b) and (d), the distributions of v_3 and v_4 for *P. canadensis* and *D. quadriceps*, respectively, are depicted. Red areas represent the selected genes, with adjusted *P*-values, determined using Eq. (1), lower than 0.01 (41 and 145 genes for *P. canadensis* and *D. quadriceps*, respectively).

To determine whether the expression of selected genes differs between queens and workers, three statistical tests were carried out (Table 3). The majority (five of six) of the applied tests confirmed that expression of the selected genes differs between queens and workers of *P. canadensis* and *D. quadriceps*, in line with the findings reported elsewhere [8].

PCA-based unsupervised FE applied to the dataset provided by Ferreira et al. [9]

PCA-based unsupervised FE was performed on the expression profiles reported in another study [9]. In

Table 2 Statistical tests for differences in the methylation rates of selected genes between queens and workers

	t	Wilcox	KS
P. canadensis	8.59×10^{-5}	3.10×10^{-3}	1.83×10^{-4}
D. quadriceps	1.11×10^{-2}	5.88×10^{-3}	1.75×10^{-3}

t: the *t* test, Wilcox: the Wilcoxon rank sum test, KS: the Kolmogorov–Sinai test, all two-sided

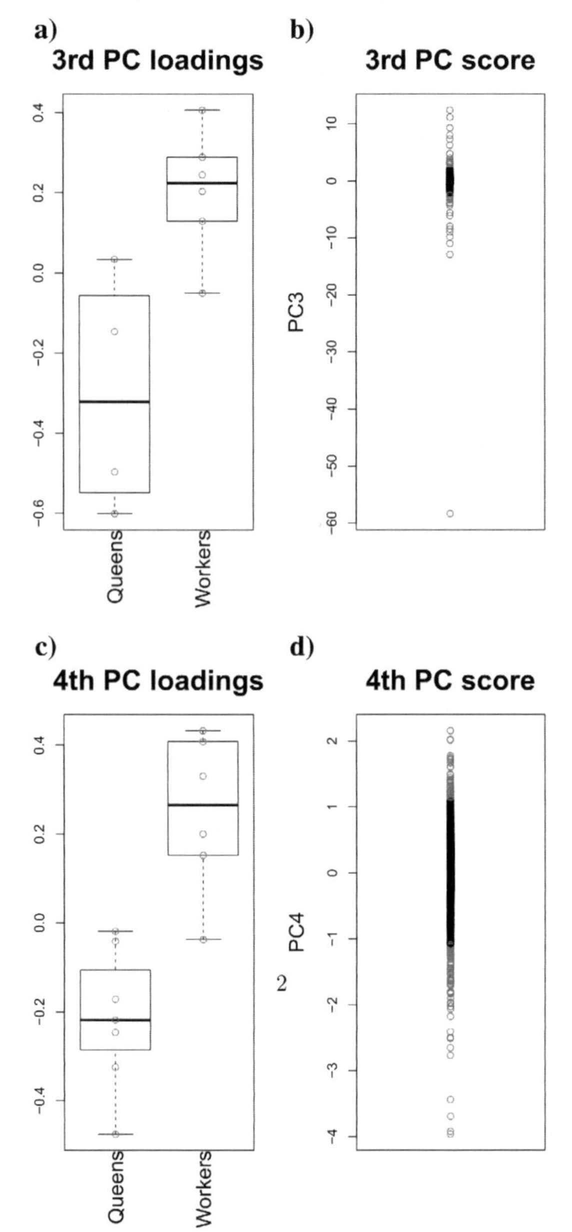

Fig. 4 Results of gene expression PCA. PCA-based unsupervised FE was applied to the gene expression profiles. **a** and **b**: *P. canadensis*. **c** and **d**: *D. quadriceps*. **a** and **c**: A boxplot of PC loadings v_1, (**b**) and (**d**): A distribution of PC scores u_1. Red areas in (**b**) and (**d**) represent the selected genes

Table 3 Statistical tests for differences in expression of the selected genes between queens and workers

	t	Wilcox	KS
P. canadensis	4.37×10^{-4}	0.07	6.45×10^{-3}
D. quadriceps	1.73×10^{-12}	2.24×10^{-12}	5.26×10^{-12}

t: the *t* test, Wilcox: the Wilcoxon rank sum test, KS: the Kolmogorov–Sinai test; all two-sided

Fig. 5(a), the third PC loadings, v_3, are presented, which demonstrate the most significant class dependence based on categorical regression (ANOVA). In Fig. 5(b), the distribution of the third PC score, u_3, is shown, where the red areas represent the selected genes (120 genes associated with the adjusted *P*-values, as determined by means of Eq. (1), lower than 0.01). Because this situation corresponds to $q > 0.99$ in the analysis provided by Ferreira et al., we demonstrated that we were able to identify a set of genes more significant than those identified in the other study, where q was > 0.6.

To test whether the selected genes reflect the class-specific up- and downregulation of expression, three statistical tests were conducted (Table 4). We separated genes into two classes based on the sign of u_{3i}, to identify the upregulation or downregulation of these genes for further comparisons. As shown in Table 4, 120 selected genes were found to be significantly upregulated or downregulated, excluding the upregulation of genes in Foudress, with the smallest number of genes identified in the other study. Therefore, we successfully identified genes that manifest class-specific upregulation or downregulation. Furthermore, the genes identified here are common for

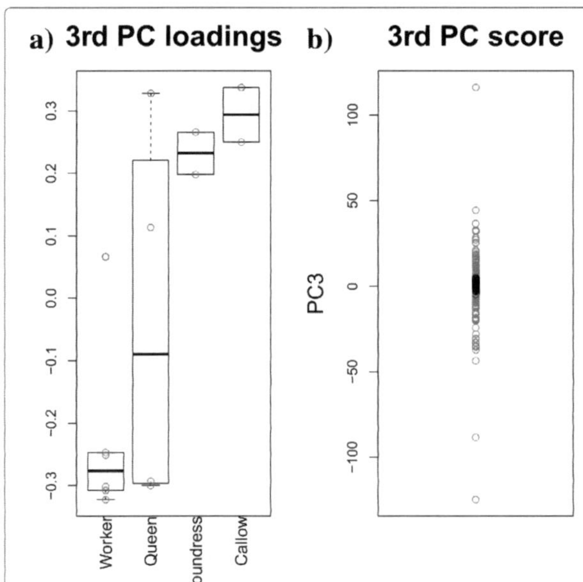

Fig. 5 The results of PCA applied to gene expression profiles. PCA-based unsupervised FE was performed on the gene expression profiles obtained from the study by Ferreira et al. **a**: A boxplot of PC loadings v_1. **b**: A distribution of PC scores u_1. Red areas in panel (**b**) represent 120 selected genes. PC3> 0: 77 genes, PC3< 0: 43 genes

Table 4 Statistical tests for upregulation and downregulation of gene expression in four categories vs. others

PC3			t	Wilcox	KS
Worker vs others	+	down	3.96×10^{-5}	*	*
	−	up	0.22	*	*
Queen vs others	+	up	0.36	0.10	8.87×10^{-4}
	−	down	0.32	0.69	4.69×10^{-3}
Foundress vs others	+	up	0.05	0.13	0.17
	−	down	0.11	3.49×10^{-10}	6.27×10^{-11}
Callow vs others	+	up	0.99	*	*
	−	down	1.14×10^{-6}	2.77×10^{-7}	3.34×10^{-6}

*:< 2.2×10^{-16}. t: the t test, Wilcox: the Wilcoxon rank sum test, KS: the Kolmogorov–Sinai test, all are two-sided

all four classes, while those identified in the other study differed between the comparisons.

TD-based unsupervised FE applied to the integrated analysis of gene expression and methylation profiles

By PCA-based unsupervised FE, we successfully identified genes associated with gene expression and methylation profiles showing significant differences between castes. Genes found by means of gene expression profiles and methylation profiles did not overlap. Fisher's test results yielded an odds ratio lower than 1.0 (data not shown because of negative results). This finding indicated that the genes identified using different datasets are quite distinct. Therefore, it is difficult to understand the mechanisms by which the epigenetic modifications affect gene expression and regulate phenotype development.

One may wonder if a gene expression alteration must not always be associated with altered methylation. Nevertheless, many authors have employed the strategy where genes associated with both altered gene expression and methylation are sought, to identify biologically important genes. Heng et al. [43, 44] have tried to find genes associated with both altered gene expression and methylation to discover genes crucial for breast cancers. Li et al. [45] have attempted to find genes associated with both altered gene expression and methylation to identify key genes in severe oligozoospermia. Mallik et al. [46] have looked for genes associated with both altered gene expression and methylation for tumor prediction. These are only a few examples of studies that involve the association of altered gene expression and methylation. Thus, altered gene expression and promoter methylation may have to be considered together to identify genes specific for castes of social insects, too.

To explore the possibility of correlating gene expression and methylation profiles via our strategy, we applied TD-based unsupervised FE to tensor $x_{ij\ell} = x_{ij}x_{i\ell}$, where i represents a gene, x_{ij} is the gene expression of the jth sample, and $x_{i\ell}$ denotes methylation of the ℓth sample. In Fig. 6(a), the first sample singular value vector for gene expression, u_{k_2j} (at $k_2 = 1$) for *P. canadensis* is shown, which has the most significant dependence upon class labels based upon categorical regression. Because the presented results are similar to those shown in Fig. 3(a), TD-based unsupervised FE was demonstrated to successfully generate biologically relevant singular value vectors for gene expression, u_{k_2j} (at $k_2 = 1$). Similarly, results depicted in Fig. 6(b) represent the third sample singular value vectors for the methylation profiles of *P. canadensis*, $u_{k_3\ell}$ (with $k_3 = 3$), which were also found to be similar to those presented in Fig. 4(a). This finding indicates that TD-based unsupervised FE can successfully generate biologically relevant sample singular value vectors.

Next, we aimed to identify core tensor $G(k_1, k_2, k_3)$ (at $(k_2, k_3) = (1, 3)$) associated with the larger absolute values, to select k_1s used for the gene selection based on *P. canadensis* profiles (Table 5). Given that $G(k_1, k_2, k_3)$ (with $9 \leq k_1 \leq 10, (k_2, k_3) = (1, 3)$) were shown to be top-ranked, the ninth and 10th singular value vectors, u_{k_1i} (with $9 \leq k_1 \leq 10$) were used for the selection of 133 genes, associated with the adjusted P-values, determined via Eq. (1), lower than 0.01 (Fig. 6(c)).

To determine whether the selected genes show differential expression and methylation between workers and queens, three statistical tests were applied (Table 6). We observed that 133 selected genes have significant differences in expression and methylation between the samples under study. Therefore, using TD-based unsupervised FE, we successfully identified a set of genes that have both differential gene expression and methylation between queens and workers; this accomplishment was not possible with the PCA-based unsupervised FE performed individually on gene expression and methylation profiles.

In Fig. 6(d), the first sample singular value vector for gene expression, u_{k_2j} (with $k_2 = 1$) for *D. quadriceps* is presented, which was shown to have the most significant dependence upon class labels based on categorical regression. Because these results were shown to be similar to those presented in Fig. 3(c), TD-based unsupervised FE was demonstrated to successfully generate biologically relevant sample singular value vectors of gene expression, u_{k_2j} (with $k_2 = 1$). Similarly, in Fig. 6(e), the fifth sample singular value vectors for the methylation profile of *D. quadriceps* are depicted, $u_{k_3\ell}$ (with $k_3 = 5$), which were found to be similar to those in Fig. 4(c). This finding indicates that TD-based unsupervised FE can lead

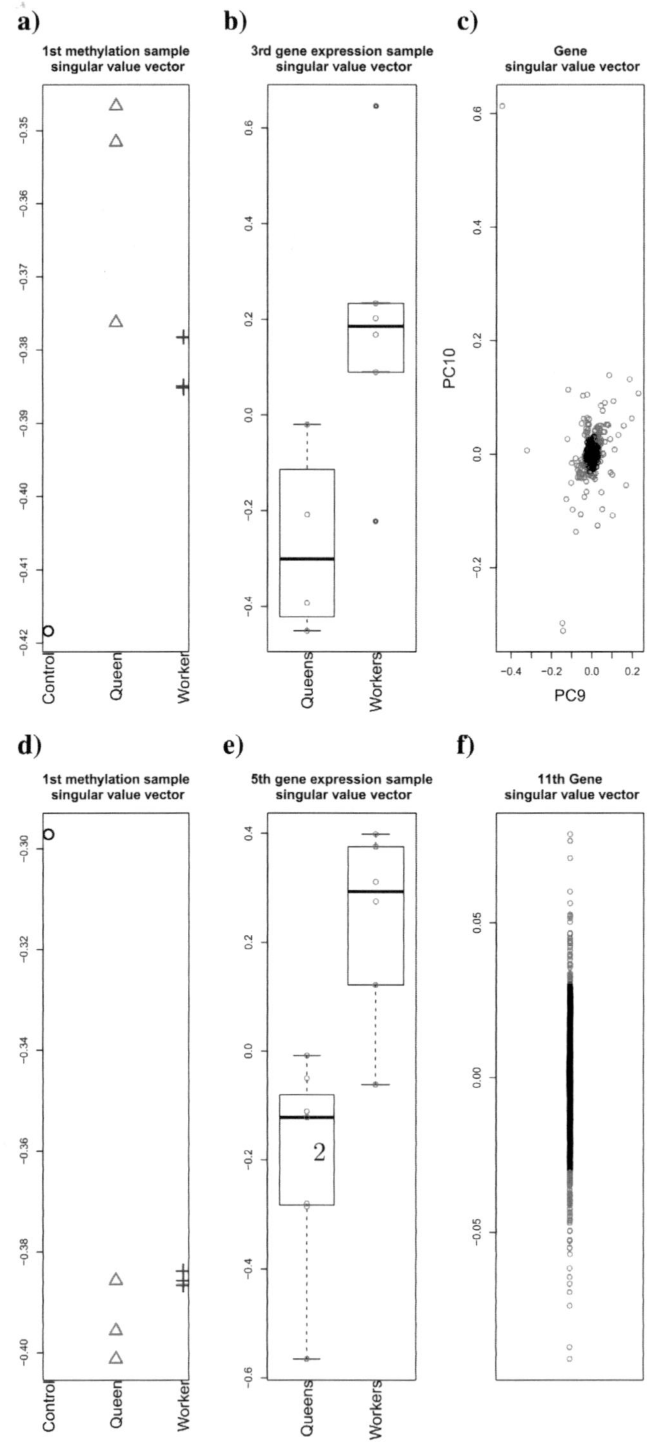

Fig. 6 TD analysis of gene expression profiles. TD-based unsupervised FE was applied to gene expression and methylation profiles of *P. canadensis* (**a**, **b**, **c**) and *D. quadriceps* (**d**, **e**, **f**). **a**: The first sample singular value vector for methylation profiles $u_{k_3\ell}$ (with $k_3 = 1$), (**b**) A boxplot of the third sample singular value vectors for gene expression profiles, u_{k_2j} (with $k_2 = 3$). **c** The ninth and 10th gene singular value vector, u_{k_1i} (with $9 \leq k_1 \leq 10$). Red areas in (**c**) represent 133 selected genes. **d**: The first sample singular value vector for methylation profiles, $u_{k_3\ell}$ (at $k_3 = 1$). **e** A boxplot of the fifth sample singular value vectors for gene expression profiles, u_{k_2j} (at $k_2 = 5$). **f** The 11th gene singular value vector, u_{k_1i} (at $k_1 = 11$). Red areas in (**f**) represent 128 selected genes

Table 5 The top 10 core tensors, G, with large absolute values

P. canadensis		*D. quadriceps*	
k_1	$G(k_1, k_2, k_3)$ $(k_2, k_3) = (1, 3)$	k_1	$G(k_1, k_2, k_3)$ $(k_2, k_3) = (1, 5)$
9	−79.8	11	−54.8
10	75.4	12	4.1
7	−61.4	25	3.4
11	38.4	2	−2.9
5	−23.4	23	2.8
4	−16.0	9	2.4
12	−11.9	20	−2.2
1	−5.4	8	2.2
13	5.4	10	−1.7
6	−4.5	22	−1.4

to successful generation of biologically relevant sample singular value vectors.

Furthermore, we aimed to identify the core tensor $G(k_1, k_2, k_3)$ (at $(k_2, k_3) = (1, 5)$) associated with the increased absolute values, to identify the k_1s used for the gene selection in *D. quadriceps* datasets (Table 5). Because $G(k_1, k_2, k_3)$ (with $(k_1, k_2, k_3) = (11, 1, 5)$) is top-ranked, the 11th gene singular value vectors, $u_{k_1=11,i}$, were employed for the selection of 128 genes associated with the adjusted P-values, determined using Eq. (1), lower than 0.01 (Fig. 6(f)).

To confirm that the selected genes are associated with differential gene expression and methylation between workers and queens, three statistical tests were applied (Table 6). We demonstrated that the 128 selected genes are associated with significant differences in gene expression but not methylation between queens and workers. Therefore, by TD-based unsupervised FE, we successfully found a set of genes associated with the differential gene expression but not methylation between queens and workers, suggesting that this analysis was not successful in the case of *D. quadriceps*.

Table 6 Statistical tests of the differences (between queens and workers) in gene expression and methylation

		t	Wilcox	KS
P. canadensis	gene expression	1.71×10^{-3}	1.89×10^{-2}	0.08
	methylation	1.74×10^{-4}	5.06×10^{-3}	1.02×10^{-3}
D. quadriceps	gene expression	2.73×10^{-12}	9.05×10^{-12}	4.41×10^{-11}
	methylation	0.3757	0.7163	0.4413

The genes identified by TD-based unsupervised FE were analyzed by *t* (the *t* test), Wilcox (the Wilcoxon rank sum test), and KS (the Kolmogorov–Sinai test), all two-sided

GO enrichment analysis

We demonstrated that PCA- and TD-based unsupervised FE can be used for the successful identification of genes associated with differential gene expression and methylation between workers and queens, but these results may be improved by showing that these sets of genes are biologically relevant as well. GO enrichment analysis was performed on three sets of genes selected on the basis of the gene expression profiles obtained from other studies (two *P. canadensis* datasets and one *D. quadriceps* dataset [8, 9], Table 7).

In contrast to the results of the other study on *P. canadensis* gene expression profiles [8], which showed no enrichment data, by means of the same dataset, we identified two enriched GO terms using the results obtained by PCA-based unsupervised FE. In the TD analysis, the number of enriched GO terms increased to three. These results indicate that we successfully performed the integrated analysis via TD-based unsupervised FE on *P. canadensis* datasets.

For *D. quadriceps* profiles, both genes identified by Patalano et al. [8] and those selected by PCA-based unsupervised FE were found to be associated with the same number of enriched GO terms, five, although the identified terms were not identical. Nevertheless, in the TD analysis, we did not observe any enrichment, and this result coincides with the fact that TD-based unsupervised FE failed to identify genes associated with differences in methylation profiles between queens and workers (Table 6).

Finally, in the analysis of genes identified by PCA-based unsupervised FE in the dataset provided by Ferreira et al. [9], five enriched GO terms were identified as well. Therefore, PCA- and TD-based unsupervised FE methods were shown to successfully identify biologically relevant sets of genes associated with significant enrichment in GO terms.

Discussion

Biological importance of the obtained results

We observed some instances of enrichment of GO terms, but the biological importance of our results should be examined further. In *P. canadensis* analysis, GO terms related to lipid transport were found to be enriched. Recently, Ament et al. [47] reported that worker honey bees undergo a socially regulated, highly stable lipid loss as part of their behavioral maturation. Given that *P. canadensis* is a bee species as well, the observed GO term enrichment of genes with differential gene expression and methylation profiles may be promising. Altered methylation of these genes may induce changes in gene

Table 7 GO enrichment analysis of the genes selected using gene expression profiles

P. canadensis		
Dataset provided by Patalano et al. [8]		
Results obtained in this study (PCA)		
GO:0005319		Lipid transporter activity
GO:0006869		Lipid transport
Results obtained in this study (TD)		
GO:0005319		Lipid transporter activity
GO:0005811		Lipid particle
GO:0006869		Lipid transport
Patalano et al. [8]		
	No enrichments	
Dataset provided by Ferreira et al. [9]		
Results obtained in this study (PCA)		
GO:0004129		Cytochrome-c oxidase activity
GO:0003735		Structural constituent of ribosome
GO:0006412		Translation
GO:0005743		Mitochondrial inner membrane
GO:0008137		NADH dehydrogenase (ubiquinone) activity
D. quadriceps		
Dataset provided by Patalano et al. [8]		
Results obtained in this study (PCA)		
GO:0005506		Iron ion binding
GO:0009055		Electron carrier activity
GO:0016705		Oxidoreductase activity, acting on paired donors, with incorporation or reduction of molecular oxygen
GO:0020037		heme binding
GO:0055114		Oxidation-reduction process
Results obtained in this study (TD)		
	No enrichments	
Patalano et al. [8]		
GO:0003735		Structural constituent of ribosome
GO:0005622		Intracellular
GO:0005840		Ribosome
GO:0005842		Cytosolic large ribosomal subunit
GO:0006412		Translation

expression that result in a highly stable lipid loss. This arrangement may enable the coexistence of multiple phenotypes.

In contrast, oxidation-reduction processes, regulated by genes expressed differently between queens and workers of *D. quadriceps*, have been reported to be upregulated in the queens of multiple ant species [48]. Because *D. quadriceps* is an ant species, this means that we correctly identified DEGs between workers and queens, via the proposed strategy. In addition to the heme-binding–associated genes, shown to be differently expressed between queens and workers of *D. quadriceps*, they were associated with aberrant methylation in termites [49], another species of social insects.

Furthermore, oxidoreductase activity–related genes, found to be differently expressed between queens and workers of *D. quadriceps*, have been reported to be expressed in the queens of honey bees too [50].

Alaux et al. [51] determined genes associated with electron carrier activity—shown to be differentially expressed between the queens and workers of *D. quadriceps* in this study as well—in another study, which analyzed the relation between aggressiveness and behavioral evolution in honey bees.

When genes of *D. quadriceps*, identified by Patalano et al. were analyzed, more general GO terms were obtained, e.g., those related to translation and ribosomes, which are unlikely to be related to social-insect–specific features. In contrast, identification of more specific GO terms and the results of other studies point to biological importance of the analyses presented here.

Future directions

This paper shows the importance of integrated analyses of gene expression and promoter methylation for finding genes that might link castes of social insects to the epigenome. At the moment, only two species were investigated, but because castes of social insects have been examined in multiple species that do not belong to even the same family from the genetic point of view [52], this approach should be extended other species. Inclusion of more species may clarify how castes of social insects have evolved and have been maintained from the standpoint of epigenetics.

Conclusions

Here, we tested newly developed PCA- and TD-based unsupervised FE for the analysis of gene expression and methylation profiles of *P. canadensis* and *D. quadriceps*. The issues observed in other studies were solved as follows:

1. GO enrichment analysis was performed successfully on *P. canadensis* gene expression profiles [8] with a strict criterion of FDR less than 0.01 (Table 7);
2. Genes found to be differentially expressed among four castes [9] were analyzed by means of a strict criterion of FDR less than 0.01 (Table 6);
3. Genes associated with differential methylation between queens and workers of *P. canadensis* were analyzed successfully [8] (Table 2).

Therefore, PCA- and TD-based unsupervised FE methods were successfully performed on 'omics datasets comprising gene expression and methylation profiles. The obtained sets of genes will help us understand how development of caste phenotypes is regulated epigenetically.

Additional file

Additional file 1: Genes selected by PCA- and TD-based unsupervised FE. (XLSX 16 kb)

Abbreviations

FE: Feature extraction; PC: Principal component; PCA: Principal component analysis; TD: Tensor decomposition

Acknowledgements

Not applicable.

Funding

This study was supported by KAKENHI 17K00417 and a Chuo University Specific Topics grant. Funding for the publication of this article was provided by the above KAKENHI.

Availability of data and materials

All data used in this study was obtained from GEO.

About this supplement

This article has been published as part of *BMC Bioinformatics Volume* 19 Supplement 4, 2018: Selected articles from the 16th Asia Pacific Bioinformatics Conference (APBC 2018): bioinformatics. The full contents of the supplement are available online at https://bmcbioinformatics.biomedcentral.com/articles/supplements/volume-19-supplement-4.

Authors' contributions

YHT planned the project, performed all the analyses, and wrote the manuscript.

Ethics approval and consent to participate

Not applicable.

Consent for publication

Not applicable.

Competing interests

The author declares that he has no competing interests.

Publisher's Note

Springer Nature remains neutral with regard to jurisdictional claims in published maps and institutional affiliations.

Published: 8 May 2018

References

1. Corona M, Libbrecht R, Wheeler DE. Molecular mechanisms of phenotypic plasticity in social insects. Current Opinion Insect Sci. 2016;13:55–60. https://doi.org/10.1016/j.cois.2015.12.003.
2. Lockett GA, Kucharski R, Maleszka R. DNA methylation changes elicited by social stimuli in the brains of worker honey bees. Genes Brain Behavior. 2011;11(2):235–242. https://doi.org/10.1111/j.1601-183x.2011.00751.x.
3. Becker N, Kucharski R, Rössler W, Maleszka R. Age-dependent transcriptional and epigenomic responses to light exposure in the honey bee brain. FEBS Open Bio. 2016;6(7):622–639. https://doi.org/10.1002/2211-5463.12084.
4. Peaston AE, Whitelaw E. Epigenetics and phenotypic variation in mammals. Mammalian Genome. 2006;17(5):365–374. https://doi.org/10.1007/s00335-005-0180-2.
5. Mohtat D, Susztak K. Fine tuning gene expression: The epigenome. Semin Nephrol. 2010;30(5):468–476. https://doi.org/10.1016/j.semnephrol.2010.07.004.
6. Triantaphyllopoulos KA, Ikonomopoulos I, Bannister AJ. Epigenetics and inheritance of phenotype variation in livestock. Epigenetics & Chromatin. 2016;9(1):. https://doi.org/10.1186/s13072-016-0081-5.

7. Duncan EJ, Gluckman PD, Dearden PK. J Exp Zool Part B: Mol Dev Evol. 2014;322(4):208–220. https://doi.org/10.1002/jez.b.22571.
8. Patalano S, Vlasova A, Wyatt C, Ewels P, Camara F, Ferreira PG, Asher CL, Jurkowski TP, Segonds-Pichon A, Bachman M, González-Navarrete I, Minoche AE, Krueger F, Lowy E, Marcet-Houben M, Rodriguez-Ales JL, Nascimento FS, Balasubramanian S, Gabaldon T, Tarver JE, Andrews S, Himmelbauer H, Hughes WOH, Guigó R, Reik W, Sumner S. Molecular signatures of plastic phenotypes in two eusocial insect species with simple societies. Proc Natl Acad Sci. 2015;112(45):13970–13975. https://doi.org/10.1073/pnas.1515937112.
9. Ferreira PG, Patalano S, Chauhan R, Ffrench-Constant R, Gabaldón T, Guigó R, Sumner S. Transcriptome analyses of primitively eusocial wasps reveal novel insights into the evolution of sociality and the origin of alternative phenotypes. Genome Biol. 2013;14(2):20. https://doi.org/10.1186/gb-2013-14-2-r20.
10. Taguchi Y-h, Iwadate M, Umeyama H, Murakami Y. Principal component analysis based unsupervised feature extraction applied to bioinformatics analysis. In: Tsai JJP, Ng K-L, editors. Computational Methods with Applications in Bioinformatics Analysis. Singapore: World Scientific; 2017. p. 153–182. Chap. 8 https://doi.org/10.1142/9789813207981_0008.
11. Taguchi Y-h. microrna-mrna interaction identification in wilms tumor using principal component analysis based unsupervised feature extraction. In: 2016 IEEE 16th International Conference on Bioinformatics and Bioengineering (BIBE); 2016. p. 71–78. https://doi.org/10.1109/BIBE.2016.14.
12. Taguchi Y-h. Principal Components Analysis Based Unsupervised Feature Extraction Applied to Gene Expression Analysis of Blood from Dengue Haemorrhagic Fever Patients. Sci Rep. 2017;7:44016.
13. Taguchi Y-h. Principal component analysis based unsupervised feature extraction applied to publicly available gene expression profiles provides new insights into the mechanisms of action of histone deacetylase inhibitors. Neuroepigenetics. 2016;8:1–18. https://doi.org/10.1016/j.nepig.2016.10.001.
14. Taguchi YH, Iwadate M, Umeyama H. Principal component analysis-based unsupervised feature extraction applied to in silico drug discovery for posttraumatic stress disorder-mediated heart disease. BMC Bioinforma. 2015;16:139.
15. Taguchi Y-h, Okamoto A. Principal component analysis for bacterial proteomic analysis. In: Shibuya T, Kashima H, Sese J, Ahmad S, editors. Pattern Recognition in Bioinformatics. LNCS. Heidelberg: Springer; 2012. p. 141–152.
16. Ishida S, Umeyama H, Iwadate M, Taguchi Y-h. Bioinformatic Screening of Autoimmune Disease Genes and Protein Structure Prediction with FAMS for Drug Discovery. Protein Pept Lett. 2014;21(8):828–39.
17. Kinoshita R, Iwadate M, Umeyama H, Taguchi Y-h. Genes associated with genotype-specific DNA methylation in squamous cell carcinoma as candidate drug targets. BMC Syst Biol. 2014;8 Suppl 1:4.
18. Taguchi Y-h, Murakami Y. Principal component analysis based feature extraction approach to identify circulating microRNA biomarkers. PLoS ONE. 2013;8(6):66714.
19. Taguchi Y-h, Murakami Y. Universal disease biomarker: can a fixed set of blood microRNAs diagnose multiple diseases? BMC Res Notes. 2014;7:581.
20. Murakami Y, Toyoda H, Tanahashi T, Tanaka J, Kumada T, Yoshioka Y, Kosaka N, Ochiya T, Taguchi Y-h. Comprehensive miRNA expression analysis in peripheral blood can diagnose liver disease. PLoS ONE. 2012;7(10):48366.
21. Murakami Y, Tanahashi T, Okada R, Toyoda H, Kumada T, Enomoto M, Tamori A, Kawada N, Taguchi Y-h, Azuma T. Comparison of Hepatocellular Carcinoma miRNA Expression Profiling as Evaluated by Next Generation Sequencing and Microarray. PLoS ONE. 2014;9(9):106314.
22. Murakami Y, Kubo S, Tamori A, Itami S, Kawamura E, Iwaisako K, Ikeda K, Kawada N, Ochiya T, Taguchi Y-h. Comprehensive analysis of transcriptome and metabolome analysis in Intrahepatic Cholangiocarcinoma and Hepatocellular Carcinoma. Sci Rep. 2015;5:16294.
23. Umeyama H, Iwadate M, Taguchi Y-h. TINAGL1 and B3GALNT1 are potential therapy target genes to suppress metastasis in non-small cell lung cancer. BMC Genomics. 2014;15 Suppl 9:2.
24. Taguchi Y-h, Iwadate M, Umeyama H. Heuristic principal component analysis-based unsupervised feature extraction and its application to gene expression analysis of amyotrophic lateral sclerosis data sets. In: Computational Intelligence in Bioinformatics and Computational Biology (CIBCB), 2015 IEEE Conference On; 2015. p. 1–10. https://doi.org/10.1109/CIBCB.2015.7300274.
25. Taguchi Y-h, Iwadate M, Umeyama H, Murakami Y, Okamoto A. Heuristic principal component analysis-aased unsupervised feature extraction and its application to bioinformatics. In: Wang B, Li R, Perrizo W, editors. Big Data Analytics in Bioinformatics and Healthcare; 2015. p. 138–162.
26. Taguchi Y-h. Integrative analysis of gene expression and promoter methylation during reprogramming of a non-small-cell lung cancer cell line using principal component analysis-based unsupervised feature extraction. In: Huang D-S, Han K, Gromiha M, editors. Intelligent Computing in Bioinformatics. LNCS. Heidelberg: Springer; 2014. p. 445–455.
27. Taguchi Y-h. Identification of aberrant gene expression associated with aberrant promoter methylation in primordial germ cells between E13 and E16 rat F3 generation vinclozolin lineage. BMC Bioinformatics. 2015;16 Suppl 18:16.
28. Taguchi Y-h. Identification of More Feasible MicroRNA-mRNA Interactions within Multiple Cancers Using Principal Component Analysis Based Unsupervised Feature Extraction. Int J Mol Sci. 2016;17(5):696.
29. Taguchi Y-h. Principal component analysis based unsupervised feature extraction applied to budding yeast temporally periodic gene expression. BioData Min. 2016;9:22.
30. Taguchi Y-h, et al. SFRP00001 is a possible candidate for epigenetic therapy in non-small cell lung cancer. BMC Med Genomics. 2016;9 Suppl 1:28.
31. Taguchi Y-h, Wang H. Genetic association between amyotrophic lateral sclerosis and cancer. Genes. 2017;8(10):243. https://doi.org/10.3390/genes8100243.
32. Taguchi Y-h. Tensor decomposition-based unsupervised feature extraction identifies candidate genes that induce post-traumatic stress disorder-mediated heart diseases. BMC Medical Genomics. in press.
33. Taguchi Y-h. Tensor decomposition based unsupervised feature extraction identified universal nature of sequence-non-specific off-target regulation of mrna mediated by microrna transfection. BMC Med Genomics. 2017. in press.
34. Tensor decomposition/principal component analysis based unsupervised feature extraction applied to brain gene expression and methylation profiles of social insects with multiple castes. BMC Bioinformtics. 2018. in press.
35. Taguchi Y-h. Identification of candidate drugs for heart failure using tensor decomposition-based unsupervised feature extraction applied to integrated analysis of gene expression between heart failure and DrugMatrix datasets. In: Intelligent Computing Theories and Application. Heidelberg: Springer; 2017. p. 517–528. https://doi.org/10.1007/978-3-319-63312-1_45.
36. Taguchi Y-h. Tensor decomposition-based unsupervised feature extraction applied to matrix products for multi-view data processing. Plos ONE. 2017;12(8):0183933. https://doi.org/10.1371/journal.pone.0183933.
37. Taguchi Y-h. Identification of candidate drugs using tensor-decomposition-based unsupervised feature extraction in integrated analysis of gene expression between diseases and drugmatrix datasets. Scientific Report. 2017. in press.
38. Benjamini Y, Hochberg Y. Controlling the false discovery rate: A practical and powerful approach to multiple testing. Journal of the Royal Statistical Society. Series B (Methodological). 1995;57(1):289–300.
39. Lathauwer LD, Moor BD, Vandewalle J. A multilinear singular value decomposition. SIAM Journal on Matrix Analysis and Applications. 2000;21(4):1253–1278. https://doi.org/10.1137/s0895479896305696.
40. Luo D, Ding C, Huang H. Are tensor decomposition solutions unique? on the global convergence HOSVD and ParaFac algorithms. In: Advances in Knowledge Discovery and Data Mining. Springer; 2011. p. 148–159. https://doi.org/10.1007/978-3-642-20841-6_13.
41. Paper Wasp and Denosaur Ant Project. Accessed 18 Nov. 2017. http://wasp.crg.eu/download.html.
42. Uniprot. Accessed 18 Nov. 2017. http://www.uniprot.org/uploadlists/.
43. Heng J, Guo X, Wu W, Wang Y, Li G, Chen M, Peng L, Wang S, Dai L, Tang L, Wang J. Integrated analysis of promoter mutation, methylation and expression of AKT1 gene in chinese breast cancer patients. PLOS ONE. 2017;12(3):0174022. https://doi.org/10.1371/journal.pone.0174022.
44. Heng J, Zhang F, Guo X, Tang L, Peng L, Luo X, Xu X, Wang S, Dai L, Wang J. Integrated analysis of promoter methylation and expression of

telomere related genes in breast cancer. Oncotarget. 2017. https://doi.org/10.18632/oncotarget.16036.
45. Li Z, Zhuang X, Zeng J, Tzeng C-M. Integrated analysis of DNA methylation and mRNA expression profiles to identify key genes in severe oligozoospermia. Frontiers in Physiology. 2017;8:. https://doi.org/10.3389/fphys.2017.00261.
46. Mallik S, Mukhopadhyay A, Maulik U, Bandyopadhyay S. Integrated analysis of gene expression and genome-wide DNA methylation for tumor prediction: An association rule mining-based approach. In: 2013 IEEE Symposium on Computational Intelligence in Bioinformatics and Computational Biology (CIBCB). IEEE; 2013. https://doi.org/10.1109/cibcb.2013.6595397.
47. Ament SA, Chan QW, Wheeler MM, Nixon SE, Johnson SP, Rodriguez-Zas SL, Foster LJ, Robinson GE. Mechanisms of stable lipid loss in a social insect. Journal of Experimental Biology. 2011;214(22):3808–3821. https://doi.org/10.1242/jeb.060244.
48. Warner MR, Mikheyev AS, Linksvayer TA. Genomic signature of kin selection in an ant with obligately sterile workers. Molecular Biology and Evolution. 2017;34(7):1780–1787. https://doi.org/10.1093/molbev/msx123.
49. Glastad KM, Gokhale K, Liebig J, Goodisman MAD. The caste- and sex-specific DNA methylome of the termite zootermopsis nevadensis. Scientific Reports. 2016;6(1):. https://doi.org/10.1038/srep37110.
50. Cristino AS, Nunes FMF, Lobo CH, Bitondi MMG, Simões ZLP, da Fontoura Costa L, Lattorff HMG, Moritz RFA, Evans JD, Hartfelder K. Caste development and reproduction: a genome-wide analysis of hallmarks of insect eusociality. Insect Molecular Biology. 2006;15(5):703–714. https://doi.org/10.1111/j.1365-2583.2006.00696.x.
51. Alaux C, Sinha S, Hasadsri L, Hunt GJ, Guzman-Novoa E, DeGrandi-Hoffman G, Uribe-Rubio JL, Southey BR, Rodriguez-Zas S, Robinson GE. Honey bee aggression supports a link between gene regulation and behavioral evolution. Proceedings of the National Academy of Sciences. 2009;106(36):15400–15405. https://doi.org/10.1073/pnas.0907043106.
52. Fischman BJ, Woodard SH, Robinson GE. Molecular evolutionary analyses of insect societies. Proceedings of the National Academy of Sciences. 2011;108(Supplement_2):10847–10854. https://doi.org/10.1073/pnas.1100301108.

Chen *et al. BMC Bioinformatics* 2018, **19**(Suppl 4):154
https://doi.org/10.1186/s12859-018-2069-6

BMC Bioinformatics

RESEARCH Open Access

A gene profiling deconvolution approach to estimating immune cell composition from complex tissues

Shu-Hwa Chen[1*†], Wen-Yu Kuo[2†], Sheng-Yao Su[1,3,4], Wei-Chun Chung[1], Jen-Ming Ho[1], Henry Horng-Shing Lu[2*] and Chung-Yen Lin[1,5,6*]

From The Sixteenth Asia Pacific Bioinformatics Conference
Yokohama, Japan. 15-17 January 2018

Abstract

Background: A new emerged cancer treatment utilizes intrinsic immune surveillance mechanism that is silenced by those malicious cells. Hence, studies of tumor infiltrating lymphocyte populations (TILs) are key to the success of advanced treatments. In addition to laboratory methods such as immunohistochemistry and flow cytometry, in silico gene expression deconvolution methods are available for analyses of relative proportions of immune cell types.

Results: Herein, we used microarray data from the public domain to profile gene expression pattern of twenty-two immune cell types. Initially, outliers were detected based on the consistency of gene profiling clustering results and the original cell phenotype notation. Subsequently, we filtered out genes that are expressed in non-hematopoietic normal tissues and cancer cells. For every pair of immune cell types, we ran t-tests for each gene, and defined differentially expressed genes (DEGs) from this comparison. Equal numbers of DEGs were then collected as candidate lists and numbers of conditions and minimal values for building signature matrixes were calculated. Finally, we used ν -Support Vector Regression to construct a deconvolution model. The performance of our system was finally evaluated using blood biopsies from 20 adults, in which 9 immune cell types were identified using flow cytometry. The present computations performed better than current state-of-the-art deconvolution methods.

Conclusions: Finally, we implemented the proposed method into R and tested extensibility and usability on Windows, MacOS, and Linux operating systems. The method, MySort, is wrapped as the Galaxy platform pluggable tool and usage details are available at https://testtoolshed.g2.bx.psu.edu/view/moneycat/mysort/e3afe097e80a.

Background

Cancers comprise a group of diseases that are characterized by uncontrolled growth of abnormal immortalized cells that can spread to other parts of the body and finally deplete resources. Hanahan and Weinberg suggested biological capabilities and hallmarks of a multistep process toward the development of human tumors [1, 2]. Among these, tumors have strategies for evading immune destruction. In contrast, the immune system sometimes over-functions and incorrectly identifies normal cells, leading to convalescence and autoimmune disease. Hence, as regulators of immunity, various immune checkpoints achieve an equilibrium of system responses. Recently, cancer cells were shown to interact with immune checkpoints and inhibit T cell activation as an immune evasion strategy [3].

Immunotherapy is a novel strategy in which immune checkpoint components are manipulated, in particular leading to blockade of T cell activation. Interactions between immune cell receptors and ligands are essential targets of immune checkpoint blockade which involves many immune cell types and distinct pathways that are

* Correspondence: sophia@iis.sinica.edu.tw; hslu@stat.nctu.edu.tw; cylin@iis.sinica.edu.tw
†Equal contributors
[1]Institute of Information Science, Academia Sinica, 128 Academia Road, Section 2, Nankang, Taipei 115, Taiwan
[2]Institute of Statistics, National Chiao Tung University, Assembly Building I, 1001 Ta Hsueh Road, Hsinchu 30010, Taiwan
Full list of author information is available at the end of the article

incompletely understood. Therefore, understanding the composition of immune cells in tissues is central to studies of immune checkpoints with the ultimate aim of developing immunotherapy. Flow cytometry and immunohistochemistry analyses have long been developed to define cell compositions but can't be easily applied to resolve immune cell types from all kind of cancers. In contrast, whole transcriptome profiling methods such as microarray and next generation sequencing can be developed as in silico methods for defining compositions of a panel of cell types that are defined using mixed gene profiling of cell-type specific genes.

Gene expression deconvolution methods have been developed for several years and most methods regard deconvolution as a linear problem. Strictly speaking, the expression level of gene i in mixture j is the sum of its expression in r cell types, as indicated by the following equation:

$$M_{ij} = \Sigma_{k=1}^{r} S_{ik} F_{kj} \quad (1)$$

where S_{ik} is specific gene expression in cell type k and F_{kj} is the proportion of cell type k in mixture j. The general model can be expressed as an approximate matrix problem as follows: $M \approx S \times F$, and the problem can be solved from the view of an equation-solving question.

Abbas et al. [4] and Clarke et al. [5] were the first to propose deconvolution methods, and Gong et al. [6] and Zhong et al. [7] used quadratic programming as an optimization tool to solve the problem. In further studies, Gaujoux et al. [8] applied non-negative matrix factorization to deconvolution and Qiao et al. [9] and Liebner et al. [10] introduced statistics to solve the deconvolution problem. Finally, Newman et al. [11] developed the novel strategy CIBERSORT to solve the linear equation, and comparisons with the six deconvolution methods listed above by benchmarking on mixture expression data showed that the method agreed with flow cytometry data and outperformed other methods.

Tumor infiltrating lymphocytes (TILs) include various cell types in variable proportions, and leave the bloodstream to infiltrate into cancer tissues. TILs are associated with tumor growth, cancer progression, and patient prognoses [12]. Thus, assessments of the composition of TILs are fundamental for developing effective immunotherapies. In this study we analyzed the composition of TILs from mixed tissue gene profiling data and revised analyses for better performance.

Methods

The deconvolution method in this study is mainly adjusted from the strategy of CIBERSORT. Table 1 describes the datasets of 22 immune cell types collected by Newman et al. [11] from 11 major leukocyte types. At least two replicates for each cell types are included in this study.

Table 1 A brief of the 113 microarrays used as the deconvolution dataset

Major leukocyte types	Subtypes	Number of replicates	
B cells	B cells naïve	7	*
	B cells memory	8	*
PCs	Plasma cells	7	
CD8 T cells	T cells CD8	4	*
CD4 T cells	T cells CD4 naïve	3	*
	T cells CD4 memory resting	3	*
	T cells CD4 memory activated	3	*
	T cells follicular helper	3	
	T cells regulatory (Tregs)	2	
Gamma delta T cells	T cells gamma delta	2	*
NK cells	NK cells resting	4	**
	NK cells activated	11	**
Monocytes and Macrophages	Monocytes	11	*
	Macrophages M0	12	
	Macrophages M1	3	
	Macrophages M2	3	
Dendritic cells	Dendritic cells resting	6	
	Dendritic cells activated	6	
Mast cells	Mast cells resting	2	
	Mast cells activated	2	
Eos	Eosinophils	2	
PMNs	Neutrophils	8	

Remarks: * and ** are cell types that used in the validation datasets (PBMC from 20 adults). Both activated and resting NK cells are counted as NK cells in the prediction on the benchmark experiment

Implement resources

We develop the algorithms using R (version 3.1.1) and the following packages: preprocessCore (version 1.34.0), limma (version 3.28.31), geneplotter (version 1.50.0), qvalue (version 2.4.2), genefilter (version 1.54.2), plyr (version 1.8.4), and e1071 (version 1.6–7). The algorithms are then exam on Windows 10 and Ubuntu Linux 16.04 server. Note that the version of R and associated packages may be varied between operating systems.

To obtain usability and extensibility for a customized workflow, we adopt the proposed algorithms to a pluggable tool, MySort, and integrate the tool into the Galaxy platform.

Research design

As mentioned previously, the deconvolution model can be presented as a matrix function. We denote the transcriptome expression level of a cell type consists of i genes. The transcription profile of a cancer biopsy will be the summation of each gene expression level cross the proportion of all cell types in the biopsy. Since we are concerning only in the composition of a particular panel of cell types, we pick t genes, a subset of genes that are only expressed in the given cell panel, to build a subset of gene expression matrix, namely signature matrix, S. Thus, S is a matrix with dimension of t rows and r columns where r represents the number of cell types in the given cell panel, the immune cells in this study. When S is defined, the vector F, the proportion of each cell type in the given panel, can be solved out. Hence, deconvolution can be divided into two parts: constructing a signature matrix and solving linear equation for F (Fig. 1).

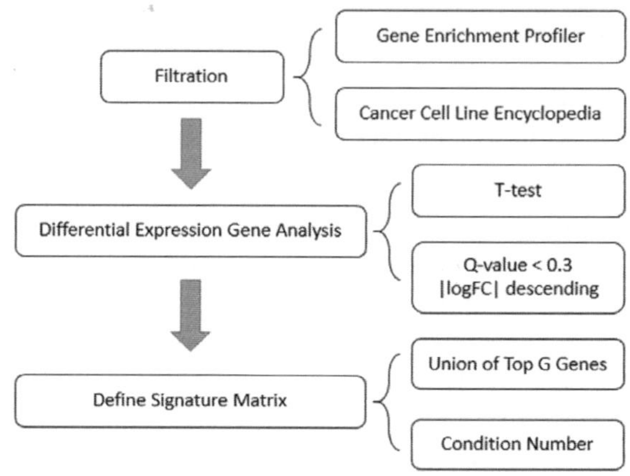

Fig. 2 Steps of constructing signature matrix

The signature matrix

The main strategy for choosing cell-type-specific genes is to do differential gene expression analysis. The following diagram explains the method of constructing signature matrix step by step (Fig. 2).

Data from the 22 cell types (the signature set, 113 arrays) are quantile normalized before detecting differentially expressed genes. In order to prevent the datasets containing bias experiment result, clustering on gene profiling is applied. The clustering method is complete linkage and in the Euclidean distance. The inconsistency of gene profiling clusters and cell type labels is further analyzed for an advanced outlier judgment. Correlation between arrays in the same cell type is calculated using Pearson correlation. For each array of a certain cell type, we judge the problematic cluster if any single array shows low gene expression correlations ($R < 0.85$) to the others in the same cell phenotype (i.e., more than 2/3 of the related array-to-array pairs within a cluster).

To define a set of feature genes for deconvolution, we first eliminate genes that are unsuitable for building model. Two methods were adopted. We use datasets and enrichment score (ES) described in Benita et al. [13] to define genes that expressed in normal tissues by the criteria of $ES > 0$ in more than 5% of observed tissue types. Besides, genes that expressed in cancer cells are collected from cancer cell line encyclopedia [14] in the criteria of log2 transformed expression level > 7. The two lists are used as black lists to remove genes expressed in normal tissues and cancer cells from the candidate list.

To select genes that can be the representative features of a cell type, we run the statistical analysis between each two immune cell type pairs. Firstly, the differentially expressed genes (DEGs) are detected using two-sided unequal variance T-test with a significant criterion in q-value < 0.3. Secondly, we sort the DEGs of each comparing pair by the absolute value of log fold change of gene expression level in descending order. Third, a top G ($G = 5$ to 100) ranked DEGs are selected from each pair to build a union set of a signature gene list (top G signature gene list) and to derive top G signature matrix, the expression profiles for each top G signature gene list. Condition number [15] which is associated with the linear equation is introduced to define the choice of G and calculated with "kappa" function in R.

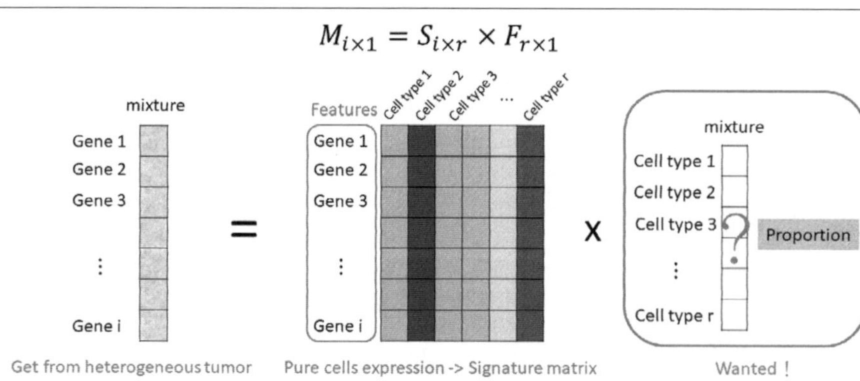

Fig. 1 The deconvolution model

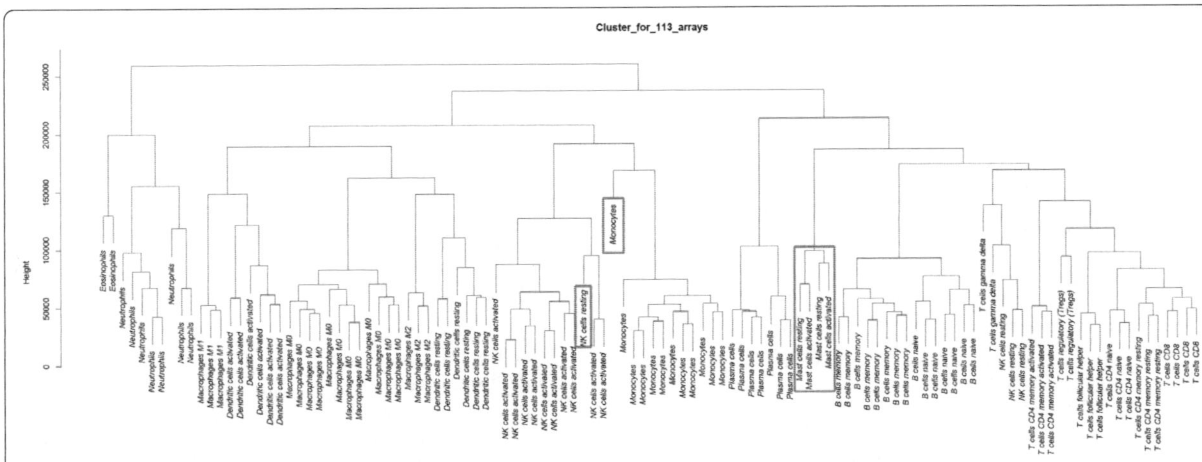

Fig. 3 Hierarchical clustering for 113 arrays to detect the consistency between cell phenotypes and gene expression pattern. Arrays with problematic grouping are boxed in red

ν-support vector regression

First, gene expression profiles from both the signature set and the validation set (described in benchmark method) are quantile normalized. To avoid inconsistency of data range, a preprocessing procedure is introduced. We use a standardized score by converting the data to z score (mean = 0, variance = 1). Furthermore, genes in the signature matrix may be not always included in the mixture data set. We use the overlaps of genes as the final features for building the model.

The strategy to solve this regression problem is a machine learning method called *ν – support vector regression* (*ν*-SVR). In this study, support vectors represent a particular subset of genes in the signature matrix. *υ*-SVR is adopted in this study by "svm" function of R in package, "e1071", with linear kernel. Furthermore, *F* have the best result with $\upsilon = \{0.25, 0.5, 0.75\}$, where the evaluation method is to estimate the lowest root mean square error between M and S × F. Negative coefficients for regression solved by SVR are set to be 0, and then all coefficients are normalized to be summed to 1. Finally, the estimation of relative proportions for cell types is solved out, and correlations and root mean squares between M and S × F for each sample are evaluated.

Benchmark method

The deconvolution method was benchmarked with flow cytometry results of twenty adult blood biopsies used in CIBERSORT study [11]. Since there are only nine cell types were labeled, we extract the prediction

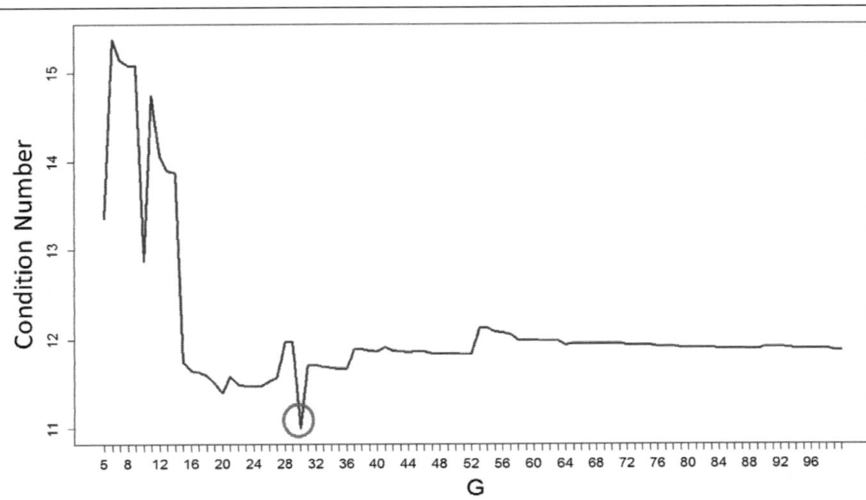

Fig. 4 The condition number of iteration for G

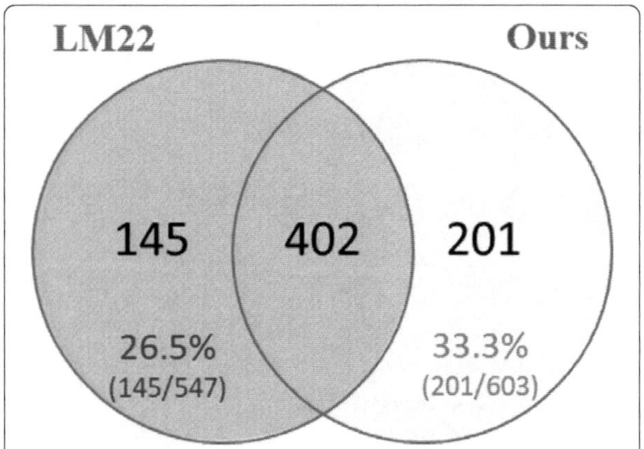

Fig. 5 Venn-diagram for a number of genes of two signature gene sets. (LM22 is signature matrix of CIBERSORT)

of these types and rescale the sum of the nine prediction value to 1. We use the cell type frequencies determined by flow cytometry as the standard and calculate correlation (Pearson correlation) and root mean square error (RMSE) of the derived expression level of signature genes to the true value to evaluate the performance of prediction.

Results

The agreement of cell phenotypes to gene expression profile clustering

Initially, we retrieved data for building a deconvolution model. After a data preprocessing step, we clustered the deconvolution dataset (113 arrays) according to gene expression profiles. Although replicates of each cell type were expected to be clustered into the same group, one array of resting NK cells was inserted into the activated NK cells group and one array of monocytes is far from its group (Fig. 3), and arrays of resting and activated mast cells were arranged without clear segregation. Therefore, we performed Pearson correlations between arrays of cell types to identify outlying arrays. One array from monocytes and another from resting NK cells were excluded according to weaker correlations than those between others within the group. In addition, we decided to merge two cell types, resting and activated mast cells, into a single mast cell category due to the sparse of evidence on segregation.

Construct signature matrix

After filtration processes, pairwise comparisons of cell types were performed using t-tests. The top G ranked DEGs ($G = 5\sim 100$) were then joined to give the top G gene list for each G value. Subsequently, we calculated the numbers of conditions for each top G gene list and defined the final gene list to build the signature matrix as the list with the lowest number of conditions. The final results were $G = 30$, number of conditions = 10.99, and number of signature genes = 603 (Fig. 4).

The top G gene list ($G = 30$) was then compared to the LM22 as defined using CIBERSORT. In the Venn-diagram for numbers of genes in Fig. 5, two-thirds of our signature genes overlapped with LM22, although 201 genes were uniquely defined by us and 145 were uniquely defined by CIBERSORT.

Benchmark method

We used cell composition data from reported peripheral blood mononuclear cells as a validation to benchmark the performance of deconvolution methods, and as mentioned in the methods, nine cell types were resolved by flow

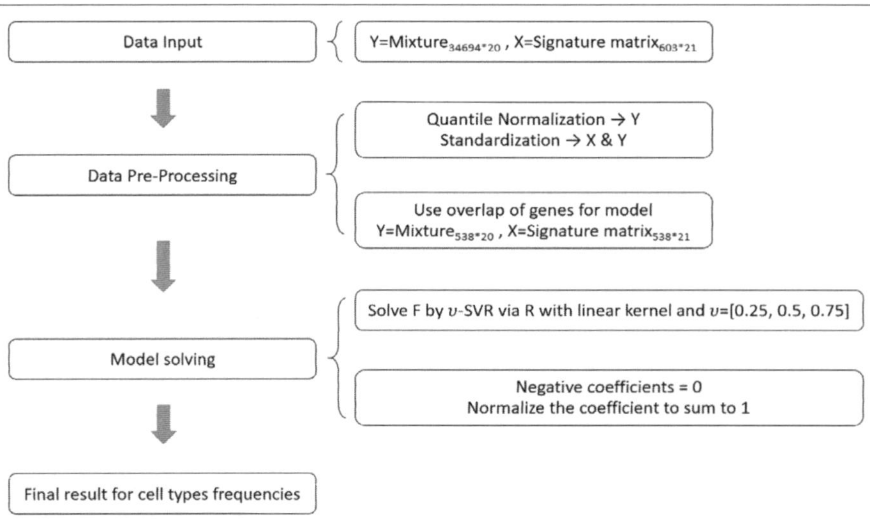

Fig. 6 Workflow of the deconvolution method

Table 2 The performance of our deconvolution method and in CIBERSORT, evaluated by RMSE of signature genes

RMSE	Naïve B cells	Memory B cells	CD8 T cells	Naïve CD4 T cells	Resting memory CD4 T cells	Activated memory CD4 T cells	NK cells	Monocytes
Our result	3.690576	2.739405	7.618429	17.603077	7.602287	2.301891	3.971373	14.821731
CIBERSORT	3.750000	1.420000	8.040000	18.25000	8.38000	2.880000	4.390000	20.29000

Method in better performance (lower RMSE) is marked yellow

cytometry in the validation dataset. We then recalculated relative proportions of all cell types, but recalculated the relative portion of nine relevant types to a sum of 1. Similarly, neither mast cells nor subtyped resting and activated mast cells from CIBERSORT were typed in the validation dataset. The process is illustrated in Fig. 6.

We compared our prediction with the result provided in CIBERSORT and estimated performance according to RMSE (Table 2) and Pearson correlations (Table 3) for each cell type. These comparisons indicate that the present deconvolution method outperforms CIBERSORT. The differences between our approach and CIBERSORT are summarized in Table 4.

Galaxy plugin

Our validation experiments show that the present workflow outperforms previous models for most cell types. To improve access to the research community interesting in detecting infiltrating immune cells, we have implemented the present algorithms MySort in R and wrapped as a Galaxy platform pluggable tool and provide usage details in the Galaxy toolshed https://testtoolshed.g2.bx.psu.edu/ view/moneycat/mysort/e3afe097e80a (Fig. 7). Using mixture expression profiling data (uploaded by the user, indexed in gene symbols with samples arranged by column) and the signature matrix (provided by this study) as inputs, MySort generates the immune cell compositions for each expression profile in csv format and a combined diagram to present the resolved cell proportion in a bar chart plot and a hierarchical clustering plot for relatedness among submitted samples.

Discussions

Signature genes play important roles in gene expression deconvolution computations. Additionally, outlier detection, gene list filtration, and support vector regression were central to the positive outcomes of our model. We also revised the deconvolution process and discovered issues that were not properly dealt with previously, including uncertain accuracy of representative data matrixes for each immune cell type and unequal contributions of DEG pairs that are used to build signature matrixes.

Removing the outliers is an important preliminary step for any in silico model. We revised the cell phenotype to the profiling clustering result and identified possible outliers and improper typing groups. Subsequently, we excluded two arrays and merged two classes in the signature set. In addition, we identified limitations of data replicates in some cell types as a cause of decreased confidence in DEGs with weak statistical power. Further technical difficulties, such as inconsistency of cell type definitions in different laboratories may introduce additional problems for data analysis. For example, gamma delta T cells are apparently difficult to identify with certainty, and as stated in the CIBERSORT study, few microarrays of gamma delta T cells contain more than two replicates.

We assumed that the present deconvolution model is linear. Thus, to realize and simplify the model, we selected features as genes that are specifically expressed in certain cell types. In addition, we used support vector regression (SVR) to resolve the regression problem of cell compositions. SVR was developed from a support vector machine as an optimization approach for binary

Table 3 The performance of our deconvolution method and in CIBERSORT, evaluated by Pearson correlation of cell proportion

Correlation	Naïve B cells	Memory B cells	CD8 T cells	Naïve CD4 T cells	Resting memory CD4 T cells	Activated memory CD4 T cells	NK cells	Monocytes
Our result	0.588169	0.736077	0.883668	0.698011	0.532620	0.637819	0.809337	0.790502
CIBERSORT	0.580000	0.6300000	0.8500000	0.6500000	0.5000000	0.6100000	0.7600000	0.7400000

Method in better performance (higher correlation) is marked yellow

Table 4 Comparison the difference of our approach with CIBERSORT. Major differences are marked gray and underlined

MySort approach	CIBERSORT approach
1. Data Preparation i. Collecting array data from 22 immune cell types for deconvolution ii. Normalization iii. Clustering iv. Pearson correlation v. Remove outliers and conclude to run deconvolution in 21 immune cell types	1. Data Preparation i. Collecting array data from 22 immune cell types for deconvolution ii. Normalization
2. Construct signature matrix i. Filtration on all gene candidates ii. T-test, log FC iii. G=30 with best Condition number=10.99 and conclude signature genes=603	2. Construct signature matrix i. T-test, logFC ii. G=102 with best Condition number=11.4 iii. Filtration on selected signature genes and conclude the signature genes (LM22, # of genes: 547)
3. Solve model by υ-SVR	3. Solve model by υ-SVR

Major differences are marked gray and underlined

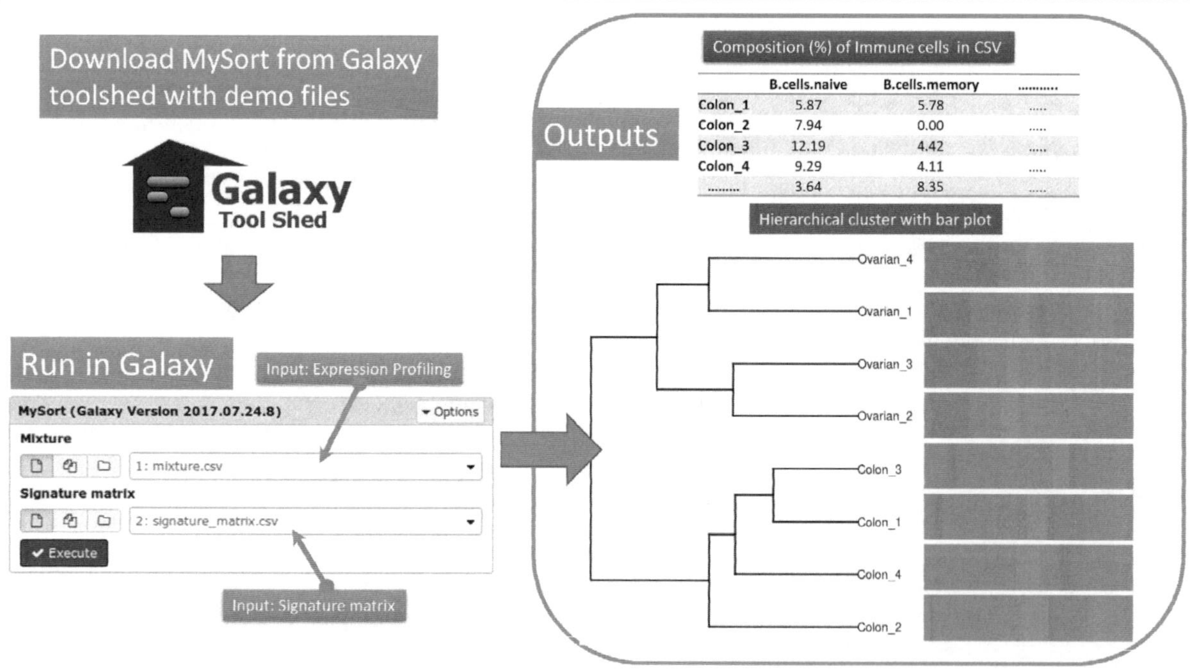

Fig. 7 The implementation of galaxy plugin for the deconvolution method

classification problems. This method defines hyperplanes that separate classes with the largest possible margin by maximizing the distance from the hyperplane to the nearest data point. In contrast, SVR seeks a hyperplane that fits the data points in a tube of width 2ε. Hence, in ε-based SVR (ε-SVR), data points at most ε values deviate from the hyperplane. Consequently, ε-SVR does not focus on data points with the ε-tube, points out of the tube are support vectors (SVs), and distances from SV's to the boundary of the tube are evaluated by the loss function. Similar to ε-SVR,υ-SVR uses υ to provide a more convenient control over the number of SVs and training errors. However, the required SV chosen by the SVR model are some important genes are selected for solving the regression function. These properties of SVR are key to the enhanced performance of the present model in comparison with other traditional methods for solving regression problems.

Gene list filtration directly introduces a black list of unrelated genes that are either expressed in cell types other than those of interest or may interfere with the deconvolution strategy. Although this is an effective strategy for building deconvolution models, the timing of filtration can alter the selection of signature genes greatly. Thus, we used the filtration step before selecting top G ranked DEGs, whereas CIBERSORT filtered genes in the black list after selecting top G ranked DEGs. Consequently, our model achieved G = 30 with 10.99 best conditions and 603 union gene sets, whereas CIBERSORT achieved G = 102 with 11.4 best conditions and 547 union gene sets. Moreover, the processes in CIBERSORT lead to the use of unequal numbers of genes to distinguish cell types.

Conclusions

Gene expression deconvolution methods can be used to reveal defined cell types from transcriptomes of samples with mixed cell types, and are demonstrably powerful strategies for identifying TILs in cancer tissues through reanalyzes of accumulated microarray databanks. Due to the heavy task of reanalysis, we implemented the algorithm, improved its performance, and then packed it as a portable application.

The present version in MySort regards the deconvolution model as a linear model. However, nonlinear methods may be applied to gene expression deconvolution, and machine learning has recently been shown to have good predictive performance. Furthermore, deep learning is an advanced method for nonlinear problems. Hence, further studies are required to develop machine learning and deep learning methods to decipher large databases, and to train the model using the data to make good predictions in gene expression deconvolution.

Finally, new high-throughput technologies such as next generation sequencing and single cell technologies are considered advanced techniques for gene profiling. Theoretically, all cell components could be resolved with knowledge of expression profiles of all cell types. Moreover, greater numbers of profiled cell types will necessitate strategies for classifying them. Subsequently, resolving gene profiling at the single cell level will drive deconvolution methods to a new level if more comprehensive and accurate cell type information can be included during model building.

Acknowledgements
We thank Dr. Su-Fang Lin from National Health Research Institutes, Taiwan for her suggestions on the methodology, and anonymous reviewers for the critical reading and suggestions throughout the whole article.

Funding
The authors thank Ministry of Science and Technology (MOST), Taiwan, for financially supporting this research and publication through MOST 103–2311-B-001 -033 -MY3, MOST 106-3114-B-038 -001 - and MOST106–2313-B-001-010 to CYL and MOST 102–2811-B-001-046 to SHC. Meanwhile, this study was also financial supported by flagship program from Institute of Information Science, Academia Sinica, Taiwan to CYL and JMH.

Availability of data and materials
The implemented method in this study, the signature matrix, and a test/demo set, are available at galaxy plug-in tool mySort through https://test-toolshed.g2.bx.psu.edu/view/moneycat/mysort/e3afe097e80a.

About this supplement
This article has been published as part of *BMC Bioinformatics* Volume 19 Supplement 4, 2018: Selected articles from the 16th Asia Pacific Bioinformatics Conference (APBC 2018): bioinformatics. The full contents of the supplement are available online at https://bmcbioinformatics.biomedcentral.com/articles/supplements/volume-19-supplement-4.

Authors' contributions
SHC, WYK and CYL, composed the whole infrastructure, conducted the experiments, and drafted the manuscript together with HSL. SHC, WYK and CYL designed and implemented the algorithm to decipher the composition of infiltrated immune cells. WCC, SYS and JMH worked on the construction of workflow and implementation of this tool on Galaxy for data visualization and deep analysis. All authors read and approved the final manuscript.

Ethics approval and consent to participate
Not applicable

Consent for publication
Not applicable.

Competing interests
The authors declare that they have no competing interests.

Publisher's Note

Author details
[1]Institute of Information Science, Academia Sinica, 128 Academia Road, Section 2, Nankang, Taipei 115, Taiwan. [2]Institute of Statistics, National Chiao Tung University, Assembly Building I, 1001 Ta Hsueh Road, Hsinchu 30010, Taiwan. [3]Bioinformatics Program, Taiwan International Graduate Program, Institute of Information Science, Academia Sinica, 128 Academia Road, Section 2, Nankang, Taipei 115, Taiwan. [4]Institute of Biomedical Informatics, National Yang-Ming University, No. 155, Sec. 2, Linong St., Beitou District, Taipei City 112, Taiwan. [5]Division of Biostatistics and Bioinformatics, Institute

of Population Health Sciences, National Health Research Institutes, 35 Keyan Road, Zhunan, Miaoli County 35053, Taiwan. [6]Institute of Fisheries Science, College of Life Science, National Taiwan University, No. 1, Sec. 4, Roosevelt Rd, Taipei 10617, Taiwan.

Published: 8 May 2018

References

1. Hanahan D, Weinberg RA. The hallmarks of cancer. Cell. 2000;100(1):57–70.
2. Hanahan D, Weinberg RA. Hallmarks of cancer: the next generation. Cell. 2011;144(5):646–74.
3. Sharma P, Allison JP. The future of immune checkpoint therapy. Science. 2015;348(6230):56–61.
4. Abbas AR, et al. Deconvolution of blood microarray data identifies cellular activation patterns in systemic lupus erythematosus. PLoS One. 2009;4(7): e6098.
5. Clarke J, Seo P, Clarke B. Statistical expression deconvolution from mixed tissue samples. Bioinformatics. 2010;26(8):1043–9.
6. Gong T, et al. Optimal deconvolution of transcriptional profiling data using quadratic programming with application to complex clinical blood samples. PLoS One. 2011;6(11):e27156.
7. Zhong Y, et al. Digital sorting of complex tissues for cell type-specific gene expression profiles. BMC Bioinformatics. 2013;14:89.
8. Gaujoux R, Seoighe C. Semi-supervised nonnegative matrix factorization for gene expression deconvolution: a case study. Infect Genet Evol. 2012;12(5): 913–21.
9. Qiao W, et al. PERT: a method for expression deconvolution of human blood samples from varied microenvironmental and developmental conditions. PLoS Comput Biol. 2012;8(12):e1002838.
10. Liebner DA, Huang K, Parvin JD. MMAD: microarray microdissection with analysis of differences is a computational tool for deconvoluting cell type-specific contributions from tissue samples. Bioinformatics. 2014;30(5):682–9.
11. Newman AM, et al. Robust enumeration of cell subsets from tissue expression profiles. Nat Meth. 2015;12(5):453–7.
12. Gentles AJ, et al. The prognostic landscape of genes and infiltrating immune cells across human cancers. Nat Med. 2015;21(8):938–45.
13. Benita Y, et al. Gene enrichment profiles reveal T-cell development, differentiation, and lineage-specific transcription factors including ZBTB25 as a novel NF-AT repressor. Blood. 2010;115(26):5376–84.
14. Barretina J, et al. The cancer cell line encyclopedia enables predictive modeling of anticancer drug sensitivity. Nature. 2012;483(7391):603–307.
15. Belsley DA, Kuh E, Welsch RE. The Condition Number, in Regression Diagnostics: Identifying Influential Data and Sources of Collinearity. New York: John Wiley & Sons; 1980. p. 100–4.

Kim *et al. BMC Bioinformatics* 2018, **19**(Suppl 4):75
https://doi.org/10.1186/s12859-018-2070-0

BMC Bioinformatics

RESEARCH

Open Access

Hierarchical structural component modeling of microRNA-mRNA integration analysis

Yongkang Kim[1], Sungyoung Lee[2], Sungkyoung Choi[2], Jin-Young Jang[3] and Taesung Park[1,2*]

From The Sixteenth Asia Pacific Bioinformatics Conference
Yokohama, Japan. 15-17 January 2018

Abstract

Background: Identification of multi-markers is one of the most challenging issues in personalized medicine era. Nowadays, many different types of omics data are generated from the same subject. Although many methods endeavor to identify candidate markers, for each type of omics data, few or none can facilitate such identification.

Results: It is well known that microRNAs affect phenotypes only indirectly, through regulating mRNA expression and/or protein translation. Toward addressing this issue, we suggest a *hi*erarchical *s*tructured *com*ponent analysis of *mi*croRNA-*m*RNA *i*ntegration ("HisCoM-mimi") model that accounts for this biological relationship, to efficiently study and identify such integrated markers. In simulation studies, HisCoM-mimi showed the better performance than the other three methods. Also, in real data analysis, HisCoM-mimi successfully identified more gives more informative miRNA-mRNA integration sets relationships for pancreatic ductal adenocarcinoma (PDAC) diagnosis, compared to the other methods.

Conclusion: As exemplified by an application to pancreatic cancer data, our proposed model effectively identified integrated miRNA/target mRNA pairs as markers for early diagnosis, providing a much broader biological interpretation.

Keywords: miRNA, mRNA, Integration analysis, Generalized Structured Component Analysis (GSCA), Hierarchical structured component analysis of miRNA-mRNA integration (HisCoM-mimi)

Background

Presently, numerous types of "omics" data are generated by many accurate and cost-effective methods. For instance, next-generation sequencing (NGS) technology is used to find DNA or RNA variations, bisulfite sequencing is used to find DNA-methylated variants, and multiple reaction monitoring (MRM) is applied to measure protein abundances [1–3]. These efficient omics data platforms allow researchers to use multi-omics data, obtained from the same subjects, for analyzing huge numbers of variants. As a result, efficient multi-omics data analysis is becoming more important in integrating large-scale data sets, making it possible to interpret fundamental biological systems [4].

MicroRNAs (miRNAs) are noncoding RNAs having a length less than 25 base pairs, regulating the expression of specific genes by mRNA degradation or blocking translation by binding to the 3′ regions of their "target" mRNAs. Many recent studies have now implicated miRNAs in the pathogenesis of cancer, including triggering cancer initiation and progression. MiRNAs have been shown to have tissue-specific and disease-specific expression patterns [5–8]. Intensive investigation is now underway for using applying miRNAs' inhibitory information to mRNAs. For example, Nam et al. developed "miRNA and mRNA integrated analysis" (MMIA) to examine biological functions of miRNA expression [9]. Moreover, Buffa et al. used pathway information to independently validate miRNAs significant for breast cancer [10],

* Correspondence: tspark@stats.snu.ac.kr
[1]Department of Statistics, Seoul National University, Seoul, Korea
[2]Interdisciplinary program in Bioinformatics, Seoul National University, Seoul, Korea
Full list of author information is available at the end of the article

while Cho et al. performed network analysis, and hierarchical clustering, to find biological "signatures" of interstitial lung diseases [11]. Most miRNA and mRNA integration analyses focus on first identifying miRNAs significantly associated with the phenotype of interest, and then experimentally validating those miRNAs' phenotype involvement by inhibiting or ectopically overregulating their expression [9–11]. Although these approaches are effective at validating significant miRNAs, they do not provide information on how they regulate expression of their target mRNAs, as relevant to the pathway level.

In this work, we propose a structured component-based analysis, for integrating omics data for identifying multiple accurate biomarkers. It is well known that miRNAs affect phenotypes indirectly, by regulating mRNA expression or protein translation [8]. Herein, we propose **hi**erarchical **s**tructured **com**ponent analysis of **mi**RNA-**m**RNA **i**ntegration (HisCoM-mimi) analysis, which models biological relationships as structured components, to efficiently yield integrated markers. Our proposed model is based on generalized structured component analysis (GSCA), which tests hypothesized relationships between observed and latent variables [12]. GSCA is a component-based method whereby each component represents a latent variable. Extending GSCA, we previously developed Pathway-based approach using hierarchical components of collapsed rare variants (PHARAOH) [13]. PHARAOH uses a hierarchial structure of rare variants, genes, and pathways. The advantage of such hierarchical structural component models is their generation of (unobservable) latent variables, such as genes and pathways, which are inferred by observed variables, such as rare variants. Using latent variables, we can collapse unstructured data into a structured form, providing less ambiguous biological explanations of the results. In this current work, mRNAs, inhibited by miRNAs, can be merged into latent variables.

Accordingly, our proposed HisCoM-mimi model can efficiently account for biological relationships between miRNA and mRNA, in the structured component, and effectively provide integrated (e.g., miRNA-to-target-mRNA) markers. As an illustration, we tried HisCoM-mimi for identifying biomarkers for the early diagnosis of pancreatic cancer (PC). Note that PC is one of the most fatal diseases in the world, having a mere 8% five-year survival rate in the USA and a 9.4% survival rate in the Republic of Korea [14–16]. In particular, the tumor heterogeneity in PC patients' tumors makes early diagnosis harder than cancers of most other organs [17]. To adjust for heterogeneity among tumor cells, we need a more robust and complex statistical model which can interpret and integrate several causes of cancer altogether. Although many bioinformatics research studies have been performed to find diagnostic markers for PC, to date, no clinically approved prognostic markers exist [18].

Here, we applied HisCoM-mimi to computationally identify diagnostic markers of pancreatic ductal adenocarcinoma (PDAC), the most common type of PC. By applying the HisCoM-mimi approach to miRNA and mRNA microarray data from PDAC patients, at Seoul National University Hospital (SNUH), we identified numerous cognate miRNA-mRNA partners, as markers for diagnosis of PDAC. Finally, our HisCoM-mimi provided integrated marker sets, with more biological and intuitive interpretation, than other existing methods.

Methods

Pancreatic ductal adenocarcinoma (PDAC) samples

Between the years 2009 and 2012, 200 pancreatic ductal adenocarcinoma (PDAC) samples were collected by the Department of Hepatobiliary and Pancreas Surgery of Seoul National University Hospital. The study protocol was approved by the Institutional Review Board of Seoul National University Hospital (IRB H-0901-010-267) and written, informed consent was obtained from each patient or legally authorized representative.

Of the 200 tumors, 96 were excluded because of RNA degradation or insufficient RNA content, leaving 104 samples valid for microarray analysis. After quality control, 97 PDAC samples remained for microarray assessment. The PDAC patients' average age was 64.3 years (standard deviation (SD): 9.7). Twenty-nine patients were male, and 31 female. For the normal groups, 17 benign pancreatic tissues were used. Subsequently, we built and implemented our mini model, using the 97 PDAC and 17 normal tissues, respectively.

HisCoM-mimi model

To perform the integration analysis of miRNA and mRNA data, we developed and implemented our HisCoM-mimi approach. This model analyzes multiple subnetworks simultaneously, with specific regard to inverse correlations between mRNA and miRNA. Figure 1 shows the flowchart of the method. First, for a given miRNA, a miRNA-mRNA subnetwork, consisting of one miRNA and multiple potential target mRNAs, is constructed if the following two conditions are satisfied: (i) the mRNAs are reported as target of the miRNA by TargetScan 7.1 (targetscan.org) [19], and the negative correlation coefficients between the mRNA and miRNAs are significant (p-value < 0.05). Second, for all entities deemed significant, we derived our hierarchical structural component model by using all miRNA-mRNA subnetworks.

As shown in Fig. 2, there are three structures to consider: miRNA-mRNA structure, miRNA integration latent structure, and phenotype-latent structure. Each

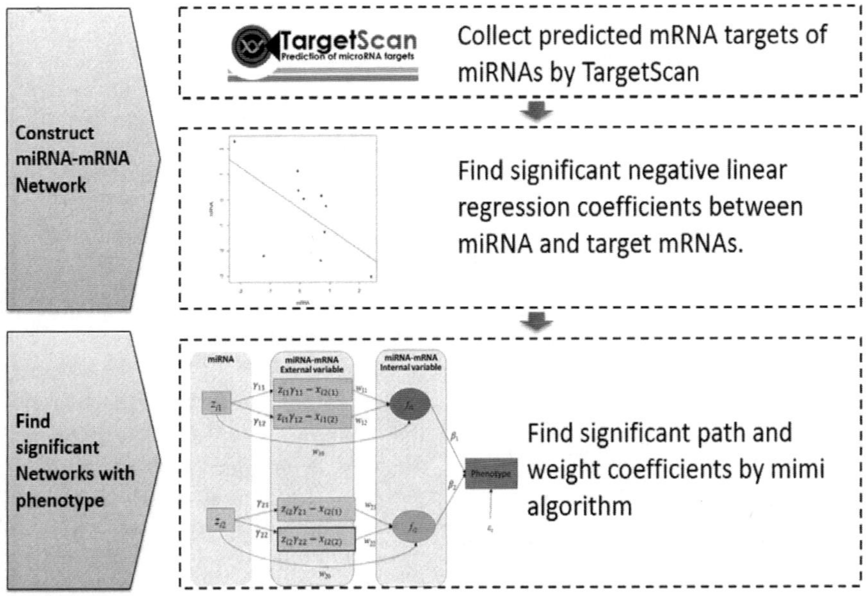

Fig. 1 Flow chart for analyzing mRNA-miRNA integration

structure can be represented as a generalized linear model, similar to PHARAOH [13].

miRNA-mRNA structure

$$\hat{X}_{ijk} = x_{ijk} - \gamma_{jk} z_{ij}, j = 1, ..., G_j, \quad (1)$$

Equation (1) shows how to obtain mRNA expression before inhibition by miRNA, subscript i means i th individual, x_{ijk} represents the mRNA expression of the kth gene related with j th miRNA, z_j the j th miRNA expression, γ_{jk} the inhibition coefficient for the j th miRNA for the k th gene, and G_j is the number of inhibited mRNAs by the j th miRNA. By estimating the coefficients γ_{jk}, mRNA expression after removing the inhibition effect of miRNA can be obtained.

miRNA latent structure

$$f_{ij} = \gamma_{j0} z_j + \sum_{k=1}^{G_j} \hat{X}_{ijk} w_{jk} \quad (2)$$

The miRNA latent variable is defined in Eq. (2). The miRNA latent variable is built by linearly combining miRNA expression values. While γ_{j0} denotes the direct

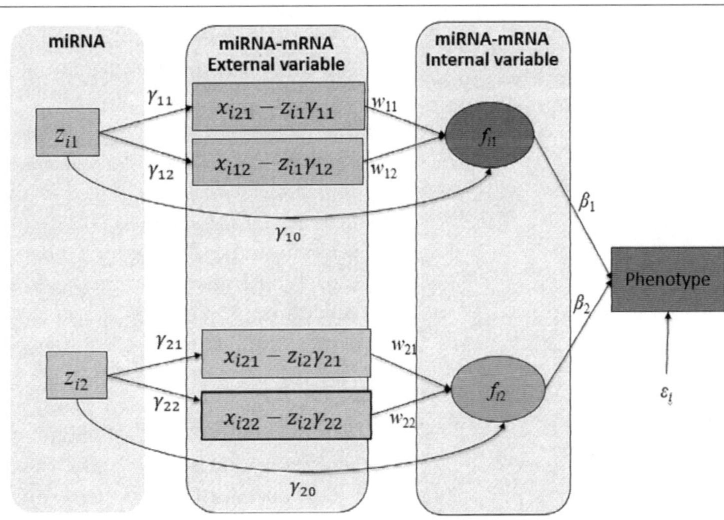

Fig. 2 Network Diagram for HisCoM-mimi model

effect of the miRNA on the phenotype. Then, the latent variable f_{ij} represents the global effect of the miRNA's activity through its inhibited mRNAs.

Phenotype-latent structure

$$logit(\pi_i) = \beta_0 + \sum_{j=1}^{J}\left[\sum_{k=1}^{G_j} x_{ij(k)}^{gene} w_{jk}\right]\beta_j = \beta_0 + \sum_{j=1}^{J} f_{ij}\beta_j \quad (3)$$

Let the phenotype variable y_i be a binary variable, distinguishing PDAC from normal tissues. Let π_i be the probability of $y_i = 1$ (PDAC). $logit(\pi_i)$ is the logit link function, β_j represents the effect of f_{ij} on the phenotype, as interpreted as a log-odds ratio.

Fitting the HisCoM-mimi algorithm

To estimate the parameters for HisCoM-mimi, we adopted our previously developed PHARAOH algorithm [13], which is based on the alternating least squares algorithm for the penalized log-likelihood function, with ridge parameters. Then, the objective function to maximize is given as follows:

$$logit(\pi_i) = \beta_0 + \sum_{j=1}^{J}\left[\sum_{k=1}^{G_j} x_{ij(k)}^{gene} w_{jk}\right]\beta_j = \beta_0 + \sum_{j=1}^{J} f_{ij}\beta_j, \quad (4)$$

$$\phi_1 = \sum_{i=1}^{n} \log p\left(y_i; \beta_j, \delta\right) - \frac{1}{2}\lambda_m \sum_{j=1}^{J}\sum_{k=1}^{G_j} w_{jk}^2 - \frac{1}{2}\lambda_{mm}\sum_{j=0}^{J}\beta_j^2 \quad (5)$$

where $p(y_i; \gamma_i, \delta)$ is the probability distribution for the phenotype of the *ith* individual. λ_m and λ_{mm} are ridge parameters for miRNA-mRNA pairs of interest, representing the integrated latent components.

To maximize the objective function, ϕ_1, the iterative reweighted least squares (IRWLS) algorithm is used. Note that when using IRWLS, maximizing ϕ_1 is equivalent to minimizing the object function ϕ_2.

$$\phi_2 = \sum_{i=1}^{n} v_i \left(z_i - \sum_{j=1}^{J} f_{ij}\beta_j\right)^2 - \frac{1}{2}\lambda_m \sum_{j=1}^{J}\sum_{k=1}^{G_j} w_{jk}^2 - \frac{1}{2}\lambda_{mm}\sum_{j=0}^{J}\beta_j^2 \quad (6)$$

Comparative models

To compare the results of HisCoM-mimi with other methods, we considered several alternative regression-based methods.

$$logit(\pi_i) = \beta_0 + \sum_{j=1}^{J}\theta_j z_{ij} + \sum_{k=1}^{K}\rho_k x_{ijk}, j = 1, \ldots, J \quad (7)$$

$$\phi_{LR}(\beta_0, \theta, \rho, \delta; X, Z) = \sum_{i=1}^{n} \log p(y_i; \beta_0, \theta, \rho) - \delta P_\alpha(\theta, \rho), j = 1, \ldots, J \quad (8)$$

Firstly, we considered the ordinary penalized logistic regression (LR) methods such as lasso or elastic-net (EN) [20, 21]. Equation 7 shows the LR model, where θ_j and ρ_k represent the effect of the *j*th miRNA and the *k*th mRNA, respectively. Equation 8 is the objective function to maximize for finding optimal parameters with the penalty function $P_\alpha(\theta, \rho)$. When lasso is used, $P_\alpha(\theta, \rho) = \Sigma_k \mid \rho_k \mid + \Sigma_j \mid \theta_j \mid$.

If EN is used, $P_\alpha(\theta, \rho) = \alpha(\sum_k |\rho_k| + \sum_j |\theta_j|) + (1-\alpha)(\sum_k \rho_k^2 + \sum_j \theta_j^2)$. Lasso or EN can then select the miRNAs and/or mRNAs of interest. However, these methods cannot use group information. Thus, ordinarily penalized LR methods cannot adequately account for the biological structure of miRNA-mRNA.

Secondly, we considered LR with a group lasso penalty (GL) [22], which has the benefit of using group information among the miRNAs and mRNAs of interest. In our analysis, a group can be defined as a set of one miRNA and its corresponding inhibited target mRNAs. GL uses the same LR in (8) with a different penalty function $P(\theta, \rho) = \sum_{j=1}^{J}\sqrt{\theta_j^2 + \sum_{k=1}^{G_j}|\rho_k|}$. Via this penalty function, miRNA integration set can be selected together. However, the GL approach does not easily provide *p*-values for each set of independent variables.

To fit the penalized LR models, we first performed 3-fold cross-validation to find the optimal tuning parameter, δ. after which we fitted the models with all the data sets.

Simulation study

To compare HisCoM-mimi to the other three methods, we performed simulation studies and computed type I errors and power, simulating data from the same miRNA and mRNA data structure in our pancreatic cancer dataset. That is, we selected miRNA and mRNA data from the pancreatic cancer dataset, and then generated phenotype data iteratively from the LR model. We then considered two simulation scenarios. Scenario 1 assumed that a true causal integration set contains two mRNAs, with the same effect size. Scenario 2 assumed that a true causal integration set contains five mRNAs, with the same effect size. For each scenario, we randomly selected one causal miRNA-mRNA subnetwork, and then randomly selected another 9 miRNA-mRNA subnetworks, for which the number of inhibited mRNAs was less than 10. The selected miRNA-mRNA subnetworks for Scenario 1 are summarized in Table 1 and for Scenario 2 are in Table 2.

For Scenario 1, we used miR-217 as a true causal miRNA. To generate phenotypes, we considered the following LR model.

Table 1 List of used miRNAs and mRNAs for simulation Scenario 1

miRNA	Role in simulation	Inhibited mRNA
miR-217	Causal	ITGBL1, ATP10A
miR-215	Non-Causal	CDC6, CTH, DNAJC19, DPP10, ELP4, FUNDC2, GLP1R, B3GALNT2, SLC39A8
miR-485	Non-Causal	CDX1, CTDNEP1, GPR3, HDAC5, KCNJ11, RASL10A, SLC39A14
miR-195	Non-Causal	CNDP2, SLC45A2, SLC7A2
miR-381	Non-Causal	DKK3, IGFBP5, LAMA4, OSBPL3, BAMBI
miR-132	Non-Causal	GLRB, GMPR, ARX, SALL3
miR-363	Non-Causal	SOSTDC1
miR-1	Non-Causal	FAM150B
miR-28	Non-Causal	SRPRB
miR-200	Non-Causal	NRG3

$$logit(\pi) = \beta_{miRNA}z_1 + \beta_1 x_1 + \beta_2 x_2, \quad (9)$$

where π is the probability of observing a disease ($Y = 1$), z_1 represents the true causal miRNA expression, and x_1 and x_2 represent two causal mRNA expression values. For type I error evaluation, we assumed $\beta_{miRNA} = \beta_1 = \beta_2 = 0$. For power comparison, we generated simulation data sets under the assumption that $\beta_{miRNA} = \beta_1 = 0.2, \ 0.25, 0.3, 0.35$. For the given 114 (97 PDAC and 17 normal tissues) values of (z_1, x_1, x_2), from our pancreatic cancer dataset, we simulated 1000 datasets.

For Scenario 2, we assumed that a true causal integration set contains five mRNAs, with the same effect size. In our dataset, miR-381 was the only miRNA having five inhibited target mRNAs. To generate phenotypes, we considered the following LR model:

$$logit(\pi) = \beta_{miRNA}z_1 + \beta_1 x_1 + \beta_2 x_2 + \beta_3 x_3 + \beta_4 x_4 + \beta_5 x_5, \quad (10)$$

where x_1, ..., x_5 represent five causal mRNA expression values. As in Scenario 1, we assumed $\beta_{miRNA} = \beta_1 = \beta_2 = \beta_3 = \beta_4 = \beta_5 = 0$, for type I error evaluation, and $\beta_{miRNA} = \beta_1 = \beta_2 = \beta_3 = \beta_4 = \beta_5 = 0.2$, 0.25, 0.3, 0.35, for power comparison. For the given 114 values of (z_1, x_1, x_2, x_3, x_4, x_5) from the pancreatic cancer dataset, 1000 simulation datasets were generated. We used the significance level $\alpha = 0.05$ for HisCoM-mimi, as an false positive rate (FPR) criterion. For lasso, EN, and group-lasso, we selected a threshold T which provides a comparable FPR to the type I error 0.05. T was determined by calculating the FPR for simulation settings such that a miRNA-mRNA subnetwork is selected when $\beta_{miRNA} \neq 0$ and $K(= \sum_{l=1}^{L} I(\beta_l \neq 0))$ exceeded the threshold T. Here, L is the number of inhibited mRNAs for true causal miRNA for each scenario: $L = 2$ for Scenario 1, and $L = 5$ for Scenario 2.

Table 2 List of used miRNAs and mRNAs for simulation Scenario 2

miRNA	Role in simulation	inhibited mRNA
miR-381	Causal	DKK3, IGFBP5, LAMA4, OSBPL3, BAMBI
miR-215	Non-Causal	CDC6, CTH, DNAJC19, DPP10, ELP4, FUNDC2, GLP1, B3GALNT2, SLC39A8
miR-32	Non-Causal	COL1A2, BGN
miR-195	Non-Causal	CNDP2, SLC45A2, SLC7A2
miR-501	Non-Causal	PARM1, SLC32A1
miR-1	Non-Causal	FAM150B
miR-212	Non-Causal	KCNK2
miR-204	Non-Causal	CDH11
miR-200	Non-Causal	NRG3
miR-363	Non-Causal	SOSTDC1

Results

Simulation results

For our analyses, we first determined the false positive error rates (FPRs) of each method, and chose the threshold values of T to make each penalized method provide (hold) FPRs close to 0.05. In Scenario 1, the type I error rate of HisCoM-mimi was 0.048 when $\alpha = 0.05$. The FPRs of lasso were 0.054, when T was 1, and that of EN was 0.064, when T was 1. Since type I error rates of lasso and EN were nearly 0.05 when T = 1, we set T = 1 to evaluate power of those two methods. The FPR of GL, when choosing a causal miRNA integration set, 0.064.

For Scenario 2, Table 3 shows the FPRs for lasso and EN, when varying the threshold T. For this result, we found that the type I error of lasso and EN were similar to 0.05, when T = 1 and 2, respectively. The type I error rate of HisCoM-mimi was 0.054. On the other hand, GL did not select a causal miRNA integration set at all, such that the type I error rate was 0. Secondly, we compared the powers of each method for Scenarios 1 and 2. Figure 3 shows bar plots of powers for scenario 1, where the x-axis shows the effect sizes (i.e., beta coefficients), and the y-axis shows the power. HisCoM-mimi showed the highest power, while EN was second, Lasso was third, and GL was last. The same tendency is shown in Fig. 4, for Scenario 2. Figure 5 shows that the differences of power between HisCoM-mimi and the others were much larger than those of Scenario 1. Consequently, GL could not find any significant miRNA-mRNA integration sets

Table 3 False positive rate when varying the number of selected mRNAs for lasso and EN

T	5	4	3	2	1
Lasso	0	0	0.007	0.022	0.053
EN	0	0.002	0.014	0.055	0.204

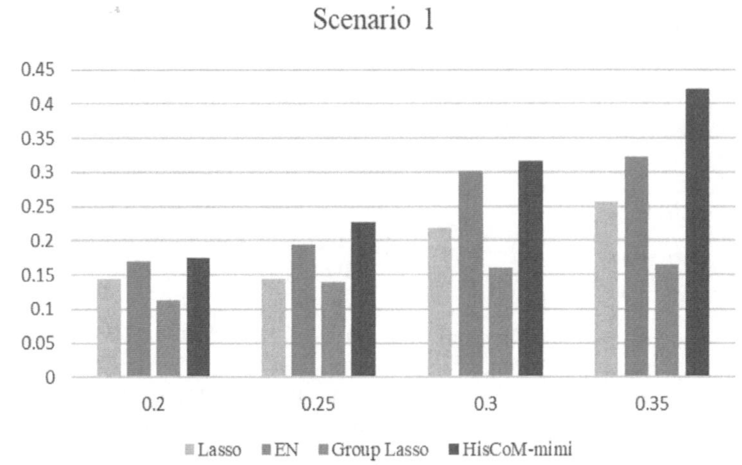

Fig. 3 Power comparison for scenario 1

under Scenario 1, due to its GL's penalty being too strict for many mRNAs, whose beta values were small.

Constructing miRNA-mRNA subnetworks

To use human mRNA and miRNA probes, we first filtered out non-annotated mRNA probes and non-human miRNA probes. After filtering, there were 22,077 mRNA probes and 3391 miRNA probes. To construct miRNA-mRNA subnetworks, we checked predicted target mRNAs, for each miRNA, from TargetScan 7.1 (targetscan.org) [19, 23]. Among predicted targets, we only selected mRNAs having significant Pearson correlation coefficients with a specific miRNA. After filtering, there were 55 miRNAs, and 2411 edges connected with mRNAs.

Integration analysis for the PDAC data

Table 4 shows the top significant weights of miRNA-mRNA integrations derived from HisCoM-mimi. To perform multiple comparison, we used false discovery rate (FDR) q-values summarized in the 7th column [24]. We could only find 12 miRNAs having q-values below 0.05. Tables 5 and 6 show the lists of the selected markers by lasso and EN, respectively. Since lasso and EN select markers without any group information, they selected miRNA and mRNA markers independently. There were no miRNAs selected by lasso or EN directly, with lasso yielding only two significant mRNAs, both related to miR-326. Other mRNAs were independently selected from different miRNAs. Consequently, there were only 12 markers selected by lasso. For EN, 58 mRNAs

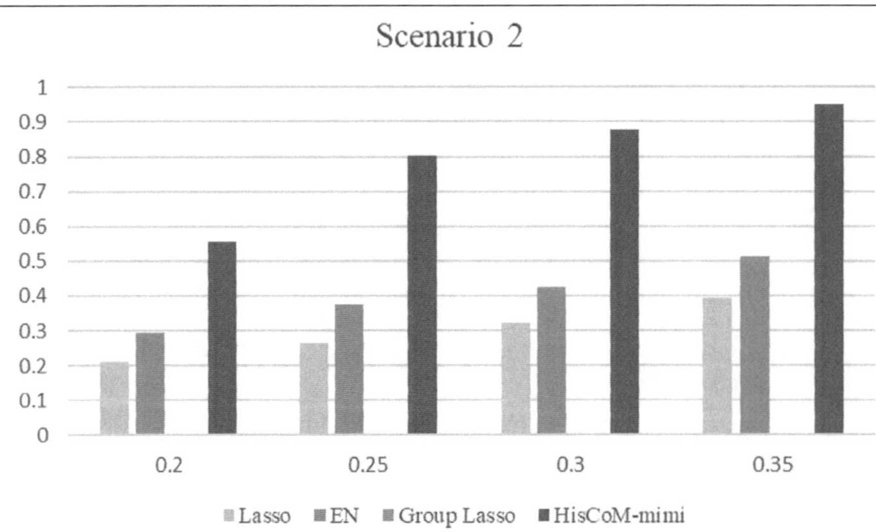

Fig. 4 Power comparison for scenario 2

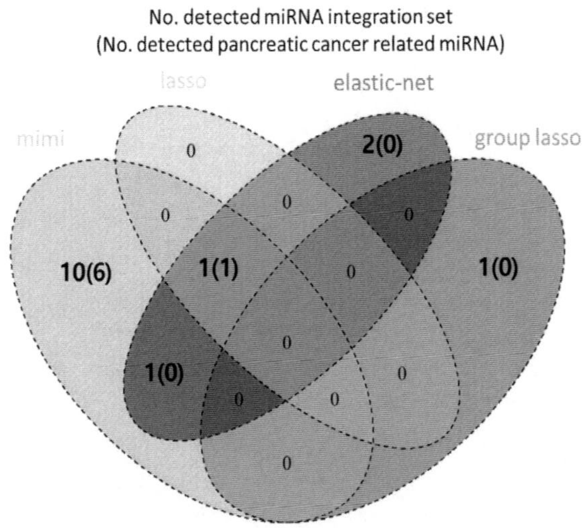

Fig. 5 Venn Diagram for number of detected miRNAs for each method

were selected. Similar to the lasso result, there were no selected miRNAs, although four miRNAs (miR-206, miR-3064, miR-222, and miR-326) connected to more than three mRNAs. Figure 5 shows a Venn diagram of the number of miRNAs selected by each method. Each number represents the total number of detected miRNAs and one in the parenthesis does the number of detected miRNAs whose relationship with pancreatic cancer were reported. HisCoM-mimi selected larger number of unique miRNAs and the majority of them were already were reported.

For the lasso group only one miRNA (miR-32) and whose related two mRNA (COL1A2, and BGN) were selected. Although miR-32 is not reported as pancreatic cancer marker, there were some reports that miR-32 is related with other cancers [25, 26].

Table 4 Significant miRNAs produced by HisCoM-mimi

Order	miRNA	Number of inhibited mRNAs	Number of significant mRNAs	β_{mimi}	$P_{HisCoM\text{-}mimi}$	$q_{HisCoM\text{-}mimi}$
1	miR-133b	81	29	0.319	0.0008	0.0126
2	miR-141	105	57	0.638	0.0008	0.0126
3	miR-222	127	70	0.587	0.0010	0.0126
4	miR-532	11	0	0.190	0.0010	0.0126
5	miR-93	80	36	−0.573	0.0014	0.0126
6	miR-219	26	3	0.278	0.0016	0.0126
7	miR-590	24	4	−0.183	0.0016	0.0126
8	miR-326	13	0	0.172	0.0022	0.0151
9	miR-203	65	11	−0.261	0.0026	0.0159
10	miR-132	4	0	−0.204	0.0034	0.0187
11	miR-96	109	42	0.701	0.0038	0.0190
12	miR-708	43	3	−0.181	0.0102	0.0468

Table 5 Selected markers by lasso. Twelve markers (12 mRNAs) were selected. No miRNAs were selected

Selected marker	Beta	Connected miRNA	Selected marker	Beta	Connected miRNA
NSD1	− 0.704	miR-206	PLCE1	0.129	miR-1271
EMX2	−0.336	miR-222	TFCP2	0.112	miR-497
BBC3	0.329	miR-222	AKAP7	−0.017	miR-1297
GSG1	0.005	miR-3064	MAMDC2	1.044	miR-670
ZRANB3	−0.414	miR-326	DRGX	0.393	miR-96
MLEC	0.051	miR-362	FBXL2	−0.187	miR-133b

Table 7 summarizes miRNAs detected by HisCoM-mimi, lasso, EN, or GL. Previously, miR-93, miR-219, miR-141, miR-222, miR-203, miR-132, miR-96, and miR-206 were reported to be pancreatic cancer-related markers [27–35]. Although other miRNAs detected by HisCoM-mimi, lasso, EN, or GL have not been reported for pancreatic cancer relation, miR-532, miR-590, miR-133b, miR-326, miR-708, miR-3064, and miR-32 were reported to associate with other cancer types [25, 36–42].

Table 8 shows the cross-validation (CV) results for comparing prediction performance for marker-sets selected by HisCoM-mimi, Lasso, EN, and Group Lasso. The first column indicates methods used to construct prediction model and the second column does the method to select marker sets. The third column shows the area under the Receiver Operating Characteristic curve (AUC) results performed by leave-one-out cross validation (LOOCV). This setting is from the previous study of Kwon et al. [23]. The fourth column indicates the average AUC values performed by four-fold CV with a hundred iterations. Here, we used four-fold and eight-fold CV to balance the number of samples in CV datasets. The fifth column indicates the average AUC values performed by eight-fold CV with a hundred iterations. For all selected marker-sets, all prediction models built by HisCoM-mimi showed the best performances yielding AUC values higher than 0.9 except the marker-set selected by Group lasso in which the number of markers is less than five and one path coefficient exists.

Discussion and conclusion

In this paper, we proposed and developed a novel method, **hi**erarchical **s**tructured **com**ponent analysis of **mi**croRNA-**m**RNA **i**ntegration ("HisCoM-mimi"), to construct a component model to identifying significantly integrated miRNA-target-mRNA cognate pairs. Since HisCoM-mimi could use subgroup information, it yelded more results, as related to phenotypes (e.g. cancer, metabolic syndrome, and etc.), than those of other existing methods that lack network information.

Table 6 Markers selected by EN

Selected mRNA	Beta	Connected miRNA	Selected mRNA	Beta	Connected miRNA	Selected mRNA	Beta	Connected miRNA
NSD1	− 0.340	miR-206	NUP214	−0.103	miR-3064	TFCP2	0.216	miR-497
FRS2	−0.046	miR-206	TCP11	−0.077	miR-3064	KDM5B	0.040	miR-524
MGAT4A	0.004	miR-206	BCL2L13	−0.022	miR-3064	RNASEH2C	−0.043	miR-670
SLC8A1	0.022	miR-206	SLC16A10	−0.016	miR-3064	MAP3K10	0.163	miR-670
PI4KA	0.027	miR-206	GSG1	0.034	miR-3064	MAMDC2	0.395	miR-670
MATR3	0.034	miR-206	LRRC34	−0.159	miR-326	TCEB3	−0.286	miR-93
OSBPL8	0.088	miR-206	ZRANB3	−0.127	miR-326	RASL11B	0.036	miR-93
EMX2	−0.275	miR-222	AQP2	−0.037	miR-326	KIAA0087	0.182	miR-96
KIAA0430	−0.039	miR-222	CTRC	−0.007	miR-326	DRGX	0.249	miR-96
AXIN2	0.003	miR-222	MLEC	0.034	miR-362	HS3ST2	0.016	miR-100
PRUNE	0.013	miR-222	NOTCH1	0.003	miR-367	SYDE2	0.098	miR-107
SHISA9	0.016	miR-222	SH3PXD2A	0.014	miR-367	AKAP7	−0.207	miR-1297
SHC3	0.031	miR-222	PTDSS1	0.017	miR-372	FBXL2	−0.373	miR-133b
RBL1	0.044	miR-222	CATSPER4	0.002	miR-378	CLIP2	0.005	miR-141
SOCS1	0.053	miR-222	TRIM55	0.071	miR-378	LYPD3	0.188	miR-152
SH3BP4	0.057	miR-222	SLC35E2B	−0.128	miR-488	PAQR9	0.308	miR-152
BBC3	0.074	miR-222	SALL4	−0.080	miR-1271	SCN1A	0.017	miR-203
SEC23IP	0.077	miR-222	MAGI3	0.009	miR-1271	CCPG1	0.070	miR-211
ESR1	0.085	miR-222	PLCE1 ,	0.198	miR-1271	BGN	−0.161	miR-32
DGKI	−0.003	miR-330-5p						

In simulation studies, we compared the performances of HisCoM-mimi, lasso, EN, and GL. From that comparison, HisCoM-mimi showed better performance than the other three methods. Controlling type I error, by HisCoM-mimi, was easier for controlling FPRs than other methods, because HisCoM-mimi uses permutation based *p*-values. In particular, HisCoM-mimi could identify miRNA-mRNA integration sets in a much more flexible way, due to better use of a standard multiple testing framework, as compared to the other methods. In real data analysis, HisCoM-mimi succesfully identified more miRNA-mRNA integration sets for pancreatic ductal adenocarcinoma (PDAC) diagnosis, compared to the other methods. Among 12 miRNAs, whose q-values were below 0.05 by HisCoM-mimi, 7 miRNAs were previously reported to associate with a panreatic cancer [27–35]. EN found two miRNAs (miR-222, and miR-206) [30, 34]. Among two miRNAs selected by lasso, only miR-222 was reported to associate with pancreatic cancer.

Although HisCoM-mimi worked well for the PDAC data sets, further biological verification of those results are needed. In future studies, we will perform additional simulation analyses to evaluate the performance of HisCoM-mimi, under numerous conditions. Furthermore,

Table 7 Cancer related miRNAs detected by methods

Method	miRNA	Number of used mRNA	Reported cancer relationship	Method	miRNA	Number of used mRNA	Reported cancer relationship
HisCoM-mimi	miR-93	80	Pancreas	HisCoM-mimi	miR-132	4	Pancreas
HisCoM-mimi	miR-219	26	Pancreas	HisCoM-mimi	miR-96	109	Pancreas
HisCoM-mimi	miR-532	11	Other	HisCoM-mimi	miR-708	43	Other
HisCoM-mimi	miR-590	24	Other	Lasso	miR-222	2	Pancreas
HisCoM-mimi	miR-141	105	Pancreas	EN	miR-206	7	Pancreas
HisCoM-mimi	miR-133b	81	Other	EN	miR-222	12	Pancreas
HisCoM-mimi	miR-222	127	Pancreas	EN	miR-3064	5	Other
HisCoM-mimi	miR-203	65	Pancreas	EN	miR-326	4	Other
HisCoM-mimi	miR-326	13	Other cancer	GL	miR-32	2	Other

Table 8 Evaluation of Prediction performance for marker set selected by HisCoM-mimi, Lasso, EN, or Group Lasso in PDAC samples

Marker set	Method	AUC-loocv	AUC-4-fold CV	AUC-8-fold CV
HisCoM-mimi	HisCoM-mimi	**0.997**	**0.996**	**0.997**
	Lasso	0.948	0.947	0.948
	EN	0.975	0.969	0.971
	Group Lasso	0.889	0.888	0.895
Lasso	HisCoM-mimi	0.976	0.975	0.976
	Lasso	0.938	0.928	0.939
	EN	0.970	0.953	0.963
	Group Lasso	0.910	0.910	0.918
EN	HisCoM-mimi	0.976	0.976	0.976
	Lasso	0.939	0.927	0.935
	EN	0.969	0.957	0.965
	Group Lasso	0.911	0.912	0.915

HisCoM-mimi can be extended in many ways, for other types of phenotypes, such as time to event. Second, it can be easily applied to other cancer studies to identify miRNA-mRNA integration sets for early diagnosis and prognosis. Third, it can be extended to combine other types of omics data such as genomics, epignomics, and proteomics data. It is now established that dysregulated miRNAs play substantial roles in a myriad of diseases [43]. We firmly believe that these methods for miRNA identification and their target transcripts could yield effective biomarkers and therapeutic targets, in addition to providing better understanding of disease mechanisms and etiology.

Abbreviations
AUC: Area under the receiver operating characteristic curve; CV: Cross-validation; EN: Elastic-net; FPR: False positive rate; GL: Group lasso; GSCA: Generalized structured component analysis; HisCoM-mimi: Hierarchical structured component analysis of microRNA-mRNA integration; IRWLS: Iterative reweighted least squares; JJ: Jin-Young Jang; LR: Logistic regression; PDAC: Pancreatic ductal adenocarcinoma; PHARAOH: Pathway-based approach using hierarchical components of collapsed rare variants; SC: Sungkyoung Choi; SL: Sungyoung Lee; SNUH: Seoul National University Hospital; TP: Taesung Park; YK: Yongkang Kim

Funding
This research was supported by a grant of the Korea Health Technology R&D Project through the Korea Health Industry Development Institute (KHIDI), funded by the Ministry of Health & Welfare, Republic of Korea (grant number: HI16C2037010016) and Bio-Synergy Research Project of the Ministry of Science, ICT and Future Planning through the National Research Foundation (grant number: 2013M3A9C4078158). Publication of this article was sponsored by the Bio-Synergy Research Project (grant number: 2013M3A9C4078158).

Availability of data and materials
An implementation of HisCoM-mimi, and normalized intensity microarray data can be downloaded from the website (http://statgen.snu.ac.kr/software/hiscom-mimi).

About this supplement
This article has been published as part of *BMC Bioinformatics* Volume 19 Supplement 4, 2018: Selected articles from the 16th Asia Pacific Bioinformatics Conference (APBC 2018): bioinformatics. The full contents of the supplement are available online at https://bmcbioinformatics.biomedcentral.com/articles/supplements/volume-19-supplement-4.

Authors' contributions
YK performed all analyses and developed the software implementation. YK and TP wrote the manuscript and developed the methodology. SL developed the software implementation. SC helped the analysis. JJ provided clinical interpretation of analysis results. All of the authors have read and approved of the final manuscript.

Ethics approval and consent to participate
Between the years 2009 and 2012, 200 pancreatic ductal adenocarcinoma (PDAC) samples were collected by the Department of Hepatobiliary and Pancreas Surgery of Seoul National University Hospital. The study protocol was approved by the Institutional Review Board of Seoul National University Hospital (IRB H-0901-010-267) and written, informed consent was obtained from each patient or legally authorized representative.

Consent for publication
Not applicable.

Competing interests
The authors declare that they have no competing interests.

Publisher's Note

Author details
[1]Department of Statistics, Seoul National University, Seoul, Korea. [2]Interdisciplinary program in Bioinformatics, Seoul National University, Seoul, Korea. [3]Department of Surgery and Cancer Research Institute, Seoul National University College of Medicine, Seoul, Korea.

Published: 8 May 2018

References
1. Reis-Filho JS. Next-generation sequencing. Breast Cancer Res. 2009;11(Suppl 3):S12.
2. Bock C, Reither S, Mikeska T, Paulsen M, Walter J, Lengauer T. BiQ analyzer: visualization and quality control for DNA methylation data from bisulfite sequencing. Bioinformatics. 2005;21(21):4067–8.
3. Anderson L, Hunter CL. Quantitative mass spectrometric multiple reaction monitoring assays for major plasma proteins. Mol Cell Proteomics. 2006; 5(4):573–88.
4. Joyce AR, Palsson BO. The model organism as a system: integrating 'omics' data sets. Nat Rev Mol Cell Bio. 2006;7(3):198–210.
5. Farazi TA, Hoell JI, Morozov P, Tuschl T. MicroRNAs in human cancer. Adv Exp Med Biol. 2013;774:1–20.
6. Kang SM, Lee HJ. MicroRNAs in human lung cancer. Exp Biol Med. 2014; 239(11):1505–13.
7. Navarro A, Monzo M. MicroRNAs in human embryonic and cancer stem cells. Yonsei Med J. 2010;51(5):622–32.
8. Negrini M, Ferracin M, Sabbioni S, Croce CM. MicroRNAs in human cancer: from research to therapy. J Cell Sci. 2007;120(11):1833–40.
9. Nam S, Li M, Choi K, Balch C, Kim S, Nephew KP. MicroRNA and mRNA integrated analysis (MMIA): a web tool for examining biological functions of microRNA expression. Nucleic Acids Res. 2009;37(Web Server issue):W356–62.
10. Buffa FM, Camps C, Winchester L, Snell CE, Gee HE, Sheldon H, Taylor M, Harris AL, Ragoussis J. microRNA-associated progression pathways and potential therapeutic targets identified by integrated mRNA and microRNA expression profiling in breast cancer. Cancer Res. 2011;71(17):5635–45.
11. Cho JH, Gelinas R, Wang K, Etheridge A, Piper MG, Batte K, Dakhallah D, Price J, Bornman D, Zhang S, et al. Systems biology of interstitial lung diseases: integration of mRNA and microRNA expression changes. BMC Med Genet. 2011;4:8.
12. Hwang HS, Takane Y. Generalized structured component analysis. Psychometrika. 2004;69(1):81–99.
13. Lee S, Choi S, Kim YJ, Kim BJ, Consortium T-G, Hwang H, Park T. Pathway-based approach using hierarchical components of collapsed rare variants. Bioinformatics. 2016;32(17):i586–94.
14. Siegel R, Naishadham D, Jemal A. Cancer statistics, 2013. Ca Cancer J Clin. 2013;63(1):11–30.
15. Rahib L, Smith BD, Aizenberg R, Rosenzweig AB, Fleshman JM, Matrisian LM. Projecting cancer incidence and deaths to 2030: the unexpected burden of

thyroid, liver, and pancreas cancers in the United States. Cancer Res. 2014; 74(11):2913–21.
16. Oh CM, Won YJ, Jung KW, Kong HJ, Cho H, Lee JK, Lee DH, Lee KH, Community of Population-Based Regional Cancer R. Cancer statistics in Korea: incidence, mortality, survival, and prevalence in 2013. Cancer Res Treat. 2016;48(2):436–50.
17. Meacham CE, Morrison SJ. Tumour heterogeneity and cancer cell plasticity. Nature. 2013;501(7467):328–37.
18. Namkung J, Kwon W, Choi Y, Yi SG, Han S, Kang MJ, Kim SW, Park T, Jang JY. Molecular subtypes of pancreatic cancer based on miRNA expression profiles have independent prognostic value. J Gastroenterol Hepatol. 2016;31(6):1160–7.
19. Agarwal V, Bell GW, Nam JW, Bartel DP. Predicting effective microRNA target sites in mammalian mRNAs. elife. 2015;4:e05005.
20. Tibshirani R. Regression shrinkage and selection via the Lasso. J Roy Stat Soc B Met. 1996;58(1):267–88.
21. Zou H, Hastie T. Regularization and variable selection via the elastic net. J R Stat Soc B. 2005;67:301–20.
22. Meier L, van de Geer SA, Buhlmann P. The group lasso for logistic regression. J R Stat Soc B. 2008;70:53–71.
23. Kwon MS, Kim Y, Lee S, Namkung J, Yun T, Yi SG, Han S, Kang M, Kim SW, Jang JY, et al. Integrative analysis of multi-omics data for identifying multi-markers for diagnosing pancreatic cancer. BMC Genomics. 2015;16(Suppl 9):S4.
24. Benjamini Y, Drai D, Elmer G, Kafkafi N, Golani I. Controlling the false discovery rate in behavior genetics research. Behav Brain Res. 2001;125(1–2):279–84.
25. Yang D, Ma M, Zhou W, Yang B, Xiao C. Inhibition of miR-32 activity promoted EMT induced by PM2.5 exposure through the modulation of the Smad1-mediated signaling pathways in lung cancer cells. Chemosphere. 2017;184:289–98.
26. Chen R, Liao JY, Huang J, Chen WL, Ma XJ, Luo XD. Downregulation of SRC kinase signaling inhibitor 1 (SRCIN1) expression by MicroRNA-32 promotes proliferation and epithelial-mesenchymal transition in human liver cancer cells. Oncol Res; 2017. https://doi.org/10.3727/096504017X14954923820137.
27. Chen S, Chen X, Sun KX, Xiu YL, Liu BL, Feng MX, Sang XB, Zhao Y. MicroRNA-93 promotes epithelial-mesenchymal transition of endometrial carcinoma cells. PLoS One. 2016;11(11):e0165776.
28. Lahdaoui F, Delpu Y, Vincent A, Renaud F, Messager M, Duchene B, Leteurtre E, Mariette C, Torrisani J, Jonckheere N, et al. miR-219-1-3p is a negative regulator of the mucin MUC4 expression and is a tumor suppressor in pancreatic cancer. Oncogene. 2015;34(6):780–8.
29. Xu L, Li Q, Xu D, Wang Q, An Y, Du Q, Zhang J, Zhu Y, Miao Y. hsa-miR-141 downregulates TM4SF1 to inhibit pancreatic cancer cell invasion and migration. Int J Oncol. 2014;44(2):459–66.
30. Greither T, Grochola LF, Udelnow A, Lautenschlager C, Wurl P, Taubert H. Elevated expression of microRNAs 155, 203, 210 and 222 in pancreatic tumors is associated with poorer survival. Int J Cancer. 2010;126(1):73–80.
31. Lee CL, He H, Jiang YJ, Di Y, Yang F, Li J, Jin C, Fu DL. Elevated expression of tumor miR-222 in pancreatic cancer is associated with Ki67 and poor prognosis. Med Oncol. 2013;30(4):700.
32. Park JK, Henry JC, Jiang J, Esau C, Gusev Y, Lerner MR, Postier RG, Brackett DJ, Schmittgen TD. miR-132 and miR-212 are increased in pancreatic cancer and target the retinoblastoma tumor suppressor. Biochem Biophys Res Commun. 2011;406(4):518–23.
33. Zhang S, Hao J, Xie F, Hu X, Liu C, Tong J, Zhou J, Wu J, Shao C. Downregulation of miR-132 by promoter methylation contributes to pancreatic cancer development. Carcinogenesis. 2011;32(8):1183–9.
34. Feng J, Yu J, Pan X, Li Z, Chen Z, Zhang W, Wang B, Yang L, Xu H, Zhang G, et al. HERG1 functions as an oncogene in pancreatic cancer and is downregulated by miR-96. Oncotarget. 2014;5(14):5832–44.
35. Keklikoglou I, Hosaka K, Bender C, Bott A, Koerner C, Mitra D, Will R, Woerner A, Muenstermann E, Wilhelm H, et al. MicroRNA-206 functions as a pleiotropic modulator of cell proliferation, invasion and lymphangiogenesis in pancreatic adenocarcinoma by targeting ANXA2 and KRAS genes. Oncogene. 2015;34(37):4867–78.
36. Hu S, Zheng Q, Wu H, Wang C, Liu T, Zhou W. miR-532 promoted gastric cancer migration and invasion by targeting NKD1. Life Sci. 2017;177:15–9.
37. Bai L, Wang H, Wang AH, Zhang LY, Bai J. MicroRNA-532 and microRNA-3064 inhibit cell proliferation and invasion by acting as direct regulators of human telomerase reverse transcriptase in ovarian cancer. PLoS One. 2017; 12(3):e0173912.
38. Sheikholeslami A, Nabiuni M, Arefian E. Suppressing the molecular signaling pathways involved in inflammation and cancer in breast cancer cell lines MDA-MB-231 and MCF-7 by miR-590. Tumour Biol. 2017;39(4): 1010428317697570.
39. Yang D, Zhao D, Chen X. MiR-133b inhibits proliferation and invasion of gastric cancer cells by up-regulating FBN1 expression. Cancer Biomark. 2017;19:425.
40. Li D, Xia L, Chen M, Lin C, Wu H, Zhang Y, Pan S, Li X. miR-133b, a particular member of myomiRs, coming into playing its unique pathological role in human cancer. Oncotarget. 2017;8:50193.
41. Wu H, Wang Y, Wu C, Yang P, Li H, Li Z. Resveratrol induces cancer cell apoptosis through MiR-326/PKM2-mediated ER stress and mitochondrial fission. J Agric Food Chem. 2016;64(49):9356–67.
42. Ji S, Zhang B, Kong Y, Ma F, Hua Y. MiR-326 inhibits gastric cancer cell growth through down regulating NOB1. Oncol Res. 2017;25(6):53–86.
43. Esteller M. Non-coding RNAs in human disease. Nat Rev Genet. 2011;12(12): 861–74.

Yang et al. BMC Bioinformatics 2018, 19(Suppl 4):98
https://doi.org/10.1186/s12859-018-2071-z

BMC Bioinformatics

RESEARCH Open Access

cmFSM: a scalable CPU-MIC coordinated drug-finding tool by frequent subgraph mining

Shunyun Yang[2†], Runxin Guo[2†], Rui Liu[3†], Xiangke Liao[2], Quan Zou[4*], Benyun Shi[5*] and Shaoliang Peng[1,2*]

From The Sixteenth Asia Pacific Bioinformatics Conference
Yokohama, Japan. 15-17 January 2018

Abstract

Background: Frequent subgraphs mining is a significant problem in many practical domains. The solution of this kind of problem can particularly used in some large-scale drug molecular or biological libraries to help us find drugs or core biological structures rapidly and predict toxicity of some unknown compounds. The main challenge is its efficiency, as (i) it is computationally intensive to test for graph isomorphisms, and (ii) the graph collection to be mined and mining results can be very large. Existing solutions often require days to derive mining results from biological networks even with relative low support threshold. Also, the whole mining results always cannot be stored in single node memory.

Results: In this paper, we implement a parallel acceleration tool for classical frequent subgraph mining algorithm called cmFSM. The core idea is to employ parallel techniques to parallelize extension tasks, so as to reduce computation time. On the other hand, we employ multi-node strategy to solve the problem of memory constraints. The parallel optimization of cmFSM is carried out on three different levels, including the fine-grained OpenMP parallelization on single node, multi-node multi-process parallel acceleration and CPU-MIC collaborated parallel optimization.

Conclusions: Evaluation results show that cmFSM clearly outperforms the existing state-of-the-art miners even if we only hold a few parallel computing resources. It means that cmFSM provides a practical solution to frequent subgraph mining problem with huge number of mining results. Specifically, our solution is up to one order of magnitude faster than the best CPU-based approach on single node and presents a promising scalability of massive mining tasks in multi-node scenario. More source code are available at:Source Code: https://github.com/ysycloud/cmFSM.

Keywords: Frequent subgraph mining, Bioinformatics, Memory constraints, Isomorphism, Many integrated Core (MIC)

Background

Frequent subgraphs mining

Frequent subgraphs mining in a collection of graph objects is a very difficult challenge, especially in the important application area "Bioinformatics" where it can help finding new drugs in pharmacological compound databases or core functional structures in biological networks. Subgraph mining problem is difficult to solve because arbitrary graph structures must be generated and matched. As isomorphism testing is a hard problem [1], subgraph miners are exponential in memory consumption and execution time.

Lin [2] have summarized that the problem of frequent subgraph mining mainly consists of two categories: (i) frequent subgraph patterns ought to be found in different regions of one large graph of massive scale; (ii) frequent subgraph patterns should be found within a large-scale collection of middle-sized graphs. The first case is usually adapted to social network domain, and the

* Correspondence: zouquan@nclab.net; benyunshi@outlook.com; pengshaoliang@nudt.edu.cn
†Equal contributors
[4]School of Computer Science and Technology, Tianjin University, Tianjin 300072, China
[5]School of Cyberspace, Hangzhou Dianzi University, Hangzhou 310018, China
[1]College of Computer Science and Electronic Engineering & National Supercomputer Centre in Changsha, Hunan University, Changsha 410082, China
Full list of author information is available at the end of the article

second case is usually adapted to the areas of computational pharmacology and bioinformatics. Both categories share several common challenges. For example, large data input size with relative low support threshold can lead to huge number of mining results, which may exceed the memory of a single machine, and require vast amounts of runtime. Given these characteristics, parallel techniques are presented as a promising solution to solve these challenges.

The objective of this problem is to find subgraphs that occur with support higher than a threshold θ, i.e., $0 \ll \theta \ll 1$. Several solutions have been put up with for the first case in either serial CPU-based techniques [3–5] or parallel computing (MapReduce, MPI, Spark) framework [6–9] and GPU [10]. However, we mainly focus on the second case, which is more practical in the field of bioinformatics and known as transaction setting [11].

Related work

In the transaction scenario, miner/frequent subgraph mining algorithm recursively generates all possible refinement extensions from empty graph by adding edges and nodes to already generated refinements. Then, isomorphism test will be performed of each new possible refinement to determine if it appears frequently. Early miner/frequent subgraph mining algorithms generated refinements in a Breadth First Search (BFS) way, e.g., AGM [12], and FSG [13]. However, the Depth First Search (DFS) approaches need less memory and almost show better performance. [14] have summarized three main subproblems (i.e. Purposive refinement Efficient enumeration and Focused isomorphism testing) have to be solved of efficient miners and made a quantitative and detailed comparison of some typical dfs-algorithms, e.g., MoFa [15], FFSM [16], gSpan [17] and Gaston [18], and some special extensions of them, e.g., CloseGraph [19], showing them attacking the three subproblems quite differently. Thus, the dfs-algorithms will be mainly used for comparative analysis with our work. A follow-up work [20] is more focus on accelerating the mining process.

All solutions discussed above are single-core serial version. When they come to large-scale mining problems, they may be difficult to meet time requirement. To solve this problem, SUBDUE [21] develops a shared-memory parallel approach by using the minimum description length (MDL) principle and embodies in the SUBDUE system. Further, [22] proposes parallel mining strategy in a multi-core system and partitions the mining tasks among multiple shared-memory processors. To some extent, these studies make full use of the machine resources on single node to accelerate the mining process.

Another problem is all these approaches are memory-based, and assume that the collection, intermediate data and mining results fit in memory. However, as data size increases and especially support threshold decreases (the scale of mining results grows exponentially), the assumption no longer reach. To address this limitation, some disk-based solutions have been proposed, e. g. ADI-Mine [23]. However, these approaches face significant overhead of accessing the data. The same as a disk-based solution, PGM [24] propose a data partition approach.

The work in IFSM [11] is relatively early to employ MapReduce [25] framework for mining frequent subgraphs. Specifically, it first determines local support for all possible single-edge subgraphs by mapping a part of the graph collection to each worker. Secondly, a reduction phase will determine the global support for each subgraph, and discard those candidates that do not reach the global support threshold. The solution continues to deal with the double-edge subgraphs, triple-edge subgraphs and so on. Similar with IFSM [11], FSM-H [26] and mrFSM [27] are also developed on MapReduce framework by an iterative method. Relatively, [27] pay more attention to the load balancing in each iteration. However, because MapReduce is not suitable for iterative computing, which may result in a lot of IO and serialization overhead, these approaches on MapReduce still create significant performance problem.

The more outstanding work so far on MapReduce framework is MRFSM [28]. It does not adopt iterative method, but the whole process is divided into two MapReduce stages: filter and refinement. The filter stage prunes based on the probability of support and outputs local frequent subgraphs in the local dataset which is divided to each machine. The refinement phase intelligently translates the local support which is gotten from the filter phase into a global support to integrate the final results. Because there are no iterations, it presents a better performance than IFSM [11], FSM-H [26] and mrFSM [27]. However, the implementation code of MRFSM is not purely native java program, most program is coding by C++, and then use Hadoop Streaming to adapt to MapReduce framework to complete distributed mining. As a result, the performance of the MRFSM will be severely restricted because of limited data exchange capacity using standard I/O and redundant data type conversions. Moreover, when the support threshold is low enough, the massive mining results tend to far exceed the single-machine memory. In this respect, because the refinement stage distributes all candidate subgraphs to all machines, this stage can easily cause severe memory pressure on each machine when a large number of candidate subgraphs cannot be filtered due to the low support threshold. Thus, MRFSM [28] may not be able to cope with scenarios with massive mining results due to low support threshold.

In contrast, our tool is implemented by native C++ program with several efficient parallel techniques, e.g., MPI and OpenMP, to maximize performance. Also, every node will not hold all candidate subgraphs but carry out its own mining process to get local final results so as to easily handle the scenarios with massive mining results.

Methods

cmFSM realizes parallelism of multiple levels and multiple granularities and utilizes MIC as accelerator. Multithreading is implemented using OpenMP aimed at hotspots of mining process. Four kinds of static task dividing strategy and a supervisor-based dynamic task dividing strategy are implemented by MPI to achieve best load balancing. Further, we used MICs in offload mode only to transfer double-edge frequent subgraphs and back up complex data structures redundantly to avoid the bottlenecks caused by excessive transmission. By making full use of the multi-core computing capacity of MIC, we can achieve a desired effect of execution speed in the scenario of CPU and MIC collaboration.

OpenMP parallelization on single node

1) The strategy of parallelization

The general dfs-algorithm of frequent subgraph mining uses a recursive approach to deal with the hotspot of mining process, which is very difficult to control the parallel granularity. Also, a simple function call can continuously find out a large number of mining results because we cannot pick out or predict the depth of recursive process. This is bound to lead to load unbalance among different mining tasks.

In order to solve this problem to achieve better effect of parallelization by OpenMP, we adopt a fine-grained parallel strategy. Specifically, we translate the common recursive mining process into a BFS loop mining process by one-edge growth of several layers, so as to implement the parallelization on the granularity of one-edge growth. The operation consists of two specific parts: (i) judgment of minimum DFS code and (ii) right-most extension. Overall, the computing scale of this operation is not too small, so that there is no possibility that most CPU resources is used in thread scheduling because the parallel granularity is not big enough. At the same time, the tasks of the two parts are specific and similar so as to easily achieve a good load balancing by dynamic scheduling strategy of OpenMP. Moreover, there is no need of system to help us manage the stack because recursive processes are replaced by loops, which may lead to additional acceleration.

Take gSpan [15] as an example, the following pseudocode compares original Algorithm 1 with new parallel Algorithm 2.

Algorithm 1. Old Algorithm.

```
Frequent_Subgraph_Mining(GS, S)
1: sort labels of the vertices and edges in GS by their frequency;
2: remove infrequent vertices and edges;
3: relabel the remaining vertices and edges in descending frequency for GS;
4: S^l ←all frequent 1-edges graphs in GS;
5: sort S^l in DFS lexicographic order;
6: S←S^l;
7: for edge e in S^l do
8:     initialize 1-edge graph g with e;
9:     Mining(GS, S, g);
10:    GS←GS − e;
11:    if |GS| < minSup then
12:        break;
13:    end if
14:end for

Mining(GS, S, g)
1:  if g ≠ min_DFS_code(g) then
2:        return;
3:  S←S ∪ {g};
4:  generate all g' potential children gs with one edge growth in GS;
5:  for graph c in gs do
6:        if support(c) ≥ minSup then
7:              g←c;
8:              Mining(GS, S, g);
9:        end if
10:end for
```

Algorithm 2. New Parallel Algorithm.

```
Frequent_Subgraph_Mining(GS, S)
1-6: same with old algorithm1-6;
7:  for edge e in S^l do
8:      initialize 1-edge graph g with e;
9:      One_edge_growth(GS, S, g, children);
10:     len←size(children);
11:     while len ≠ 0 do
12:        #pragma omp parallel
13:           #pragma omp parallel for
14:           for graph c in children
15:               One_edge_growth(GS, S, c, tchildren);
16:               lchildren←lchildren ∪ tchildren;
17:               clear(tchildren);
18:           end for
19:           #pragma omp critical
20:               cchildren←cchildren ∪ lchildren;
21:           #end parallel
22:           swap(children, cchildren);
23:           clear(cchildren);
24:           len←size(children);
25:     end while
26-29: same with old algorithm10-13;
30:end for

One_edge_growth(GS, S, g, children)
1-2: same with old algorithm1-2;
3:  #pragma omp critical
4:  S←S ∪ {g};
5:  generate all g' potential children gs with one edge growth in GS;
6:  for graph c in gs do
7:        if support(c) ≥ minSup then
8:              g←c;
9:              children←children ∪ {g};
10:       end if
11:end for
```

As the pseudocode shown above, in order to complete the parallelization, we apply for four new categories of buffer: children, tchildren, lchildren, cchildren. The children are used to record the set of subgraphs obtained from each level extension where the subgraphs in same level have same number of edges. When children is not empty, the next level mining process will be carried out. It is a global variable and will be used sequentially. The tchildren is a local variable within single-thread tasks, recording the subgraphs obtained from one-edge growth of each subgraph. The lchildren is also a local variable within single-thread tasks, but it is a summary of the results of all one-edge growths for each thread and gotten from the union of tchildren in every iteration. The cchildren also records the set of subgraphs obtained from each level extension. At the end of single-thread tasks, the lchildren will be summarized to cchildren in the critical area. Also, the cchildren and children will be exchanged out of the parallel area in order to carry out the next level iterative mining. It is worth noting that the meaning of the existence of cchildren is that we cannot directly summarize lchildren to children, because at the parallel computing scenario we cannot make sure all extension tasks in every thread are over at the same time. The thread not yet complete tasks will continue to use the data of children, which may lead to failures.

2) 2) Memory Management Deep Optimization

The main challenge of frequent subgraph mining is the memory constraint. In order to achieve the purpose of memory reuse and the efficient utilization of memory space, we adopt the strategy named "apply dynamically & store pointers". Specifically, when the subgraph is extended, the program applies edges dynamically and stores edge pointers rather than actual edge objects in the graph code structure, so that the new frequent graphs share most edges with their ancestors, which will lead to a significant saving in memory space. The schematic diagram is shown in Fig. 1. It can be easily seen that only the edge pointers are stored in the graph code and each edge has only one instance in memory, so as to achieve the purpose of memory reuse.

Multi-node multi-process parallel acceleration

1) The strategies of task division

The biggest challenge of multi-node program is the communication overhead. To solve this problem, a coarse-grained parallel strategy is used among processes through dividing single-edge frequent subgraphs by MPI. The main tasks of each process are completely independent to avoid a large amount of communication overhead, and then each processes will write out its corresponding results in its own node to avoid the single-node memory pressure causing by massive mining results. We should notice that all output files can be easily merged to form the whole results. Further, combining with the multi-thread work on single node, our implementation allows to generate a second level of parallelization by creating multiple threads per MPI process to achieve a good performance.

However, this coarse-grained strategy is not conducive to load balancing, which is easily lead to data tilt and cannot make full use of system resources to achieve the best performance. Therefore, we design and provide four kinds of static task dividing (i.e. equality, single, increment, circle) strategies and a supervisor-based dynamic task dividing strategy based on different characteristics of datasets to achieve load balancing as much as possible.

a) static division___equality

A simple strategy is to divide the single-edge frequent subgraphs equally. However, we found that the load is extremely unbalanced after the experiment, and it is easy to achieve bottlenecks. Also, a lot of mining tasks are concentrated on the front of the nodes, and they will also pick out most results. This is because these single-edge frequent subgraphs will be sorted in descending order in the pretreatment stage. The more front they sorted, the more frequent they are and also the more frequent subgraphs they may pick out. And those sorted behind will be closer to the support threshold, they may no longer a frequent subgraph after one-edge growth even one time, so that they may pick out a little results and stop tasks early. Moreover, we find that the scale of results is decreased exponentially as the pretreatment order. Therefore, this intuitive strategy in most cases are difficult to achieve load balancing.

b) static division___increment

In order to solve the problem of equal division, we propose an incremental task allocation strategy. Specifically, the first node gets only one graph, the second node gets two graphs, the third node gets three graphs, and so on, and the last node gets all remaining graphs.

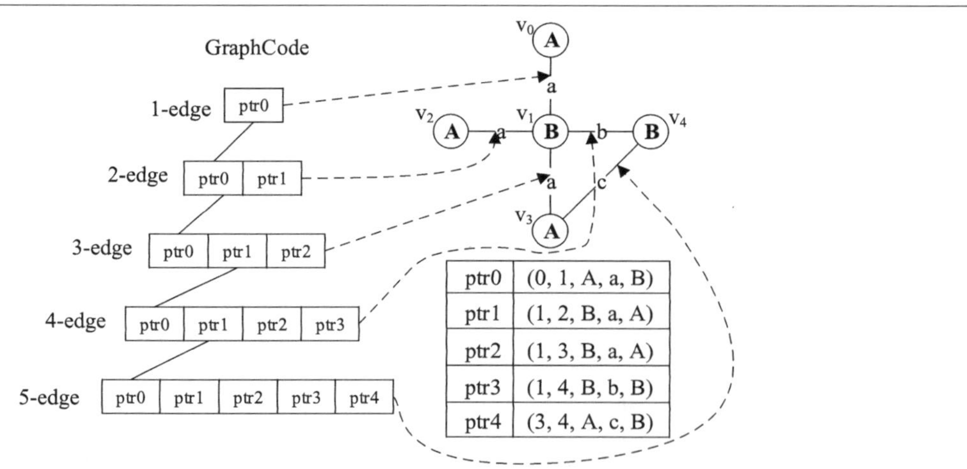

Fig. 1 Memory Reuse. The left side shows the frequent subgraph extension process in the form of graph code. The upper right subfigure is the original subgraph. Below it lists the codes of each edge and their corresponding pointers

Implementation of this strategy actually improves the performance and achieve a better load balancing.

c) static division___single

Actually, when the dataset is big enough, although the tasks of single-edge subgraphs ranking behind will drastically reduce, the gap among the tasks of top-ranked single-edge subgraphs is not obvious. In this case, incremental strategy will lead the relative top nodes to undertake more tasks, so that the load is still not balance. For this scenario, we proposed a single task allocation strategy, in which we assign all preceding nodes only one single-edge frequent subgraph, and the remaining subgraphs are allocated to the last node. This strategy sometimes can achieve a better performance.

d) static division___circle

Considering the single strategy may make a large number of tasks focused on the last node when the dataset is big enough but the degree of parallelism is not particularly high, we proposed a circle task allocation strategy, in which we first allocate all nodes a round of single-edge subgraphs in turn, and then we allocate all nodes a round of graphs in reverse order, and then we allocate all nodes a round of graphs in turn again, and so on until all single-edge frequent subgraphs are allocated to their corresponding nodes. This strategy is expected in most cases to achieve better load balancing.

e) dynamic division___supervisor-based

In fact, because the mining process is very complex and it is difficult for us to predict or measure the scale of tasks, these static task division strategies certainly cannot adapt to all practical scenarios. A more ideal method is to use dynamic division strategy based on task queue, in which we first allocate a round of single-edge subgraphs in turn to all nodes, and then one of the remaining graphs will be allocated to the node that finishes tasks earliest to carry out mining process until the end of all tasks. In theory, this strategy can always achieve a better load balancing compare to static division strategies.

In order to implement this dynamic strategy, we treat process0 as a supervisor, which manages all tasks uniformly. When other processes finish their current tasks, they first ask process0 for a remaining single-edge frequent subgraph. The process0 will search its task queue and reply to them. When the task queue is not empty, the process0 will allocate a new single-edge frequent subgraph to other processes, otherwise, it replies − 1 and counts. When the count reaches the number of process, the process0 will end its work. On the other hand, when other processes receive − 1 once, they will also end their work.

The Fig. 2 shows an example of five division strategies. The dynamic strategy can always achieve a better load balancing than static strategies, but the overall performance is not necessarily more optimal because of other operations such as request, wait, communication and synchronization. Thus, users can choose all these strategies. However, in most cases it is recommended to adopt dynamic strategy.

2) Remove multi-node redundant results

Another problem in multi-node scenario is the redundant results. From Algorithm 1, it is not difficult to find that the original single-edge frequent subgraph must be deleted after its extension tasks. This is to avoid to consider the high-frequency single-edge subgraphs which have been used in the following mining process, which may lead to redundant results. This possibility can result in some difficulties in multi-node scenario. Because each process can only handle its own single-edge frequent subgraphs in current mining process, the high-frequency single-edge subgraphs in other nodes which should not be considered will not be deleted.

We extend the parallel algorithm to multi-node scenario and show it in the following Algorithm 3.

In order to extend the parallel algorithm, we notice that the single-edge subgraphs which are allocated to every node are also be handled in turn. This characteristic make it possible for us to remove high-frequency single-edge subgraphs which priori to current object before the mining process.

Algorithm 3. New Parallel Algorithm.
Every_Process_FSM(*GS*, *S*)

```
1-5: same with Algorithm 2:1-5;
6:  if this is root process then
7:      S←S^1;
8:  end if
9:  use proper strategy to split S^1 and get the task
queue Q in current process;
10: pre←0;
11: for edge e in Q do
12:     for edge t between S^1[pre] and e do
13:         GS←GS − t;
14:     end for
15-32: same with Algorithm 2:8-25;
33-35: same with Algorithm 2:27-29;
36: pre← the index of e in S^1;
37: end for
```

CPU-MIC collaborated parallel optimization

1) Collaborated Parallelization of cmFSM

The collaboration among CPUs and MICs we employ a medium-grained parallel strategy. In detail, we divide

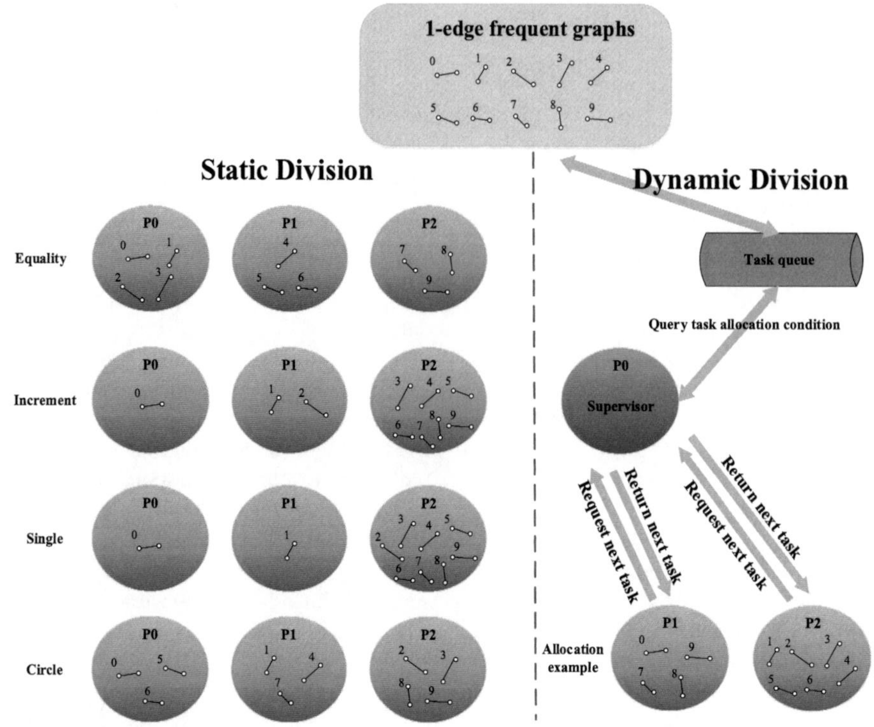

Fig. 2 Different kinds of task dividing strategies. Note that the equality division is not absolute, but each process holds the number of single-edge frequent subgraphs will be the same or keep the difference that no more than one. At the same time, the example of dynamic division in this figure shows only a possible result which is not necessary in this case

the double-edge frequent subgraphs among CPUs and MICs, which are gotten from one-edge growth of initial single-edge frequent subgraphs in every process. Also, we adopt offload mode to transfer double-edge frequent subgraphs from host to MICs. By tolerating appropriate communication overhead and making full use of the multi-core computing capacity of MIC, we can achieve a desired effect of load balancing and computing speed.

The Fig. 3 shows interactive process between CPU and MIC to achieve the ideal collaboration. It is worth noting that the coarse-grained strategy is not used because it is difficult to effectively reach load balancing under this strategy. Also, the single-edge frequent subgraphs are allocated to each process are not definitely able to be divided reasonably. For example, only one single-edge frequent subgraph may be allocated to some processes in many cases. Coupled with the truth that there is a difference of computing capacity between CPU and MIC, load balancing will be a great challenge. On the other hand, the fine-grained strategy also should not be considered, because it is not shared memory between CPU and MIC. There must be a huge scale of communication overhead to transfer and divide the graphs by offload mode. Thus, this strategy is also not conducive to improve the overall performance.

2) Memory Reuse

The memory of one MIC card is approximately 5 Gigabytes, which cannot be matched to the general node memory. Also, the speed of memory allocation and release is still slower than CPU. Experimental tests show that allocating 1 Gigabyte of memory on MIC takes approximately 5 s. Therefore, you must reduce the frequency of memory allocation and release on MIC and maximize possibility of memory reuse.

In addition to the usage of "apply dynamically & store pointers" strategy we introduced before, cmFSM reduces the memory allocation time on MIC by memory reuse. We create a counter JobCount to record the job number. If count = 1, it uses "alloc_if(1) free_if(0)" to allocate memory for the array and object listed in the offload segment. When count > 1, it employs "alloc_if(0) free_if(0)" to reuse the memory. Until the last time, it adopts "alloc_if(0) free_if(1)" to release the memory after the operation is completed. By this way, we can minimize the frequency of memory allocation and release on MIC.

On the other hand, when the dataset is relatively large and the mining process is deep enough, even if a whole extension process of one single-edge frequent subgraph can use up the memory on MIC. In this scenario, it is not

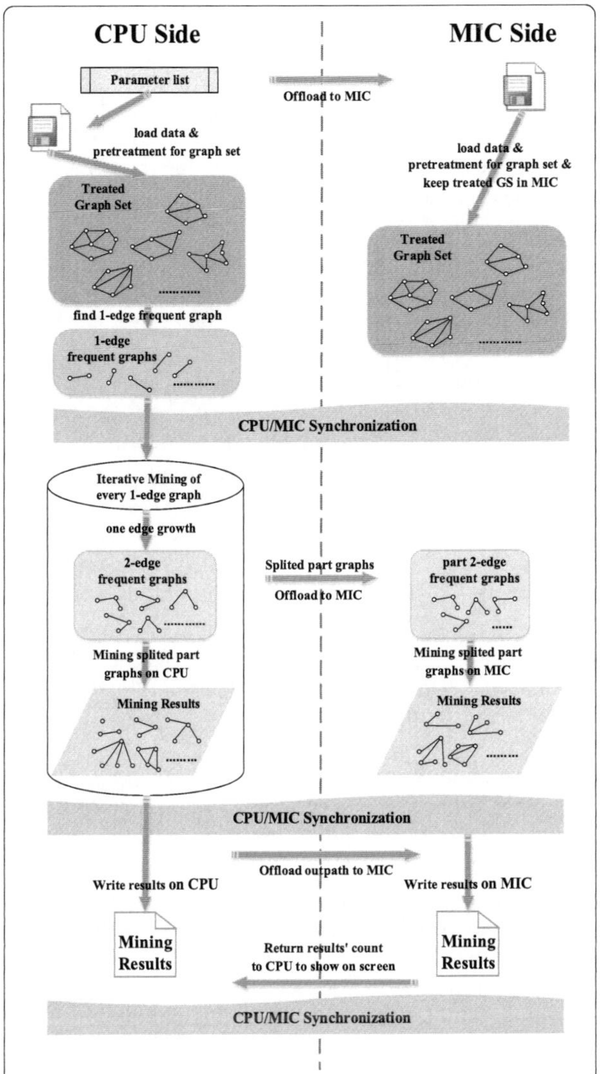

Fig. 3 Framework of CPU/MIC collaboration. There are mainly three stages of MIC process: (i) Data loading and pretreatment (ii) Iterative mining process and (iii) Output the results. All these tasks are collaborated with CPU at the same time. Moreover, at the end of each stage, the two devices will synchronize their tasks to ensure the correctness and efficiency of the program

suitable to transfer all the data to be mined once and then mining their results. Therefore, instead of uploading all the double-edge frequent graphs at one time, we adopt an iterative method to upload only a part of the graphs obtained from the same single-edge frequent graph at one time to facilitate data compression and save the memory space on MIC, which can be clearly seen from Fig. 3.

3) Data Transmission Optimization

Although the C++ STL container and class is supported by MIC, the ICC compiler does not support the use of offload mode to transfer these structures. It can only support the basic datatype and array without pointers. Thus, we adopt the strategy named "dismantle & restore" to transfer the objects. The Fig. 4 shows the format of n double-edge frequent subgraphs that are supposed to be transfered in an iteration.

We first dismantle these objects and integrate their elements to make those of same kind are stored in the same successive buffer. Then, we transfer these buffers to MIC by offload mode and allocate memory of original objects on MIC. Finally, we restore these objects on MIC by filling corresponding elements from these buffers. This is no doubt a troublesome process, but fortunately we only need to transfer some double-edge frequent subgraphs with the same first edge in an iteration according to the above idea.

From the Fig. 4, we can find that there are seven buffers are used in transmission process. Because the format of edge code is a quintuple (ix, iy, x, a, y), the first five buffers are used to transfer edges. The elements with subscript 0 represent the first common edge. The following n elements represent the second edges of every subgraph. The next buffer represents the number of nodes of n subgraphs. The gs buffer is the numbered lists of original graphs where this subgraph appeared in dataset. Because the number begins from 0, we simply employ -1 to separate these lists. From this formation, we can organize and transfer data between CPU and MIC efficiently.

On the other hand, some complex data structures will be used in the whole mining process, such as original graph set after the pretreatment, which may lead to unbearable transmission overhead and memory allocation overhead. Therefore, we back up complex data structures redundantly which are reusable and difficult to transfer to maximize performance. In detail, at the beginning we only transfer the analytical parameters to MIC, the coprocessor can read data and construct graph set by itself based on these parameters, which is not the hotspot of calculation and can be quickly completed on

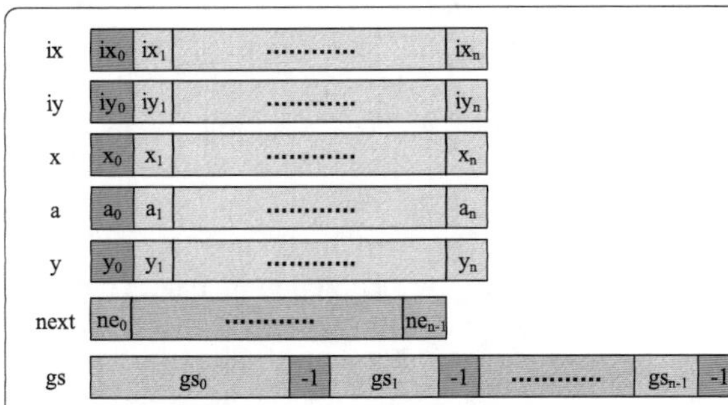

Fig. 4 Data structure used in offload mode. The format of n double-edge frequent subgraphs that are supposed to be transfered in an iteration

the MIC. Also, in many cases the mining results is too big to transfer by offload mode. Thus, at the end, instead of returning all mining results to CPU, we only return the count of results to show the overall consequence on CPU. Specific mining results will be directly written on MIC. These files can be easily merged to get the whole results. The two stages are clearly illustrated in Fig. 3.

4) Load balancing and data division among CPU/MICs

According to the previous strategy, only the double-edge frequent subgraphs obtained from the one-edge growth of the same single-edge frequent graph will be transferred to MIC in an iteration. If the process continues to mine from this layer, the scale of calculation is supposed to be greatly closer than from single-edge subgraphs. Considering the computing capacities of CPU and MIC are still close in our environment after tests, we simply adopt a static strategy using interval division and make the host device with slightly stronger calculation ability start first, which is because the front double-edge subgraphs theoretically still have more potential to extend, to achieve an efficient load balancing among CPUs and MICs.

The data division and CPU/MIC collaborating mining process in an iteration is shown in the Fig. 5. Taking into account the truth that there are three MICs on single node of Tianhe-2, we take three MICs as an example in Fig. 5. Multiple CPUs in single node share memory, we can manage their computing resources uniformly and call them Host. The method of interval division can be clearly seen from Fig. 5. At the same time, the mining depth or scale of calculation on each device cannot be sure, but all their processes will end until there are no more new frequent subgraphs after one-edge growths. In the multi-node scenario, we just allocate every node different single-edge frequent subgraphs to form different task queues. There are no more other differences.

Results

We have evaluated the performance of cmFSM under five aspects: (i) parallelization on single node, (ii) multi-node division strategy, (iii) efficiency of multi-node multi-thread acceleration, (iv) CPU/MIC collaboration and (v) multi-node CPU/MIC collaboration.

Setup and dataset

The cmFSM was implemented in C++ using the STL and compiled with the –O2 compilation flag. The first experiment is performed on a high performance server which consists of 8 Xeon E7–8800 v3 18 core CPU processors with 2 Xeon Phi 3120 57 core coprocessors and 2 K40 M GPU. It has a 2 TB memory in total and the operation system is Red Hat Enterprise Linux Server release 7.1. The next four experiments are performed on the Tianhe-2 supercomputer. The configuration is listed in the following Table 1.

A comprehensive performance study was conducted in our experiments on both real molecular and synthetic datasets. The first real dataset we tested is the Predictive Toxicology dataset (PTE). It is sparse and contains 340 chemical compounds, 24 different atoms, 66 atom types, and 4 types of bonds. We use the type of atoms and bonds as labels. The second real dataset is the AIDS antiviral screen compound dataset from Developmental Therapeutics Program in NCI/NIH. It contains 43,905 chemical compounds. The results of the screening tests can be categorized into three classes: CA: confirmed active; CM: confirmed moderately active; and CI: confirmed inactive. We only use CA class in our tests which consists of 422 molecules (dataset DTP).

The synthetic graph dataset is using a synthetic data generator similar to that described in [12]. A user can set parameters to decide the number of graphs and their average size. We generate three datasets (S1, S2 and S3) for our tests, which consist of 10,000 graphs, 20,000 graphs and 100,000 graphs respectively. More information of these four datasets in shown in Table 2.

Parallelization on single node

We try to compare our tool with a wide range of functionally comparable frequent structure miners, such as FSM [12], FFSM [14], gSpan [15] and Gaston [17]. We should note that some of these miners had restrictions regarding the number of labels or were restricted to molecular database. For these algorithm we only publish limited results.

In this part, we have used first three datasets for analysis to show that cmFSM can easily present a better performance than any other famous miners in a relatively low level of parallelization. Table 3 compares results and runtimes among cmFSM and other miners on PTE dataset.

From Table 3, it is not difficult to find that cmFSM presents a significant performance advantage. The last three lines in Table 3 represent that cmFSM starts 2 threads, 8 threads and 32 threads respectively. It can be seen that even the serial version, the runtime of cmFSM is less than gSpan. In addition, as long as we start more than 8 threads, the runtime of cmFSM is less than all other tools. This proves that cmFSM can show better performance than other state-of-the-art miners even if we only hold a few parallel computing resources. Moreover, the consistency of the mining results also demonstrates that our parallel optimization does not affect the correctness of the miners.

Fig. 6 reflects the mining conditions on DTP datasets. From Fig. 6, it is not difficult to see that cmFSM can also

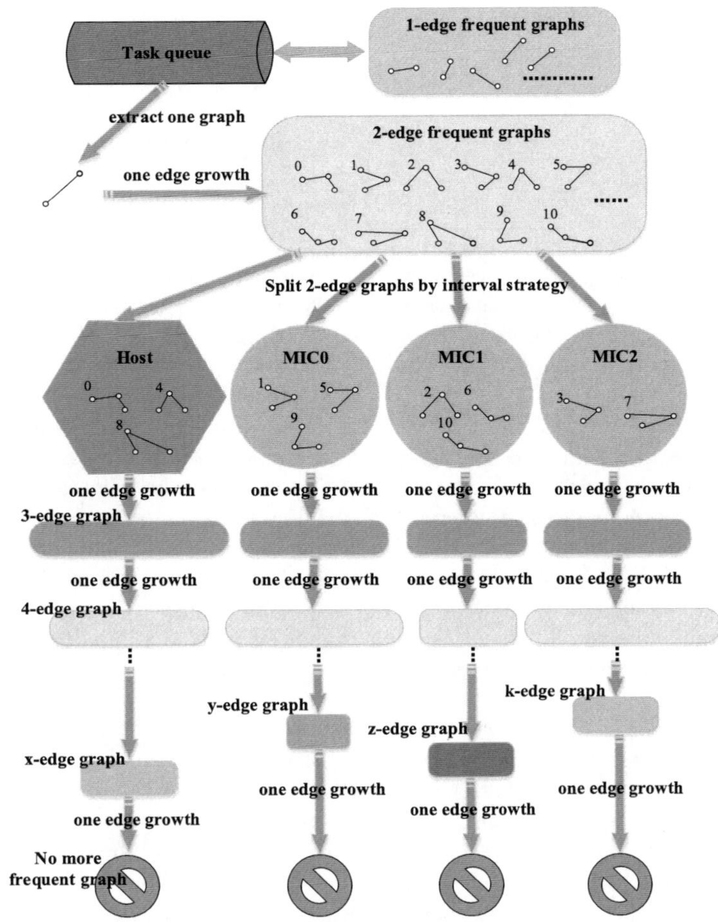

Fig. 5 Data division among CPU/MICs and Mining Schematic. The data division and CPU/MIC collaborating mining process in an iteration

achieve better performance than any other state-of-the-art miners with a small number of threads on DTP dataset. Also, the mining scale will drastically decrease with the increase in support threshold.

The experiments on S1 dataset is shown on Fig. 7a, which reflects the excellent parallel acceleration effect of cmFSM. We have set 1%, 2%, 3% and 4% as support thresholds respectively to form different scale of mining process. Basically, doubling the number of threads, the speedup is doubling too, which is close to the linear speedup. Also, the smaller the support threshold is, the larger the mining scale will be, where cmFSM presents a better parallel efficiency. This means it can be well applied to large-scale mining scenes.

Table 1 Tianhe-2 supercomputer Configuration

Hardware	Index
Node number	16,000
Computing node	2 Xeon E5 CPUs and 3 Xeon Phi coprocessors
CPU memory	64 GB
Xeon Phi memory	8 GB
Communication system	High-speed interconnection network
File system	Lustre file system
Operating system	Kylin 2.6.32–431.TH.x86_64

Multi-node division strategy

In order to compare the advantages and disadvantages of the five division strategies under different computing scenarios, we have experimented on DTP, PTE and S2 datasets, and set 4%, 2% and 1% as support thresholds respectively. Also, in order to eliminate the impact of multi-thread parallelization, we have only started one thread of each process. The Table 4 shows the results on DTP dataset.

From Table 4, it is not difficult to see that the runtime did not be significantly reduced among anyone of these five strategies with the number of process increases. This is because the first single-edge frequent subgraph will pick out 80% of the results on DTP dataset. Thus, the first process will become the bottleneck. Moreover, the

Table 2 Dataset information

Dataset	Molecules Or Graphs	Average #edges	Largest #edges	Average #vertices	Largest #vertices
PTE	340	28	214	27	214
DTP	422	42	196	40	188
S1	10,000	29	276	26	225
S2	20,000	32	214	30	197
S3	100,000	45	321	38	278

equality strategy is undoubtedly the worst strategy. The single and increment strategies are relatively faster, because the first single-edge frequent subgraph will be uniquely allocated to first process. Although the dynamic strategy also has such a division result, it has additional communication scheduling overhead, so that the performance is slightly lower than these two strategies. However, no matter how much the processes are started, the mining time of first single-edge frequent subgraph is always bottleneck on such a special dataset.

Figure 7b and c reflects the multi-node runtimes on PTE and S2 datasets. The condition that more than half of the results are picked out by the first single-edge frequent subgraph will not happen again on these normal datasets, such as PTE and S2.

From Fig. 7b, we can find the equality is also the worst strategy. The single and increment strategies present close performance. The circle strategy gradually shows the advantages of performance with the number of process increases. The dynamic strategy is the best at first, but with the increase in the number of process, its performance begin to slowly weaker than circle. This is due to the fact that the mining process of single-edge frequent graphs can be completed quickly on this dataset, so that as the number of process increases, there are frequent task requests and competitions. Coupled with the communication costs, the dynamic strategy shows weakness compared to circle. Fortunately, this is not an obvious weakness.

Table 3 Results for PTE dataset

MinSup %	2	4	6	8	10	20	30
MinSup Abs	7	14	20	27	34	68	102
Freq graphs	136,949	5935	2326	1323	844	190	68
Runtime in Seconds							
FSM	312.21	11.22	4.12	2.51	1.69	0.66	0.31
FFSM	78.12	5.21	2.01	1.03	0.75	0.58	0.29
gSpan	101.12	7.21	2.31	1.21	0.83	0.65	0.33
Gaston	36.53	2.13	1.01	0.66	0.38	0.12	0.06
cmFSM	97.32	3.28	1.25	0.71	0.49	0.15	0.08
cmFSM_2t	68.23	2.98	1.01	0.63	0.47	0.17	0.09
cmFSM_8t	21.51	1.03	0.42	0.27	0.21	0.08	0.07
cmFSM_32t	6.32	0.44	0.21	0.16	0.14	0.06	0.06

From Fig. 7c, we can find that the equality is still the worst strategy on average. At the beginning, single and increment are weaker than equality, which is because a large number of tasks are allocated to the final process when the number of process is small so as to form a bottleneck. Also, it is not difficult to find that the circle is an optimal choice among static strategies when the mining scale is big enough. The single and increment strategies still present close performance and the dynamic strategy is always better than all these static ones. Therefore, it is more recommended to use dynamic strategy, especially in the face of large-scale mining tasks.

Efficiency of multi-node multi-thread acceleration

In order to evaluate the efficiency of multi-node acceleration, we have experimented on S1, S2 and S3 datasets, and set 1%, 1% and 2% as support thresholds respectively.

The Table 5 shows the result of multi-node scalability. We should note that we have always adopted dynamic strategy in the following experiments because this division strategy can achieve an average best performance among all these five division strategies. By this way, we can simplify following tests.

Figure 7d indicates the comparison of parallel efficiency on different datasets which we can get from Table 5. we can easily find that the parallel efficiency will be better maintained with the increase in the number of cores for larger datasets, which also means cmFSM can be well applied to large-scale mining scenes.

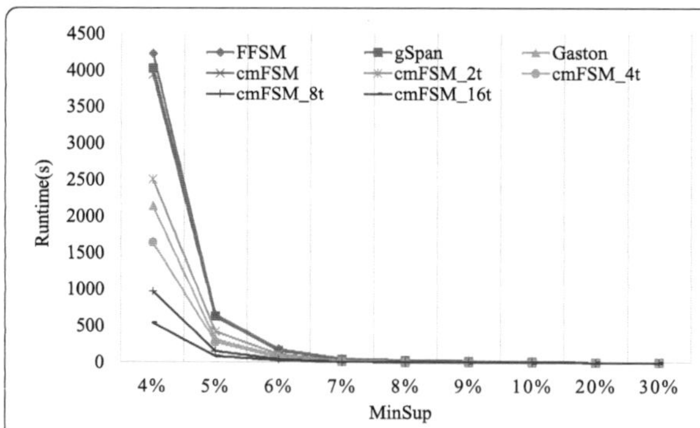

Fig. 6 Comparison of some miners for DTP dataset

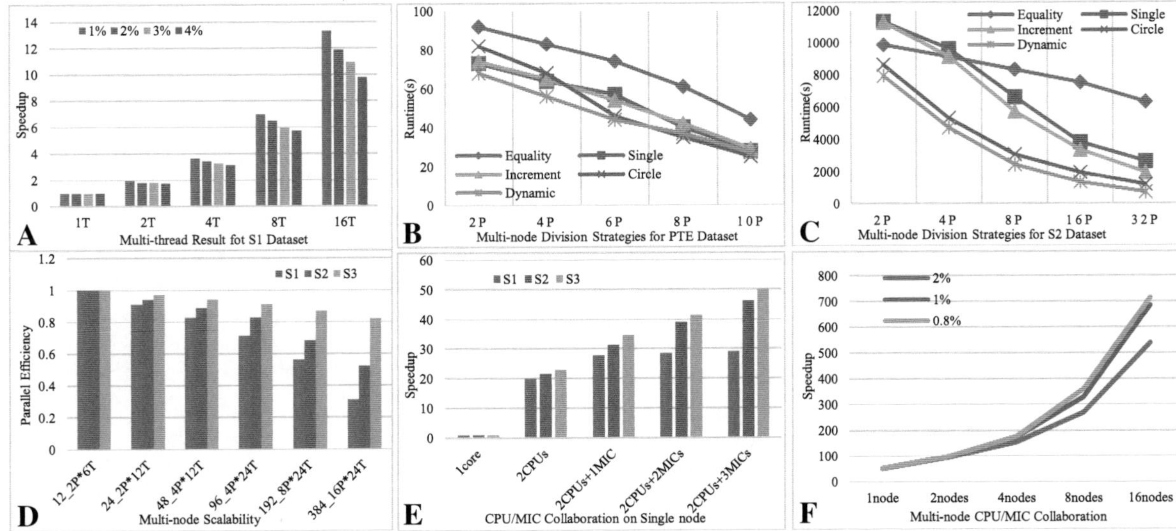

Fig. 7 Experiments of parallel optimization. There are six pictures labeled as **a**,**b**,**c**,**d**,**e** and **f** respectively. The "T" in coordinate axis means the number of thread we have started. The "P" in coordinate axis means the number of process we have started. The "1Core" means we have only started one process with single thread to carry out these experiments. The "2CPUs" means we have started one process with 24 threads to carry out these experiments. This is because there are two CPUs with 12 cores on single node of Tianhe-2 supercomputer

CPU/MIC collaboration

We have also used the last three datasets with 1%, 1% and 2% as support thresholds respectively to evaluate the effect of CPU/MIC collaboration on single node. The following Table 6 shows the results. In general, the computing capacity of 2 CPUs should be close to 3 MICs on single node. Therefore, we expect to be 24 times faster with 2 CPUs and achieve 48 times speedup with 2 CPUs and 3 MICs to make full use of computing resources on single node.

Figure 7e reflects the comparison of speedup on different datasets with different CPU/MIC collaboration modes which we can get from Table 6. We can easily find that the better accelerating effect can be achieved with larger scale mining tasks. Moreover, we obtained more than 50 times speedup finally on S3 dataset, which is better than we expect to get. This is because of a series of optimization means we adopt, such as memory reuse, data transmission optimization and vectorization. In addition, the experiment on S1 dataset quickly reached the bottleneck and there is no obvious difference between 2 MICs mode and 3 MICs mode. This should also be caused by the characteristics of the dataset itself. On this dataset, a large number of tasks are always concentrated in the Host and first MIC coprocessor. However, the accelerating effect of CPU/MIC is still nice in most large-scale mining scenarios.

Multi-node CPU/MIC collaboration

We have employed the biggest dataset S3 with 2%, 1% and 0.8% as support thresholds respectively to evaluate the effect of CPU/MIC collaboration. The following Table 7 shows the results where each node have made full use of 2 CPUs and 3 MICs.

Figure 7f reflects the comparison of multi-node speedup on S3 with different support thresholds which we can get from Table 7. We can also easily find that the better multi-node accelerating effect can be achieved with larger scale mining tasks. In overall, all these experiments present weaker and weaker speedup than linear speedup, which is mainly caused by multi-node communication, process competition and synchronization. However, this condition does not affect the excellent scalability of cmFSM under large-scale mining tasks.

Table 4 Multi-node results for DTP dataset

Process number	Equality	Single	Increment	Circle	Dynamic
2	3646	3183	3192	3532	3203
3	3584	3174	3185	3493	3213
4	3567	3178	3191	3468	3208
5	3552	3185	3188	3452	3216

Table 5 Multi-node scalability

	12Core 2P*6 T	24Core 2P*12 T	48Core 4P*12 T	96Core 4P*24 T	192Core 8P*24 T	384Core 16P*24 T
S1	742	408	223	131	83	74
S2	1732	921	487	261	161	104
S3	22,821	11,763	6069	3135	1640	871

Table 6 CPU/MIC collaboration on single node

	1Core	2CPUs	2CPUs &1MIC	2CPUs &2MICs	2CPUs &3MICs
S1	7590	382	274	234	208
S2	18,944	875	607	486	410
S3	266,794	11,625	7749	6457	5320

Conclusions

cmFSM is a scalable parallel frequent subgraph mining tool using CPUs and MICs in collaboration. It realizes parallelism of multiple levels and multiple granularities. We first adopt a fine-grained parallel strategy among threads by translating the common recursive mining process into a BFS loop mining process on single node. In addition to some special datasets, cmFSM can obtain near-linear speedup. Second, a coarse-grained parallel strategy is used among processes by dividing single-edge frequent subgraphs. Four kinds of static task dividing (i.e. equality, single, increment, circle) strategies and a supervisor-based dynamic task dividing strategy are implemented to achieve load balancing as much as possible. Some experiments show the dynamic strategy mostly presents better performance than all these static ones, especially in the face of large-scale mining tasks. Also, combining with the multi-thread work on single node, our implementation allows to generate a second level of parallelization by creating multiple threads per MPI process, which shows a promising scalability of massive mining tasks. Third, the collaboration among CPUs and MICs we employ a medium-grained parallel strategy by dividing the double-edge frequent subgraphs which is gotten from one-edge growth of initial single-edge frequent subgraphs. We also back up complex data structures redundantly to avoid the bottlenecks caused by excessive transmission. By memory reuse and making full of the multi-core computing capacity of MIC, we can obtained more than 50 times speedup on single node for some datasets. Also, the multi-node CPU/MIC collaboration presents an excellent scalability under large-scale mining tasks.

Moreover, Experimental evaluation results on several real compound molecular datasets show that cmFSM clearly outperforms the existing state-of-the-art miners even if we only hold a few parallel computing resources, which sufficiently demonstrates the effectiveness of our tool in the field of bioinformatics. However, on some special datasets, which will concentrate most of the mining tasks on a few front single-edge subgraphs, cmFSM will show a great limitation. In this scenario, it will quickly reach the bottleneck, which needs to further our work to solve in the future.

Table 7 Multi-node CPU/MIC collaboration

MinSup	1node	2nodes	4nodes	8nodes
2%	5320	2816	1736	993
1%	18,329	9802	5508	2918
0.8%	43,253	23,026	12,847	6295

Abbreviations

FSM: Frequent subgraphs mining; GPU: Graphics Processing Unit; MIC: Many Integrated Core

Acknowledgments

We would like to thank Xiaoyu Zhang, Kaiwen Huang and Wenhe Su for helpful discussions, comments and solidarity.

Funding

The authors would like to acknowledge that the publication of this article was sponsored by National Key R&D Program of China 2017YFB0202600, 2016YFC1302500, 2016YFB0200400, 2017YFB0202104; NSFC Grants 61772543, U1435222, 61625202, 61272056; and Guangdong Provincial Department of Science and Technology under grant No. 2016B090918122, National Natural Science Foundation of China (Grant Nos 81402760) and the Natural Science Foundation of Jiangsu Province, China (Grant No. BK20161563).

Availability of data and materials

Not applicable.

About this supplement

This article has been published as part of *BMC Bioinformatics* Volume 19 Supplement 4, 2018: Selected articles from the 16th Asia Pacific Bioinformatics Conference (APBC 2018): bioinformatics. The full contents of the supplement are available online at https://bmcbioinformatics.biomedcentral.com/articles/supplements/volume-19-supplement-4.

Authors' contributions

SY designed the method, wrote the software, do some experiments and wrote the paper. RG conceived the study, wrote the software and wrote the paper. XL and QZ give important comments on algorithm design and writing. SP designed the method and algorithm. All of the authors have read and approve of the final manuscript.

Ethics approval and consent to participate

Not applicable.

Consent for publication

Not applicable.

Competing interests

The authors declare that they have no competing interests.

Publisher's Note

Author details

[1]College of Computer Science and Electronic Engineering & National Supercomputer Centre in Changsha, Hunan University, Changsha 410082, China. [2]School of Computer Science, National University of Defense Technology, Changsha 410073, China. [3]Department of Oncology, The Second Xiangya Hospital of Central South University, Changsha 410011, China. [4]School of Computer Science and Technology, Tianjin University, Tianjin 300072, China. [5]School of Cyberspace, Hangzhou Dianzi University, Hangzhou 310018, China.

Published: 8 May 2018

References

1. Mckay BD. Practical graph isomorphism. J Symb Comput. 2013;60(1):94–112.
2. Lin W. Efficient techniques for subgraph mining and query processing. Singapore: School of Computer Engineering, Nanyang Technological University; 2015.

3. Huan J, Wang W, Prins J, et al. SPIN: mining maximal frequent subgraphs from graph databases. Tenth ACM SIGKDD International Conference on Knowledge Discovery and Data Mining. Seattle, Washington, Usa: August, DBLP; 2004. p. 581–6.
4. Jiang X, Xiong H, Wang C, et al. Mining globally distributed frequent subgraphs in a single labeled graph. Data Knowl Eng. 2009;68(10):1034–58.
5. Kuramochi M, Karypis G. Finding frequent patterns in a large sparse graph. Data Min Knowl Disc. 2005;11(3):243–71.
6. Kang U, Tsourakakis CE, Faloutsos C. PEGASUS: mining peta-scale graphs. Knowl Inf Syst. 2011;27(2):303–25.
7. Reinhardt S, Karypis G. A Multi-Level Parallel Implementation of a Program for Finding Frequent Patterns in a Large Sparse Graph. Parallel and Distributed Processing Symposium, 2007. IPDPS 2007. IEEE international. IEEE; 2007. p. 1–8.
8. Wu B, Bai YL. An efficient distributed subgraph mining algorithm in extreme large graphs. Artificial Intelligence and Computational Intelligence - International Conference, Aici 2010, DBLP, Sanya, China, 2010:107–115 October 23–24, 2010, Proceedings, .
9. Yan Y, Dong Y, He X, et al. FSMBUS: a frequent subgraph mining algorithm in single large-scale graph using spark. J Comput Res Dev. 2015;52(8):1768–83.
10. Lin W, Xiao X, Xie X, et al. Network motif discovery: a GPU approach. IEEE, International Conference on Data Engineering IEEE. 2015. p. 831–42.
11. Hill S, Srichandan B, Sunderraman R. An iterative MapReduce approach to frequent subgraph mining in biological datasets. ACM Conference on Bioinformatics, Computational Biology and Biomedicine ACM. 2012. p. 661–6.
12. Inokuchi A, Washio T, Motoda H. An Apriori-based algorithm for mining frequent substructures from graph data. European conference on principles of data mining and knowledge discovery, London, UK. 2000;1910(1):13–23.
13. Kuramochi M, Karypis G. Frequent subgraph discovery. IEEE International Conference on Data Mining IEEE Xplore. 2001:313–20.
14. Meinl T, Fischer I, Philippsen M. A quantitative comparison of the subgraph miners mofa, gspan, FFSM, and gaston. European Conference on Principles and Practice of Knowledge Discovery in Databases Springer-Verlag. 2005: 392–403.
15. Borgelt C, Berthold MR. Mining Molecular Fragments: Finding Relevant Substructures of Molecules. IEEE International Conference on Data Mining, 2002. ICDM 2002. Proceedings. IEEE Xplore. 2002:51–8.
16. Huan J, Wang W, Prins J. Efficient mining of frequent subgraphs in the presence of isomorphism. IEEE International Conference on Data Mining IEEE. 2003:549–52.
17. Yan X, Han J. gSpan: graph-based substructure pattern mining. IEEE International Conference on Data Mining, 2002. ICDM 2002. Proceedings. IEEE Xplore. 2002:721–4.
18. Nijssen S. A quickstart in frequent structure mining can make a difference. Tenth ACM SIGKDD International Conference on Knowledge Discovery and Data Mining. Seattle, Washington, Usa: August, DBLP; 2004. p. 647–52.
19. Yan X, Han J. CloseGraph: mining closed frequent graph patterns. ACM SIGKDD International Conference on Knowledge Discovery and Data Mining: Washington, Dc, Usa, August DBLP; 2003. p. 286–95.
20. Maunz A, Helma C, Kramer S. Large-scale graph mining using backbone refinement classes. ACM SIGKDD International Conference on Knowledge Discovery and Data Mining, Paris, France, DBLP, 2009:617–626 June 28 - July.
21. Cook DJ, Holder LB, Galal G, et al. Approaches to parallel graph-based knowledge discovery. J Parallel & Distributed Comput. 2001;61(3):427–46.
22. Buehrer G, Parthasarathy S, Chen YK. Adaptive parallel graph mining for CMP architectures. International Conference on Data Mining IEEE Computer Society. 2006:97–106.
23. Wang C, Wang W, Pei J, et al. Scalable mining of large disk-based graph databases. Tenth ACM SIGKDD International Conference on Knowledge Discovery and Data Mining. Seattle, Washington, Usa: August DBLP; 2004. p. 316–25.
24. Nguyen SN, Orlowska ME, Li X. Graph mining based on a data partitioning approach. Nineteenth Conference on Australasian Database: Australian Computer Society, Inc; 2008. p. 31–7.
25. Dean J, Ghemawat S. MapReduce: simplified data processing on large clusters. Conference on Symposium on Opearting Systems Design & Implementation; 2004. p. 107–13.
26. Bhuiyan MA, Al HM. An iterative MapReduce based frequent subgraph mining algorithm. IEEE Transactions on Knowledge & Data Engineering. 2013;27(3):608–20.
27. Lu W, Chen G, Tung AKH, et al. Efficiently extracting frequent subgraphs using MapReduce. IEEE International Conference on Big Data IEEE. 2013: 639–47.
28. Lin W, Xiao X, Ghinita G. Large-scale frequent subgraph mining in MapReduce. IEEE, International Conference on Data Engineering. IEEE. 2014:844–55.

Takenaka *et al. BMC Bioinformatics* 2018, **19**(Suppl 4):89
https://doi.org/10.1186/s12859-018-2072-y

BMC Bioinformatics

RESEARCH Open Access

Automated transition analysis of activated gene regulation during diauxic nutrient shift in *Escherichia coli* and adipocyte differentiation in mouse cells

Yoichi Takenaka[1,2,3*], Kazuma Mikami[4], Shigeto Seno[2] and Hideo Matsuda[2]

From The Sixteenth Asia Pacific Bioinformatics Conference
Yokohama, Japan. 15-17 January 2018

Abstract

Background: Comprehensively understanding the dynamics of biological systems is among the biggest current challenges in biology and medicine. To acquire this understanding, researchers have measured the time-series expression profiles of cell lines of various organisms. Biological technologies have also drastically improved, providing a huge amount of information with support from bioinformatics and systems biology. However, the transitions between the activation and inactivation of gene regulations, at the temporal resolution of single time points, are difficult to extract from time-course gene expression profiles.

Results: Our proposed method reports the activation period of each gene regulation from gene expression profiles and a gene regulatory network. The correctness and effectiveness of the method were validated by analyzing the diauxic shift from glucose to lactose in *Escherichia coli*. The method completely detected the three periods of the shift; 1) consumption of glucose as nutrient source, 2) the period of seeking another nutrient source and 3) consumption of lactose as nutrient source. We then applied the method to mouse adipocyte differentiation data. Cell differentiation into adipocytes is known to involve two waves of the gene regulation cascade, and sub-waves are predicted. From the gene expression profiles of the cell differentiation process from ES to adipose cells (62 time points), our method acquired four periods; three periods covering the two known waves of the cascade, and a final period of gene regulations when the differentiation to adipocytes was completed.

Conclusions: Our proposed method identifies the transitions of gene regulations from time-series gene expression profiles. Dynamic analyses are essential for deep understanding of biological systems and for identifying the causes of the onset of diseases such as diabetes and osteoporosis. The proposed method can greatly contribute to the progress of biology and medicine.

Keywords: Gene regulatory network, Network dynamics, Time course, Cell differentiation, Adipocyte

*Correspondence: takenaka@kansai-u.ac.jp
[1]Faculty of Informatics, Kansai University, Ryousenji 2-1-1, Takatsuki, Osaka, Japan
[2]Graduate School of Information Science and Technology, Osaka University, Yamadaoka 1-5, Suita, Osaka, Japan
Full list of author information is available at the end of the article

Background

Acquiring a comprehensive understanding of biological systems dynamics is among the biggest challenges in biology and medicine. The processes of all organisms (including humans) are time-variant. Cells cycle and divide, change their internal states and differentiate into other cell types when stimulated from the outside. Abnormal cell differentiation causes diseases and cancer. Dynamic systems are universal in organisms.

Biological systems dynamics can now be understood through advanced biological technologies. On-chip hybridization and DNA sequencing has enabled the simultaneous measurement of the expression levels of many genes. Many of the accumulated data are registered in public databases such as GEO (Gene Expression Omnibus) [1, 2], Array Express [3] and Sequence Read Archive [4]. Special-purpose databases are also accessible. The references of mammalian tissue gene expressions are stored in RefEx (Reference Expression Dataset) [5], whereas the data from human tissues, cell lines and cancers are distributed through the Human Protein Atlas [6]. Mouse data are archived in the specialized database GXD (Gene expression database) of the Mouse Genome Informatics resource [7]. These databases are freely accessible to all users.

The above databases store the time-course movements of the gene expression levels in multiple organisms. For example, a search for "*time-course*" (2017/7/12) returned 913 datasets in GEO and 7623 entries in SRA. Based on these data, researchers have revealed the biological mechanisms of processes such as cell cycling [8] and diauxic shifts [9]. The adipogeneses of mice and humans have also been compared using time-course profiles [10]. These studies cover a broad range, from biology, through medicine, to pharmacology.

Numerous systems operate in living cells. Examples are metabolism, genetic and environmental information processing, cellular processes, and organismal systems [11]. Concurrent functioning of these systems in vivo maintains the life activities of the cell. Systems have been extensively researched and summarized as pathways in the KEGG (Kyoto Encyclopedia of Genes and Genomes) database [11]. Control of gene expression levels is another life-sustaining system. Gene regulation defines the relation between a controlling agent and its target gene. The gene regulatory system can be represented as a network of nodes (genes) connected by directed edges (regulation of a downstream gene by an upstream gene). The gene regulatory networks of model organisms are available in databases such as RegulonDB [12, 13], which contains the nearly-complete gene regulation of *Escherichia coli*, and the yeast network databases YEASTRACT [14–17]. The FANTOM projects have attempted to exhaustively reveal the relations between mouse and human in many tissues [18–22]. Gene regulatory networks are also provided in commercial databases such as TransFac [23, 24] and Quiagen's Ingenuity Pathway Analysis [25]. The regulatory relations between genes, at least in the model organisms, will be revealed exhaustively in the near future. However, although databases provide numerous gene regulations, the relationships are basically static. Each database entry supplies a gene regulatory relation and the experimental environment in which the relation was activated. Very few entries include the chronological order of the activations, despite our increasing knowledge of signaling pathway cascades. When the chronological order is given, the experimental environment includes the cell cycles and diauxic shift of the nutrient source. Adipocyte differentiation is accompanied by a rough sketch of the transition of active expression controls [26] (see Fig. 1). The differentiation involves a single transition point and two waves of activated gene regulations, although more than two waves have been suggested [26]. The dynamics of gene-expression control is a new research topic, with many unknowns at present.

Gene regulation activity can be dynamically analyzed from gene expression profiles. Given time-course data, researchers have sought the transformations in the gene regulatory network through models, such as ordinary differential equations and graphical models. Ordinary differential equations can explain the activity level of a gene regulation. The explanatory and objective variables are the expression levels of the upstream genes and the downstream genes, respectively [27–29]. Graphical models describe the control relations in the gene expression as a network of nodes (genes) with directed edges (controls). The dynamics of biological systems can be analyzed by a graphical Gaussian model with a Markov property [30, 31], which measures the independence of any two genes using partial correlations.

Other graphical models are the gene relevance network and the Bayesian network. The former generates a gene regulatory network from the mutual information, and measures the degree of association between gene expression levels [32–34]. This model can also manage time-course gene expression profiles [35].

In the Bayesian network, a directed edge represents the probabilistic relationship between two genes by information-theoretic approaches. The plausibility of the network is measured by the Bayesian information criterion and the minimum description length [36, 37]. Bayesian networks have been extended to time-course data by introducing Markov properties into the joint probability distribution. The time-dependent Bayesian network is called a dynamic Bayesian network [38, 39].

The above-described methods have analyzed not only the gene expression profiles, but also general data. Researchers have obtained meaningful results from all

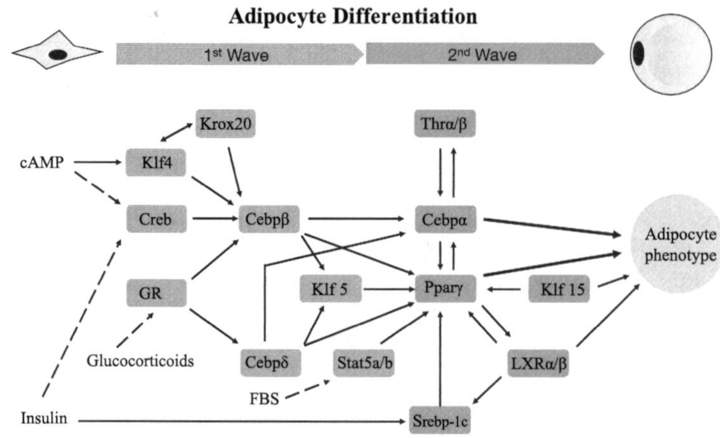

Fig. 1 The gene regulatory network of murine cell differentiation into adipocytes occurs in two waves (distinguished by different colors) [26]

data except time-course gene expression profiles, which are most severely limited by the small number of time points. Figure 2 shows the distribution of time points in the GEO database. There are 913 datasets with time-series expression profiles. The average, mean and standard deviation of the time points are 4.66, 4, and 2.82, respectively. The number of time points is far below the numbers of other common data. To clarify this fact, Table 1 lists the number of data points in benchmark problems for inferring causal relationship networks [37, 40–44]. All of these benchmarks are included in bnlearn of the R package [44]. Comparing Fig. 2 and Table 1, we observe that the number of time points is two orders of magnitude lower in the time-series expression profile than in the benchmarks. This data limitation not only prevents the estimation of correct relationships, but also stifles the dynamic analyses.

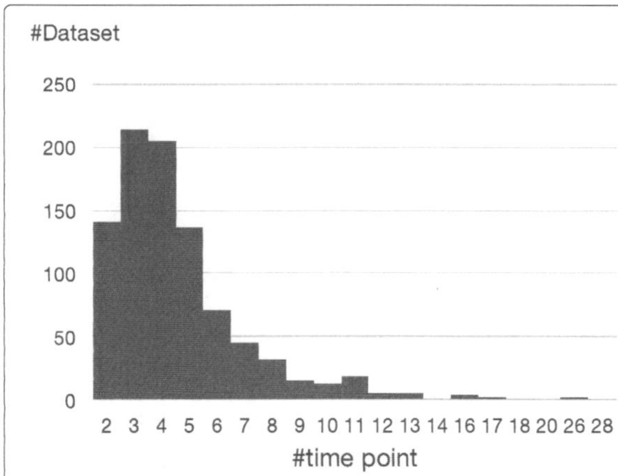

Fig. 2 Number of time points in datasets returned by a "time course or time-course" query in the Gene Expression Omnibus. The average, mean and standard deviation of the distribution are 4.66, 4 and 2.82, respectively

Takenaka et al. [45] proposed a method that captures the dynamic changes in gene expression control from time-course expression profiles with a small number of time points. Their method quantifies the intensity at which genes control the expression levels of other genes. The temporal resolution of intensity is one time point (the observation unit of the time-course gene expression profile). Single time-point resolution is finer than the resolutions of other methods [46–49]. Assuming a profile with N time points, the method generates N sub-profiles, each omitting one time point. The effectiveness of the method was confirmed in the diauxic shift of *E.coli* [9] and the adipocyte cell differentiation in mouse tissue [45, 50, 51]. The results detected the time of the diauxic shift in *E. coli*, and two waves of active regulations during adipocyte cell differentiation (Fig. 3). The authors concluded that the method effectively reveals the dynamics of gene regulations. But the method does not have any procedure how to determine the number of breakpoints and when the breaks happen. Therefore, finding the breakpoint of active gene expression control by this approach is somewhat subjective, and is undeniably influenced by already known biological findings. In this research, we propose a mechanical method that eliminates this subjectivity.

Table 1 Numbers of samples and variables in benchmark problems for inferring causal relationship networks

Benchmark	#Samples	#Variables
Learning.test	5000	6
Cigaussian.test	5000	8
Asia	5000	8
Alarm	20000	37
Insurance	20000	27
Hailfinder	20000	56

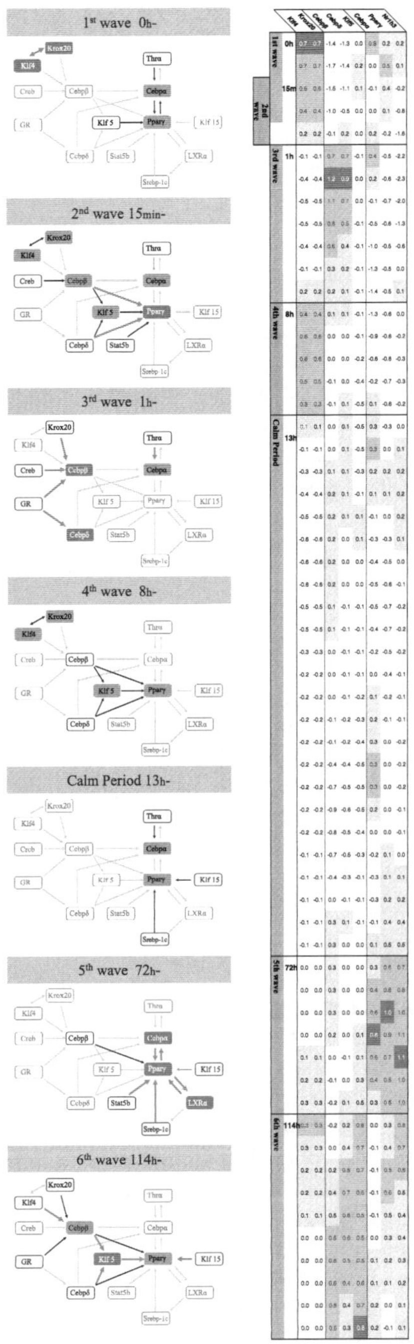

Fig. 3 Six waves and one calm period during adipocyte cell differentiation. The tables show the strengths of the gene regulations calculated by [45]. Eight genes are regulated by other genes, and the strength changes between 0 and 196 h (white and red cells denote low and high strengths of the gene regulation, respectively). The period of each wave is also shown in the table. The left side of the figure presents the active gene regulations in each wave. The waves before and after the calm period correspond to the two waves reported in [26]. This analysis was performed by Takenaka and presented at Recomb 2016 [51]

Method

We propose a method that analyzes the dynamics of gene regulations from time-series gene expression profiles. Here, the dynamics of gene regulations take two meanings. The first interpretation considers the dynamics of each gene regulation; whether the control of gene expression level by other genes is activated or inactivated at each time point in the time-series data. The second considers the dynamics of the gene regulatory network; the periods and pairs of their start and end times. The periods (sometimes called states, waves, or phases) are often biologically meaningful. The most well-known period is the cell cycle, which has two states; interphase and cell division. The interphase is divided into three sub-states called phases (Gap1, Synthesis, and Gap2), each with sub-phases. The cell cycle has four phases in total. Under these dynamics, the number of periods in a cell cycle is not fixed. The same situation occurs in other biological processes. Therefore, determining the number of periods in a biological process is a difficult task, and the dynamics of the gene regulatory network are difficult to analyze, especially in a manual analysis.

Notations

To analyze the dynamics of gene regulations in the two interpretations by an automated approach, we formulate the analysis as an optimization problem. The notations of the formulation are listed below.

Exp : time-course gene expression profiles with $|T|$ time points and $|G|$ genes.

T : a set of time points in *Exp*. If $i < j$, then $t_i \in T$ is earlier than $t_j \in T$.

G : a set of genes in *Exp*.

M : a two-dimensional matrix $|G| \times |T|$, where M_{ij} represents the intensity at which the expression of gene $g_i \in G$ is controlled by others at time $t_j \in T$. M is calculated from *Exp* by the method proposed in [45].

In our method, the activity of gene regulation is represented by a binary number, with 1 and 0 denoting active and inactive, respectively. The notations of the gene-regulation dynamics are explained below.

a_g : a binary vector of length $|T|$ representing the activity of gene $g \in G$ at time $t_j \in T$.

Act : a binary matrix of size $|G| \times |T|$, where Act_i equals a_{g_i}.

Tra : a subset of $|T|$ representing the breaks between periods. For example, if $Tra = \{t_3\}$, the system has two periods; the first ending at t_3 and the second starting at t_4. If *Tra* is an empty set, no transition of the gene regulation occurs throughout the period.

Score function

As previously mentioned, we analyze the dynamics of gene regulations by solving the optimization problem; that is, by finding the most plausible dynamic model of the gene regulations. This model is conditional on the activity of each gene and the model complexity. Regarding the first condition, a positive M_{ij} indicates high likelihood of activating the gene g_i at time point t_j. The second conditional expression describes the penalty term of the Bayesian information criterion. Here, the complexity of the model is defined as the number of periods in the model, which deals with the trade-off between the goodness of fit to the model and the tractability of the model.

To find a plausible model for M, we must optimize the following score:

$$\text{Score}(Act) = \sum_{i=1}^{|G|}\sum_{j=1}^{|T|} M_{ij}\cdot Act_{ij} - |Tra|\cdot\ln(|G|\cdot|T|), \quad (1)$$

where Tra is calculated from Act. A time point $t_i \in T$ is a member of Tra if $\exists j Act_{ij} \neq Act_{i(j+1)}$.

Algorithm

The algorithm that finds the most plausible dynamic model of gene regulation is described below.

Input a Matrix M calculated from the gene expression profile and a gene regulatory network by the method in [45].
Output a Matrix Act, a set Tra, and a $score$.
Measure Maximize the score function Score(Act).
Variables

k: The integer k represents the number of breaks between the periods in the model.

V: A set of binary vectors of length $|T|$. Each element of V represents the activity of a gene regulation.

step 1 Initialize $k \leftarrow 0$ and $Tra \leftarrow \{\}$
step 2 Generate a set of vectors V by the following sub-steps.

(1) add a zero vector to V
(2) add a one vector to V

step 3 $score_0 \leftarrow 0$
step 4 for each $g_i \in G$, select a vector $v \in V$ that maximizes the g_i score in the following sub-steps.

(1) select a vector v in V that maximizes $M_i \cdot v^t$, where M_i is the i-th row of M.
(2) $score_0 \mathrel{+}= M_i \cdot v^t$
(3) insert vector v into the i-th row of Act; $Act_i \leftarrow v$.

step 5 $score_{best} \leftarrow score_0$; $Act_{best} \leftarrow Act$; $Tra_{best} \leftarrow Tra$

step 6 $k \mathrel{+}= 1$; If $k = |T|$, goto **Step 9** .
step 7 for each $Tra = \{t|t \in T\}$, where $|Tra| = k$,

step 7-1 generate a set of vectors V as follows:

(1) for each period i of Tra, generate a set V_i with two elements; a zero vector and a unit vector with lengths equal to that of the period.
(2) generate V as the direct product of all V_i.

step 7-2 $score_k \leftarrow -2 \cdot k \cdot \ln(|T| \cdot |G|)$
step 7-3 for each $g_i \in G$, select a vector $v \in V$ that maximizes the score by the following sub-steps.

(1) select a vector v in V that maximizes $M_i \cdot v^t$, where M_i is the i-th row of M.
(2) $score_k \mathrel{+}= M_i \cdot v^t$
(3) insert vector v into the i-th column of Act; $Act_i \leftarrow v$.

step 7-4 if $score_k > score_{best}$, $score_{best} \leftarrow score_k$; $Act_{best} \leftarrow Act$; $Tra_{best} \leftarrow Tra$.

step 8 if $score_k > score_{k-1}$, goto **Step 6** .
step 9 output Act_{best}, Tra_{best} and $score_{best}$.

The above algorithm is a brute-force search with poor computational efficiency. However, as the gene expression profile contains very few time points, the above algorithm satisfies our requirements.

For larger datasets, the calculation time can be shortened by using a hash function from a gene and a period to the sub-score. Meanwhile, the computational complexity can be reduced by dynamic programming.

Results and discussion

The effectiveness of the proposed method was tested on two biological datasets. The results of our method were also compared with a manual analysis of the periods in the time series. The first dataset contains the time-course gene expression profiles during the diauxic shift from glucose to lactose metabolism in *E. coli*. In this test, the method was assessed by its ability to detect the time of the diauxic shift.

The second dataset reports the adipocyte cell differentiation in the house mouse *Mus musculus*. Two periods (called waves in [26]) of gene regulation have been identified during this differentiation, and more waves are expected. This test evaluated whether the method can detect three or more plausible periods.

Diauxic shift

Material

The first dataset is the time-course gene expression profile GSE7265 in the GEO dataset. The diauxic shift was observed in *E. coli* MG1655 and isogenetic mutants cultured in a medium containing glucose and lactose [9]. The expression levels of the genes were measured in oligonucleotide microarrays containing 70-base oligonucleotide probes. The wild-type profile contains 17 time points from 780 min. to 1089 min. after inoculation. At each time point, the triplicated expression values were averaged to obtain the representative expression level.

The time points are divided into two phases; the first lasting from 780 min. to 939 min., the second from 969 *min.to* 1089 min. The diauxic shift from glucose to lactose has been recognized in the first half, but information on the second half is completely lacking. Although the GS2765 gene expression profile contains 17 time points, the original paper [9] trimmed the profile to 10 time points (830-939 min. post-inoculation), and reported the growth ratio in the first half only. Therefore, the full dynamics of the diauxic shift can only be surmised.

The gene regulatory network of the diauxic shift comprises 31 genes and 50 gene regulations [45]. Fourteen enzymes are related to the glycolytic and lactose metabolic pathways. According to RegulonDB [12, 13], the expression levels of the enzymes are controlled by 14 transcription factors (TFs), and 50 TF–enzyme interactions are known. The present study adopts the gene list and the gene regulatory network used in [45].

The proposed method was implemented in R, and the matrix M representing the intensities of the gene regulations was calculated by an R package from GitHub https://github.com/takenakayoichi/tacs. The computed matrix M was input to the proposed method.

Results

The proposed method detected five periods during the diauxic shift of wild type *E. coli*. The maximum score during each period is tabulated in Table 2. Although the proposed algorithm identified 6 periods, the maximum score was recorded over 8 periods for reference. The score peaks during Period 5 and the between-period breakpoints are 878 min., 908 min., 999 min. and 1049 min. The first, second and third periods belong to the first half of the diauxic shift; the remainder belong to the second half. The periods are also indicated by the vertical lines in Fig. 4.

Table 2 Maximum scores in each period of diauxic shift in *E. coli* (central column) and differentiation to murine adipocytes (right column) (maximum scores are in bold)

#Period	Diauxic shift	Adipocyte
1	16.42	1.50E-13
2	23.57	50.38
3	26.23	70.30
4	28.00	**74.76**
5	**28.44**	73.41
6	26.81	72.02
7	25.02	69.25
8	23.14	67.87

Takenaka et al. [45] manually detected three periods in the GS2765 gene expression profile, with breakpoints at approximately 888 − 908 min. and 1070 − 1089 min. The breakpoints were expressed as ranges because they were difficult to pinpoint by the manual approach. In their analysis, the first and second periods belonged to the first half of the diauxic shift, and the third period belonged to the second half.

First half

Our method divided the first half of the experiment into three periods. Figure 4 shows the growth ratios in each period. During the first and third periods, the *E. coli* exist in the logarithmic growth phase and are consuming glucose and lactose, respectively. The second period is consistent with cessation of cell proliferation.

In comparison, the manual curation identified only two periods in the first half. The division point was exactly centered in the growth arrest period. From the difference between our results and the manually curated results, we can identify whether a mechanism that regulates the gene expression level exists independently in the logarithmic and growth-arrest phases.

Which of these methods expresses more realistic biological properties? Manual curation suggests that lactose consumption begins immediately after glucose depletion. On the other hand, the proposed method suggests that once the glucose is depleted, the cells cease growing while seeking a new nutrient source, and resume growth when an alternative nutrient is found. We consider that the proposed method presents a more plausible dynamic model than manual curation. One expects that the *E. coli* cells cannot immediately access a new nutrition source once the original source is depleted. If this were possible, logarithmic growth would continue without interruption.

Second half

The second half of the experiment was continuous in the manual curation, but divided into two periods in the proposed method. Once the *E. coli* had consumed the glucose and lactose in their nutrient medium, they began seeking the next nutrition source. When no further source was found, the gene expression regulations changed sufficiently for detection by both methods. However, as the expression profile provides the only available information in the second half, the actual number of periods in the second half of the experiment is difficult to judge.

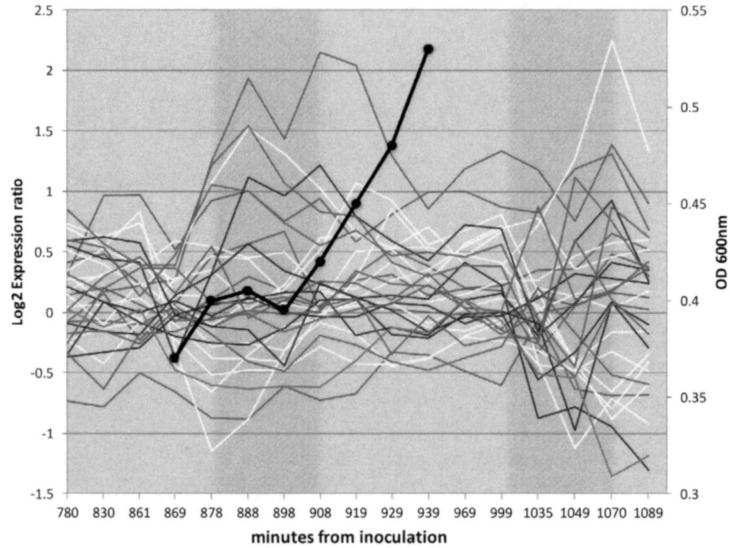

Fig. 4 Expression levels of 31 genes (colored plots) and growth ratio (solid black line) during diauxic shift in *E. coli*. Vertical lines delineate the periods of the dynamic model detected by the proposed method

Adipocyte differentiation

Materials

The second dataset was a time-course gene expression profile of mouse adipocyte differentiation. The RNAs in this dataset were collected from mouse ST2 bone marrow stroma cell-derived stem cells (RCB0224) obtained from the RIKEN BioResource Center (BRC, Tsukuba, Japan). ST2 cells were induced by changing the medium from RMPI1640 to DMEM supplemented with 10% FBS, 0.5 mM 3-isobutyl-1-methylxanthine (MIX), 0.25 μM DEX, and with insulin-transferrin-selenium-X supplement containing 5μg/ml of insulin and 1 μM rosiglitazone. After 48 hours, the differentiation medium was replaced with DMEM supplemented with 10% FBS. The RNAs collected at 62 time points during adipocyte differentiation were measured using an Affymetrix GeneChip Mouse Genome 430 2.0 Array. The time points were 0, 5, 15, 30 and 45 min., hourly from 1 to 30 h, and 6-hourly from 36 to 192 h after adipogenesis induction. Each datum was background-subtracted and normalized in the robust

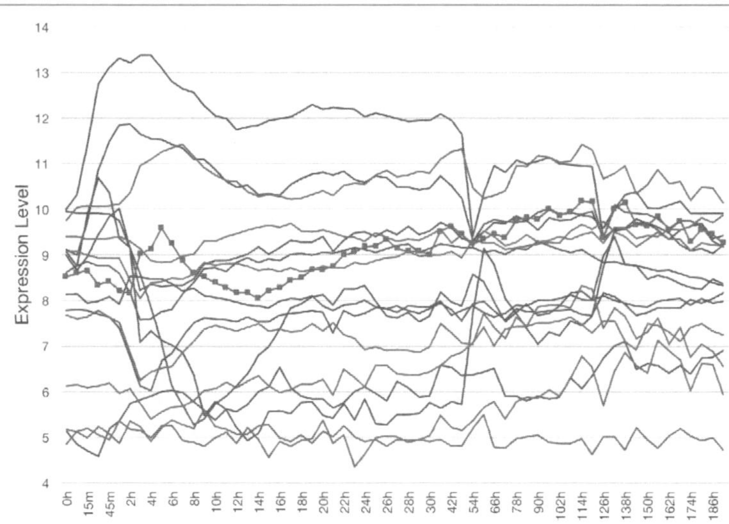

Fig. 5 Expression Levels of 14 genes during murine adipocyte differentiation. Blue and red lines indicate the expression levels of genes belonging to the first and second waves in Fig. 1, respectively. The red line with square markers plots the expression levels of pparg, a marker gene of adipose cells

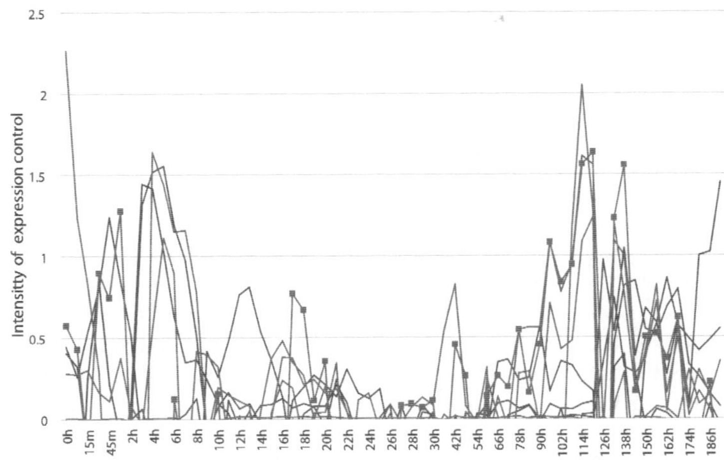

Fig. 6 Intensity of genes whose expression levels are controlled by other genes. Blue and red lines show the intensities of genes belonging to the first and second waves in Fig. 1, respectively. The red line with square markers plots the intensity of pparg, a marker gene of adipose cells

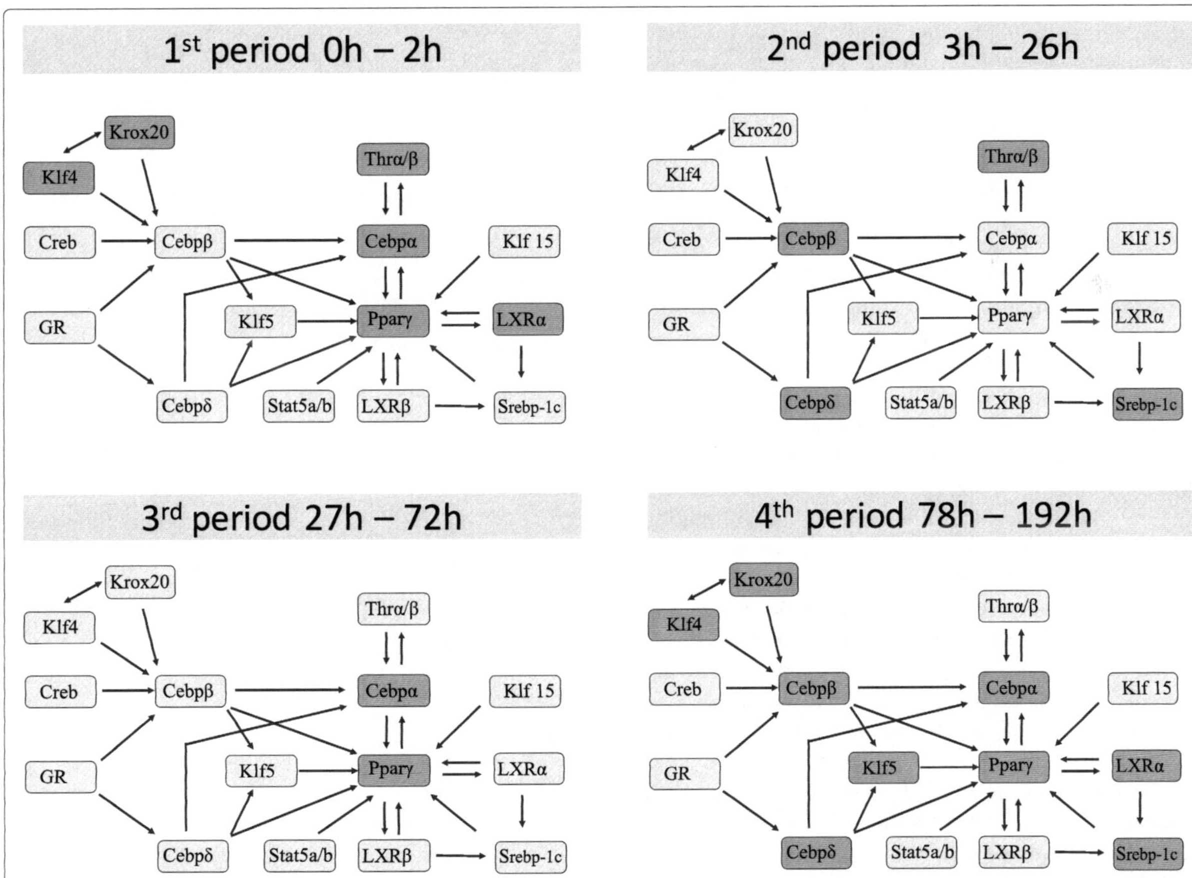

Fig. 7 The four periods of differentiation into murine adipocytes detected by the proposed method. Dark- and light-shaded boxes enclose genes whose expression level is controlled and not controlled by upstream genes, respectively

multi-array analysis (RMA). The gene expression profiles are available from the Genome Network Platform [50].

Previous research has unveiled the important regulators of adipocyte development and a two-wave cascade in this process [26] (see Fig. 1). The network includes 24 regulations between 14 genes, but undetected sub-waves have been predicted [26].

The expression levels of the 14 genes are shown in Fig. 5. Among them, the line with square indicates pparg, a known marker of adipocyte cells. The expression level of pparg increases from 2 to 6 h and then decreases. The decline is followed by a gradual rise after 14 h.

Figure 6 shows the intensities of the expression levels controlled by other genes, which constitute the intensity matrix M of our algorithm. The intensities were calculated from the gene expression profiles in Fig. 5 and the gene regulatory network in Fig. 1.

Result

The proposed method detected four periods of adipocyte differentiation. The maximum score in each period is listed in Table 2. The algorithm terminated after five periods, but 8 periods are included for referencing purposes. The score peaks during period 4, and the between-period breakpoints are 2 h, 26 h and 72 h. Figure 7 shows the dynamics of the gene regulatory network. In the figure, we classify the gene whose expression level during a period is controlled by upstream genes or not by the average of the intensity during the period is larger than zero or not.

The two waves of the known cascade in Fig. 1 appear in the four periods detected by our algorithm. During the first period of adipocyte differentiation, the expression controls of the most upstream genes (Klf4 and Knox20) are activated. During the second period, the expressions of Cebpβ and Cebpδ are controlled and the regulations of the second wave (Thrα/β and Srebp-1c) begin activating. The pparg and cebpα genes are regulated during the third period, and the most downstream genes are regulated during the fourth period. The four-period differentiation identified by our method includes the known waves, but appears to be more precise.

The four periods detected by our method differ from the two-wave differentiation in two ways. First, the regulations of Pparg and Cebpα are activated twice. Once at the first period, and re-activated in the second wave composed of third and fourth periods. Pparg is a marker of adipose cells and a known regulator of adipocyte differentiation. The two activations of the pparg gene are confirmed in Fig. 5. The expression level of Pparg peaks at 6 h and 114 h. Most of the gene expression regulations are activated in the fourth period. By the start time of this period (78 h), the cells will have produced lipid droplets [52]. Therefore, we consider that the differentiation into adipocytes was completed during the fourth period, and was stabilized thereafter. It appears that many regulations of the adipocyte-related gene expression levels are activated in the fourth period.

Conclusions

We proposed a method that detects the dynamics of gene regulatory networks at the temporal resolution of single time points in the underlying gene expression profiles. The dynamics are modeled as periods of activated regulations. The plausibility of the model was quantified using the Bayesian information criterion. The problem constitutes a combinatorial optimization problem that find the highest-scoring model. The algorithm inputs the gene expression profiles and a gene regulatory network, and returns the activated regulations, divided into periods.

The effectiveness of the method was validated in two datasets; the diauxic shift from glucose to lactose in *E. coli* and adipocyte differentiation in the mouse. In both datasets, the proposed method detected more plausible dynamic models than existing models. We believe that the proposed method can precisely reveal the dynamics of biological systems.

Acknowledgements
This work was partially supported by JSPS KAKENHI Grant Numbers 15K00402 and 16H03261.

Funding
Funding for the publication of this article was provided by JSPS KAKENHI Grant Number 15K00402.

Availability of data and materials
The E.coli data analyzed in this research was available as GSE7265 Gene Data Set in Gene Expression Omnibus. The mouse data analyzed in this research was downloaded from Genome Network Platform [50]. The code used in this study is available in https://github.com/takenakayoichi/tacs.

About this supplement
This article has been published as part of *BMC Bioinformatics* Volume 19 Supplement 4, 2018: Selected articles from the 16th Asia Pacific Bioinformatics Conference (APBC 2018): bioinformatics. The full contents of the supplement are available online at https://bmcbioinformatics.biomedcentral.com/articles/supplements/volume-19-supplement-4.

Authors' contributions
YT conceived and designed the algorithm and the experiments and wrote the manuscript. YT and KM implemented the algorithm and performed the analyses. SS and HM contributed by providing suggestions. All authors have read and approve the final manuscript.

Ethics approval and consent to participate
Not applicable.

Consent for publication
Not applicable.

Competing interests
The authors declare that they have no competing interests.

Publishers Note
Springer Nature remains neutral with regard to jurisdictional claims in published maps and institutional affiliations.

Author details
[1]Faculty of Informatics, Kansai University, Ryousenji 2-1-1, Takatsuki, Osaka, Japan. [2]Graduate School of Information Science and Technology, Osaka University, Yamadaoka 1-5, Suita, Osaka, Japan. [3]Graduate School of Medicine, Osaka University, Yamadaoka 2, Suita, Osaka, Japan. [4]Recruit Holdings Co. Ltd., Marunouchi 1-9-2, Chiyoda, Tokyo, Japan.

Published: 8 May 2018

References

1. Edgar R, Domrachev M, Lash AE. Gene expression omnibus: Ncbi gene expression and hybridization array data repository. Nucleic Acids Res. 2002;30(1):207–10. https://doi.org/10.1093/nar/30.1.207.
2. Barrett T, Wilhite SE, Ledoux P, Evangelista C, Kim IF, Tomashevsky M, Marshall KA, Phillippy KH, Sherman PM, Holko M, Yefanov A, Lee H, Zhang N, Robertson CL, Serova N, Davis S, Soboleva A. Ncbi geo: archive for functional genomics data sets-update. Nucleic Acids Res. 2013;41(D1):991–5. https://doi.org/10.1093/nar/gks1193.
3. Kolesnikov N, Hastings E, Keays M, Melnichuk O, Tang YA, Williams E, Dylag M, Kurbatova N, Brandizi M, Burdett T, Megy K, Pilicheva E, Rustici G, Tikhonov A, Parkinson H, Petryszak R, Sarkans U, Brazma A. Arrayexpress update–simplifying data submissions. Nucleic Acids Res. 2015;43(Database issue):1113–6.
4. Kodama Y, Shumway M, Leinonen R. The sequence read archive: Explosive growth of sequencing data. Nucleic Acids Res. 2012;40(D1): 2011–3. https://doi.org/10.1093/nar/gkr854.
5. Reference Expression Dataset. http://refex.dbcls.jp. Accessed 13 July 2017.
6. Uhlen M, Fagerberg L, Hallstrom BM, Lindskog C, Oksvold P, Mardinoglu A, Sivertsson A, Kampf C, Sjostedt E, Asplund A, Olsson I, Edlund K, Lundberg E, Navani S, Szigyarto CA-K, Odeberg J, Djureinovic D, Takanen JO, Hober S, Alm T, Edqvist PH, Berling H, Tegel H, Mulder J, Rockberg J, Nilsson P, Schwenk JM, Hamsten M, von Feilitzen K, Forsberg M, Persson L, Johansson F, Zwahlen M, von Heijne G, Nielsen J, Ponten F. Tissue-based map of the human proteome. Science. 2015;347:6220. https://doi.org/10.1126/science.1260419.
7. Ringwald M, Eppig JT, Begley DA, Corradi JP, McCright IJ, Hayamizu TF, Hill DP, Kadin JA, Richardson JE. The mouse gene expression database (gxd). Nucleic Acids Res. 2001;29(1):98–101. https://doi.org/10.1093/nar/29.1.98.
8. Bähler J. Cell-Cycle Control of Gene Expression in Budding and Fission Yeast. Annu Rev Genet. 2005;39(1):69–94. https://doi.org/10.1146/annurev.genet.39.110304.095808.
9. Traxler MF, Chang DE, Conway T. Proc Natl Acad Sci of the United States of America. 2006;103(7):2374–9.
10. Mikkelsen TS, Xu Z, Zhang X, Wang L, Gimble JM, Lander ES, Rosen ED. Comparative epigenomic analysis of murine and human adipogenesis. Cell. 2010;143(1):156–69. https://doi.org/10.1016/j.cell.2010.09.006.
11. Kanehisa M, Goto S. Kegg: Kyoto encyclopedia of genes and genomes. Nucleic Acids Res. 2000;28(1):27–30. https://doi.org/10.1093/nar/28.1.27.
12. Huerta AM, Salgado H, Thieffry D, Collado-Vides J. Regulondb: A database on transcriptional regulation in escherichia coli. Nucleic Acids Res. 1998;26(1):55–9. https://doi.org/10.1093/nar/26.1.55.
13. Salgado H, Peralta-Gil M, Gama-Castro S, Santos-Zavaleta A, Muniz-Rascado L, Garcia-Sotelo JS, Weiss V, Solano-Lira H, Martinez-Flores I, Medina-Rivera A, Salgado-Osorio G, Alquicira-Hernandez S, Alquicira-Hernandez K, Lopez-Fuentes A, Porron-Sotelo L, Huerta AM, Bonavides-Martinez C, Balderas-Martinez YI, Pannier L, Olvera M, Labastida A, Jimenez-Jacinto V, Vega-Alvarado L, del Moral-Chavez V, Hernandez-Alvarez A, Morett E, Collado-Vides J. Regulondb v8.0: omics data sets, evolutionary conservation, regulatory phrases, cross-validated gold standards and more. Nucleic Acids Res. 2013;41(D1):203–13. https://doi.org/10.1093/nar/gks1201.
14. Teixeira MC, Monteiro P, Jain P, Tenreiro S, Fernandes AR, Mira NP, Alenquer M, Freitas AT, Oliveira AL, Sá-Correia I. The yeastract database: a tool for the analysis of transcription regulatory associations in saccharomyces cerevisiae. Nucleic Acids Res. 2006;34(suppl 1):446. https://doi.org/10.1093/nar/gkj013.
15. Monteiro PT, Mendes ND, Teixeira MC, d'Orey S, Tenreiro S, Mira NP, Pais H, Francisco AP, Carvalho AM, Lourenço AB, Sá-Correia I, Oliveira AL, Freitas AT. Yeastract-discoverer: new tools to improve the analysis of transcriptional regulatory associations in saccharomyces cerevisiae. Nucleic Acids Res. 2008;36(suppl 1):132. https://doi.org/10.1093/nar/gkm976.
16. Abdulrehman D, Monteiro PT, Teixeira MC, Mira NP, Lourenço AB, dos Santos SC, Cabrito TR, Francisco AP, Madeira SC, Aires RS, Oliveira AL, Sá-Correia I, Freitas AT. Yeastract: providing a programmatic access to curated transcriptional regulatory associations in saccharomyces cerevisiae through a web services interface. Nucleic Acids Res. 2011;39(suppl 1):136. https://doi.org/10.1093/nar/gkq964.
17. Teixeira MC, Monteiro PT, Guerreiro JF, Gonçalves JP, Mira NP, dos Santos SC, Cabrito TR, Palma M, Costa C, Francisco AP, Madeira SC, Oliveira AL, Freitas AT, Sá-Correia I. The yeastract database: an upgraded information system for the analysis of gene and genomic transcription regulation in saccharomyces cerevisiae. Nucleic Acids Res. 2014;42(D1): 161. https://doi.org/10.1093/nar/gkt1015.
18. Ravasi T, Suzuki H, Cannistraci CV, Katayama S, Bajic VB, Tan K, Akalin A, Schmeier S, Kanamori-Katayama M, Bertin N, et al. An atlas of combinatorial transcriptional regulation in mouse and man. Cell. 2010;140(5):744–52.
19. Suzuki H, Forrest AR, van Nimwegen E, Daub CO, Balwierz PJ, Irvine KM, Lassmann T, Ravasi T, Hasegawa Y, de Hoon MJ, et al. The transcriptional network that controls growth arrest and differentiation in a human myeloid leukemia cell line. Nat Genet. 2009;41(5):553–62.
20. Forrest A, Kawaji H, Rehli M, Baillie J, de HM, Haberle V, Lassmann T, Kulakovskiy I, Lizio M, Itoh M, Andersson R, Mungall C, Meehan T, Schmeier S, Bertin N, Jorgensen M, Dimont E, Arner E, Schmidl C, Schaefer U, Medvedeva Y, Plessy C, Vitezic M, Severin J, Semple C, Ishizu Y, Young R, Francescatto M, Alam I, Albanese D, Altschuler G, Arakawa T, Archer J, Arner P, Babina M, Rennie S, Balwierz P, Beckhouse A, Pradhan-Bhatt S, Blake J, Blumenthal A, Bodega B, Bonetti A, Briggs J, Brombacher F, Burroughs A, Califano A, Cannistraci C, Carbajo D, Chen Y, Chierici M, Ciani Y, Clevers H, Dalla E, Davis C, Detmar M, Diehl A, Dohi T, Drablos F, Edge A, Edinger M, Ekwall K, Endoh M, Enomoto H, Fagiolini M, Fairb. A promoter-level mammalian expression atlas. Nature. 2014;507:462. https://doi.org/10.1038/nature13182.
21. Andersson R, Gebhard C, Miguel-Escalada I, Hoof I, Bornholdt J, Boyd M, Chen Y, Zhao X, Schmidl C, Suzuki T, et al. An atlas of active enhancers across human cell types and tissues. Nature. 2014;507(7493):455–61.
22. Arner E, Daub CO, Vitting-Seerup K, Andersson R, Lilje B, Drabløs F, Lennartsson A, Rönnerblad M, Hrydziuszko O, Vitezic M, et al. Transcribed enhancers lead waves of coordinated transcription in transitioning mammalian cells. Science. 2015;347(6225):1010–4.
23. Wingender E, Chen X, Hehl R, Karas H, Liebich I, Matys V, Meinhardt T, Prüß M, Reuter I, Schacherer F. Transfac: an integrated system for gene expression regulation. Nucleic Acids Res. 2000;28(1):316–9.
24. Wingender E. The transfac project as an example of framework technology that supports the analysis of genomic regulation. Brief Bioinform. 2008;9(4):326–32.
25. QUIAGEN's Ingenuity Pathway Analysis. https://www.qiagenbioinformatics.com/. Accessed 14 July 2017.
26. Siersbæk R, Nielsen R, Mandrup S. Transcriptional networks and chromatin remodeling controlling adipogenesis. Trends Endocrinol Metab. 2012;23(2):56–64.
27. Tegner J, Yeung MS, Hasty J, Collins JJ. Reverse engineering gene networks: integrating genetic perturbations with dynamical modeling. Proc Natl Acad Sci. 2003;100(10):5944–9.
28. Gardner TS, Di Bernardo D, Lorenz D, Collins JJ. Inferring genetic networks and identifying compound mode of action via expression profiling. Science. 2003;301(5629):102–5.
29. Kim S, Kim J, Cho KH. Inferring gene regulatory networks from temporal expression profiles under time-delay and noise. Comput Biol Chem. 2007;31(4):239–45.
30. Schäfer J, Strimmer K. An empirical bayes approach to inferring large-scale gene association networks. Bioinformatics. 2005;21(6):754–64.
31. Stark E, Drori R, Abeles M. Partial cross-correlation analysis resolves ambiguity in the encoding of multiple movement features. J Neurophys. 2006;95(3):1966–75.
32. Butte AJ, Kohane IS. Mutual information relevance networks: functional genomic clustering using pairwise entropy measurements. In: Pac Symp Biocomput. Singapore: World Scientific Pub Co Inc.; 2000. p. 418–29.

33. Hausser J, Strimmer K. Entropy inference and the james-stein estimator, with application to nonlinear gene association networks. J Mach Learn Res. 2009;10:1469–84.
34. Margolin AA, Nemenman I, Basso K, Wiggins C, Stolovitzky G, Favera RD, Califano A. Aracne: an algorithm for the reconstruction of gene regulatory networks in a mammalian cellular context. BMC Bioinformatics. 2006;7(Suppl 1):7.
35. Zoppoli P, Morganella S, Ceccarelli M. Timedelay-aracne: Reverse engineering of gene networks from time-course data by an information theoretic approach. Bmc Bioinformatics. 2010;11(1):154.
36. Jensen FV, Nielsen TD. Bayesian Networks and Decision Graphs, 2nd edn. New York: Springer; 2007.
37. Neapolitan RE. Learning Bayesian Networks. Upper Saddle River: Prentice-Hall, Inc.; 2003.
38. Perrin BE, Ralaivola L, Mazurie A, Bottani S, Mallet J, d' Alche–Buc F. Gene networks inference using dynamic bayesian networks. Bioinformatics. 2003;19(suppl 2):138–48.
39. Zou M, Conzen SD. A new dynamic bayesian network (dbn) approach for identifying gene regulatory networks from time course microarray data. Bioinformatics. 2005;21(1):71–9.
40. Agresti A. Categorical Data Analysis, 3rd edn. Wiley Series in Probability and Statistics. 111 River Street: Wiley-Interscience; 2012.
41. Lauritzen SL, Spiegelhalter DJ. Local Computation with Probabilities on Graphical Structures and their Application to Expert Systems (with discussion). J R Stat Soc Ser B Stat Methodol. 1988;50(2):157–224.
42. Beinlich IA, Suermondt HJ, Chavez RM, Cooper GF. The alarm monitoring system: A case study with two probabilistic inference techniques for belief networks. In: Proceedings of the 2nd European Conference on Artificial Intelligence in Medicine. Berlin: Springer-Verlag; 1989. p. 247–56.
43. Binder J, Koller D, Russell S, Kanazawa K. Adaptive Probabilistic Networks with Hidden Variables. Mach Learn. 1997;29(2–3):213–44.
44. Scutari M. Learning Bayesian Networks with the bnlearn R Package. J Stat Softw. 2010;35(3):1–22. Accessed 2 Aug 2017.
45. Takenaka Y, Seno S, Matsuda H. Detecting shifts in gene regulatory networks during time-course experiments at single-time-point temporal resolution. J Bioinforma Comput Biol. 2015;13(05):1543002. https://doi.org/10.1142/S0219720015430027.
46. Jia Y, Huan J. Constructing non-stationary dynamic bayesian networks with a flexible lag choosing mechanism. BMC Bioinformatics. 2010;11(Suppl 6):27.
47. Nakayama T, Daiyasu H, Seno S, Takenaka Y, Matsuda H. Bioinformatics and Computational Biology: The 2013 WorldComp International Conference Proceedings. In: Arabnia HR, Tran Q-N, editors. Herndon: Mercury Learning and Information; 2014. p. 375–9.
48. Song L, Kolar M, Xing EP. Time-varying dynamic bayesian networks. In: Advances in Neural Information Processing Systems. NY: Curran Associates, Inc.; 2009. p. 1732–40.
49. Tamada Y, Araki H, Imoto S, Nagasaki M, Doi A, Nakanishi Y, Tomiyasu Y, Yasuda K, Dunmore B, Sanders D, Humphreys S, Print CG, Charnock-Jones SD, Tashiro K, Kuhara S, Miyano S. Unraveling dynamic activities of autocrine pathways that control drug-response transcriptome networks. In: Altman RB, Dunker AK, Hunter L, Murray T, Klein TE, editors. Pacific Symposium on Biocomputing. Singapore: World Scientific Pub Co Inc.; 2009. p. 251–63.
50. Adipocyte Differentiation of Mouse, Genome Network Plaform. ST2 stem cell was induced DMEM with 10% FBS, Affymetrix GeneChip 430 2.0 Array, 62 time points. http://genomenetwork.nig.ac.jp/index_e.html. Accessed 2 Aug 2017.
51. Takenaka Y, Seno S, Matsuda H. Dynamic Analysis of Gene Regulations 734 using Leaving-One-Out Expression Profile. Proceeding of The 20th Annual International Conference on Research in Computational Molecular Biology, RECOMB2016, P48, 2016.
52. Qian SW, Li X, Zhang YY, Huang HY, Liu Y, Sun X, Tang QQ. Characterization of adipocyte differentiation from human mesenchymal stem cells in bone marrow,. BMC Dev Biol. 2010;10:47. https://doi.org/10.1186/1471-213X-10-47.

Hayashi et al. BMC Bioinformatics 2018, 19(Suppl 4):62
https://doi.org/10.1186/s12859-018-2073-x

BMC Bioinformatics

DATABASE Open Access

MEGADOCK-Web: an integrated database of high-throughput structure-based protein-protein interaction predictions

Takanori Hayashi[1,2], Yuri Matsuzaki[2], Keisuke Yanagisawa[1,2], Masahito Ohue[1,3,4*] and Yutaka Akiyama[1,2,3,4,5*]

From The Sixteenth Asia Pacific Bioinformatics Conference
Yokohama, Japan. 15-17 January 2018

Abstract

Background: Protein-protein interactions (PPIs) play several roles in living cells, and computational PPI prediction is a major focus of many researchers. The three-dimensional (3D) structure and binding surface are important for the design of PPI inhibitors. Therefore, rigid body protein-protein docking calculations for two protein structures are expected to allow elucidation of PPIs different from known complexes in terms of 3D structures because known PPI information is not explicitly required. We have developed rapid PPI prediction software based on protein-protein docking, called MEGADOCK. In order to fully utilize the benefits of computational PPI predictions, it is necessary to construct a comprehensive database to gather prediction results and their predicted 3D complex structures and to make them easily accessible. Although several databases exist that provide predicted PPIs, the previous databases do not contain a sufficient number of entries for the purpose of discovering novel PPIs.

Results: In this study, we constructed an integrated database of MEGADOCK PPI predictions, named MEGADOCK-Web. MEGADOCK-Web provides more than 10 times the number of PPI predictions than previous databases and enables users to conduct PPI predictions that cannot be found in conventional PPI prediction databases. In MEGADOCK-Web, there are 7528 protein chains and 28,331,628 predicted PPIs from all possible combinations of those proteins. Each protein structure is annotated with PDB ID, chain ID, UniProt AC, related KEGG pathway IDs, and known PPI pairs. Additionally, MEGADOCK-Web provides four powerful functions: 1) searching precalculated PPI predictions, 2) providing annotations for each predicted protein pair with an experimentally known PPI, 3) visualizing candidates that may interact with the query protein on biochemical pathways, and 4) visualizing predicted complex structures through a 3D molecular viewer.

Conclusion: MEGADOCK-Web provides a huge amount of comprehensive PPI predictions based on docking calculations with biochemical pathways and enables users to easily and quickly assess PPI feasibilities by archiving PPI predictions. MEGADOCK-Web also promotes the discovery of new PPIs and protein functions and is freely available for use at http://www.bi.cs.titech.ac.jp/megadock-web/.

Keywords: Protein-protein interaction, Protein-protein docking, Predicted protein-protein interaction database, MEGADOCK, MEGADOCK-Web

* Correspondence: ohue@c.titech.ac.jp; akiyama@c.titech.ac.jp
[1]Department of Computer Science, School of Computing, Tokyo Institute of Technology, 2-12-1 W8-76 Ookayama, Meguro-ku, Tokyo 152-8550, Japan
Full list of author information is available at the end of the article

Background

Many proteins interact with each other in nature [1]; these interactions are called protein-protein interactions (PPIs). PPIs play crucial roles in living cells, including signal transduction and regulation of metabolic pathways. In recent years, PPI inhibitors and candidate compounds have also been developed [2, 3] in drug discovery to treat PPI-related diseases. Therefore, understanding of PPIs is important not only for elucidation of biological phenomena but also for pharmaceutical research. PPIs are identified by biochemical experiments, such as the yeast two-hybrid method [4], mass spectrometry [5], and cocrystallization of protein complexes [6]. Although many proteins and their functions have been discovered each year, the discovery of PPIs by biochemical assays is time-consuming and expensive. In addition, several PPIs are involved in protein signaling networks, and many attempts have been made to elucidate the exhaustive interactions among groups of proteins within a specific cellular function category through experiments and calculations [7–9]. In order to improve our understanding of biological processes, computational prediction of PPIs, which is less time-consuming and less expensive than experimental approaches, is needed to discover proteins with a high likelihood of involvement in PPIs [10, 11].

Computational PPI prediction methods can be categorized into three types: 1) those utilizing the amino acid sequence information of the protein [12–14], 2) those based on co-evolution information [15], and 3) those using the three-dimensional (3D) structure of the protein [16–22]. The third method, based on the protein 3D structure, has been shown to be promising because elucidation of the 3D structure and binding surface localization of the protein can facilitate the design of PPI inhibitors [23, 24]. Methods using the 3D structure of proteins are further divided into two types: 3-a) methods based on sequence homology and structural similarity with known complex structures [16–18], and 3-b) methods based on rigid body docking of a pair of protein structures [19–22]. Since rigid body docking for two protein structures does not require a priori PPI information, PPIs whose interfaces are not necessarily similar to known complexes in terms of 3D structures or of amino acid compositions may be found by rigid body docking. To predict such novel PPIs, we developed a structure-based PPI prediction software program named MEGADOCK [25]. MEGADOCK uses a template-free (i.e., no reference to experimentally determined cocrystal structures) and fast Fourier transform (FFT)-based rigid body protein-protein docking method. Since the evaluation function of MEGADOCK is computable at high speed, MEGADOCK is faster than other template-free docking tools; for example, MEGADOCK is ~ 10 times faster than ZDOCK [25, 26]. Additionally, its implementation of multithreading, GPU acceleration, and MPI parallelization enables 30,000 PPI predictions in 2 h with 70-node GPU-equipped computing clusters [27].

However, in order to utilize PPI predictions (including predicted 3D complex structures), it is necessary to archive prediction results and their 3D structures and make this information easily accessible. For example, some existing databases, such as PrePPI [16], HOMCOS [17], PRISM [28], and Interactome3D [29], provide PPI predictions. PrePPI provides highly confident PPI predictions by combining structural, functional, evolutionary, and expression information. HOMCOS can predict PPIs on demand using query protein homology. PRISM can predict PPIs on demand based on known template protein-protein interfaces and provides a small database that includes PPI predictions. Interactome3D provides PPI predictions within PPI networks. Unfortunately, it is difficult for such systems to identify PPIs having 3D structures that differ from those of known complexes because they are based on a known PPI template. For example, it is difficult to search the predicted complex structures of adenomatous polyposis coli (APC) protein and Axin protein, a well-known PPI pair [30, 31], using these databases. In the PrePPI database, no predicted structures of the APC and Axin complex are available because PrePPI shows only high-prediction-scored complex structures. In the case of HOMCOS, it is impossible to predict the PPI because homologous proteins consisting of heterodimers have not been reported. In PRISM, the PPI prediction between AOC and Axin is not provided because of the small size of the database. Although PRISM can build predicted complex structure models on-demand, it takes about 1 h to obtain the model. In the case of Interactome3D, a predicted complex structure model is not obtained because it provides only a part of the experimentally determined structure, even if the full-length structure of APC exists Therefore, a novel database that contains exhaustive PPI predictions from the protein structure is needed to discover PPIs that have structures different from those of known PPIs.

In this study, we developed a database containing comprehensive PPI predictions by MEGADOCK among 7528 representative structures of human protein chains registered in the Protein Data Bank (PDB) [32] (a total of 28,331,628 PPI predictions) and a web interface for the database, named MEGADOCK-Web. MEGADOCK-Web provides the following four powerful functions in order to make the predicted PPI information more useful: 1) searching PPI predictions, 2) providing annotations for each predicted protein pair with known PPIs, 3) visualizing candidates of interactions with query proteins in biochemical pathways, and 4) visualizing predicted complex structures using a web browser-based 3D molecular viewer. These functions will allow users to easily access comprehensive PPI predictions from a web browser and facilitate discovery of unknown PPIs.

Construction and contents

In this section, we describe the construction of MEGADOCK-Web and the contents stored in MEGADOCK-Web. Figure 1 shows the database schema of MEGADOCK-Web. The database of MEGADOCK-Web consists of three models: 1) Protein, 2) Docking, and 3) Pathway. These attributes provide protein structure, predicted PPIs annotated with experimentally known PPI information, and biochemical pathway information in detail.

System architecture

We developed MEGADOCK-Web using Play Framework [33], which is a Java and Scala's web application framework including components and application programming interfaces (APIs) necessary for web application development. Figure 2 depicts the system architecture of MEGADOCK-Web. MEGADOCK-Web adopts a client-server model in the same way as a general web application. Requests from clients are classified into requests to static resources (such as protein PDB files and docking result files) and requests to the web pages. In the former case, Apache directly returns the corresponding resources to clients. In the latter case, Apache redirects the requests to the web application running on the local port. For server-side implementation, we stored information in the relational database H2 Database Engine [34] with MySQL [35] wrapped around Play Framework. Receiving requests from users invokes the web application, which generates HTML dynamically to refer to the database. The server-side implementation is multithreaded to respond immediately. Javascript and jQuery are adopted as client-side implementation tools to control web pages.

Protein structures

This section explains the protein structures registered in MEGADOCK-Web. Since it is important for supporting drug discovery to integrate data for *Homo sapiens*, 3775 PDB entries were obtained from the PDB based on the following criteria: protein (not nucleic acids or complexes with nucleic acids); taxonomy of protein is *H. sapiens*; resolution of 2.5 Å or less; and nonredundant so that the sequence similarity is less than 90% using the PDB search function, which utilizes NCBI BLASTClust [36].

These PDB entries were divided into chain units, and we excluded those with less than 25 residues or with mutations. As a result, 7528 protein chain structures were subjected to PPI prediction. In addition, we annotated each protein chain structure with the ID of UniProt [37] (UniProt AC), the ID of Kyoto Encyclopedia of Genes and Genomes (KEGG) [38], organism name

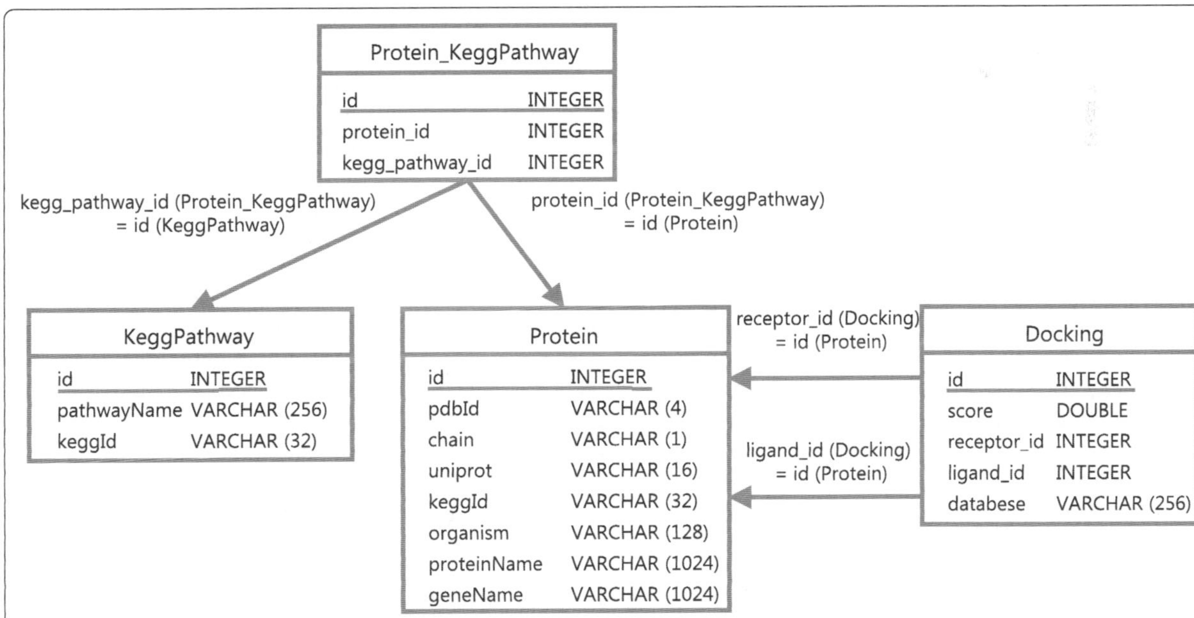

Fig. 1 Database schema. Each protein has a PDB ID, chain ID, UniProt AC, protein name, gene name, source organism, KEGG pathways to which it belongs, and docking information. Docking describes the results of the prediction and has information on the receptor and ligand for the protein, the databases to which the protein is registered, and *PPIScore*. The red underline in the database columns indicates the primary key in each database table. Note that "column *X*1 (table *T*1) = column *X*2 (table *T*2)" for each arrow indicates that column *X*1 in database table *T*1 matches column *X*2 in database table *T*2, and these columns are the foreign key

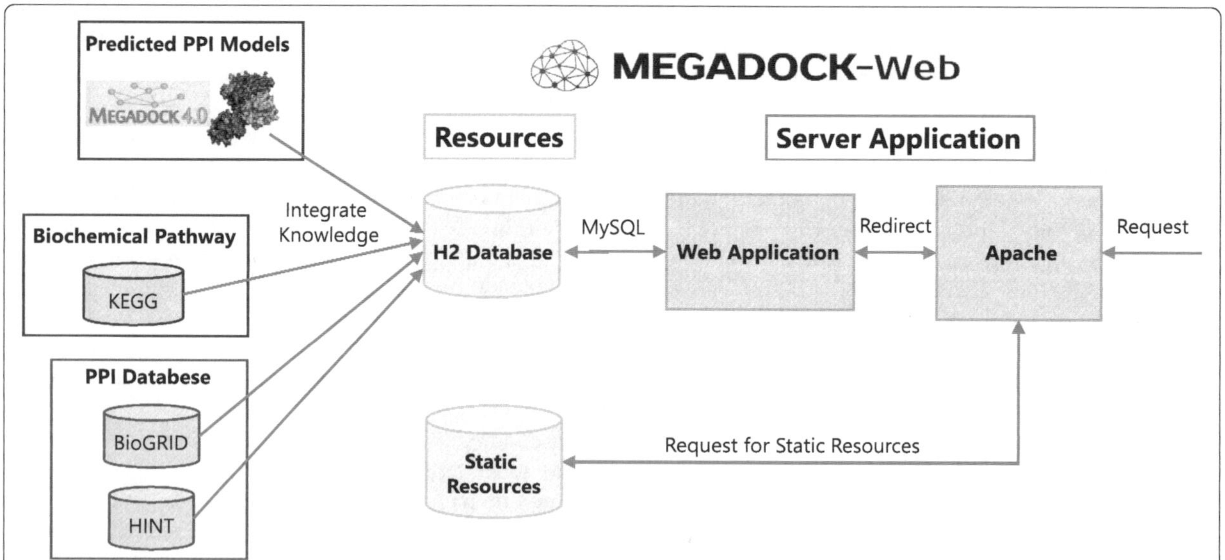

Fig. 2 Archtecture of MEGADOCK-Web. We integrated KEGG, BioGRID, HINT, and predicted PPI information in H2 Database in advance. When a GET or POST request comes, Apache distributes the request depending on whether access to static resources or access to web pages. In the latter case, the web application returns the appropriate page referring to DB with MySQL

(currently only *H. sapiens*), protein name, and gene name using Bioservices [39], a package for Python.

Known PPI pairs

MEGADOCK-Web shows known PPI pairs as well as vast number of prediction results. HINT [40] (acquired July 13, 2016) and BioGRID [41] (acquired July 13, 2016) are used as databases of known PPI pairs. MEGADOCK-Web shows known PPI information based on UniProt AC because these databases describe known PPIs based on UniProtAC.

Biochemical pathway information

MEGADOCK-Web uses KEGG pathways as biochemical pathways in order to help understand protein functions. The KEGG pathway is a collection of manually drawn pathway maps representing knowledge on molecular interaction, reaction, and relation networks. We obtained KEGG pathway information for each protein via the web API upon startup of the server just once.

Predicted PPI information

MEGADOCK-Web integrates exhaustive PPI predictions among 7528 protein chain structures using MEGADOCK. All-to-all PPI predictions (a total of 28,331,628) are stored in MEGADOCK-Web.

The procedure for PPI predictions by MEGADOCK is as follows. First, hydrogens and small molecules in HETATM rows in PDB files, such as cofactors and small ligands, were excluded in the predocking process. Secondly, predicted protein-protein conformations were created with MEGADOCK. MEGADOCK outputs the three best translational positions that provide the best value of docking evaluation function (*DockingScore*) per specific protein rotation, and 3600 patterns of protein rotations are computed, resulting in 10,800 predicted conformations in total. To calculate the *DockingScore* for each position, we used methods described in our previous papers [25, 42]. MEGADOCK uses FFT to enable an efficient global docking search on a 3D grid, and calculates shape complementarity, electrostatic interactions, and desolvation free energy [42]. Finally, from each *DockingScore*, MEGADOCK calculates the *PPIScore*, which describes the possibility of a PPI. *PPIScore* is defined as $PPIScore = (S_{top} - \mu)/\sigma$, where S_{top} is the highest *DockingScore* in the protein pair, μ is the average of 10,800 *DockingScore*s, and σ is the standard deviation of these scores [24]. Higher *PPIScores* indicate a higher possibility of a PPI. Although *DockingScore* cannot be compared between different pairs of proteins, the *PPIScore*, which is normalized *DockingScore*, represents how likely the PPI is to occur. The sensitivity of PPI prediction can be adjusted by tuning the threshold value of the *PPIScore*. Additional file 1: Figure S1 illustrates the precision and recall of MEGADOCK for each threshold. We also reported the ability of MEGADOCK using real pathway datasets in our previous studies [9, 43, 44].

As a result, we calculated $7528^2 = 56,670,784$ predictions since the number of registered 3D protein chain structures was 7528. Reversing the receptor file and the ligand file generally allows predictions to be performed on the same protein pair. Notably, FFT-based docking, such as MEGADOCK, usually shows different results (*DockingScores*) when the order of PDB input pair is switched. In order to avoid such redundancy for users, we stored the protein pairs that had higher *PPIScores* for prediction between the same PDB files. As a result, MEGADOCK-Web displays ${}_{7528}\mathbf{C}_2 + 7528$ (homo dimers) = 28,331,628 PPI predictions. The total computation time was approximately 500 CPU years.

Utility

Figure 3 provides the page transition diagram of MEGADOCK-Web. In this section, we explain the utilities of MEGADOCK-Web in three situations: 1) searching for PPI candidates of a query protein, 2) searching for PPI candidates on a specific pathway, and 3) assessing the possibility of a PPI for a pair of proteins.

Searching PPI candidates of a query protein

In this section, we describe the use of MEGADOCK-Web to search PPI candidates of a query protein. First, a user can search with a query word (for example, protein name, PDB ID, UniProt AC, etc.) typed in the "Search Single Protein" window in the Top page (Additional file 1: Figure S2), resulting in transition to the protein selection page (Additional file 1: Figure S3) showing the list of search results. In this page, protein structure hits with the query are listed accompanying identifiable protein information for each protein (PDB ID, chain, UniProt AC, protein name, and gene name). Clicking the "View" button takes users to the PPI prediction list page (Additional file 1: Figure S4). In the PPI prediction list page, proteins are shown in descending order of *PPIScore* and are paged per 10 items. The list contains information on the following three points for all the proteins in the database: 1) identification information, 2) *PPIScores* with the query protein, and 3) known PPIs for the query protein. Each row also has a "View" button to transit to the predicted complexes page (Additional file 1: Figure S5) between the query protein structure and the corresponding structure within the row.

On the predicted complexes page, 10 predicted complexes based on the results of the MEGADOCK dockings between the two proteins are displayed with 3D molecule viewers. From the "Download Complex in this page" link, 10 predicted complex structures can be downloaded as PDB files together in zip format. The table includes the following four columns: 1) rank (up to 1000) of the *DockingScore*, 2) *DockingScore*, 3) a molecule viewer displaying the 3D structure of the predicted complex, and 4) a PDB file download link for the complex structure. Molmil [45] is adopted as a molecular viewer because of its lightness and stability; the program enables users to rotate and enlarge the structure easily. It is also possible to download the docking results as a text file describing the translation distance, rotation angle of the ligand protein, and *DockingScore* (ZDOCK 3.0.1 and MEGADOCK output style).

Searching PPI candidates on a specific pathway

Assuming that a user has a protein of interest known to affect a certain biochemical pathway with unknown interactions, this section describes how to discover PPI candidates in a specific pathway.

First, as in the previous section, the user can search a query by typing in the "Search Single Protein" window at the top of the page (Additional file 1: Figure S2) and select the protein structure in the protein selection page (Additional file 1: Figure S3). Next, in the PPI prediction list page (Additional file 1: Figure S4), the user can move to the pathway selection page (Additional file 1: Figure S6) by clicking the "Show Pathway" button. The pathway selection page displays a list of pathways containing at least one predicted binder. The predicted binder is defined as a protein with a *PPIScore* of greater than or equal to the user-defined threshold.

The threshold of *PPIScore* can vary from 6.0 to 12.0 with steps of 1.0 (default value is 8.0). Lowering the threshold increases the sensitivity and decreases the specificity. In the pathway selection page, links to the KEGG pathways to which the predicted binders belong are displayed. Note that the query protein does not always exist on the KEGG pathway; obtaining many predicted binders on the pathway without the query protein indicates a possibility of unknown interactions of the query

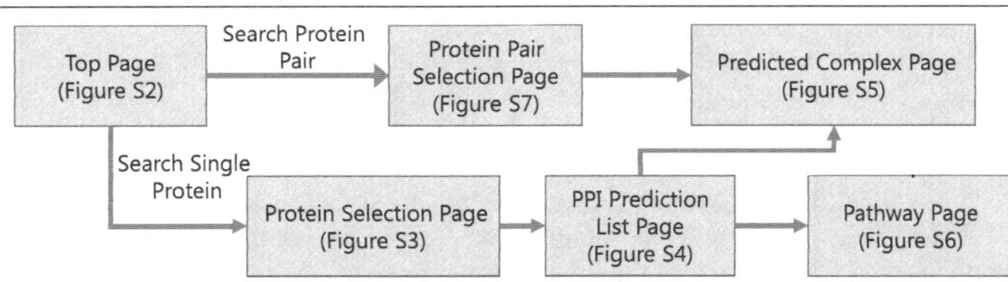

Fig. 3 Page transition diagram of MEGADOCK-Web. For a single query, users can transit to the PPI prediction list page via the protein selection page. From this page, it is possible to transit to the pathway selection page to which the predicted binder belongs and to the prediction complex display page. For two queries, it is possible to transit to the prediction complex display page through the protein pair selection page

protein in the pathway. Furthermore, in order to display only pathways with specific functions, it is possible to filter the pathways by inputting a word or phrase in the "Pathway Filtering" window at the top of the pathway selection page. Clicking the link takes the user to the corresponding KEGG page, in which the query protein is colored in blue, and predicted binders are in red.

Assessing the possibility of PPIs for a pair of proteins

MEGADOCK-Web provides a functionality to assess the possibility of interactions with two proteins from functional and structural points of view. First, the user can search for the two proteins in the "Search Protein Pairs" window at the top of the page, resulting in transition to the protein pair selection page (Additional file 1: Figure S7). Selection of protein structures for each query is shown at the upper part of the page, whereas *PPIScore*s of all combinations of pairs of those protein structures are shown in descending order in the lower part of the page. The upper table has information similar to that of the protein selection page. If the structures of proteins to be used have already been specified, the predicted complexes page (Additional file 1: Figure S5) can be accessed by choosing the individual structure of each protein in the upper table.

The lower part of the page is useful if the structures of proteins to be used have not yet been determined. Since pairs of protein structures are presented in descending order of *PPIScores*, higher *PPIScore* structures (structures with higher PPI possibility) can be found easily. From the "View" link, it is also possible to transit to the predicted complex page (Additional file 1: Figure S5) of the corresponding proteins.

Discussion

How to utilize MEGADOCK-Web: Concrete examples

In this section, three case studies with specific proteins are explained to provide examples of the use of MEGADOCK-Web.

Case 1: APC and Axin (recalling a known PPI)

Cancer is one of the most serious diseases in humans. APC is a cancer-related protein; thus, PPI candidates with APC are likely to be related to cancer. In this section, we describe a search for PPI candidates of APC using MEGADOCK-Web.

First, when APC is specified as a query, structure information of APC is displayed as search results. After selecting the structure of 1DEB_A (chain A of 1DEB) from several structures, users are taken to the PPI prediction list page (Fig. 4a). On this page, the pathway selection page with the threshold set to 8.0 is shown in order to view the KEGG pathways to which the predicted binders belong. Pathway filtering can be used to find cancer-related pathways. After filtering with the word "cancer", the user can select breast cancer (KEGG ID: hsa05224). In the identified pathway, Axin (colored in red as a predicted binder) is present next to APC (colored in blue as the query protein; Fig. 4b). In order to more deeply assess the possibility of an interaction between APC and Axin, the user can use the pair query function and show the PPI prediction in detail. When searching APC and Axin as a pair query, some structures are found for both proteins. Selecting 1DEB_A and 1DK8_A results in transition to the predicted complex page (Fig. 4c). In the predicted complex page, predicted complexes are browsed, and the possible interactions can be assessed.

By following the above procedure, MEGADOCK-Web can facilitate the assessment of a specific PPI. In fact, the interaction between APC and Axin has been revealed experimentally [31].

Case 2: Sons of Sevenless (SOS) and SH2B adaptor protein 2 (SH2B2) (exploring a novel PPI)

The potential for identification of PPIs was described in the previous section. Here, we demonstrate the ability to identify a feasible new PPI using MEGADOCK-Web. We explain how to search unknown PPIs against SOS protein, which is also associated with cancer, as an example.

First, the pathway selection page is chosen, as described in the previous section, and the list includes the neurotrophin signaling pathway (hsa04722). In this pathway, an adaptor protein containing Pleckstrin homology (PH) and Src homology-2 (SH2) domains (APS), also known as SH2B2, is adjacent to SOS (Fig. 5a). Second, in order to search the possibility of an interaction between these proteins, searching SOS and SH2B2 as a pair query brings the user to the protein pair selection page, after which the user selects the predicted complex of 1DBH_A and 1Q2H_B with the highest score in the table (Fig. 5b). On the predicted complexes page (Fig. 5c), the possibility of an interaction is assessed. The *PPIScore* between 1DBH_A and 1Q2H_B was 8.2, indicating a possibility of a PPI.

The PPI database STRING DB stores experimental information and literature-based predictions [46]. Analysis using the database also supported the possibility of an interaction between SOS and SH2B2 when searching their UniProt ACs Q07889 and O14492, respectively. These results suggested that it may be useful to further investigate the interaction between SOS and SH2B2 (APS).

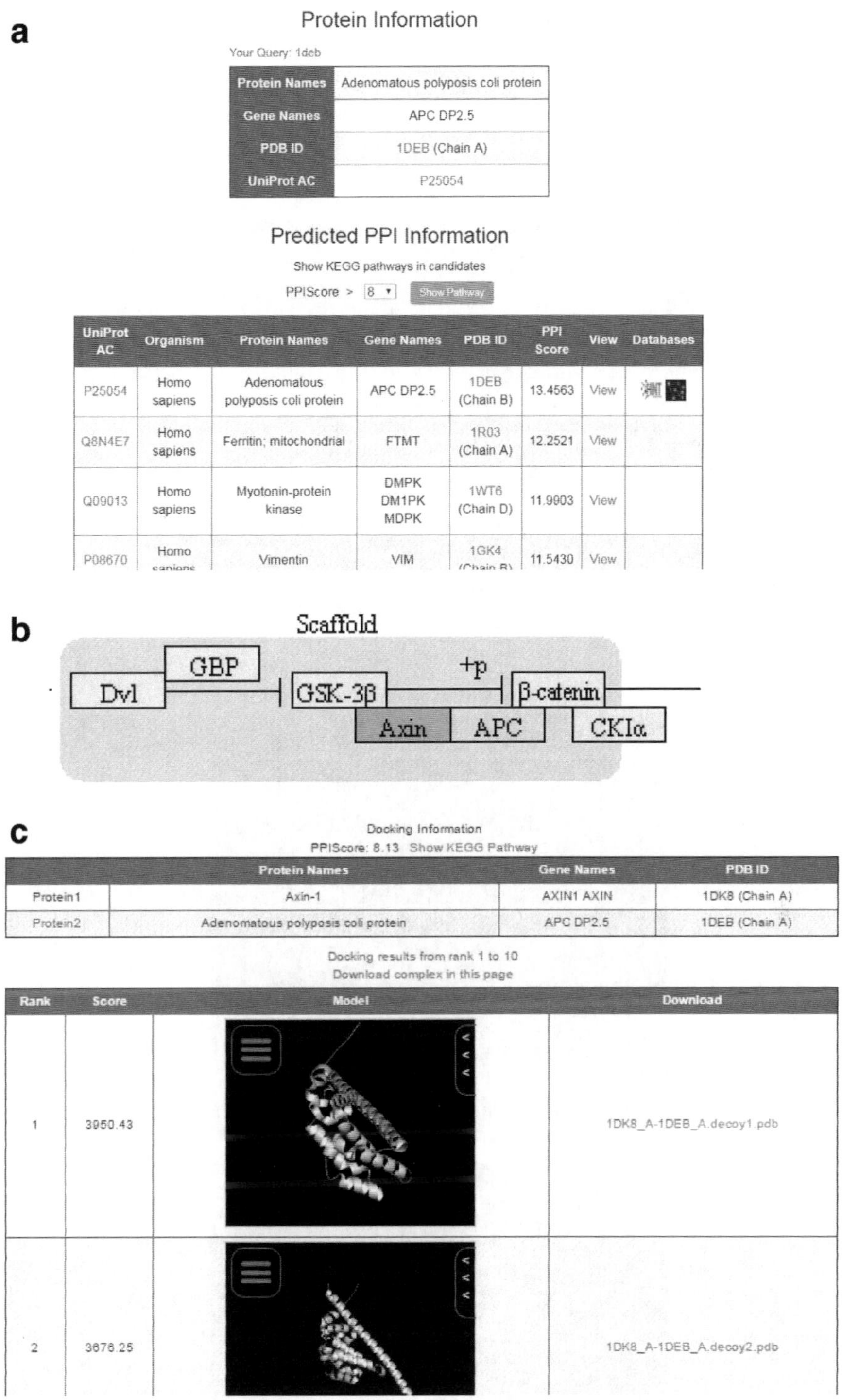

Fig. 4 An example of page transition when searching for protein interaction candidates and APC. **a** The PPI prediction list page of APC. **b** Part of the KEGG pathway (breast cancer, hsa05224). The query protein (APC) is colored in blue, and the predicted binders (Axin) are in red. **c** The predicted complex page of 1DEB_A (APC) and 1DK8_A (Axin)

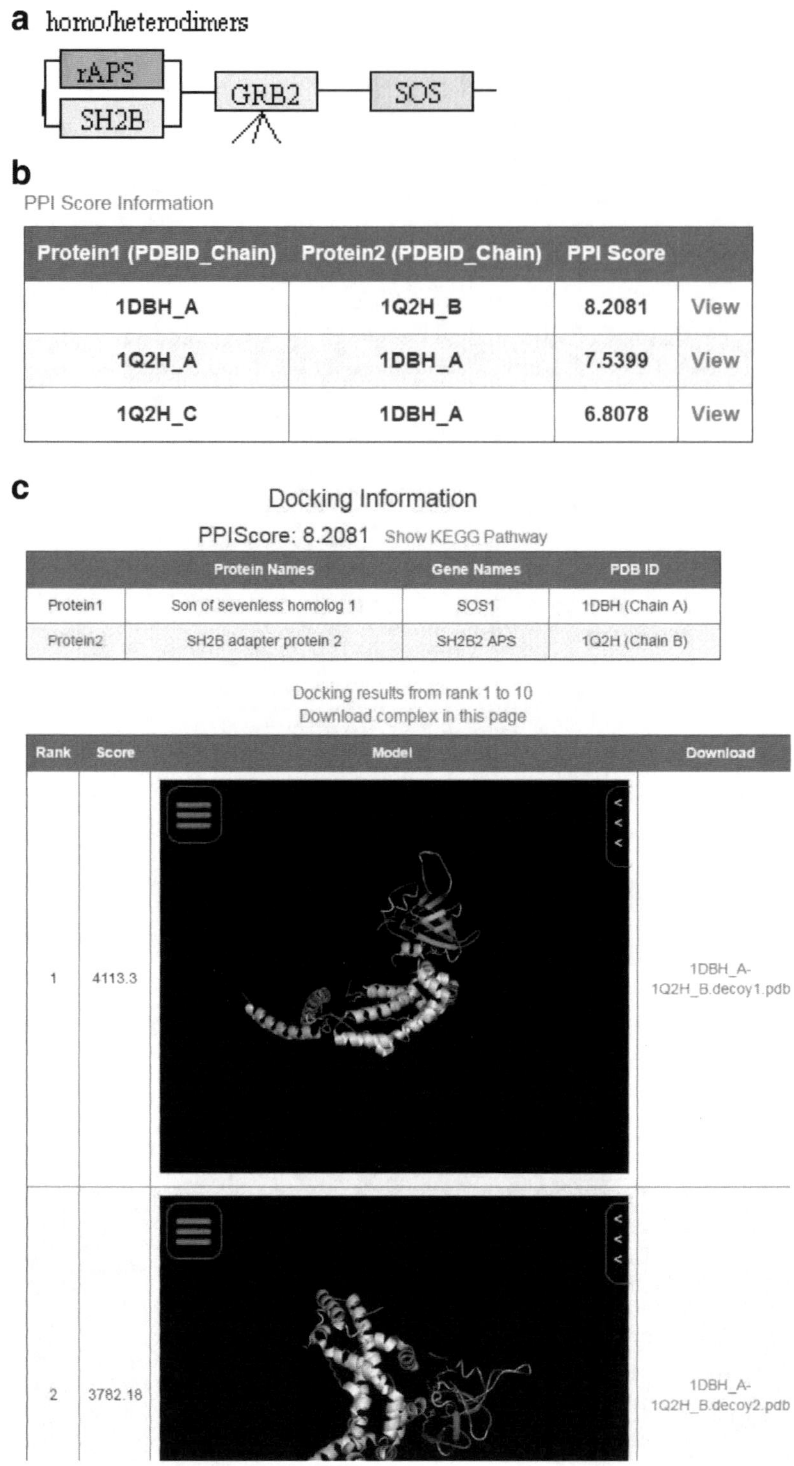

Fig. 5 An example of page transition when searching for SOS protein interaction candidates. **a** Part of the KEGG pathway (neurotrophin signaling pathway, hsa04722). The query protein (SOS) is colored in blue, and the predicted binder (APS) is in red. Note that the notation of rAPS (rat ortholog of human APS) is considered to be a mistake in APS of KEGG. **b** The table in the lower part of the protein pair selection page for SOS and SH2B2 (APS). **c** The predicted complex page for 1DBH_A (SOS) and 1Q2H_B (SH2B2)

Case 3: Membrane Metallo-endopeptidase (MME) and methylenetetrahydrofolate dehydrogenase 1 (MTHFD1) (predicting the complex structure of a known PPI)

MME and MTHFD1 are a pair of proteins that interact in prostate cancer cells [47]; however, the complex structure of these proteins has not been determined. MEGADOCK-Web can provide predicted complex structures for these proteins.

When MME and MTHFD1 are searched as queries, candidate structures are obtained. Here, 1R1H [48] describes the complex structure of MME and an inhibitor, and 1A4I [49] describes a homodimer complex structure of MTHFD1. Thus, the user may select 1R1H_A and 1A4I_B, and the predicted complexes are displayed in the predicted complex page. Figure 6a shows the predicted complex with Rank 1, and Fig. 6b shows that with Rank 2. The predicted complexes are created for different interaction surfaces. Therefore, users are able to discuss the possibility of PPIs on various interaction surface candidates.

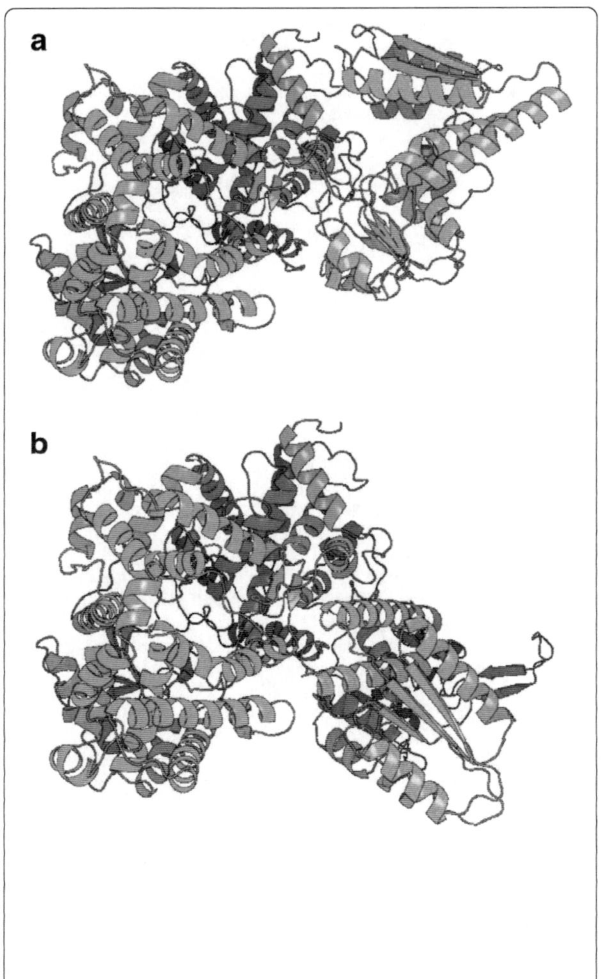

Comparison with other databases

Some existing databases contain 3D structural information similar to MEGADOCK-Web, such as PrePPI [16], HOMCOS [17], PRISM [28], and Interactome3D [29]. Table 1 shows a comparison of these databases and MEGADOCK-Web. MEGADOCK-Web was found to be superior in the following points: 1) provides many PPI predictions, 2) allows the user to search for pair proteins, 3) has a shorter response time, 4) permits viewing of some prediction complexes, and 5) allows mapping of proteins onto biochemical pathways.

First, MEGADOCK-Web contains a larger number of PPI predictions than other databases because it uses MEGADOCK, a rapid rigid docking tool. For example, MEGADOCK-Web provides 10 times more PPI predictions than PrePPI and 100 times more PPI predictions than Interactome3D. Second, MEGADOCK-Web supports both single protein and paired protein queries. Thus, it is possible to conduct searches from various perspectives. Furthermore, since PrePPI displays only one prediction model, users cannot discuss the predicted complexes on various interaction surfaces. In contrast, MEGADOCK-Web enables users to investigate the complexes because up to 1000 candidate complexes can be shown in the predicted complexes page. Finally, since proteins are mapped on the biochemical pathways (KEGG pathways), MEGADOCK-Web helps users to discuss the possibility of PPIs based on the biological roles of the predicted interactions.

Future plans

MEGADOCK-Web is expected to help to annotate proteins with unknown functions by providing comprehensive PPI predictions, irrespective of species. This program can be expected to help researchers discover new PPIs more broadly if we expand the database to include other species (e.g., *Saccharomyces cerevisiae, Escherichia coli* and *Mus musculus*). Although the current version of MEGADOCK-Web provides only human proteins, we will expand MEGADOCK-Web to include other species in the near future.

MEGADOCK-Web provides PPI predictions for proteins registered in the database. It will be useful to expand this system to perform docking calculations on-demand and allow browsing of PPI prediction results. With ordinary computers and implementations, long computation times are required for on-demand rigid body docking calculations; however, our system operates at multiple GPU computing nodes at high speeds. In addition, we have already completed a porting of MEGADOCK onto a public cloud computing resource (Microsoft Azure) that can use computing resources according to demand (Ohue M, et al:

Table 1 Comparison with other PPI prediction databases

	PrePPI [16]	HOMCOS [17]	PRISM [28]	Interactome3D [29]	MEGADOCK-Web
Number of PPI predictions	1,350,000	–	30,475	227,815	28,331,628
Comprehensive PPI prediction					✓
Search a single protein query	✓		✓	✓	✓
Search a pair protein query	✓	✓	✓	✓	✓
Viewing of multiple modeled complex structures		✓	✓	✓	✓
Mapping biochemical pathways				✓	✓
Template-free PPI prediction					✓

The numbers of archived PPI predictions as of June 1, 2017 are shown. Because HOMCOS is an on-demand prediction server, the number of PPI predictions is shown as '-'. '✓' and blank cells indicate enable and disable

MEGADOCK-Azure: cloud-based high-performance protein-protein interaction prediction system on Azure HPC, in preperation). We will release a service for on-demand PPI prediction in the near future.

Conclusions

In this paper, we described our new database, named MEGADOCK-Web, which is an integrated database of high-throughput structure-based PPI predictions. MEGADOCK-Web is freely available for use at http://www.bi.cs.titech.ac.jp/megadock-web/.

MEGADOCK-Web provides PPI predictions to users quickly by archiving exhaustive PPI predictions based on docking calculations. MEGADOCK-Web provides the following powerful functions: 1) searching PPI predictions by MEGADOCK, 2) providing annotations for each predicted protein pair with experimentally known PPIs, 3) visualizing candidate interactions with query proteins in biochemical pathways, and 4) visualizing predicted complex structures using a 3D molecular viewer. MEGADOCK-Web also gives comprehensive PPI prediction results in cooperation with biochemical pathways and enables users to easily and rapidly examine PPI possibilities. MEGADOCK-Web is expected to facilitate the discovery of new PPIs and protein functions.

Availability and requirements

MEGADOCK-Web is freely available for use at http://www.bi.cs.titech.ac.jp/megadock-web/.

Operating systems

MEGADOCK-Web is accessed from a web browser; therefore, it is platform-independent.

Browsers

MEGADOCK-Web has been extensively tested with Google Chrome, Safari, Mozilla Firefox, and Microsoft Edge.

Any restrictions for use

MEGADOCK-Web is free to use by everyone; there are no restrictions.

Additional files

Additional file 1: Supplementary Figures. PDF file containing various figures with detailed MEGADOCK performance and MEGADOCK-Web screenshots. (PDF 819 kb)

Abbreviations

3D: Three-dimensional; APC: Adenomatous Polyposis Coli; API: Application Programming Interface; APS: an adaptor molecule containing PH and SH2 domains; FFT: Fast Fourier Transform; GPU: Graphics Processing Unit; HTML: HyperText Markup Language; KEGG: Kyoto Encyclopedia of Genes and Genomes; MME: Membrane Metallo-endopeptidase; MTHFD1: Methylenetetrahydrofolate Dehydrogenase 1; PDB: Protein Data Bank; PPI: Protein-Protein Interaction; SH2: Src homology-2; SH2B2: SH2B adaptor protein 2; SOS: Son of Sevenless

Acknowledgements

We thank Mr. Kazuki Nagasawa for helping to implement the MEGADOCK-Web database system and Dr. Nobuyuki Uchikoga for testing the MEGADOCK-Web service.

Funding

This work was partially supported by the Core Research for Evolutional Science and Technology (CREST) "Extreme Big Data" (grant number JPMJCR1303) from the Japan Science and Technology Agency (JST), the Research Complex Program "Wellbeing Research Campus: Creating new values through technological and social innovation" from the Japan Science and Technology Agency (JST), KAKENHI (grant numbers 15K16081, 16K00388, and 17H01814) from the Japan Society for the Promotion of Science (JSPS), Platform Project for Supporting Drug Discovery and Life Science Research (Basis for Supporting Innovative Drug Discovery and Life Science Research) from Japan Agency for Medical Research and Development (AMED), the Regional Innovation and Ecosystem Formation Program "Program to Industrialize an Innovative Middle Molecule Drug Discovery Flow through Fusion of Computational Drug Design and Chemical Synthesis Technology" from the Japanese Ministry of Education, Culture, Sports, Science and Technology (MEXT), Leave a Nest Grant from Leave a Nest, Co., Ltd., and Microsoft Japan, Co., Ltd. The publication charges for this article were funded by JST (CREST JMPJCR1303 and the Research Complex Program).

Availability of data and materials

Not applicable.

About this supplement

This article has been published as part of *BMC Bioinformatics* Volume 19 Supplement 4, 2018: Selected articles from the 16th Asia Pacific Bioinformatics Conference (APBC 2018): bioinformatics. The full contents of the supplement are available online at https://bmcbioinformatics.biomedcentral.com/articles/supplements/volume-18-supplement-17.

Authors' contributions

TH, YM, and MO integrated the external resources and participated in the design of the MEGADOCK-Web database and website. MO and YA coordinated the project and provided guidance for system implementations. TH constructed the database and developed the website. TH, YM, and KY designed the use cases of MEGADOCK-Web. TH, KY, MO, and YA wrote the manuscript. All authors read and approved the final manuscript.

Ethics approval and consent to participate

Not applicable.

Consent for publication

Not applicable.

Competing interests

The authors declare that they have no competing interests.

Publisher's Note

Springer Nature remains neutral with regard to jurisdictional claims in published maps and institutional affiliations.

Author details

[1]Department of Computer Science, School of Computing, Tokyo Institute of Technology, 2-12-1 W8-76 Ookayama, Meguro-ku, Tokyo 152-8550, Japan. [2]Education Academy of Computational Life Sciences, Tokyo Institute of Technology, 2-12-1 W8-93 Ookayama, Meguro-ku, Tokyo 152-8550, Japan. [3]Advanced Computational Drug Discovery Unit (ACDD), Institute of Innovative Research, Tokyo Institute of Technology, 4259 Nagatsutacho, Midori-ku, Yokohama, Kanagawa 226-8501, Japan. [4]AIST-Tokyo Tech Real World Big-Data Computation Open Innovation Laboratory (RWBC-OIL), National Institute of Advanced Industrial Science and Technology (AIST), 1-1-1 Umezono, Tsukuba, Ibaraki 305-8560, Japan. [5]Molecular Profiling Research Center for Drug Discovery (molprof), National Institute of Advanced Industrial Science and Technology (AIST), 2-4-7 Aomi, Koto-ku, Tokyo 135-0064, Japan.

Published: 8 May 2018

References

1. Stelzl U, Worm U, Lalowski M, Haenig C, Brembeck FH, Goehler H, et al. A human protein-protein interaction network: a resource for annotating the proteome. Cell. 2005;122:957–68.
2. Oltersdorf T, Elmore SW, Shoemaker AR, Armstrong RC, Augeri DJ, Belli BA, et al. An inhibitor of Bcl-2 family proteins induces regression of solid tumours. Nature. 2005;435:677–81.
3. Popowicz GM, Czarna A, Wolf S, Wang K, Wang W, Dömling A, et al. Structures of low molecular weight inhibitors bound to MDMX and MDM2 reveal new approaches for p53-MDMX/MDM2 antagonist drug discovery. Cell Cycle. 2010;9:1104–11.
4. Ito T, Chiba T, Ozawa R, Yoshida M, Hattori M, Sakaki Y. A comprehensive two-hybrid analysis to explore the yeast protein interactome. Proc Natl Acad Sci U S A. 2001;98:4569–74.
5. Ho Y, Gruhler A, Heilbut A, Bader GD, Moore L, Adams S, et al. Systematic identification of protein complexes in Saccharomyces Cerevisiae by mass spectrometry. Nature. 2002;415:180–3.
6. Gamble TR, Vajdos FF, Yoo S, Worthylake DK, Houseweart M, Sundquist WI, et al. Crystal structure of human cyclophilin a bound to the amino-terminal domain of HIV-1 capsid. Cell. 1996;87:1285–94.
7. Yachie N, Petsalaki E, Mellor JC, Weile J, Jacob Y, Verby M, et al. Pooled-matrix protein interaction screens using barcode fusion genetics. Mol Syst Biol. 2016;12:863.
8. Stein A, Mosca R, Aloy P. Three-dimensional modeling of protein interactions and complexes is going 'omics. Curr Opin Struct Biol. 2011;21: 200–8.
9. Matsuzaki Y, Matsuzaki Y, Sato T, Akiyama Y. In silico screening of protein-protein interactions with all-to-all rigid docking and clustering: an application to pathway analysis. J Bioinforma Comput Biol. 2009;7:991–1012.
10. Skrabanek L, Saini HK, Bader GD, Enright AJ. Computational prediction of protein-protein interactions. Mol Biotechnol. 2008;38:1–17.
11. Matsuzaki Y, Uchikoga N, Ohue M, Akiyama Y. Rigid-docking approaches to explore protein-protein interaction space. Adv Biochem Eng Biotech. 2017; 160:33–55.
12. Murakami Y, Mizuguchi K. Homology-based prediction of interactions between proteins using averaged one-dependence estimators. BMC Bioinform. 2014;15:213.
13. Shen J, Zhang J, Luo X, Zhu W, Yu K, Chen K, et al. Predicting protein-protein interactions based only on sequences information. Proc Natl Acad Sci U S A. 2007;104:4337–41.
14. You Z, Lei Y, Zhu L, Xia J, Wang B. Prediction of protein-protein interactions from amino acid sequences with ensemble extreme learning machines and principal component analysis. BMC Bioinform. 2013;14(Suppl 8):S10.
15. Zhou H, Jakobsson E. Predicting protein-protein interaction by the mirrortree method: possibilities and limitations. PLoS One. 2013;8:e81100.
16. Zhang QC, Petrey D, Garzón JI, Deng L, Honig B. PrePPI: a structure-informed database of protein-protein interactions. Nucleic Acids Res. 2013; 41:D828–33.
17. Fukuhara N, Kawabata T. HOMCOS: a server to predict interacting protein pairs and interacting sites by homology modeling of complex structures. Nucleic Acids Res. 2008;36:185–9.
18. Ghoorah AW, Devignes MD, Smaïl-Tabbone M, Ritchie DW. Spatial clustering of protein binding sites for template based protein docking. Bioinformatics. 2011;27:2820–7.
19. Chen R, Li L, Weng Z. ZDOCK: an initial-stage protein-docking algorithm. Proteins. 2003;52:80–7.
20. Zhang C, Lai L. SDOCK: a global protein-protein docking program using stepwise force-field potentials. J Comput Chem. 2011;32:2598–612.
21. Ritchie DW, Venkatraman V. Ultra-fast FFT protein docking on graphics processors. Bioinformatics. 2010;26:2398–405.
22. Kozakov D, Beglov D, Bohnuud T, Mottarella SE, Xia B, Hall DR, et al. How good is automated protein docking? Proteins. 2013;81:2159–66.
23. Bourgeas R, Basse MJ, Morelli X, Roche P. Atomic analysis of protein-protein interfaces with known inhibitors: the 2P2I database. PLoS One. 2010;5:e9598.
24. Fuller JC, Burgoyne NJ, Jackson RM. Predicting druggable binding sites at the protein-protein interface. Drug Discov Today. 2009:155–61.
25. Ohue M, Matsuzaki Y, Uchikoga N, Ishida T, Akiyama Y. MEGADOCK: An all-to-all protein-protein interaction prediction system using tertiary structure data. Protein Pept Lett. 2014;21:766–78.
26. Mintseris J, Pierce B, Wiehe K, Anderson R, Chen R, Weng Z. Integrating statistical pair potentials into protein complex prediction. Proteins. 2007;69: 511–20.
27. Ohue M, Shimoda T, Suzuki S, Matsuzaki Y, Ishida T, Akiyama Y. MEGADOCK 4.0: an ultra-high-performance protein-protein docking software for heterogeneous supercomputers. Bioinformatics. 2014;30:3281–3.
28. Tuncbag N, Gursoy A, Nussinov R, Keskin O. Predicting protein-protein interactions on a proteome scale by matching evolutionary and structural similarities at interfaces using PRISM. Nat Protoc. 2011;6:1341–54.
29. Mosca R, Céol A, Aloy P. Interactome3D: adding structural details to protein networks. Nat Methods. 2012;10:47–53.
30. Behrens J, Jerchow BA, Würtele M, Grimm J, Asbrand C, Wirtz R, et al. Functional interaction of an axin homolog, conductin, with beta-catenin, APC, and GSK3beta. Science. 1998;280:596–9.
31. Nakamura T, Hamada F, Ishidate T, Anai K, Kawahara K, Toyoshima K, et al. Axin, an inhibitor of the Wnt signalling pathway, interacts with beta-catenin, GSK-3beta and APC and reduces the beta-catenin level. Genes Cells. 1998;3: 395–403.
32. Rose PW, Prlić A, Bi C, Bluhm WF, Christie CH, Dutta S, et al. The RCSB protein data Bank: views of structural biology for basic and applied research and education. Nucleic Acids Res. 2015;43:D345–56.
33. Play Framework. https://www.playframework.com/. Accessed 1 June 2017.
34. H2 Database Engine. http://www.h2database.com/html/main.html.. Accessed 1 June 2017.
35. MySQL. https://www.mysql.com/. Accessed 1 June 2017.
36. Altschul SF, Madden TL, Schäffer AA, et al. Gapped BLAST and PSI-BLAST: a new generation of protein database search programs. Nucleic Acids Res. 1997;25:3389–402.
37. Wasmuth EV, Lima CD. UniProt: the universal protein knowledgebase. Nucleic Acids Res. 2016;45:1–12.

38. Kanehisa M, Furumichi M, Tanabe M, Sato Y, Morishima K. KEGG: new perspectives on genomes, pathways, diseases and drugs. Nucleic Acids Res. 2017;45:D353–61.
39. Cokelaer T, Pultz D, Harder LM, Serra-Musach J, Saez-Rodriguez J. BioServices: a common python package to access biological web services programmatically. Bioinformatics. 2013;29:3241–2.
40. Das J, Yu H. HINT: high-quality protein interactomes and their applications in understanding human disease. BMC Syst Biol. 2012;6:92.
41. Chatr-Aryamontri A, Oughtred R, Boucher L, Rust J, Chang C, Kolas NK, et al. The BioGRID interaction database: 2017 update. Nucleic Acids Res. 2017;45: D369–79.
42. Ohue M, Matsuzaki Y, Ishida T, Akiyama Y. Improvement of the protein-protein docking prediction by introducing a simple hydrophobic interaction model: an application to interaction pathway analysis. Lecture Notes in Bioinform. 2012;7632:178–87.
43. Ohue M, Matsuzaki Y, Shimoda T, Ishida T, Akiyama Y. Highly precise protein-protein interaction prediction based on consensus between template-based and de novo docking methods. BMC Proc. 2013;7:S6.
44. Matsuzaki Y, Ohue M, Uchikoga N, Akiyama Y. Protein-protein interaction network prediction by using rigid-body docking tools: application to bacterial chemotaxis. Protein Pept Lett. 2014;21(8):790–8.
45. Bekker GJ, Nakamura H, Kinjo AR. Molmil: a molecular viewer for the PDB and beyond. J Cheminform. 2016;8:42.
46. Szklarczyk D, Morris JH, Cook H, Kuhn M, Wyder S, Simonovic M, et al. The STRING database in 2017: quality-controlled protein-protein association networks, made broadly accessible. Nucleic Acids Res. 2017;45:D362–8.
47. Dall'Era MA, Oudes A, Martin DB, Liu AY. HSP27 and HSP70 interact with CD10 in C4-2 prostate cancer cells. Prostate. 2007;67:714–21.
48. Oefner C, Roques BP, Fournie-Zaluski MC, Dale GE. Structural analysis of neprilysin with various specific and potent inhibitors. Acta Crystallogr Sect D Biol Crystallogr. 2004;60:392–6.
49. Allaire M, Li Y, MacKenzie RE, Cygler M. The 3-D structure of a folate-dependent dehydrogenase/cyclohydrolase bifunctional enzyme at 1.5 å resolution. Structure. 1998;6:173–82.

Gao *et al. BMC Bioinformatics* 2018, **19**(Suppl 4):100
https://doi.org/10.1186/s12859-018-2065-x

BMC Bioinformatics

RESEARCH Open Access

RaptorX-Angle: real-value prediction of protein backbone dihedral angles through a hybrid method of clustering and deep learning

Yujuan Gao[1,2], Sheng Wang[2], Minghua Deng[1,3,4*] and Jinbo Xu[2*]

From The Sixteenth Asia Pacific Bioinformatics Conference
Yokohama, Japan. 15-17 January 2018

Abstract

Background: Protein dihedral angles provide a detailed description of protein local conformation. Predicted dihedral angles can be used to narrow down the conformational space of the whole polypeptide chain significantly, thus aiding protein tertiary structure prediction. However, direct angle prediction from sequence alone is challenging.

Results: In this article, we present a novel method (named RaptorX-Angle) to predict real-valued angles by combining clustering and deep learning. Tested on a subset of PDB25 and the targets in the latest two Critical Assessment of protein Structure Prediction (CASP), our method outperforms the existing state-of-art method SPIDER2 in terms of Pearson Correlation Coefficient (PCC) and Mean Absolute Error (MAE). Our result also shows approximately linear relationship between the real prediction errors and our estimated bounds. That is, the real prediction error can be well approximated by our estimated bounds.

Conclusions: Our study provides an alternative and more accurate prediction of dihedral angles, which may facilitate protein structure prediction and functional study.

Keywords: Dihedral angle prediction, Protein structure prediction, Clustering, Residual network, Deep learning

Background

It has been shown that sequences contain rich information for protein tertiary structure prediction as well as functional study [1, 2]. But it is challenging to directly predict tertiary structure from primary sequence, so the hierarchical approach has been widely accepted as one of the most efficient methods. That means to transform the ultimate goal into several sub-problems, such as secondary structure prediction, solvent accessibility prediction, residue-residue contact prediction, etc. [3] reviewed the progress in the field of intermediate state or one-dimensional property prediction. It has been shown that predicted secondary structure is useful in the prediction of disordered and flexible regions, fold recognition and function prediction. However, secondary structure states are described as discrete classes and there is no clear boundary between coil and helical/strand states. It is a significant step towards establishing the structure and function of a protein to predict local conformation of the polypeptide chain. The local structural bias information restricts the possible conformations of a sequence segment and therefore narrows down the conformation space of the whole polypeptide chain significantly. Thus, prediction of dihedral angles is especially useful for protein tertiary structure prediction.

On the whole, dihedral angle prediction may benefit protein tertiary structure prediction in several aspects. Firstly, dihedral angle prediction may act as substitute or supplement for secondary structure prediction [4–6]. Secondly, It can be used in generation of sequence/structure alignment. For one thing, it can be directly applied to

*Correspondence: dengmh@pku.edu.cn; dengmh@math.pku.edu.cn; jinbo.xu@gmail.com;
[1]Center for Quantitative Biology, Peking University, Beijing, China
[2]Toyota Technological Institute at Chicago, 6045 S Kenwood Ave., Chicago, USA
Full list of author information is available at the end of the article

structure alignment methods based on dihedral angles [7, 8] and may aid refinement of target-template structure alignment. For another, considering predicted angles to refine multiple sequence alignment may narrow the gap between sequence and structure alignment, thus aiding de novo prediction of structural properties. In addition, dihedral angle prediction may also find applications in protein structure prediction that includes but not limits to fold recognition approaches [9, 10], fragment-free tertiary structure prediction [11], tertiary structure refinement and structure quality assessment [12] and functional study, such as ligand-binding site prediction [13].

There are mainly two kinds of problems in dihedral angle prediction: angle region prediction and real value prediction, which corresponds to two different representations of protein backbone local structural bias.

Initially, Ramachandran basin is an intuitive description of local structural bias [14]. A Ramachandran basin is a specific region of a Ramachandran plot and illustrates the preference of torsion angle values. Each angle pair can be assigned a basin label. With more basins, the assignment would be harder but the representation would be more accurate and vice versa. Colubri et al. tested the ability to recover the native structure from a given basin assignment for each residue to investigate the level of representation required to simulate folding and predict structure, resulting in five basins [15]. Gong et al. partitioned ϕ, ψ-space into a uniform grid of 36 squares, each $60^\circ \times 60^\circ$, thus resulting in 36 basins, and showed that they successfully reconstructed six proteins solely from their mesostate (basin label) sequences [16]. There are also some other methods to define basins and do angle region prediction with different definitions of basins [17–20]. Although it is vital to determine the proper number of regions and clearly define the boundary, a universal algorithm to generate Ramachandran basins and assign basin labels remains to be developed. In our study, k-means clustering serves as the basin generator and label assigner.

While Ramachandran basin provides an overall description of conformation, it is a coarse-grained representation and lacks statistical explanations describing the torsion angle distributions of each basin. In consideration of the circular nature of angles, traditional parametric or non-parametric density estimation methods cannot work properly to approximate Ramachandran distributions. Fortunately, directional distributions such as von Mises distribution could solve the problem [21]. Bivariate von Mises distribution (mixtures) has been used to model protein dihedral angle distribution [22, 23], which removes arbitrariness in defining the boundary between discrete states. In this study, we assume angle pairs in each basin follow a bivariate von Mises distribution to derive the log-likelihood of each clustering.

Thanks to the rapid growth of Protein Data Bank and computational and algorithmic development in machine learning (especially deep learning), several supervised machine learning methods have been proposed to predict real values of dihedral angles. As ϕ values in α-helices and β-sheets are quite similar, ψ seems more informative. Wood et al. first developed a method DESTRUCT for prediction of real-valued dihedral angle ψ and used this information for prediction of the protein secondary structure with high accuracy [4]. Wu et al.proposed a composite machine-learning algorithm called ANGLOR to predict real-value protein backbone torsion angles from protein sequences [24]. The input features of ANGLOR include sequence profiles, predicted secondary structure and solvent accessibility. The mean absolute error (MAE) of the ϕ/ψ prediction was reported to be $28^\circ/\ 46^\circ$. Later Song et al. developed TANGLE based on a two-level support vector regression approach using a variety of features derived from amino acid sequences, including the evolutionary profiles and natively disordered region as well as other global sequence features [25]. The MAE of the ϕ/ψ was $27.8^\circ/\ 44.6^\circ$. Xue et al. established a neural network method called Real-SPINE, with sequence profiles generated from multiple sequence alignment and predicted secondary structures as inputs [26]. In 2015, they presented SPIDER2 [27] by improving SPIDER [28] through iterative learning, which used a deep artificial neural network (ANN) with three hidden layers of 150 nodes. They fed the predicted torsion angles of last layer as the input to the following generation and reported 19° and 30° for mean absolute errors of backbone ϕ and ψ angles, respectively. As it is impossible to introduce all methods here, interested readers can refer to excellent reviews [29, 30].

Although there has been tremendous development, their performance is still limited by their shallow architectures. Inspired by the excellent performance of convolution neural network in predicting secondary structure [31] and order/disorder regions [32] and also the success of residual framework to do contact prediction [33], we adopt the ultra deep residual framework of convolutional neural network to do k-means basin label probability prediction.

However, even though a protein backbone conformation can be highly accurately rebuilt from its respective native dihedral angles, accumulation of errors in predicted angles can lead to large deviation in three-dimensional structures, which prevents angle prediction from its direct use in building protein structures [27]. It is of great significance to produce the corresponding confidence scores for the real value predictions, i.e., we need to know the confidence level of the predictions. Otherwise the effect of predicted dihedral angles as restraints for three dimensional structure prediction would be limited [34]. Zhou et al. had developed SPIDER2 [27] to predict real-valued angles and then separately SPIDER2-Delta [35] to predict error of those predicted structural properties. Here we describe a simple hybrid technique to predict angles and confidence scores simultaneously.

Another problem that need to be considered is the periodicity of angles. For example, if an angle $\theta = 179°$ is predicted to be $-179°$, the error would be treated as 358° instead of 2°. There are some approaches proposed to reduce the impact of cyclic nature of angles. One was angle shifting to reduce confusion at 0° and 360° (or $-180°$ and 180°), e.g., shifting ψ by 100° and ϕ by $-10°$ [26] or adding 100° to the angles between $-100°$ and 180° and adding 460° to the angles between $-180°$ and $-100°$ [34]. But the improvement was limited and strongly depended on the angle range. For amino acids such as alanine that had minimal residues in the affected range, angle shifting made little difference [29]. A better choice was to take advantage of the inherent angle periodicity of trigonometric functions, that is, mapping the angles to their sine and cosine values [27], which has achieved best performance so far. Inspired by this, we deal with equivalent trigonometric representations of dihedral angle pairs, rather than real value angles.

Considering dihedral angles share similar patterns in alpha helix and beta strand, the acceptable (ϕ, ψ) patterns are limited. Moreover, it is much easier to do classification than regression. Also indebted to mixture models and Expectation-maximization algorithm, we develop a hybrid method of k-means clustering and deep learning to do angle prediction, combining advantages of discrete and continuous representation of dihedral angles. Specifically, we firstly generate a set of clusters of (ϕ, ψ) from training data, in which we could get the distribution of each cluster; then we use deep learning methods to predict discrete labels; lastly we predict real value angles by mixing empirical clusters with their predicted probabilities. We employ a residual framework of convolutional neuron network in RaptorX-Angle to predict the cluster label probabilities. We test our method on filtered PDB25 dataset as well as CASP (Critical Assessment of protein Structure Prediction) targets and compare with other three state-of-art methods. Tested on the subset of PDB25, our method gains about 0.5°and 1.4°for ϕ and ψ better MAE than SPIDER2, currently among the best backbone angle predictors. Our method also performs better than SPIDER2 on the CASP11 and CASP12 test targets. The advantage is even more obvious when looking into detailed secondary structural regions.

Methods

K-means clustering of angle vectors

Genearating k-means "centers" from angle vectors

For a dihedral angle pair (ϕ, ψ), we can equivalently denote it by an angle vector

$$\mathbf{v} = (\cos(\phi), \sin(\phi), \cos(\psi), \sin(\psi)) .$$

Conversely, given the vector representation $\mathbf{v}$, we can easily derive the corresponding angles ϕ and ψ (Additional file 1: S1.1). We run k-means on angle vectors to cluster dihedral angle pairs in training set into $K = 10, 20, \ldots, 100$ clusters. Then we normalize the K centres $\{\mathbf{C}_k\}_{k=1}^K$ and get the final "centers" $\left\{\widetilde{\mathbf{C}}_k = (\widetilde{c}_{k0}, \widetilde{c}_{k1}, \widetilde{c}_{k2}, \widetilde{c}_{k3})\right\}_{k=1}^K$, so that each "centre" $\widetilde{\mathbf{C}}_k$ is a valid representation for some angle pair (Additional file 1: S1.2).

Predicting "true" labels from k-means

Given the K normalised vector "centres" $\left\{\widetilde{\mathbf{C}}_k\right\}_{k=1}^K$, we could assign the "true" label for each dihedral angle pair as the one whose corresponding normalised centre was closest to its respective vector representation. Then the "true" labels can be used as the training labels to build a deep learning model as a classifier to predict labels for testing data.

Deep learning model details

Deep Convolutional Neural Network (DCNN)

DCNN consists of multiple convolutional blocks. A convolutional block is a neural network that implements a composite of linear convolution and nonlinear activation transformation. Convolution is used in place of general matrix multiplication, which can better capture local dependency. It has been widely accepted that protein torsion angles strongly depend on neighbour residues [36–38]. So DCNN is ideal to abstract angle information from sequence.

Residual Network (ResNet)

DCNN can integrate features in hierarchical levels and some work has shown the significance of depth [39]. However, with the depth increasing, accuracy gets saturated and even degraded. That is because adding more layers may lead to higher training error as identity mapping is difficult to fit with a stack of nonlinear layers [40]. ResNet was proposed as a residual learning framework to ease the training of substantially deeper networks [41]. Figure 1 demonstrates the basic architecture of ResNet in RaptorX-Angle. Figure 1a is a residual block, which consists of 2 convolution layers and 2 activation layers, and the ResNet consists of stacked residual blocks (Fig. 1b). The activation layer conducts a simple nonlinear transformation of its input depending on the activation function with no additional parameters. In this work, we used the ReLU activation function [42].

Logistic regression layer

DCNN and ResNet can capture information from data and output abstract features. To do classification for

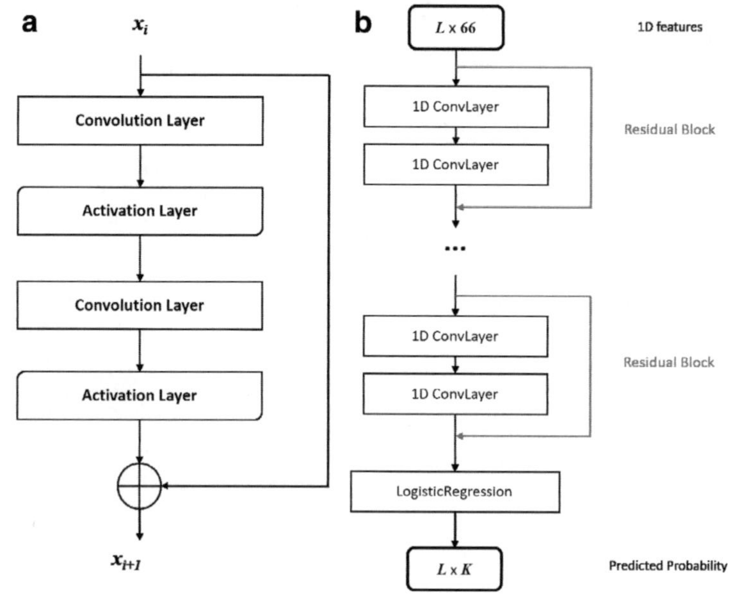

Fig. 1 Illustration of the ResNet model in RaptorX-Angle. **a** A building block of ResNet with x_i and x_{i+1} being input and output, respectively. **b** The ResNet model architecture as a classifier with stacked residual blocks and a logistic regression layer. Here L is the sequence length of the protein or total number of residues under prediction and K is the number of clusters

residues, a logistic regression layer is added as the final layer in RaptorX-Angle, which could output the marginal probability of K labels (Fig. 1b).

Loss function

We train model parameters through maximizing the probability of angle pairs belong to the "true" labels. Naturally, the loss function is defined as the negative log-likelihood averaged over all residues of the training proteins.

Regularization and optimization

As is widely used in machine learning, the log-likelihood objective function is penalized with a L_2-norm of the model parameters to prevent overfitting. Thus, the final objective function has two items: loss function and regularization item, with a regularization factor λ to balance the two items. That is, the final objective function is:

$$\max_{\theta} \quad \log P_\theta(Y|X) - \lambda \|\theta\|^2$$

where X is the input features, Y is the output labels, θ is the model parameters and λ is the regularization factor used to balance the log likelihood and regularization. We use Adam [43] to minimize the objective function, which usually can converge within 20 epochs. The whole algorithm has been implemented by Theano [44] and mainly run on a GPU card.

Input features

For each residue in each protein sequence, we generate a total of 66 input features, of which 20 from position specific scoring matrix(PSSM) of PSI-BLAST [45], 20 from position-specific frequency matrix (PSFM) of HHpred [46, 47], 20 from primary sequence, 3 from predicted solvent accessibility (ACC) and 3 from predicted secondary structure(SS) probabilities (Additional file 1: S1.3).

Predicting real-value angles from predicted marginal probability

From the last logistic regression layer of the deep learning model, we could predict the marginal probability $\mathbf{P} = (p_1, p_2, \ldots, p_K)$ of an angle pair for each label. We use the marginal probability rather than the single predicted label to reduce bias. Concretely, we calculate the weighted mean by:

$$\widehat{\mathbf{v}} = (v_0, v_1, v_2, v_3) = \sum_{k=1}^{K} p_k \widetilde{\mathbf{C_k}},$$

Finally, we normalise $\widehat{\mathbf{v}}$ to get

$$\widehat{\cos(\phi)} = \frac{v_0}{\sqrt{{v_0}^2 + {v_1}^2}}, \widehat{\sin(\phi)} = \frac{v_1}{\sqrt{{v_0}^2 + {v_1}^2}},$$

$$\widehat{\cos(\psi)} = \frac{v_2}{\sqrt{{v_2}^2 + {v_3}^2}}, \widehat{\sin(\psi)} = \frac{v_3}{\sqrt{{v_2}^2 + {v_3}^2}}.$$

and we could derive the predicted real values $\widehat{\phi}, \widehat{\psi}$ from this angle vector (Additional file 1: S1.1). We also tried to predict real-value angles from labels with top $R(R < K)$ probabilities when K is well chosen (Additional file 1: S2.3).

Programs to compare and evaluation metrics

We compare our method with three available standalone softwares SPIDER2 [27], SPINE X [11], and ANGLOR [24]. All the programs are run with parameters suggested in their respective papers.

We evaluate the performance by Pearson Correlation Coefficient (PCC) and Mean Absolute Error (MAE) as described by [48], for assessing the prediction of ϕ/ψ angles. Considering the periodicity of angles, PCC is calculated between the cosine (sine) values of predicted and experimentally determined angles. MAE is the average absolute difference between predicted and experimentally determined angles. The periodicity of an angle has been taken care of by utilizing the smaller value of the absolute difference $d\left(= |\theta_{pred} - \theta_{exp}|\right)$ and $360-d$ for average, where θ_{pred} is the predicted angle and θ_{exp} is the true angle value.

Results

Datasets

We use the targets from PDB25 updated in February, 2016. The set consists of 10820 non-redundant protein chains, in which any two chains share no more than 25% sequence identity. To remove impact of disordered regions, we filter out proteins with internal disordered regions by DSSP [49]. Finally we get 7604 proteins. We then randomly select 5070 proteins as the candidate training set, 1267 as validation set (VL1267, see Additional file 2) and the remaining 1267 as test set (TS1267, see Additional file 3). We also test on 85 CASP11 targets (see Additional file 4) and the latest 40 CASP12 targets (see Additional file 5) with publicly released native structures. To remove redundancy between training proteins and CASP targets, we run MMseqs2 [50], which is similar but more sensitive and faster than BLAST (PSI-BLAST) for protein sequence homology search, with *seqID* cutoff 0.25 and also E-value cutoff 0.001 to filter 5070 the candidate training proteins, resulting in 5046 training proteins (TR5046, see Additional file 6).

Choosing a proper number of clusters

A vital problem is how to select the number of clusters, which can be reduced to defining measures for clustering evaluation. Here we adopt two measures: (i) entropy loss based on discrete distribution; (ii) loglikelihood based on continuous distribution to evaluate 10 different clusterings ($K = 10, 20, \ldots, 100$). Firstly, we do k-means clustering on TR5046 and get K empirical clusters. Secondly, we train the deep learning models and do classification on VL1267, then we can obtain the predicted marginal probability of the K clusters $\mathbf{P_i} = (p_{i1}, p_{i2}, \ldots, p_{iK}), i = 1, 2, \ldots, N$, where i is the index of residue and N is the total number of residues in VL1267.

Entropy loss

Entropy $H(\cdot)$ is always used to measure the information of a distribution. From k-means clustering on TR5046, the background distribution among clusters $\mathbf{P_0} = (p_{01}, p_{02}, \ldots, p_{0K})$ could be derived. Then the entropy loss of this clustering on VL1267 can be calculated as the mean difference between entropy of background distribution and predicted marginal distribution:

$$
\begin{aligned}
EL &= \frac{1}{N}\sum_{i=1}^{N}\left(H\left(\mathbf{P_0}\right) - H\left(\mathbf{P_i}\right)\right) \\
&= \frac{1}{N}\sum_{i=1}^{N}\left(\sum_{k=1}^{K} p_{0k}\log\left(p_{0k}\right) - \sum_{k=1}^{K} p_{ik}\log\left(p_{ik}\right)\right)
\end{aligned}
$$

which can roughly evaluate the information gain from the clustering. Here N is the number of residues in VL1267.

Loglikelihood

To demonstrate the detailed information of each cluster, we need a continuous angular(circular) distribution defined on the torus. Mixture bivariate von Mises distributions are successfully used to describe the local bias of torsion angle pair (ϕ, ψ) [21–23], we assume that angle pairs belong to the same cluster k obey a common bivariate von Mises distribution f_k with parameters $\Theta_k = \left(\kappa_1^k, \kappa_2^k, \kappa_3^k, \mu^k, \nu^k\right)$. Here,

$$
\begin{aligned}
f_k(\phi, \psi) = c\left(\kappa_1^k, \kappa_2^k, \kappa_3^k\right)\exp\Big\{&\kappa_1^k\cos\left(\phi - \mu^k\right) \\
&+\kappa_2^k\cos\left(\psi - \nu^k\right) + \kappa_3^k\cos\left(\phi - \mu^k - \psi + \nu^k\right)\Big\}
\end{aligned}
$$

where μ^k and ν^k are the mean value of ϕ and ψ, respectively; κ_1^k, κ_2^k are the concentrations, κ_3^k allows for the dependency between the two angles and $c\left(\kappa_1^k, \kappa_2^k, \kappa_3^k\right)$ is a normalization constant:

$$
c\left(\kappa_1^k, \kappa_2^k, \kappa_3^k\right) = (2\pi)^2\left\{I_0\left(\kappa_1^k\right)I_0\left(\kappa_2^k\right)I_0\left(\kappa_3^k\right) + 2\sum_{p=1}^{\infty} I_p\left(\kappa_1^k\right)I_p\left(\kappa_2^k\right)I_p\left(\kappa_3^k\right)\right\}
$$

in which $I_p(\kappa)$ is the modified Bessel function of the first kind and order p. Parameters $\left\{\Theta_k = \left(\kappa_1^k, \kappa_2^k, \kappa_3^k, \mu^k, \nu^k\right)\right\}_{k=1}^{K}$ can be intuitively estimated from the empirical clusters $\{(\phi, \psi)_k\}_{k=1}^{K}$ [51]. Then the density function for the torsion angle pair (ϕ, ψ) can be approximately described as:

$$
f(\phi, \psi) = \sum_{k=1}^{K} p_k f_k(\phi, \psi)
$$

where p_k is the predicted marginal probability of (ϕ, ψ) belongs to cluster k. Then the log-likelihhod for the VL1267 can be calculated as:

$$LL = \frac{1}{N}\sum_{i=1}^{N} \log f(\phi_i, \psi_i) = \frac{1}{N}\sum_{i=1}^{N} \log \sum_{k=1}^{K} p_{ik} f_k(\phi_i, \psi_i)$$

Selecting proper K

Figure 2 shows the result of entropy loss and loglikelihood with respect to the number of clusters. As expected, the loglikelihood increases along with K, which means it can better describe the data with more clusters. But when K goes larger than 30, there is an obvious decrease in entropy loss. Maybe that is because the more clusters are used, the more challenging it would be to do angle prediction. As there is a soaring information gain when K goes from 10 to 20 and little difference when K increases from 20 to 30, we test every single clustering between 20 and 30 and there is no significant benefit with more clusters. So we just choose $K = 20$ to do following studies.

Feature contribution study

The features can be divided into three categories: sequence information including amino acid (aa) and profile, predicted secondary structure (SS) and solvent accessibility (ACC). Sequence profile information are generated from PSI-BLAST (PSSM) and HHpred (PSFM) (See Additional file 1: S1.3 for more details). To test the impact of different feature combinations, we design six experiments: (1) basic1 = 20 PSSM + 20 aa; (2) basic2 = 20 PSFM + 20 aa; (3) basic = 20 PSSM + 20 PSFM + 20 aa; (4) basic + 3 ACC; (5) basic + 3 SS; (6) basic + 3 ACC + 3 SS. The network architecture is fixed as $N_{layers} = 5, N_{nodes} = 100, halfWinSize = 3$ (ResNet 5-100-3), and the regularization factor is fixed to be 0.0001.

Table 1 shows the MAE performance of different feature combinations on TS1267. From the first three experiments with only sequence information involved, there is little performance difference between PSSM and PSFM, and the combination of PSSM and PSFM gains the best accuracy. So PSSM and PSFM are complementary and both unignorable. ACC and SS both contribute significantly and also the combination gain the best accuracy. Finally we use the whole set of features.

Overall PCC performance of cosine values compared with other methods

To tune proper regularization factor and also network architectures, we perform 5-fold cross validation on TR5046 (Additional file 1: S2.1 and S2.2). Finally, we choose an ensemble of 6 networks (Additional file 1: S2.2). We test our method on TS1267 and also the popular CASP targets, including 85 CASP11 targets and 40 CASP12 targets. Table 2 shows the PCC performance of cosine values on the three benchmarks. RaptorX-Angle has gained the highest PCC on all datasets. We also evaluate PCC performance of sine values (See Additional file 1: S2.4) and get similar results.

Overall MAE performance compared with other methods

Table 3 shows the MAE performance on the three benchmarks in different secondary structural regions of our RaptorX-Angle comparing with other three methods. All methods have larger MAE on CASP targets than on TS1267. It is reasonable since CASP targets are usually hard to predict. It can be seen that RaptorX-Angle performs the best on all benchmarks, with about 0.5° and 1.4° for ϕ and ψ better MAE on both TS1267 and CASP12 and slightly better performance on CASP11 than the second best method SPIDER2. We perform Student's t test of absolute errors between RaptorX-Angle and SPIDER2. As a result, the p-values for ϕ/ψ are $8.65e-12/2.79e-33$, $5.13e-2/8.36e-2$ and $1.28e-5/2.59e-8$ on TS1267, CASP11 and CASP12, respectively. That is , the advantage of RaptorX-Angle over SPIDER2 on TS1267 and CASP12 is statistically more significant than on CASP11. These results demonstrate the rationality of representing the Ramachandran plot with a limited number of clusters, say

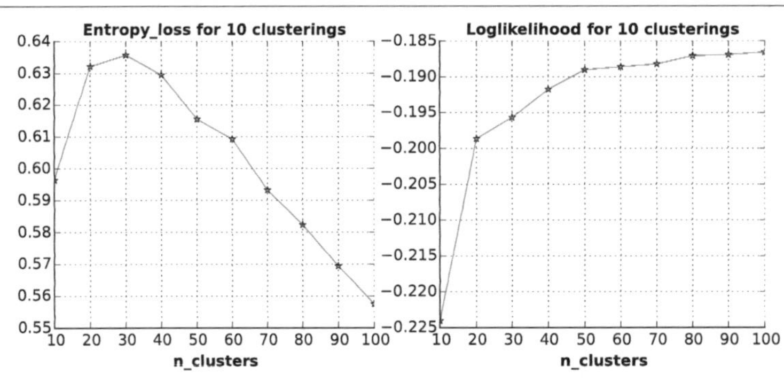

Fig. 2 Selecting proper number of clusters K. Left: relationship between entropy loss of discrete label probabilities and number of clusters K; Right: relationship between loglikelihood of mixture bivariate von Mises distribution and number of clusters K

Table 1 The mean absolute error of different feature combinations with ResNet 5-100-3 on TS1267

Feature combination	Phi	Psi	Phi_H	Psi_H	Phi_E	Psi_E	Phi_C	Psi_C
basic1=PSSM+aa	19.97	31.97	9.82	17.57	20.70	26.97	29.84	49.66
basic2=PSFM+aa	20.02	31.78	9.86	17.68	20.46	26.38	30.10	49.39
basic=PSSM+PSFM+aa	19.27	30.04	9.11	15.58	19.70	24.64	29.35	48.02
basic+ACC	19.08	29.30	9.07	15.44	19.36	23.18	29.11	47.10
basic+SS	19.19	28.73	8.56	13.76	19.29	22.43	31.00	47.95
basic+ACC+SS	18.58	27.98	8.45	13.37	19.03	22.14	28.61	46.21

Phi and Psi denote MAE for all residues
Phi_H and Psi_H denote MAE for residues in helix region
Phi_E and Psi_E denote MAE for residues in beta strand region
Phi_C and Psi_C denote MAE for residues in coil region

20 clusters, and also reflect the power of deep learning methods.

Mean absolute error performance study in VL1267

In methodology, the conversion from angle pair to trigonometric vector is nonlinear, the prediction error may depend on angles. And in biology, prediction error may differ for different amino acids with different microscopic biochemical properties, and also for different protein classes with different macroscopic structures. So we perform detailed studies on prediction error in VL1267.

Study mean absolute error performance for different clusters

As each cluster corresponds to a certain angle region, we calculate the MAE for each cluster in VL1267. We observe that the 20 clusters are well consistent with Ramachandran plot and also the two peaks for ϕ and ψ [11] (Fig. 3 Left). And the prediction errors differ a lot between clusters. It turns out that clusters with more residues in coil region tend to result in larger prediction errors. Moreover, prediction error for ϕ is smaller than for ψ. But there are three uncommon clusters with larger MAE for ϕ, i.e., 5, 6 and 10 (Fig. 3 Right). Clusters 5 and 6 are totally in one of the peak areas in Ramachandran plot, which may indicate some interesting biological discoveries.

Mean absolute error performance for each amino acid type

As different amino acids have different stereochemical and physiochemical properties, they are anticipated to have different degrees of difficulty for the torsion angle prediction. In Table 4, we examine the MAE performance for each of 20 amino acid types. Glycine, with no side-chain atom except for a proton, has least steric restriction to backbone dihedral angle motions. As a result, it has the largest prediction error (43.32° / 39.59° for ϕ/ψ). In contrast, Proline has the least MAE (8.84°) for ϕ but has an unusually large MAE (33.00°) for ψ prediction due to its special side-chain structure, which is consistent with [24]. In addition, three of the amino acids (Ile, Leu and Val) with the smallest MAE are all hydrophobic.

Table 2 Pearson correlation coefficient of cosine values between predicted and true angles

	TS1267	CASP11	CASP12
	$\cos(\phi)/\cos(\psi)$	$\cos(\phi)/\cos(\psi)$	$\cos(\phi)/\cos(\psi)$
RaptorX-Angle	**0.7111/0.7576**	**0.6585/0.7103**	**0.6539/0.6979**
SPIDER2	0.6893/0.7427	0.6485/0.7095	0.6299/0.6761
SPINE X	0.6410/0.6543	0.5015/0.4891	0.4990/0.5039
ANGLOR	0.4775/0.6226	0.4437/0.5868	0.4431/0.5772

Mean absolute error performance for different protein classes

After studying on MAE performance in microcosmic view, we intend to study the performance for different macroscopical structures. We abstract 99, 117, 171, 117 proteins from VL1267 (resulted in 17696, 24874, 47304 and 19645 residues) in all α, all β, α/β and $\alpha + \beta$ classes, respectively. We calculate the absolute error for every residue in each class. Figure 4 shows the violin plot of prediction error for ϕ (Left) and ψ (Right). A violin plot is similar to box plot except that it also shows the probability density of the data. We can see although the MAE for ϕ are smaller for all protein classes, prediction errors belong to each protein class have their own distribution pattern and the pattern is similar between ϕ and ψ. Overall, prediction errors are smallest in all α proteins and largest in all β for both ϕ and ψ predictions.

Estimating confidence score of predicted angles

Generally, variance σ^2 includes variance within cluster σ_w^2 and variance between cluster σ_b^2. To produce the confidence score of our predicted angles, we calculate the standard deviation from variances within a cluster. Specifically, for each cluster k, we can get the in-cluster variance $\sigma_k^2(\theta)$ from training data, where $\theta = \phi$ or ψ. Then we derive the variance of prediction by:

Table 3 Mean absolute error of four methods for different secondary structural regions on three benchmarks: TS1267, 85 CASP11 targets and 40 CASP12 targets

(°)	Phi	Psi	Phi_H	Psi_H	Phi_E	Psi_E	Phi_C	Psi_C
TS1267								
RaptorX-Angle	**18.08**	**26.68**	**8.35**	**12.98**	**18.24**	**20.94**	**27.88**	**44.11**
SPIDER2	18.57	28.02	8.59	14.52	19.28	23.09	28.28	44.73
SPINE X	20.31	34.05	9.32	16.69	22.23	31.23	30.32	53.42
ANGLOR	24.01	43.59	9.29	26.41	27.47	40.88	36.89	62.72
CASP11								
RaptorX-Angle	**20.00**	**30.14**	**9.49**	**15.65**	**18.82**	**23.58**	**29.87**	46.89
SPIDER2	20.18	30.32	9.53	16.05	19.77	24.50	29.88	**46.84**
SPINE X	24.85	46.58	13.57	29.65	26.25	43.65	33.88	63.49
ANGLOR	25.69	46.17	9.99	27.72	28.08	43.85	37.96	64.03
CASP12								
RaptorX-Angle	**20.69**	**32.73**	9.28	**16.73**	**19.94**	**26.06**	**31.22**	**51.02**
SPIDER2	21.13	34.17	**9.13**	17.19	21.35	28.56	31.95	52.76
SPINE X	24.85	46.57	11.52	26.34	26.98	46.04	35.85	65.33
ANGLOR	25.79	47.37	9.69	28.81	29.11	44.79	38.65	65.74

Same notations with Table 1

$$var(\theta) = \sigma^2(\theta) = \sum_{k=1}^{K} p_k \sigma_k^2(\theta)$$

Figure 5 shows the mean standard deviation for ϕ and ψ in different regions. As expected, the smallest variance appears in helix region, and then strand and lastly coil region. The standard deviation in disordered regions are rather large and quite similar to coil regions, which is consistent with our prior knowledge that disordered region resembles loop region and is rather flexible.

Figure 6 demonstrates the relationship between MAE and mean standard deviation for ϕ and ψ in different regions on VL1267. Roughly, the relationship is linear ($R^2 = 0.8911$). So the MAE can be bounded well by the standard deviation. We predict the error for each residue in each target from TS1267 and calculate corresponding Pearson and Spearman correlation coefficients (*PCC* and *SCC*) between prediction errors and true errors, and also the mean absolute error for prediction errors (MAEPE). Finally, we obtain $PCC = 0.3109, SSC = 0.5427, MAEPE = 13.94$ for ϕ and $PCC = 0.2597, SCC = 0.4751, MAEPE = 26.21$ for ψ. We also try to fit two linear models for ϕ and ψ separately on the all data points in VL1267 and get similar testing results. This indicates that the mean for different secondary structural regions almost contains enough information about the relationship between the estimated standard deviation and prediction error (Additional file 1: S2.6).

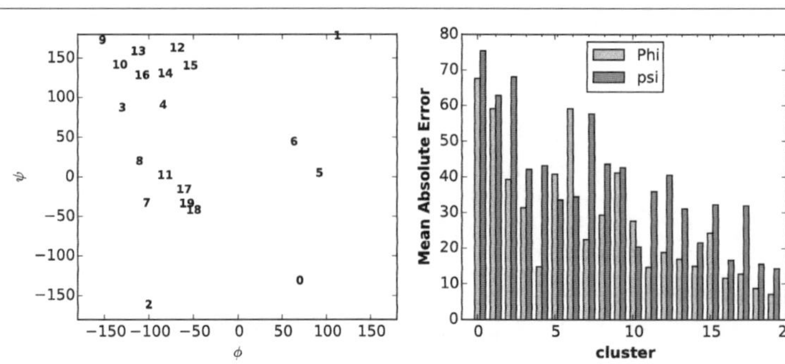

Fig. 3 Mean absolute error performance for different clusters in VL1267. Left: visualization of 20 cluster centers on the Ramachandran plot with smaller number indicating smaller size. Right: mean absolute error for different clusters

Table 4 Mean absolute error performance for each amino acid type in VL1267

Amino acids	Abundance	Frequency(%)	$\phi(^\circ)$	$\psi(^\circ)$
A (Ala)	22527	8.46	13.87	22.92
C (Cys)	3151	1.18	20.50	28.66
D (Asp)	15946	5.99	20.71	30.80
E (Glu)	18326	6.89	14.75	23.97
F (Phe)	10812	4.06	18.13	26.10
G (Gly)	19133	7.19	43.32	39.59
H (His)	5989	2.25	22.04	31.12
I (Ile)	15302	5.75	12.79	20.12
K (Lys)	15299	5.75	16.71	25.83
L (Leu)	24731	9.29	12.49	21.37
M (Met)	5833	2.19	16.71	24.86
N (Asn)	11383	4.28	27.38	32.04
P (Pro)	11977	4.50	8.84	33.00
Q (Gln)	10163	3.82	15.96	24.72
R (Arg)	13529	5.08	16.81	25.45
S (Ser)	15991	6.01	20.83	33.92
T (Thr)	14309	5.38	17.12	30.92
V (Val)	18612	6.99	13.70	20.94
W (Trp)	3854	1.45	18.05	27.61
Y (Tyr)	9287	3.49	18.83	27.02
Total	266154	100	18.32	27.15

Computational cost analysis

All mentioned methods could do angle prediction target by target, so the computational cost is bounded by the longest protein (i.e., protein with the largest number of residues). To generate angle predictions for 1xdoA, the largest protein in TS1267 with 685 residues, it takes 726s, 123s, 370s and 524s for ANGLOR, spineX, SPIDER2 and RaptorX-Angle, respectively.

As far as we see, the computational cost is mainly determined by method outline, network complexity, feature engineering and technical resources. ANGLOR is a composite method and the technology was not so developed at that time, it needs the most time. While spineX just adopted a simple network, SPIDER2 used more features iteratively in a more complex network and it takes longer than spineX.

Compared with the second best SPIDER2, RaptorX-Angle used much deeper networks and also adopted profile information from hhblits (PSFM), besides PSSM from PSI-BLAST harnessed by spineX and SPIDER2. As a result, it takes SPIDER2 360s to generate features with 4 CPUs and 20s to predict angles using a CPU, while it takes RaptorX-Angle 385s to generate features with 4 CPUs, and 200s to predict angles from the features using a GPU card.

However, we can integrate the features of a total batch of proteins and run them all at once. Actually, it just takes 750s to do angle prediction for all proteins in TS1267, while other methods needs many CPUs in parallel. Overall, our method is faster for prediction of many proteins and has gained better prediction accuracy.

Discussion

We have transformed the hard real-valued prediction problem into a discrete label assignment problem, which has simplified the problem and also gained better results. Overall, our RaptorX-Angle gains the best PCC in terms of cosine and sine of angles on all datasets. It has about 0.5° and 1.4° for ϕ and ψ better MAE than the second best method SPIDER2 on a subset of PDB25. We have also calculated the two-state accuracy to see how much improvement there would be in large angle errors. RaptorX-Angle performs the best and has about 0.15 and 1 percent improvement over SPIDER2 for ϕ and ψ on TS1267(See Additional file 1: S2.5). Our method also works very well on the CASP targets.

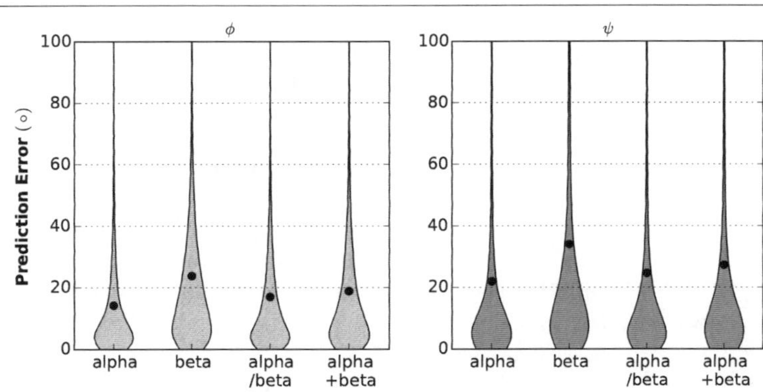

Fig. 4 Mean absolute error performance for different protein classes in VL1267. Left: for ϕ prediction. Right: for ψ prediction

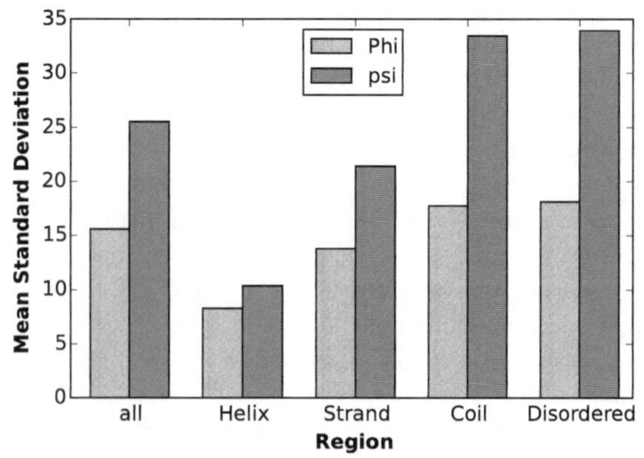

Fig. 5 Mean standard deviation for different secondary structural regions in TS1267

Moreover, we have estimated the prediction errors at each residue by a mixture of the clusters with their predicted probabilities. It has been shown that there is approximately linear relationship between the real prediction error and in-cluster standard deviation. That is a unique feature of our method. In addition, we check the prediction for disordered regions. As there is no angle information, we just analyze the standard deviation and get quite large values and similar patterns to coil region. It is consistent with our prior knowledge that disordered region is rather flexible and resembles loop region. We also do comprehensive studies on prediction performance in VL1267, both in microscopic and macroscopic view.

This simple technique has gained better performance than other state-of-art methods. It demonstrates that for protein structures, the 20 clusters contain enough information for (ϕ, ψ), which is an efficient compression of information. The idea that to predict dihedral angles from clustering has turned out to be successful due to three aspects. The first is the continuous growth of the solved structures [52], so we have enough training data. The second is the novel idea to predict real-value angles by mixing a set of clusters with their respective predicted probabilities. Conversely, such good performance demonstrated that the distribution of protein backbone dihedral angles can be described through a set of clusters. Last but not the least, the everlasting development of deep learning models and optimization methods proves to be a powerful tool to promote new ideas and exploit new methods.

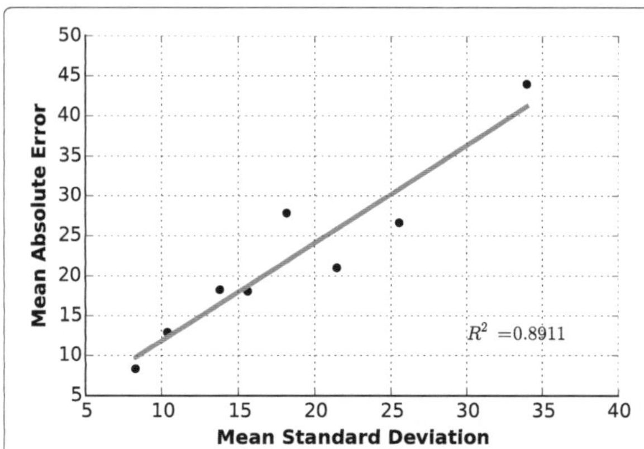

Fig. 6 Relationship betwee n prediction error and standard deviation. Eight points are for two kinds of angles (ϕ, ψ) in four secondary structural regions (total, helix, strand, coil)

But there is still room for improvement. RaptorX-Angle just used one-dimensional features and adopted 1D CNN. It cannot extract information of long range interaction. Heffernan et al. has developed more accurate SPIDER3 employing Long Short-Term Memory (LSTM) Bidirectional Recurrent Neural Networks (BRNNs), which are capable of capturing long range interactions [53]. That is, considering pairwise interaction can further increase prediction accuracy. We will include two-dimensional features and exploit 2D CNN to see how much improvement could be achieved.

Moreover, as mentioned before, accumulation of prediction errors has buried the usefulness of torsion angles to construct 3D models. There is a great demand to develop a proper technique to deal with the errors. A general pipeline to add angle restraints and confidence to improve protein tertiary structure prediction need to be developed.

Conclusions

In conclusion, this study has made a more reliable prediction of dihedral angles and may facilitate protein structure prediction and functional study. In the near future, we can use the angle restraints to do tertiary structure prediction, which should be considered carefully to deal with errors and flexibility. We can also adopt the angle prediction to aid structure alignment and fold recognition.

Additional files

Additional file 1: Supplementary Material. This additional file contains two parts: S1. supplementary methods for vector representation and normalisation of angles, and also feature generation; S2. supplementary results for determining regularization factor, testing different network architectures, testing different number of labels to use for real value prediction and performance comparison. (PDF 250 kb)

Additional file 2: This additional file contains target list for the validation set from PDB25 consisting of 1267 proteins. (TXT 7.42 kb)

Additional file 3: This additional file contains target list for the testing set from PDB25 consisting of 1267 proteins. (TXT 7.42 kb)

Additional file 4: This additional file contains target list for the testing set from CASP11 consisting of 85 proteins. (TXT 0.51 kb)

Additional file 5: This additional file contains target list for the testing set from CASP12 consisting of 40 proteins. (TXT 0.24 kb)

Additional file 6: This additional file contains target list for the training set from PDB25 consisting of 5046 proteins. (TXT 29.5 kb)

Abbreviations
ACC: Solvent accessibility; CASP: Critical assessment of protein structure prediction; DCNN: Deep convolutional neural network; MAE: Mean absolute error; PCC: Pearson correlation coefficient; ResNet: Residual network; SS: Secondary structure

Acknowledgements
None.

Funding
This work has been partly supported by the National Key Research and Development Program of China(No.2016YFA0502303), the National Key Basic Research Project of China (No. 2015CB910303), the National Natural Science Foundation of China (No.31471246) and China Scholarship Council.
It is also supported by National Institutes of Health grant R01GM089753 to JX and National Science Foundation grant DBI-1564955 to JX. The authors are also grateful to the support of Nvidia Inc. The publication costs of this manuscript were financially supported by National Institutes of Health grant R01GM089753 to JX and National Science Foundation grant DBI-1564955 to JX.

Availability of data and materials
The software package for our method, including feature generation and angle prediction, is open source and freely available from https://github.com/lacus2009/RaptorX-Angle.git. The target lists of datasets supporting the conclusions of this article are included as additional files. All the data are available upon request.

About this supplement
This article has been published as part of *BMC Bioinformatics* Volume 19 Supplement 4, 2018: Selected articles from the 16th Asia Pacific Bioinformatics Conference (APBC 2018): bioinformatics. The full contents of the supplement are available online at https://bmcbioinformatics.biomedcentral.com/articles/supplements/volume-19-supplement-4.

Authors' contributions
YG carried out the experiments and drafted the manuscript. SW participated in the feature generation and data integration. YG and JX implemented the package. JX and MD participated in the design of the study. All authors read and approved the final manuscript.

Ethics approval and consent to participate
Not applicable.

Consent for publication
Not applicable.

Competing interests
The authors declare that they have no competing interests.

Publisher's Note

Author details
[1]Center for Quantitative Biology, Peking University, Beijing, China. [2]Toyota Technological Institute at Chicago, 6045 S Kenwood Ave., Chicago, USA. [3]School of Mathematical Sciences, Beijing, China. [4]Center for Statistical Sciences, Beijing, China.

Published: 8 May 2018

References
1. Marks DS, Hopf TA, Sander C. Protein structure prediction from sequence variation. Nat Biotechnol. 2012;30(11):1072–80.
2. De Juan D, Pazos F, Valencia A. Emerging methods in protein co-evolution. Nat Rev Genet. 2013;14(4):249–61.
3. Kurgan L, Miri Disfani F. Structural protein descriptors in 1-dimension and their sequence-based predictions. Curr Protein Peptide Sci. 2011;12(6): 470–89.
4. Wood MJ, Hirst JD. Protein secondary structure prediction with dihedral angles. PROTEINS: Struct Funct Bioinform. 2005;59(3):476–81.
5. Kountouris P, Hirst JD. Predicting β-turns and their types using predicted backbone dihedral angles and secondary structures. BMC Bioinformatics. 2010;11(1):407.
6. Faraggi E, Zhang T, Yang Y, Kurgan L, Zhou Y. Spine x: improving protein secondary structure prediction by multistep learning coupled with prediction of solvent accessible surface area and backbone torsion angles. J Comput Chem. 2012;33(3):259–67.
7. Miao X, Waddell PJ, Valafar H. Tali: local alignment of protein structures using backbone torsion angles. J Bioinform Comput Biol. 2008;6(01): 163–81.
8. Jung S, Bae S, Son H. Validity of protein structure alignment method based on backbone torsion angles. J Proteomics Bioinform. 2011;4: 218–26.
9. Zhang C, Hou J, Kim SH. Fold prediction of helical proteins using torsion angle dynamics and predicted restraints. Proc Natl Acad Sci. 2002;99(6): 3581–5.
10. Zhang W, Liu S, Zhou Y. Sp 5: improving protein fold recognition by using torsion angle profiles and profile-based gap penalty model. PloS ONE. 2008;3(6):2325.
11. Faraggi E, Yang Y, Zhang S, Zhou Y. Predicting continuous local structure and the effect of its substitution for secondary structure in fragment-free protein structure prediction. Structure. 2009;17(11):1515–27.
12. Sims GE, Kim SH. A method for evaluating the structural quality of protein models by using higher-order φ–ψ pairs scoring. Proc Natl Acad Sci USA. 2006;103(12):4428–32.
13. Cao C, Xu S. Improving the performance of the plb index for ligand-binding site prediction using dihedral angles and the solvent-accessible surface area. Sci Rep. 2016;6:33232.
14. Ramachandran GT, Sasisekharan V. Conformation of polypeptides and proteins. Adv Protein Chem. 1968;23:283–437.
15. Colubri A, Jha AK, Shen M-Y, Sali A, Berry RS, Sosnick TR, Freed KF. Minimalist representations and the importance of nearest neighbor effects in protein folding simulations. J Mol Biol. 2006;363(4):835–57.
16. Gong H, Fleming PJ, Rose GD. Building native protein conformation from highly approximate backbone torsion angles. Proc Natl Acad Sci USA. 2005;102(45):16227–32.
17. Dowe DL, Allison L, Dix TI, Hunter L, Wallace CS, et al. Circular clustering of protein dihedral angles by minimum message length. Proc. 1st Pacific Symp. Biocomput. 1996.
18. Kuang R, Leslie CS, Yang AS. Protein backbone angle prediction with machine learning approaches. Bioinformatics. 2004;20(10):1612–21.
19. Zimmermann O, Hansmann UH. Support vector machines for prediction of dihedral angle regions. Bioinformatics. 2006;22(24):3009–15.
20. Zhang S, Jin S, Xue B. Accurate prediction of protein dihedral angles through conditional random field. Front Biol. 2013;8(3):353–61.
21. Singh H, Hnizdo V, Demchuk E. Probabilistic model for two dependent circular variables. Biometrik. 2002;89:719–23.
22. Mardia KV, Taylor CC, Subramaniam GK. Protein bioinformatics and mixtures of bivariate von mises distributions for angular data. Biometrics. 2007;63(2):505–12.
23. Li SC, Bu D, Xu J, Li M. Fragment-hmm: A new approach to protein structure prediction. Protein Sci. 2008;17(11):1925–34.
24. Wu S, Zhang Y. Anglor: a composite machine-learning algorithm for protein backbone torsion angle prediction. PLoS ONE. 2008;3(10):3400.
25. Song J, Tan H, Wang M, Webb GI, Akutsu T. Tangle: two-level support vector regression approach for protein backbone torsion angle prediction from primary sequences. PloS ONE. 2012;7(2):30361.
26. Xue B, Dor O, Faraggi E, Zhou Y. Real-value prediction of backbone torsion angles. Proteins Struct Funct Bioinform. 2008;72(1):427–33.
27. Heffernan R, Paliwal K, Lyons J, Dehzangi A, Sharma A, Wang J, Sattar A, Yang Y, Zhou Y. Improving prediction of secondary structure, local backbone angles, and solvent accessible surface area of proteins by iterative deep learning. Sci Rep. 2015;5:11476.

28. Lyons J, Dehzangi A, Heffernan R, Sharma A, Paliwal K, Sattar A, Zhou Y, Yang Y. Predicting backbone $c\alpha$ angles and dihedrals from protein sequences by stacked sparse auto-encoder deep neural network. J Comput Chem. 2014;35(28):2040–6.
29. Singh H, Singh S, Raghava GP. Evaluation of protein dihedral angle prediction methods. PloS ONE. 2014;9(8):105667.
30. Zimmermann O. Backbone dihedral angle prediction. In: Prediction of Protein Secondary Structure. New York: Humana Press; 2017. p. 65–82.
31. Wang S, Peng J, Ma J, Xu J. Protein secondary structure prediction using deep convolutional neural fields. Sci Rep. 2016;6.
32. Wang S, Weng S, Ma J, Tang Q. Deepcnf-d: predicting protein order/disorder regions by weighted deep convolutional neural fields. Int J Mol Sci. 2015;16(8):17315–30.
33. Wang S, Sun S, Li Z, Zhang R, Xu J. Accurate de novo prediction of protein contact map by ultra-deep learning model. PLOS Comput Biol. 2017;13(1):1005324.
34. Faraggi E, Xue B, Zhou Y. Improving the prediction accuracy of residue solvent accessibility and real-value backbone torsion angles of proteins by guided-learning through a two-layer neural network. Protein Struct Funct Bioinformatics. 2009;74(4):847–56.
35. Gao J, Yang Y, Zhou Y. Predicting the errors of predicted local backbone angles and non-local solvent-accessibilities of proteins by deep neural networks. Bioinformatics. 2016;32(24):3768–73.
36. Betancourt MR, Skolnick J. Local propensities and statistical potentials of backbone dihedral angles in proteins. J Mol Biol. 2004;342(2):635–49.
37. Keskin O, Yuret D, Gursoy A, Turkay M, Erman B. Relationships between amino acid sequence and backbone torsion angle preferences. Proteins Struct Funct Bioinform. 2004;55(4):992–8.
38. Jha AK, Colubri A, Zaman MH, Koide S, Sosnick TR, Freed KF. Helix, sheet, and polyproline ii frequencies and strong nearest neighbor effects in a restricted coil library. Biochemistry. 2005;44(28):9691–702.
39. Szegedy C, Liu W, Jia Y, Sermanet P, Reed S, Anguelov D, Erhan D, Vanhoucke V, Rabinovich A. Going deeper with convolutions. In: Proceedings of the IEEE Conference on Computer Vision and Pattern Recognition. CVPR; 2015. p. 1–9.
40. Srivastava RK, Greff K, Schmidhuber J. Training very deep networks. In: Advances in Neural Information Processing Systems. NIPS; 2015. p. 2377–85.
41. He K, Zhang X, Ren S, Sun J. Deep residual learning for image recognition. In: Proceedings of the IEEE Conference on Computer Vision and Pattern Recognition. CVPR; 2016. p. 770–8.
42. Nair V, Hinton GE. Rectified linear units improve restricted boltzmann machines. In: Proceedings of the 27th International Conference on Machine Learning (ICML-10). ICML; 2010. p. 807–14.
43. Kinga D, Adam JB. A method for stochastic optimization. In: International Conference on Learning Representations (ICLR). ICLR; 2015.
44. Bergstra J, Breuleux O, Bastien F, Lamblin P, Pascanu R, Desjardins G, Turian J, Warde-Farley D, Bengio Y. Theano: A cpu and gpu math compiler in python. In: Proc. 9th Python in Science Conf. Scipy; 2010. p. 1–7.
45. Camacho C, Coulouris G, Avagyan V, Ma N, Papadopoulos J, Bealer K, Madden TL. Blast+: architecture and applications. BMC Bioinformatics. 2009;10(1):421.
46. Söding J. Protein homology detection by hmm–hmm comparison. Bioinformatics. 2004;21(7):951–60.
47. Remmert M, Biegert A, Hauser A, Söding J. Hhblits: lightning-fast iterative protein sequence searching by hmm-hmm alignment. Nat Methods. 2012;9(2):173–5.
48. Kountouris P, Hirst JD. Prediction of backbone dihedral angles and protein secondary structure using support vector machines. BMC Bioinformatics. 2009;10(1):437.
49. Kabsch W, Sander C. Dictionary of protein secondary structure: pattern recognition of hydrogen-bonded and geometrical features. Biopolymers. 1983;22(12):2577–637.
50. Steinegger M, Söding J. Sensitive protein sequence searching for the analysis of massive data sets. bioRxiv. 2016. p. 079681.
51. Hamelryck T, Mardia K, Ferkinghoff-Borg J. Bayesian Methods in Structural Bioinformatics. Cambridge: Springer; 2012, pp. 159–78.
52. Berman HM, Westbrook J, Feng Z, Gilliland G, Bhat TN, Weissig H, Shindyalov IN, Bourne PE. The protein data bank. Nucleic Acids Res. 2000;28(1):235–42.
53. Heffernan R, Yang Y, Paliwal K, Zhou Y. Capturing non-local interactions by long short term memory bidirectional recurrent neural networks for improving prediction of protein secondary structure, backbone angles, contact numbers, and solvent accessibility. Bioinformatics. 2017;33:218.

Lee et al. BMC Bioinformatics 2018, 19(Suppl 4):79
https://doi.org/10.1186/s12859-018-2066-9

BMC Bioinformatics

RESEARCH

Open Access

CrossMark

Pathway-based approach using hierarchical components of rare variants to analyze multiple phenotypes

Sungyoung Lee[1], Yongkang Kim[2], Sungkyoung Choi[1], Heungsun Hwang[3] and Taesung Park[1,2*]

From The Sixteenth Asia Pacific Bioinformatics Conference
Yokohama, Japan. 15-17 January 2018

Abstract

Background: As one possible solution to the "missing heritability" problem, many methods have been proposed that apply pathway-based analyses, using rare variants that are detected by next generation sequencing technology. However, while a number of methods for pathway-based rare-variant analysis of multiple phenotypes have been proposed, no method considers a unified model that incorporate multiple pathways.

Results: Simulation studies successfully demonstrated advantages of multivariate analysis, compared to univariate analysis, and comparison studies showed the proposed approach to outperform existing methods. Moreover, real data analysis of six type 2 diabetes-related traits, using large-scale whole exome sequencing data, identified significant pathways that were not found by univariate analysis. Furthermore, strong relationships between the identified pathways, and their associated metabolic disorder risk factors, were found via literature search, and one of the identified pathway, was successfully replicated by an analysis with an independent dataset.

Conclusions: Herein, we present a powerful, pathway-based approach to investigate associations between multiple pathways and multiple phenotypes. By reflecting the natural hierarchy of biological behavior, and considering correlation between pathways and phenotypes, the proposed method is capable of analyzing multiple phenotypes and multiple pathways simultaneously.

Keywords: Pathway-based analysis, Next-generation sequencing data, Multivariate analysis, Generalized structured component analysis, Hierarchical analysis

Background

In the past decade, genome-wide association studies (GWAS) have played a key role in identifying genetic associations between Single Nucleotide Variants (SNVs) and many complex biological pathologies, including type 2 diabetes (T2D), heart disease, and schizophrenia [1–3]. However, large-scale genetic analyses continue to suffer from incomplete association, of single nucleotide variants (SNVs), with distinct phenotypes ("missing heritability"), and difficulties of biological interpretation [4].

Among many proposed solutions to solve the missing heritability problem, many researchers have focused on "rare variants". Methods for rare variants analysis arose from extending individual variant-level approaches to those at the gene-level [5, 6], and extending those at the gene level, to multiple phenotypes [7–9].

As the number of publicly available biological resources is increasing, recent methods for analyzing rare variants utilize pathway knowledge as a priori information. Since most biological behaviors manifest from a complex interaction of biological pathways [10, 11], analyzing pathway information for identifying rare variants has several advantages. In contrast to variant-level analysis, the number of

* Correspondence: tspark@stats.snu.ac.kr
[1]Interdisciplinary Program in Bioinformatics, Seoul National University, Seoul, South Korea
[2]Department of Statistics, Seoul National University, 1 Gwanak-ro Gwanak-gu, Seoul 08826, Korea
Full list of author information is available at the end of the article

statistical tests is substantially smaller in pathway analysis, resulting in less strict multiple testing corrections. Moreover, since pathways explain curated biological behaviors with multiple genes, it is easier to interpret statistically significant pathways than variant- or gene-level analyses. In this respect, many pathway-based approaches have been proposed especially using the pathway databases, which resulted in improvement of the interpretation of discoveries [12, 13].

Another effort to enhance the power of rare variants is to develop multivariate analysis methods. In general, many complex diseases arise from multiply correlated traits. For example, according to American Diabetes Association guidelines, diabetic status is diagnosed based on four traits: fasting glucose, two hours after plasma glucose, random plasma glucose, and HbA1c [14]. In that regard, simultaneous analysis of those correlated traits offer two substantial advantages over univariate analysis. First, multivariate analysis can elevate statistical power to identify additional causal biomarkers, which are not discovered by single phenotype analysis. Second, by analyzing multiple traits at once, the required number of statistical tests can be reduced, compared to those of univariate analysis. Those advantages have been well documented in past studies of large-scale sequencing datasets [15, 16].

There have now been many applications of multivariate analysis to large-scale datasets. In particular, for variant- and gene-level analysis, many multivariate methods, for common and rare variants, have been proposed [8, 15]. Despite those efforts, only a number of pathway-based multivariate analyses have been deemed feasible. Recently, three multivariate approaches, for region-level analyses, were proposed: MARV, aSPU, and MURAT. MARV [17] uses a statistical approach, reverse regression, to investigate associations between genetic regions and multiple phenotypes, by treating phenotypes as independent variables, hence enabling rapid multivariate analysis of large-scale datasets. On the other hand, aSPU [18], extends an original concept, data-adaptive sum of powered score test, to multivariate analysis, using summary statistics from single SNVs. For multivariate extension of powerful gene-based tests, MURAT (Multivariate Rare-variant Association Test) extended the original SKAT (sequence kernel association test) method to multiple phenotypes [19]. However, it might not be adequate to apply SKAT-based methods to pathway-based analysis, as we have previously demonstrated [20]. Moreover, none of the above methods are available for multivariate pathway-based association tests for rare variants with multiple pathways. Since the established pathway databases have substantial overlap among their pathways, they may ignore significant correlations between pathways, leading to misleading biological interpretations [21, 22].

In this report, we introduce a new method, "PHARAOH-multi" (**P**athway-based approach using **H**ier**A**rchical component of collapsed **RA**re variants **O**f **H**igh-throughput sequencing data), for analyzing **multi**ple phenotypes. Previously, we proposed a component-based hierarchical model for analysis of multiple pathways with a single model [20]. Here, while keeping the advantages of our previous approach, we extend it to enable analysis of multiple traits using hierarchical components of genetic variants. In addition, the proposed model can identify associations between multiple phenotypes and multiple pathways, with a single model, in the presence of subsequent genes within pathways, as a hierarchy.

Methods

Exome sequencing dataset for discovery study

To demonstrate the validity of the proposed method for examining large-scale datasets with multiple phenotypes, in real (biological) data analysis, we analyzed whole-exome sequencing (WES) data from a Korean population study. In brief, the dataset consists of next generation sequencing of 1087 individuals' genomes, using the Illumina HiSeq2000 platform (Illumina, Inc., San Diego, CA), selected by the Korean Association REsource (KARE) study [23], as a part of the T2D-GENES consortium. For pathway-gene mapping, we retrieved pathway information from MSigDB [24], and mapped the genes to 217, 186 and 674 pathways extracted from the Biocarta, KEGG [25] and Reactome [26], respectively.

Exome chip dataset for replication study

For replication of the identified pathways from the discover study, an independent cohort from Koreans, the Health Examinee shared control study (HEXA), was used. HEXA is a part of the KoGES population based cohort, initiated in 2001 [27]. In total, genotypes of 3445 individuals were acquired using the HumanExome BeadChip v1.1 (Illumina, Inc., San Diego, CA). With same quality control criteria, 24,474 rare variants were used in the analysis.

PHARAOH-multi method

Our ultimate goal was to find an association between Q phenotypes and K pathways, each of whose number of genes was $T_1, \ldots, T_K$, under the presence of distinct parameters for ridge penalization. The proposed method is based on Generalized Structural Component Analysis (GSCA) [28], and an exemplary structure of the model is shown in Fig. 1.

Let $\mathbf{Y} = [y_{11} \ldots y_{1Q}; \ldots; y_{N1} \ldots y_{NQ}]$ be the matrix of phenotypes for N samples, where y_{iq} is the observation of the ith sample on the qth phenotype, and let $\mathbf{X}$ be the matrix of gene-level collapsed variables generated by summing rare variants according to their gene variant-gene mapping. Let $g_{ij} \in \{0, 1, 2\}$ be the number of minor alleles for the jth genetic variant of the ith sample. Regarding the elements of X, x_{ikt} is a gene-level summary of rare variants which is defined as weighted sum of the ith sample's rare variants in the tth gene of the kth pathway,

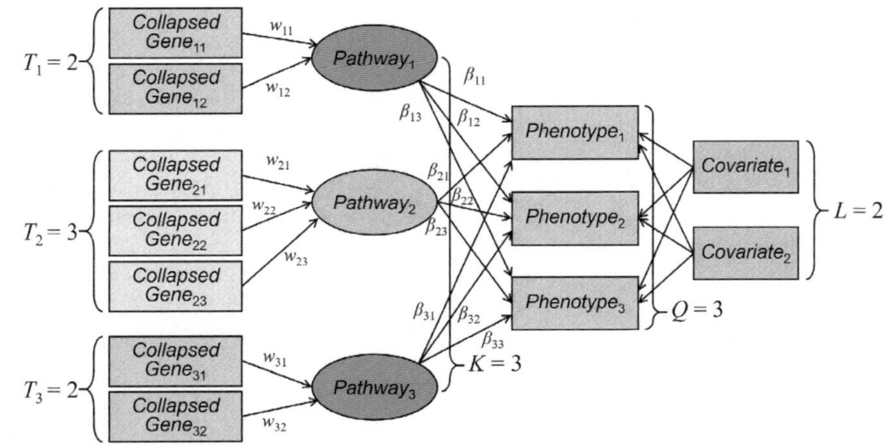

Fig. 1 A graphical representation of PHARAOH-multi. The exemplary model is described with the number of pathways $K = 3$, the number of phenotypes $Q = 3$, the number of covariates $L = 2$, and the number of genes for each pathway T_1, T_2 and T_3 are 2, 3 and 2, respectively. Variable w_{kt} denote the weights assigned to the collapsed genes, and β_{ik} are coefficients on the pathway latent variables. Residual terms were omitted

denoted by $x_{ikt} = \Sigma_{j \in M_{kt}} \omega_j g_{ij}$, where M_{kt} is an index set that defines which rare variants are mapped onto the tth gene in the kth pathway. Several weighting parameters, ω_j, can be used, as previously described in [20]. By imposing two penalty parameters λ_G and λ_P on the genes-pathway and pathways-phenotype, we sought to address potential multicollinearity problems, in both genes and pathways, in the proposed method. Such problems may adversely affect the estimation of weights and coefficients. The proposed model assumes that the phenotype, y_{iq}, arises from the multivariate normal distribution with mean μ and covariance Σ ($i = 1,\ldots,Q$ and $j = 1,\ldots,N$). Then we define the proposed PHARAOH-multi model as.

$$\begin{aligned} y_{iq} &= \beta_{0q} + \sum_{k=1}^{K} \left(\sum_{t=1}^{T_k} x_{ikt} w_{tk} \right) \beta_{kq} + \tilde{\epsilon}_i \\ &= \beta_{0q} + \sum_{k=1}^{K} f_{ik} \beta_{kq} + \tilde{\epsilon}_i = F_i \tilde{\beta}_q + \tilde{\epsilon}_{i.} \end{aligned} \tag{1}$$

Here, $f_{ik} = \sum_{t=1}^{T_k} x_{ikt} w_{tk}$ and F_i indicate the ith observation's score of the kth pathway, and its matrix form across Q phenotypes, respectively. Moreover, $\tilde{\beta}_q = [\beta_{0q}\ \beta_{1q} \cdots \beta_{Kq}]$ is a vector of coefficients for the qth phenotype, and $\tilde{\epsilon}_i = [\epsilon_{i1} \cdots \epsilon_{iQ}]$ is a vector of residuals for the ith sample.

Parameter estimation

The proposed model seeks to associate pathways and phenotypes. The effect of the kth pathway, on multiple phenotypes, can be determined by testing all coefficients of the pathways simultaneously ($H_0 : \beta_{k1} = \ldots = \beta_{kQ} = 0$).

Moreover, by its nature, the proposed method can further assess three more associations: (1) the effect of a gene on multiple phenotypes conditioned by a given pathway; (2) the effect of a gene on a phenotype conditioned by the pathway; and (3) the effect of a pathway on a phenotype. Detailed characteristics of the proposed model (PHARAOH-multi), including relationships and coefficients, are shown in Table 1.

Let B is a matrix of $\tilde{\beta}_1, \cdots, \tilde{\beta}_Q$. From the above model, we seek to maximize the penalized log-likelihood function, to estimate the parameters w_{tk} and β_{kq}, subject to the conventional scaling constraint $\sum_{i=1}^{N} f_{ik}^2 = N$ [29]. The penalized log-likelihood function is expressed to

$$\begin{aligned} \ell(B\ ,\ W, \Sigma | Y_i, X) = &-\frac{NQ}{2} \log \pi - \frac{N}{2} \log \det \Sigma \\ &-\frac{1}{2} \sum_{i=1}^{N} \left(Y_i - B' F_i \right)' \Sigma^{-1} \left(Y_i - B' F_i \right) \\ &-\frac{1}{2} \lambda_G \sum_{k=1}^{K} \sum_{t=1}^{T_k} \| w_{tk} \|_2 - \frac{1}{2} \lambda_G \sum_{q=1}^{Q} \sum_{k=1}^{k} \left\| \beta_{kq} \right\|_2 \end{aligned} \tag{2}$$

where λ_G and λ_P are the penalty parameters for each

Table 1 Parameters related to specific relationships for the proposed model

		Coefficients			Coefficients
Relationship	$P_k \rightarrow Y_*$	$\beta_{k1}, \ldots, \beta_{kQ}$	Relationship	$G_{tk} \rightarrow Y_*$	$w_{tk}\beta_{k1}, \ldots, w_{tk}\beta_{kQ}$
	$P_k \rightarrow Y_q$	β_{kq}		$G_{tk} \rightarrow Y_q$	$w_{tk}\beta_{kq}$

P_k indicates the kth pathway, Y_q is the qth phenotype, Y_* indicates all phenotypes, and G_{tk} indicates the tth gene in the kth pathway specific gene and pathway, respectively, and $\|w_{tk}\|_2$ and $\|\beta_{kq}\|_2$ are the ridge penalty function.

We previously introduced an iteratively reweighted least square (IRLS) method to minimize an univariate version of (2) under the presence of ridge penalties [20], which is similar to the alternating regularized least-squares algorithm [30]. Here we extend the previous algorithm to multivariate analysis. Let R_i be a "column-trimmed" matrix of GSCA [30], defined by $F_i \otimes I_K$, where $\otimes$ is Kronecker product, and I_K is $K \times K$ identity matrix. Maximization of (2) in respect of B and W is equivalent to minimizing the following least-square functions:

$$\begin{aligned}\phi_B &= \sum_{i=1}^{N}\left(Y_i - B'F_i\right)'\Sigma^{-1}\left(Y_i - B'F_i\right) + \lambda_P\sum_{q=1}^{Q}\sum_{k=1}^{K}\|\beta_{kq}\|_2\\ &= \sum_{i=1}^{N}(Y_i - R_i vec(B))'\Sigma^{-1}(Y_i - R_i vec(B)) + \lambda_P vec(B)'vec(B)\\ &= (vec(Y) - Rvec(B))'(vec(Y) - Rvec(B)) + \lambda_P vec(B)'vec(B)\end{aligned} \tag{3}$$

$$\begin{aligned}\phi_w &= \sum_{i=1}^{N}\left(Y_i - B'X_iW\right)'\Sigma^{-1}\left(Y_i - B'X_iW\right) + \lambda_G\sum_{K=1}^{K}\sum_{t=1}^{T_k}\|w_{tk}\|_2\\ &= \sum_{i=1}^{N}(Y_i - \Phi_i W)'\Sigma^{-1}(Y_i - \Phi_i W) + \lambda_G\sum_{k=1}^{K}w_k'w_k\\ &= (vec(Y) - \Phi W)'(vec(Y) - \Phi W) + \lambda_G\sum_{k=1}^{K}w_k'w_k\end{aligned} \tag{4}$$

These least-square functions are subject to diag($R'R$) = NI_{NQ}, where Φ_i is a column-trimmed matrix of $B' \otimes X_i$ [30], and $vec(\cdot)$ is a vectorization operator. Then, it can be easily shown that the covariance matrix Σ is not related to the above equations since the PHARAOH-multi model uses multivariate linear model. An estimation of Σ can be done after convergence of B and W, by minimizing the first derivate of (2) with respect to Σ, as:

$$\hat{\Sigma} = \frac{1}{N}(Y - Rvec(B))'(Y - Rvec(B)) \tag{5}$$

Similarly, B and W can be updated by equating (3) and (4) to zero. This then gives the estimating equation of B and W as:

$$vec(\hat{B}) = (R'R)^{-1}R'vec(Y) \tag{6}$$

$$vec(\hat{w}) = (\Phi'\Phi)^{-1}\Phi'vec(Y) \tag{7}$$

Taken together, the overall procedure of the proposed algorithm is as follows:

1. Let $t = 1$.
2. Assign random initial values to W, which are then represented by $W_{(0)}$.
3. Calculate $F_{(t)}$, using $W_{(t-1)}$.
4. Update $B_{(t)}$, using $F_{(t)}$.
5. Update $W_{(t)}$, using $F_{(t)}$ and $B_{(t)}$.
6. Repeat until the sum of the differences $|W_{(t)} - W_{(t-1)}| + |B_{(t)} - B_{(t-1)}|$ converges the threshold.

Finally, we determine the values of λ_G and λ_P, before applying the parameter estimation procedure. To that end, we can implement k-fold cross-validation (CV) to determine the values of λ_G and λ_P. First, we construct a two-dimensional grid of different λ_G and λ_P values. Then we compute the deviance of each model with the given λ_G and λ_P, for all CV fold values. Finally, λ_G and λ_P are selected by their average deviance, which is minimized.

Significance testing

To assess the significance of genes or pathways, resampling methods can be used to test the statistical significance of the estimated effects of all pathways on the phenotype. In the proposed method, we utilize a permutation test to obtain p-values. By permuting the given phenotype, our method first generates null distributions for both pathways and gene coefficients. By computing the quantile of estimated pathway and gene coefficients, from the non-permuted dataset in each empirical null distribution, we can obtain an empirical p-value for any specific pathway and gene.

The testing of joint effects, between multiple phenotypes, is crucial. As shown in Table 1, PHARAOH-multi provides the individual effects of a pathway on each phenotype through β_{k1}, ..., β_{kQ}. The global effect of a pathway, on all phenotypes, can be evaluated by jointly testing β_{k1}, ..., β_{kQ}. Here, we introduce two different schema for determining a joint p-value for the kth pathway, from multiple phenotypes.

Our first approach was to combine the individual p-values (referred as "P_K"). Since there are considerations among the estimated coefficients β_{k1}, ..., β_{kQ}, these correlations should be accounted for combining multiple p-values. Let the p-values from the kth pathway be denoted by P_{k1}, ..., P_{kQ}. The simplest way to combine those p-values is to use Fisher's method, which is denoted by $\Psi_k = -2\sum_{i=1}^{Q}\log P_{ik}$ under the independence assumption. Then, the statistic, Ψ_k, follows the χ^2 distribution, with the degrees of freedom, $2Q$, under the null hypothesis. An extended version of Fisher's method, Brown's method, can combine dependent p-values using a rescaled χ^2distribution and covariance of p-values [31]. However, an analytical computation of the covariance is not feasible for large-scale datasets, due to their computational complexity. A solution for this problem [32] introduced an approximation using a third-order polynomial for the covariance, denoted by $\mathrm{cov}(-2\ \log P_i, -2\ \log P_j) \approx 3.263\rho_{ij} + 0.71\rho_{ij}^2 + 0.027\rho_{ij}^3$. To that end, Kost's approach has been shown to be one of the best working methods for combining p-values [33]. Here, we adopt Kost's method by substituting ρ to the empirical correlation of estimated coefficients, $\beta_{k1}, \ldots, \beta_{kQ}$, and derive the statistic for joint effect between the kth pathway and multiple phenotypes, as follows:

$$P_{kost,k} = 1 - \Phi_{2d_k}(\Psi_k / c_k), \tag{8}$$

where c_k, d_k and Φ_{2d_k} are the scale parameter, the rescaled degree of freedom, and the cumulative distribution function of χ^2, with the degree of freedom $2d_k$ for the kth pathway, respectively [32].

Our second approach was to construct a single statistic that combines all Q coefficients (referred as "P_M"). Here, we define a Wald-type statistic, $\boldsymbol{T}$, as below, and utilize $\boldsymbol{T}$ for the following permutation testing scheme:

$$T = \tilde{\beta}_k' \ \mathrm{cov}^{-1}\left(\tilde{\beta}_k\right)\tilde{\beta}_k \tag{9}$$

Then, the estimated covariance $\widehat{cov}(\tilde{\beta}_k)$ can be directly estimated using (6) with equation $\widehat{cov}(\tilde{\beta}) = (F'F + \lambda_P I)^{-1} F'F(F'F + \lambda_P I)^{-1} \otimes \hat{\Sigma}$ [34], or can be altered by calculating sample covariance of $\tilde{\beta}_k$, from permutations.

Multiple testing correction

Since the number of pathways is far less than those of genes or genetic variants, the "multiple testing problem" remains. While Bonferroni correction can be a straightforward approach for adjusting for multiple testing, it may impose an adjustment that is too stringent, especially for correlated results [35]. To overcome this issue, we applied two types of multiple testing corrections.

First, PHARAOH-multi corrects p-values using the Westfall & Young permutation procedure [36], which can be easily adopted, since PHARAOH-multi already uses a permutation scheme. Let $T_{(0)}$ be a vector of the statistics calculated using observed, unpermuted phenotypes, and let $T_{(j)}$ be those from the jth permutation. First, we rank the values of $T_{(0)}$ in ascending order, and let the rank of the kth pathway, and the kth index, be r_k and $r_{(k)}$, respectfully. Then, for each permutation $j = 0, 1, \cdots, J$, let $T'_{(j)}$ be $T_{(j)r_{(1)}}, \cdots, T_{(j)r_{(K)}}$, to define $T^M_{(j)}$ as a cumulative maximum of $T'_{(j)}$. Let $I_{j,k}$ be an indicator function that resolves to 1.0, if $T'_{(0)r_k} < T^M_{(j)r_k}$, or 0.0, if that condition does not hold. The adjusted p-value for the kth pathway, by the Westfall & Young procedure, is then defined as:

$$P_k^{adj} = \frac{1 + \sum_{j=1}^{J} I_{j,k}}{1 + J} \tag{10}$$

Second, PHARAOH-multi provides False Discovery Rate (FDR) adjustment, by calculating q-values [37]. Here, we first obtain K as the number of permutation p-values, and from those, we can derive q-values, using the Benjamini-Hochberg step-up procedure.

Simulation study

To evaluate the performance of the proposed method, we conducted simulation studies, under various scenarios. For generating rare variants, we first produced a pool of genetic variants, using SimRare [38], a rare variant simulator with well-established genetic assumptions. A pool was then generated, with default settings and gene lengths of 1Kbp. Next, we generated a simulation dataset of 10 pathways, with 1000 samples, for each replicate. All simulation scenarios were evaluated, using 1000 replicates. Based on the genotypes, the simulated phenotypes were generated by the following model, with an assumption that only the first pathway is causal to the phenotypes:

$$\begin{aligned} y_{iq} &= \beta_{1q}\tilde{f}_{i1} + \epsilon_{iq} = \beta_{1q}\sum_{t=1}^{H_1} w_{1t}x_{i1t} + \epsilon_{iq} \\ &= \beta_{1q}\sum_{t=1}^{H_1}\left(w_{1t}\sum_{j=1}^{M_{1t}}\gamma_{1tj}g_{i1tj}\right) + \epsilon_{iq} \end{aligned} \tag{11}$$

This is then subject to $\mathrm{diag}(F'F) = NI_K$, where H_1 is the number of causal genes in the first pathway, and M_{1t} is the number of rare variants in the tth gene of the first pathway (i.e., causal pathway).

In the above model (eq. 11), γ_{1tj} denotes the effect of the jth genetic variant, of the tth gene, set to $|\log_{10}$-$MAF_{tj}|$, such that ϵ_{iq} denotes the residual and follows $\mathrm{MVN}(0, \Sigma)$. In our simulation, the settings $q = 1{,}2$, and $H_1 = 1, 2, 5, 10$, were used. For each replicate, all rare variants were collapsed into genes.

Results

For the simulation and the analysis of the real dataset, a workstation system with two Intel Xeon E5–2620 CPUs, with a combined RAM of 128GiB, were used. Note that the aSPU and MARV analyses were performed using the default settings, except that aSPU was performed without "genetic variant pruning" capability, as we observed that aSPU raises "unrecoverable error" with that capability. For our proposed method and aSPU, the number of permutations was 5000, to prevent possible lower bound limitation.

Simulation study

For PHARAOH-multi, we selected the tuning parameters λ_G and λ_B based on three-fold CV for each replicate, using two-dimensional grids of λ_G and λ_B with six different starting points of ridge parameters, ranging from 10^1 to 10^6 (on a logarithmic base 10 scale). In the simulation, we considered the following conditions: number of effective genes in the causal pathway (H_1), gene-level effect (w_{t1}), pathway-level effect (β_{1q}), and residual correlation (ρ). In our simulation, H_1, w_{t1}, β, and ρ were assumed to be 1, 2, and 5; 0.1 and 0.2; 0.1, 0.15, and 0.2; and 0, 0.25, 0.5, respectively, with evaluation of their exhaustive combinations. Other parameters, Q, K and T_K, were fixed to 2, 10, and 10, respectively.

We first compared the type 1 error rates of the proposed method vs. the traditional methods. Here, type 1 error rate was computed as a proportion of p-values for the pathways with no effect, and was less than the significance level, across 1000 replicates of permuted phenotypes. As shown in Fig. 2, we evaluated the type 1 errors using two significance levels, 0.01 and 0.05. As a result, type 1 errors were controlled well in the traditional methods, but PHARAOH-multi showed a moderately deflated type 1 error rate (P_M), while the inflated rate is P_K, when $\rho = 0$. In contrast, the quantile-quantile (Q-Q) plots in Fig. 3 show no inflation or deflation pattern, in all the methods, except for P_K, with no correlation between phenotypes.

It was also worthwhile to assess the gain of power in the multivariate analysis, as compared to univariate analysis. In this respect, our simulation study was conducted to compare the power gain from multivariate methods, and between multivariate and univariate analyses.

First, we checked whether PHARAOH-multi with multiple phenotypes boosts power compared to PHARAOH with a single phenotype, under the same scenarios of the power simulation. As a result, we observed that the power of PHAROH-multi was at least 2.52 times larger than PHARAOH, and this difference becomes even larger, as w and β increase (data not shown).

Second, we assessed the statistical power of PHARAOH-multi and the compared methods, defined as the proportion of the adjusted q-value of the simulated causal pathway (the first pathway) being less than the significance threshold, e.g., 0.05. Despite the proposed method supporting the Westfall-Young permutation procedure, it was not considered in the simulation study, due to the absence of corresponding adjustments in the compared methods.

Figures 4 and 5 show comparison results of statistical power simulation from 1000 replications. Each row in the grid of plots represents the same settings of w and β, with different numbers of causal genes in the causal pathway, and each column represents the same number of causal genes, with different effect settings.

In most scenario comparisons, the two proposed statistics obtained by PHARAOH-multi (P_M) and p-value

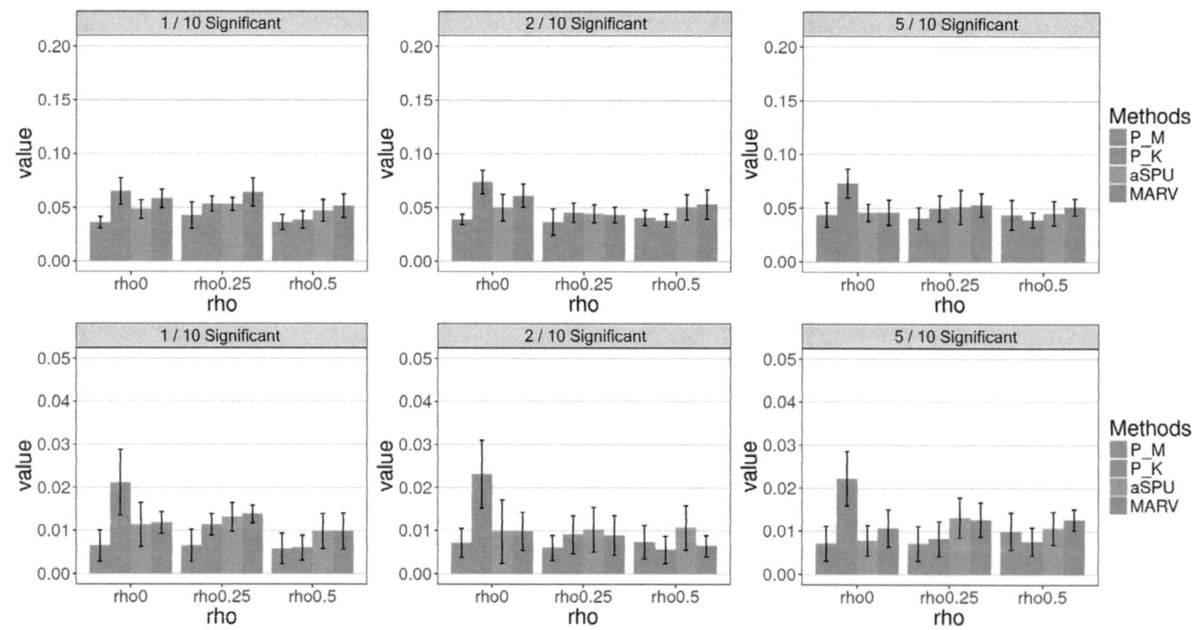

Fig. 2 Type 1 error simulation result. Plots in the first and second row represent the type 1 error at $\alpha = 0.05$ and $\alpha = 0.01$, respectively. Each bar is the mean and the error bars represent SD

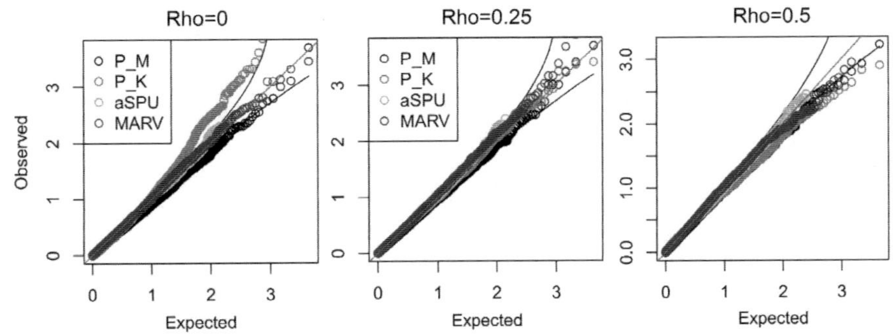

Fig. 3 Quantile-Quantile plots of type 1 error evaluation, without multiple testing adjustment

aggregation (P_K) showed greater power than the other two approaches, aSPU and MARV. However, this did not hold when 50% of the genes were causal for a specific pathway, with effect sizes of $w = 0.1$ and $\beta = 0.1$. In order to investigate whether or not there are significant differences among powers, we performed paired *t*-tests between a pair of methods. In Fig. 4 for the case of $w = 0.1$, the *p*-values were 3×10^{-7} for comparing powers of P_M and aSPU, and 4.2×10^{-6} for comparing those of P_M and MARV. In Fig. 5 for the case of $w = 0.2$, the same pairwise comparison

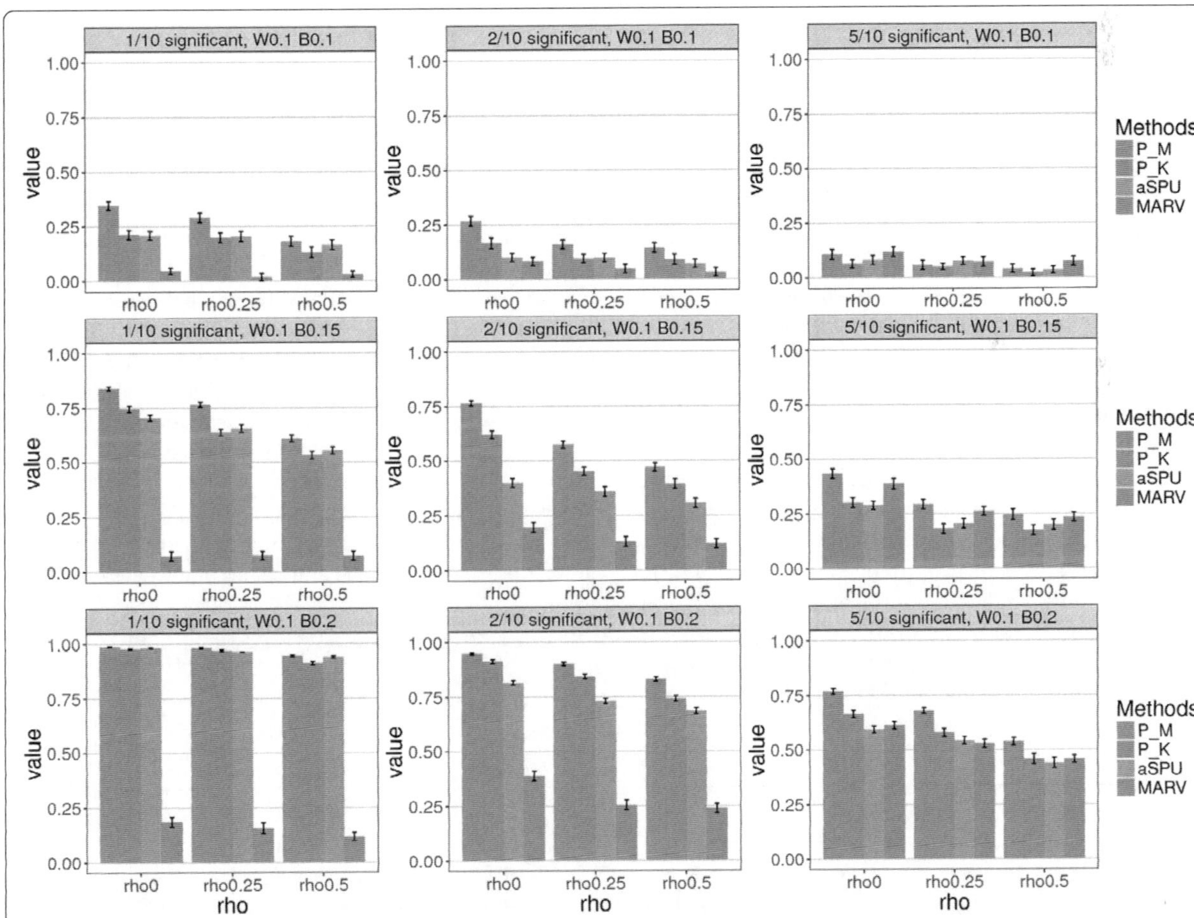

Fig. 4 Comparison of simulation results of statistical power from various methods of multiple testing adjustment. The value of $w = 0.1$. Red and green bars represent results obtained by the PHARAOH-multi ("P_M", using joint testing) and *p*-value aggregation ("P_K") methods, respectively. Teal and purple bars represent the aSPU and MARV methods, respectively. Powers were calculated by the proportion of the causal pathway's *q*-value, obtained by the Benjamini-Hochberg procedure (< 0.05). Paired *t*-test *p*-values are 7.3×10^{-7} and 4.2×10^{-6} for P_M vs. aSPU and P_M vs. MARV, respectively

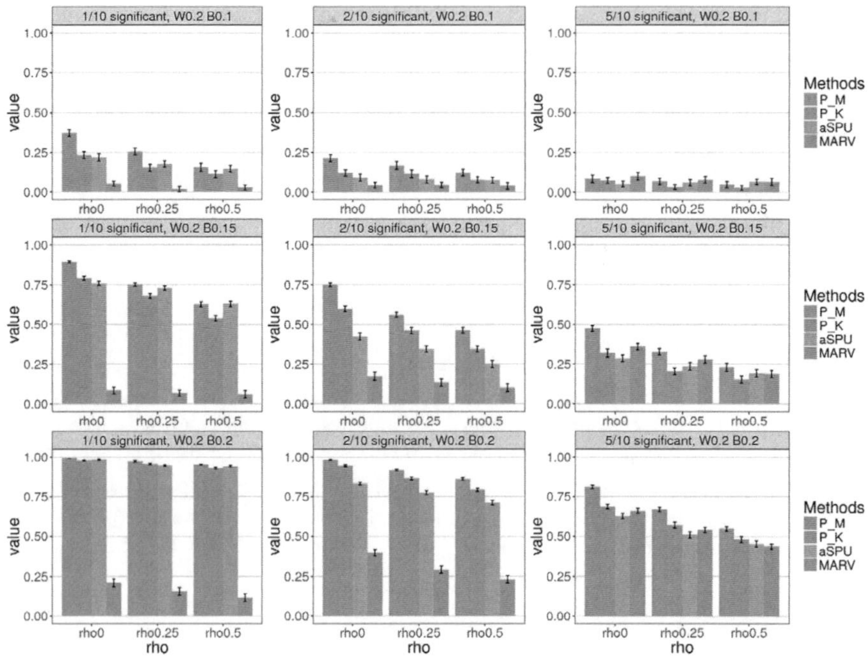

Fig. 5 Comparison of simulation results for statistical power using multiple testing adjustment at $w = 0.2$. Red and green bars represent results obtained by the PHARAOH-multi ("P_M," using joint testing) and p-value aggregation ("P_K") methods, respectively. Teal and purple bars represent the aSPU and MARV methods, respectively. Powers were calculated by the proportion of the causal pathway's q-value obtained by the Benjamini-Hochberg procedure (< 0.05). Paired t-test p-values are 3.4×10^{-6} and 1.8×10^{-6} for P_M vs. aSPU and P_M vs. MARV, respectively

for the p-values were 7.3×10^{-7} and 4.2×10^{-6}, respectively. In overall scenarios, powers of P_M were larger up to 18%p compared to aSPU in $H_1 = 5$, $w = 0.2$ and $\beta = 0.2$, and were larger up to 83%p compared to MARV. Generally, P_K exhibited smaller power than P_M, and showed comparable or slightly smaller power, than aSPU.

Here, we observed three interesting patterns in the results. First, the proposed P_M and aSPU methods showed lower power, when the proportion of causal genes increase, compared to MARV. Second, the power rapidly increased, as β increased, as shown in Fig. 4. Third, the contribution of w to the power was relatively moderate, compared to β, as shown in corresponding scenarios of Figs. 4 and 5. The reason for the occurrence of those two patterns is that the model(s) generate phenotypes for power simulation, and eq. (11) requires the constraint of the so-called "latent variable," in GSCA (see Methods). While both PHARAOH-multi and aSPU construct hierarchies of genes and pathways, MARV essentially treats a pathway as a large set of SNVs, since the motivation of MARV is for region- vs. pathway-based tests. The simulation setting and its overall effect on phenotypes is summarized, first at the gene-level, and then by the expression of a linear combination of those genes. In this respect, the results of PHARAOH-multi and aSPU were more plausible than those of MARV, because those two methods more properly reflected the simulation settings.

To confirm the above hypothesis, we performed an additional comparison using the same dataset, except that the phenotypes were generated without the constraint. As shown in Fig. 6, PHARAOH-multi ("P_M") showed larger power than in the previous simulation, while the power of MARV increased as the number of causal genes increased. However, in contrast to the previous results, the powers of PHARAOH-multi and aSPU also increased.

Finally, we investigated whether or not the statistical power changes by M_{kt}. For simplicity, we split 1000 simulation datasets into two groups: the first group where the number of variants is small and the second where it is large.

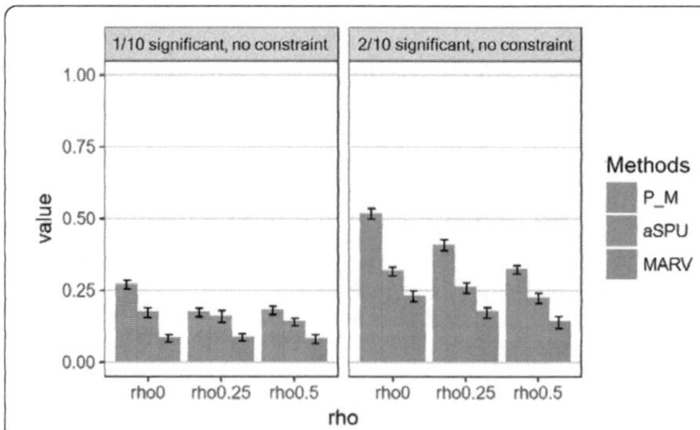

Fig. 6 Result of additional power simulation, without constraint of manifest variables

Then, we compared the power of each method between two groups using *t*-test. As a result, the *p*-values were 0.097 for aSPU, 0.684 for MARV, and 0.825 for PHARAOH-multi. Thus, we concluded that M_{kt} is unlikely to affect the simulation result regardless of the methods.

Real data discovery from whole-exome sequencing dataset

To evaluate the practical performance of PHARAOH-multi, we conducted a discovery study using a large-scale sequencing dataset. Many studies suggest that the major underlying risk factors for metabolic disorders include high density lipoprotein (HDL), blood pressure (SBP, DBP), waist circumference (WAISTC), fasting glucose (FAST_GLU), and triglycerides (TG). In this regard, we conducted a multivariate analysis of metabolism-related traits, using a large-scale sequencing dataset, obtained from the Type 2 Diabetes Genetic Exploration by Next-generation sequencing in multi-Ethnic Samples (T2D-GENES) Consortium, comparing our proposed (PHARAOH-multi) and other common methods. In detail, we analyzed a dataset consisting of 1086 samples selected from the Korean Association REsource (KARE) study [23].

After removal of samples with any missing observations of the aforementioned six phenotypes, we included 1085 samples for analysis. The quality controls with genotype call rates were < 95%, or for Hardy-Weinberg Equilibrium (HWE) test $P < 10^{-5}$, the minor allele frequency was < 5% and the minor allele count was > 2, resulted in 198,761 variants. The final dataset was then mapped to genes, using the human genome-19 (hg19) reference genome coordinates, with 10Kbp flanking regions. The gene range of hg19 reference, was extracted from RefSeq track of UCSC Table Browser, as of October 2014. Finally, the gene-level collapsed variable was generated using Workbench for Integrated Superfast Association study with Related Data (WISARD), with beta-transformation weighting, as suggested in [5], with the number of genes being 4388.

Next, we compared our multivariate and univariate analysis results, using PHARAOH-multi and PHARAOH. As shown in Figs. 7 and 8, Q-Q plots of the results showed no substantial inflation or deflation pattern for either the multivariate or univariate results. However, with regard to pathway discovery, the results did show significant differences.

These comparisons clearly support the one advantage of multivariate analysis that we discussed above: elevation of statistical power. As with multivariate analysis, we calculated *q*-values for each univariate result. Interestingly, no univariate analysis identified significant pathways, except for SBP with KEGG, which identified three pathways (drug metabolism cytochrome P450, glutathione metabolism and progesterone-mediated oocyte maturation). As shown in Table 2, only one pathway, glutathione metabolism, was identified in the univariate analysis, and the *q*-values of univariate analyses for pathways identified by multivariate analysis, were not significant.

Second, we compared the result of multivariate analyses, using PHARAOH-multi, aSPU and MARV. As shown in Fig. 8, PHARAOH-multi exhibited generally acceptable *p*-value trends, despite the result from KEGG being modestly deflated, due to the optimization of lambda. The Q-Q plots of MARV look similar to PHARAOH-multi. In contrast, aSPU showed unacceptably inflated patterns of Q-Q plots, regardless of the pathway databases, which were not used in the simulation study. This could possibly be due to substantial overlap of existing pathway databases.

As shown in Table 2, the multivariate analysis successfully identified eight pathways from three pathway databases, with Benjamini-Hochberg *q*-value < 0.1. Interestingly, PHARAOH-multi identified glutathione-related pathways in both KEGG and Reactome pathway databases, which supports the result of PHRAOH-multi. As shown in Fig. 8, the quantile-quantile plots of aSPU for the real dataset are highly inflated (i.e., their *p*-values are very small). As a result, 57.7% (Reactome), 29.5% (Biocarta) and 71.5% (KEGG) of the tested pathways by aSPU were statistically significant (*q*-value < 0.1). Unfortunately, these pathways are highly false positives. In this respect, we included the results of significant pathways identified by either PHARAOH-multi or MARV.

The identified pathways suggested evident relationships with metabolic syndrome. Since the peroxisome pathway elucidates peroxisome biogenesis, which contributes to fatty acid oxidation and biosynthesis of ether lipids, many studies have discussed interrelationship between peroxisomes and metabolic processes [39, 40]. Likewise, identification of the GABA pathway can also be explained by the relationship between GABA and peroxidation, and putative relationship of obesity [41, 42]. Moreover, identification of glutathione metabolism, and its conjugation, explain that PHARAOH-multi successfully captured a key process of metabolic disorders [43]. Finally, another report suggested a putative role of adhesion molecules in metabolic diseases, as explained by "cell2cell" pathway [44]. For the two pathways identified by MARV, we found that Caspase pathway has been known to be related to metabolic stress or perturbation [45], but no evidence for D4GDI pathway was found.

Replication study using independent exome chip dataset

We conducted a replication study using exome chip dataset from an independent cohort, using the identified pathways in the discovery study. Despite the insufficiency of detected variants in the exome chip dataset, as a result, we successfully replicated two pathways with *p*-value < 0.1, the peroxisome pathway in KEGG ($p = 0.059$) and cell2cell pathway in Biocarta ($p = 0.093$). As shown in the literature search, the two pathways we replicated have strong relationships with metabolic disorders.

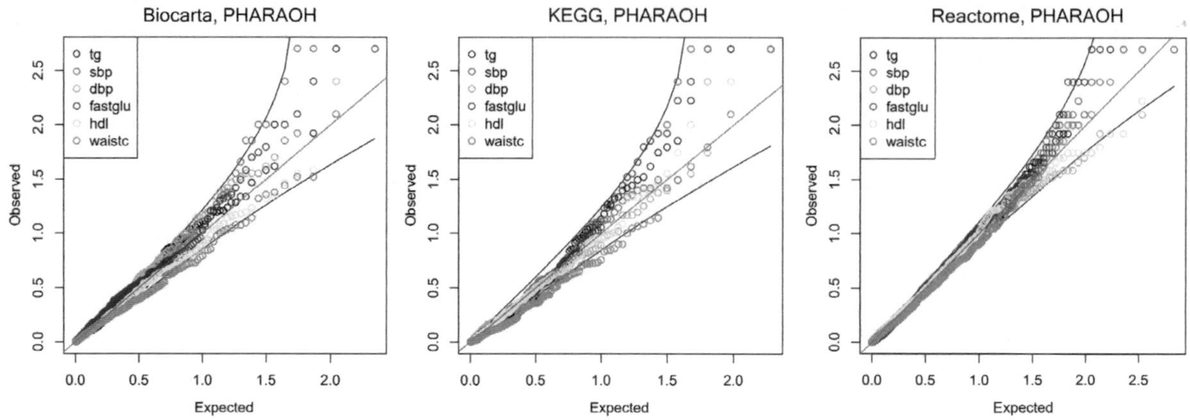

Fig. 7 Q-Q plots of univariate PHARAOH for each pathway database

Discussion

Compared to univariate approaches, which analyze each phenotype individually, our real data analysis successfully demonstrated that the multivariate approach could identify pathways commonly associated with specific traits. It is important to construct a systematic analysis that considers the correlation between complex diseases and their underlying biological traits. In addition, our results from two well-established pathway databases were strongly supported by many existing

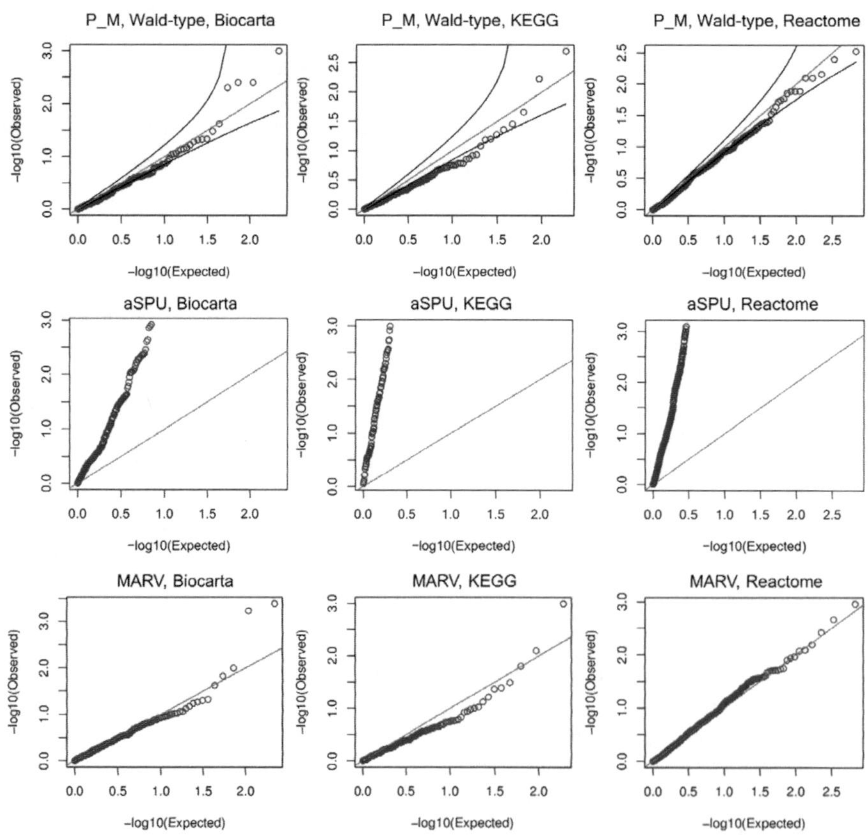

Fig. 8 Q-Q plots of discovery study using multivariate methods. The first to third rows show the results of P_M, aSPU and MARV, respectively. The plots were drawn by unadjusted *p*-values. For each method, three pathway databases (Biocarta, KEGG and Reactome) were used. Many *p*-values of aSPU are not appear in the Q-Q plots since aSPU reports many zero *p*-value that cannot be drawn in log scale

Table 2 Significant pathways of PHARAOH-multi and MARV, and their *q*-values of multivariate and univariate analyses

DB	Pathway	# of variants	Multivariate *q*-value			Univariate *q*-value (PHARAOH)					
			P_M	aSPU	MARV	tg	sbp	dbp	fastglu	hdl	waistc
KEGG											
	Peroxisome	421	**0.0396**	0	0.9826	0.6886	0.7	0.9138	0.9899	0.9942	1
	Glutathione metabolism	187	**0.044**	0.0076	0.9826	0.999	**0.0939**	0.9138	0.993	0.9942	1
Biocarta											
	CDMAC pathway	63	**0.0858**	0.5739	0.9638	0.9817	0.1094	0.5743	0.953	0.9967	0.9962
	Cell2cell pathway	112	**0.0208**	0	0.9638	0.7293	0.3063	0.5722	0.8234	0.9967	0.9962
	GABA pathway	46	**0.0497**	0.0085	0.8134	0.9783	0.1094	0.5722	0.8234	0.9967	0.9962
	MPR pathway	179	**0.0208**	0	0.9638	0.9783	0.1094	0.2188	0.8234	0.9845	0.9962
	Caspase pathway	649	0.8358	0.1741	**0.0634**	0.997	0.8863	0.6418	0.7584	0.999	0.9928
	D4GDI pathway	422	0.7626	0.3727	**0.0634**	0.997	0.9867	0.6418	0.4377	0.999	0.995
Reactome											
	Glutathione conjugation	99	**0.0979**	0.0567	0.9979	0.9813	0.3859	0.9254	0.999	0.9793	0.979
	Phase II conjugation	270	**0.0571**	0	0.9979	0.9813	0.3859	0.9254	0.999	0.9793	0.99

Bold numbers are the *q*-values below the significance threshold 0.1. P_M, aSPU and MARV indicate *q*-values from the joint testing method of multiple phenotypes, and univariate *q*-values indicate the *q*-values of PHARAOH analysis for each phenotype

publications, thus demonstrating the advantage of our proposed approach.

Compared to existing multivariate analysis methods, PHARAOH-multi features several advantages. Firstly, by constructing a hierarchical structure of genes-pathways-phenotypes, four types of associations (gene-single phenotype, gene-multiple phenotypes, pathway-single phenotype, and pathway-multiple phenotypes) can be estimated simultaneously. Compared to our proposed method, existing methods of multivariate analysis were limited to gene-level analysis, and hence, the combinatorial effect of multiple genes, via biological pathways, was impossible to estimate. In addition, the proposed method considers the correlation between genes, pathways, and phenotypes, by imposing penalty parameters on the estimation procedure.

Secondly, PHARAOH-multi provides multiple options for correcting for the multiple testing issue. Although Bonferroni correction is simple, and powerfully controls type 1 error, it is a well-known fact that the Bonferroni correction often results in controls that are too stringent, when the tests are correlated. Under such conditions, application of the Westfall-Young permutation procedure can be an appropriate alternative, since its asymptotic optimality under dependence is known [35]. In this respect, the proposed method has the advantage of identifying causal pathways, by considering correlation among pathways.

For analysis times of both simulation and real datasets, MARV was the fastest among all the methods, while PHARAOH-multi ran slightly faster than aSPU. For example, in the analysis of simulation dataset of 100 genes with 1000 samples, the running times of MARV, PHARAOH-multi, and SPU were 13, 67 and 235 s, respectively. The trends of execution time were consistent regardless of simulation parameters or datasets. However, PHARAOH-multi can be further accelerated with multithreading which is not supported by MARV and aSPU. With multithreading of 8 threads, the analysis time of PHARAOH-multi was reduced to 12 s.

At this point, there are a number of subjects we can consider for future research. Our current analysis is limited only to Korean population. In our future study, we apply our method to the whole data of 13,000 WES dataset of T2D-GENES consortium [46] which contains our KARE samples. It would be a challenging work to identify novel pathways across multiple populations. For the methodological aspect, our approach uses gene-level collapsing of multiple rare variants. Although the collapsing method has the advantage that the analysis of very rare variants is possible, it cancels out the effects of variants with opposite direction (e.g, gene upregulation vs. downregulation). Despite such limitations, our method showed great potential in identifying causal genetic structure in the real data analysis. However, further research, on a more sophisticated approach that can consider the effect direction of variants, is needed. Moreover, we plan to improve our proposed multivariate analysis by applying Generalized Estimating Equations (GEE) or Linear Mixed Model (LMM). Our method can be extended to prediction models, rather than association tests, using other types of penalization, such as LASSO or SCAD [47]. Lastly, our method can also be extended to pathway interaction analysis that has been commonly performed in gene expression data analysis [48].

Conclusion

In this study, we proposed a novel statistical approach for multivariate pathway-based analysis of rare variants, from large-scale sequencing datasets. Analyses of multiple phenotypes have been successful in analyzing various complex diseases, including type-2 diabetes (T2D) or hypertension. In general, curated guidelines suggest diagnosing T2D according to traits observed in the individual. Consequently, incorporating multiple correlated traits, to be investigated for association with specific diseases, via multivariate analysis, elevates the statistical power. In this respect, our simulation study reflects the relationship between diseases and their related traits. Throughout the simulation study, PHARAOH-multi outperformed existing multivariate methods. In addition, our proposed method successfully demonstrated several advantages of multivariate analysis, including significantly improving the detection power of causal pathways, as compared to univariate analysis, while also retaining detection power for the individual phenotype. Moreover, we successfully demonstrated that the proposed method is capable of identifying plausible pathways in the real dataset, by identifying eight pathways in the discovery study, and replicating two pathways in the replication study. We firmly believe that the proposed method will assist researchers in understanding the genetic structures that underlie many complex diseases.

Acknowledgements
Not applicable.

Funding
This research was supported by grants of the Korea Health Technology R&D Project through the Korea Health Industry Development Institute (KHIDI), funded by the Ministry of Health & Welfare, Republic of Korea (HI15C2165, HI16C2037), supported by the Bio-Synergy Research Project (2013M3A9C4078158) of the Ministry of Science, ICT and Future Planning through the National Research Foundation. The publication of this article was sponsored by the Bio-Synergy Research Project.

Availability of data and materials
An implementation of PHARAOH-multi can be downloaded from the website (http://statgen.snu.ac.kr/software/pharaoh-multi). The KARE exome sequencing dataset is a part of T2D-GENES consortium, and is available upon approval of T2D-GENES project committee. The HEXA exome chip dataset is a part of KoGES, and is available upon approval of the genome center in Korea National Institute of Health.

About this supplement
This article has been published as part of *BMC Bioinformatics* Volume 19 Supplement 4, 2018: Selected articles from the 16th Asia Pacific Bioinformatics Conference (APBC 2018): bioinformatics. The full contents of the supplement are available online at https://bmcbioinformatics.biomedcentral.com/articles/supplements/volume-18-supplement-17.

Authors' contributions
SL performed all analyses and developed the software implementation. SL and TP conducted the entire study, developed the methodology, and wrote the manuscript. YK and SC helped with the performing of analyses. HH helped developing the methodology. All of the authors have read and approved of the final manuscript.

Ethics approval and consent to participate
KARE and HEXA dataset are a part of Korean Genome Epidemiology Study (KoGES). In this study, the exome sequencing data from 1037 samples of KARE study and the exome chip data from 3445 samples of HEXA study were used. KARE dataset was used under the partnership of T2D-GENES. All participants of KARE and HEXA studies provided written informed consent. The study using KARE and HEXA samples was approved by two independent institutional review boards at Seoul National University.

Consent for publication
Not applicable.

Competing interests
The authors declare that they have no competing interests.

Publisher's Note
Springer Nature remains neutral with regard to jurisdictional claims in published maps and institutional affiliations.

Author details
[1]Interdisciplinary Program in Bioinformatics, Seoul National University, Seoul, South Korea. [2]Department of Statistics, Seoul National University, 1 Gwanak-ro Gwanak-gu, Seoul 08826, Korea. [3]Department of Psychology, McGill University, Montreal, Canada.

Published: 8 May 2018

References
1. Schizophrenia Working Group of the Psychiatric Genomics C. Biological insights from 108 schizophrenia-associated genetic loci. Nature. 2014; 511(7510):421–7.
2. Lettre G, Palmer CD, Young T, Ejebe KG, Allayee H, Benjamin EJ, Bennett F, Bowden DW, Chakravarti A, Dreisbach A, et al. Genome-wide association study of coronary heart disease and its risk factors in 8,090 African Americans: the NHLBI CARe project. PLoS Genet. 2011;7(2):e1001300.
3. McCarthy MI, Zeggini E. Genome-wide association studies in type 2 diabetes. Curr Diab Rep. 2009;9(2):164–71.
4. Manolio TA, Collins FS, Cox NJ, Goldstein DB, Hindorff LA, Hunter DJ, McCarthy MI, Ramos EM, Cardon LR, Chakravarti A, et al. Finding the missing heritability of complex diseases. Nature. 2009;461(7265):747–53.
5. Wu MC, Lee S, Cai T, Li Y, Boehnke M, Lin X. Rare-variant association testing for sequencing data with the sequence kernel association test. Am J Hum Genet. 2011;89(1):82–93.
6. Choi S, Lee S, Cichon S, Nothen MM, Lange C, Park T, Won S. FARVAT: a family-based rare variant association test. Bioinformatics. 2014;30(22):3197–205.
7. Wang L, Lee S, Gim J, Qiao D, Cho M, Elston RC, Silverman EK, Won S. Family-based rare variant association analysis: a fast and efficient method of multivariate phenotype association analysis. Genet Epidemiol. 2016;40(6):502–11.
8. Zhou X, Stephens M. Efficient multivariate linear mixed model algorithms for genome-wide association studies. Nat Methods. 2014;11(4):407–9.
9. Lee S, Won S, Kim YJ, Kim Y, Consortium TD-G, Kim BJ, Park T. Rare variant association test with multiple phenotypes. Genet Epidemiol. 2016;
10. Costanzo M, Baryshnikova A, Bellay J, Kim Y, Spear ED, Sevier CS, Ding H, Koh JL, Toufighi K, Mostafavi S, et al. The genetic landscape of a cell. Science. 2010;327(5964):425–31.
11. Hirschhorn JN. Genomewide association studies–illuminating biologic pathways. N Engl J Med. 2009;360(17):1699–701.
12. Lee JH, Zhao XM, Yoon I, Lee JY, Kwon NH, Wang YY, Lee KM, Lee MJ, Kim J, Moon HG, et al. Integrative analysis of mutational and transcriptional profiles reveals driver mutations of metastatic breast cancers. Cell Discov. 2016;2:16025.
13. Zhao XM, Liu KQ, Zhu G, He F, Duval B, Richer JM, Huang DS, Jiang CJ, Hao JK, Chen L. Identifying cancer-related microRNAs based on gene expression data. Bioinformatics. 2015;31(8):1226–34.
14. American Diabetes A. Diagnosis and classification of diabetes mellitus. Diabetes Care. 2014;37(Suppl 1):S81–90.
15. O'Reilly PF, Hoggart CJ, Pomyen Y, Calboli FC, Elliott P, Jarvelin MR, Coin LJ. MultiPhen: joint model of multiple phenotypes can increase discovery in GWAS. PLoS One. 2012;7(5):e34861.
16. Yang Q, Wang Y. Methods for analyzing multivariate phenotypes in genetic association studies. J Probab Stat. 2012;2012:652569.
17. Kaakinen M, Magi R, Fischer K, Heikkinen J, Jarvelin MR, Morris AP, Prokopenko I. A rare-variant test for high-dimensional data. Eur J Hum Genet. 2017;
18. Kwak IY, Pan W. Adaptive gene- and pathway-trait association testing with GWAS summary statistics. Bioinformatics. 2016;32(8):1178–84.
19. Sun J, Oualkacha K, Forgetta V, Zheng HF, Brent Richards J, Ciampi A, Greenwood CM, Consortium UK. A method for analyzing multiple

continuous phenotypes in rare variant association studies allowing for flexible correlations in variant effects. Eur J Hum Genet. 2016;24(9):1344–51.
20. Lee S, Choi S, Kim YJ, Kim BJ, Consortium Td-G, Hwang H, Park T: Pathway-based approach using hierarchical components of collapsed rare variants. Bioinformatics 2016, 32(17):i586-i594.
21. Alexa A, Rahnenfuhrer J, Lengauer T. Improved scoring of functional groups from gene expression data by decorrelating GO graph structure. Bioinformatics. 2006;22(13):1600–7.
22. Skarman A, Shariati M, Jans L, Jiang L, Sorensen P. A Bayesian variable selection procedure to rank overlapping gene sets. BMC bioinformatics. 2012;13:73.
23. Cho YS, Go MJ, Kim YJ, Heo JY, Oh JH, Ban HJ, Yoon D, Lee MH, Kim DJ, Park M, et al. A large-scale genome-wide association study of Asian populations uncovers genetic factors influencing eight quantitative traits. Nat Genet. 2009;41(5):527–34.
24. Liberzon A, Subramanian A, Pinchback R, Thorvaldsdottir H, Tamayo P, Mesirov JP. Molecular signatures database (MSigDB) 3.0. Bioinformatics. 2011;27(12):1739–40.
25. Kanehisa M, Goto S, Sato Y, Furumichi M, Tanabe M. KEGG for integration and interpretation of large-scale molecular data sets. Nucleic Acids Res. 2012;40(Database issue):D109–14.
26. Fabregat A, Sidiropoulos K, Garapati P, Gillespie M, Hausmann K, Haw R, Jassal B, Jupe S, Korninger F, McKay S, et al. The Reactome pathway knowledgebase. Nucleic Acids Res. 2016;44(D1):D481–7.
27. Kim YJ, Go MJ, Hu C, Hong CB, Kim YK, Lee JY, Hwang JY, Oh JH, Kim DJ, Kim NH, et al. Large-scale genome-wide association studies in east Asians identify new genetic loci influencing metabolic traits. Nat Genet. 2011; 43(10):990–5.
28. Hwang H, Takane Y. Generalized structured component analysis. Psychometrika. 2004;69(1):81–99.
29. Takane Y, Hwang H. An extended redundancy analysis and its applications to two practical examples. Computational Statistics & Data Analysis. 2005; 49(3):785–808.
30. Hwang H. Regularized generalized structured component analysis. Psychometrika. 2009;74(3):517–30.
31. Brown MB. 400: a method for combining non-independent, one-sided tests of significance. Biometrics. 1975;31(4):987–92.
32. Kost JT, McDermott MP. Combining dependent P-values. Stat Probabil Lett. 2002;60(2):183–90.
33. Alves G, Yu YK. Accuracy evaluation of the unified P-value from combining correlated P-values. PLoS One. 2014;9(3):e91225.
34. Hoerl AE, Kennard RW. Ridge Regression - Biased Estimation for Nonorthogonal Problems. Technometrics. 1970;12(1):55&.
35. Meinshausen N, Maathuis MH, Bühlmann P. Asymptotic optimality of the Westfall–young permutation procedure for multiple testing under dependence. Ann Stat. 2011;39(6):3369–91.
36. Westfall PH, Young SS. Resampling-based multiple testing : examples and methods for P-value adjustment. New York: Wiley. 1993;
37. Benjamini Y, Hochberg Y. Controlling the false discovery rate: a practical and powerful approach to multiple testing. J R Stat Soc Ser B Methodol. 1995;57(1):289–300.
38. Li B, Wang G, Leal SM. SimRare: a program to generate and analyze sequence-based data for association studies of quantitative and qualitative traits. Bioinformatics. 2012;28(20):2703–4.
39. Azhar S. Peroxisome proliferator-activated receptors, metabolic syndrome and cardiovascular disease. Futur Cardiol. 2010;6(5):657–91.
40. Hall D, Poussin C, Velagapudi VR, Empsen C, Joffraud M, Beckmann JS, Geerts AE, Ravussin Y, Ibberson M, Oresic M, et al. Peroxisomal and microsomal lipid pathways associated with resistance to hepatic steatosis and reduced pro-inflammatory state. J Biol Chem. 2010;285(40):31011–23.
41. Deng Y, Xu L, Zeng X, Li Z, Qin B, He N. New perspective of GABA as an inhibitor of formation of advanced lipoxidation end-products: it's interaction with malondiadehyde. J Biomed Nanotechnol. 2010;6(4):318–24.
42. Ma YH, Hu JH, Zhou XG, Zeng RW, Mei ZT, Fei J, Guo LH. Transgenic mice overexpressing gamma-aminobutyric acid transporter subtype I develop obesity. Cell Res. 2000;10(4):303–10.
43. Wu G, Fang YZ, Yang S, Lupton JR, Turner ND. Glutathione metabolism and its implications for health. J Nutr. 2004;134(3):489–92.
44. Wagner O, Jilma B. Putative role of adhesion molecules in metabolic disorders. Horm Metab Res. 1997;29(12):627–30.
45. McIlwain DR, Berger T, Mak TW. Caspase functions in cell death and disease. Cold Spring Harb Perspect Biol. 2015;7(4)
46. Fuchsberger C, Flannick J, Teslovich TM, Mahajan A, Agarwala V, Gaulton KJ, Ma C, Fontanillas P, Moutsianas L, McCarthy DJ, et al. The genetic architecture of type 2 diabetes. Nature. 2016;536(7614):41–7.
47. Fan J, Li R. Variable selection via nonconcave penalized likelihood and its Oracle properties. J Am Stat Assoc. 2001;96(456):1348–60.
48. Liu KQ, Liu ZP, Hao JK, Chen L, Zhao XM. Identifying dysregulated pathways in cancers from pathway interaction networks. BMC Bioinformatics. 2012;13:126.

Zhou et al. BMC Bioinformatics 2018, 19(Suppl 4):60
https://doi.org/10.1186/s12859-018-2067-8

BMC Bioinformatics

RESEARCH Open Access

CNNH_PSS: protein 8-class secondary structure prediction by convolutional neural network with highway

Jiyun Zhou[1,2], Hongpeng Wang[1], Zhishan Zhao[1], Ruifeng Xu[1*] and Qin Lu[2]

From The Sixteenth Asia Pacific Bioinformatics Conference
Yokohama, Japan. 15-17 January 2018

Abstract

Background: Protein secondary structure is the three dimensional form of local segments of proteins and its prediction is an important problem in protein tertiary structure prediction. Developing computational approaches for protein secondary structure prediction is becoming increasingly urgent.

Results: We present a novel deep learning based model, referred to as CNNH_PSS, by using multi-scale CNN with highway. In CNNH_PSS, any two neighbor convolutional layers have a highway to deliver information from current layer to the output of the next one to keep local contexts. As lower layers extract local context while higher layers extract long-range interdependencies, the highways between neighbor layers allow CNNH_PSS to have ability to extract both local contexts and long-range interdependencies. We evaluate CNNH_PSS on two commonly used datasets: CB6133 and CB513. CNNH_PSS outperforms the multi-scale CNN without highway by at least 0.010 Q8 accuracy and also performs better than CNF, DeepCNF and SSpro8, which cannot extract long-range interdependencies, by at least 0.020 Q8 accuracy, demonstrating that both local contexts and long-range interdependencies are indeed useful for prediction. Furthermore, CNNH_PSS also performs better than GSM and DCRNN which need extra complex model to extract long-range interdependencies. It demonstrates that CNNH_PSS not only cost less computer resource, but also achieves better predicting performance.

Conclusion: CNNH_PSS have ability to extracts both local contexts and long-range interdependencies by combing multi-scale CNN and highway network. The evaluations on common datasets and comparisons with state-of-the-art methods indicate that CNNH_PSS is an useful and efficient tool for protein secondary structure prediction.

Keywords: Protein secondary structure, Convolutional neural network, Highway, Local context, Long-range interdependency

Background

The concept of secondary structure was first introduced by Linderstrøm-Lang at Stanford in 1952 [1, 2] to represent the three dimensional form of local segments of proteins. Protein secondary structure is defined by the pattern of hydrogen bonds between the amine hydrogen and carbonyl oxygen. There are two ways used for the classification of protein secondary structures: three-category classification(Q3) and eight-category classification(Q8). Q3 classifies target amino acid residues into helix(H), strand(E) and coil(C) while Q8 classifies target amino acid residues into (1) 3-turn helix(G), (2) 4-turn helix(H), (3) 5-turn helix(I), (4) hydrogen bonded turn(T), (5) extended strand in parallel and/or anti-parallel β-sheet conformation(E), (6) residue in isolated β-bridge (B), (7) bend(S) and (8) coil(C) [3–5]. Most protein secondary structure prediction studies have been focused Q3 prediction. Q8 prediction is more challenging and can reveal more structural details [6, 7], so we focus the Q8 prediction in this study.

* Correspondence: xuruifeng@hit.edu.cn
[1]School Computer Science and Technology, Harbin Institute of Technology Shenzhen Graduate School, HIT Campus Shenzhen University Town, Xili, Shenzhen, Guangdong 518055, China
Full list of author information is available at the end of the article

Protein secondary structure prediction is secondary structure inference of protein fragments based on their amino acid sequence. In bioinformatics and theoretical chemistry, protein secondary structure prediction is very important for medicine and biotechnology, for example drug design [8] and the design of novel enzymes. Since secondary structure can be used to find distant relationship for proteins with unalignable primary structures, incorporating both secondary structure information and simple sequence information can improve the accuracy of their alignment [9]. Finally, protein secondary structure prediction also plays an important role in protein tertiary structure prediction. Protein secondary structure can determine the structure types of protein local fragments, so the freedom degree of protein local fragments in the tertiary structure can be reduced. Therefore accurate secondary structure prediction is potential for improving the accuracy of protein tertiary structure prediction [4, 7, 10].

Three experimental methods were proposed to determine secondary structures for proteins: far-ultraviolet circular dichroism, infrared spectroscopy and NMR spectrum. Far-ultraviolet circular dichroism predict pronounced double minimum at 208 and 222 nm as α-helical structure and single minimum at 204 nm or 217 nm as random-coil or β-sheet structure, respectively [11]. Infrared spectroscopy uses the differences in the bond oscillations of amide groups for prediction [12] while NMR spectrum predict protein secondary structure by using the estimated chemical shifts [12]. As experimental methods are costly and the proteins with known sequence continue to outnumber the experimentally determined secondary structures, developing computational approaches for protein secondary structure prediction becomes increasingly urgent. Existing computational approaches for protein secondary structure prediction can be divided into 3 categories. The first category is statistical model based methods, which can date back to 1970s. Early, this category uses statistical models to analyze the probability of secondary structure elements for individual amino acid residue [13]. Next, the statistical models were applied for the prediction of segments of 9–21 amino acids. For example, the GOR method [14] used amino acid segment to predict the structure of its central residue. However, the performances (< 60% Q3 accuracy) of this category of methods are far from practical application due to inadequate features.

Due to the lacking of inadequate features for the statistical model based methods, evolutionary information based methods have been proposed. These methods usually used the evolutionary information of proteins from a same structural family [15] extracted by multiple-sequence alignment or position-specific scoring matrices (PSSM) [16] from PSI-BLAST for prediction. An earlier evolutionary information based method was developed based on a two-layered feed-forward neural network, for which the evolutionary information in the form of multiple sequence alignment is used as input instead of single sequences [15]. As SVM [17] is significantly better than neural network in a wide range of pattern recognition problems [18–21], Hua and Sun first proposed a SVM classifier for protein secondary structure prediction [22]. The input for SVM is evolutionary information in the form of multiple sequence alignment. Unbalanced data is a challenging problem in protein secondary structure prediction and existing methods lack the ability to handle it [23, 24]. So Kim and Park proposed a new protein secondary structure prediction method, SVMpsi, by an improved SVM, which reduces the influence of imbalanced data by using different penalty parameters in the improved SVM [23]. By using different penalty parameters, SVMpsi resolved the situation where the recall value of the smaller class is too small. Another SVM based method is PMSVM which was proposed by using dual-layer support vector machine (SVM) and evolutionary information in form of PSSMs [25].

Protein sequence usually contains two types of sequence information: local context and long-range interdependencies [4, 26, 27]. Local contexts denote the correlations between residues with distance less than or equal to a predefined threshold while long-range correlation are the correlations between residues with distance more the threshold. Inspired by the success of convolutional neural networks (CNN) [28] for local context extraction in natural language processing tasks [29, 30], multi-scale CNN has been used to capture local contexts for protein secondary structure prediction [31]. For example, Wang et al. proposed conditional neural fields (CNF) [7] to extract local contexts for prediction by integrating a windows-based neural network with conditional random field (CRF). In addition to local contexts, long-range interdependencies also are important for protein secondary structure prediction(BRNN) [6, 32–34]. In order to extract both local contexts and long-range interdependencies for prediction, Zhou and Troyanskaya proposed GSN [4] by using convolutional architecture and supervised generative stochastic network, which is a recently proposed deep learning technique [26]. In addition to GSN, a novel deep convolutional and recurrent neural network (DCRNN) also has been proposed by Li and Yu [27] for protein secondary structure prediction by extracting both local contexts and long-range interdependencies.

In summary, the statistical model based methods and evolutionary information based methods cannot extract local contexts and long-range interdependencies for prediction. For the deep learning based methods, some methods cannot extract both local contexts and long-range interdependencies for prediction. Although several methods can extract both local contexts and long-range interdependencies, such as GSN and DCRNN, they need extra complex models to extract long-range interdependencies, which are complex and time-consuming. In this paper, we propose a novel method, referred to as CNNH_PSS, by combining multi-scale CNN with highway network, which has ability to extract both local contexts and long-range interdependencies without needing extra models. CNNH_PSS consists of two parts: multi-scale CNN and fully connected and softmax layer. In the multi-scale CNN, any two neighbor convolutional layer contains a highway to deliver information from current layer to the output of the next one to keep local contexts. As the convolutional kernels in higher layer can extract long-range interdependencies by using the local contexts extracted by lower layers, thus with the layer number increasing, CNNH_PSS can extract long-range interdependencies covering more remote residues while keeping local contexts extracted by lower layers by using highway. So CNNH_PSS can extract both local contexts and long-range interdependencies covering very remote residues for prediction. The source code of our proposed method CNNH_PSS is provided for free access to the biological research community at http://hlt.hitsz.edu.cn/CNNH_PSS/ and http://119.23.18.63:8080/CNNH_PSS/.

Methods

As shown by many recently published works [35–37], a complete prediction model in bioinformatics should contain the following four components: validation benchmark dataset(s), an effective feature extraction procedure, an efficient predicting algorithm, a set of fair evaluation criteria. In the following text, we will describe the four components of our proposed CNNH_PSS in details.

Datasets

Two publicly available datasets: CB6133 and CB513 were used to evaluate the performance of our proposed method CNNH_PSS and compare with state-of-the-art methods.

CB6133

CB6133 was produced by PISCES CullPDB [38] and is a larger non-homologous protein dataset with known secondary structure for every protein. It contains 6128 proteins, in which 5600 proteins are training samples, 256 proteins are validation samples and 272 proteins are testing samples. This dataset is publicly available from literature [4].

CB513

CB513 is a public testing dataset and can be freely obtained from [4, 39]. For the testing on CB513, CB6133 is used as the training dataset. As there exists redundancy between CB513 and CB6133, CB6133 is filtered by removing sequences having over 25% sequence similarity with sequences in CB513. After filtering, 5534 proteins left in CB6133 are used as training samples. Since GSN [4] and DCRNN [27] as well as other state-of-the-art methods [31, 40] performed a validation on CB513 to get their best performance, we also perform a validation on CB513 to get the best performance of our method CNNH_PSS to make fair comparisons with them.

Feature representation

Given a protein with L amino acid residues as $X = x_1, x_2, x_3, \cdots, x_L$, where $x_i(\in\mathbb{R}^m)$ is the m-dimensional feature vector of the ith residue, the secondary structure prediction for this protein is formulated as determining $S = s_1, s_2, s_3, \cdots, s_L$ for X where s_i is a Q8 secondary structure label. In this study,x_i is encoded by both sequence features and evolutionary information. Sequence features are used to specify the identity of the target residue. Two methods are used to encode sequence features: one hot and residue embedding. One hot encodes sequence features of each residue by a 21-dimension one-hot vector, in which only one element equals to 1 and the remaining elements are set to 0, where 21 denotes the 20 standard types of residues and one extra residue type which represents all non-standard residue types. However, one-hot vector is a sparse representation and unsuitable for measuring relation between different residues. In order to get dense representation of sequence features, an embedding technique in natural language processing is used to transform 21-dimensional one-hot vector to a 21-dimensional denser representation [41]. The embedding technique maps words or phrases from the vocabulary to vectors of real numbers. Specifically, it maps words from a space with one dimension per word to a continuous vector space with much lower dimension. So the embedding technique provides a real value for every dimension. As the dimension of amino acid representation is already low, we only calculate a real value for every dimension by embedding technique and don't decrease the dimension. The residue embedding in this paper is implemented by a feedforward neural network layer before multi-scale CNN in CNNH_PSS [42].

Evolutionary information such as position-specific scoring matrix (PSSM) is considered as informative features for predicting secondary structure by previous research [16]. PSSM is a common representation for evolutionary information and has been used in many bioinformatics studies including protein functionality annotation and protein structure prediction [43–47]. In this study, PSSM is calculated by PSI-BLAST [48] against the UniRef90 database with E-value threshold 0.001 and 3 iterations. UniRef90 database contains the known protein sequences with sequence identity less than 90 from almost all known species. So the PSSM calculated from UniRef90 database contains the common sequence information among the known protein sequences of different species. Overtime, scientists have reached a consensus that a protein's structure primarily depends on its amino acid sequence and concluded that the local and long-range interaction are a cause of protein second and tertiary structure. Based on this hypothesis, we can deduce that proteins with similar amino acid sequence tend to have similar secondary structure sequence. Therefore, the common sequence information contained by PSSM can contribute to the secondary structure prediction. For a protein with length L, PSSM is usually represented as a matrix with $L \times 21$ dimensions where 21 denotes the 20 standard types of residues and one extra residue type which represents all non-standard residue types. Before PSSMs are used inputs for CNNH_PSS, they need to be transformed to 0–1 range by the sigmoid function. By concatenating sequence features and evolutional information, each residue in protein sequences can be encoded by a feature vector with dimension of 42.

Multi-scale CNN with highway between neighbor layers

In CNN model, a kernel can examine a local patch in input sequence and extract interdependence among the residues contained in the local patch. With stacking of convolutional layers, the kernels for deep layers have ability to cover correlations among more spread-out residues in the input sequence. So, CNN model with more number of convolutional layers have the ability to extracted long-range interdependencies between residues with more large distance. However, with the number of layers increasing, CNN model will lose the local contexts extracted by lower layers. In this paper, we propose a novel method, referred to as CNNH_PSS, to resolve this problem. CNNH_PSS contains a highway between any two neighbor convolutional layers in multi-scale CNN. As the number of convolutional layers increases, CNNH_PSS can not only extract long-range interdependencies by higher layers, but also obtain the local contexts extracted by lower layers through highway. The frame of CNNH_PSS is shown in Fig. 1. Figure 1 shows that CNNH_PSS contains three parts: input section, multi-scale CNN with highway and output section. In the input section, $x_i \in R^m$ denotes the feature vector of the ith residue in protein, which is the concatenation of sequence features and evolutional information. Thus a protein of length L is encoded as a $L \times m$ matrix $x_{1:L} = [x_1, x_2, \cdots, x_L]^T$, where L and m denote the length of protein and the number features used to encode residues, respectively. In this study, m equals to 42. In order to keep the output of convolutional layer have the same height with the input, we need to pad $\lfloor h/2 \rfloor$ and $\lfloor (h-1)/2 \rfloor$ m-dimensional zero vectors to the head and the tail of the input $x_{1:L}$, respectively, where h is the length of convolutional kernels in the convolutional layer. The second section contains two parts: multi-scale CNN and highway, where the multi-scale CNN contains n convolutional layers. In the $(t-1)^{th}$ layer, the convolution operation of the kth kernel $w_k^{t-1} \in R^{h \times m}$ executed on protein fragment $x_{i:i+h-1}$ is expressed as

$$c_{k,i}^{t-1} = f^{t-1}\left(w_k^{t-1} \cdot x_{i:i+h-1} + b_k^{t-1}\right) \quad (1)$$

where h is the length of convolution kernel,b_k^{t-1} is the bias of the kth kernel, f is activation function and $x_{i:i+h-1}$ denotes the protein fragment x_i, x_{i+1}, x_{i+2}, $\cdots$, x_{i+h-1}. Through executing convolution operation of the kth kernel on all fragments with length h of the padded input, we get a novel feature vector

$$c_k^{t-1} = \left[c_{k,1}^{t-1}, c_{k,2}^{t-1}, c_{k,3}^{t-1}, \cdots, c_{k,L}^{t-1}\right]^T \quad (2)$$

Suppose we have d kernels in the convolutional layer, thus we can get d novel features vectors. By concatenating the d novel feature vectors, we can get a novel feature matrix with dimension $L \times d$

$$c^{t-1} = \left[c_1^{t-1}, c_2^{t-1}, c_3^{t-1}, \cdots, c_L^{t-1}\right] \quad (3)$$

This novel feature matrix is used as the input of the next convolutional layer. If there are n convolutional layers and θ_t is used to denote the kernels and the bias of the tth convolutional layer, then the output of the nth convolutional layer is

$$c^n = f_{\theta_n}^n\left(f_{\theta_{n-1}}^{n-1}\left(\cdots f_{\theta_1}^1(x_{1:L})\right)\right) \quad (4)$$

Finally, the output of the nth convolutional layer is used as the input of the fully connected softmax layer for prediction

$$y_i = argmax(w \cdot c_i^n + b) \quad (5)$$

where w and b is the weight and bias of the fully connected softmax layer, respectively. c_i^n is the feature vector of the ith outputted by the nth convolutional layer and y_i is its predicted secondary structure.

CNN has achieved huge progress in many tasks of image processing filed, one common sense is that the successes of CNN are attributed to the multiple convolutional layers in CNN, because CNN with more number of layers can extract correlations covering more residues. However, with the increasing of number of layers in CNN, the information communication between layers will become more difficult and the gradient will disappear [49]. Furthermore, the local contexts extracted by lower layers also will lose. Srivastava et al. [49] have proposed highway network to resolve these problems. So

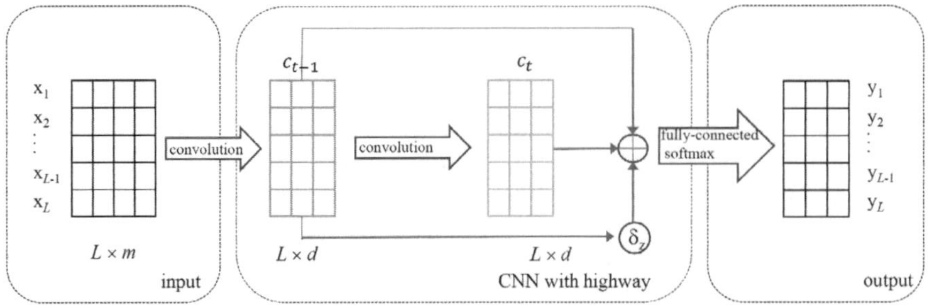

Fig. 1 The frame of CNNH_PSS

CNNH_PSS incorporates highway network and multi-scale CNN to extract both local contexts and long-range interdependencies for secondary structure prediction.

In CNNH_PSS, each convolutional layer except the last layer has three accesses to the next layer (shown in Fig. 1). Two accesses are used to deliver information from the current layer to the output and convolution kernels of the next layer, respectively. The other one is a weight used to determine the share of information in for information from highway. So the output c^t of the tth convolutional layer is the weighted sum of the information delivered by highway from last layer and that outputted by the convolution kernels of current layer

$$z_t = \delta\left(w^z c^{t-1}\right) \quad (6)$$

$$c^t = (1-z_t) \times f^t_{\theta_t}\left(c^{t-1}\right) + z_t \times c^{t-1} \quad (7)$$

where $\delta(\cdot)$ is *sigmoid* function,z_t is the weight of the highway and $f^t_{\theta_t}(\cdot)$ is the convolution operation of current convolutional layer. So the output of thet^{th}convolutional layer contains two portion: information from the $(t-1)^{th}$convolutional layer delivered by highway and that outputted by the convolution kernels of current layer.

Results

The purpose of the evaluation is to examine the effectiveness of our proposed CNNH_PSS over other methods. Four sets of evaluations are conducted here. The first experiment evaluates the performance of multi-scale CNN on CB6133 and CB513. The second experiment evaluates our proposed method CNNH_PSS on CB6133 and CB513. The third experiment compares CNNH_PSS with state-of-the-art methods. Finally, based on CB6133, we analyze the local contexts and long-range interdependencies learned by CNNH_PSS. As we mainly focus the Q8 prediction of protein secondary structure in this study, the performances of prediction methods are measured by Q8 accuracy [4, 27]. The Q8 accuracy is the percentage of the amino acid residues for which the predicted secondary structure labels are correct. The source code of our proposed method CNNH_PSS is provide for free access at http://hlt.hitsz.edu.cn/CNNH_PSS/.

The performance of multi-scale CNN model

In this section, multi-scale CNN is used to predict secondary structure for proteins. The hyper-parameters of the multi-scale CNN for protein secondary structure prediction in this study are listed in Table 1. Note that three kernel lengths are used in the multi-scale CNN model and 80 kernels are used for each kernel length. To conveniently encode and process protein sequences, the length of all protein sequences are normalized to 700. When sequences are shorter than 700, they will be padded with zero vectors. And when sequences are longer than 700, they will be truncated. In order to get best performance, we need to determine how many convolutional layers the multi-scale CNN should contains. We conduct experiments to evaluate the performances of the multi-scale CNNs with different number of convolutional layers on CB513. The performances are shown in Fig. 2, where the *x*-axis is the number of epochs used to train multi-scale CNN and the *y*-axis is Q8 accuracy. Fig. 2 shows the performances for models with number of convolutional layers from 1 to 5. From this figure, we see that the model with 3 convolutional layers gets the best accuracy. When the number of convolutional layers is increased to 4 or 5, the accuracy is decreased obviously. The main reason for this phenomenon may be the loss of extracted local contexts with the increasing of the

Table 1 Hyper-parameters of multi-scale CNN

Layer	Hyper-parameter	Value
Multi-scale CNN	Kernel length	[7, 9, 11]
	Number of kernels	80 for each kernel length
	Batch size	50
	Learning rate	2e-3
	Regularizer	5e-5
	Decay rate	0.05
	Activation function	ReLU

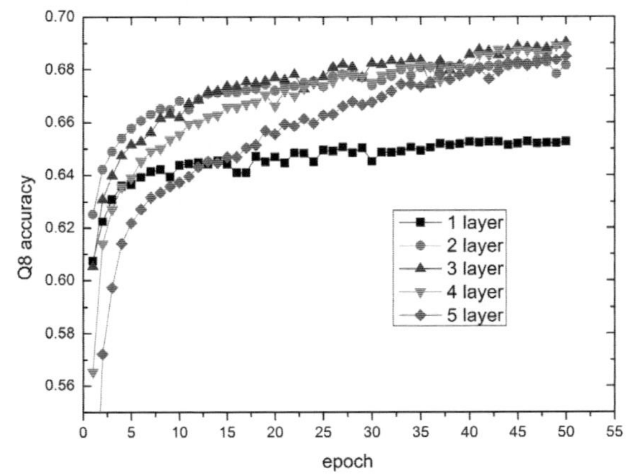

Fig. 2 The performance of multi-scale CNN with different number of convolutional layers

number of convolutional layers in CNN. With the increasing of the number of convolutional layers in CNN, the correlations extracted by higher can cover more residues so that they contains interdependencies between more remote residues. When the number of convolutional layers is increased to 3, the CNN may achieve both local contexts and long-range interdependencies, which is validated by that the CNN with 3 convolutional layers gets the best accuracy in our problem. However, when the number of convolutional layers is more than 3, most local contexts extracted by lower layers are lost in the transport processes of information cross layers, causing the relationships outputted by the last layer in CNN contains less and less local contexts. So the predicting accuracy starts to decrease when the number of convolutional layers is more than 3.

The performances of multi-scale CNN with 3 convolutional layers on CB6133 and CB513 are shown in Table 2, where two sequence features encoding methods for residues are evaluated: one hot and residue embedding. Table 2 shows that residue embedding outperforms one hot on both CB6133 and CB513 by at 0.004 Q8 accuracy, indicating that residue embedding is a better encoding method for sequence features of residues. In the following text, we will use residue embedding method to encode sequence features in the multi-scale CNN model and our proposed method CNNH_PSS.

Table 2 The Q8 accuracy of Multi-scale CNN with 3 convolutional layers

datasets	CB6133	CB513
Multi-scale CNN(one hot)	0.721	0.689
Multi-scale CNN(embedding)	*0.729*	*0.693*

The data in italic denote the best performance

The performance of CNNH_PSS

Local contexts are the relationships among residues at close range while long-range interdependencies are the relationships among remote residues. As there is no strict bounds between local contexts and long-range interdependencies, we specify the information extracted by the first convolutional layers as local contexts and that extracted by all other layers as long-range interdependencies in this study. In CNNH_PSS, any two neighbor convolutional layers have a highway to deliver information from current convolutional layer to the output of the next one, so it can make sure that the output of each layer contains a portion of the local contexts. Furthermore, the convolution kernels of each layer except the first one can extract long-range interdependencies by using the information from previous layer. With the increasing of the number of convolutional layers, CNNH_PSS can extract long-range interdependencies between more remote residues. Therefore, the output the last convolutional layer in CNNH_PSS contains two portion of information: local contexts extracted by the first layer and long-range interdependencies extracted by all other layers.

Similarly, we evaluate the performances of our proposed method CNNH_PSSs with different number of convolutional layers on CB513. The performances are shown in Fig. 3. Figure 3 shows that CNNH_PSS achieves the best performance when the number of convolutional layers is 5. When number of convolutional layers is more than 5, the performance of our method starts to decrease. So CNNH_PSS with 5 convolutional layers is used in the following text. Comparing the multi-scale CNN model with 3 convolutional layers described in above section, our proposed method CNNH_PSS not only contains a highway between any two neigbouring layers, but also have more number of

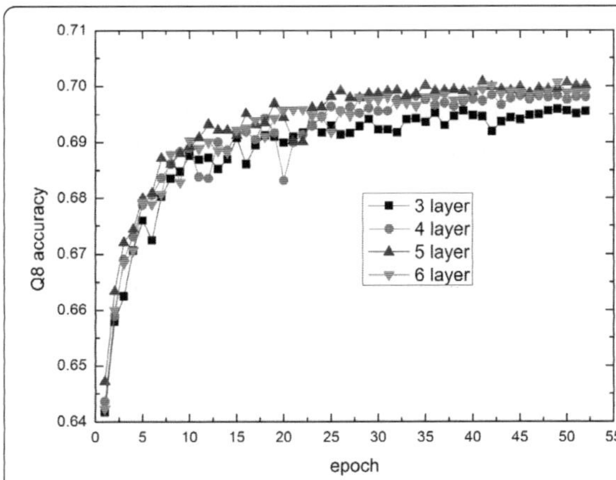

Fig. 3 The performance of CNNH_PSS with different number of convolutional layers

convolutional layers. It makes sure that CNNH_PSS with 5 layers can not only extract local contexts, but also capture long-range interdependencies between more remote residues than the multi-scale CNN model with 3 layers. The performances of CNNH_PSS and the multi-scale CNN model on CB6133 and CB513 are shown in Table 3. Table 3 shows that our proposed method CNNH_PSS outperforms the multi-scale CNN model by 0.011 Q8 accuracy on CB6133 and 0.010 Q8 accuracy on CB513. The outperformance of CNNH_PSS over the multi-scale CNN model on both CB6133 and CB513 validates that the highway in CNN indeed are useful for protein secondary structure prediction.

Comparison with state-of-the-art methods

Protein secondary structure prediction is an important problem in bioinformatics and critical for analyzing protein function and applications like drug design. So many state-of-the-art methods have been proposed for the prediction. SSpro8 is a prediction method proposed by Pollastri et al. [6] by combining bidirectional recurrent neural networks (RNN) and PSI-BLAST-derived profiles. CNF is a Conditional Neural Fields based method which was proposed by Wang et al. [40], which can not only extract relationships between sequence features of residues and their secondary structures, but also capture local contexts [40]. Later, an extension version of CNF (DeepCNF) was proposed by Wang et al. [31] using deep learning extension of conditional neural fields, which is an integration of conditional neural fields and shallow neural networks. It can extract both complex sequence-structure relationship and interdependency between adjacent SS labels. These three methods can only extract local contexts for prediction. Furthermore, GSN is a prediction method proposed by Zhou and Troyanskaya [4] using supervised generative stochastic network and convolutional architectures. Supervised generative stochastic network is a recently proposed deep learning technique [26], which is well suitable for extracting local contexts and also can capture some long-range interdependencies. Finally, DCRNN is the best performing method up to now, which was recently proposed by Li and Yu [27] using multi-scale CNN and three staked bidirectional gate recurrent units (BGRUs) [50]. GSN and DCRNN can extract both local contexts and long-range interdependencies. We first compare our proposed method CNNH_PSS with the three state-of-the-art methods which only can extract local contexts on CB513. The performances of these three methods and our method on CB513 are listed in Table 4. Table 4 shows that CNNH_PSS outperforms the three methods by at least 0.020 Q8 accuracy. The outperformance of CNNH_PSS over the three state-of-the-art methods which only can extract local contexts indicates that the long-range interdependencies extracted by CNNH_PSS are indeed useful for protein secondary structure prediction.

Table 3 The Q8 accuracy of CNNH_PSS

Method	CB6133	CB513
Multi-scale CNN	0.729	0.693
CNNH_PSS	*0.740*	*0.703*

The data in italic denote the best performance

Then, we compare our method CNNH_PSS to GSN and DCRNN by both CB6133 and CB513, which also can extract both local contexts and long-range interdependencies. The performances of these two methods and our method on CB6133 and CB513 are listed in Table 5. The table shows that CNNH_PSS performs better than GSN and DCRNN by at least 0.008 Q8 accuracy on CB6133 and 0.009 Q8 accuracy on CB513. GSN and DCRNN consist of CNN for local context extraction and extra models for long-range interdependencies extraction. As the extra models of the two methods are complex and time-consuming, these two methods need consume much computer resource. Trained on GTX TITANX GPU, CNNH_PSS tends to converge after only a half hour while DCRNN needs more than 24 h to converge [27]. So CNNH_PSS is almost 50 times faster than DCRNN. Although we do not known the exact running time for GSN, we know that GSN needs to be trained for 300 epochs [4] while CNNH_PSS tends to converge after training for less than 35 epochs shown by Fig. 3. It means that CNNH_PSS is almost 9 times faster then GSN. Therefore, the outperformance of our method over GSN and DCRNN further demonstrates that CNNH_PSS can not only cost less computer resource but also achieves better predicting performance.

Discussion

The advantage of our proposed method CNNH_PSS over state-of-the-art methods is that it can extract both local contexts and long-range interdependencies by using multi-scale CNN with highway. In CNNH_PSS, any two neighbor convolutional layers have a highway to deliver information from current convolutional layer to the output of the next one and each layer except the first one have convolution kernels to extract long-range interdependencies by using the information from

Table 4 The Q8 accuracy of CNNH_PSS and state-of-the-art methods containing only local contexts

Method	CB513
SSpro8	0.511
CNF	0.633
DeepCNF	0.683
CNNH_PSS	*0.703*

The data in italic denote the best performance

Table 5 The Q8 accuracy of CNNH_PSS and state-of-the-art methods containing both local contexts and long-range interdependencies

Method	CB6133	CB513
GSN	0.721	0.664
DCRNN	0.732	0.694
CNNH_PSS	*0.740*	*0.703*

The data in italic denote the best performance

previous layer. In this section, we use CNNH_PSS with 5 convolutional layers and the kernel length of 11 to introduce the process for local contexts and long-range interdependencies extraction, which is shown in Fig. 4. First, the target protein are inputted to the first layer and the convolution kernels in the first layer extract local contexts from the inputted protein. So the output of first layer contains local contexts among 11 residues. Then the information in the output of the first layer are delivered to the output of the second layer by two ways: highway between them and the convolution kernels in the second layer. Finally, the output of the second layer is the weighted sum of the information transmitted by the two way. As the convolution kernels in the secondary layer can extract relationships among 21 residues, the output of the second layer contains both local contexts among 11 residues and long-range interdependencies among 21 residues. And so on, the output of the fifth layer contains two parts. One part is the information from the output of the fourth layer by the highway between them, which contains local contexts among 11

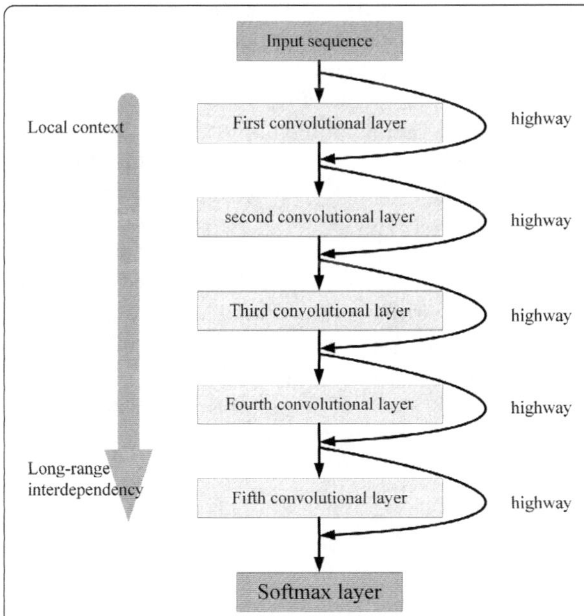

Fig. 4 Extraction process for local contexts and long-range interdependencies

and long-range interdependencies among 21, 31 and 41 residues. The other part is the information outputted by the convolutional kernels of current layer, which contains long-range interdependencies among 51 residues. Therefore CNNH_PSS can output local contexts among 11 and long-range interdependencies among 21, 31, 41 and 51 residues while the multi-scale CNN with the same number of convolutional layer as CNNH_PSS outputs only long-range interdependencies among 51 residues.

In order to demonstrate the importance of learned local contexts and long-range interdependencies in protein secondary structure prediction, we show the local contexts and the long-range interdependencies learned in a representative protein PDB 154 L [51], which obtained from the publicly available protein data bank [52]. The learned local contexts and long-range interdependencies by CNNH_PSS in protein PDB 154 L are shown in Fig. 5. In Fig. 5, the five rows correspond to the predicted results by CNNH_PSS with 5 layers, that by CNNH_PSS with 3 layers, that by the multi-scale CNN with 5 layers, real secondary structures and protein sequence, respectively. The reason for why CNNH_PSS with 5 layers, CNNH_PSS with 3 layers and the multi-scale CNN with 5 layers are selected for comparison is that CNNH_PSS with 5 layers can extracted local contexts and long-range interdependencies among up to 51 residues while CNNH_PSS with 3 layers cannot extracted long-range interdependencies among more than 41 residues and the multi-scale CNN with 5 layers cannot extract local contexts. Fig. 5 shows three instances for long-range interdependencies: (1) interdependencies among 24th, 25th and 60th amino acid; (2) that between 60th and 100th and (3) that between 85th and 131th amino acid. As the number of residues covered by these three learned interdependencies is more than 31 and less than 51 residues, both CNNH_PSS with 5 layers and the multi-scale CNN with 5 layers can extract them for correct prediction them while CNNH_PSS with 3 layers cannot capture them. So both CNNH_PSS with 5 layers and the multi-scale CNN with 5 layers make correct prediction for the 24th, 25th, 85th, 100th and 131th residues while CNNH_PSS with 3 layers cannot make correct predictions for them. It validates that CNNH_PSS with more layers indeed can extract long-range interdependencies between more remote residues.

Furthermore, Fig. 4 also shows 4 instances for learned local contexts: (1) contexts from 31th to 35th residues; (2) that from 111th to 115th residues; (3) that from 146th to 149th residues and (4) that from 158th to 163th residues. Both CNNH_PSS with 3 layers and that with 5 layers can learn these four contexts so that the secondary structures of all the residues in the learned contexts can be correctly predicted. However, the multi-

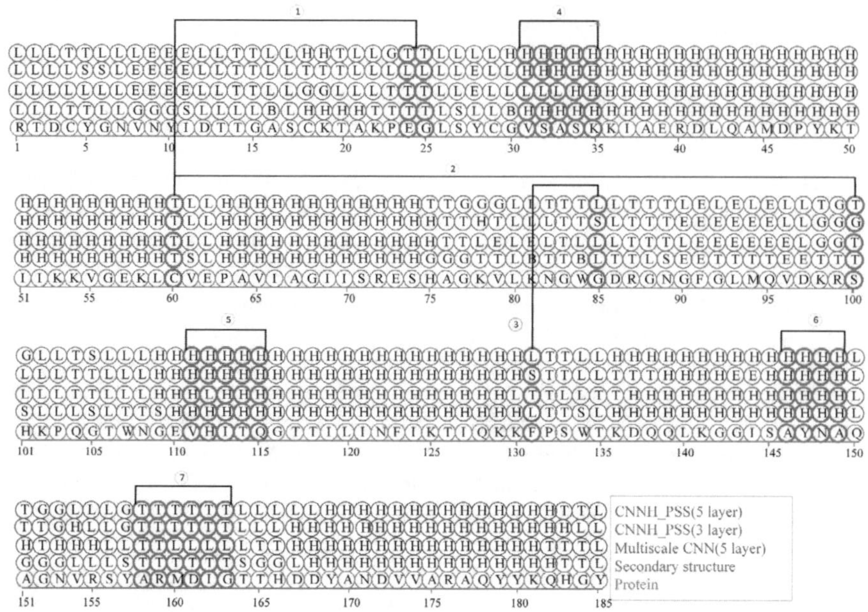

Fig. 5 Prediction results of 154 L by CNNH_PSS

scale CNN with 5 layers cannot learn these four contexts. So it cannot predict the secondary structures correctly for these residues. It validates that the highways in the CNNH_PSS indeed can be used to extract local contexts for prediction.

Conclusion

Protein secondary structure prediction is an important problem in bioinformatics and critical for analyzing protein function and applications like drug design. Several experimental methods have been proposed to determined the secondary structures for proteins, such as far-ultraviolet circular dichroism, infrared spectroscopy and NMR spectrum. However, experimental methods usually are costly and time-consuming. And the proteins with known sequence continues to outnumber the experimentally determined secondary structures. So developing computational approaches that can accurately handle large amount of data becomes increasingly urgent. However, most of these proposed methods cannot extract either local contexts or long-range interdependencies. Although GSM and DCRNN can extract both of them, they are build by combing CNN architecture and extra complex models. Yet CNNH_PSS is developed by only multi-scale CNN with highway. So comparing to GSM and DCRNN, CNNH_PSS may cost less computer resource. We evaluate CNNH_PSS on two commonly used datasets: CB6133 and CB513. CNNH_PSS outperforms the multi-scale CNN without highway on both datasets, which demonstrates that the extracted local contexts through highways are indeed useful for protein secondary structure prediction. CNNH_PSS also outperforms CNF and DeepCNF as well as SSpro8 on CB513, which cannot extract long-range interdependencies. It indicates that long-range interdependencies extracted by CNNH_PSS are useful for protein secondary structure prediction. Furthermore, CNNH_PSS performs better than GSM and DCRNN, demonstrating that CNNH_PSS can not only cost less computer resource but also achieves better predicting performance than state-of-the-art methods. We also analyze the local contexts and long-range interdependencies learned by CNNH_PSS in protein PDB 154 L and show their roles in protein secondary structure prediction. X-ray diffraction crystallography and NMR can measure the structures of proteins, so these methods can be used to calculate the distances between any two residues in a protein sequence. By analyzing the second structures of long-range residues but with short distance in space and short-range residues, we can further demonstrate the importance of long-range interdependencies and local contexts for second structure prediction. Therefore, our future works will validate the conclusions achieved in this paper by using these experimental methods.

Abbreviations

B: residue in isolated β-bridge; C: coil; CNN: convolutional neural network; CRF: conditional random field; E: extended strand in parallel and/or anti-parallel β-sheet conformation; G: 3-turn helix; GSN: generative stochastic network; H: 4-turn helix; I: 5-turn helix; NMR: nuclear magnetic resonance; PDB: Protein Data Bank; PSSM: Position Specific Score Matrix; Q3: three-category classification; Q8: eight-category classification; RNN: recurrent neural networks; S: bend; SVM: Support Vector Machine; T: hydrogen bonded turn

Acknowledgements
Not applicable.

Funding
Publication of this article was supported by the National Natural Science Foundation of China U1636103, 61632011 and Shenzhen Foundational Research Funding 20170307150024907.

Availability of data and materials
All data generated or analysed during this study are included in this published article. The datasets analysed during this study and the source code for CNNH_PSS can be accessed from our webserver of this paper http://hlt.hitsz.edu.cn/CNNH_PSS/ and http://119.23.18.63:8080/CNNH_PSS/.

About this supplement
This article has been published as part of *BMC Bioinformatics* Volume 19 Supplement 4, 2018: Selected articles from the 16th Asia Pacific Bioinformatics Conference (APBC 2018): bioinformatics. The full contents of the supplement are available online at https://bmcbioinformatics.biomedcentral.com/articles/supplements/volume-18-supplement-17.

Authors' contributions
JZ initiated and designed the study. RX made substantial contributions to acquisition of data, analysis and interpretation of data. JZ drafted the manuscript. RX and QL involved in drafting the manuscript or revising it. ZZ and HW provided valuable insights on biomolecular interactions and systems biology modeling, participated in result interpretation and manuscript preparation. All authors read and approved the final manuscript.

Competing interest
The authors declare that they have no conflicting interests.

Ethics approval and consent to participate
Not applicable.

Consent for publication
Not applicable.

Publisher's Note

Author details
[1]School Computer Science and Technology, Harbin Institute of Technology Shenzhen Graduate School, HIT Campus Shenzhen University Town, Xili, Shenzhen, Guangdong 518055, China. [2]Department of Computing, the Hong Kong Polytechnic University, Hung Hom, Hong Kong.

Published: 8 May 2018

References
1. Linderstrøm-Lang KU. Lane medical lectures: proteins and enzymes. California: Stanford University Press; 1952. p. 115.
2. Schellman JA, Schellman CG. Kaj Ulrik Linderstrøm-Lang (1896-1959). Protein Sci. 1997;6(5):1092–100.
3. Kabsch W, Sander C. Dictionary of protein secondary structure: pattern recognition of hydrogen-bonded and geometrical features. Biopolymers. 1983;22(12):2577–637.
4. Zhou J, Troyanskaya O. Deep supervised and convolutional generative stochastic network for protein secondary structure prediction. In: Proceedings of the 31st international conference on machine learning (ICML-14); 2014. p. 745–53.
5. Yaseen A, Li Y. Template-based c8-scorpion: a protein 8-state secondary structure prediction method using structural information and context-based features. BMC Bioinformatics. 2014;15(Suppl 8):S3.
6. Pollastri G, Przybylski D, Rost B, Baldi P. Improving the prediction of protein secondary structure in three and eight classes using recurrent neural networks and profiles. Proteins: Structure, Function, and Bioinformatics. 2002;47(2):228–35.
7. Wang Z, Zhao F, Peng J, Xu J. Protein 8-class secondary structure prediction using conditional neural fields. Proteomics. 2011;11(19):3786–92.
8. Noble ME, Endicott JA, Johnson LN. Protein kinase inhibitors: insights into drug design from structure. Science. 2004;303(5665):1800–5.
9. Simossis VA, Heringa J. Integrating protein secondary structure prediction and multiple sequence alignment. Curr Protein Pept Sci. 2004;5(4):249–66.
10. Ashraf Y, Yaohang L. Context-based features enhance protein secondary structure prediction accuracy. Journal of chemical information and modeling. J Chem Inf Model. 2014;54(3):992–1002.
11. Pelton JT, McLean LR. Spectroscopic methods for analysis of protein secondary structure. Anal Biochem. 2000;277(2):167–76.
12. Meiler J, Baker D. Rapid protein fold determination using unassigned NMR data. Proc Natl Acad Sci U S A. 2003;100(26):15404–9.
13. Chou PY, Fasman GD. Prediction of protein conformation. Biochemistry. 1974;13(2):222–45.
14. Gascuel O, Golmard JL. A simple method for predicting the secondary structure of globular proteins: implications and accuracy. Computer Appl Biosci. 1988;4(3):357–65.
15. Rost B, Sander C. Prediction of protein secondary structure at better than 70% accuracy. J Mol Biol. 1993;232(2):584–99.
16. Jones TD. Protein secondary structure prediction based on position-specific scoring matrices. J Mol Biol. 1999;292(2):195–202.
17. Cortes C, Vapnik V. Support vector networks. Mach Learn. 1995;20:273–93.
18. Scholkopf B, Burges C, Vapnik V. Extracting support data for a given task. In: Proceedings, first international conference on knowledge discovery and data mining. Menlo Park, CA: AAAI Press; 1995. p. 252–7.
19. Roobaert D, Hulle MM. View-based 3D object recognition with support vector machines. In: Proceedings of the IEEE neural networks for signal processing workshop. NJ: IEEE Press; 1999. p. 77–84.
20. Schmidt M, Grish H. Speaker identification via support vector classifiers. In: The proceedings of the international conference on acoustics, speech and signal processing, 1996. Long Beach, CA: IEEE Press; 1996. p. 105–8.
21. Drucker H, Wu D, Vapnik V. Support vector machines for spam categorization. IEEE Trans Neural Netw. 1999;10:1048–54.
22. Hua S, Sun Z. A novel method of protein secondary structure prediction with high segment overlap measure: support vector machine approach. J Mol Biol. 2001;308(2):397–407.
23. Kim H, Park H. Protein secondary structure prediction based on an improved support vector machines approach. Protein Eng Des Sel. 2003;16(8):553–60.
24. Zhou J, Lu Q, Xu R, He Y, Wang H. EL_PSSM-RT: DNA-binding residue prediction by integrating ensemble learning with PSSM relation transformation. BMC Bioinformatics. 2017;18:379.
25. Guo J, Chen H, Sun Z, Lin Y. A novel method for protein secondary structure prediction using dual-layer SVM and profiles. Proteins: Structure, Function, and Bioinformatics. 2004;54(4):738–43.
26. Bengio Y, Thibodeau-Laufer É, Alain G, Yosinski J, preprint arXiv:.1091. Deep generative stochastic networks trainable by backprop. Computer Sci. 2013;2:226–34.
27. Li Z, Yu Y: Protein secondary structure prediction using cascaded convolutional and recurrent neural networks. 2016.
28. Lawrence S, Giles CL, Tsoi AC, Back AD. Face recognition: a convolutional neural-network approach. IEEE Trans Neural Netw. 1997;8(1):98–113.
29. Yih W, Toutanova K, Platt JC, Meek C. Learning discriminative projections for text similarity measures. In: Proceedings of the fifteenth conference on computational natural language learning; 2011. p. 247–56.
30. Zhang X, Zhao J, LeCun Y. Character-level convolutional networks for text classification. In: Advances in neural information processing systems; 2015. p. 649–57.
31. Wang S, Peng J, Ma J, Xu J. Protein secondary structure prediction using deep convolutional neural fields. Sci Rep. 2016;6:18962.
32. Baldi P, Brunak S, Frasconi P, Soda G, Pollastri G. Exploiting the past and the future in protein secondary structure prediction. Bioinformatics. 1999;15(11):937–46.
33. Schmidler SC, Liu JS, Brutlag DL. Bayesian segmentation of protein secondary structure. J Comput Biol. 2000;7(1–2):233–48.
34. Chu W, Ghahramani Z, Wild DL. A graphical model for protein secondary structure prediction. In: Proceedings of the twenty-first international conference conference on machine learning (ICML); 2004. p. 161–8.
35. Xu R, Zhou J, Liu B, He Y, Zou Q, Wang X, Chou KC. Identification of DNA-binding proteins by incorporating evolutionary information into pseudo amino acid composition via the top-n-gram approach. J Biomol Struct Dyn. 2015;33(8):1720.

36. Xu R, Zhou J, Wang H, He Y, Wang X, Liu B. Identifying DNA-binding proteins by combining support vector machine and PSSM distance transformation. BMC Syst Biol. 2015;9(S1):1–12.
37. Zhou J, Xu R, He Y, Lu Q, Wang H, Kong B. PDNAsite: identification of DNA-binding site from protein sequence by incorporating spatial and sequence context. Sci Rep. 2016;6:27653.
38. Wang G, Jr DR. Pisces: a protein sequence culling server. Bioinformatics. 2003;19(12):1589–91.
39. Cuff JA, Barton GJ. Evaluation and improvement of multiple sequence methods for protein secondary structure prediction. Proteins: Structure, Function, and Bioinformatics. 1999;34(4):508–19.
40. Wang Z, Zhao F, Peng J, Xu J. Protein 8class secondary structure prediction using conditional neural fields. IEEE Int Conf Bioinformatics Biomed. 2011; 11(19):3786–92.
41. Mesnil G, Dauphin Y, Yao K, Bengio Y, Deng L, Hakkani-Tur D, He X, Heck L, Tar G, Yu D, et al. Using recurrent neural networks for slot filling in spoken language understanding. IEEE/ACM Trans Audio Speech Lang Process. 2015; 23(3):530–9.
42. Mikolov T, Sutskever I, Chen K, Corrado G, Dean J. Distributed representations of words and phrases and their compositionality. In: Advances in neural information processing systems; 2013.
43. Kumar M, Gromiha M, Raghava G. Identification of DNA-binding proteins using support vector machines and evolutionary profiles. BMC Bioinformatics. 2007;8(1):563.
44. Jones DT. Improving the accuracy of transmembrane protein topology prediction using evolutionary information. Bioinformatics. 2007;23:538–44.
45. Biswas AK, Noman N, Sikder AR. Machine learning approach to predict protein phosphorylation sites by incorporating evolutionary information. BMC Bioinformatics. 2010;11(1):273.
46. Ruchi V, Grish CV, Raghava GPS. Prediction of mitochondrial proteins of malaria parasite using split amino acid composition and PSSM profile. Amino Acids. 2010;39:101–10.
47. Zhao XW, Li XT, Ma ZQ, Yin MH. Prediction of lysine ubiquitylation with ensemble classifier and feature selection. Int J Mol Sci. 2011;12:8347–61.
48. Schaffer AA, Aravind L, Madden TL, Shavirin S, Spouge JL, Wolf YI, Koonin EV, Altschul SF. Improving the accuracy of PSI-BLAST protein database searches with composition-based statistics and other refinements. Nucleic Acids Res. 2001;29(14):2994–3005.
49. Srivastava RK, Greff K, Schmidhuber J. Training very deep networks. In: Cortes C, Lawrence ND, Lee DD, Sugiyama M, Garnett R, editors. Advances in neural information processing systems, vol. 28; 2015. p. 2377–85.
50. Cho K, Merrienboer BV, Gulcehre C, Bahdanau D, Bougares F, Schwenk H, Bengio Y. Learning phrase representations using RNN encoder-decoder for statistical machine translation. Computer Sci. 2014;
51. Simpson RJ, Morgan FJ. Complete amino acid sequence of embden goose (anser anser) egg-white lysozyme. Biochimica et Biophysica Acta (BBA)-Protein Structure and Molecular Enzymology. 1983;744(3):349–51.
52. Berman HM, Westbrook J, Feng Z, Gilliland G, Bhat TN, Weissig H, Shindyalov IN, Bourne PE. The protein data bank. Nucleic Acids Res. 2000;28:235–42.

Gong et al. BMC Bioinformatics 2018, 19(Suppl 4):162
https://doi.org/10.1186/s12859-018-2064-y

BMC Bioinformatics

RESEARCH

Open Access

A new method to measure the semantic similarity from query phenotypic abnormalities to diseases based on the human phenotype ontology

Xiaofeng Gong, Jianping Jiang, Zhongqu Duan and Hui Lu*

From The Sixteenth Asia Pacific Bioinformatics Conference
Yokohama, Japan. 15-17 January 2018

Abstract

Background: Although rapid developed sequencing technologies make it possible for genotype data to be used in clinical diagnosis, it is still challenging for clinicians to understand the results of sequencing and make correct judgement based on them. Before this, diagnosis based on clinical features held a leading position. With the establishment of the Human Phenotype Ontology (HPO) and the enrichment of phenotype-disease annotations, there throws much more attention to the improvement of phenotype-based diagnosis.

Results: In this study, we presented a novel method called RelativeBestPair to measure similarity from the query terms to hereditary diseases based on HPO and then rank the candidate diseases. To evaluate the performance, we simulated a set of patients based on 44 complex diseases. Besides, by adding noise or imprecision or both, cases closer to real clinical conditions were generated. Thus, four simulated datasets were used to make comparison among RelativeBestPair and seven existing semantic similarity measures. RelativeBestPair ranked the underlying disease as top 1 on 93.73% of the simulated dataset without noise and imprecision, 93.64% of the simulated dataset with noise and without imprecision, 39.82% of the simulated dataset without noise and with imprecision, and 33.64% of the simulated dataset with both noise and imprecision.

Conclusion: Compared with the seven existing semantic similarity measures, RelativeBestPair showed similar performance in two datasets without imprecision. While RelativeBestPair appeared to be equal to Resnik and better than other six methods in the simulated dataset without noise and with imprecision, it significantly outperformed all other seven methods in the simulated dataset with both noise and imprecision. It can be indicated that RelativeBestPair might be of great help in clinical setting.

Keywords: Human phenotype ontology (HPO), Semantic similarity, Disease, Diagnosis

* Correspondence: huilu@sjtu.edu.cn
Department of Bioinformatics and Biostatistics, SJTU-Yale Joint Center for Biostatistics, Shanghai Jiao Tong University, Shanghai, China

Background

Correct diagnosis based on the observed clinical features of patients is a quite important task for physicians, especially in the field of rare genetic diseases, where different diseases often share some features. Recently, with the rapid development of sequencing technology, it becomes possible to improve diagnosis by providing physicians with patients' genotype data in a short time [1]. While techniques like whole genome sequencing and whole exome sequencing allows a patient's genotype data to be used to detect mutations, the relative high expense and the ability to identify disease-causing variants make it difficult to be put into practical clinical use. However, back to the beginning, if the performance of diagnosis based on clinical features can be improved, it will be of great help to the clinicians.

Thus, to make full use of clinical features or phenotypic information, many databases have been established to record and reorganize phenotypic data of diseases, such as OMIM [2] and Orphanet [3]. Furthermore, the Human Phenotype Ontology (HPO) [4–6] was constructed to describe human phenotype abnormalities in a structured and controlled vocabulary and has been widely used in research.

Recently, HPO has been widely applied in various fields. A web application called the *Phenimizer* provides ontology similarity search based on HPO to assist the clinical diagnosis workflow [7]. *PhenoTips*, a deep phenotyping tool and database, is developed to collect phenotypic information of patients with genetic disorders using HPO and suggest additional clinical investigations and possible disorders in Online Mendelian Inheritance in Man (OMIM) [8]. *PhenoDB*, a Web-based portal which can store and analyze phenotypic information using mapped HPO terms as well as other clinical information, is also developed [9]. Besides, several methods or tools have been introduced to combine phenotypic information based on HPO and genotypic data with other information available to make variant or gene prioritization, including eXtasy [10], Phen-Gen [11], an initial study using semantic similarity [12], PHIVE/Exomiser [13], Phevor [14], PhenoVar [15], PhenIx [16] and OMIM Explorer [17]. Despite the short history of HPO, it has drawn much attention from researchers and scientists and been broadly used in scientific researches.

In this article, we focus on using similarity between observed phenotypes of a patient and the annotated phenotypes of diseases to rank the candidate diseases of the patient. From this point of view, several methods and tools [7, 12, 18] has been presented to exploit HPO-based semantic similarity borrowing ideas from semantic similarity measures used in Gene Ontology (GO), which have been widely studied and broadly used during the last decade. Most of them utilized information content (IC) to calculate the semantic similarity. Although those approaches have been used in clinical research, the results are still uncertain and can be further imporved. Here we present a new method called RelativeBestPair. RelativeBestPair takes the ideas from information content and the best pair method. Our work shows better diagnosis using the RelativeBestPair method over other methods.

Methods

Human phenotype ontology (HPO)

An ontology is a knowledge-based structured system, which consists of a rich, standardized vocabulary to describe entities and the semantic relationships between them. The Human Phenotype Ontology (HPO) provides a standardized vocabulary of phenotypic abnormalities encountered in human disease. Terms in HPO, representing different phenotypic abnormalities, are related to their parent terms by "is a" relationship in a relaxed hierarchy which allows a term to possibly have multiple parent terms (Fig. 1). With HPO terms corresponding to phenotypic abnormalities, diseases can be described in a detailed and organized way. The HPO (version 1.2 releases/2017–2-14) currently contains approximately 12,000 terms (still growing) and over 120,000 phenotype-disease annotations. Here we concentrate on annotations about 6918 diseases listed in Online Mendelian Inheritance in Man (OMIM) to calculate the semantic similarity scores.

RelativeBestPair method

Based on the HPO structure and annotations, the information content of a term t in HPO is defined as follows:

$$\mathrm{IC}(t) = -\log\frac{N_t}{N} \tag{1}$$

where N is the total number of annotated diseases and N_t is the number of diseases annotated by term t and all its descendants. When comparing the similarity between two sets of phenotypes, the best pair method just simply counts the number of same terms in both two sets, which does not take the semantic inheritance structure of HPO and the different importance of the terms into consideration.

Thus we propose RelativeBestPair, a new semantic similarity measure based on the information content and the best pair method. Inspired by the idea of information content, we collect diseases annotated by a phenotype t and its descendants to measure the different importance of terms. RealtiveBestPair is described as follows.

A. For a given term t, we denote $D(t)$ as the set of diseases annotated by term t and all its descendants and N_t as the size of $D(t)$. Then, the sccn term t is defined as

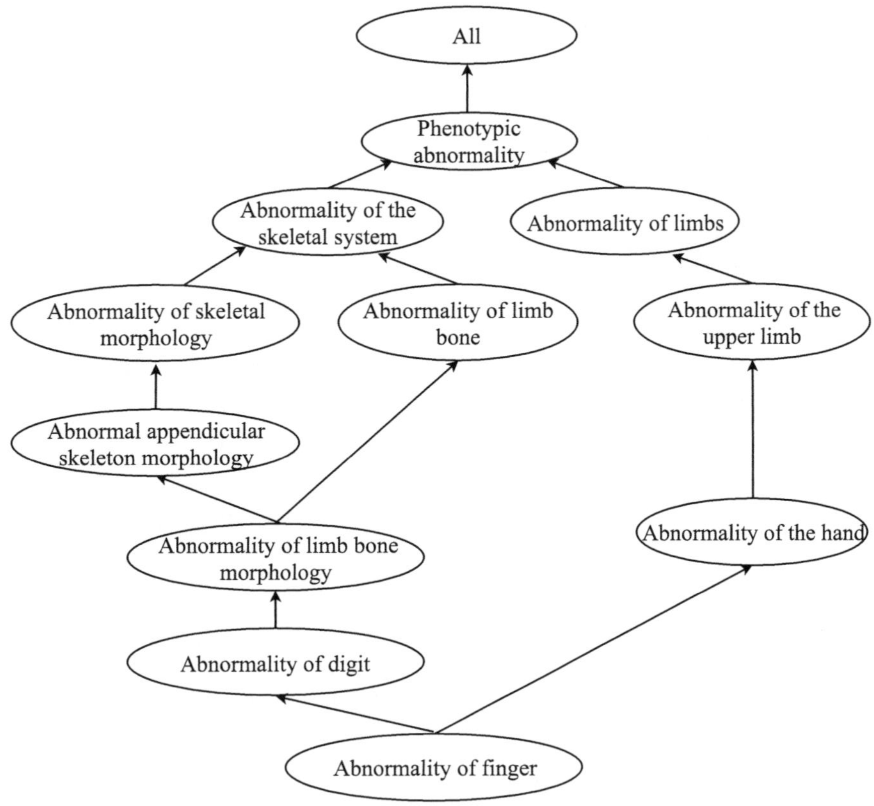

Fig. 1 Example of the structure of HPO. Term Abnormality of finger (HP:0001167) and all its ancestors are shown. Each term, representing a phenotypic abnormality, is related to parents terms by "is a" relationship

$$S(D|t) = \begin{cases} 1/N_t, & D \in d(t) \\ 0, & otherwise \end{cases} \quad (2)$$

B. Then we can get all the scores of being each disease given each term. For a sets of phenotypes $\{t_1, t_2, \ldots, t_n\}$ and a disease D_k, the semantic similarity score can be calculated as

$$Sim(D_k|t_1, t_2, \ldots, t_n) = \sum_{i=1}^{n} \min(\alpha, S(D_k|t_i)) \quad (3)$$

where α is a given threshold.

The threshold α is introduced to control the contribution of a single term. If only several diseases are annotated by a single term, then the score of being one of those diseases given this term will be so large that it may dominate the semantic similarity score and ignore the contributions of other terms. For example, we observed a patient with ten terms $\{t_1, t_2, \ldots, t_{10}\}$. If the score of being D_1 given each of $\{t_1, t_2, \ldots, t_9\}$ is suitable like 0.005 while the score of being D_2 given t_{10} is quite large, for example 0.1, the semantic similarity score between the patient and D_2 will be larger than that between the patient and D_1. Thus we use the threshold α to avoid the such extreme cases. Although the choice of α may affect the performance, generally we set it to be 0.01.

Disease diagnosis based on RelaitveBestPair can be summarized as followed (Fig. 2). With the input of HPO and its annotations, the ontology and the database (containing the scores of being each Disease D given each term t using Eq. (2)) are constructed first. Then given a query set of phenotype terms, the similarity scores from query terms to each disease can be calculated with Eq. (3). Finally, diseases are ranked according to these scores from the largest to the smallest.

Existing semantic similarity measures

We compared the performance of RelativeBestPair with other seven existing approaches summarized in HPOsim [19]. Among those, six approaches are based on information content. The Resnik measure [20], the Lin measure [21], the Jiang-Conrath measure [22], the Relevance measure [23], the information coefficient measure [24] and the graph IC measure [25] define the similarity between two terms as follows:

$$sim_{Resnik}(t_1, t_2) = IC(t_{MICA}) \quad (4)$$

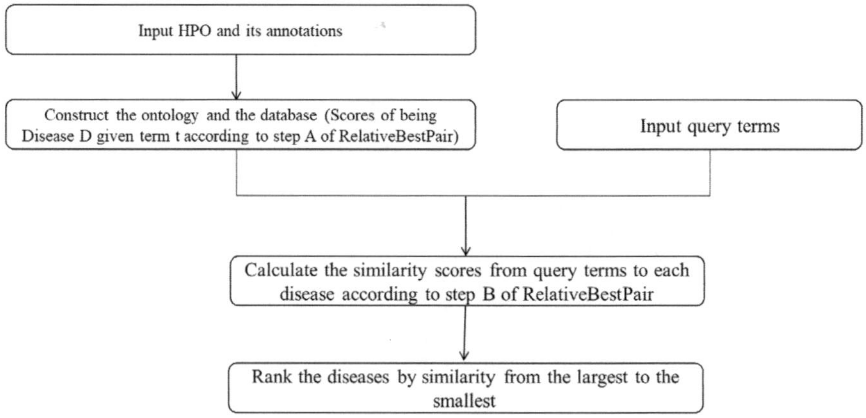

Fig. 2 The workflow of disease diagnosis based on RelativeBestPair

$$sim_{Lin}(t_1, t_2) = \frac{2 \times IC(t_{MICA})}{IC(t_1) + IC(t_2)} \quad (5)$$

$$sim_{JC}(t_1, t_2) = \frac{1}{(1 + IC(t_1) + IC(t_2) - 2 \times IC(t_{MICA}))} \quad (6)$$

$$sim_{Rel}(t_1, t_2) = sim_{Lin}(t_1, t_2) \times (1 - p(t_{MICA})) \quad (7)$$

$$sim_{IC}(t_1, t_2) = sim_{Lin}(t_1, t_2) \times \left(1 - \frac{1}{1 + IC(t_{MICA})}\right) \quad (8)$$

$$sim_{GraphIC}(t_1, t_2) = \frac{\sum_{t \in (A(t_1) \cap A(t_2))} IC(t)}{\sum_{t \in (A(t_1) \cap A(t_2))} IC(t)} \quad (9)$$

Where IC is defined as (1), t_{MICA} is the most informative common ancestors, $p(t_{MICA})$ is the proportion of diseases annotated by t_{MICA} and $A(t)$ is the set of the ancestors of term t in HPO.

Besides, the Wang measure [26] is based the structure of ontology. For a given term t, $DAG_t = (t, T_t, E_t)$ represents the subgraph made up of term t and its ancestors, where T_t is the set of the ancestors of t and E_t is the corresponding set of edges is DAG_t. In DAG_t, $S_t(n)$ is defined as:

$$\begin{cases} S_t(t) = 1 \\ S_t(n) = \max\{w_e * S_t(n') | n' \in children\ of(n)\} if\ t \neq n \end{cases} \quad (10)$$

here we choose w_e equal to 0.8. Therefore the similarity between two terms is defined as:

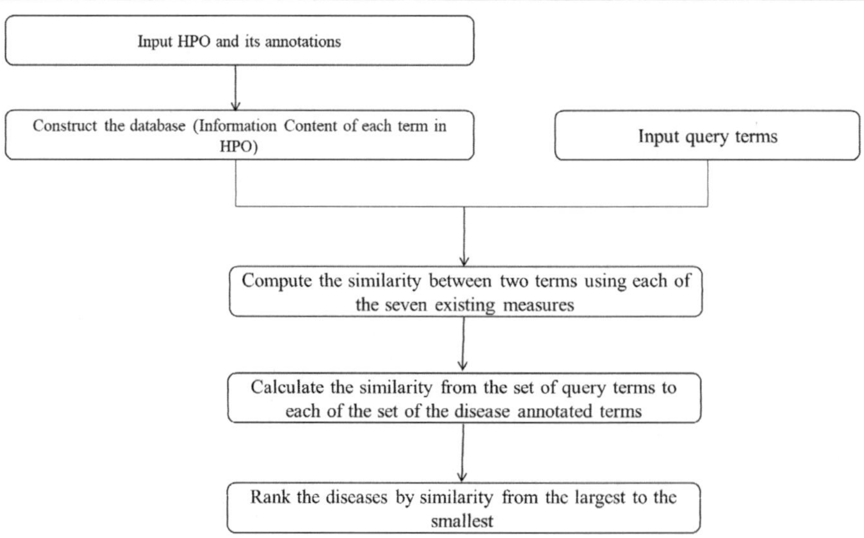

Fig. 3 The workflow of disease diagnosis based on the seven existing methods

$$sim_{Wang}(t_1, t_2) = \frac{\sum_{t \in T_{t1} \cap T t_2} S_{t_1}(t) + S_{t_2}(t)}{SV(t_1) + SV(t_2)} \quad (11)$$

where $SV(t)$ is the sum of $S_t(n)$ for n in DAG_t.

In order to get the similarity between the query set of terms and the set of disease associated terms, we used the one-sided search algorithm as it was showed to be superior to the symmetric version in [7]. The one-sided search algorithm is defined as:

$$sim_{one-sided}(Q \rightarrow D) = avg\left[\sum_{t_1 \in Q} \max_{t_2 \in D} sim(t_1, t_2)\right] \quad (12)$$

where Q is the set of the query terms (observed phenotypes of the patient), D is the set of terms annotated with a given disease, and $sim(t_1, t_2)$ can be one of the seven approaches.

Disease diagnosis based the seven semantic similarity measures is quite similar with that based on.

RelativeBestPair (Fig. 3). Firstly, the ontology and the database (containing information content of each term in HPO) are constructed based on HPO and its annotation files. Secondly, given a query set of phenotype terms, the similarity score from these query terms to each disease are calculated with term-term similarity based on each of the seven methods and then one-sided search algorithm. Finally, diseases are also ranked from the largest score to the smallest score.

Performance evaluation and generation of simulated patients

Since it is difficult to get clinical features about a large number of patients, we used similar method and same data in [7] to generate simulated patients. In the data used in [7], 44 complex dysmorphology syndromes were identified with detailed frequency of phenotypes. The simulation process is as follows. First, we assigned a disease to each patient. Second, for each phenotype associated with the assigned disease, a random integer between 0 and 100 was generated. If the number was smaller than the relative occurrence in 100 patients (frequency*100), the corresponding phenotype was kept. For each of the 44 diseases, we generated 25 patients with at least three phenotypes. Finally, we got a dataset of 1100 simulated patients. To make the simulation more realistic, three more datasets were also generated just as what was done in [7, 12]. We generated a dataset with 'noise' by adding half as many noise terms, unrelated with the underlying disorder, to the present terms, a dataset with 'imprecision' by randomly substituting each of the present phenotypes with one of its ancestors in HPO, and also a dataset with both 'imprecision' and 'noise' by imprecision step first and then noise step. With the four simulated datasets, we evaluated the performance of semantic similarity measure by the ranks of the true disease and adopted the criterion from [12, 19].

Table 1 Summary results of different methods on the four simulated datasets

	Resnik	Lin	JC	Rel	IC	GraphIC	Wang	RBP
Dataset 1(Noise:-, Imprecision:-)								
Top 1	1027	1016	1029	1018	1021	1029	1023	1031
Top 5	1087	1071	1082	1071	1075	1079	1078	1091
Top 10	1089	1077	1088	1077	1079	1081	1081	1095
Top 20	1092	1078	1092	1078	1080	1083	1081	1096
Dataset 2(Noise:+, Imprecision:-)								
Top 1	992	997	1036	996	1006	1031	1001	1030
Top 5	1074	1059	1081	1063	1070	1077	1071	1089
Top 10	1081	1069	1086	1071	1077	1080	1078	1094
Top 20	1087	1074	1089	1076	1078	1083	1079	1095
Dataset 3(Noise:-, Imprecision:+)								
Top 1	434	243	104	302	336	120	172	438
Top 5	767	502	261	583	603	341	446	765
Top 10	866	613	342	685	707	482	604	863
Top 20	926	714	440	785	797	620	725	926
Dataset 4(Noise:+, Imprecision:+)								
Top 1	183	130	97	143	162	73	77	370
Top 5	453	327	239	383	406	252	263	694
Top 10	579	452	319	509	533	393	384	786
Top 20	703	570	420	640	657	540	535	860

Resnik the Resnik measure, *Lin* the Lin measure, *JC* the Jiang-Conrath measure, *Rel* the Relevance measure, *IC* the information coefficient measure, *GraphIC* the graph IC measure, *Wang* the Wang measure, *RBP* RelativeBestPair method
The seven existing measures are all implemented with one-sided search algorithm. The numbers represent the number of patients in 1100 cases that the true diseases are ranked within top 1, top 5, top 10 or top 20

Results

We evaluated the performance of the seven existing approaches and RealtiveBestPair method in the four simulated datasets respectively. We denoted the dataset without noise and imprecision, the dataset with noise and without imprecision, the dataset without noise and with imprecision, and the dataset with both noise and imprecision as "Dataset 1(Noise:-, Imprecision:-)", "Dataset 2(Noise:+, Imprecision:-)", "Dataset 3(Noise:-, Imprecision:+)", and "Dataset 4(Noise:+, Imprecision:+)". As we moved on from Dataset 1 to Dataset 4, it became more difficult to make the correct diagnosis. It would show us the real abilities of those methods to identify the true underlying disease.

For a given patient, we calculated the similarity score from the patient to each of the 6918 OMIM diseases using one kind of semantic similarity measure, and then rank all the diseases by their similarity scores (from the largest to the smallest). In case that some diseases received the same score, the average rank was returned to make it more reasonable. The results of all the eight methods on the four datasets are shown in Table 1 and Figs. 1, 2, 3 and 4.

It can be seen that in the seven existing semantic similarity measures, the Resnik measure has a modest advantage over other six approaches, similar to the results in [7]. The RelativeBestPair method shows the almost the best performance in all four datasets (Table 1). Although in Dataset 1 and Dataset 2, two datasets that do not include "imprecision", all methods reveal good results by ranking the true diseases as top 1 on over 90% of the patients and within top 20 on over 95% of the patients (Table 1, Figs. 4 and 5), their performances deteriorate with different extents in Dataset 3 and Dataset 4. In Dataset 3 with imprecision, RelativeBestPair method, along with the Resnik measure, tends to be superior with the underlying diseases being ranked within top 1, top 5, top 10, top 20 on 39.82%, 69.55%, 78.45%, 84.18% of the cases for RealtiveBestPair and 39.45%, 69.73%, 78.73%, 84.18% for Resnik (Table 1, Fig. 6). The corresponding percentages using other measures are much smaller. In Dataset 4, a more real situation by both introducing unrelated phenotypic noise and using terms that are more general, RelativeBestPair achieves the best performance among the eight methods (Table 1, Fig. 7). On 33.64% of the patients, their underlying diseases are ranked the highest when applying RelativeBestPair. In comparison, the percentages using Resnik, Lin, Jiang-Conrath, Relevance, information coefficients, Graph IC and Wang measures are only 16.64%, 11.82%, 8.82%, 13%, 14.73%, 6.64% and 7% respectively. Even if a higher rank threshold is employed to give out a candidate list, RelativeBestPair still turns out to be significant better than other methods (Fig. 4). In total, it indicates that RelativeBestPair has the potential to provide a candidate disease/disease list for clinician to improve the diagnosis efficiency as well as accuracy.

Discussion and conclusion

Recently, the rapid development of sequencing technology makes it possible to get personal genotype data for clinical use, which may be helpful in disease diagnosis. However, the relative high cost and low ability to identify the disease-related causal variants prevent it from being widely used in real cases. While lots of effort and money have been paid to study the relationship between diseases and genetic mutations, to speed up the process of sequencing and to promote the accuracy of sequencing results, in this article we focus on the improvement in the field of phenotypic diagnosis. Compared with genotypic data, it is much easier to get phenotypic data from patients. With the construction and development of the Human Phenotype Ontology and the enrichment and completeness of disease-phenotype annotations, the observed phenotypes of a particular patient can provide more information about the underlying disease he/she might suffer.

Here we proposed a novel method called RelativeBestPair to measure the semantic similarity from a given set

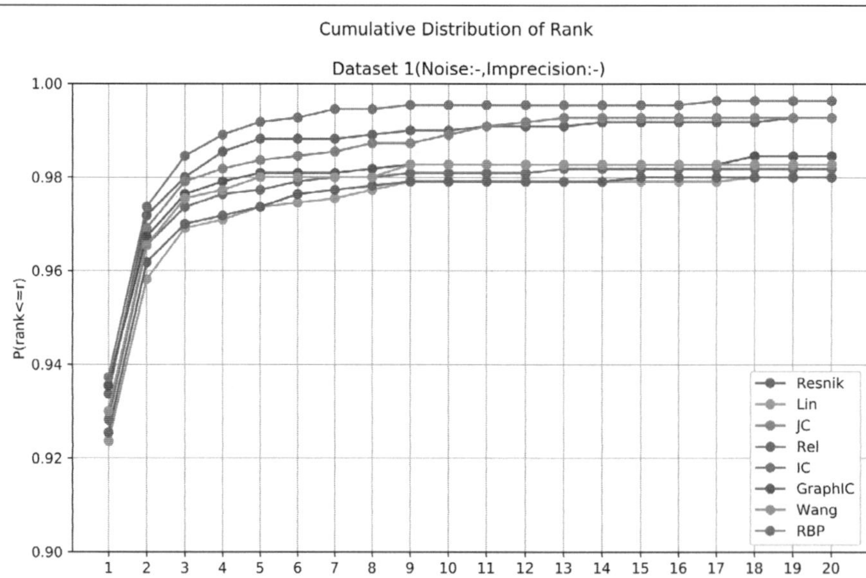

Fig. 4 Cumulative Distribution of the rank of the underlying diseases on the simulated dataset without noise and imprecision. The horizontal axis is the threshold for the disease rank. The vertical axis is the corresponding ratio of patients satisfying the ranking threshold

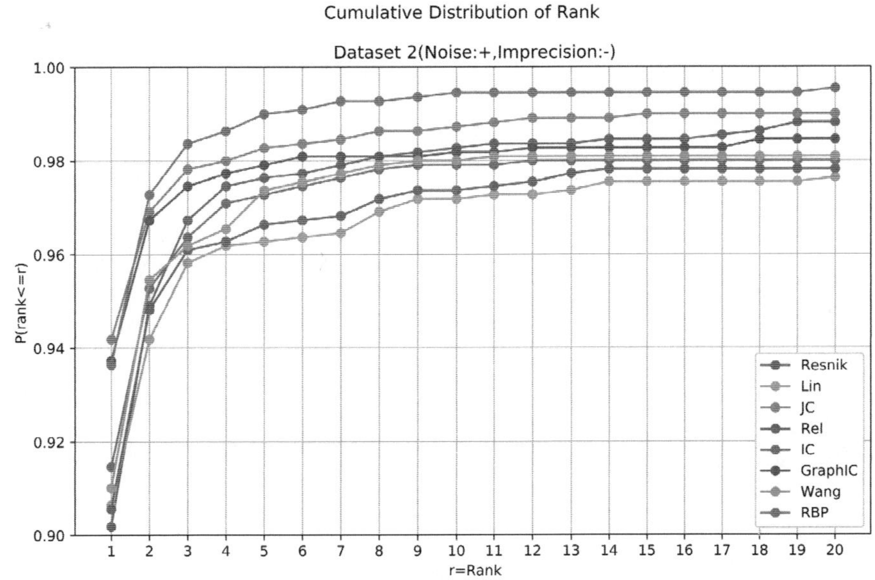

Fig. 5 Cumulative Distribution of the rank of the underlying diseases on the simulated dataset with noise and without imprecision. The horizontal axis is the threshold for the disease rank. The vertical axis is the corresponding ratio of patients satisfying the ranking threshold

of phenotypes to a disease. Different from those existing approaches that calculate the similarity from the query set to a certain disease based on term-term comparison, we directly define the contribution of one phenotype term to the certain disease. To evaluate the performance of RelativeBestPair and seven existing methods, we adopted the procedure similar to that in [7, 12] to generate four kinds of simulated patients from the easiest situation to the most difficult situation. In order to be adapted to the scenario of disease diagnosis, the one-sided search algorithm, which showed better performance than symmetric version in [7], was chosen for the seven existing methods. The results on the simulated datasets demonstrated that RelativeBestPair outperformed other methods in all situations especially when "noise" and "imprecision" were added, typical in the clinical setting.

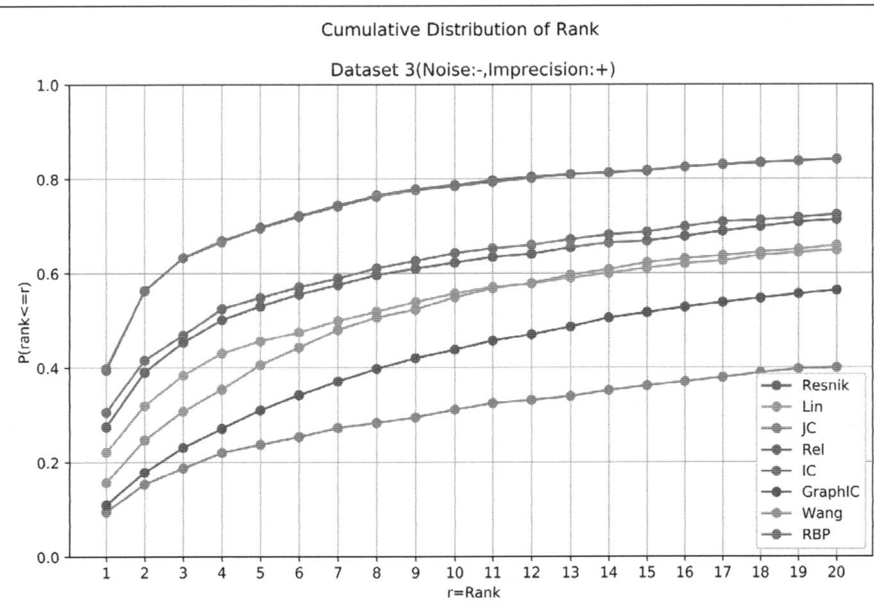

Fig. 6 Cumulative Distribution of the rank of the underlying diseases on the simulated dataset without noise and with imprecision. The horizontal axis is the threshold for the disease rank. The vertical axis is the corresponding ratio of patients satisfying the ranking threshold

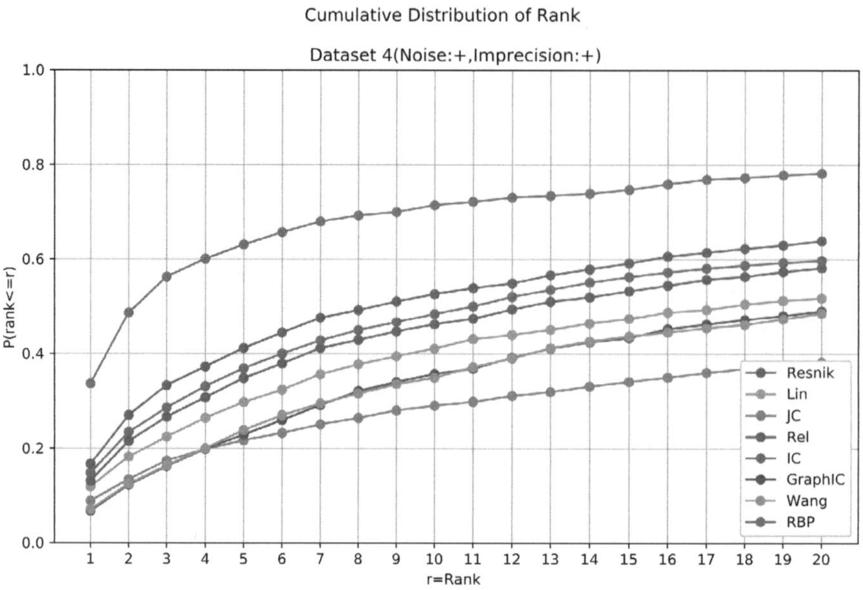

Fig. 7 Cumulative Distribution of the rank of the underlying diseases on the simulated dataset with both noise and imprecision. The horizontal axis is the threshold for the disease rank. The vertical axis is the corresponding ratio of patients satisfying the ranking threshold

Despite the well performance in simulation, there still remains much for RelativeBestPair to take into consideration. Firstly, the optimal value for α requires further discussion. The introduction of threshold α played a key role in the performance of RelativeBestPair since we found poor results when the threshold α was not employed. Therefore, the choice of threshold α would substantially affect the performance. Other than 0.01, we also tested other values for α including 0.001–0.005, 0.015, 0.02, 0.025 and 0.03. Although those results showed some minor difference (data not shown), considering the fact that on average one term annotates about 150 diseases which indicates that average score of being the given disease is $1/150 \approx 0.0067$, empirically the choice of 0.01 for α might be enough to make sure that the contribution of one single term won't be too large. Other choices are also welcomed as long as α is neither too big nor too small. Secondly, unlike the seven existing approaches, RelativeBestPair cannot be used to compute the similarity between two phenotype terms. The usage of RelativeBestPair might be limited in disease diagnosis and its expansion to other biomedical ontologies and other usages may be uncertain. Finally, without thousands of real cases, the true ability of RelativeBestPair as well as other semantic similarity measures in disease diagnosis is still unknown. As mentioned before, all the simulations are based on 44 complex diseases with detailed frequencies of phenotypes [7]. Then, we cannot assert the performance in any cases. However, from the simulation results, RelativeBestPair might have a large potential to identity the true underlying diseases of patients.

In conclusion, we have presented a new method, RelativeBestPair, that calculates the semantic similarity from the given query terms to each disease. Our method has the advantage of pay special attention to the fields of disease diagnosis. This approach can be applied to the real clinical setting by providing clinicians with a candidate disease list. We have shown that RelativeBestPair achieved a better performance of identifying the true disease as top-ranked diseases against other methods in four simulated dataset, mimic to the real cases.

Acknowledgements
We would like to thank Center for HPC, Shanghai Jiao Tong University and Children Hospital of Shanghai for providing computation help.

Funding
This work and the publication charges for this article was supported by the SJTU-Yale Seed Fund for Research collaborations and by the "Neil Shen's SJTU Medical Research Fund".

Availability of data and materials
The data used to generate simulated patients was reported by Köhler et al. in [7] and is available as supplemental data in http://www.cell.com/ajhg/fulltext/S0002-9297(09)00399-1.

About this supplement
This article has been published as part of *BMC Bioinformatics* Volume 19 Supplement 4, 2018: Selected articles from the 16th Asia Pacific Bioinformatics Conference (APBC 2018): bioinformatics. The full contents of the supplement are available online at https://bmcbioinformatics.biomedcentral.com/articles/supplements/volume-18-supplement-17.

Authors' contributions
XFG designed and implemented the method. JPJ and ZQD provided valuable advice on the method. XFG prepared the data and analyzed the results. XFG wrote the manuscript. HL provided guidance and revised the manuscript. All authors read and approved the final manuscript.

Ethics approval and consent to participate
Not applicable.

Consent for publication
Not applicable.

Competing interests
The authors declare that they have no competing interests.

Publisher's Note

Springer Nature remains neutral with regard to jurisdictional claims in published maps and institutional affiliations.

Published: 8 May 2018

References

1. Smedley D, Jacobsen JO, Jäger M, Köhler S, Holtgrewe M, Schubach M, et al. Next-generation diagnostics and disease-gene discovery with the exomiser. Nat Protoc. 2015;10(12):2004-2015.
2. Hamosh A, Scott AF, Amberger JS, Bocchini CA, McKusick VA. Online Mendelian inheritance in man (OMIM), a knowledgebase of human genes and genetic disorders. Nucleic Acids Res. 2005;33(suppl 1):D514–7.
3. Aymé S. Orphanet, an information site on rare diseases. Soins. 2003;672:46.
4. Robinson PN, Köhler S, Bauer S, Seelow D, Horn D, Mundlos S. The human phenotype ontology: a tool for annotating and analyzing human hereditary disease. Am J Hum Genet. 2008;83(5):610–5.
5. Köhler S, Doelken SC, Mungall CJ, Bauer S, Firth HV, Bailleul-Forestier I, et al. The human phenotype ontology project: linking molecular biology and disease through phenotype data. Nucleic Acids Res. 2014;42(D1):D966–74.
6. Köhler S, Vasilevsky NA, Engelstad M, Foster E, McMurry J, Aymé S, et al. The human phenotype ontology in 2017. Nucleic Acids Res. 2017; 45(D1):D865–76.
7. Köhler S, Schulz MH, Krawitz P, Bauer S, Dölken S, Ott CE, et al. Clinical diagnostics in human genetics with semantic similarity searches in ontologies. Am J Hum Genet. 2009;85(4):457–64.
8. Girdea M, Dumitriu S, Fiume M, Bowdin S, Boycott KM, Chénier S, et al. PhenoTips: patient phenotyping software for clinical and research use. Hum Mutat. 2013;34(8):1057–65.
9. Hamosh A, Sobreira N, Hoover-Fong J, Sutton VR, Boehm C, Schiettecatte F, Valle D. PhenoDB: a new web-based tool for the collection, storage, and analysis of phenotypic features. Hum Mutat. 2013;34(4):566–71.
10. Sifrim A, Popovic D, Tranchevent LC, Ardeshirdavani A, Sakai R, Konings P, et al. eXtasy: variant prioritization by genomic data fusion. Nat Methods. 2013; 10(11):1083–4.
11. Javed A, Agrawal S, Ng PC. Phen-gen: combining phenotype and genotype to analyze rare disorders. Nat Methods. 2014;11(9):935–7.
12. Masino AJ, Dechene ET, Dulik MC, Wilkens A, Spinner NB, Krantz ID, et al. Clinical phenotype-based gene prioritization: an initial study using semantic similarity and the human phenotype ontology. BMC Bioinform. 2014;15(1):248.
13. Robinson PN, Köhler S, Oellrich A, Wang K, Mungall CJ, Lewis SE, et al. Improved exome prioritization of disease genes through cross-species phenotype comparison. Genome Res. 2014;24(2):340–8.
14. Singleton MV, Guthery SL, Voelkerding KV, Chen K, Kennedy B, Margraf RL, et al. Phevor combines multiple biomedical ontologies for accurate identification of disease-causing alleles in single individuals and small nuclear families. Am J Hum Genet. 2014;94(4):599–610.
15. Trakadis YJ, Buote C, Therriault JF, Jacques PÉ, Larochelle H, Lévesque S. PhenoVar: a phenotype-driven approach in clinical genomics for the diagnosis of polymalformative syndromes. BMC Med Genet. 2014;7(1):22.
16. Zemojtel T, Köhler S, Mackenroth L, Jäger M, Hecht J, Krawitz P, et al. Effective diagnosis of genetic disease by computational phenotype analysis of the disease-associated genome. Sci Transl Med. 2014;6(252):252ra123-252ra123.
17. James RA, Campbell IM, Chen ES, Boone PM, Rao MA, Bainbridge MN, et al. A visual and curatorial approach to clinical variant prioritization and disease gene discovery in genome-wide diagnostics. Genome Med. 2016;8(1):13.
18. Deng Y, Gao L, Wang B, Guo X. HPOSim: an R package for phenotypic similarity measure and enrichment analysis based on the human phenotype ontology. PLoS One. 2015;10(2):e0115692.
19. Peng J, Xue H, Shao Y, Shang X, Wang Y, Chen J. Measuring phenotype semantic similarity using human phenotype ontology. In bioinformatics and biomedicine (BIBM), 2016 IEEE international conference on Shenzhen: IEEE; 2016. pp. 763-6.
20. Resnik P. Using information content to evaluate semantic similarity in a taxonomy. In: Proceedings of the 14th international joint conference on artificial intelligence (IJCAI-95); 1995.
21. Lin D. An information-theoretic definition of similarity. In: ICML, vol. Vol. 98, no. 1998; 1998. p. 296–304.
22. Jiang JJ, Conrath DW. Semantic similarity based on corpus statistics and lexical taxonomy. In: Proc of 10th international conference on research in computational linguistics, ROCLING'97; 1997.
23. Schlicker A, Domingues FS, Rahnenführer J, Lengauer T. A new measure for functional similarity of gene products based on gene ontology. BMC Bioinforma. 2006;7(1):302.
24. Li, B., Wang, J. Z., Feltus, F. A., Zhou, J., & Luo, F. (2010). Effectively integrating information content and structural relationship to improve the GO-based similarity measure between proteins. arXiv preprint arXiv: 1001.0958.
25. Pesquita C, Faria D, Bastos H, Falcao A, Couto F. Evaluating GO-based semantic similarity measures. In: Proc. 10th annual bio-ontologies meeting, vol. Vol. 37, no. 40; 2007. p. 38.
26. Wang JZ, Du Z, Payattakool R, Yu PS, Chen CF. A new method to measure the semantic similarity of GO terms. Bioinformatics. 2007;23(10):1274–81.

AUTHOR INDEX

AUTHOR INDEX

BioMed Central Ltd
236 Gray's Inn Road
London WC1X 8HB
UK

ISBN 978-1-5108-6019-3

27th Annual Conference of the International Society of Exposure Science (ISES 2017)

Integrating Exposure Science Across Diverse Communities

Book of Abstracts

Research Triangle Park, North Carolina, USA
15 - 19 October 2017

International Society
of Exposure Science
Science for Better Environmental Health